THE ORGAN MOUNTAIN RANGE

ITS HISTORY AND ITS ORCHIDS

Cattleya dormaniana Rchb.

PESSANHA 2003

Cattleya dormaniana
A botanical illustration by Álvaro Pessanha

THE ORGAN MOUNTAIN RANGE

ITS HISTORY AND ITS ORCHIDS

David Miller, Richard Warren,
Izabel Moura Miller and Helmut Seehawer

The Organ Mountain Range, Its History and Its Orchids

David Miller, Richard Warren, Izabel Moura Miller and Helmut Seehawer

Cover:

Huntleya meleagris - Izabel M. Miller

Art: Daniel Scart and Inghra Ursula Scart

ISBN **978-85-60217-01-4**

1st **Text Preparation:** Marcelo França Dragoud

1st **Text Revision:** Richard Warren

2nd **Text Revision:** David Miller

Final Revision: David Miller and John Makant

Technical Coordinator: Rafael Pereira Franco

Digital File Finishing: Fábio Herdy

Photography: Izabel Moura Miller (except where indicated otherwise)

Botanic Illustrations (Aquarelles): Álvaro Pessanha

Orchid Lip line drawings: David Miller and Izabel M. Miller

Coloured Botanical Drawings: Helmut Seehawer

Layout: Daniel Scart e Inghra Ursula Scart

Published by Editora Scart

Nova Friburgo / RJ - Brazil

Collaborators:

John & Monica Miller
The Bromley Trust
The Rio Atlantic Forest Trust
Reserva Ecológica de Guapiaçu
Sheila Grey
Chris Dobson
Maren Talbot
San Diego Orchid Society
Thames Valley Orchid Society
Maria do Rosário de Almeida Braga
Timothy Moulton
Álvaro Pessanha
Cheshire and North Wales Orchid Society

Laelia crispa
A botanical illustration by Álvaro Pessanha

This book is dedicated to
Luiz Buarque de Hollanda
(in memorium)

Laelia cinnabarina, Batem. HCinna N.F.

PESSANHA 95

Laelia cinnabarina
A botanical illustration by Álvaro Pessanha

ACKNOWLEDGMENTS

We have found It essential to live for many years within the bounds of the Organ Mountain Range Pluvial Atlantic Forest, in order to understand the dynamics of this forest. In addition and due to the complexity and the interdependence of all the forest parts, we had the good fortune to count on the help and support of many friends and institutions. Within these we high light the Rio de Janeiro Botanic Garden, its executives and more than sixty botanists who participated in the "Atlantic Rain Forest Project" in our property in Macaé de Cima – N.F. during the years 1986 - 1995.

In addition we would like to give special thanks to Chris Dobson who participated in much of the field reserach as well as discovering unusual and rich sources giving more information about George Gardner and the English traders, the Harrison Family, in Liverpool and London – UK. To all we offer our heart felt thanks.

Álvaro Pessanha (Botanic Artist and Architect); Athayde Tonhasco Jr., PhD (UENF – Apoidea "Euglossa"- Campos-R.J.); Bengt Janér (BIODOMUS Institute – Silva Jardim); Bock I. (AL); Carlos Minc (State Deputy -RJ); Cláudio Belmonte de Athayde Bohrer, PhD (Geography/Ecology -UFF-RJ); Cláudio Brandão Azambuja, Dr. (lawyer-N.F.); Darcilio Fernandes, PhD (Water Insects FIOCRUZ-RJ); Dina Lerner, Dra. (INEPAC –RJ); Elizabeth Garlipp (*In memorium* – President of Soc. Macaé de Cima); Everardo e Lenora Grossi (FRILAB – N.F.); Fritz Johann (NF); Geraldo Antonio R. de Coutinho, Dr. (Public Prosecutor – N.F.); Gustavo Martinelli, PhD (Bromeliaceae – JB-RJ); Haroldo Lima, PhD (Leguminosae – JB-RJ); Jackei Blackmer, PhD (UENF- Apoidea "Euglossa" – Campos-RJ); James Ratter, PhD (Royal Botanic Garden, Edinburgh); Julio Brandão Azambuja, Dr. (Lawyer -NF); Leiner P. Rezende, Dr. (Civil Engineer-NF); Liliana Couto (Agência de Viagens e Turismo Clássico Ltda); Lynn Clark, PhD (Gramineae – American Bamboos – Iowa State Univerty-USA); Melissa Bocayuva, Dr. (Jardim Botânico-RJ); Marcelo França Dragaud (Digitação); Marcus Nadruz, Msc. (Philodendron Schott – JB-RJ); Marta Moraes, Msc. (Orquidário – JB-RJ); Milan (Kronokoma Foto Ltda – RJ); Nicholas Locke (REGUA-Guapiaçu-RJ); Oscar V. Sachs Jr, Paulo Costa (TV Serra Mar-NF); Peter Tobias (San Diego Orchid Society-USA); RAFT – Rio Atlantic Forest Trust – UK, Renato Pineschi (Ornithology); Rodrigo Singer, PhD (Orchid Pollination –UNICAMP-SP); Rosário Almeida Braga e Tim Moulton (Petrópolis –RJ); Samantha Koehler, MSc.(Bifrenaria e Maxillaria – UNICAMP – SP); Solange Pessoa, MSc. (Passiflora-JB-RJ); Tania Muniz (FEEMA-RJ); Thames Valley Orchid Society-UK; Victor Rogério Magalhães de Sá e Melo, Dr. (*in memorium* - Public Prosecutor-NF). **Field Work Help**: Alex Lima Abreu (Santa Maria Madalena); Álvaro Pessanha – R.J.; Andy (Pedra Branca – Cachoeiras de Macacu); Carlos Alberto(garrucha – Sumidouro - NF); Chris Dobson (UK); Gabriel Dreyfus Cattan(Petropolis); Girson Overney (Silva Jardim); Jair José da Silva, Dr. (President of The Leopoldina Railroad Engineers' Association); Lauro de Freitas Moreira, Dr. (Nova Friburgo); Luiz Antonio Bueno (Santa Maria Madalena); Nicholas Locke (Guapiaçu); Pascoal Grossi (Coleóptero - Nova Friburgo); Paulo Roberto Dutra da Silva (Nova Friburgo); Rafael Dutra da Silva (Nova Friburgo); Ricardo Ferreira Nines (Santa Maria Madalena); Rita Gripp (Macaé-Glicério); Rosário Almeida Braga (Petrópolis); Vanusa da Silva Rocha (Conceição de Macabu).

Cleistes vinosa
A botanical illustration by Álvaro Pessanha

SUMMARY

Sophronitis coccinea
A botanical illustration by Álvaro Pessanha

PROLEGOMENONA

There is a critical absence of information on the state of both what is left of original forest and small remaining fragments in the Organ Mountain Range, and so in 1997 we decided to prepare this book. We have, since the start, been deeply concerned with the conservation of original forest and its remnants and our objectives have been to preserve this and to stimulate the regeneration of native forest in this centre of endemism that is the Organ Mountain Range.

Geographically the Organ Mountain Range extends for about two hundred and fifty km. in length and between thirty and seventy km. in width. It is situated between latitudes 41° 30' and 43° 30' W. and between 21° 50' and 22° 45' S. For the length of the range, the mountains are divided between scarp and anticline as defined by the watershed. The scarps are the steep rocky slopes facing the Atlantic Ocean, while the slopes of the anticline descend more gradually and irregularly from the ridges of the scarp to the west in the direction of the River Paraiba do Sul. For practical purposes we have divided the range transversely into eight sections - four on the scarp and four on the anticline.

The arrival of botanists and plant collectors in Brazil and specifically into the Organ Mountains, can be said truly to have begun in earnest between 1815 and 1860, a period that coincided with a frenzy for cultivating tropical orchid species in Europe. We chose George Gardner and the Harrison family as the most representative of the scientists and plant collectors of this period as they have left a well-documented trail of their activities.

We visited every single section at various times and many several times over, during different seasons. During the nine years of the investigation, we determined the state of the forest and identified, described, photographed and illustrated the orchids that we found. Sections V and VI, as we explain later, received more attention for two reasons: firstly, we live in these areas and secondly because they are the least devastated of the entire range.

Our approach to the project was to use the orchid family as a touchstone for the state of the forest. Everyone has at some time or another encountered orchids. They evoke images of beauty, luxury and mystery, while the trees, - fundamental parts for the survival of this unique, complex and interdependent ecosystem - are, well....just trees. The orchids are one of the largest flowering plant families (Angiosperms) with approaching thirty thousand species and tens of thousands of magnificent hybrids. But most importantly from our viewpoint is their existence, or their absence, as an indication of the state of health and the richness of the biodiversity of the tropical humid forest, which in pre-colonial times completely covered the Organ Mountain Range.

The forested slopes of the scarp contain significant fragments of original forest and on occasion show considerable forest regeneration. In contrast, the anticline has an almost total absence of original forest and less healthy regeneration. Widespread continuing clearance for sugar and coffee cycles, coupled with seventy years of clearance for and damage from the railways, managed to transform a substantial part of the region into a semi-desert.

Original forest is sustained by a thick layer of humus and organic matter. The roots, which run mainly horizontally, are intimately interwoven with this organic layer. In particular, an intense system of branched capillary roots collect and retain rainwater like a sponge, releasing it gradually to descend the slopes. This store of water is of vital importance for the months of low rainfall. Pastures, on the other hand, release water as the rain falls, discharging it directly into rivers, functioning basically like a flush toilet.

The effect of deforestation on the slopes and valleys of the scarp, together with the deforestation and drainage of the coastal plain is already creating problems for the supply of water to the cities of the plains and the coast. These problems will become critical with every further natural period of six months without rain, if these slopes and valleys are not soon reforested. In other words, reforestation and conservation of surviving native forest are not simply the concern of 'greens' and ecologists but are of vital economic importance to the urban populations, especially in those rapidly expanding cities on the coastal plain.

The regeneration of native forest in these mountains is a very complex subject. Success depends on various factors: there are four or five development phases and these are influenced by the location, the situation, the altitude, etc. The single most practical and economical system for areas larger than one hectare is natural regeneration. Neighbouring fragments of original forest are indispensable for the success of the project, as is the exclusion of domestic herbivores - cows, horses, mules, goats - from the area. For the great majority of projects, the starting state of affairs will be degraded pasture with only a layer of subsoil. Invading pioneer tree species arrive as wind-borne seeds and rapidly colonise the area creating a band of humus. These pioneers enjoy a short life but, as the humus layer becomes thicker, a new wave of secondary pioneers appears. These are taller and longer lived. The organic soil and humus layers thicken. A sub-forest forms and when soil, vegetation and humidity are relatively constant, the colonization and growth of more noble trees from original forest can begin to happen. It is a long process. Time is needed for each phase, and success depends on both the location and many other factors. Above all, the major question is, does the slope face north or south? Slopes facing the south receive limited sunlight for half the year, but soon create a constant humidity, a humus layer and vegetation much more readily than a slope that faces north. Natural regeneration will take time but the only expense will be rolls of barbed wire to exclude domestic animals. We have found that the capacity of the humus to sequester and store water begins after five or six years.

The reintroduction of orchid species is a theme as complex as it is controversial. One thing is unarguable: a forest regenerating naturally for about forty years will only show pioneer orchid species. Even if the regrowth is right beside original forest, we have discovered that no more than 10% of the species found in that original forest fragment are able to colonize the area of regenerating forest. One interesting exception is regrowth around a river. But here, of course, there is the benefit of the high humidity generated by water.

The introduction of mature orchid species rescued from an original forest, into another forest at the same altitude is, in our experience, always successful. Therefore, when forest has been felled, usually to build a road or a reservoir in the region, relocation of plant species seems us to be a laudable, even if occasionally questioned, project.

We have had several successes introducing plants cultivated in vitro into areas which had previously been seriously depleted through collection. However, a good deal of care must be taken in the weaning of seedlings after they are removed from the flasks and before they are replanted on the final substrate. It is, of course, also vital that recolonized areas are protected from further orchid collectors.

In summary, we found over 620 species of orchid from 110 genera in the Organ Mountain Range. Astonishingly, this represents 25% of all Brazilian orchid species and 50% of Brazilian genera, all found in less than 1% of the Brazilian land mass.

These totals can be broken down as follows:

√ 65% epiphytes, 23% terrestrials, and 12% lithophytes.

√ 38% common, 37% occasional and 25% rare.

√ 58% in sections V & VI, 24% in sections I and IV and 18% in sections VII & VIII

√ 72 % on scarp slopes and 28% on the anticline

√ Also, 26% were found between 0-800 M. and 73% between 800-1600M.

These figures reflect the altitudes where the largest fragments of original forest are found.

√ 73% have their flowering time between November and April.

We came to a sad conclusion. Five hundred years ago, the Organ Mountain Range was completely covered with dense, moist forest, on both the scarp and the anticline. It is reasonable to assume that there was continuity of epiphytes, particularly orchid species, with only minor endemic exceptions. Back then, the humid mists in the underforest must have been the same on both sides of the watershed. The rivers would have been permanently full (unlike today). The dense garden of epiphytic bromeliads created a huge suspended lake because of the water kept in their rosettes, functioning as a regulator of humidity and temperature during dry periods.

Consider the contrast now between the species numbers of epiphytes on the scarp slopes and the anticline. We estimate that about 325 species of epiphytic orchids have been lost from the anticline or are extremely rare, and many of the survivors no longer have the benefit of their preferred habitat.

Without doubt, the most distinctive characteristic of the humid forest of the Organ Mountain Range is the complex interdependent relationships of all the forms of tropical life which make up the whole: the matrix. Many millions of years of inexhaustible speciation, at times intense, at times slow, but always present, have created what we are trying so hard to conserve today.

One of the most curious and obvious examples of such interdependence is the relationship of bromeliads with the climate and with other species. Five hundred or so years ago, bromeliads in the original forest canopy would have held up to 18,000 litres of water per hectare in their rosettes. They formed, effectively, a suspended sea, evenly distributed, which controlled humidity and climate throughout the whole forest.

Today on the mountain ridges, bromeliads grow as an immense humid terrestrial garden. By accumulating rainwater, mist and dew, they enable the survival of numerous epiphytic orchid species, which germinate and grow ten metres above in the stunted, twisted trees of the elfin forest.

In regrowth forest, the bromeliads grow as a terrestrial garden and rapidly create a constant cloud of humidity formed by the evaporation of water trapped in their rosettes and guaranteeing the rapid development of the forest and the introduction of epiphytic species.

And finally, large populations of bromeliads, several containing up to 20 litres of water and which are stuck to sheer rock faces, create a permanent vertical lake controlling the humidity and permitting the germination of seeds of other plants in this otherwise hostile environment.

During the Proceedings of the Second International Orchid Conservation Congress in 2004, Stuart Pimm asked in his keynote speech (Selbyana, 26, pp. 5-13): *'Do you know where your orchids are?'* and, *'Do forest fragments smaller than 100 hectares have conservation value for orchids?'* In this book lies both answers. Along with its precursor Orchids of the High Mountain Atlantic Rainforest in Southeastern Brazil, it contains the information and experience of thirty years of work. These two volumes now provide in a clear and accessible way hitherto unavailable information about the orchids which still exist in the Organ Mountain Range, along with essential details of their habitats, altitudes, frequency and prospects for survival. As Professor Pimm so astutely observed, if you are conserving, it is essential to know what you are conserving. This book fills that gap.

What we hope to do with this volume is convince people of the vital importance of this forest, not simply for its orchids and the amazing biodiversity, but for its practical role in the all-important issues of climate control and water conservation. Preserve the forest and you help control the climate and conserve water. The orchids will then look after themselves.

Zygopetalum maxillare
A botanical illustration by Álvaro Pessanha

1. Introduction

1.   INTRODUCTION

We picked up the European members of our team at the Rio de Janeiro international airport at 6.30 AM off a very punctual Varig flight from London. Richard Warren a large, bow-tied botanist, over six foot tall and arguably the best producer of *in vitro* orchid species from seeds in the UK, must have passed through hell sitting in a peasant-class seat for 12 hours. Chris Dobson, a dapper man from Cheshire, as always impeccably dressed, wearing a dusty blue sports jacket, grey well-pressed slacks and a perfect light red tie. He is not only a brilliant orchid grower but a competent researcher into historical orchid personalities. Finally Helmut Seehawer, just retired from 35 years of driving Boeing 747s around the world for Lufthansa and has made himself into an accomplished botanical artist and also an out-and-out expert on the Organ Mountain Range Pleurothallidineae.

We are, Izabel Moura Miller who has become a fine orchid photographer with a falcon's eye for spotting orchids in the forest, and David Miller, an environmentalist who together with Izabel, has bought large tracts of original forest in the municipality of Nova Friburgo with a view to conservation.

Luggage aboard we set off, due west, for the Organ Mountain Range, on the historic road to Petrópolis, and thence across the plain, driving parallel to the mountains.

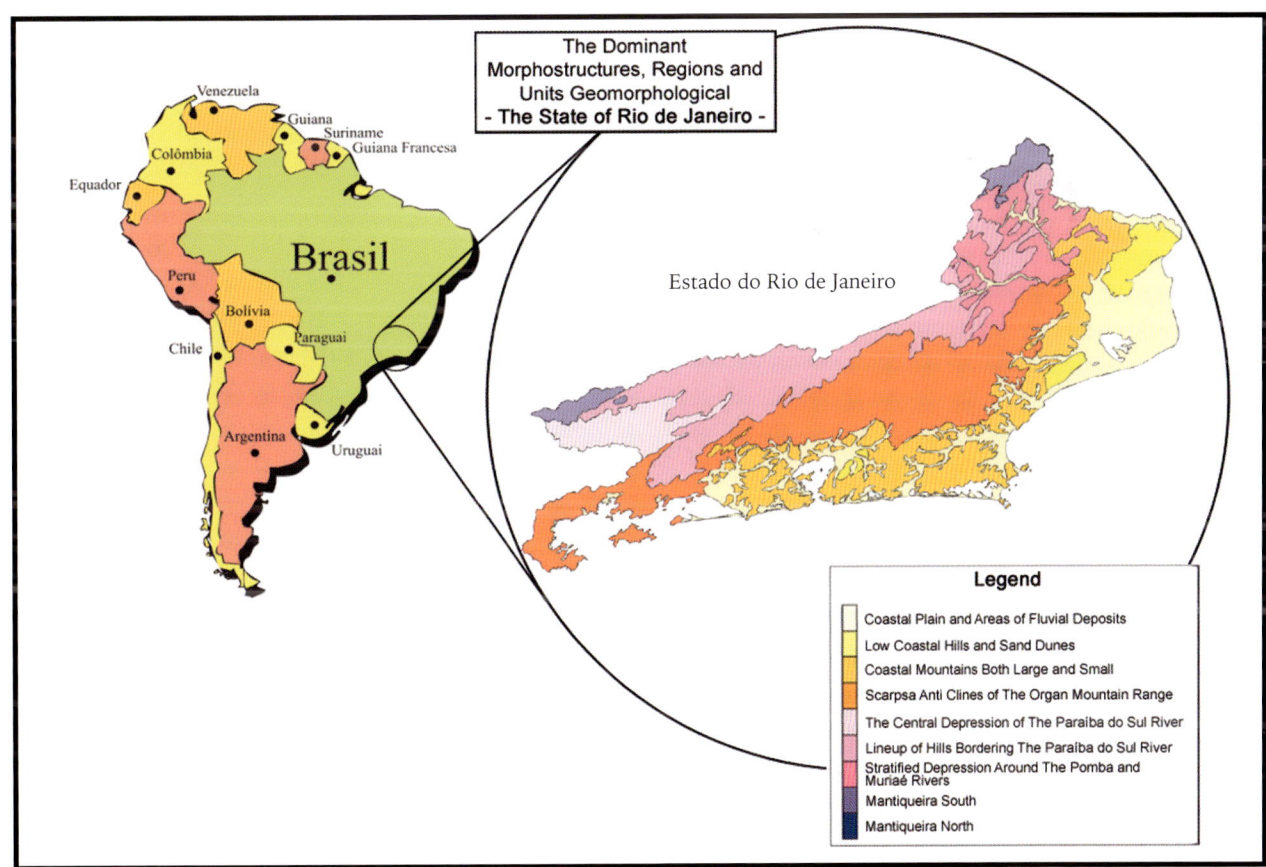

Source Project Radam Brazil

THE POLITICAL AND ADMINISTRATIVE DIVISIONS OF RIO DE JANEIRO STATE - 1998
WITH THE ORGAN MOUNTAIN RANGE AND OUR RESEARCH SEGMENTS SUPERIMPOSED

THE ORGAN MOUNTAIN RANGE

THE DIVISION: SCARP / ANTICLINE

I a VIII   SEGMENTS RESEARCHED: SCARP / ANTICLINE

Source: CIDE - The Rio State Centre For Information and Data (as adapted to show additions above)

The mountains loom as we near them. This was the vision that must have amazed the first Portuguese sailors into Guanabara Bay, and made George Gardner, our botanist hero, mad with anticipation as he did his six month apprenticeship in 1836 on the Tijuca massif in Rio de Janeiro, familiarizing himself with the flora and learning the Portuguese language.

George March

We turn north almost at the base of these mountains, the old but sturdy Kombi keeping up a constant 80 km an hour on the turnpike while all others pass us at 100. The early morning sun strikes this mountain wall. This palisade, apparently impossible to penetrate, has a collar of mist which the sun will soon dissipate. It appears vertical, but mostly forest covered. The Organ Mountain Group, so named by some long forgotten Portuguese adventurer who imagined them as pipes of the organ in his cathedral back home. We pass through Guapimirim, somewhat amazed by the apparent richness of the foothill forest, but the heavy incidence of silver-leafed *Cecropia*, and purple-flowered Melastomataceae species show us that this must be regrowth dating from the age when coffee was planted here, over 70 years ago.

On through Magé which George Gardner visited on his way to George March's establishment, with a mule train on Christmas day 1836. George March was the richest and most successful English trader in Rio de Janeiro in the first four decades of the 19th century. There still is a trail, albeit more like the dry bed of a mountain torrent, leading to Teresópolis.

Just a while later we arrive at Cachoeiras de Macacu where the Leopoldina Railway Co. had a maintenance and marshalling yard; here they prepared their compositions to climb the steep mountain pass at Teodoro de Oliveira and thence to Nova Friburgo and the interior.

We enter the mountains on a very steep, winding road. The old railroad is at the other side of the valley on a gradient much less steep (one day not very far off, it will certainly be used to form part of a dual carriageway). The forest around, from about 600 M.asl appears to be, and is, original. Across the forest-covered valley, where the railroad track ran, the forest shows the scars both of having been burnt by sparks from the locomotives during the dry season, also from the trails where timber was cut for sleepers and fuel.

Our base in Macaé de Cima.

We reach the top of the pass at Teodoro de Oliveira some 1075 M.asl, drive gently down to Mury where we leave the asphalt and climb a dirt road, and we arrive at our HQ, 1450 M.asl, in the middle of original montane rain forest not three hours away from sophisticated Rio de Janeiro.

This was to be the final month-long field-work expedition; the cut-off point where we leave the forests and mountains and get on with writing this book.

Nine years ago, emboldened by the success of a book we had published,

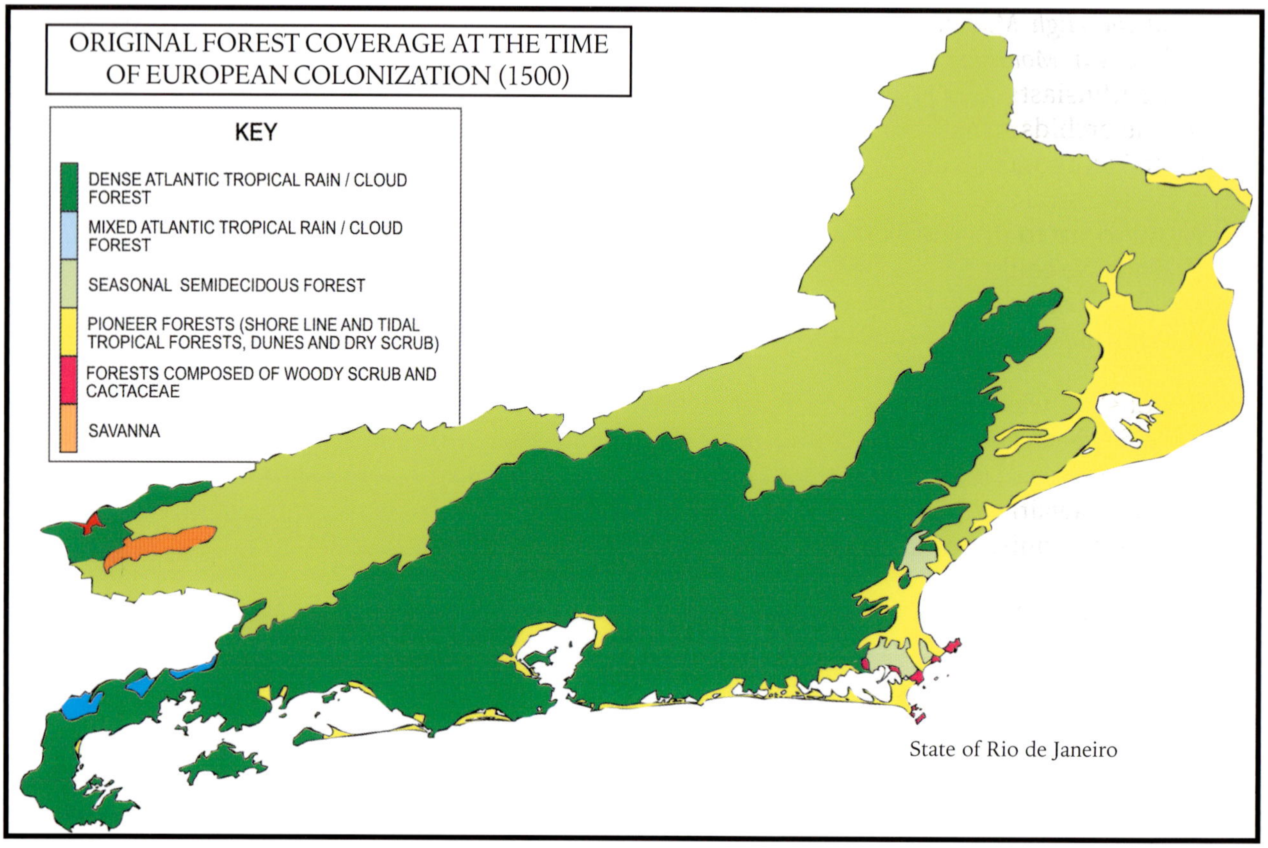

ORIGINAL FOREST COVERAGE AT THE TIME OF EUROPEAN COLONIZATION (1500)

KEY

- DENSE ATLANTIC TROPICAL RAIN / CLOUD FOREST
- MIXED ATLANTIC TROPICAL RAIN / CLOUD FOREST
- SEASONAL SEMIDECIDOUS FOREST
- PIONEER FORESTS (SHORE LINE AND TIDAL TROPICAL FORESTS, DUNES AND DRY SCRUB)
- FORESTS COMPOSED OF WOODY SCRUB AND CACTACEAE
- SAVANNA

State of Rio de Janeiro

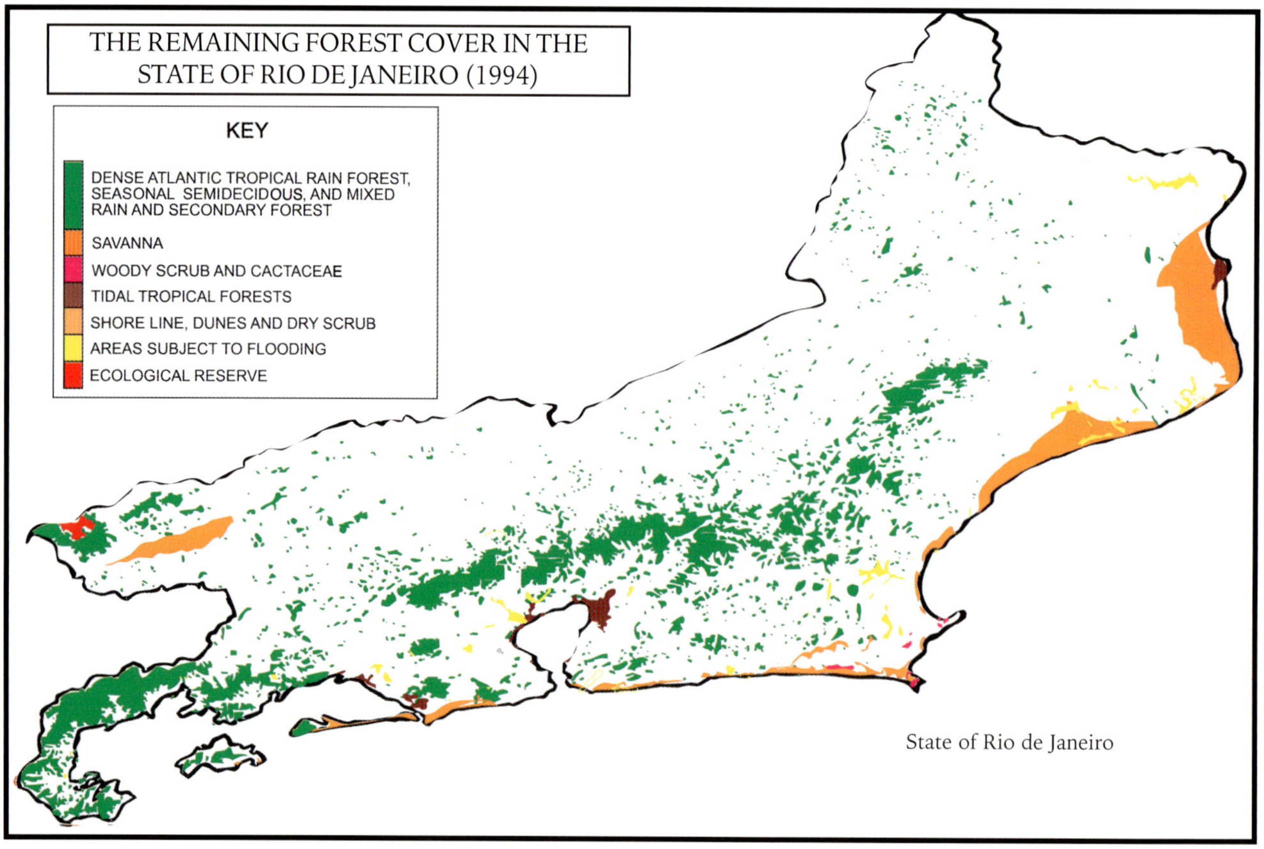

THE REMAINING FOREST COVER IN THE STATE OF RIO DE JANEIRO (1994)

KEY

- DENSE ATLANTIC TROPICAL RAIN FOREST, SEASONAL SEMIDECIDOUS, AND MIXED RAIN AND SECONDARY FOREST
- SAVANNA
- WOODY SCRUB AND CACTACEAE
- TIDAL TROPICAL FORESTS
- SHORE LINE, DUNES AND DRY SCRUB
- AREAS SUBJECT TO FLOODING
- ECOLOGICAL RESERVE

State of Rio de Janeiro

Source: Project Radam Brazil

| 26 |

Orchids of the High Mountain Atlantic Rain Forest in Southeastern Brazil, we decided to produce this book, *The Organ Mountain Range, Its History and Its Orchids,* a somewhat ambitious project for a small group of enthusiasts who had to earn their wherewithal while undertaking the project. With a view to place the orchids species into their context of a large defined tropical mountain forest complex we decided to incorporate a fairly large introduction dealing with the geological history of this mountain range, the botanical evolution, human occupation and exploitation in current, colonial and pre-colonial times, and to try to produce some evidence as how and why the remains should be conserved, and how perhaps we could generate ideas on the regeneration of some of this forest without treading too much on the corns of local politicians, real estate agents, subsistence farmers and more or less everybody else.

Obviously it was impossible for us to examine every inch of the Organ Mountain Range. In fact this proved essentially unnecessary as most of the anticline, the area invaded by coffee, has been turned into a virtual desert populated by grass and ants with flocks of vultures floating "in the blue Brazilian sky" awaiting the death of starved cattle during the long dry winter season. So comprehensive exploration was quite unnecessary.

The various maps that accompany this treatise will show how this mountain range is the spine of Rio de Janeiro state, dominating it in a way that few other small subtropical areas are dominated and showing such a botanical richness, containing 50% of Brazil's orchid genera, 25% of Brazil's orchid species and almost as high a number of vascular plants as the Amazon forest in an area only 0.25% of the country's landmass.

We have divided this mountain range into two principal parts; the scarp, that is the steep palisade facing the South Atlantic, running from Tinguá in the south east to Santa Maria Madalena in the north east, and the anticline sloping gradually and irregularly down to the Paraíba do Sul River from the ridge of the scarp. This somewhat vague and irregular division is, however, emphasized by rainfall. The scarp slope receives an average annual rainfall of between 2.0-2.5 metres per annum, whereas the anticline receives between 1.0-1.5 metres. This of course has a profound effect on the flora, particularly in the presence and diversity of characteristic species.

The scarp slope is still fairly intensely forested and from 600 metres to the summit there are significant tracts of original forest which, though disturbed, essentially are examples of the original montane rain forest eco-system. The reasons are clear: these steep slopes are of difficult access, impossible for any type of agriculture, quite useless for pasture and being intensely humid are difficult to burn either spontaneously or deliberately.

The anticline forests on the other hand, growing in valleys and on smoother slopes

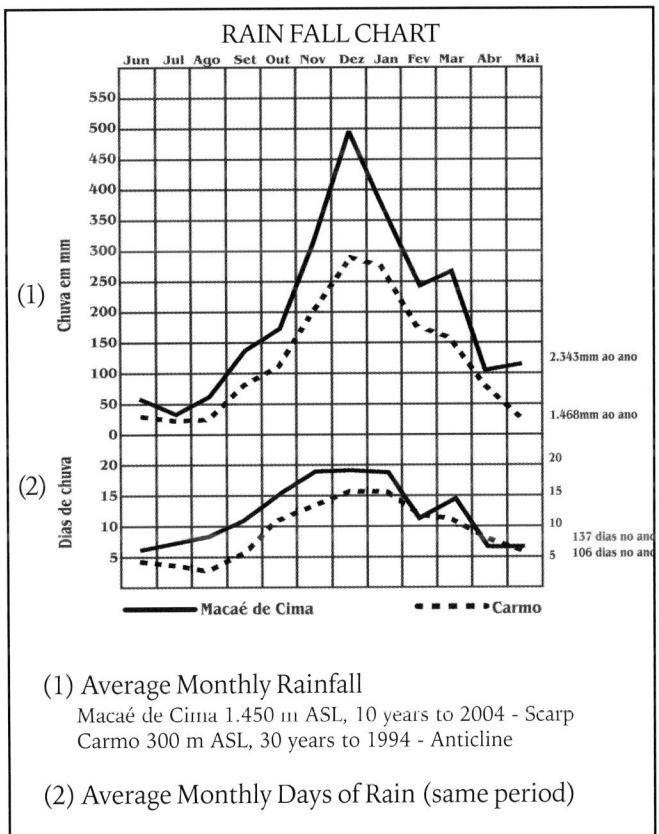

(1) Average Monthly Rainfall
  Macaé de Cima 1.450 m ASL, 10 years to 2004 - Scarp
  Carmo 300 m ASL, 30 years to 1994 - Anticline

(2) Average Monthly Days of Rain (same period)

(nick-named half-oranges in Paraíba do Sul) were both accessible, capable of sustaining coffee plantations, albeit for one cycle only, and with a climate that was almost perfect for coffee's flowering, fruit development, harvesting and drying. There is little, if any, left of the original forest in the whole region and this puts the epiphyte hunter somewhat at a loss. In addition the humidity levels throughout the region in regrowth forest are not consistent, and a constant high humidity is a *sine qua non* for the germination of most epiphytic species.

Notwithstanding this apparently insurmountable obstacle and after climbing fruitlessly through many mountainous regrowth coppices, we adopted a different approach, only examining trees along permanent water courses, while always searching for an "Oasis" or a huge relict tree, standing alone in almost desert-like pastures. The "Oases" were found generally high up between two very close sugar-loaf-type mountains, impossible to cultivate and too remote to make even fire-wood collection worthwhile. We found such profitable areas throughout the anticline, sparsely and *per se* precarious, but giving us clues to the original orchid species content of these long forgotten and extensive forests. Surprisingly the anticline rock faces show a rich crop of lithophytes not found on the scarp, and their vegetation is fairly pristine as these slopes are of no use to man or beast (except lithophyte collectors).

Having made the critical division between scarp and anticline, we then sectioned off both areas into convenient geographical segments. The reasons for doing this are in the first place practical. We take a transect on both the scarp side and its opposite on the anticline and then, depending on forest availability, explore both linearly and sometimes extensively, within each segment. The second reason is that in no way do we wish to identify the exact location where such or such a plant was found and thus call in a host of collectors. However, we do want our readers to have a very good idea of what kind of conditions the plants grow in, at what altitude and conditions, so that they will then be able to work out the best schemes for cultivation.

We have sampled fairly well all segments of the Organ Mountain Range, with particular emphasis on section VI and V, for two reasons. The first being that we live in the middle of these segments, and the second that they are by far the richest and least devastated areas in the Organ Mountain Range. The other segments we have visited at least several times at different seasons. The one exception is the Tinguá forest in section II. This is a forest reserve, very difficult to get permission to visit, and is also dangerous as it is surrounded by shanty towns where rival gangs are reputed to dump their murdered victims. In this segment we relied on research undertaken in the summer of 2002, by friends from the Rio de Janeiro Botanic Garden.

2. The Origins of the Organ Mountain Range and Its Rain Forests

2. THE ORIGINS OF THE ORGAN MOUNTAIN RANGE AND ITS RAIN FORESTS

The "Blue Planet" was formed as a relatively solid object some 4.5 billion years ago. However it needed 1.5 billion or so more years before the first living organisms appeared and a soup of primitive algae and bacteria began to spread through the seas. Very gradually these organisms began to invade the land mass. It was only 345 million years ago that the first forests evolved. Primitive ferns and pine-like plants became abundant during this next lengthy period, known as the Carboniferous. Over a period of 100 million years, huge swamps accumulated the organic detritus from these forests and formed the coal and oil deposits which we exploit today.

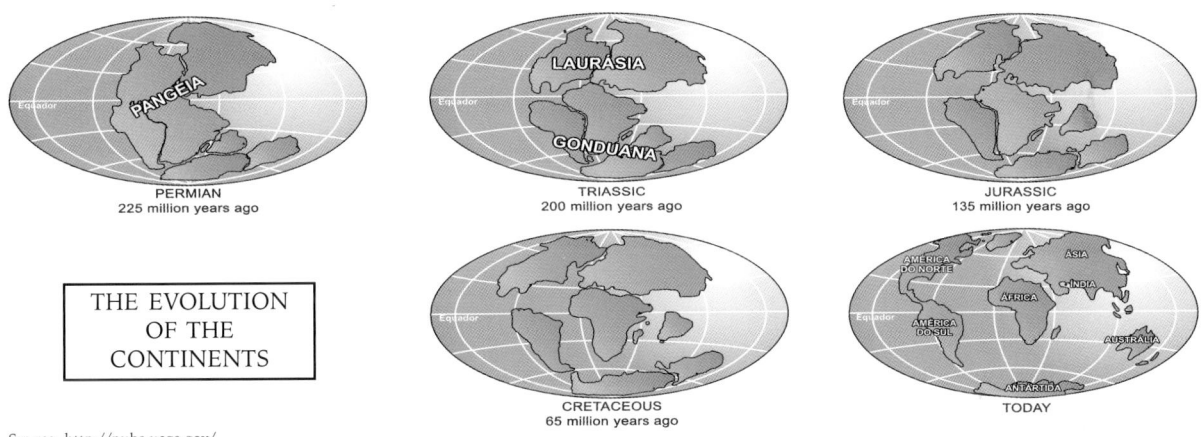

PERMIAN
225 million years ago

TRIASSIC
200 million years ago

JURASSIC
135 million years ago

THE EVOLUTION OF THE CONTINENTS

CRETACEOUS
65 million years ago

TODAY

Source: http://pubs.usgs.gov/

Up to 200 million years ago the planet's total land mass was held in the undivided super continent of Pangaea which had slowly been formed in the continental lithospheric crust.

The Lithosphere is the one directly accessible to observation and is relatively cold and rigid. It comprises both the crust and the upper part of the mantle. The oceanic Lithosphere crust varies in thickness from 6-8 km. at the mid-ocean ridges, to about 100 km thick in the oldest parts of the oceans, whereas the continental Lithosphere is probably between 20-90 km, and on average, about 35 km, thick. The Lithosphere floats over the Asthenosphere, which is hotter, less rigid and capable of being deformed over long periods of time.

At the end of the Triassic, 200 million years ago, the Pangaea super-continent was still largely intact but just about to start to break-up. By 160 million years ago (the mid Jurassic), ocean areas in the Caribbean and Central Atlantic to the east had opened a continuous equatorial sea-way between Laurasia (North America and Eurasia) and Gondwana (Africa, South America and other southern landmasses). By the late Cretaceous (65 million years ago) Africa and South America had separated, had detached from Antarctica and the Atlantic Ocean was a continuum up to the North Pole. India and Australasia had also separated from Antarctica and were relatively speaking, racing North Eastwards. North and South America were tenuously joined by the Panamanian isthmus but this later broke and left South America isolated for some 50 million years before reforming some 3 million years ago. By about 20 million years ago all the land masses were near their present day relative positions.

The original fault line between the African and South American continents is the divisor between two major tectonic plates. These tectonic plates are rather like a very simple jig saw puzzle where the Lithosphere is divided up into seven major pieces and a number of smaller ones. The large pieces are the South American, African, Antarctic, North American, Pacific, Eurasian and Indian, which touch one piece or another on all sides, and the whole drifts at a rate of one or maybe 20 cm a year in whatever direction a pressure area forces them. These confrontations can be direct; a good example would be the South and Northern American plates being forced against the Pacific plate. The force behind the two former could not budge the inertia of the latter so they climbed over and forced under the Pacific plate forming the Andes and the Rocky Mountains.

ERA	PERIODS OR SYSTEMS	BEGINNING	DURATION	END
TABLE SHOWING GEOLOGICAL TIME, ERAS AND PERIODS (PART)				
Quaternary	Recent	25 thou	25 thou	Now
Quaternary	Pleistocene	600 thou	575 thou	25 thou
Tertiary	Pliocene	10 mil	9.4 mil	600 thou
Tertiary	Miocene	25 mil	15 mil	10 mil
Tertiary	Oligocene	35 mil	10 mil	55 mil
Tertiary	Eocene	55 mil	20 mil	35 mil
Tertiary	Paleocene	65 mil	10 mil	55 mil
Cretaceous		135 mil	70 mil	65 mil
Jurassic		200 mil	65 mil	135 mil
Triassic		225 mil	25 mil	200 mil
Permian		250 mil	55 mil	225 mil
Carboniferous	Pennsylvanian	310 mil	30 mil	280 mil
Carboniferous	Mississipian	345 mil	35 mil	310 mil
	Devonian	405 mil	60 mil	345 mil
	Silurian	425 mil	25 mil	405 mil
FORMATION		4.500 bil	4.075 bil	425 mil

The primitive and current force behind the South American plate came from the original fault line that was created long before Pangaea split up and South America and Africa separated. Probably, the mountains of the Brazilian coastal range were formed before the separation by the upsurge of magma through the Asthenosphere. This continues up to the present day, forming the mid Atlantic ridge.

At the end of the Jurassic and just before the break up of the continents, it is probable that flowering and seed bearing plants started to evolve. This coincided with a surge in insect life which was vital to the proliferation of these "modern" plant families, which ultimately through speciation, adaptability and mobility, took over the neo-tropical forests from the primitive ferns and pines.

Two major events took place during this immense period. Probably an asteroid collision wiped out most earth life around 230 million years ago, and yet another had a similar effect about 65 million years ago, terminating the dinosaurs' domination of the planet at the end of the Cretaceous.

The southern continents also evolved their own unique flora and fauna.

In the case of flora, many plant families exist on the southern continents that are common to all, but with rare exceptions the genera are different (this shows a common evolutionary base with speciation occurring after the separation of the land masses). South American mammals which evolved during the long isolation of the continent during the tertiary are represented by marsupial opossums, edentates, sloths, armadillos and anteaters, *platyrrhine* monkeys, and some rodent groups. But these were joined by immigrants from North and Central America during the so-called 'Great American Interchange' which occurred when the Panama Land Bridge reformed towards the end of the Pliocene. Examples of such immigrants include Jaguars (and all other S. American cats), tapirs, deer, llamas, peccaries etc.

The east coast Brazilian Mountain Range, of which the Organ Mountain Range forms a central and significant part, is very old and was formed long before Gondwana broke up and drifted northwards, probably in the Cambrian, which lasted from 600 million to 500 million years ago.

Periods of intense volcanic activity produced the now recognizable "Serra do Mar". The mountain chain must been around twice its present altitude and it may have developed along, or even created, a fault line many thousand kilometres long which itself, later facilitated the fateful separation of the South American and African continents. Be that as it may, it is almost certain that while still part of Gondwana, this region must have suffered several prolonged ice ages which would have started the grinding down of the mountain range towards its present size and shape. Proof of the more current erosion by rain and winds since the continent was formed, is shown by the thick silt deposits in the Paraíba do Sul river valley and the coastal plain, both on and offshore, between Rio de Janeiro and Campos.

Once the South American continent had been more or less settled in its current position relative to other land masses, its climate began to establish a pattern, influenced by ocean currents, water temperatures and the consequent air movement.

During the Quaternary period, 600 thousand years ago up to the present-day, the world experienced many cold and warm periods. It seems that the cold periods, when the glaciers advanced in the temperate parts of the world, lasted much longer than the warm periods. During these cold phases vast quantities of water were held in ice covering much of the present day temperate land-masses, and furthermore the lowering of temperature reduced the volume of the water so that the oceans retreated from their warm period shore lines. This in turn would have pushed the warm Brazil current away from the Organ Mountain Range. This current arises in the mid-Atlantic round about latitude 10° south, hits the Brazilian eastern coast line at 15° south, branches at 25° south, the main stream moving out to sea and a smaller stream continuing down the coast line until it meets the cold Falklands current and dissipates at 28° south around the gulf of Santa Catarina. The relative difference in mid-ocean water temperatures would have been minimized; the wind patterns would have changed; the air moisture content would have been seriously reduced, so during the glaciation millennia, rainfall would have been sparse over the cold dry mountain barrier. And thus the Organ Mountain Range forests would have retreated into deep valleys still with permanent watercourses and the plants that survived would have adapted to these harsh conditions. Many species would have been wiped out, but a significant nucleus remained.

It is probable that the Organ Mountain Range suffered no glaciation after the continent separated from **Gondwana** and drifted to its present position and thus, there has been an evolutionary and speciation process only arrested by the dry cold periods which resulted in this region becoming a classic centre of endemism.

During the following warm wet periods glaciation retreated world wide, the oceans warmed up and regained their old shore lines, the warm Brazil current would have moved inshore, the western trade winds would have grown stronger with a higher moisture content and consistent rains would have started to fall again on the mountains.

The forests and their plants from this centre of endemism, crouching in their retreats, would have started to repopulate the vast areas abandoned thousands of years before and occupied by species adapted to a much warmer climate. Moreover, the isolated and reduced populations within the refugia evolved to create new species. They certainly created adaptable species; witness the number of *Cactaceae* and other succulents currently found in these now wet forests. Not to forget the orchid genera like *Scuticaria*, *Leptotes* and a host of *Maxillaria* species and Pleurothallidinae which have terete, pine-needle like leaves, capable of retaining moisture. And adaptations like the flowers of many of the *Zygopetalum* alliance which currently populate the high mountain fields, and whose flowers demonstrate a clear origin from the deep dark forests by their capacity to develop chloroplasts in their tepals to help in the photosynthesis process during seed development in a dark low light. Many scientists have pondered that this is the warmest and wettest period the quaternary world has known. Certainly it was the most propitious for the spread of neotropical rain forests, and the result over the ensuing millennia was to cover the whole Organ Mountain Range with a forest of such grandiosity, complexity, variety and intrinsic beauty never before seen on the "Blue Planet".

3. Human Contact with the Organ Mountain Range

3.   HUMAN CONTACT WITH THE ORGAN MOUNTAIN RANGE

We can now leave the Organ Mountain Range forests to expand, luxuriate and effectively climax during the next 25 millennia.

The central, strategically placed forest nuclei in the Organ Mountain Range would have crept out of their valley refuges, climbed mountains, spread along increasingly stronger and more permanent rivers until they covered most of the State of Rio and crossed the Paraíba river into Minas Gerais, as far as the annual rainfall permitted.

Albert Eckhout
Tapui Dance
Oil on wood, 168x294 cm
The Denmark National Museum, Copenhaguen
This work is also known as Tarairiu-Dancers

It was probably just after the middle of this period, some 12 thousand years ago, that the first humans arrived, descendants of the pioneers who crossed the Bering Strait from east Asia, spreading through North America, down along the Panamanian land bridge until they arrived in present day Brazil, probably from Peru, Bolivia and the Northeast South American Coast.

These peoples formed several amazing civilizations in Mexico, Central America and Peru, but their forerunners, who thinly populated Brazil, remained at least partially hunter-gatherers until the first European colonists arrived in 1500. Partly nomadic and living principally on the coastal plain and river courses, they lived off shellfish from the extensive mangrove swamps, and fish from the rivers and lakes. They also practiced slash-and-burn agriculture, planted manioc, perhaps other root crops and even some Leguminosae. They also knew how to stockpile some basic foods; smoked and dried fish, roasted manioc meal and so on, and of course would rely on many species of wild animals and birds for protein, as well as an abundance of wild fruit. In spite of this, the area occupied by a tribe, sub-tribe or family was limited to the area that could sustain the unit for a certain amount of time.

It is quite possible that these people, the first human occupants of Brazil, were in fact moving towards an Inca, Aztec or Mayan-type civilization. However they had not yet invented the tools; their axes were of stone, they had no beasts of burden and no domesticated animals and they apparently did not trade amongst themselves or with other tribes to any significant degree.

They were at a stage that Arnold Toynbee, the historian, would have called "an arrested civilization", and as a consequence had no power to affect the virgin forests as the Eskimos had no way of developing and advancing civilization in the Arctic. They could only overcome the natural environment by fire and had they so desired could still only have nibbled at the forest edges. In fact they only hunted for wild animals and birds, and passed through the forests on nomadic wanderings in search of further sustenance.

Jean Baptiste Debret
A manual machine to extract sugar cane juice. Rio de Janeiro, 1822 - Water color on paper, 17,9 x 24,9 cm
The Castro Maya Museum/IBPC. Rio de Janeiro, Brasil

Pedro Álvares Cabral, captain of a fleet of Portuguese caravels, *en route* to the East Indies by way of the Cape of Good Hope, used the easterly trade winds to tack across to South America before taking the westerlies to continue his voyage. The fleet anchored at Porto Seguro, on the 22 of April 1500, now one of the most desirable pieces of real estate on the Bahian coastline. The interchange with the occupants of the land was by all accounts cordial, even friendly. This was the first step of a small European nation claiming a foothold on a continent which had been divided between Spain and Portugal some six years before by the Treaty of Tordesilhas.

During the following 250 years of colonization, the virgin forests of the Organ Mountain Range suffered but peripheral damage. The sugar cane plantations on the coastal plain in the North East were sufficiently extensive, starting a dent into the mountain forest, supplying wood to fuel the sugar refining process. The need for firewood in the slowly growing city of Rio de Janeiro began to make small inroads, as did the need to clear land for crop planting and pasture.

The gold and diamond discoveries in Minas Gerais set off a rush of immigrants, colonists and adventurers never before seen in South America and lasting for most of the 18th century. To support the increase in population brought about by these discoveries of alluvial gold required land for pasture and for planting food crops which, together with the physical destruction of forest to get at the gold, obliterated much of the rainforest in South East Minas Gerais. The freighting of supplies in and gold out required the use of an enormous number of mules and significant pasture lands and maize fields to sustain them. There were two or three official routes to carry these supplies and gold to and from Rio de Janeiro: through a road from Rio, via Tinguá around what is now Petrópolis, through Paraíba do Sul, to Ouro Preto in Minas Gerais, and from Rio via what are now Nova Friburgo and Cantagalo, to the Serra do Espinhaço in Minas Gerais. But there were other trails such as one through what is now Teresópolis and it is estimated that through these, half or more of the gold and diamonds were smuggled and did not appear in official records. During this period the Organ Mountain Range forests were breached: penetrated but still not seriously affected by all the activity.

In 1720 or so a Brazilian army officer, part of a commission to define the boundary between French Guiana and Brazil, took a pocket full of seeds from a coffee bush he found in the Cayenne Botanic Gardens. The French, like the English and the Portuguese, were avid collectors of useful plants, transferring them from one tropical continent to another, particularly in their own possessions. The officer brought them back to Belém in Pará where they were planted. The plants in Belém, apparently, prospered only modestly and not surprisingly since the plant's natural habitat is as a forest under-storey bush from the high Ethiopian or Yemen rain or cloud forests. In 1762, plants, or seeds were brought from Belém to Rio de Janeiro. Captain James Cook on his round-the-world trip in 1768, commented that coffee in Rio de Janeiro mostly came from Portugal. In 1772 a Dutchman who looked after the Academia Fluminense gardens in Rio, distributed coffee seeds and advice on their planting and care, and yet in 1779 it was still a relatively unimportant crop, cultivated in the coastal lowlands and on the Tijuca massif.

The essential differences between the expansion of territory by Stone Age tribes of semi-nomadic hunter-gatherers at one extreme, and an eighteenth century mercantile society at the other, are very distinct. The first exploits what it can to sustain its members in a given area, and then moves on. Generally speaking, wild animals and fish are the main prey, incidentally adding wild fruits, roots and seeds as dietary supplements. These semi-nomads tend not to destroy the environment because they have neither the need, nor in fact, the wherewithal, to do so. To say that they live, by choice, in harmony with their natural surroundings is obviously untrue. That they are limited by their ability to capture and gather food in their natural surroundings is much more to the point.

Washing for Gold, near to Itacolomi Mountain.
30,5 x 26,2 cm
Engravers: A. Joly e É. Wattier (figures)
Casa litográfica: Thierry Frères, succ.rs de Engelmann et C.ie, Paris

At the other extreme is the Eurasian mercantile society, with thousands of years of interaction along an east-west axis stretching from China through southern and northern Europe, where written languages were evolved and disseminated over five millennia and where every invention including intensive agriculture and animal husbandry evolved; where trade in produce and ideas between peoples became commonplace, each endeavouring to improve their respective life styles or to dominate their neighbours and so accumulate wealth.

During most of the 18th century, Portugal had become heavily dependent on the gold and diamond revenues from the interior of Minas Gerais and São Paulo, but by the last quarter of the century this source of income was slowing down. Miners from central east Minas Gerais moved southwards crossing the Paraíba do Sul river to the very base of the Organ Mountain Range anticline, where they settled on pasture lands which had previously been cleared for cattle and sugar. At the time of their arrival it was becoming clear that these lands and the climate were ideal for the growing of coffee. Further, the infrastructure for freighting this crop to Rio and receiving supplies in return, already existed; mule trains and passable trails, albeit grossly inefficient, had been moving goods through the route Paraíba do Sul - Rio de Janeiro for over half a century.

Incipient coffee plantations were established in the 1780's and by 1790 had consolidated into real money-spinners, not only for the ex-miners from Minas Gerais but also for ambitious adventurers from Rio de Janeiro.

Here then is the contrast between the subsistence-living semi-nomads of the 15th century and the traders, the exploiters and wealth accumulators and conquerors from 18th century Europe. They had iron tools, firearms, horses and mules for transport, a capacity to organize and were enslavers to an astonishing degree. They were also virtually unbound by law save that of the gun and having to give only token allegiance to a weak central authority.

The expansion of coffee culture continued apace in the 1790's up to the first decade of the 19th century. There were extraordinary happenings in Europe, a permanently squabbling, war-like and war-torn Europe, with a half dozen established colonial and potentially colonialist countries warring against each other and arranging and breaking alliances. Napoleon and his French armies were dominating the continent and the French and the British were at serious loggerheads after Napoleon ordered all countries to cease trading with Britain. For a long period Britain had more or less controlled the Portuguese trade, particularly in fortified wines, principally Port and Madeira amongst other items and of course, selling their products from the accelerating Industrial Revolution. The British fleet was *inter alia* protecting this trade, when an exasperated Napoleon decided to invade Portugal with his great army. He did, but before he got to Lisbon, Dom João (at this stage prince regent of Portugal) with a retinue of at least 15,000, boarded both their own ships and ships of the British fleet and sailed off to Rio de Janeiro. It was, of course, much more complex than this but in the final analysis our interest is the Organ Mountain Range forests and not in European "real politik".

The fleet reached Rio de Janeiro in 1808 and Rio became the seat of the Portuguese government; one of Dom João´s first acts was to open Brazilian ports to foreign trade. No doubt this was partly a *quid pro quo* to the British for escorting the Portuguese court over to Brazil. This new era was not just for trade, but also an invitation for European scientists to visit these grandiose, complex and intrinsically beautiful virgin forests.

4. The Metamorphosis of the Organ Mountain Range

4. THE METAMORPHOSIS OF THE ORGAN MOUNTAIN RANGE

THE COFFEE INVASION

The under-storey bush from Ethiopia or Yemen, is a member of the Rubiaceae, a large family of around 500 genera and over 7000 species, mainly concentrated in tropical and sub-tropical parts of the world. In the tropics and sub-

Café.
Jean Baptiste Debret
Water color on paper. 24 x 18,9 cm.
Castro Maya Museum/IBPC, Rio de Janeiro, Brasil

or shrubs while in the temperate herbaceous. Coffee is this family's However, Quinine is extracted Rubiaceae species, and until the the only alleviator of malarial fever. herbaceous plants are used as tropics, most members are trees regions its representatives are most important product. from the bark of another latter half of the 20th century was Many of the family's bushes and ornamentals in gardens.

The coffee bush, in nature, like has a lateral root system, endures end of the Asian monsoon, and a spring, and so was brilliantly suited the Organ Mountain Range. The experience that on the scarp slopes Mountain Range, the summer winter period far too humid to dry sun. Consequently even today we most mountain rain forest plants, a heavy rainy summer at the tail long dry autumn, winter and to the climate of the anticline of coffee growers discovered by and the summits of the Organ rainfall was too high and the the coffee beans efficiently in the can find significant tracts of disturbed but original forest on the steeper eastern-facing scarp. The rainfall in section VI of the Organ Mountain Range averages 2.6 metres annually (see our 8 year rainfall readings - and our readings are taken from the second tier of mountains, page 27). The scant readings taken by friends on the first tier indicate an annual rainfall of over three metres, mainly between late September and early April. Rainfall on other ocean-facing sections of the Organ Mountain Range is surely similar. When one crosses the rough dividing line between scarp and anticline, into the odd-numbered sections, the annual rainfall drops dramatically and even more so when one moves westward towards the Paraíba do Sul river. In this instance the interior anticline rainfall in all sections is below 1.5 metres annually, with a shorter wet period and consequently a longer dry autumn, winter and spring. One can only speculate that current annual rainfall reflects a forested past. It is certain that the pattern now is the same but the recycled transpiration from the intact forest would certainly have given greater humidity, less evaporation loss, heavier local rains and certainly a far lower average annual temperature within.

The valley of the river Paraíba do Sul was the focal point for the subsequent enormous coffee plantations. Although the Afro-Asian coffee bush was a sub-storey plant, the Brazilians ignored this facet of the plant's origins.

The topographic configuration of the anticline is a serried but haphazard rank of mounds or mountains declining in altitude as one moves westward. Northern slopes obviously receive more sunlight than those facing south, and so it was on these that the techniques for massive deforestation were first applied. These involved half cutting all the trees from the valley base upwards until the rounded peak was reached, work not given to slaves but to a group of skilled woodsmen, as it was extremely dangerous. At the climax, the topmost trees were simultaneously felled, creating a domino effect on those trees below, resulting in the whole mountainside forest crashing down in one fell swoop. If this succeeded, and it sometimes

An old coffee farm house in the municipality of Trajano de Moraes, RJ.

did, the foliage was left to dry out and, at the height of the dry season, fires were started at the valley base. Hopefully only the trunks remained after this ordeal by fire, since to burn the humus below would have been disastrous. In the 1850's and 1860's the smoke from these fires was so thick and intense that it created a grey haze as far away as Rio de Janeiro - just like present day pyrotechnics achieve in Mato Grosso and the Amazon region, burning both forest, scrub and pasture. As an aside we can imagine the plant collectors passing through these felled forests before the fire, taking their pick of exotic orchids and other plant species and shipping them back to Europe, to be accused, much later, of burning forests down to exclude their competitors from such bounty!

In principle, all tree trunks remaining faced downwards which we suppose, dictated that the coffee bushes were planted vertically in lines, between the trunks. This had a certain logic for to rotate the trunks horizontally and plant on the contours would have been extremely dangerous and very labour intensive. As a side benefit, vertical planting made the supervision of slaves much easier, but it also meant soil erosion of disastrous proportions. Contour planting was a technique already well known in Europe and for thousands of years well known in Asia, but these plantation owners were not farmers and only interested in fast fortunes. When one plantation became unviable, uneconomic, they simply moved on to the next tract of virgin forest to repeat the same process over again. We have seen the same technique used in original forest in the municipality of Silva Jardim by subsistence farmers not 30 years ago.

Coincident with the spread of the coffee plantations on the anticline of the Organ Mountain Range, was the need by Dom João VI, now king of the Portuguese dominions, to keep his 15,000 imported government officials and their hangers-on happy, not to mention a somewhat rebellious

Photo: Archives, The Ecological Reserve of Guapiaçu (REGUA)

populace. Gold and diamond revenues were down to a trickle, so coffee revenues became the obvious substitute. From 1500 to 1808 the Portuguese had kept foreign scientists out of their Brazilian territories. The gold and diamond revenues gave them the wherewithal to enforce this ban during the turbulent 18th century when the European enlightenment was in full swing, fearing the dissemination of dangerous liberal ideas and attitudes in such an unstable environment. This policy was clearly

Ex-slaves.
Cachoeiras de Macacu, RJ, 1927.

shown in the 1790's, when Humbolt and Bonpland, two of the century's greatest scientists were barred entry to Amazonian Brazil from Venezuela via the Rio Negro. To a large extent this prohibition also applied to foreign traders who were kept out of the ports by high taxes, the hostility of the old Portuguese colonial regime and the rules dictated by Lisbon.

There were no universities or even printing presses in Brazil, at that time. The intelligent young and rich had to go to Portugal to obtain a higher education. To a great degree this explains, or reflects, the lack of interest in the country's natural resources, other than gold and precious stones. It was only in 1772 that the University of Coimbra (Portugal) introduced natural sciences into its curriculum. In summary, Portugal, and consequently, Brazil, were at the heel end of the European enlightenment, and any Brazilian initiatives were stifled by Portugal's mercantile and extractive attitudes towards its colony. Ignorance then, contributed significantly to the 19th century destruction of the Organ Mountain Range forests.

Photo: Archives, REGUA.

Cutting of Sugar Cane. Cachoeiras de Macacu, RJ, 1927.

After 1808 both traders and scientists poured into Brazil. There were over 60 English trading houses in Rio de Janeiro alone by 1825, and the English traders sold everything that the industrial revolution could produce, cheaply and in quantity, to a completely supply starved population: cloth, glass, iron and steel products of all sorts, plumbing, water closets, baths, whisky, porcelain, butter, an endless list of articles which mostly came through the port of Liverpool and contributed strongly to that city's prosperity and created great fortunes there amongst its merchants. The trade was pretty much one-sided. In 1825, goods valued at 6.4 million pounds were exported to Brazil (at today's value; around US$ 400 million) which in three years, by 1828, had risen to 28 million pounds (today this would be around US$ 1.25 billion). Brazil's exports of coffee, hides and tallow were worth only 10 million pounds (today US$ 600 million). So Brazil entered into the foreign debt spiral which even to this day is still significantly subsidised by exploiting non-renewable natural resources. During the 16th, 17th and 18th centuries, this included the enslavement of the first inhabitants, and in the 18th and 19th, the unfortunate slaves from Africa.

Coffee bearers on their way to the city.
Jean Baptiste Debret. Rio de Janeiro, 1826.
Watercolor on paper. 15,9 x 22 cm.
Castro Maya Museum/IBPC, Rio de Janeiro, Brasil.

By 1820 probably 15% of the Organ Mountain Range anticline had been deforested to plant coffee bushes. The system had not changed from that described earlier and was mostly concentrated in the upper Paraíba do Sul river valley. In 1815 Dom João declared Brazil a kingdom co-equal with Portugal. In 1820 he returned to Portugal and his eldest son, Pedro I declared Brazil independent, with particular support from the coffee barons, and in no time at all, he proclaimed himself Emperor.

1820 saw one of Dom João's tentative efforts to populate the Organ Mountain Range with solid yeomen peasants from a depressed Europe. This resulted in the influx of over 2000 Swiss-French who were to colonize an area in what is now Nova Friburgo. The concept was great. However the land allocated to the majority of the colonists was

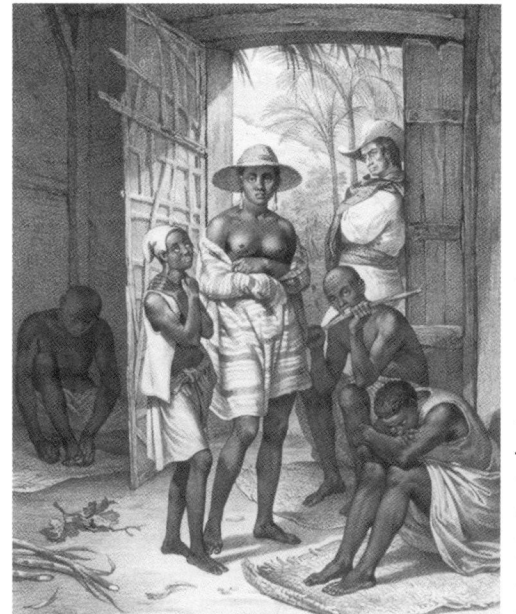

Fresh Slaves
29,4 x 24,5 cm.
Engraver: N. Maurin
Casa Litográfica: Engelmann

useless mountainsides and unsuited to the most basic of subsistence farming as it still is. A similar sized group of Germans were imported to settle around Petrópolis during the same period, in a similar type of environment with understandably similar results. These land grants illustrate the continuing operation of unscrupulous and corrupt land agents who make a living out of misrepresenting the properties they negotiate to gullible clients, in this case Central Government. Naturally a significant number of the more moneyed and industrious immigrant families moved away and populated the adjacent more promising areas - to plant coffee and sugar cane. They moved to Cantagalo, Bom Jardim, Macuco, Trajano de Moraes, Cordeiro, São Sebastião do Alto, northeast from Nova Friburgo and set about to deforest, with a will, the North Fluminense. The ex-miners from Minas Gerais were doing their bit by entering into what are now the municipalities of Itaperuna and Muriaé to join up with the ex-Friburguense much later. These areas were slower in developing and although the north western area of Rio de Janeiro State has nothing to do with the Organ Mountain Range, it did export more coffee between 1870-1880 than any other area in the State.

By 1860 it has been estimated that nearly 60% of the State of Rio de Janeiro was under coffee. This is possibly an exaggeration as most coffee was grown on the anticline of the Organ Mountain Range and this takes up only 20% of the state's land mass. In addition there still were significant tracts of original forest therein. Nevertheless, the Organ Mountain forests had been severely, if not mortally wounded.

The coffee bushes were planted in vertical lines that can still be seen as furrows and erosion channels on steep pasture configurations anywhere on the Organ Mountain Range anticline. The plants matured in four or five years. The humus top soils, built up over millennia under the forest, were very fertile and made even more so by the nutrients released by the burning of the forest cover. In theory the plants could produce fruit economically for up to 30 years, but the unselected seeds produced poor plants, and the dramatic erosion caused by the heavy tropical summer rains on the steep hillsides, cut the 30 year period down to 20 years or less, when the owners moved on to new virgin forest. Production would have been around a kilo of ripe beans per bush, per year, so a plantation of 100,000 bushes would have given a tidy annual profit. Current coffee plantations in the Organ Mountain Range anticline, and there are some extant around Bom Jardim and Cordeiro, are contour-planted with selected seeds and, using fertilizer, show a yield of over 3 kilos a bush and the plants last much longer than 20 years.

Moving to fresh virgin forests was simple. Land grants were asked for and almost always given either through bribery or pure influence with colonial or post-colonial governments. Boundaries were never defined or if occasionally they were, it was the gun and not the law that prevailed in a boundary dispute, because of a central government that had very little control over local oligarchies (a situation which is more or less the norm at present in the Amazon territories).

A good question would be why such environmental destruction was allowed to take place. We think that the answer is fairly clear:

➢ the planters in general had no agricultural background and understood neither topsoil nor subsoil dynamics;

➢ nor had they any, or only minimal education;

➢ the central government was self-serving and basically formed by the very oligarchies which it should have been controlling;

➢ the local authorities were thin on the ground and owed their allegiance to local landowners;

A sad end to the Coffee Cycle. Visconde de Imbé, RJ.

➢ the central authority's law enforcers were even thinner on the ground, corrupt (often to save their own lives) and consequently ineffective;

➢ planters were "miners" of the topsoil not "farmers". The concept of *"live as if you will die tomorrow, farm as if you will live forever"* was quite absent from their lexicon;

➢ the inheritance laws did not help; large monocultural units became uneconomic when divided up amongst a large number of heirs;

➢ and finally the central government required revenues at any cost to keep itself in business, and the exploitation of natural resources whether renewable or not was then the only way open to them.

Gustavo Schuch, Baron of Capanema was a lone voice at this time warning of the dangers to and possible destruction of the Atlantic forests. But in many, if not most ways, the destroyers were little different then from the other European exploiters of the Old and New Worlds. First of all let us remember

Erosion caused by the Coffee Cycle. Municipality of Carmo, RJ.

that every square metre of exploitable Europe has been exploited many times to the detriment of the environment and to the flora and fauna found therein. The English invaded North America and immigrants created dust bowls by ploughing up the central prairies, eliminated the millions of buffalo, passenger pigeon, and the other human occupants, the Indians, from these regions. Currently the Americans and the Canadians are eliminating the forests for pulp wood to make newspapers, magazines, houses and books. They kid themselves and others that they are replanting forests to maintain the continuum. A professor from Iowa State University told us, only the other day, that Iowa now has about 2% of its original coverage and she suspects that since the onset of the European settlers that the overall humus level in that state has dropped significantly since intensive agriculture

A modern coffee plantation, respecting the contours. Municipality of Bom Jardim, RJ.

started. Iowa does not have Brazilian torrential rains on mountain territory. It is essentially flat or rolling so erosion would be a minor hazard. The Americans had no problem in attempting to defoliate large tracts of the Vietnam tropical rain forests to expose the Vietcong. China has no problem in helping to turn Mongolia into a dust bowl in order to provide yak products for its ever increasing population and has already destroyed much of its own natural environment. The Australians, having made great inroads in eliminating the original occupants of their captured territories had no problem whatever in turning the "pleasant island" Nauru, in the Pacific into a wasteland to extract valuable phosphates. We quote *The Economist*, December 22, 2001 *"greed, phosphate and gross incompetence in a tropical setting. The history of Nauru is stranger than fiction"*. We have almost eliminated the great whales, we are within a spit of eliminating many fish stocks, herring, cod and God knows what else. Pretty soon we will be trawling the krill in the Antarctic to feed pigs and incidentally eliminate the whole eco-chain. No doubt at this stage Norway, Japan and probably Spain will try to uphold their inalienable rights to continue the depredation. In other words the coffee barons were doing what comes naturally, no different to anybody else.

The tragic problem is that boreal or sub-boreal forests are relatively simple and can be reproduced with ease. They contain comparatively few plant and insect species, whose common characteristics are an ability to survive harsh post glacial conditions. The Neotropical forests of the Organ Mountain Range, however, are incredibly rich in interdependent plant, insect, animal and fungal species, and it is virtually impossible to reproduce the original forest once the base has been devastated.

THE SCIENTISTS, COLLECTORS AND TRADERS

Source: http://bioweb.cs.earlham.edu

Beagle

The scientists that visited Brazil after 1808 were all European. Their almost common denominator was that the majority set off on their expeditions from Rio de Janeiro through the Organ Mountain Range into the interior. The first real Brazilian effort was in fact an initiative undertaken by Archduchess Leopoldina of Austria, married to Pedro, the son of king João, who included amongst her followers a number of outstanding Austrian and German scientists. This group included Johann Pohl, Johann Von Spix and above all Karl von Martius. These three travelled throughout Brazil collecting and naming plants, animals, reptiles et al. which were then sent to many European academic institutions. Von Martius (1794-1868), changed the face of Brazilian botany with his Flora Brasiliensis. Publication was completed 66 years after the start, 38 years after his death and included 648 species of orchids described by Cogniaux and with illustrations copied from Barbosa Rodrigues.

The descriptions of their journeys and those of August Saint Hilaire, a Frenchman and Heinrich Schott, a German, who collected mainly around the state of Rio de Janeiro, attracted a great deal of attention not only in European scientific circles but also from horticulturists, avidly collecting and introducing new exotic plants into gardens and glasshouses, especially in England and Belgium.

August S. Hilaire

Source: http://www.jbrj.gov.br/index.html

Barbosa Rodrigues

Source: Museu Paulista, SP.

Empress Leopoldina with her children
A. Failutti
Oil on canvas.

Source: http://en.wikipedia.org/wiki/

Charles Darwin

Source: http://www.botanik.biologie.uni-muenchen.de/botsamml/

Martius

Such is the discipline of scientists in general and botanists in particular that they concentrate their attentions on the specific and largely ignore the general context. This is understandable. If one has been in the forests all day collecting plants and then spent the evening drying and pressing them in primitive conditions, probably with poor food, and sleeping under a makeshift shelter, one tends not to pay too much attention to describing eloquently, the grandiosity of the forest around. And this was the case of all but a few of these scientists. One of the few was Charles Darwin. In his book *The Voyage of the Beagle* he describes his 1832 journey from Botafogo, Rio de Janeiro to Macaé on the state's north eastern coastal plain in detail. However, he did not visit the Organ Mountain Range but he did give us some clues to the remaining coastal plain forests and the coffee plantations therein.

"A few miles North of Rio de Janeiro we entered a forest which in the grandeur of all its parts could not be exceeded", while two days riding on, *"we again entered the forest - the trees were very lofty and remarkable compared with those of Europe from the whiteness of their trunks...wonderful and beautiful flowering parasites invariably struck me as the most novel object in these grand scenes"*.

He also described within a large farm near to Macaé probably near Glicerio *"A large pile of coffee beans drying in the sun"*, and he states that *"in this part of the country the chief produce is coffee"* and further that *"the cultivated ground was surrounded on every side by a wall of dark green luxuriant forest."*

In other words, in 1832, there were large tracts of virgin forest on the north east Rio de Janeiro plain, and coffee was an important cash crop for this area.

A young farmer, the grandson of a farmer who had cultivated a huge tract of the scarp foothills of the Organ Mountain Range in section IV early in the 20th century, told us that the cycle of planting in that region was, deforest, burn, plant coffee vertically, followed by manioc and finally bananas. After the banana plantations gave a low yield, the land was left as pasture or the forest was allowed to regenerate. He believes that was the pattern of exploitation up to 600 M.asl in many areas of the scarp foot hills. Coffee planting came to an end there with the Great Depression, overproduction and consequent glut in the 1930's.

The third group that had involvement in the Organ Mountain Range were the plant collectors, contracted mainly by English and European orchid nurseries. These were generally young adventurers, probably with a smattering of basic botanical knowledge and a willingness to travel to remote areas. Very little is known of these young men. Some say that they cut down forests to get at the orchids they were hunting. They did not need to in the Organ Mountain Range. The coffee barons were doing it for them. All they would have to do was follow the expanding coffee plantations and without effort or cost remove the plants before the felled forest was torched.

As far as we can see from our research into the remnant Organ Mountain anticline forests, few if any of the most desirable orchid species to horticulture, are or were found there in such concentrations as to justify the cutting down of a significant area of forest. The Cattleyas mainly lived in the swampy marshlands of the coastal plain and these could be plucked off gnarled trees without even climbing them. The Oncidiums do not form large colonies and thus the cost: benefit ratio rules out the cutting down of large trees to obtain a few plants. *Laelia crispa* does form large groups in the first bifurcation of large trees but this plant, due to its short flowering, was never popular. *Sophronitis* species are found in large numbers on montane forest scrub trees over 1300 M.asl, mostly at eye level while *Brassavola* species can most easily be collected on accessible rock faces. *Maxillaria* and

A gigantic tree in the brazilian tropical forest, 1830.
Rugendas

Oil on canvas. 46,5 x 38,5 cm.
State Palaces and Gardens, Potsdam-Sanssouci - Alemanha

Bifrenaria species are found at any level in original forest while *Miltonia* species do in fact occupy difficult niches in original forest. Sending small boys up trees would obviate the necessity of cutting them down and be much cheaper. *Cyrtopodium* and *Zygopetalum* species are terrestrial on high mountain fields. In summary, forests were not cut down on the Organ Mountain Range to obtain exotic epiphytes.

There was one who did visit the Organ Mountain Range many times and who also wrote a book in which he made this very point. The book was *Travels in the Interior of Brazil*, which Gilberto Freyre, perhaps Brazil's most important socio-historian, calls *"one of the most interesting of all books written in the English language (about Brazil) as it deals not only with plants and animals but also Brazilian customs"*. This man was George Gardner, a young Scots surgeon turned botanist, who ranks as one of the most competent of all the scientist-collectors to visit the Organ Mountain Range and who also formed strong links with the English trading fraternity and local Brazilian oligarchies.

As an example, inadvertently George Gardner shows this in his first letter to his sponsors, via Sir J. Hooker of Kew Gardens in 1836.

J. Hooker

"Near the summit of the Pedra Bonita (Tijuca Massif), *there is a small Fazenda, or farm, the proprietor of which was then clearing away the forest which covers it, converting the larger trees into charcoal. From the massive trunks of some of them which had just been felled we obtained some very pretty Orchidaceous plants..."*

Some of the traders were also collectors of exotic plants, notably the Harrison brothers from Liverpool, and many of these traders' descendants subsequently became distinguished Brazilian families whose surnames are still found today amongst the population. One of the most successful, a Mr. March, obtained a grant of many square leagues of land on which is now the municipality of Teresópolis. He could well be called the father of this city as he opened its first hotel there and received the elite fleeing from the summer heat and fever epidemics in Rio de Janeiro; amongst his guests were George Gardner and Harrison Co. employees who accompanied him on collecting trips. March was so rich that only the British consul's wife, Lady Amherst's establishment in Rio de Janeiro was more magnificent. Lady Amherst later had her name joined to that of Earl Grey of Groby, a munificent patron of horticulture, by the great John Lindley of Kew Gardens, who named a striking orchid from the Organ Mountain Range, *Grobya amherstiae*. Gardner was indignant at the price Mr. March charged at his establishment for a monthly stay, - over US$ 300 at present values. But the rich get rich by making money and not by giving it away!

George Gardner had qualified as a surgeon physician at Glasgow University, in Scotland. His father had been gardener to several of the most notable of Scottish exotic plant collectors and he spent most of his limited spare time with his father in these gardens or with W. J. Hooker, then curator of the Glasgow Botanic Gardens. Hooker, later a director of Kew Gardens, helped the boy, who clearly showed a bent for botany; a short time later, after Gardner had produced a definitive thesis on British mosses, Hooker arranged a collecting trip to Brazil for him. His was a short life; he died of a stroke at 39 years old in Ceylon. Even reading his book *Travels in the Interior of Brazil*, one can not quite get to grips with the man. He was clearly charming with an enormous capacity to integrate into any level of society, a workaholic, but a complete enigma, revealing nothing one could identify with in a warm way; sometimes sardonic, describing brilliantly and intimately the interior of Brazil but giving away nothing of his own.

He was 23 years old when he and Sir W. J. Hooker persuaded 24 backers to finance a plant collecting trip to Brazil. The principal sponsor was the Duke of Bedford and the backers agreed to buy the dried and live specimens collected. Gardner arrived in Rio de Janeiro in 1836 and immersed himself in the Rio de Janeiro rain forest, exploring the Tijuca massif around which Rio de Janeiro has now spread. His letters to Sir W. J. Hooker, which form the first chapter of his book, are so crammed with detail, description and eloquence as to make one want to follow his collecting trips around these mountains.

Having learnt Portuguese and got comfortable with the amazing flora in the remaining virgin forest on the Tijuca massif he felt himself sufficiently prepared to confront the Organ Mountains.

The Organ Mountain Range is the spine of Rio de Janeiro State (see map on page 23). It's central position is in the Atlantic coastal chain, itself nearly 3,000 km long and stretching from Rio Grande do Sul in the South almost to Rio Grande do Norte in the North. It is probable that it suffered no glaciation during the past three million years, but passed through enormously long cold and dry periods, interspersed with shorter (but still long) warm and wet periods. Evolution and speciation was slowed down but never halted and consequently the current situation of incredible richness, diversification, interaction and inter-dependence of flora, fauna, insects and fungal life evolved. It is a brilliantly strong and inter-active ecosystem, but one which depends on the sum of its parts. Take one of its major supports away and the whole structure collapses. Such was the case with the coffee cycle which took away the principal support - the forest - and left behind it a world of "ants and

The Organ Mountains
Engravers: J. M. Rugendas e A. Joly. 22,6 x 29,9 cm. Casa Litográfica: Engelmann.

A view of Guanabara Bay taken from Russel Beach. Rio de Janeiro, 1850.
C. J. Martin.
Oil on canvas. 68 x 105 cm. From the Sérgio Fadel Collection, Rio de Janeiro, Brasil.

grass" (to paraphrase the author of an article in *The New Yorker* Magazine, which depicted the world after a nuclear war) and this is how much of the mountainous Organ Range anticline appears today.

George Gardner was very fortunate. He arrived at the beginning of the end of the forests and as such he was able to observe the forest and its parts which had been opened up, as it were, like a corpse on the pathologist's slab, showing its vital organs, quite clearly defined, while the green mould of the coffee plantations was spreading throughout the body.

"...we started together, on the 24th of Dec., along with two or three English merchants, who were going up to spend the Christmas holidays with their families. It was mid-day before we could leave the city, and, under the influence of a strong sea-breeze, we (crossed Guanabara Bay) *reached Piedade, the landing-place* (some 30 km away at the head of the bay).... *At Piedade, mules from Mr. March's Fazenda were waiting for us, and our luggage, and, after a short stay for the arrangement of the latter, we began the land part of our journey...."* From Magé to Frechal, the place *where we slept for the night, the distance is about 14 miles. The road still continued flat, but wound round many low hills, the sides of which are covered with plantations of Mandioca. We met several troops of mules coming down from the interior, loaded with produce..." "Loaded mules start daily from Rio, Piedade, and Porto d'Estrella, to make journeys into the interior of from five hundred to two thousand miles and upwards. They seldom travel above twelve or sixteen miles a day, and the load allowed to each varies from six to eight arrobas of thirty-two pounds each....*

From thence to Mr. March's Fazenda, which stands at an elevation of upwards of 3,000 feet above the level of the sea, is twelve miles. During the whole way the road is very bad, and in many places so steep, that it is with considerable difficulty the mules make their way up it. Indeed, to one unused to travel on such paths, which have more the appearance of the bed of a mountain torrent than a road for beasts of burden, many parts of it appear impassable; but he is soon undeceived by the slow yet sure manner in which the mules pass along the worst portion of it, especially if left entirely to themselves. The whole length of the road is through one dense forest, the magnificence of which cannot be imagined by those who have never seen it, or penetrated into its recesses. Those remnants of the virgin forest which still stand in the vicinity of the capital, although they appear grand to the eye of a newly-arrived European, become insignificant when compared with the mass of giant vegetation which clothes the sides of the Organ Mountains....Many of the trees are of immense size, and have their trunks and branches covered with myriads of those plants which are usually called parasites, but are not so in reality, consisting of Orchideae, Bromeliaceae, Ferns, Peperomiae &c....*Many of the trees have their trunks encircled with twiners, the stems of which are often thicker than those they surround. This is particularly the case with a kind of wild fig, called by the Brazilians, Cipo Matadór."* (whose seeds germinate on the upper branches of its host tree, sending out roots that eventually embrace the tree, grow thicker and finally strangle its host and substitutes it in the forest canopy with its gross roots, planting it solid in the humus around.)

And what more did our 24 year-old surgeon, physician, pathologist and naturalist see on the 20 km up to Mr. March's property, nearly half the size of the present municipality of Teresópolis? He saw immense Cedros (*Cedrela odorata*), Jequitibá rosa (*Cariniana legalis* Mart.), the largest tree in the Organ Mountain Range, up to 50 metres tall and with a trunk so huge that 15 men with arms outstretched could not circle the bole, and Lauraceae like Canela tapinoã (*Mezilaurus crassiramea*) so useful for ship-building that it was named Canela naval which the Portuguese reserved for their own fleets. Other Lauraceae such as Canela preta (*Nectandra megapotamica*), the huge Canela murici *Vochysia* sp. with its canopy garden of bright yellow flowers, not to forget the thick and tall bamboos, the cecropias and the purple flowered Melastomataceae which enter rapidly to patch up wounds in the forests.

The party arrived on Christmas day at Mr. March's headquarters. The property was 64 square leagues in extent; the major part covered in virgin forest and included the spectacular, church-organ-like peaks.

The entrance to the present Organ Mountains State park trail, leading to Petrópolis, some 50 km to the west, surely was the region that George Gardner described:

"My first journey of any length into the virgin forest here, was made in company with M. Lomonosof, the Russian minister at the court of Brazil, and Mr. Heath. M. Lomonosof was desirous to witness a tapir hunt, that animal being very common on this range.... For the first mile and a half we had a tolerable path, leading through a forest of fine trees, with very little underwood except young palms, hundreds of which were cut down by the blacks who were clearing the way for us. In going up the valley we crossed and recrossed a small river, called the Imbuhy, several times, on the

banks of which I added largely to my botanical collections.....Having accomplished this, we came upon the old track of a tapir. It was about two feet broad, well beaten, and had foot marks of the animal on it... While Mr. Heath was endeavouring to get the dogs upon a recent track, I occupied myself in collecting a number of curious plants which grew on a sloping bank by the side of the stream. It was now nearly four o'clock in the afternoon, and the rain was beginning to fall heavily; we therefore sought for a place where we might encamp for the night, as we were ten miles distant from the Fazenda, and M. Lomonosof was too much fatigued to be able to return. The place we selected was under the shade of some large trees, near which grew abundance of the small cabbage palm (Euterpe edulis, Mart.), the terminal bud of which is so much made use of as a vegetable by the Brazilians. A hut was soon erected, and thickly thatched over with leaves of this palm. At first we were dreadfully annoyed by mosquitoes and a little sand-fly, but the kindling of a large fire in front of our hut soon dispersed them. Palm leaves were spread upon the floor for our bed, and we had a small log of wood for a pillow. It rained heavily all night, but we did not suffer from it. We got up next morning by break of day, and prepared to return home, as it still continued to rain....After a slight breakfast we commenced our journey homewards; but before getting out of the forest, M. Lomonosof, little accustomed to a hunter's life, became so exhausted from fatigue, that it was with difficulty he reached the place where horses had been ordered to be sent to await our return."

If you can imagine a plant collecting trip lasting for several months under such conditions you will not be surprised that few coherent, descriptive and readable travelogues were prepared by the doughty scientists and collectors that swarmed throughout Brazil during the 19th century.

A scene in the brazilian tropical forest, 1830. Rugendas
Oil on canvas. 62,5 x 49,5 cm. State Palaces and Gardens, Potsdam-Sanssouci - Alemanha

Poor M. Lomonosof, who was probably fat, over 50 and sedentary, came under the 24 years old Gardner's somewhat amused disdain.... We had a similar experience in the same mountains in the same month 165 years later. The conditions were different; four of us trying to find a rare variety of *Sophronitis coccinea* 1000 metres above and 10 km. away from our base, in searing, windless heat accompanied by two botanists from Rio de Janeiro Botanic Garden in their 20s. We got three quarters of the way up before collapsing from fatigue and exhaustion. The two 24 year olds bounded onwards and upwards. We staggered and stumbled back to base and were sucking on cold beers with our feet in a cold stream and licking our wounds when the triumphant young botanists floated back. They had found the plant. The look of amused disdain on their faces could only have been equalled by that of George Gardner when he watched "...very unsuited to a hunter's life, M. Lomonosof staggering and stumbling towards the horses". A 40 year age gap makes much difference in these conditions, and the young are unforgiving.

Gardner lists the animals he came across during the following six months and we show these below with their probable current status.

The current status is assessed from our own observations over 30 years and from discussions with hunters throughout the Organ Mountain Range. The data obviously apply to the niches wherein the animals were or are slotted; if there are no trees you won't find monkeys, sloths or squirrels; with no water or fish you won't find otters.

George Gardner´s brief was plant collecting, live where possible, but also seeds and dried plants. However his natural curiosity knew no bounds and he

Gardner's List 1836	Name in Latin	Condition 1836	Condition 2002
Jaguar	*Panthera onça*	Rare	Extinct
Black Jaguar	*Panthera onça*	Very rare	Extinct
Puma	*Puma concolor*	Common	Rare
Howler Monkey	*Alouatta guariba*	Common	Rare
Several Monkey Species		Common	Occasional
Woolly Spider Monkey	*Brachyteles arachnoides*	Occasional	Very rare
Sloth	*Bradypus variegatus*	Common	Rare
Otter	*Lutra longicaulis*	Common	Rare
Capybara	*Hydrochaeris hydrochaeris*	Occasional	Common
Collared Peccary	*Tayassu tajacu*	Common	Rare
White Lipped Peccary	*Tayassu pecari*	Common	Rare
Deer	*Mazama americana*	Common	Rare
Armadillo	*Dasypus septemcinctus*	Common	Common
Tapir	*Tapirus terrestris*	Common	Extinct
Ant-Eater	*Tamandua tetradactyla*	Common	Extinct
Wild Dog	*Cerdocyon thous*	Common	Rare
Porcupine	*Coendou prehensilis*	Common	Common
Squirrels	*Sciurus aestuans*	Common	Common

must have eaten meat from quite a number of those animals listed, which would have been absolutely normal for the guides and muleteers who looked after the pack animals on long trips. In any event he noted that the wild pig, peccary, armadillo and tapir meat was much prized and again like his Brazilian contemporaries he gave little attention to the possibility that wild animal protein was a finite resource.

As he was a surgeon and physician he was more than welcome at the remote settlements and plantations he passed through, and while at Mr. March's establishment was called on frequently to tend the sick. The first such call came on Christmas night on the evening of his arrival. A 32 year old female African had been bitten on her hand by a poisonous snake, *Jararaca fonseca*.

"She was then moved to the Fazenda (after having been bled of one litre of blood as her temperature was very high) and had two grains of calomel administered to her, and about an hour after a large dose of castor oil. Next day a number of little vesicles made their appearance of the back of the hand and a little above the wrist, which, when opened, discharged a watery fluid. For the next two days she continued to suffer much pain, to relieve which poultices were constantly applied. More vesicles formed, and the cuticle began to peel off in the vicinity of the bite. On the morning of the 29th, that is, on the fourth day after the accident, when the poultice was removed, she complained of no pain at all in her hand, and on careful examination I found that gangrene had taken place, all below the wrist being dead. From the state of the arm, there was every appearance of the mortification extending....She was now very weak, the pulse 136, small and feeble, and she appeared to be fast sinking. Amputation being the only means that seemed to offer her a chance of recovery, I decided at once to take off the arm. As the crepitation extended to a few inches above the elbow, and the swelling itself to the shoulder, I determined to take it off as close to the latter as possible. As there was no room for the application of the tourniquet, I got Mr. Heath to apply pressure with a padded key over the artery where it passes under the clavicle, and Mr. March held the arm while I performed the operation. A good deal of blood was lost before I could secure the artery, which had to be done before the bone was sawn through. In a fortnight after, the stump had nearly healed up, and she was walking around the room. Four years afterwards I

again saw her, and her general health had not suffered in the least, but she had become extremely irritable and ill-tempered."

George Gardner was indeed a man of many parts; he had the capacity to integrate himself rapidly into the community, to learn the language, to familiarize himself immediately with the almost unknown exotic flora and fauna of a relatively closed continent and further to use his surgeon's skills for the benefit of those he met on his travels. Scotland should be proud of him, but he seems long forgotten. We could find no bust or even portrait in Edinburgh or

Jararaca fonseca.
Macaé de Cima, RJ.

Photo by: Klaus Seehawer.

London. Only a monument in the botanic gardens he managed in Sri Lanka. Perhaps the fact that his voyages in Brazil and Ceylon took up all his short working life explains why he made no inroads into the British botanical hierarchy.

By contrast, Bernardino Gomes, a Portuguese naval surgeon, who collected plants around Rio de Janeiro at the beginning of the 19th century has his statue erected in the Lisbon Botanic Garden, in a country which at that time paid little interest or attention to botany. Gardner's next trip was to be to the high mountain peaks of the Teresópolis massif.

"As the Organ Mountains rise to an elevation of about four thousand feet above Mr. March's house, I had long been desirous to spend a few days among the high peaks, for the purpose of making collections of their vegetable productions. The only botanists who had visited Mr. March's estate before me were Langsdorff, the celebrated voyager, at that time Consul-General for Russia in Brazil, Burchell, the African traveller and a German of the name of Lhotsky....None of them had botanized higher than the level of Mr. March's house, and the knowledge of this fact made me the more anxious to explore a field which promised so much novelty."

At the beginning of April he planned these ascents but heavy rains intervened and the trip was postponed until May, the beginning of the dry season. On the 6th, accompanied by four Africans, one of them, father Felipe, a sixty year old was the guide. This person, by now his friend, was the most active of his age that Gardner had ever met - maybe the young are not totally disdainful of the aged: only the fat, unfit ones. He was accustomed to the forest since childhood and was one of the best hunters in the region. They entered into the virgin forest two km. to the North of Mr. March's headquarters and then struck due west. Even though the old woodsman had cut a trail two years previously it was hard going, for the high mountain bamboos had half closed the trail over, so progress was slow.

Many beautiful begonias were seen on the trunks of Melastomataceae, Myrtaceae and Rubiaceae, bromeliads, orchids and gesneriads abounded together with *Zygocactus* and *Rhipsalis*. One low hill he found was covered with orchids.

"...but with the exception of the beautiful little Sophronites grandiflora, which was then in flower, all had been previously met with at a lower elevation...At 4

o'clock P.M., we reached a place by the side of a small stream, where I determined to remain for the night; and, while the blacks were occupied in cutting wood for a fire and in preparing dinner, I took a walk up the course of the little stream. As I estimated this spot to be at an elevation of about 4,500 feet, I naturally expected a vegetation different from that in the valley below. The first plant that attracted my attention was what I imagined to be a fine individual of Cereus truncatus (now Zygocactus truncatus), in full flower, hanging from the under side of the trunk of a large tree that was bent over the stream, but on getting possession of it, it proved to be a new, and, perhaps, a still more beautiful species. I have named it Cereus Russellianus, (now Zygocactus russellianus), in honour of His Grace the late Duke of Bedford, one of the most liberal supporters of my mission to Brazil".

The following day, Gardner and his party reached the areas of the high mountain fields. These are meadows with grasses and low scrub with no large trees, partly created by altitude and the consequent harsh and extreme climate, and partly because of the poor and thin soils. To a great extent they show a unique flora, only replicated to some degree temporarily in burnt recovering forests at lower altitudes, and in the dry scrub forests some hundreds of km to the west. It is possible that some, at least, of this flora had withdrawn during the cold dry periods of world glacial maxima to small isolated refugia which had retained moisture, to move out only during the inter-glacial warm and wet periods and spread both outwards and upwards, to areas which the rapidly developing rain forest could not penetrate. The plants are almost all slow growing, short and scrubby which makes one wonder if, as such, components of this flora could be relics of that present many thousands of years ago.

Many are succulents or semi-succulent, all require full light, abhor shade and cannot withstand fast growing competition. This is why a sharp transitional line exists between this flora and the lower slope montane forests. It would be interesting to know if any insects are also endemic to this ecosystem.

George Gardner was probably the first botanist to visit these regions, discovering and naming a number of new species, to his evident delight. One of these was *Prepusa connata* Gard., a fleshy creeper with a wine-red cup found again by Glaziou in 1869 and by Gustavo Martinelli in 1974. He also found a new scarlet *Salvia, S. benthamiana* Gard., named in honour of George Bentham a director of Kew Gardens who later took part in producing Martius's *Flora Brasiliensis*. He found many other new species in this region but was unable to visit other similar areas around 2000 M.asl as his guides refused to sleep near the peak since there was no water near at hand.

"The summit of the peak on which we now were, was quite a little flower garden; A pretty Fuchsia, in full flower, was trailing over the bare rocks; in their clefts grew a handsome Amaryllis, and on all sides numerous flowering shrubs. The coolness of the air and the stillness were quite refreshing; not a sound was to be heard; and the only animals to be seen were a few small birds, so tame that they allowed us to come quite close to them...In the six months of my stay in these mountains, (Mr. March's establishment) was always replete with visitors and it was a rare night without some diversion with one or another group... and even today I recall these few months as the happiest period of my life for besides those diversions I was occupied daily in my favourite task - that is exploring almost unvisited virgin forest."

For the next four years Gardner made many long trips through Bahia, Pernambuco, Alagoas, the valley of the San Francisco river, Piaui, North Minas, Goias and down through Diamantina to Ouro Preto, and finally, Barbacena in south east Minas Gerais on his way back to Rio de Janeiro. From there, he travelled due east and crossed the Paraíba do Sul river probably near Três Rios in 1840.

"Three days after we passed the Rio Parahyba, we arrived at a large fazenda called Padre Correia, the distance between the two places being about seven leagues; the road in many places was very bad, and the country still continued hilly and densely covered with virgin forests.

The fazenda of Padre Correa is situated in a hollow surrounded by bare hills....An extensive manufactory of horse-shoes, and such iron implements (iron horse shoes were vital for the mule trains, quite as vital as good pasture) *as are used in the country, is carried on in this place. Our next journey brought me once more in sight of the sea. The road between Padre Correa and the pass of the Serra d'Estrella which is a continuation of the Organ mountains, was then under repair; the workmen were Germans, who lived in a small village by themselves...* (Petrópolis, the second abortive attempt, after Nova Friburgo, to colonize the state of Rio with docile yet sturdy European farmer yeomen). *The country very much resembles that I have elsewhere described, between the Organ Mountains and the Swiss colony of Nova Friburgo, being very hilly and covered with magnificent virgin forests. From the top of the pass, there is a fine view of the country around Rio de Janeiro, and of the bay with its numerous verdant islands. On reaching this spot, I stood for a long time admiring the scene of my first labours in Brazil. My feelings, on looking down on the magnificent view before me, were such as would be experienced after returning to my native country, for everything brought to my recollection the remembrance of past times, and of kind friends; the Sugar Loaf, the Corcovado, the Gavea, and the Peak of Tijuca, were rearing high their cloudless summits, as if to welcome me back to a place of civilization"*.

The road, in good condition, led back to the Porto de Estrela about six km. and at its base he passed the farm where the late M. Langsdorff, ambassador to the Russias and celebrated explorer, who had given lodging to many scientists on the first stage of their expeditions into the interior, had lived…. *"the farm was now a gunpowder factory and belonged to the government"*.

He left Walker, his travelling companion of several years to look after his collections and bring them along afterwards, while he took a boat to Rio to find a house to live and work in. All we know now of Walker is a *Cattleya, C. walkeriana* named in his honour.

"As the land breeze was very faint, the boat had to be rowed nearly all the way, and on this account it was about four o'clock in the morning before we reached the city. Not wishing to disturb any of my friends at so early an hour, I remained in the boat till six o'clock, when I went to the home of Messrs. William Harrison & Co., and received from my old friends there a kind and hearty welcome back to Rio de Janeiro after an absence of more than three years…."

Poor Mr. Walker arrived with the collections two days later. Our hero had rented a house where his collections of plants could be prepared and organized. It took three months of intense work to prepare the 60,000 specimens from 3,000 species to be shipped to England, identified wherever possible, catalogued and even put in the correct packets for the individual original subscribers to his voyage. The Harrison brothers duly shipped the collections which were very well received. Amusingly, then as now, merchant exporters were up to all sorts of tricks to pay less tax. Gardner had given a value for each plant within the individual packets. The value for excise tax stamped on the outside was much lower. He was very worried that Sir W. J. Hooker, when he examined the packets at Kew gardens, would mistakenly give the lower price to the subscribers.

There are between 25-30,000 orchid species found throughout the world but only a few have become all-time favourites, grown by successive generations of fanciers. Ease of cultivation coupled with the reward of regularly produced and beautiful flowers, are traits which guarantee popularity. Three of the genera which fulfil these criteria are *Cattleya, Oncidium* and *Bifrenaria* and these are almost always to be found in even the smallest of orchid collections. Where did they come from and who introduced them into European cultivation? The answer to the first question, is South America, principally Brazil and significantly the Organ Mountain Range. The answer to the second question is more complex because many people were involved, but the name of Harrison is honoured in three species from these genera, *Cattleya harrisoniana, Oncidium harrisonianum* and *Bifrenaria harrisoniae*, all from the Organ Mountains.

We have found little information in Brazil about the Harrison Brothers outside of several references by Gilberto Freyre in his superb book *The English in Brazil*.

> "…Another fact that should be recorded is that it was an English Trader, resident in Rio de Janeiro during the first half of the 19th century - one Mr. William Harrison - who contributed to make Brazilian orchids well known in Europe. The residence of his brother Richard in Liverpool became the Mecca for English orchid enthusiasts" … "In spite of the phrase of Sir W. J. Hooker, director of the Royal Botanic Gardens at Kew *near* London, *that England was the graveyard of all tropical orchids brought thereto* (stove houses, at extreme heat, using stinking manure which the English imagined was a tropical environment) *the cultivation of orchids in that country became a splendid victory for the technique of creating artificial environments and this victory was won with the cooperation of the more or less romantic naturalists, but also with the English traders established in Brazil of a type, apparently prosaic, like Mr. William Harrison."*

We also found two references in the same book to the firm of Harrison Latham & Co. in Recife and Salvador, referring to their arrogance and that of other English traders. Probably the firm in Rio de Janeiro was a sole proprietorship owned by Richard Harrison. However our doughty researcher at the Liverpool end, Chris Dobson, uncovered a plethora of information through research in the library of the Liverpool Botanic Gardens and in discussions with descendants of Richard and William Harrison.

> *"There are two collections found in this country (England) that are pre-eminently rich in the plants of Brazil, that of Mrs. Arnold Harrison and of Richard Harrison, both of Aigburth, near Liverpool. Their connection with Rio de Janeiro and the circumstance of opportunity of collecting the vegetable treasures of that country for them, have been the means of their introducing to Britain some of the choicest productions in our stoves."*

(CURTIS BOTANICAL MAGAZINE, AUGUST, 1827)

William Harrison, (1769-1812), was a Justice of the Peace and a Manchester cotton merchant, a single man until his brother Richard died circa 1790 and left him with seven boys to look after. This responsibility prompted him to marry, and the union resulted in three more children, a daughter Elizabeth and two sons William and Henry. Henry was born in 1795, in due course was sent to Eton, a school for sons of the elite then as it still is today, and placed under the care of the famous Dr. Keate by whom he had the *"honour of been flogged"* for either poaching or breaking bounds. Apart from a little Latin, he himself said that he *"only learned to poach"* there. He was a wild one. For example, on one occasion he fought Lord Derby, the future prime minister over an insult to his father. On another, as he later pointed out an old jail to his cousin, Arnold Harrison, where he had spent an uncomfortable night awaiting sentencing by a Magistrate for some sin committed the previous day. When he was 17 years old his father died and he was placed in the offices of the family firm of Harrison Latham and Co., Brazilian merchants, in Liverpool, to learn the business. But he admitted that he would rather be catching rats beneath the street market stalls than sitting behind an office desk.

Shortly afterwards he sailed to Brazil to work in that branch of the business. However, being adventurous and really more interested in the outdoors, he spent a great deal of his time hunting, fishing and riding in the environs of Rio de Janeiro. Henry Harrison collected butterflies amongst many other diversions. His family in Liverpool kept this collection for many years. A family anecdote suggests that he could also be inconsequent and thoughtless: *"One day Henry was walking in a coffee plantation and killed a snake and turning a corner came suddenly upon a slave. As a joke he threw the snake at her and it coiled around her neck. She was so terrified that she fell down and died"*.

It was probably in the 1820's when the very successful Harrison Brothers started to export plants to England. The Harrisons were in the normal run of traders, making a substantial profit both in Liverpool and in Brazil on whatever they bought or sold. What separates them from the norm is the whole family's interest in exotic plants. In this they were very lucky to have two close relatives in Liverpool intensely interested in tropical plants. Elizabeth Eliza Harrison (1793-1834), one of William's three children, was a more than accomplished botanical artist as her paintings in oil and water colour clearly show. That she managed to draw and paint so well while not only maintaining an enormous establishment with extensive gardens and greenhouses but also bearing eight children, puts her on the early Victorian plinth for female fortitude and determination. Many of her paintings of hitherto unknown plants from Brazil were sent to the curator of the botanical gardens at Kew, London.

ELIZABETH HARRISON. 1743 1834

Source: Records of the Harrison familly unearthed, by Chris Dobson.

Up to the laying of the first Atlantic telegraph cable, ship cargo manifests were mostly sent on in fast frigates before the actual cargo. This way the owner knew beforehand the cargo he could expect. The manifest would then be compared with the manifest carried by the merchantman's captain, a copy of which was registered by H. M. Customs and Excise who duly levied the appropriate taxes. It should be remembered that in these days, before Income Tax was firmly established, these levies were a main source of revenue. All ships' manifests were filed, and apparently still are filed in date order, in the Liverpool customs house. Our doughty and competent researcher was advised not to waste his time hunting through these millions of files, for three reasons: (i) most are illegible in hand-written, faded ink, (ii) countless plants arrived rotten and would not have been registered, and (iii) most importantly, shipments of plants for influential people would have been carried in the Captain's day cabin, not in the slimy, wet cargo decks, and these would not have been registered at all. This would explain why Gardner's collections arrived in perfect condition and why there seems to be no record of the Harrison & Co. brothers' importations. It seems that the wars between the collectors of exotic plants continued on the high seas as well as on land!

"The interesting plant I had the pleasure of seeing in September 1826 in the fine collection of stove plants at the Liverpool Gardens. It was introduced from Brazil by Mrs. Harrison of Aigburth and, unable to refer it to any described genus, I am anxious to dedicate it to that lady, who has been the means of adorning our gardens with so many new plants and who has cultivated them with eminent success in her own collection."

(CURTIS BOTANICAL MAGAZINE, 1826. Referring to *Harrisonia loniceroides*)

Her brother William had engineered an effective greenhouse which was no small achievement in an age before electricity. We remember one in our place in Ireland many years ago that had a coke-fired furnace servicing hot water pipes which ran throughout the stove house, and in winter this had to be stoked day and night by the unfortunate establishment employees.

We know that a large quantity and a wide assortment of orchids were shipped by the Harrisons to Liverpool in 1826. What these orchids were remains unknown, but we do know that *Brassavola perrinii* was flowered by Richard Harrison in 1828 and that the first flowering in England of *Epidendrum oncidioides* was achieved by Richard in 1832 at the same time *Gongora quinquenervis* had its first English flowering under the care of his gardener Mr. Perrin. In recognition of his skill, a newly discovered orchid, *Laelia perrinii* was dedicated to him, as well as the aforementioned *Brassavola*. Yet another orchid in the Organ Mountains collected by William Harrison was named *Oncidium harrisonianum* by John Lindley in 1832. In 1836 he sent back to England the *Cattleya* species that also bears his name, *C. harrisoniana*. In 1826 Eliza Harrison, flowered a species of orchid belonging to an undescribed genus collected in Botafogo Bay, Rio de Janeiro by William Harrison. John Lindley, the most prolific and competent orchid taxonomist of his day, named the new genus *Sophronitis* from the Greek Sophron meaning 'modest', an unfortunately inappropriate name for this plant and the other beautiful species from this genus. These are mostly found in the Organ Mountains and have helped to make millions upon millions of US Dollars for hybridists, through introducing their brilliant scarlet colours into other plants in the *Cattleya* and *Laelia* Alliances.

Though there appears to be no direct evidence, we strongly suspect that the Liverpool Harrisons made serious profit from this trade in Brazilian plants between 1810-1840 and, by 1840, the known forests of Rio de Janeiro were running out of commercially attractive new plants. The arrival of George Gardner, with his brief to find as many new species as possible, was a godsend to their business, so they befriended him, probably facilitated his financing and as a *quid pro quo* accompanied him on a number of his plant collecting expeditions, finding new species later to collect for themselves for export.

George Gardner, despite his mere 24 years, was a canny Scot. He probably balanced carefully the plusses and minuses to come out of this friendship, and finding on the whole that the benefit was mutual, proceeded with it.

At 40 years old, in 1840 or so, Henry Harrison returned to Liverpool to marry and settle down, leaving the Rio de Janeiro business end to William. He was rumoured to have speculated heavily in some enterprise, more or less destroying the business. Again we suspect that other countries had benefited from the British initiated industrial revolution and were competing more efficiently with the traditional, and by many accounts, arrogant British traders. In addition, the exotic plant business was drying up as few new glamorous species were being discovered from the Organ Mountain Range.

The firm Harrison & Co. closed in 1865 when William Harrison declared bankruptcy.

After George Gardner had completed despatching his plants to England; *"...I resolved to make another journey to the Organ Mountains, being desirous of devoting more time to the investigation of the botany of the higher regions of that chain* (high mountain fields) *than I had been able to do during my former residence there".*

To this end he left Rio de Janeiro in March 1841 and for a month busied himself once again with expeditions from Mr. March's headquarters. *"...and having been joined by Mr. George Hockin from the house of Messrs. Harrison and Co., who had frequently accompanied me in previous excursions in the neighbourhood of Rio..."*

The weather was very changeable, and it was only on the 9th April 1841 at 8 o'clock that they set off for the high mountain fields.

> *"...taking with us three blacks, besides my own servant; my old guide, Pai Felipe, was now too infirm to undertake such a journey, but his place was filled by one of his sons. Following the path I had made four years before, we reached, about four o'clock, the highest point I had attained on my former visit, and at this place, under the ledge of a rock, we slept for the night; this being a very convenient and well sheltered spot, we decided to make it our head-quarters during the few days we remained in the mountains..."*

On the way up, besides the many species he had already found, he discovered new ones, and two of the most notable were a *Fuchsia* (*F. alpestris*. Gard.) and an extraordinary *Utricularia* (*U. nelumbifolia* Gard.) that lives *inter alia* in the rosette of a giant bromeliad, now *Alcantara imperialis*, at over 1,800 M.asl. The following morning they scaled the highest peak, which only six weeks earlier had been visited by an English gardener, a Mr. Lobb, sent by a plant nursery to collect seeds and live plants. When they arrived at this peak, they unfolded a flag to show the people at Mr. March's establishment, down below, that they had arrived safe and sound. However they found that there was an even higher peak to the east which they decided to scale on the following day. Here Gardner found a plant, *Hemitelia capesis*, first discovered on the Cape of Good Hope in Southernmost Africa (which brings us back to Gondwana and the separation of the continents) and another new genus of the 'Compostas' (Asteraceae) which he gave the name in honour of his friend J. E. Bowman of Manchester.

The following morning, on leaving their campsite, the explorers crossed a virgin forested valley in high montane forest whose floor was covered by *Alstroemeria nemorosa*, a lovely plant of the Amaryllidaceae, as common now as it was then. Curiously, it has become so popular during the last century in Europe, interbred and hybridized, that we recently received a frantic request for seeds from the original wild plants so that the breeders could introduce some cold tolerance.

The trip up to this new height was much more strenuous than that of the previous day as they had to cut a trail where tapir tracks could not help them. On passing this forest they entered into montane forest with much lower and gnarled trees; *"...where I was rather surprised to observe that while the stem and branches of almost every tree were covered with the beautiful little Sophronitis grandiflora, no other orchidaceous plant was to be seen;..."* as it is today, but however they found *Prepusa hookeriana* in quantity and the following day they returned to Mr. March's establishment.

> *"In order to gratify my desire of examining the* (anticline) *virgin forests which exist on the banks of the Rio Parahyba I determined to make a hurried visit there previous to my return to Rio de Janeiro.... On this expedition I was again accompanied by Mr. Hockin* (of Harrison and Co.) *and was glad to have so excellent a companion.*

We left the fazenda on the 24th of March, and after a journey of seven leagues, arrived at a farm called Serra do Capim....By far the greater part of the country through which we travelled was in a state of nature, being covered with virgin forests, abounding in tree-ferns and palms."

They dined well and it transpired that the manager, a kind old man had been an apothecary in Minas Gerais and that, as in many Brazilian farms, he acted as doctor to the estate's hospital. He was very grateful to be able to consult George Gardner on the majority of the patients under his care and would not allow them to leave until after lunch on the following day. When they left the farm they passed through some of the most beautiful forests they had seen in the State. After about seven leagues (42 km.), they reached a large coffee plantation called Monte Café. The plantation was in its initial stages, but even so it was expected to produce around 180,000 kilos of coffee beans that year. The region was composed of low rounded hills, some still covered by forest but the majority in coffee.

"There were about 200 slaves on the estate, 70 of whom only were employed as field-labourers, the others being occupied at various trades, such as cabinet-makers, carpenters, masons, blacksmiths &c. A few days previous to our arrival, about twenty young negro boys, recently imported, were brought up from Rio; they appeared to be between ten and fifteen years of age, and none could yet speak Portuguese. They were all active, healthy little fellows, running about laughing, playing and seemingly happy and unconscious of the circumstances in which they were placed. In justice, however, to the Brazilians, I must say of them, after an experience of five years, that they are far from being hard task-masters, and that with very few exceptions, I found them kind and considerate to their slaves".

On the 28th of April 1841 they left Monte Café for the Paraíba river, a mere 9 km. away, at a small village called Sapucaia, now the capital of the eponymous and treeless municipality. The road ran along the river bank through a splendid forest formed of very large trees, some over 40 metres tall, and on their trunks they saw an abundance of orchid species of which *Cattleya labiata* was the most beautiful They found lovely examples of *Huntleya meleagris,* (which we bet is now extinct over the whole anticline. They also saw a number of howler monkeys.

A shop on Valongo street
Jean Baptiste Debret. Litograph. The Nacional Library, Rio de Janeiro, Brasil.

"In riding along (the river Paraíba) *I could not help feeling deep regret, that in these regions many square leagues of such forests were being cut down and burned, in order to make room for plantations of coffee…"*

As they could not cross the river at Sapucaia, though a bridge was under construction, they travelled a short way up river to Porto D'anta in the same municipality, now only named Anta where they managed to do so. They visited two large coffee plantations on the Minas Gerais side of the river producing between them some 315.000 kilos of coffee beans annually. On the morning of the 31 March, there is a mix up of dates here, they departed the Minas Gerais coffee plantations and

Bridge over the river Paraíba do Sul. Resende, RJ.

" in the evening reached Porto da Cunha, which is six leagues further down the river; we had to travel more than eight leagues (48 km.), having mistaken our road. (not a great problem driving in a Land Rover but pure hell on horseback) Some parts of the country through which we travelled were very romantic, particularly by the side of the river, the banks of which were often rocky and well-wooded; the forests are, indeed, the most magnificent that it is possible to imagine".

They departed Porto do Cunha in the now treeless municipality of Carmo were they had serious problems in finding maize for their mules and food and lodging for themselves and went east to the small town of Cantagalo, some time ago renowned for its alluvial gold.

"The villa of Cantagallo is situated in a narrow valley, bounded on each side by rather high hills; it consists principally of one long street, and a large square, of which only two sides are completed; the houses are mostly well built, and on the whole it has a neat and clean appearance. Formerly there were many gold washings in the neighbourhood, but now scarcely any one occupies himself in searching for this metal. The great article of produce is coffee, with which immense tracts are planted: it is conveyed by mules to the head of the bay, and then shipped for Rio".

On the second morning, after having rested in Cantagalo they recommenced their journey and soon reached the Swiss colony of Nova Friburgo, some 48 km. away. The first part of the way was through well cultivated flat lands but afterwards it became very mountainous and in particular the last 12 km. from Bom Jardim to Nova Friburgo took them through a deep, wild and romantic valley.

"The town of Novo Friburgo called also Morro Queimado, is built in the form of a square, with the houses nearly all of one story; it is inhabited principally by natives of Switzerland, who emigrated to Brazil many years ago…The greater number of the colonists, however, are scattered in the country for several miles around. They are very poor, having been placed by the Brazilian government in one of the worst possible places for the exertion of their industry, the situation being elevated more than 3,000 feet above the level of the sea, with a bad soil, and a climate quite unfitted for the cultivation of either coffee or sugar. Their principal crops are Indian corn,

and a few European vegetables; they also make a little butter. The climate being very agreeable during the summer months, many families, both foreign and Brazilian, come here to escape the great heat of the city".

They left Nova Friburgo, journyed due west to Teresópolis, through very mountainous country totally covered by forest, and under heavy rains, noticing enormous quantities of scarlet flowered *Gesneria bulbosa* and abundant orchids, the most beautiful in flower being *Oncidium forbesii*, until they reached Mr. March's establishment and from there to Rio de Janeiro.

"Notwithstanding that this visit to the Organ mountains was made at the same season as my previous one, the variety of the vegetation is so great, that I added to my collection several hundred plants I had not formerly met with".

A view of Nova Friburgo (The Swiss Colony, at Burnt Mountain). Aquatint designed by J. Steimann and engraved by Fr. Salaté. Kastro digital files. Nova Friburgo, RJ, Brasil.

THE TRANSPORT TRANSFORMATION

The Train Station of the Cantagallo Railroad, later the Leopoldina Railway.
It has been modified. The Barão de Nova Friburgo's Palace now houses the Executive Branch of the municipal government.
Kastro Digital Files. Nova Friburgo - RJ, Brasil.

It is interesting to try to determine which system of transport and freight contributed more towards the deforestation of the Organ Mountain Range. Thousands of mules organized in trains of thirty or more, carrying up to 3000 kilos per train and needing up to 10 days from the coffee plantations to the ports would have required a lot of cleared land for pasture and maize to complete their journey. This system would have been continuous, a sort of moving staircase of mule trains, until the years' harvest of coffee beans had reached the ports. Take the example of the farm George Gardner visited in the Paraíba do Sul river valley that produced 180,000 kilos of coffee beans a year. This would require 60 such mule trains or 18,000 mule-days of fodder and the cleared land to supply this. The advantage of mules was that they required a low, up front, capital investment. Furthermore, if and when they became surplus, a good profit could be made from selling them to warring armies in Europe, as happened during the Franco-Prussian war and even during the First World War: Europe had no traction animals to match their staying power in logistical support.

The railroads on the other hand required a very large starting capital and to obtain this, the potential financiers required the Brazilian government to guarantee a return, and further the financiers and the government quite obviously would want to be reasonably sure that there would

The Baroness.

The Leopoldina Railway network around the Organ Mountain Range
(1965)

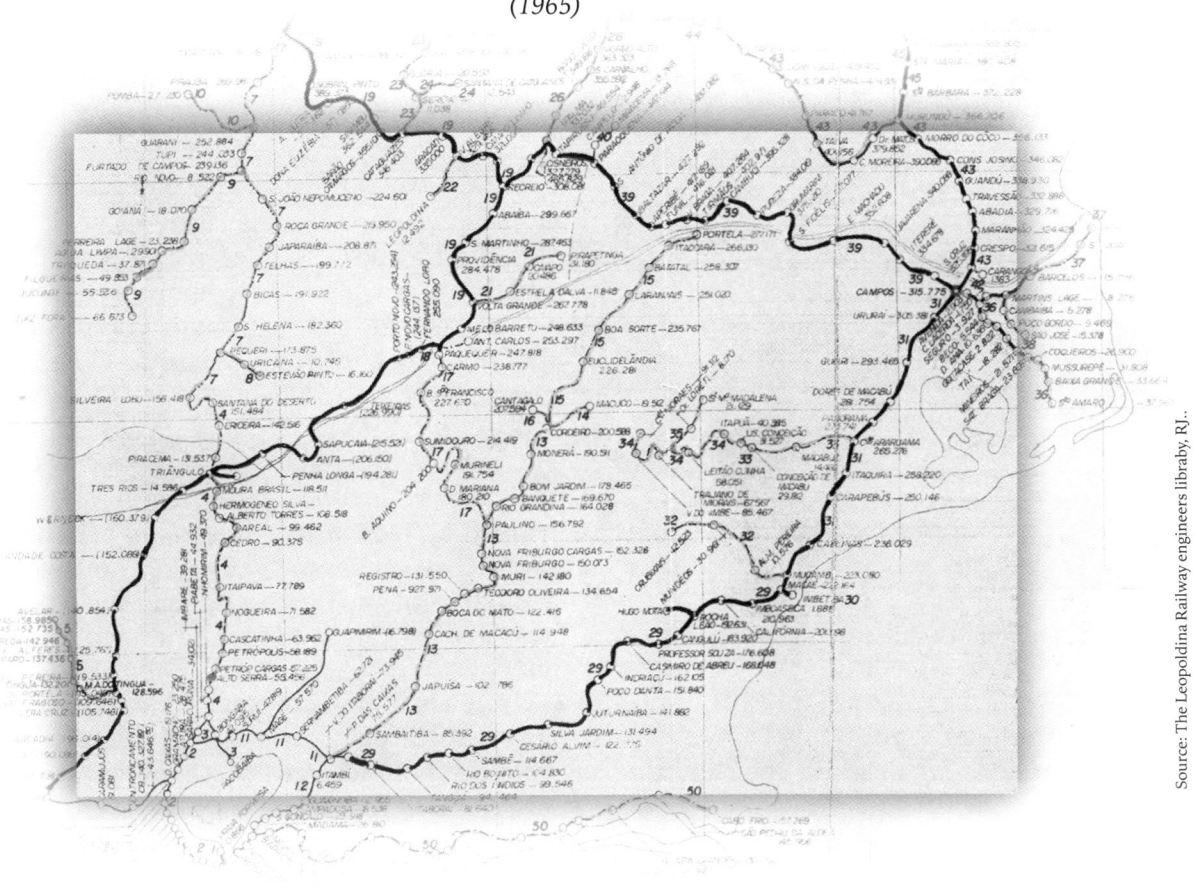

be a growing market of profitable freight and passenger traffic.

One suspects that by the 1860's it was becoming at least half clear that *après le café, le déluge,* to paraphrase Luis XV, that when a coffee plantation became uneconomic, there was no other freightable crop to compensate. The mass of plantation workers were slaves who did not participate in the money economy and anyway they would have been moved to the new plantations. Perhaps this reluctance to invest is reflected in the slow development of the railroads in and around the Organ Mountain Range.

In 1835 the Brazilian government prepared the first legislation to regulate railroad concessions throughout the country. It took 17 more years before the Barão de Mauá inaugurated the first concession. This was a short run from Rio de Janeiro to the foot of the Organ Mountain Range on the Serra de Estrela where 12 years earlier George Gardner had ended his various incursions into these mountains. The English-built locomotive, The Baronesa, so called in honour of the Barão's wife, has recently been restored as part of a plan to reopen this line for historical and tourism purposes.

The scarp slopes of the Organ Mountain Range appear as an almost vertical wall, with very few passes, even for mules, and because of this it took 39 years more to construct a successful railroad. At that time, freight and passengers detrained, climbed the mountain pass by road to Petrópolis where they again entrained to proceed on their journey to Paraíba do Sul and thence to Minas Gerais. In 1859 a concession was prosecuted from Niterói to Cachoeiras de Macacú and between 1870-1873 the scarp slope was breached from Cachoeiras to Nova Friburgo.

An old train station at the foot of the mountains.
Boca do Mato, RJ.

Within the anticline other concessions had been granted. From 1879 to 1884 the connection was made from Nova Friburgo right through to the river Paraíba at Itaocara with branch lines running to Cantagalo and Macuco. In 1887 the spur line from Nova Friburgo through Sumidouro to Carmo and Sapucaia was opened, while other lines were laid from Santa Maria Madalena and Manoel de Moraes through Trajano de Moraes to Conceição de Macabu and thence to Quissama, to hook up with the coastal line from Rio de Janeiro through Campos and finally the line was built from Macaé to Glicerio, where Darwin visited in 1832.

Most of these railroads were grossly inefficient and badly maintained. However, they were cheaper and more effective than mule trains and further they carried passengers. They were, however, useful enough to join up the coastal plain sugar plantations with the incipient coffee plantations in the north west of the Organ Mountain Range.

In 1897 the Leopoldina Railway Company was formed in London to standardise the gauge of the railroads in the State of Rio de Janeiro, buy out the railroad segments and swap the debentures in these operations for shares in the new company, and this was achieved successfully.

By this time the original coffee plantations in the south west Organ Mountain Range anticline from Resende through to Paraíba do Sul were in total decline. There was no more virgin forest to cut and the plantations had migrated to São Paulo and Minas Gerais States. The belated official, end of slavery also played its part, for the São Paulo plantations were manned by European immigrants, principally Italians. However, there were still virgin forests in the middle and north of Rio de Janeiro State, even though these areas were steeper and more mountainous and with a far lower percentage of suitable land for coffee than in the Paraíba do Sul section.

And the consolidated railroads played an important part in the elimination of the remaining virgin forests. The trains were mainly fuelled by wood, a very inefficient fuel but much cheaper than imported coal. Each station had its team of wood cutters and most concessions allowed the railroads to cut trees in a 15 kilometre swathe alongside the track. In most cases the track and its sleepers were laid directly on the sandy soil, not as today and then, in other countries, on a layer of stones. In addition the sleepers were not treated with a preservative. The net result was that Leopoldina Railway sleepers had an average life of five years, whereas sleepers, well treated, in the USA, had then a useful life of over 30 years. A fairly large number of locomotives and compositions were permanently collecting wood for fuel, sleepers and for the sugar and alcohol producing process. Furthermore in the dry season, sparks from the locomotives caused numerous forest fires.

The Leopoldina Railway Company faced an enormous task in standardising these concessions, building acceptable stations, replacing, standardising and re-laying track, upgrading rolling stock and so on. But the main problem remained that the rolling stock and locomotives that climbed and descended the scarp, were different from those used on the flat lands of the coastal plain or on the relatively smoother slopes of the interior. The Petrópolis scarp track was crenelated, that is to say, it was a cog railroad with a load capacity of only 32 tons per train, and, even when cargo was transhipped at the top of the scarp to larger wagons, or more wagons were added to the trains, great care had to be observed as the original concessions used different criteria for minimum curve radius and on this depended the length of wagons that could be used. Apart from these inhibitions the various lines were consolidated by 1907 and other spur lines added.

A tansformed railway bridge built 1895 on the line Friburgo/ Sumidouro, RJ.

By the time the railroads really got underway in a consolidated form, the south west of the Organ Mountain Range had been virtually deforested and the Leopoldina Railroad Co. had contributed but little to this devastation except as a mopping up operation using the remaining forests for fuel. In the area to the north, north-east and east of Nova Friburgo there were still large forested areas suitable for both coffee and sugar plantations. Through this region the Leopoldina Railway Co. consolidated nine different concessions, constructed six spur lines and built or rebuilt forty-two stations. Here the line that penetrated the scarp for 12 km up to the pass to 1,075 M.asl from Boca do Mato to Teodoro de Oliveira had a load capacity of 42 tons per train and a central rail to which both locomotive and wagons were attached by a scissor-like device for braking.

From almost the date that the Leopoldina Railway Co. was formed, the company began to run into difficulties. To start with the consolidation cost much more than the management had bargained for. Secondly the prime reason for the north-eastern and north-western rail network was to freight coffee and sugar which gave profitable freight per ton rates. The first crop, coffee, had a finite life span of about 20 years, due to the way the land was prepared and the bushes planted and, when the plantation wore out, there was no freightable substitute to replace it. Without entering into too many tedious statistics, coffee freight as part of a group that the railroad considered profitable was as shown in the table below.

The First World War was problematic for the company. Coal became unavailable and materials for all but minor repairs were unobtainable, so productivity and profits dropped. This period was followed by the great depression 1929-1937 and the overproduction of coffee which caused prices in the world market to collapse. Stockpiling was used to force prices up and there was bureaucratic chaos in releasing and receiving stocks. Price controls of freight, particularly sugar cane, meant losses on every wagon filled. Control of passenger fares and new labour legislation raised costs causing serious finance and administrative complications, all adding to the problems. Finally the Second World War resulted in all the problems of the First with some more new ones.

An abandoned bridge, on the line. Sumidouro / Carmo, RJ.

The owners of the Leopoldina Railway Company finally threw in the towel in 1948 after heavy losses in 1947 coupled with very poor results during the previous decade.

The railroad continued to operate, however, in government hands, until 1965, when the lines

An Analysis of the Composition of Profitable Cargoes					
	Coffee	Sugar/Alcohol	Construction Materials	Other	Animals
1895 – 1905	33%	34%	10%	32%	1%
1906 – 1915	23%	32%	19%	23%	1%
1916 – 1925	20%	30%	22%	27%	1%
1926 – 1935	22%	27%	23%	26%	2%
1936 – 1945	11%	28%	33%	26%	3%
Average Price of Transports Cargo/ton	100%	39%	15%	56%	51%

Source: The Leopoldina Railway Engineers' Library.

within the depopulating Organ Mountain Range were closed down. There was little worthwhile freight left to carry. The road network had developed to a point where it was more convenient for freight to be carried in trucks and people in buses or their own cars.

The railroad cycle came to an end for the Rio de Janeiro interior in 1965, not with a bang but a whimper as had the coffee cycle, although rather more slowly, with the result that the northwest anticline of the Organ Mountain Range became depopulated. As yet no substitute has been found to give adequate employment for the remaining population in these once fairly prosperous municipal capitals.

One can conclude in fact that railroads seriously contributed to the destruction of the virgin forests in the centre and the north-west Organ Mountain Range. Such destruction had been achieved without significant help from them in the south-west where the mule trains had flourished, first up to the Serra do Espinhaço and on to Diamantina and then to freight coffee from around Paraíba do Sul to the Rio de Janeiro ports.

The railroad owners had bet that the coffee frontiers were limitless in Rio de Janeiro and further did not realize that the mindless coffee plantation expansion in São Paulo and Paraná was so reckless

A sad end to the railroads in the interior of the state of Rio. Conceição de Macabu, RJ.

that it would destroy prices for everyone by overproduction in such proportions as to cause this collapse.

Railroads are different. The steam trains in England were the first mechanical form of land transport invented and in their early days made speeds of over 40 km an hour. This terrified an enormous number of people as well as spreading rumours that these speeds could cause heart attacks and brainstorms. However their speed and efficiency in the United Kingdom by the 1870's was quite on a par with that achieved today. They united the country as never before, made inter-city travel commonplace and united all to the great metropolis of London.

While railroads were in no way as extensive on the anticline of the Organ Mountains, they achieved some of the same psychological effects on the inhabitants of the remote and isolated communities in this region. The train service to and from Rio de Janeiro must have given them a feeling of being part of, or at least firmly connected to the metropolis. It carried mail, newspapers, necessary merchandise and passengers on a regular timetable.

Very gradually automobiles came on the scene and equally very slowly the highway system improved. It was not until the early 1950's that truck transport became a viable, although a still somewhat precarious, alternative to the train in the Rio de Janeiro State interior and only in the early 1960's, when the Brazilian automobile industry got seriously underway, was the writing on the wall for the railways. For many years after, inter-municipal bus services were woefully inadequate, only becoming really efficient in the last decade of the last century.

Small wonder then that the older inhabitants of the scattered municipalities still feel nostalgia for the railroads. In some small towns the old train stations have almost become a shrine, in others left to rot, while in a few they were knocked down to give place to more modern buildings. The main station in Rio de Janeiro for example is a monument to Victorian grandiosity and may yet become a shopping centre. Only at the very end of the last century, with the help of a non state-owned telephone system, world-wide television, efficient inter-municipal bus services and better roads, have remote populations once more recovered the sense that railroads gave them that they were once again part of Brazil.

A substitute for the train. 2005, RJ.

CURRENT DANGERS

The maps on page 26 show the original forest cover and the satellite image of existing forest. Neither shows a true picture, as in 1500 for example the occupants of the country had no doubt nibbled at the more accessible lower mountain areas with slash and burn agriculture. Similarly the satellite photo does not differentiate between original and regrowth forest. In reality it is only by inspection on the ground that one can be reasonably sure of things, and even then mistakes can be made, as the forest is not static but in a state of constant dynamic renewal.

One thing is certain, however; there is no accessible undisturbed virgin forest remaining in the Organ Mountain Range. Almost every metre has seen hunters and their dogs. Hunting is prohibited in these mountains and yet it is, after football, perhaps the favourite hobby of many rural inhabitants. Along animal trails in the remotest of forest, we have found traps, of many types, the most lethal being a well fixed gun with a hair-trigger, aimed about 30 cm. high across the trail and triggered by a taut fine nylon tripwire set across the trail at a similar height. Many people have lost legs or have been badly injured by such apparatus. These and other types of trap have been even found within

The Macaé de Cima Ecological Reserve, RJ.

Photo: Bengt Janer.

both the State and Federal forest reserves. Yet another popular hobby, also an economic activity and not permitted by law, is the trapping of song birds to cage them. The cages are hung around the house, or the birds are sold to "distributors".

Many apparently original forests have in fact had the most valuable trees such as *Meliaceae* and *Lauraceae* (cedros and canelas) removed while under storey trees such as *Euterpe edulis* (palm hearts) whose new shoot George Gardner likened to asparagus when cooked, have been mercilessly cut for sale over the centuries, as they still are. Fire, mainly deliberate, is another factor which can change the character of an apparently original forest remnant or eliminate an isolated remnant altogether. The "fire-up" time in Rio de Janeiro State is June to September. During this period poor pasture is burnt, principally on the anticline, to eliminate poisonous herbs and, believe it or not, ticks. This burning, in general, degrades the pasture even more, but the practice seems to be universal. In July 2000 we watched the fires throughout the interior from the top of the Organ Mountain Range's highest peak, Caledônia at 2,400 M.asl (no great climbing feat as a road brings one to within 50 metres of the top). To the south-west, west and north-west there was a layer of smoke, as far as one could see, rather like the smoggy inversion layer sometimes seen over Rio de Janeiro on a windless late autumn morning.

June and July are the dry, winter months when another band of hobbyists enter into play and threaten the scarp slopes of the Organ Mountain Range. These are the balloonists. All around Guanabara Bay, clubs are formed by these enthusiasts, many of whom compete with one another as to who can produce the largest and most beautiful balloon that flies the highest and furthest. Normally at this time of year the winds are very light but with a slight tendency towards the north-west, which brings the scarp forests within their range. The hot air balloons are made of paper and depending on size, are powered by a can of kerosene fired through a gross cloth wick, or in the case of really large balloons by a natural gas bottle slung underneath. This hobby is also prohibited by law, but this does not prevent the winter sky being speckled with these artefacts. If the kerosene or gas bottle burns out, the balloon falls harmlessly. However should a wind force the balloon down with the kerosene or the gas bottle still alight then a forest fire could be started. Several such have come down in the dense forest around our headquarters, so far fortunately with no damage. They also create a serious hazard for aircraft around the Rio de Janeiro International Airport.

Natural fires are normally started by lightning associated with tropical thunder storms which fortunately are rare during the dry winter months. When they do occur in the summer months, quite frequently we have seen lightning strike a large tree which bursts into flame and only the heavy rain accompanying the storm washes the fire out after 10 minutes or so.

The current major danger to the scarp slope's original forest is urban sprawl and the building of holiday home condominiums, both on the periphery and within these large forest fragments. It is an ironic fact, that the more publicised the need to preserve these remaining forests and their natural wonders, the more desirable they become to real estate developers and potential middle class holiday home owners. The laws concerning the conservation of these forests are very specific but they are broken with impunity.

A clear case in point is an attempt to build a proposed condominium of 200 holiday homes right in the centre of the Macaé de Cima Ecological Reserve. This area is totally protected by Federal, State and Municipal law. The perpetrators apparently had no problem in persuading the local mayor to revoke the Municipal law that had created the reserve some 10 years before; a gesture that clearly indicated that local government would turn a blind eye. And as the region is at present only accessed by a 16 kilometre dirt road, they must have assumed that neither Federal nor State authorities would find out about it until it was too late and then it would be *fait accompli*. At a certain point the condominium owners would request the Municipality for permission to open a road, at their cost, following an old mule train track to the main Nova Friburgo-Rio de Janeiro road only some 6 km away, thus shortening by 10 km the journey on dirt roads and 10 km. less on the asphalt. Permission granted, the condominium would kick in the money and everybody would become rich. They had already sold options to a number of middle class clients longing to get rich when the Macaé de Cima Society, a local conservation unit, found out and informed the Federal and State authorities who confirmed the situation. The Public Prosecutor was called in. A civil and criminal legal action was brought against the perpetrators and the project was stopped in its tracks. The points of note from this example are that:

1 Municipal authorities are not concerned with conservation. They are desperate to increase their tax base;

2. Federal and State authorities will act to uphold the law but their forest agents are far too few to cover the vast areas that they are responsible for patrolling;

3. The Public Prosecutors will indeed act and rapidly but they must receive information to act upon, preferably from an association rather than an individual; and

4. this association must be able to produce documented evidence for a successful legal action to be brought. The Macaé de Cima Society had prepared or obtained, all the ingredients for the successful opposition and thus a *fait accompli* was avoided.

We suspect that such situations arise with frequency along the scarp slopes of the Organ Mountain Range but we doubt if there are many entities as qualified as the Macaé de Cima Society to evaluate and inform the authorities of such incursions into the remaining forests. Before leaving this theme we should remember that life is rough in these remote areas for subsistence farmers. Weekend cottagers will give them an additional income. Their land will take on an amenity rather than a productive value and they may see this as a way to escape from generations of poverty. As such, forest conservation retreats from their lexicon, to be replaced by the rather normal human condition of avarice. Furthermore, imagine that your father bought a large tract of forested land in the mountains many, many years ago. He bought it for a song and for its forest. He lived there, died there and his six children inherit it. All of them are in so called liberal professions. All are married and have children and all suffer the high costs of modern middle class urban living and all have no interest in forests except in an abstract way. What do you think their reaction will be when a wily real estate agent offers them half a million US$ for the property? You got it! Until the various levels of government show a political willingness seriously to conserve the incredible botanical richness that still remains and could remain for future generations, then the examples above will be the norm.

5. Our Travels Through the Organ Mountain Range

5.   OUR TRAVELS THROUGH THE ORGAN MOUNTAIN RANGE

We complained earlier on, that few scientists and plant collectors visiting the Organ Mountain Range, with the admirable exception of George Gardner, have left descriptions of the forests they explored or passed through. So saying we must now proceed to fill this gap with descriptions of our wanderings within these mountains.

Right from our initial interest in wild orchids up to today, we have only once made an expedition to search for a specific species. The 600-odd species described in this book have all resulted incidentally from the hundreds of forays into these mountains at different altitudes from 50-2400 M.asl, different latitudes, different types of forest, rock-faces and high mountain fields. Ninety percent of our trips last only a day, but we sometimes stay in a region for periods of up to a week in order to explore it thoroughly.

Our headquarters are located at 1,450 M.asl in the Macaé De Cima Ecological Reserve surrounded by some 250 hectares of montane forest. Part of the lower third at 1100-1300 M.asl was accidentally burnt some 40 years ago. However as most of this slope faces south, is humid and has never been used for agriculture or pasture, natural regeneration has been rapid and successful. We also own a larger forested area some 12 km. away at the head of the Flowers river valley. This property is covered by virtually undisturbed, original montane forest from 1000-1300 M.asl. An NGO, owned by friends of ours, has bought a 500 hectare forested property contiguous to ours for various kilometres, with disturbed original mountain forest running down from 1300-600 M.asl. Yet another friend bought

The Macaé de Cima ecological reserve, RJ.

Photo: Bengt Janer.

One way to examine the canopy!

the contiguous head water valleys of the Macaé River with over 1000 hectares of original and maturing regrowth forest, with a view to its permanent protection.

Within this complex we can oversee a thick swathe of sometime disturbed original montane forest from 1600 down to 600 M.asl in the centre scarp slopes of the Organ Mountain Range and of course, it was the area we first explored in real depth. After nearly 30 years in these mountains we have also concluded that it is the most complete and least disturbed area of wet forest in the Organ Mountain Range. Almost 100% of known tree species from the ocean-facing slopes of the whole Mountain Range, as well as some new species and species that were considered probably extinct, have been identified in this region principally by researchers from the Rio de Janeiro Botanic Garden - The Atlantic Rain Forest Project.

It took a long time, but over the years we have developed fairly sound methods of prospecting a forest for orchid species. Quite obviously, in tall mountain rain forest one can not identify satisfactorily, even if in flower, orchids growing from the first bifurcation and upwards to the canopy. So we pass through the region on a monthly or bi-monthly basis, fanning out for maybe 30 metres on either side of the trail looking for fallen branches or trees. These forests are dynamic and the turnover, not only in foliage, but also in large branches of trees and even groups of trees is remarkably high. The collapse of one tree, over-weighted by saturated epiphytes, can create a domino effect on the trees around. And when the branches, or the trees, fall we can harvest the orchids, identify, rescue and replant them in an appropriate environment. If some of these plants are unknown to us, we take them to a first class horticulturist living at a more-or-less similar altitude. And when this expert brings them to flower we can identify, photograph or draw, paint and describe them and ultimately return them to their original environment.

An expedition, during the course of which we found and identified *Huntleya meleagris*, a curious plant with stupendous flowers, in an original second storey forest, beneath immense trees such as Cedros (*Cedrela odorata*) and the impossibly huge Jequitibá (*Cariniana legalis*) was not that unusual; more mind stretching was the only expedition we made with the specific purpose of finding a rare orchid species: *Cattleya dormaniana*.

We set off at dawn with the intention of reaching a hut, half way to our goal, a totally remote virgin-forested valley some 20 kilometres away, inaccessible except on foot from our headquarters. We reached the hut fairly early and spent the remaining daylight hours examining the forest around and finding *Oncidium harrisonianum en masse*, *Oncidium trulliferum*, many Pleurothallidinae, the fan-like *Dipteranthus grandiflorus* and large colonies of the lovely terrestrial, *Warrea warriana*. It was the month for the mountain *Laelias* to flower. *Laelia crispa* showed

Photo: Bengt Janer.

Logistical suport: where we spent the night on the way to find *Cattleya dormaniana*.

up like snow in the first bifurcations of large forest trees.

One of our competent forest guides.

The following day, again at dawn, we trudged the 10 km. to where we imagined we would find *Cattleya dormaniana*. It rained, and how it rained, as we squelched through these ancient forests, passing over one ridge only by crawling. This was a scrub-covered ridge, a rarity at 600 M. asl., and was surrounded by original forest. It was of pure granite with only *Bromeliads* and stunted *Clusia* trees giving a hand or a foot hold. But the higher light hinted at the possibility of *C. dormaniana*. We groped our way through the sword-like bromeliads, without them we wouldn't have been able to scale this very steep ridge. In their midst we found colonies of *Oncidium ramosum*, *Cirrhaea loddigesii* and then, but suddenly, we were face-to-face with a colony of *Cattleya dormaniana*. It had many flowers, and a species of humming bird, *Phaethornis eurynome*, was assiduously visiting every flower. We simply sat down, smoked a cigarette, quite stunned, and just looked at this beauty. The plants were perched on a *Clusia* bush, overhanging a steep, deep valley and receiving up draughts of misty humidity in the middle of this original forest which would certainly receive more than three metres of rain a year, a fair percentage of which was falling on us as we gazed.

And there were other expeditions to section V where we found accessible rock-faces covered by *Vellozia* bushes. One particular slope at around 800-900 metres altitude was angled between 30-40 degrees. We had to climb and cut our way through 500 metres of scrub forest composed mainly of very thorny pioneer species with no orchids at all. This took nearly two hours. When we reached the granite slope base, we were all bloody from cuts and scratches. All had taken turns at being the lead 'hacker'. Coupled with this discomfort, it was nearly midday and the sauna-like heat had left us exhausted. To add further to our pain, the stink of decaying horse flesh from one of those unfortunate animals that had ventured out on the rock slopes during the previous dry months did not serve to lift our spirits.

But, nothing ventured - nothing gained. We scaled the first very steep ten metres to come face to face with a huge colony of *Cyrtopodium glutiniferum* and stretching upwards for over 500 metres were broad thickets of shrub-like *Vellozia compacta*, smothered like snow with its lily-like white flowers and with thousands of *Pseudolaelia corcovadensis* flower spikes pushing their lovely pink flowers half a metre or so above the *Vellozia* blanket. *Oncidium blanchetii*, also contributed its sulphur yellow, while around the bases of these *Vellozia* islands on the rock face, were stacks of the wine-coloured *Pleurothallis teres*. If that wasn't enough, there were also huge colonies of the dramatic and blatantly beautiful *Bifrenaria harrisoniae* and not forgetting the huge *Epidendrum robustum* with its large panicles of honey-coloured flowers. This brilliant

Unchanged in 500 years.

Chris Dobson seems worried by the possible tick swarms.

day had two further setbacks; the first was that, baking in the sun on this north-facing slope, we almost collapsed from heat and sun stroke. The second was that all of us arrived back at base covered from head to foot in thousands of minute ticks (*micuim*), the removal of which caused great hilarity amongst the spectators. We stood in line in our shorts while the non-combatants sprayed us with alcohol and bug-killer. Further incursions to this paradise were made; always arriving before 6 a.m. and always with a litre or so of bug repellent, and each visit has been a success, helping to raise the orchid score up to 40 species on this one slope.

In another situation after a half dozen incursions into the anticline regrowth forest in section V, after finding absolutely no epiphytic orchids, we came across colonies of *Laelia pumila* around a small lost mountain creek. You can not imagine the emotion generated by a find like this. Even a non-orchidophile worries about us in such a situation. The plant is meant to be long extinct in these areas and here we've found lots of it in regrowth forest, in a region that no sane botanist would waste his time even glancing at. And while we are deep in these recollections, what about the trip made in section III, again on the anticline. An Afghanistan-like mountain landscape; trees as rare as water in the desert. An off the track visit to a wet seepage revealed simply thousands of the beautiful yellow-flowered terrestrial *Epidendrum aquaticum* growing like, and looking like ripe wheat, in a fenced-off, boggy environment: a brilliant example of what can happen when grazing animals are kept away. We continued on and found another similar environment but this time with a 300 plant colony of *Phragmipedium vittatum* interspersed with *Bletia catenulata*: two more beautiful and rare plants would be hard to imagine and it would be even harder to imagine the emotions of the finders. Yet more astounding and not a half mile distant from this discovery, we found a carpet of *Stanhopea guttulata* in a ravine and the scrub forest around, somehow loosely anchored to the almost vertical rock face, was peppered with the most beautiful *Centroglossa nunes-limae*.

The habitat of *Phragmipedium vittatum*.

This is not trawling the remoter Amazonian forests, where our friend João Batista finds new orchids species at the drop of a hat. No! This is exploring the most raped area of Brazil to search for what once was there, to see if it still exists, to try to piece together what the original forest might have looked like, and to think about how we can preserve areas where the forest and its orchid species may continue to exist.

Continuing north-westwards, we found and explored one day, a forest fragment in section V with some fairly large trees at 600 M.asl. Only maybe two hectares in area, very steep, full of ravines and caves with a sub-storey forest of weak straggling coffee bushes. Not, on the face of it, a very promising environment. No preliminary diagnosis could ever have been so mistaken; large trees, mainly *Leguminosae* had stunning, orange-red colonies of *Sophronitis cernua*; other trees and rock faces bore clumps of *Maxillaria chrysantha*, *Miltonia russelliana* and *M. clowesii*, while within the ravines were the huge *Epidendrum paniculatum*, the delicate *Cochleanthus candida*, *Galeandra beyrichii*, *Xylobium variegatum* and *Cirrhaea dependens*, and the sub-storey forest sported several *Notylia* species, nests of the ethereal *Warmingia eugenii*, *Campylocentrum robustum*, *Leptotes bicolor*, *Oncidium pumilum*, and a host of Pleurothallidinae.

One can find surprises around rivers and permanent streams in the high mountains.

Again in section III we were guided to an oasis at 400 M.asl between two sugar-loaf type mountains, only about one hectare in extent at the foot of a towering, stark rock face. Here a curious forest of Leguminosae with a sub forest of Bromeliads and *Vellozia* bushes, were quite smothered by the delicate *Lanium avicula*, the semi-terrestrial *Ponera striata* and a number of lovely *Eltroplectris roseo-alba*, while the boulders and *Vellozia* bushes were respectively covered in *Schomburgkia crispa* and *Brassavola flagellaris*.

The heat was insupportable on our second visit at mid-summer. There was evidence of other visitors in the interim and the few *Laelia perrinii* plants we had found on our April visit seemed to be missing, as was an enormous clump of *Catasetum hookerii*. However, the new species we had found during our April visit, *Maxillaria* aff. *paulistana* was growing strongly. Hence our terror of letting anybody know the exact location of species that are rare or new. The syndrome - collect it, because if you don't somebody else will - is terribly prevalent amongst orchidophiles. We have likened it, in published articles, to the syndrome of a happily married man or woman, making that fantastic eye contact at a cocktail party… The desire is there to collect, but if one does the result is generally disastrous, as it indubitably is if one collects rare plants. As we said the heat was preposterous on our second visit. It was like cutting your way up through a dense regrowth forest within a sauna. The resultant exhaustion, seriously limits the amount of time one can spend in a target area some 100 km from headquarters. Therefore, we made our headquarters in the potentially rich target areas of sections VII and VIII when we moved to Santa Maria Madalena and Glicerio.

In our three-week long visits to the state park in Santa Maria Madalena in section VII, we hired local guides, and drove the faithful but battered kombi up impossible dirt roads to get to the fringes of Desengano State forest. This forest has been brutally attacked around its ever retreating fringes and it took us more than a little time to reach original but disturbed forest. A large tree cluster had fallen across a stream and the harvest was enormous. Most species were common to those found in section VI but two stood out: a *Bifrenaria*

Maxillaria rigida well anchored in a granite cleft.

and an *Oncidium*. The first turned out to be *Bifrenaria melicolor* while the second was *Oncidium kraenzlinianum*, a very rare plant which Pabst and Dungs show as Brazil *sine loco*. We made several more expeditions to these mountains, which showed signs of over-collection, and found most of the old friends from section VI. On the final day of our last trip we were taken to a small piece of remnant forest on the anticline at about 500 M.asl. And there we surely struck pay dirt in *Cattleya guttata*, *Oncidium baueri* and *Trichocentrum fuscum* not forgetting *Amblostoma tridactylum* and the unbelievable *Cycnoches pentadactylon*.

We get by with a little help from our friends.

In Glicerio, section VIII, below the Pico de Frade we were given great help by a local orchidophile who showed us *Cirrhaea saccata*, quite the most lovely of all the *Cirrhaea* species, *Encyclia bracteata* perhaps the only dramatic *Encyclia* species on these mountains, and quantities of *Laelia crispa*, *Warmingia eugenii* and a very unusual *Campylocentrum*, *C. calostachyum*. But these forests had been so obviously over collected that having found many of our old friends as a point of reference, we transferred our headquarters back to Macaé de Cima, a difficult 50 km. journey, mostly over mountain dirt roads, through several bucolic villages that time had left behind.

Area IV showed great richness on the scarp slopes particularly over 600 M.asl where without a doubt almost undisturbed forest still exists. Enormous *Jequitibá*, cedros, Lauraceae and Myrtaceae dominate these forests, but at a lower level, around the waterfalls, trees covered from stem to gudgeon with *Miltonia clowesii* give a most awesome sight that stops one in one's tracks, while *Gongora bufonia* sprawls over boulders and stunted trees, stirring the iridescent blue *Euglossine* male bees to a frenzy collecting perfume for their females back home (just like us); massive colonies of *Xylobium variegatum*, the spectacular *Warrea warreana* which grows in and around the wet areas near these waterfalls, while the low tree trunks sparkle with the flowers of *Promenaea stapelioides*. The area is a paradise, but the sauna-like heat persists; to penetrate any further, we would have to camp and as most of us hate camping, this has inhibited further exploration until now.

Pico do Frade - Glicério, Macaé, RJ.

The higher altitudes of section IV, in a more accessible area, gave a surprising crop of species not found in section VI, with *Oncidium praetextum* and *Oncidium waluewa* as examples. We spent several days in the area of this State park, which at the higher edges is essentially old regrowth forest, nevertheless quite rich in epiphytes. But one ridge at 1200 M.asl gave us a plethora of those mentioned above together with immense colonies of *Sophronitis coccinea* and of course a myriad of species already found in section VI.

Izabel changing lenses.

We could not visit Tinguá, in section II but the Rio de Janeiro Botanic Garden team substituted for us. At last they found *Miltonia spectabilis*, which had evaded us in all other scarp sections, with another escapee, the unmistakable *Hexadesmia sessilis*.

Section I gave us little new except for a rogue colony of *Cattleya bicolor*, a variety of *Sophronitis coccinea* and *Grobya galeata*, plus some terrestrials. The area was so heavily collected in the past that it is a wonder that any orchid species can still be found in regrowth forest, for except around the highest high mountain fields, no original forest seems to exist.

High mountain ridges between 1300-1600 M.asl, covered with wind tortured and twisted scrub, and trees rarely more than six metres tall, untouched by any form of human activity, are by far our favourite forested sections.

These pristine areas are common around our headquarters. At least 10 km of them border our properties alone. Mostly these ridges are very abrupt. The ridge itself is between five up to maybe 15 metres wide. Almost always the sub-forest is a tightly packed Bromeliad garden and each ridge seems to have a different dominant species. This is so striking as to repeat a remark from our ex-Boeing 747 pilot team-mate. If he should land by parachute on any ridge in this region he would immediately know where he was by the dominant bromeliad species around. This also made us wonder and worry if 747 pilots carried parachutes on passenger flights!

The ridges are often mist-covered and even during the dry winters the Bromeliad rosettes are full of water. So one can consider them collectively as a continuous pond, picking up mist, dew and rainwater which evaporates during hot dry days to give a constant humid atmosphere at all seasons.

The ridge trees at these altitudes hold an abundance of epiphytic orchids above these "ponds". Quite the most dramatic are the thousands of *Sophronitis coccinea* plants which give an incredible show of scarlet flowers on most branches in August and September, mainly at head height or a little more, allowing perfect viewing. We have found just about every known colour variety of this plant on these ridges.

Maxillaria picta, with its yellow, blotched red, flowers is also a very common epiphyte found here together with *Epidendrum addae* and many *Oncidium* species, notably *Oncidium ramosum*. Both of the latter species we found many years ago when we circled the Rio das

David and Helmut pushing through the bromeliads in the under forest.

The other side always looks more promising.

Flores property after we had contracted a trail-cutting team to enable a surveyor to define the property's boundaries.

We were conducting a youngish, (at least younger than us), Californian doctor, on a complete tour of the Atlantic rain forest in one or two days, as North Americans wish. We had completed the four km to the source of the Flowers river and thence to the abrupt ridge some 100 metres further on. There we marvelled at the swarms/-clouds/-flocks of *Sophronitis coccinea* in the low ridge forest around. The doctor was told that this was the point of no return, either to go on or to back-track was the same distance. Go on he said, and so on we went, with magnificent views all the way to Rio de Janeiro, and to Búzios, the fabled sea-coast resort, over the well cut trail, a pleasure to walk on. But even paradise has its problems. On these ridges grows a bamboo species, very dense with thin wiry stems which resist a bush knife as a dedicated ideologue resists reason and logic. However our boundary cutting team had done a superb job of making a trail for the surveyor, so we were able to pass through with ease. Mountain ridges are not regular, they go up and down. On the third up and down, when there were more than a few to go, the doctor told us that he had a heart condition.

This was more than worrisome. So we made a stretcher and carried him up the steep ups and let him walk down the steep downs. *'How near are we'* he would ask. *'Just one more up and down'* we would lie as we hauled him up several more ups! And all the time we were passing by new orchid species which we only glanced at - the heart condition had taken precedence.

Well, we arrived back at head quarters II. Izabel filled him up with heavily sugared lemon juice

from our own bitter lemons and he rapidly forgot his heart condition. It was to be five years before we repeated the trip and identified *Epidendrum addae* and *Oncidium ramosum* as well as numerous other plants. However this trip stands out as one of our most traumatic memories.

Dick going, Chris returning, both living dangerously.

David on a high mountain ridge.

We forgot, there was one other expedition to find a specific orchid. *Cattleya velutina* was the target and our guide, who had collected one large clump from the ridge years ago, still had it growing and flowering well, in the middle of an *Azalea* bush next to his house. *C. velutina* is a curious but very beautiful plant, never common or dense in any area, and only introduced into Europe by accident amongst some other *Cattleyas* in the 1870's. We visited this ridge, which divides our Rio das Flores property from that of our NGO friends. It is maybe 300 metres long, very abrupt and only one to three metres wide, falling off almost vertically on both sides. It was so rich in new species to us, that several trips to this site were needed to identify all the orchid bounty found there and yes, you guessed it: the predominant bromeliad was unique to this area, as were masses of *Encyclia flabellifera*, *Maxillaria caparaoensis*, the button orchid *Phloeophila echinantha*, over a dozen Pleurothallids, *Sophronitis wittigiana* var. *brevipedunculata*, but not a trace of *Cattleya velutina*.

We had named the ridge, Velutina ridge, more in hope than anything else. So to satisfy our illusions we managed to obtain a flask of *C. velutina* seedlings from a grower. We affixed them to a number of different trees in 1998 on Velutina ridge and in 2002 while not exuberant, about ten plants are growing reasonably well, and we hope that sometime soon will flower and be hand-pollinated with pollen from the original plant in our guide's garden, or by far ranging bumble bees which are the probable natural pollinators.

We have made hundreds of other such trips through this extraordinary mountain range, some extremely productive, but the majority, particularly on the anticline, were made more in hope than with results.

David, Dick, Izabel and Helmut

6. Regeneration of the Forests

6.   REGENERATION OF THE FORESTS

NATURAL REGENERATION

One cannot make general rules as to the time it might take for a forest to regenerate in the Organ Mountain Range. There are so many variables to be taken into account and an equal number of micro regions. However, there are a few more or less constant criteria. Most Atlantic mountain rain forest trees have a lateral root system. This is to say that their roots are sub-surface, running through a topsoil layer and rarely penetrate the mineral sub-soils below. Consequently a layer of organic topsoil must exist, or must be created, to permit regeneration. Our excavations in disturbed original forest between 1,300-1,500 M.asl show a topsoil principally composed of organic matter from 30-70 cm. deep with organic traces continuing for 10-20 cm. deeper. This topsoil is held together and secured by a hessian-like root mass and a continuous rain of falling leaves, fruits, branches and rotting fallen trees is constantly added. Herein live the insects and micro organisms essential for the breakdown of this detritus into the nutrients essential for tree growth.

The second constant is the need for a reasonably consistent topsoil humidity. This is clearly more easily obtained on the scarp slopes with high rainfall and frequent mists, particularly those facing the South Atlantic and receiving only glancing or no direct sunlight during the dry winter period. Yet another constant, particularly true on dry anticline pastures, where regeneration of forest is desirable, is the exclusion of herbivorous domestic animals that will eat anything green, especially during the long dry winter. However an expert told us that such animals are essential to add vital micro-organisms to the impoverished soil. Take your pick!

We believe that natural regeneration is the most practical and economic way to achieve and increase indigenous forest cover in the Organ Mountain Range. How this happens and how long it takes we will try to explain in a general way and then examine how the effects of different conditions in different micro environments will accelerate or slow down the process.

The regeneration process occurs in four distinct phases or waves:

1. The area is invaded by scrubby pioneer species. These will probably be short lived Compositae and Melastomataceae interspersed with ferns, tall coarse grasses, club mosses etc. These plants grow and in fact thrive on the apparently sterile subsoil. Their function is to cool the surface and begin the building of an organic topsoil. Their seeds and spores are mainly wind dispersed. These plants endure total exposure to the elements, so obviously are extremely hardy. A significant hazard at this stage is that most of these plants have uses for humans either as fuel, building material, or rot-resistant fence posts and are very susceptible to fire.

2. After a number of years, depending entirely on the location of the area, either the first stage will recycle or the second phase begins. Again Melastomataceae species are the principal actors on the stage, mainly *Tibouchina* species, Quaresmas or Lent flowers, so-called because they mainly flower

during February and March. Sometimes they are accompanied by Leguminosae species such as the yellow-flowered *Senna multijuga, (Canafístula)* which can form an almost homogeneous regrowth forest in the high mountains. One or two *Cecropia* species (*Imbaúba*) will also sometimes appear at this stage in dense stands and are quite unmistakable by the silver undersides to their large palmate leaves, visible for kilometres. If you see a forest across a valley at flowering time with a preponderance of purple or yellow and speckled with silver you can be sure that you are looking at a second stage regenerating native forest. Its undergrowth will be reasonably dense with spiny Leguminosae (*unha de gato* or cats' claws) and perhaps incipient colonies of several bamboo species. The humus layer is slowly building up. Most of these trees have a relatively short life, maybe up to 30 years or so, but a few lucky ones will survive the transition into the next phase.

3. Once again, depending on the location and the other forest fragments around, stage three will begin. This will probably be led by the introduction of seeds collected in more mature forest by birds and fruit-eating bats, defecated in the second phase forest, where these animals roost or search for other fruit. The predominant species brought in is yet another Melastomataceae, *Miconia cinnamomifolia* here called *Jacatirão*. This is a much larger and taller tree than its predecessors and with a longer life. Sometimes it can form an almost homogeneous canopy. Other taller and longer lived species appear, such as the yellow-flowered *Vochysia* species and an Euphorbiaceae, *Alchornea triplinervia*. The bamboo colonies may also have developed strongly. Under storey trees such as the palmitos, principally *Euterpe edulis* or *Juçara*, are also brought in by larger birds such as *Penelope superciliaris*, the Guan here known as *Jacu*, and Toucanettes. It is amusing to find dozens of small recently germinated palmitos in a tight cluster directly below a roost, and no mature palms within a hundred metre radius. The humus layer is now up to maybe 20 cm. thick. The reasonably consistent humidity layer now creeps up to about three metres. This is shown by the epiphytes, particularly Bromeliads and mosses, which had previously been found, in phase 2, on the ground, but are now to be found colonizing these heights. Even some hardier orchids, such as species of *Oncidium, Gomesa, Bulbophyllum* and some pleurothallids are making a tentative introduction on the low trunks of the trees.

This phase can continue for some time, perhaps for 30 years or so. If there is a fragment of original forest near, then windborne seeds or seeds brought in by birds, bats and even animals such as squirrels, will slowly be deposited in this phase three forest. As bamboo colonies are now becoming very thick and tall allowing little light to the forest floor, many and various seeds which fall to the ground beneath them may well remain dormant until the bamboo dies back and allows light in, while the surprisingly dry but thick leaf litter, moistens up allowing their germination.

Fire is the main hazard at all stages up to this point. It is very rarely spontaneous but started by adolescents or idiots. The urban equivalent is the spraying of graffiti on buildings. There seems to be no coherent explanation for either habit.

The phenomenon of the whole population of a bamboo species in a region showing synchronised flowering, fruiting and death, with characteristic intervals for different species, is remarkable. The predominant large species found between 1100-1500 M.asl mainly in regrowth forest, as far as we can calculate, enjoys a 40 year cycle. Reproduction and death of this species occurred in the Macaé de Cima valley in 2001. We are carefully watching a transitional period between phase three and the final phase. This has coincided with a heavy flowering and fruit crop of *Miconia cinnamomifolia* and an enormous influx of birds and bats from original but disturbed forest, are leaving behind seed from those areas while dropping the *Miconia* seeds in other needy parts. As an aside thousands of small finches came in to feast on the bamboo seeds, while it seemed that half the State population of Spotted wood quail, *Odontophorus capueira*, here called *Capoeira*, joined the party. One young man born and bred in the region, caught six together in one trap. Like we said, its in the blood. It's a major ineradicable hobby. How long the forest at this stage takes to return to its original pre 1500 form is impossible to say, but if it reaches stage four, at least a firm base is set.

To be rather depressing, one could imagine the scenario when the forest started creeping out of its refuge at the beginning of the world warm-up 25,000 years ago, and compare this to a forest, receiving full protection, creeping out of oases and narrow stream-fed gullies, on the anticline today. With the devastation wrought on the Organ Mountain Range anticline one imagines it would take thousands of years even if all domestic animals and humans were removed.

We can help to speed up the faster development of forests once they reach stages three and four. Perhaps this is the stage that forest specialists can really contribute. There is a significant humus layer. It took a long time but it cost nothing. Now, seedlings or saplings of the more regal tree species that dominated the original virgin forests could be selectively introduced. These species have been so mercilessly cut that natural regeneration is almost impossible as so few potential parent trees remain.

SPECIFIC REGENERATION SITUATIONS

The first human occupants of the fringes of the Organ Mountain Range practiced a type of slash and burn agriculture known as swidden farming. This practice involved cutting down trees in a given area, making a clearing in the forest. This clearing varied in size depending on the number of people that composed the family or tribal unit. The felled trees, scrub and foliage were burnt, and manioc, maize or beans planted. The clearing was probably used for two maybe three seasons and then allowed to regenerate, not to be used again for up to 20 years, if at all.

This surely must be the easiest, most rapid and most successful of all forest regeneration situations. There was still significant top soil and thus most of the essential micro-organisms remained. The clearing, by definition, was surrounded by forest and consequently would receive a constant shower of seeds and without a doubt in 20 years would have recreated a fertile patch for replanting or, if untouched, in 50 or so years would have blended with the surrounding forest and be almost unrecognisable as regrowth.

Contrast this with current practice by subsistence farmers around the mountain range. In many situations not even eight year intervals are allowed, the area is unlikely to be surrounded by mature forest and ultimately will revert to poor pasture.

The coffee barons first considered northern facing slopes as being the most desirable for the initial planting of coffee bushes, as these slopes would receive 12 months of sunshine or high light. If this was the case, we find it curious that a forest under-storey shrub would give its best in such conditions. Be that as it may, a coffee plantation in such a position, worn out and now reverted to poor pasture is the very worst situation for natural regeneration. Most essential micro-organisms will have been burnt out by the sun and most remaining topsoil eroded away by the heavy summer rains. It will take an enormous number of years to recover naturally. Again, drawing on hypothetical knowledge, we should imagine the forest in retreat during the long cold glacial periods to ravines containing permanent running water and thus a constant year round humidity. Such forests would be ready to spread out from these refuges when conditions ameliorated. Without a doubt under present conditions this type of niche can be used for a rapid regeneration if, and only if, a swathe of a number of metres is fenced off from cattle, horses and humans. And if so, rapid regeneration will be achieved with little effort. While the meadows around this pasture are obviously also the best and richest, fencing off sections of such areas can only benefit the farmer and guarantee his year round water supply.

Perhaps the second most easily regenerated area is a forest accidentally or naturally burnt, an area that was never used afterwards for any form of agriculture or animal husbandry. We have studied several such situations over a number of years and never cease to be amazed by the contrast between regeneration on the north and south facing slopes. In one such area where a long mountain ridge running east to west was totally burnt 40 years ago, the southern facing slopes have reached stage three, moving on to stage four, whereas the northern slope is barely out of stage one. We believe that the higher humidity, less direct sunlight and lower temperature ranges at ground level are the key factors creating the success on the southern slopes. Both slopes have the same access to seeds from nearby disturbed original forest.

In summary, regeneration of the Organ Mountain Range native forests is possible, but due to the characteristics and topsoil necessities of the main components of the original forests and save in exceptionally favourable circumstances, natural regeneration is the only practical and economic method. In addition cattle, sheep, goats, horses *et al.* must be excluded from the selected areas.

Some situations are very favourable for faster natural regeneration than others, while other areas are heart-breakingly difficult or virtually impossible.

A seed-bank for at least pioneer species, those species that grow well on poor or newly-exposed richer soils, must be adjacent. If seeding of pioneer species is contemplated one must make certain that these are pioneer species common to the region and altitude.

7. The Microenvironments within the Organ Mountain Range

7. THE MICROENVIRONMENTS WITHIN THE ORGAN MOUNTAIN RANGE

Early on in the introduction we discussed our division of the Organ Mountain range into two major segments and eight subdivisions or sections. The steep scarp slopes facing the Atlantic rise principally from the coastal plain to an average height of over 1000M. asl on the whole, which is over 250 km. long and between 3 and 30 km. broad, and forms the first major segment. The anticline, on the continental side, is much broader and contains a number of lesser scarp faces and anticlines and slopes irregularly, almost imperceptibly, down to the Paraiba do Sul River. The dividing line between these two segments is very irregular. The forest's components on either side and to the extent that any original forest can be found on the anticline, are different. Probably before the coffee cycle the difference was not as noticeable as it is now. Regeneration and forest fragments on the anticline show a preponderance of Leguminosae-Caesalpinoideae and many deciduous trees that lose all their leaves over the dry winter months. This phenomenon is much less frequently seen on the wetter scarp slopes.

The differences in the vegetation are caused mainly by rainfall which, on the principal scarp, not only averages twice that on the anticline but also is spread over the full 12 months, consequently creating higher and more constant humidity levels within the forest. Altitude is an important factor as far as orchids are concerned. As examples, we have only once found *Miltonia cuneata* below 1000 M. asl and never have we found *Miltonia clowesii* above this altitude. Exactly the same applies to *Promenaea xanthina* and *Promenaea stapelioides*. The sub-1000 M.asl plants and their higher growing counterparts are morphologically identical except for the flower colour and yet if transplanted to each other's environment all four plants will grow apparently happily. The third obvious differentiating factor between scarp and anticline is temperature. The anticline, being lower, is hotter. As a rule of thumb we use 0.8 of a degree Celsius for each hundred metres of altitude. Thus a temperature of 30 degrees Celsius at sea level implies one of 26 degrees at 500 M.asl, one of 22 degrees at 1000 M.asl and one of 18 degrees at 1500 M.asl. This is why we have included the altitudes where the orchids have been found in all the species descriptions. The cultivator will thus know whether the orchid he or she is caring for requires hot, intermediate or cool conditions. These examples alone, and there are many others, begin to show the complexity of the relationships of orchid species within the forest, the many specific microenvironments contained within and the unknown pollinators, mycorrhiza etc. We certainly do not know much about this micro-complexity and as far as we can see, in the overall context, neither does anyone else, or if they do their findings and conclusions don't seep out to the general public.

In any event, after 30 years of stumbling, crawling, scrambling and sometimes even walking upright in the Organ Mountain's forests and rock faces, we have formed a reasonable idea of where, how and why orchid species are to be found. This of course is where **we found them.** The original anticline forests have been practically eliminated and the climate significantly modified since it was forest-covered, so consequently in these areas, orchid species have been found in places which are not necessarily their optimal habitat but rather a niche where they are simply capable of surviving.

However this is not the situation on the scarp slope. This segment shows a sufficient area of original, as well as maturing regrowth forest for one to be able to assess that where orchids are found, they exist mainly in the microenvironment that they have inhabited for at least the last 500 years. So the information that we have shown concerning the habitat and cultivation requirements should be of great value to horticulturists, growers and conservationists.

As explained, the eight subdivisions we have made of the Organ Mountain Range are merely convenient geographical segments wherein we can locate the species found without giving away their specific locations.

THE ANTICLINE AND ITS SPECIFIC MICROENVIRONMENTS

Large secular relict trees

Deforestation of the north-eastern slopes of the anticline is relatively recent, up until the 1930's, and for various reasons, some large trees survived this holocaust. One finds them in large pastures, around the ruins of the coffee barons' great houses, along roadsides and on long closed railroad routes. Amazingly, most of these trees are repositories of an impressive number of orchid species: *Maxillaria chrysantha, Brassavola tuberculata,* various species of *Catasetum, Schomburgkia crispa, Oncidium forbesii, Encyclia oncidioides,* some of the hardier pleurothallids and a number of smaller orchid species. These relicts also have large colonies of big bromeliads which hold significant quantities of water, condensed from mists and collected from dew and rain, which alleviate the orchids' stress during the long, dry winters. The trees' exposed conditions may well permit strong winds to carry the orchid and bromeliad seeds a very long way and thus give them a remote chance to fall on a suitable host.

Steep ravines starting high in the mountains with a modicum of forest cover on almost inaccessible sides and a permanent stream at the base

These areas, not uncommon, are perhaps the most important havens for epiphytic orchids on the anticline. In an extraordinary way they remind one of how the forests retreated during the ice ages, particularly the most recent. Sadly, when these mountain torrent valleys open out to allow the establishment of even the most precarious of pastures, then they are put to the torch. When the stream becomes a river, against all common sense, trees are removed from the riverbanks and any chance of epiphytic survival is eliminated. Only when the river runs through a steep rocky gorge with scrub trees clinging to its unclimbable sides, can there be a repository for epiphytes. We have found a large number of orchids in these redoubts, many common to those from the wet scarp slopes, but also a cross-section of plants typical to drier climes and lower altitudes, amongst the most spectacular of which are *Laelia pumila, Laelia perrinii, Cattleya bicolor* and *Stanhopea guttulata.*

Oases on the anticline

These areas, which again are not uncommon, are often associated with year-round water courses some hundreds of metres apart and with intervening maturing secondary forest. In one such area only the terrestrial *Oeceoclades maculata* was found while the only epiphyte was *Catasetum hookeri*. Facing any direction, they are often tucked in between two very close, almost vertically-sided, sugar-loaf-shaped mountains. The distance from the watercourse is compensated by water run-off from the steep rock faces. This rainwater comes down in torrents, bringing with it dead and dying lithophytes which become trapped in caverns and amongst the boulders and rock falls. adding seriously to the humus layer at the upper edges of the oasis. Here the ground cover is amazingly thick with large bromeliads, which collect and hold mist and dew as water even during the driest spells. So a very deep humus layer is created, which, sheltered and nurtured by the bromeliad garden, never dries out. The atmosphere is always humid, and trees although sparse, grow fast, due to the water and nutrients brought down in the summer torrents, and thus the conditions have been created to form an oasis.

One of the necessary components or characteristics is that access is very difficult because of the boulder-strewn, cavernous and steeply inclined approach. In addition the terrain must be such that no planting of any crop would be contemplated, and the area must be far enough away from human habitation to avoid the collection of firewood.

In these microenvironments we have found the majority of epiphytic orchids still extant on the anticline: well over a dozen *Maxillaria* species, headed by the mighty *M. chrysantha* down to the tiny *M. minuta*. Shoals of *Sophronitis cernua* on the upper branches of large *Leguminosae*, several *Notylia* species, the translucent white *Warmingia eugenii* and masses of *Brassavola* species, *Xylobium variegatum*, *Schomburgkia crispa*, the delicately-flowered *Cochleanthes candida*, half a dozen *Campylocentrum*, the pretty pink-flowered *Ponera striata* and many Pleurothallidinae not found on the scarp slopes. Also some half dozen *Oncidium* species, from the regal *O. lietzii* to the puffball-like *O. pumilum*, not to forget the delicate *Lanium avicula*. There are also many terrestrial species: *Cyclopogon* and *Beadlea*, *Erythrodes*, *Sauroglossum*, all dominated by the massive canes and enormous flower clusters of *Epidendrum paniculatum* and the ethereal beauty of *Eltroplectris roseo-alba*.

The treeless rock faces of the anticline

The Organ Mountain Range anticline is studded with curious pre-Cambrian granite, whale-like or sugar-loaf-shaped mountains, the majority of which are unassailable on two sides at least, except by a dedicated alpinist. However, almost all are strongly populated by bromeliads whose seeds fall into cracks, fissures and faults in the granite rock faces, germinate and anchor the growing plant. It seems that their roots play little part in the food-gathering process but spread around, gluing the plant tenaciously to the mountainside. Nutrients are mainly collected from the air, dust and detritus falling onto the leaves, together with water from rain, mists and dew. As the plants grow, detritus, flowing down with rain, collects around the roots forming a microhabitat where other plants may take root.

Not infrequently one can come across secondary slopes, lower than those almost vertical and which are angled between 30 - 40 degrees. These areas are just too dangerous for cattle or horses to venture out on. In fact the majority of these slopes are usually fenced off to prevent such incursions by domesticated animals as almost certain death is the fate of famished beasts risking the search for fodder during the long dry winter. Bones and skeletons of such venturesome animals can frequently be found at the foot of these slopes. It is here that one finds areas that are the only accessible pristine locations on the anticline. And they are rich in lithophytes. The description of the bromeliads populating the almost vertical rock faces above, holds good for most the lithophytes populating these slopes. The terrain is less harsh, run-off water flows more smoothly, the rock faces are rougher and have more fissures and cracks.

The most important lithophytes are white- and blue-flowered *Vellozia* species, coarse grasses and sedges, mosses and lichens, which form mats, accumulate detritus around their roots and thrive, exposed to all the elements.

The orchids most commonly found in these areas are *Pseudolaelia corcovadensis*, whose seeds germinate on the *Vellozia* shrubs and are often found in large populations. Also the magnificent tall clumps of *Epidendrum robustum* with its honey-coloured panicles, quite the equal of the glorious *E. paniculatum* from the humid oases. *Oncidum batemannianum*, four *Cleistes* species, five *Habenaria* and various *Epidendrum* species. *Cyrtopodium glutiniferum*, *Zygopetalum crinitum v. major*, *Bifrenaria harrisoniae*, *Laelia cinnabarina* are also regularly found along with the curious *Pleurothallis teres*.

A rare microenvironment on the anticline

We came across one very unusual situation, although there are probably more, of a two hectare, possibly remnant, piece of forest on the banks of a large tributary of the Paraiba do Sul River. Amongst many other orchid species, this area had an abundance of *Oncidium baueri* and some plants of *Cattleya guttata*. Both of these species grow in coastal regions and we had never even heard of them growing on the anticline. We seriously doubt that either of them are typical 'hoppers' such as *Brassavola*, *Sophronitis cernua*, *Oeceoclades maculata etc.* which are found at low altitudes on the scarp slopes and then only found again on the anticline at a similar altitude. Could it be possible then that before the forests were decimated, particularly those on the low coast-facing slopes and river sides, that these plants had once extended from the coastal plains, along the rivers and even up and along their tributaries as far as the climate allowed? If this were so, then this particular fragment could be giving us clues as to the original epiphytic content of the lower altitude anticline forests in pre-colonial times. Alternatively could it be true that the so-called "hoppers" were not "hoppers" at all, but migrants from the anticline to the lower scarp slopes via the riverside forests and round to the lower scarp side forest, above the coastal plain?

The mountain tops

A number of mountain tops have been identified as holding scrub vegetation. Mostly they have been torched through uncontrolled forest burning on the slopes below. The original name for Nova Friburgo in the 1820's, was "*Morro Queimado*"- burnt hill, and all the mountain tops towards the interior show the same signs. High mountain field orchid species may be found on some; in particular *Cyrtopodium glutiniferum*, some *Catasetum* species, *Zygopetalum* species, *Cleistes* species, some *Habenaria* and various *Pelexia* species, and quite often, *Laelia cinnabarina*, together with some terrestrial *Epidendrum* species.

Hillside pastures and scrub forest away from permanent watercourses

These environments, which occupy perhaps 90% of the anticline, are virtual deserts as far as orchid species, both epiphytic and terrestrial, are concerned. The low and variable humidity eliminates the possibility of germination for most epiphytes, while grazing animals and leaf-cutter ants will eliminate any precocious terrestrials.

Roadside banks

This is an environment that is worth watching although one normally passes through it at speed in a car or bus. If the wayside banks are sufficiently steep to avoid browsing animals, a walk along any of the roads in a region can be fruitful. Many terrestrial orchid species can be found if the season is right and light or shade conditions adequate. Surprisingly, on dirt roadsides on the anticline we have found *Phragmipedium vittatum* with *Bletia catenulata*, *Prescottia nivalis*, many *Epidendrum* species and large groups of *Habenaria petalodes*, amongst many others.

THE SCARP SLOPES AND THEIR SPECIFIC MICROENVIRONMENTS

This segment's microenvironments are far easier to consider than those of the anticline for the simple reason that the majority described are original. This is in the sense that the microenvironments and the orchid species found there are as they always were and a grower may be sure that the conditions described show the conditions he or she should aim for. In addition, for anybody interested in finding a specific orchid species, it is essential to know where to look for it. There is no point in looking for *Pabstia jugosa* in a dry, open, regrowth forest and even less point of hunting for *Promenaea xanthina* below 1000 M asl, even in original forest.

Most Brazilian orchid species, in any book available to the public, are described as coming from Brazil or better eastern Brazil or even better Rio de Janeiro State, or wow! the coastal mountains in Rio de Janeiro State. The truth is that either the writer does not know, or does not want you to know, the exact habitat of the orchid in question, or does not think that this information is important.

One of the purposes of this book is to show how and where orchids grow in their natural environment and why we should protect such habitats, not only because of the orchids. Through an understanding of the complexities of tropical forest, we may even be able to make comparisons with the now almost mythical forests on the anticline and in this way reinforce our thesis that, come what may, these remaining scarp slope forests must be conserved.

Taking a typical original mountain and montane forest cross-section, from the top of one side of the valley down to the river, we will find the following microenvironments. This is not an imaginary valley but a real one. We start at 1300 M. asl on the Atlantic facing mountain ridge to the west, dropping down to the river at 1000 M. asl.

The high mountain ridges covered by wind-tortured gnarled montane forest above 1000 M. asl.

A thin stemmed resilient bush type bamboo species provides ground cover where there is high light, while all other shady areas are covered by a garden of bromeliad species. These are effectively continuous ponds along the ridge, which is narrow, five to 15 metres wide. This garden drops down some 10-15 metres on either side of the ridge as far as good light allows. The bromeliad garden, collector and repository of water, releases the precious liquid by evaporation during long dry spells, and thus creates the constant humidity so necessary for the germination and growth of orchid seeds.

In this microenvironment a number of tree-top and mid-tree orchid species often grow as terrestrials but are found mainly at head height on the stunted tree branches. *Maxillaria picta, M. phoenicanthera* are the larger *Maxillaria* species but the smaller ones are also well represented, including a half dozen species from the *M. madida* alliance along with the ever-present *M. cerifera*. *Epidendrum* species abound including the beautiful *E. addae*. But the true epiphytic inhabitants of this microenvironment are the *Sophronitis* varieties, in almost every colour form, size, and shape, in their thousands. The creeping *Gomesa glaziovii, Neogardneria murrayana, Scuticaria hadwenii, Elleanthus crinipes,* several *Bifrenaria* species and *Laelia virens* are all typical. Other specific denizens of this region are the long rhizomed *Oncidium ramosum*, which sprawls through and over the low scrub, while amongst the terrestrials typical to this region are *Zygopetalum* spp., *Cleistes* spp., and the quite beautiful *Psilochilus modestus*. Pleurothallidinae are well represented and quite the most striking are the "button orchid" *Phloeophila echinantha* and the unforgettable *Dryadella* species together with a host of *Octomeria* spp.

The low tree trunk and forest floor environments

As one leaves the mountain ridge and plunges almost vertically down the steep slope towards the river, one rapidly enters into a forest dominated by trees 20 to 25 metres tall. The light is medium to low at ground level and undergrowth is sparse. Humidity and temperature are relativity stable while air movement is low. One does not expect a quantity of orchid species in this environment.

Terrestrial species are few but interesting; *Cyclopogon* and *Prescottia* species, *Malaxis excavata*, *Erythrodes* species, and *Govenia utriculata*. On the tree trunks and lianas may be seen *Lankesterella gnomus*, *Oncidium hians*, *O. cornigerum* and *O. truncatum*. Where the humidity is higher as one nears the river, *Promenaea xanthina*, *Epidendrum armeniacum*, *Cirrhaea dependens*, *Houlletia brocklehurstiana* and *Pabstia jugosa* start to appear. Around the riverbank many orchid species are found in the under forest. *Phymatidium tillandsoides* forms football-sized clumps hanging from small branches. *Dichaea muricata* and *Epidendrum paranaense* surprise one by growing downwards, and then upwards to find light, over the fast-flowing river where *Octomeria grandiflora* and *O. tricolor* cover low tree trunks. This is the true home for pleurothallids and *Barbosella* species and most branches are quite covered by these miniature plants.

The mid-tree zone

This zone includes an area from half way up the trunk to a few metres above the first bifurcation and occupies some 10 to 20 metres from the floor. It receives filtered sunlight, a reasonably stable humidity, temperature and wind movement. The branches are thick and platforms for rooting are firm. This is the zone with the greatest concentration of epiphytes and in particular the bromeliads so vital to the stabilisation of the humidity layer. Once again one can only glimpse the richness of the orchid content from the ground and it is only when a typical forest tree falls that one can really appreciate the quality and quantity of the epiphytic garden.

Perhaps a description of a major tree fall would better illustrate these phenomena than an almost endless listing of orchids and other epiphytic plants found.

A large tree, Myrtaceae sp., fell after many days of rain and high winds. We analysed the epiphytic content upwards from the tree's first bifurcation This analysis showed some 240 bromeliad plants, mostly *Vriesia* spp. and ranging from a medium size to large. We calculated that the average bromeliad weighed 2 kilos when dry and when filled with water, held another 2 kilos/liters within their rosettes. The weight of the detritus around their roots as well as the weight of the other epiphytes, gesneriads, aroids, *Amaryllis*, cacti *Orchidaceae et. al.*, together with mosses and lichens we calculated to total some 1350 kilos of which some 600 kilos was water: the most of which was held in the bromeliads.

The majority of original forest trees on both the scarp and the anticline bear large populations of epiphytes, principally bromeliads. The relict trees on the anticline and the carpets of bromeliads in the ravines and oases show evidence of a climate much more humid than today. Assuming that there were some 30 trees, similar to the one analysed above, per hectare, then we have 18,000 litres of water suspended in the canopy of one hectare of forest. If we extrapolate this to the thousands of square kilometres of virgin forest on the anticline, we can see that what was eliminated by the coffee cycle was not only trees and humus but quite as important - a sea: a suspended and shallow sea but a sea none-the-less that regulated the humidity and buffered and controlled the ground temperatures. To quote a popular song from the northeast of Brazil, the coffee barons managed to turn the sea into a desert: 'O mar virou sertão'.

This sea, therefore, together with the underlying humidity from the forest floor, constant mists, dew falls and rain and no competition from terrestrial plants, make the mid-tree zone the ideal environment for those plants that have evolved to populate it: principally the orchids. Over two thirds of the epiphytic orchids in the Organ Mountain range scarp slope forests have a preference for this niche.

The tree top forest canopy

Little is known about this area in the high mountain rain forests because of its inaccessibility. On a still sunny day one can get a fair glimpse of it from high mountain ridges where the fall slope is very steep. This viewing will be from 20 metres or so on the horizontal but even with binoculars one can only get the feel that it is an active and unique ecosystem, replete with its own insect life.

When a mature tree falls, or even a large crown branch, and one can examine it immediately on falling, then a little more is revealed. On the very top amongst the leaves obviously there is little epiphytic growth. However, a few metres down along the branch, the epiphytes start to appear; seedling bromeliads and orchid species, many of which will only flower when the branch thickens. The really typical species in this area are *Sophronitis coccinea* (at over 1200 M.asl), *Zygostates multiflora*, *Rodrigueziopsis microphyta*, *Oncidium cogniauxianum*, *Bulbophyllum napellii*, some very small pleurothallids and even some *campylocentrum*. It may be, as some experts have speculated, that in terms of nutrients this area in fact is one the richest due to leaf exudates, constant dust falls and even nitrates, formed by lightning and carried by rain.

THE SCARP SLOPES FACING THE ATLANTIC, DROPPING DOWN TO THE COASTAL PLAIN WITHIN ORIGINAL FOREST BELOW 1000 M. asl.

The low tree trunk and the forest floor environment

We start from same point on the same ridge at 1300 M. asl whose characteristics have already been described. We only have to move some five metres to make the equally steep descent towards the east. For the first 400 metres the forest and its contents are almost a replica of those on the western slope, however, from this point on the differences start to become apparent.

The trees are thicker and much taller. The under storey forest is sparse. The terrestrial orchids at 600 M. asl are *Sarcoglottis grandiflora*, *Warrea warreana*, *Erythrodes arietina* and the saprophyte, *Wullschlaegelia aphylla*, while on boulders and tree trunks *Promenaea stapelioides*, *Xylobium variegatum*, *Cirrhaea dependens*, *C. loddigesii* and *C. longiracemosa* are found together with *Epidendrum mantiqueirianum*, and the lovely *Huntleya meleagris*.

The mid tree zone in original forest

This area starts at 15 metres or so from the ground, depending on light penetration, up to 10 metres above the first bifurcation. As with its high mountain counterpart and for the same reasons, this is the favoured region for orchids. The variety is not so great, but the fewer orchid species are found in larger concentrations. In particular, large groups of *Laelia crispa* from the first bifurcation and upwards, which look like pockets of melting snow at their February flowering time. *Miltonia clowesii* and *M. flavescens* show the same characteristics but in fact can cover even more area. The light allowed into the forest along the banks of a large mountain torrent, exposes tree trunks and permits the mid-tree orchid species to become visible. It is then that *Miltonia clowesii* and *M. flavescens* cover such trunks from top to bottom; a truly awesome sight during the flowering period. The light exposure also brings *Gongora bufonia*, *Epidendrum difforme*, a very large variety of *Octomeria grandiflora*, several *Oncidium* species, *Epidendrum hololeucum* and huge clumps of *Elleanthus brasiliensis* into clear view on the lower trunk. *Saundersia paniculata*, *Ornithidium parviflorum*, *Oncidium trulliferum*, *Maxillaria monantha*, *M. ferdinandiana* and *Binotia brasiliensis* are the most interesting occasional species. The most curious is a *Vanilla*, which starts life beside a tree trunk and climbs up to 20-25 metres before flowering.

The tree top canopy at 600 M. asl and below

The trees are much taller and their trunks straighter and thicker while the first bifurcation in general occurs at 25 metres or even higher. From the ground then, it is quite obvious that epiphytic species can hardly be seen let alone identified. Curiously it is very unusual to find a fallen forest giant or even a large branch. We have never had the luck, or rather misfortune, to accompany the felling of an original forest for planting coffee, and the few small branches we have found on the ground have had few epiphytes. Where we have found them, the most notable were *Isabelia virginalis*, *Sophronitella violacea*, *Maxillaria desvauxiana*, several *Octomeria* species and young plants of *Miltonia clowesii*.

A relect tree. Trajano de Moraes, RJ.

REGROWTH FORESTS

Natural regrowth

In the chapter dealing with regeneration we have discussed the varied stages through which a forest may regenerate naturally. During stages one and two no epiphytic orchids will appear save in exceptional circumstances. Terrestrial species such as *Sauroglossum*, several *Epidendrum* species, *Prescottia*, *Cyclopogon*, *Liparis nervosa*, and *Habenaria* species are the probable pioneers. In stage three, when humus and humidity have built up, bromeliads form a terrestrial carpet and together with mosses, are epiphytic up to three metres in height. Only then do the first epiphytic orchids appear. Subsequent colonization by orchids depends on the circumstances, such as neighbouring seed banks, location etc., but if conditions are ideal then the process can start.

Surprisingly four or five *Oncidium* species are the first pioneers, above 1000 M. asl; *O. hookeri*, *O. crispum*, *O. forbesii*, *O. marshallianum* and *O. ramosum*; while *Grobya amherstiae*, *Bulbophyllum* spp., *Gomesa crispa* and *G. recurva*, *Capanema thereziae* and a half dozen hardy Pleurothallidinae also occur. Further down the scarp to 600 M. asl, particularly in orchards or abandoned coffee plantations, *Phymatidium* species, *Comparettia coccinea*, *Rodriguezia obtusifolia*, *R. bracteata* and several different pleurothallids are typical first comers.

Assuming that the regenerating process continues or has continued, and the humus levels

First stage of regeneration.

grow while average humidity levels become more constant, then bromeliads start to appear in the mid-tree zone. Invasion by other epiphytic species is slow and depends on a series of factors. Perhaps the most important are the proximity of seed banks, prevailing winds, the tree species mix and so on. We have suffered quite a number of headaches trying to rationalise why maturing regrowth forest, up to 400 M. asl on the scarp slopes, shows such a paucity of orchid species at the mid-tree zone while original forest from 600 M. asl upwards shows the expected richness. We have tentatively concluded that the prevailing winds, from the southeast , blow over an almost treeless coastal plain with no seed banks in their passage. On the other hand we may have misread the maturity of this forest. In any event the only orchid species guaranteed to be found in this zone are *Oeceoclades maculata*, occasional *Oncidium* species and a few *Gomesa* spp.

Roadside banks

From 1200 down to 800 M. asl, roadside banks hold many interesting orchid species. This area in fact is an artificial high mountain field. The vegetation, often for many metres above the road, is cut back every year producing a man-made replica of a natural environment normally found above 1800 M. asl in the Organ Mountain range. Most of the terrestrial and lithophytic orchid species from these regions and indeed also from the high mountain ridges, seem to thrive in this artificial environment. *Oncidium blanchetii, O. batemannianum*, four *Zygopetalum* species, *Laelia cinnabarina*, many *Cleistes* and several *Habenaria* species as well as various *Epidendrum* spp. are quite common, as are *Eulophia alta, Stenorrhynchus lanceolatus* and two very hardy epiphytes, *Gomesa crispa* and *Comparettia coccinea*.

So if you are interested, get out of your fast air-conditioned car and walk a few hundred metres; you may be pleasantly surprised.

8. Reintroduction of Orchid Species

8.   REINTRODUCTION OF ORCHID SPECIES

Our observations on natural reintroduction and our experiences with artificial reintroduction may help readers make up their own minds of what can and should be done in tropical forests. The subject is divided into natural and artificial reintroduction into original or regrowth forest. Natural reintroduction needs no justification being simply our observations for 30 years. However, our own artificial reintroductions do.

We have three main objectives for artificially reintroduced plants. Firstly to re-establish populations of orchids taken from deforested areas. Secondly to boost local populations of healthy but sparse and threatened species which may have been over collected in the past. And thirdly and perhaps most importantly, to see how reintroduced orchid species react and adapt, particularly when put into regenerating forest at various stages of that forest's maturity.

NATURAL REINTRODUCTION OF ORCHIDS INTO REGROWTH FOREST

Scarp Forest

We have studied, for 30 years, areas of regrowth and watched the gradual arrival of orchid species. In the early stages of regeneration, epiphytic orchids are found growing as, and alongside terrestrial species. Species of *Encyclia*, *Bifrenaria* and *Oncidium* are typical epiphytic components with *Prescottia*, *Cranichis*, *Sauroglossum*, terrestrial *Oncidium* species and *Zygopetalum* among the terrestrials. When the canopy begins to close after 20 years or so, and the humidity within the forest rises, mosses and bromeliads begin to flourish on the trunks and seedlings from the above-mentioned epiphytes start to appear, while their terrestrially-growing parents, now in darker and damper conditions, start to degenerate.

After about 30 years, when the regrowth trees are more mature and flowering regularly, birds and bats are attracted to the flowers and fruits from their homes in the neighbouring original forest fragments. They feed and leave their droppings which contain a range of seeds from the original forest and these serve to produce a sudden increase in the diversity of the growing forest. There are also special niches for orchids, particularly on the trees that die in the overcrowded and competitive regrowth. *Grobya amherstii* and *Catasetum* spp. are mainly found associated with rotting wood often along with *Eurystyles* spp., and if tree ferns are present, they will generally have their associated *Zygopetalum maxillare* growing on the trunks.

However, even after 35 years of growth exposed to the prevailing winds from a neighbouring original forest seed bank and constant visits from birds and bats, the population range of epiphytic orchids is small, amounting to less than 10% of the orchid flora from the original forest nearby and composed of the orchids we would class as terrestrials or, if epiphytic, pioneers.

Natural regeneration of forest. Nova Friburgo, RJ.

Anticline Forest

Original forest fragments are almost non-existent or at best much rarer on the anticline and the patterns of natural reintroduction show a smaller range of species than on the scarp. What is clear is that existing seed banks are few and far between and further, that these seed banks are themselves mostly in regrowth forest. By definition then it is quite possible that their orchid content reflects a fraction of the original populations found in the now almost mythical anticline forests. As far as we can calculate, over 300 orchid species have been eliminated from the anticline by human activity in this area. The range of species that succeed and flourish is described under The Anticline and its Specific Microenvironments.

NATURAL RECOLONIZATION OF ARTIFICIAL SITES - ROADSIDE BANKS AND CUTTINGS

Scarp Slopes

Roadside cuttings and old railroad banks offer a special opportunity to study recolonization and such areas can be especially rewarding. On one 18 km. stretch of mountain dirt road on the scarp slopes in section VI between 900-1200 metres altitude, 54 species of orchid were found and identified (Miller & Warren 1994). Curiously, although these banks are fully exposed to the nearby original forest, the majority of orchids found are not from such forest but from the high mountain fields and ridges above. Of the 54 species found, nearly 75% were understandably terrestrials, including nine species of *Habenaria*, nine of *Cleistes* along with *Cranichis, Prescottia, Cyclopogon, Stenorrhynchus, Eulophia, Pelexia, Galeandra* and *Oncidium blanchetii*. The epiphytic species found were mainly associated with the forest around.

Anticline

Here the lack of original forest both as cover and seed source combined with ever present and starving grazing animals results in the sparse colonization of banks. However, there are sometimes surprising finds of spectacular species which may well have blown in from the interior. Here we have found substantial colonies of *Phragmipedium vittatum*, growing with *Bletia catenulata*; *Prescottia*, and many *Epidendrum* and *Habenaria* species, but very few epiphytes.

Photo: Inghra Scart.

Road side banks. Nova Friburgo, RJ.

ARTIFICIAL REINTRODUCTION INTO ORIGINAL FOREST

Laelia crispa virtually disappeared from a high mountain valley in section VI because of over collection and the destruction of the original forest from 800-1000M. asl around the base of the valley. This is about the top altitude for *L. crispa* in the region. Subsistence farmers had cut down some two hectares of original forest in a contiguous valley at around 700 M. asl and some 10 km. distant. We visited this site and found most of the felled trees to have large colonies of *Laelia crispa* in the high forks. So we borrowed four mules, complete with eight panniers and a muleteer, and in two trips to a dirt road, we carried hundreds of plants out, filled our pick-up truck and freighted all back to our headquarters. This was in 1976. We left thousands of plants behind to rot or be burnt.

The original but disturbed forest around our headquarters is between 1200-1400 M. asl where *Laelia crispa* is a rare plant. We fixed all the colonies in suitable light and wind conditions onto low tree trunks.. It took a week. During the last 26 years, all plants have flourished, flowered well and produced seed without artificial pollination. Visitors to the flowers include humming birds, *Eulaema*, bumble and carpenter bees. Seedlings appeared in profusion above and below the parent plants and on neighbouring trees. Their growth was slow and the attrition rate very high. The earliest seedling to flower was ten years later.

Third stage of regeneration, *Miconia cinnamomifolia.*

ARTIFICIAL REINTRODUCTION INTO REGROWTH FOREST

Terrestrial species

A technique we have used to boost populations of species that are probably sparse through over collection is to raise and replant laboratory grown seedlings. It is important that the seed used for this is collected from local plants growing at similar altitudes and in similar conditions and not from any old plant or cultivar from an orchid grower's collection. The objective is not so much to introduce the laboratory-raised plants *per se*, but rather to boost a breeding population, thus increasing the chances of seed being produced and spread naturally. This has been successfully used for both *Oncidium blanchetii* and *Laelia cinnabarina* and a second generation of seedlings from these plants has now appeared.

When dealing with laboratory-raised seedlings, there are some crucial steps. Firstly, when taken out of flask, the seedlings need protected incubation until they have produced a new and significant root system, as the roots feeding the plants in flask are often immature. Ideally, they would be grown in pots in a vented incubator in a humid atmosphere for a month or so. The incubator can be removed after a month and the plants gradually hardened-off by increased exposure to light and air movement. When they appear vigorous with new growths, they can be transplanted around the parent plants. Out of 19 *Oncidium blanchetii* seedlings raised and transplanted from one flask, 17 flowered within two years and two died due to animal activity.

The conclusion: transplanting certain terrestrial species into regrowth forest or high mountain fields presents few or no problems provided the seedlings are weaned properly.

High mountain river, near source.

Photo: Melissa Bocayuva.

Epiphytic species: *Laelia crispa*

Our experiences with rescued *Laelia crispa* plants relocated into regrowth forest have been the most illuminating. Firstly, the plants themselves are difficult to establish on their epiphytic host trees, unlike their counterparts in original forest which had more or less a total success rate. If attached to normal, healthy young trees, the plants generally failed to thrive, producing only a few roots, small shoots and no flowers. If, however, the plants were attached low enough on the trunk that the new roots soon encountered soil, then the plants grew swiftly, flowered and produced seed. Also, if the plants were attached to parts of trees that had dead wood or, indeed, if the tree itself was dead or dying, then again, the plants produced masses of root, grew and flowered swiftly, producing seed after natural pollination.

We have monitored one particular colony over 22 years and each year it has produced seeds which have been shed in their millions over the moss-covered bark and humus around. To date, not a single seedling has been found, thus illustrating that regrowth forest is either inhibitory to seedling growth or one of the vital components (humidity, fungus, pH) is lacking. This ties in with our observations in Section 1. (Natural reintroduction,) that the only epiphytes found in regrowth are pioneer species. We believe that a consistently high humidity is the missing factor and lack of this prevents the seeds of both the absent 90% of species from the nearby original forest and of the reintroduced *Laelia crispa* from germinating. In time, and given uninterrupted development, the regrowth forest should provide the correct environment and this would be demonstrated by a sudden flush of *Laelia crispa* seedlings and the influx of new species from original forest.

APPLICATIONS OF REINTRODUCTION

Recreation of a Microenvironment

Cattleya harrisoniana is a beautiful magenta-flowered plant, once common in the wetlands on the periphery at the base of the Organ mountain scarp. Drainage of the wetlands, 25 years ago, turned the area into pasture and arable land, eliminating the swamp-loving stunted trees and shrubs, host to the orchid. What plants were left on isolated trees were soon collected by small boys for roadside sale. An NGO, managing a large tract of forest and flat land has recreated a significant area of wetland to attract back the original birds and animals. The first reptiles returned were alligators. The indigenous trees and shrubs planted will soon become host to *Cattleya harrisoniana*, colonized from plants which were collected and placed on trees over an artificial lake many years ago. This is a brilliant example of recreating an original environment, complete with the orchid species. There is the added advantage that the alligators will act as a stern deterrent against collecting.

Other potential applications

Various orchid species on the anticline would benefit from the addition of laboratory-raised plants to their impoverished and over collected stock. One can think of many species careering towards endangerment and extinction where this technique, described for *Oncidium blanchetii* above, would help. *Laelia perrinii*, *Laelia pumila*, *Laelia fidelensis*, *Cattleya dormaniana* and *Cattleya velutina* are all species that fall into this category.

There are two important criteria for this type of reintroduction to work. Firstly the plants concerned must be placed in fragments of original forest for the successful spread of seedlings. Our experience in regrowth forest with *Laelia crispa* has shown that you can successfully reintroduce plants but seedling growth will only occur in original forest. And secondly, since the populations of these plants are sparse because of over collection, this work can only succeed with effective protection of the sites. This will involve the daunting task of encouraging the cooperation of landowners, environmental authorities and Municipalities.

9. Basic Orchid Morphology

9.   BASIC ORCHID MORPHOLOGY

Orchids grow on every continent except Antarctica. In temperate northern and southern forests and meadows they are terrestrials with their roots or tubers permanently submerged in soil. However, the majority of orchid genera and species are found between the tropics and are epiphytes, growing on other plants. The epiphytic state includes many and various habits.

Twig epiphytes live in the tree-tops and since life on a twig is precarious, as a percentage fall each year, their life cycle is short and they grow, flower and set seed in less than one year. E.g. *Capanemia* spp., *Phymatidium* spp. and *Thysanoglossa organensis*. Likewise the twig epiphytes such as *Rodrigueziopsis microphyta* and the exquisite small *Oncidium, O. cogniauxianum* and *O. hookeri*, which again have short lives, clinging to the topmost twigs and bleached by the sun. A proportion of these, broken off by wind, fall to the forest floor where they may flourish as lush green colonies, but flowering less frequently in this alternative life.

On the larger branches we find colonies of bigger plants like *M. picta, M. leucaimata* and *Miltonia flavescens, M. cuneata* whose fine roots form a tight network clinging to the branches. Also the larger *Oncidium (O. crispum, O. marshallianum)* whose thick roots extend for many metres either side of the plant, glued firmly to the bark.

A special case are the 'suicide plants' like *Catasetum* spp. and *Grobya amherstiae* which are almost always found growing on or near rotting wood. Here the roots penetrate, branch and proliferate, devouring the nutrients released by the decaying material until a branch may appear as a bag full of worms. Naturally, these branches soon fall but not before the plants have come rapidly to flowering and have shed millions of seeds to germinate on like sites.

On the main trunks in original forest, few species are found unless high light penetrates on the forest edge, or maybe on exposed trunks by river banks, as most orchids are growing high, above the first bifurcations. However, in the relative gloom we do find *Promenaea xanthina, Pabstia jugosa* and *Cirrhaea* spp. among others.

Although in tall deep forest most orchids are found in the upper branches and canopy, there are specific situations where plants are much more accessible and these areas are the most fruitful for orchid hunting. On high mountain ridges, the trees are small and stunted - the so-called elfin forests - and on such trees we can observe orchids at eye level. Similarly, along river banks, where the canopy descends to eye level, orchids can be easily found and studied.

A small proportion of orchid species are pioneers and able to penetrate and colonize regrowth forest, and in certain situations, like abandoned coffee plantations and citrus orchards, they may spread dramatically. We have seen huge populations of *Rodriguezia bracteata* growing in every available niche, tree or rock, in abandoned orchards, and *Warmingia eugenii* and *Notylia* spp. in profusion on old, long abandoned coffee bushes.

All orchids are exposed to climatic stresses and each has adaptations for dealing with this.

STORAGE ORGANS

Because of the climatic stress that most orchids and all epiphytes are exposed to, each species has adaptations to it's anatomy and morphology to conserve water. The most obvious orchid storage organ is a modified, thickened stem: the pseudobulb. But let us first look at the adaptations shown by those plants that lack pseudobulbs.

The many genera of pleurothallids mostly have strap-shaped leaves, covered with a thick cuticle which prevents water loss. Terete or cylindrical leaves, which have a reduced surface area, also help to conserve water but are not common in this group. Many of the epiphytic Spiranthoids like *Lankesterella* spp. and *Eurystyles* spp. which seem so slight and vulnerable, nevertheless have a water retentive cuticle and often large, fleshy roots which have a storage role. The tiny *Phymatidium* and *Thysanoglossa* species have fine, almost fibrous leaves, thick roots and these are almost always submerged under moss and lichen. Species of *Campylocentrum*, the only American *Angraecoid*, have thick stems and very fleshy roots which, in some of the leafless species, take over the function of photosynthesis.

Tropical terrestrial orchids also have anatomical adaptations to conserve water. This may be a root tuber as in *Habenaria*, and the roots themselves may be thickened as in *Cleistes*, or there may be modified underground stems adapted for storage as in the corms of *Eulophia alta*, or even with pseudobulbs such as *Oeceoclades maculata*. The rhizome itself, the length of stem that joins successive years' shoots, may also be thickened and serve as a storage organ. Above the rhizome, the shoot may be thickened into the familiar pseudobulb form. Each joint in the stem between the leaves is called an internode, and where the leaves attach is called a node.

STEMS AND LEAVES

Under this heading we include rhizomes, pseudobulbs, corms and leaves but we will not attempt to describe the extraordinary range and variety of form found in the Orchidaceae of Brazil. Most South American orchids are sympodial in their growth habit, each shoot being of limited growth and terminating in an inflorescence. A similar shoot arises during the next year from a dormant axillary bud: *Cattleya* and *Laelia* species are typical examples. This habit is believed to be primitive compared to the monopodial form found in the Vandeae, where the stem grows continuously from a single apical bud, with occasional lateral buds also developing. *Campylocentrum* is the only Brazilian monopodial genus although there are monopodial species in the genus *Zygopetalum* and others.

Rhizomes

A rhizome is a length of stem joining successive shoots. It may be underground or on the soil surface in terrestrial species and, in epiphytes, it either lies on the bark surface or hangs in a pendulous fashion – *Maxillaria loefgrenii*. In some small twig epiphytes, it may be long and wiry and straddle across branches as in *Rodrigueziopsis microphyta* or so short as to be barely visible as in *Chytroglossa*. Generally rhizomes are covered with sheaths formed from the dried up, brown bases of the old leaves. The roots are produced from the underside of the rhizome and these can anchor the plant to its support. Rhizomes can be short as in *Oncidium forbesii* or *Maxillaria desvauxiana*, causing the pseudobulbs to be clustered; they can be of medium length as in *Sophronitis coccinea* or long and strung-out as in *Maxillaria loefgrenii*. Among the mainly pseudobulbous *Zygopetalum* species there are two interesting rhizome modifications. *Z. pedicellatum* has no pseudobulbs but instead has a long, creeping rhizome which may reach some metres in length. In this case the rhizome's coating of old dried sheaths encloses the root system. In *Zygopetalum maxillare*, a species always associated with tree-fern, the rhizome lengths can vary remarkably giving the plant two growth forms, dimorphic. Seedlings that grow on a tree-fern trunk produce small pseudobulbs followed by a long stretch of rhizome and then another small pseudobulb. This process continues until the leading growth reaches the crown of the tree fern where short lengths of rhizome are produced. Then the orchid forms clusters and flowers through the tree fern crown.

Pseudobulbs and corms

Thickened sections of stems that are neither technically a bulb nor a tuber are pseudobulbs. They function as storage organs and they have an extraordinary variety of form. Pseudobulbs can be formed by the thickening of one of these internodes as in *Bifrenaria* spp. or *Gomesa crispa*, or from the thickening of a few in *Encyclia vespa* or from many as in *Catasetum, Cycnoches* or *Cyrtopodium*. Stems may have no visible thickenings at all as in the reed-stemmed *Epidendrum* or *Elleanthus* spp. Pseudobulbs may be tiny as in *Zygostates multiflora* and in some *Bulbophyllum* species, or huge ovoid footballs as in *Oncidium hydrophilum*. They may be long and cigar-shaped as in *Encyclia vespa*, involving many nodes, or angular and with squared-off corners as in *Bifrenaria*. Sometimes they are reduced to a rudiment as in the terete-leaved *Scuticaria hadwenii* or *Brassavola* spp., or they may make up the bulk of the plant as in some mule-eared *Oncidium* species. Most pseudobulbs are above the ground but in a few species such as *Galeandra beyrichii* or *Liparis nervosa*, they are buried. Internally, they are all very similar and composed of large water-storage cells. Most species, particularly of the Oncidiinae and Epidendroideae, have two opposite, incipient buds at the base of the stem or pseudobulb. Normally only one is activated and the other remains dormant as a reserve, and only grows should the first be destroyed. This is why, before it was discovered how to germinate and grow orchids from seed, the plants were simply split up and planted separately, so activating the dormant buds and forming new growing plants. Most features of the pseudobulb such as the shape, size, number of involved nodes and their position of the rhizome, are all significant when making an identification using vegetative characteristics alone.

Leaves

Orchid leaves are very varied. Most leaves are produced in two rows, opposite or alternate on the stem. This is not at all obvious when you look at a mature *Laelia perrinii* or a *Cirrhaea saccata* with a pseudobulb bearing only a single pair of leaves. But if you look at a young growth, and can bear to dissect it, you would find that the leaves are arranged alternately in two ranks on opposite sides of the stem, and at the base of each leaf is an axillary bud. Most leaves develop within the bud in one of two ways. There are the concertina-like folded leaves (plicate) that are found in *Cirrhaea*, *Stanhopea* and *Sobralia*. These tend to be thin, with many parallel veins, and often short-lived. Many other orchids have leaves that develop as a folded pair, one within the other, conduplicate, as in *Phragmipedium vittatum*. When opened up, these show the familiar central vein and flattened surface associated particularly with epiphytes. The variation in shape is great and of significance in making identifications from vegetative characters. The shapes can range from long and thin, subulate and linear, to oblong and elliptic, spoon-shaped, spathulate, or spear-shaped, lanceolate. The leaf tips can be blunt or pointed or irregular in shape and the margins too can vary. In cross-section, leaves may approach circular, terete, as in *Octomeria aloefolia*. For a fuller glossary of all the leaf shapes and their tips and margins. look at The Manual of Cultivated Orchid Species, Third Edition, page 571 (Bechtel, Cribb & Launert, 1992).

Leaf bases also have a part to play. They may completely enclose the young stem and give it support while it is weak and growing. And when the leaf falls, the sheaths may remain as a dried coat around the stem or rhizome along with the remains of small rudimentary leaves. In some species, these sheaths enclose and protect roots e.g. *Maxillaria acicularis*, while in others they collect water and act as a reservoir.

The length of time that leaves are retained on the plant is related to its vigour. The main functions of the leaves are to collect light and to exchange gases through the tiny stomatal pores. There is a point in a water-stressed environment like the tree-top, where the value of many leaves photosynthesising is outweighed by the amount of water they lose through the stomata. This is probably why, in many epiphytes living in exposed conditions, only a pair of apical leaves is retained whereas, in moister, less stressful situations a plant of *Oncidium baueri*, for instance, may retain the leaves on the previous five years' pseudobulbs. A similar comparison can be made between epiphytes and specimens of the same species growing terrestrially in regenerating forest. The terrestrial specimen which receives more moisture from decaying wood and detritus, always retains leaves for more years than its epiphytic counterpart.

ROOTS

This account will not concentrate on the details of anatomy and fine structure which may be sought in any good orchid text. Rather, we will look at those aspects of roots which we have been able to examine and explore in the plant's habitats. It is significant that the plant system which is most important to the orchid is also the system about which least is known. We rarely see what roots do inside flower pots until it is repotting time, then we cut them off! And dried herbarium specimens tell us even less. We know something about the structure of roots, but less about their functions in adhesion, absorption and the interactions with their mycorrhizal fungi.

Anatomy

The roots of many epiphytes have an outer coating of dead, white cells which may be thick, called the velamen. This contrasts with the vivid green (or sometimes red and orange) of the living, growing tips. Inside the velamen and protecting the inner living tissue, is a layer of dead, waterproof cells called the exodermis. At intervals in this barrier are small living cells called passage cells. How do these layers work?

Roots of *Campylocentrum* sp.

Function

At night, when dew falls or after the first rains, a dilute mineral solution is washed down the trunks of trees to be trapped by the orchid roots and absorbed by the dry velamen. Water and nutrients then fill the velamen and pass through to the passage cells and from there into the vascular system of the plant. Some simple measurements we made recently using conductivity meters demonstrate the importance of the first rain washings after a prolonged dry spell. We collected rain water, both out in the open and from the trunk of a tree, down which the rain had travelled. Runoff contained six times more dissolved solids than pure rain water. Since most epiphytes on healthy trees get their main supply of nutrients from dew and rain, it is interesting to see how they interrupt this downward flow. Species such as *Promenaea xanthina*, *Sophronitis cernua* and *Miltonia clowesii* have colonies whose roots often girdle the entire twig, branch or trunk so that any passing water must flow over or through the colony, the nutrients being extracted on the way. *Oncidium*, *Gomesa crispa* and some *Epidendrum* species have roots that tend to run for great lengths along the lengths of branches and trunks so that any flow of water would trickle along the length of the roots until they become saturated.

But what happens in the canopy? How do the tiny leafless *Campylocentrum* species and the diminutive *Barbosella* plants feed? Although the impression is that the run-off water gathers nutrients as it travels along the bark, this is not the whole story. Work by Tukey and coworkers (Tukey, Wittwer and Tukey, 1958) showed that you can leach practically any nutrients from leaves using a fine mist spray of water. This suggests that the dew-soaked canopy is probably the richest nutrient zone where even the smallest twigs are well supplied and it is probable, from the occurrence of fungal fruiting bodies in the canopy, that there is an entire nutrient cycle that occurs there, totally independent of the forest floor. This also suggests that the run-off collected at the base of the tree may also be relatively nutrient-poor since it will have passed over and through the colonies of epiphytes which are so efficient at extracting nutrients.

Another important aspect of orchid roots is their ability to adhere to their epiphytic or lithophytic supports and this feature, along with root size and number, gives clues to their correct identification from vegetative characteristics. Some, like *Laelia cinnabarina*, stick with their own glue to the rocky subsoil. Roots of *Oncidium* spp. stick for great lengths along branches. Little is known about the mechanism of this adhesion but it does seem to be independent of the mycorrhizal fungus since many species have roots that can stick to the glass and plastic walls of the culture vessels *in vitro*. Adhesion is related to the plant's ability to absorb because many species will not thrive until their roots have actually stuck to some substrate.

Mycorrhiza

Orchids will not germinate on their own but need the association with a fungus to start to grow. Strictly, this does not concern the roots as the first association is between the fungus and the developing orchid embryo. The fungal *hyphae* penetrate the embryo and for a while the orchid is parasitic on the fungus, although on occasion the fungus can devour the growing plant. As the successful seedling grows it controls the fungus which, however, remains in the roots of most adult plants. In some genera like **Cattleya**, an annual reinfection of the new root occurs. Fungal infection of seed seems a hit-or-miss affair and the huge quantities of seed released from a capsule would tend to confirm this. However, one of us (RW) worked many years ago on the fungi living on plant surfaces. While leaves and bud scales on temperate trees carried huge fungal populations, so did bark and among the cultures of fungi isolated were many weird and wonderful cultures - still waiting to be named. Although temperate trees have little relevance to orchid studies, the density of fungi, about 40 million / sq. cm, is a clue to what seeds, arriving on bark, may meet. Certainly, our own observations about seed germination in nature would confirm that germination is high and the high attrition rate occurs well after germination.

Root form

Roots come in a multitude of varieties but to help in making identifications from vegetative characteristics we have divided them into four gross categories:

1. Thick and many, for example *Zygopetalum crinitum*, *Epidendrum hololeucum* and *Laelia* spp;

2. Thick and few, as in *Cleistes*, *Eurystyles* and *Sophronitis*;

3. Fine and many as in *Stelis*, *Gomesa* and *Miltonia* spp;

4. Fine and few, as in *Oncidium truncatum* and *Masdevallia*.

An extreme example of 3 is found in the thick, hessian-like pad of roots formed by *Miltonia cuneata* the 'nest-epiphyte'. Then there are the text-book velamen-coated roots of *Laelia crispa* and *Campylocentrum* which, in *Epidendrum proligerum* and *Grobya amherstii* appear as grotesquely swollen bags of worms. Highly swollen roots are also found in the terrestrial *Zygopetalum* species and in some epiphytes like *Elleanthus* and *Sobralia*, while in *Isochilus* this swelling is accompanied by a reduction in the shoot. Extreme root development is found in the leafless and near leafless species of *Campylocentrum* where the shoot is much reduced or absent and the roots are large and fleshy Among the data we have collected, we have calculated the lengths of root found on typical specimens; some of these are listed below.

Root Quantity

Length of roots by pseudobulbs and by plant for the following orchid species

SPECIES	HABITAT	ROOTS/SHOOT	ROOT LENGTH	ROOT LENGTH (M/PLANT)
Neogardneria murrayana	epiphyte	5	20 cm	15 m/15 bulb plant
Encyclia inversa	epiphyte	2-4	15-25 cm	2 m/ 5 bulb colony
Laelia cinnabarina	lithophyte	10	20 cm	12 m/ 6 bulb colony
Oncidium hydrophilum	terrestrial	14 (branched)	30 cm	30 m/ 6 bulb colony
Oncidium longipes	epiphyte	20	140 cm	7 m/ 5 bulb colony
Promenaea xanthina	epiphyte	3-5	150	5.1 m/ 10 bulb colony

Ducted Root Systems

A curious system is found in some of the aciculate *Maxillaria* species (*M. madida* alliance) such as *M. acicularis* and *M. cogniauxiana*. These plants often form drooping colonies on the trunks of trees or the undersides of branches in the mid-tree region. Pabst and Dungs write of this alliance as *'plants usually pendent, fixed only by their basal root-bearing pseudobulbs, the subsequent pseudobulbs being rootless'*. The colonies, which may contain hundreds of pseudobulbs, certainly are fixed and anchored by ropes of roots which, after reaching the substrate, branch out fan-like to form a very effective anchor. Lengths of 14-20 pseudobulbs may be hanging free from the tree, joined together by a rhizome which is covered by overlapping brown scales.

However, if these scales are dissected away, quantities of white, parallel-running roots are uncovered, closely bound along the rhizome and running back along the plant for lengths of 40 cm. or more. These roots have little groups of swellings on the sides facing the rhizome which may be seen with a hand lens. On the oldest colonies and at the point of attachment to the tree or branch, the rhizome may be absent and is replaced in function by a rope of roots.

Laelia crispa with off-spring.

A similar hidden root arrangement is found in the quite unrelated *Zygopetalum pedicellatum* which has no pseudobulbs, but instead, a long scale-covered rhizome which sprawls upward through the supporting scrub for lengths of two metres or more, with its short apical tuft of leaves bending upwards towards the light. If the scales surrounding the thick rhizome are dissected away, the large white and fleshy roots are exposed, closely adpressed to the rhizome surface. Only at the point where the rhizome enters the soil do the roots appear and spread into the leaf litter where they form an extensive and penetrating network.

These two species, *Maxillaria acicularis* and *Zygopetalum pedicellatum*, although unrelated and very different in structure, have in common this specialised root ducting system which allows them to grow well away from their substrate without suffering water stress. It is also probable that water conducted by these covered roots is not all carried in the normal fashion by the vascular system in the roots, but instead it is carried on the outside of the roots by capillary action. Certainly, whenever these roots are exposed by removing the scales, the surfaces of the roots are always wet.

LEAF GATHERERS

Many tropical orchids have strategies that enable them to collect nutrients. In some plants the upright posture of the pseudobulbs and leaves provides a rake arrangement where passing leaves and twigs become entrapped and slowly decay. In *Catasetum*, the roots may be found growing vertically upwards to form another sort of leaf trapping structure, and in the next section we will see how an upwardly angled rhizome allows the freshly developed roots to trap leaf litter and fallen detritus and to benefit from their decay. This trapping is also seen in terrestrially growing specimens of plants that normally are epiphytic. *Encyclia vespa* and especially *Encyclia calamaria* produce long 'prop-roots', when growing as terrestrials, which branch when they reach the soil so both stabilizing the plant and creating a grid against which all manner of detritus becomes trapped and retained while it decays. *Pleurothallis limae*, naturally a terrestrial species, also collects leaves but using its fine root network which grows upwards into the fallen leaves. The fungal network from these roots is apparent and penetrates the surrounding fallen leaves.

Nest epiphytes

This term was coined by Hoehne to describe species that have developed a particular arrangement of roots; *Miltonia* spp. are excellent examples. These species grow at the tops of trunks, around the major bifurcation and produce hundreds of thin, white and branched roots. When a colony is stripped from a fallen tree and the reverse of it is examined, the first impression is that the roots resemble a hessian doormat or a bird's nest, so thick is the network. The erect leaves and old flower spikes at the top of the colony serve to collect all manner of detritus falling from above and this mass decays and ferments among the roots. Each night when dew falls and when it rains, the composted products are washed through the colony and the nutrients are taken up by the active roots and used to build up the massive exuberant colonies that *Miltonia* spp. produce. But it is not called a nest-epiphyte merely because the roots resemble a bird's nest. Frequently, in larger colonies you will actually find birds' nests and even on occasion, a possum's nest, and with all that added nitrogen in an otherwise impoverished environment, it is small wonder the colonies grow so fast and to such sizes.

Tussock formers

Excavating a terrestrial colony of *Zygopetalum mackayi* is astonishing because of the extent and complexity of the root system uncovered. We selected a typical colony and firstly removed the leaf litter from around the new growth during early summer and when the emerging shoot was about 2 cm high. New roots were found which measured 5-20 cm in length growing radially outwards from the colony, horizontally into the leaf litter and trapping leaves beneath them. This trapping is achieved because the new growth is on a rhizome that is slightly angled upwards. The new roots are white and brittle and their pointed tips have a yellowish tinge (Fig. ,1). The previous year's growth is seen to have a more extensive network (Fig. ,2) and this, being slightly lower than the current years growth, has roots running in the leaf mould and litter where there are recognizable leaf fragments. Three-year-old pseudobulbs are a good 4 cm below the level of the newest growth and their branching dark brown roots run into well-decayed, soggy soil-like leaf mould. Each root has about eight branches and some of these have secondary branches. (Fig. ,3) Four-year-old, leafless pseudobulbs, that you might be forgiven for thinking dormant, have roots of an average length of 60 cm, brown, 6 mm diameter and undulating. They run deep into the subsoil, branch after 50 cm length and the secondary branches from these grow upwards into the leaf litter zone. The oldest root uncovered on this plant was attached to a leafless four-year-old pseudobulb. It was 80 cm long and branched through the subsoil. Secondary branches rose into the leaf mould layer where they showed active root tips (page 133). The entire colony had a root system which covered a minimum of eight square metres and had a network of roots measuring a minimum of 50 metres long, all parts of which showed signs of activity. The conclusions are interesting. Firstly root growth appears not to be finite. Regrowth from old pseudobulbs is common as is branching and sub-branching from apparently dormant pseudobulbs. The pattern of root growth from a rising rhizome also allows the newly formed roots to trap and enclose recently fallen leaves and detritus and gradually to form a tussock. The new roots are thus in

Agroup of *Maxillaria picta*.

Photo: Melissa Bocayuva.

the richest, fastest rotting layer of leaf litter which is also penetrated by the smallest branches of the oldest roots. The overall root structure demonstrates a wide, three-dimensional branching network at all levels of the forest soil, anchoring the plant, stabilising the soil while keeping the root tips of the youngest and oldest roots in the fertile zone.

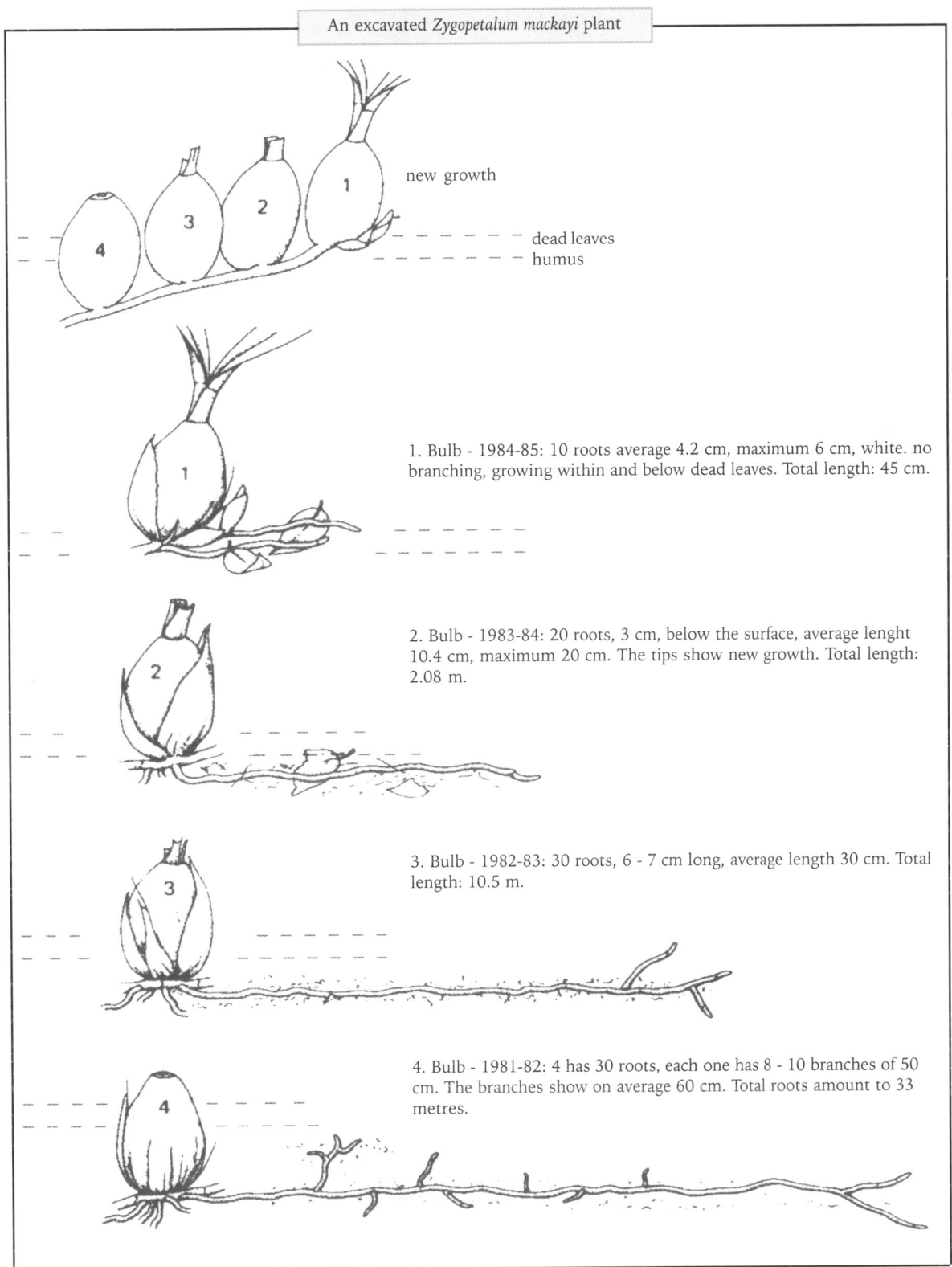

An excavated *Zygopetalum mackayi* plant

new growth

dead leaves

humus

1. Bulb - 1984-85: 10 roots average 4.2 cm, maximum 6 cm, white. no branching, growing within and below dead leaves. Total length: 45 cm.

2. Bulb - 1983-84: 20 roots, 3 cm, below the surface, average lenght 10.4 cm, maximum 20 cm. The tips show new growth. Total length: 2.08 m.

3. Bulb - 1982-83: 30 roots, 6 - 7 cm long, average length 30 cm. Total length: 10.5 m.

4. Bulb - 1981-82: 4 has 30 roots, each one has 8 - 10 branches of 50 cm. The branches show on average 60 cm. Total roots amount to 33 metres.

INFLORESCENCE AND FLOWERS

If there is one word to describe orchids and their flowers, diversity springs to mind. The flowers may be small, ugly and grotesque but they may also be supremely beautiful and exotic. There is also, as every book will tell you, a mystique about orchids. Standing in the rainforest and looking up at a magnificent group of *Miltonia clowesii* in full flower is wonderous. As far as a biological role is concerned, the only known one for an orchid is with a species of *Stanhopea* which is important in the pollination of Brazil nuts. Their role may be more subtle and complex than we realise but at present we can only consider them to be a specialised and exotic evolutionary product highly adapted both functionally to their habitat and anatomically to the form and functions of their specific pollinators. Such specialization means that they are highly vulnerable.

The *inflorescences* may be single or many-flowered and arise from the top of the pseudobulb (*Encyclia*), the base (*Oncidium*) or occasionally both, (*Gomesa*) or it may spring from the young growth (*Pabstia* and *Zygopetalum*). They may be made up of single flowers (*Maxillaria*) or many (*Epidendrum*) and they may be loosely arranged (*Gomesa glaziovii*) or compact and many-flowered as a cone (*Epidendrum armeniacum*). In their structure, the sepals are generally more or less similarly shaped.

In regrowth forest, the establishment of mature *Laelia crispa* plants depends on root growth. If this is limited the plant will not flower and may even die back.

The two petals are also similar to each other. The third petal is highly specialised to form the lip or labellum and because it frequently functions as a landing platform for pollinators, it is orientated by rotating through 180º during its development to present at the flower's lower side opposing the column. Resupinate describes any orchid flower which has the lip placed in such a way, on an erect inflorescence. In flowers with the lip uppermost, the flower bud may have rotated through a full 360º to present in that position which is called non-resupinate, as in *Encyclia inversa*.

The labellum is very distinctive, often a different size to the other petals and placed exactly opposite the fertile anther of the column. Although it is usually larger, it may also, as in some pleurothallids, be smaller; it may also be entire or lobed to varying degrees of complexity, or as in some *Campylocentrum* spp., hardly differentiated from the other petals. The lips are often decorated with outgrowths such as the distinctive calli. In *Oncidium*, for example, the calli may be hairy (*O. longipes*), warty (*O. hookeri*), like powder-puffs, (*O. robustissimum*) or nose-like (*O. crispum*) and this feature immediately assigns the plant to one or other section of that genus. The lips may bear hairs as in *Zygopetalum crinitum* or carry a powder-like *farina* or *pseudopollen* which is collected by bees, as in some species of *Maxillaria*.

Although the labellum is a feature in the asymmetric pattern of the orchid flower, such a modified petal is also found in other families. *Lobelia,* for example, has a lip-like petal arrangement. What makes the orchids unique is the structure of the sexual parts where we find the male and female units fused into a single organ called the *column*. Generally the male anther is found, covered by a cap, near to the tip of the column and contains the pollen fused into soft, mealy or hard *pollinia*. These *pollinia* are often joined by a short strap called the *stipe* or *caudicle* which has a sticky tip of quick-drying glue called the *viscidium,* which attaches to the pollinator. The column is of diagnostic importance as it may bear wings (*Oncidium*) or be extended into a column-foot where the lip attaches; this may also be extended to form a spur - *Galeandra* spp. *and Campylocentrum* spp. which functions as a nectary. The female stigma is generally a shallow depression on the underside of the column which is filled with a sticky substance to which the *pollinia* attach and within which they germinate.

Flower colour may vary within a species, for example, through the various shades of brown of *Oncidium forbesii*, or the amount of yellow in *Oncidium gardneri* to the *alba* forms that we have found in, for example, *Promenaea xanthina*. Scent may also be of diagnostic importance. For example *Zygopetalum crinitum* has a sweet scent while the similar *Z. mackayi* has a distinctly peppery aroma. The honey scent of *Maxillaria ubatubana* and the cheap perfume of *Maxillaria ochroleuca* warns you that you are nearing a colony from quite some distance away. Oddly neither plant has any apparent reward for the pollinator.

Fertilization and seed set occurs with varying frequency. Some species like *Maxillaria ubatubana* are found with seed pods every year while others like *Grobya amherstii* are only found with fruit after long, hot summers. Presumably, at higher altitudes, climate can affect the availability of pollinators such that a small drop in temperature might exclude a specific insect.

It is interesting that the inflorescences of many plants pollinated by day-flying insects, like *Miltonia* or *Rodriguezia* spp., have all their flowers pointing in one direction, often towards the morning sun. This could well be to show up the guidelines, invisible to the human eye, for the best benefit of the pollinators. Humming-bird pollinated flowers are arranged randomly (like *Sophronitis*) or in whorls (*Stenorrhynchos*). Flowers pollinated by night-flying insects like *Campylocentrum* and *Habenaria* often have heavy scents, consistent colouring but no obvious orientation.

Finally, the flowers of some species have a role in seed pod development. Some genera, related to *Zygopetalum* such as *Promenaea, Neogardneria* and *Pabstia jugosa* retain their perianth after the flowers have been fertilized and these sepals and petals turn green and presumably contribute through photosynthesis to the development of the pod. This feature is related to the amount of light available to the plant (Miller & Warren, 1992).

Laelia crispa plants placed in 30 years old regrowth forest on dead or dying trees showed good growth and flowered. However, if placed on young and healthy trees, they showed limited root growth and the plants degenerated or died.

10. Using This Text

10.   USING THIS TEXT

This section, describing the orchid species found in the Organ Mountain Range in the period 1997 to 2005, has been divided into two parts.

Section one – contains descriptions of orchid species excluding those of the Sub-tribe Pleurothallidinae. The species described, which constitute the majority of both genera and species and with the exception of some terrestrial species and some miniatures of the Oncidiinae, are familiar to most orchidophiles. In addition, many of them are far more common in cultivation than in the wild. Our descriptions of these species thus concentrate far more on:

√ The habitat and the redoubts, where appropriate, where they are found;

√ Their pollination and pollinators;

√ Their rooting processes;

√ Their cultivation,

and in a language which avoids, where possible, latinized words, so confusing to mere mortals, who object to learning a second language in order to read about their orchids.

Section two – contains descriptions of species in the sub tribe Pleurothallidinae. This division has been made for several reasons:

√ The sub tribe Pleurothallidinae contains over a quarter of known Brazilian orchid species and yet these species are the least well known of all Brazilian epiphytic orchids;

√ Almost all these species are small or very small and consequently to illustrate them with photographs would be next to useless for identification purposes;

√ As there is very little available literature concerning this sub tribe and virtually none for the average orchid grower, we decided to:

a) Be extremely specific with the plant descriptions;

b) Illustrate the plants and flowers by way of coloured drawings, all taken from living wild plants; the flowers and their parts were drawn from under a microscope;

√ The genera and species of the Pleurothallidinae live almost exclusively in very humid conditions in original forest or along forested river banks. In other words, if the Pleurothallidinae are present in serious numbers, then that forest is almost certainly original;

√ Here again we have tried to avoid using latinized words.

Taxonomy or classification is not our trade and so we will be content merely to explain the system we are using and the reasons for adopting it. Pabst&Dungs in their *Orchidaceae Brasilienses* (1975-77) adopted the systems of classification of Garay (1960) and Dressler and Dobson (1960) which were in turn based on modifications of the work by Rudolf Schlechter. Although these systems have been changed and updated (see Dressler 1993), it still suits us for the simple reason that it makes further study of these orchids easier. Not only have we adopted the classification used in *Orchidaceae Brasilienses*, but we have also added the exact numbering used by Pabst and Dungs so that cross-referencing between our book and theirs (the only available and vaguely comprehensive text on Brazilian orchids) is simplified. There are obvious areas where Pabst&Dungs system has been left behind. The sub tribe Pleurothallidinae has been recently revised by Carlyle Luer who has created a number of new genera. These are explained in detail in the Pleurothallidinae section of the book. However, we have continued to use the numbering system of *Orchidaceae Brasilienses* throughout these new genera. We believe that it is more important for the hobbyist to be able to flip from one complementary book to another than to try and keep up-to-date.

Very briefly, and in much simplified form, there are five subfamilies in the family Orchidaceae. Subfamilies have names which end in -oideae and start with a generic name in that subfamily. Firstly there is the subfamily Apostasioideae which some authors argue may not even be orchids, and because of this, many pages of text are devoted to discussing their relationships. But they do not occur in Brazil and so we can ignore them. Next comes the Cypripedioideae: their main features are the slipper-shaped lip, 2 lateral anthers and a sterile apical staminode. The pollen is in a viscous liquid or paste and it includes the genera *Phragmipedium* and *Selenipedium* in South America.

The large subfamily Neottioideae (named after the saprophytic genus ,) is composed of largely terrestrial species marked by having powdery and separable pollen grains. Included here are *Cleistes*, *Prescottia* and *Stenorrhynchos*. Although Dressler (1981) lists over 100 genera for the subfamily Orchidoideae (named after the genus Orchis) and characterised by being terrestrial and having tuberous roots, only *Habenaria* is represented in Brazil; it has, however, over 160 species.

Lastly and most significantly, as it contains most of the hobbyists' epiphytic orchids, is the Epidendroidae. Within this subfamily are two major divisions (Tribes) called the Epidendreae and the Vandeae. Soft, malleable pollinia suggest that the species are in the Epidendreae, for example *Cattleya*, *Encyclia* and *Pleurothallis*. There is also a strong tendency for members to flower from the apex of the pseudobulb. Hard cartilaginous pollinia with stipes formed from the rostellum, puts genera like *Zygopetalum*, *Oncidium* and *Maxillaria* into the Vandeae. These tend to flower from the base of the pseudobulb as in *Oncidium*, or from the new growth as in *Neogardneria* and *Zygopetalum*.

Of course, with probably over 30,000 species and new species being discovered all the time, there are always oddities and exceptions. For instance, there are occasional apical inflorescences in *Oncidium* and basal inflorescences in *Epidendrum* species. And monopodial growth is understood to be restricted to sarcanthoid orchids, represented in Brazil only by the genus *Campylocentrum*, but it does also crop up in one species of *Zygopetalum* (*Z. pedicellatum*).

What follows is a list of the orchids found in the Organ Mountain Range in the order they are described in Pabst&Dungs (1975-77). Although an alphabetical listing might seem to be more convenient, the natural or semi-natural order the species are presented in here is much more workable for the purposes of identification. The species are described in closely related groups and so the book becomes simple to use. For those wishing an alphabetical listing, this will be found in the index. We are conscious of the inadequacies of the system but feel strongly that as it is a working handbook, its usability should be paramount.

We have used some abbreviations which should be explained:

◆ V. and Var. = Varieties of plants or flowers that do not fit descriptions in books of reference that we have consulted. In most cases these varieties are not recognised;

◆ Aff. = An affinity or similarity to; in some situations this could indicate a possible new species;

◆ M. asl. = Metres, in altitude, above the average sea level;

◆ L; H; W; = Light needs; Humidity needs; Wind movement; all on a scale of 1 to 10. e. g.

   L 5 = significant shade.   H 8 = very high humidity.   W 2 = very little air movement;

◆ Al. = Alliance within a group of similar orchids.

-

11. Some Tentative Conclusions

11. SOME TENTATIVE CONCLUSIONS

We have prepared a systematic index of all the species of orchids we have found in the Organ Mountains using the system of Pabst & Dungs in *Orchidaceae Brasiliensis*. In addition we have included relevant information taken from the orchid species descriptions. This includes the following:

1. The genus;

2. The species;

3. The number used by Pabst&Dungs;

4. Habitat - <u>E</u>piphyte, <u>T</u>errestrial, <u>L</u>ithophyte;

5. Frequency - <u>C</u>ommon, <u>O</u>ccasional, <u>R</u>are;

6. The section wherein the plant was found;

7. The altitude at which the plant was found;

8. The plant's flowering period.

For example, *Oncidium marshallianum* could be classified as common in section VI above 1200 M. but apparently absent at these altitudes in similar forested areas of the Organ Mountain Range. *Oncidium concolor*, however, is very rare in section VI. but relatively common above 1200 M.asl. in the mountains to the northeast and southwest in sections IV and VIII.

The second polemic, 'the sections wherein we found the orchid species', is, on the face of it, easier to explain. We have lived and worked in sections V and VI for over 30 years and as a consequence have missed very few orchid species in these sections. Other areas have been less frequently visited but, generally speaking, always with a team of 4-6 hawkeyed people. Nevertheless, orchid species, particularly the rarer terrestrials, may well have been missed. Sections V and VI contain the best forested and the least disturbed areas on the whole mountain chain. The Rio de Janeiro Botanic Garden team used section VI above 1,100 metres when initiating the Mata Atlantica Project in the State of Rio. Not only did they find it to be the richest area in tree species but they subsequently used their findings as a standard to compare all other similar studies throughout the State.

Having made these *caveats*, we can now examine the information flowing from the summarized data.

Some species fall into all three categories: good examples would be *Zygopetalum crinitum*, *Ponera striata* and *Cochleanthes candida*. More are both lithophytic and epiphytic. Good examples are *Brassavola tuberculata*, *Laelia crispa* and *Sophronitis coccinea* , while others can be quite happily both epiphytic and terrestrial. Common species in this category are *Gomesa crispa*, *Stanhopea guttulata* and *Xylobium variegatum*. The following table, totalling 760 species shows the degree of adaptability of some of the 620 species involved. A tendency towards this type of adaptability is shown on the scrub-covered mountain ridges and, more emphatically as the high mountain fields are approached. The true light-loving terrestrials are mostly found on the roadside banks above 800 M.asl. Curiously *Habenaria*, *Cleistes* and *Eulophia* species are far more commonly found on roadside banks on the scarp side rather than the anticline. We believe this to be because the scarp banks are steeper than those on the anticline and thus more inaccessible to starving herbivorous animals which are so numerous on the smoother, accessible verges on the anticline.

The deep original forest terrestrials will hardly ever be found outside this environment and it almost goes without saying that the vast majority of the epiphytic species are found at mid- to high-tree level in original humid forest.

* The habitat of the orchid species found *

	Number of species	Percentage (%)
Epiphytes	495	65
Terrestrials	173	23
Lithophytes	92	12
	760	100 %

It is extremely difficult to identify all but the largest epiphytic orchids from ground level so we have to rely on fallen branches and trees to identify all plants in the epiphytic tree gardens. Unfortunately trees do not fall on a consistent basis throughout the forest; many die and decay while still standing or supported by their neighbours. The categories above then are not truly representative of the frequency of epiphytic orchids in original forest but rather an indication only. Mature trees and original forest are rare on the anticline, consequently it would not be unfair to categorize all epiphytes as rare or, at best, locally common, and to push this point even further, there is good reason to suggest that epiphytic orchids found on the anticline today are not necessarily found in their true habitat but rather in a habitat wherein they are capable of surviving.

Light loving terrestrials on the scarp are mostly colonizers from the high mountain fields and scrub-covered mountain ridges, whereas those found on the anticline have their origins in the scrub forests of the interior and their frequency is most probably determined by the access or otherwise of herbivorous domestic animals.

To an extent, the number of species found in sections V and VI reflect the fact that we have lived and worked in this region for 30 years or so. However, without a doubt, section VI has the largest

* Frequency of the orchid species found *

Number of species	Frequency	Percentage (%)
230	common	38
224	occasional	37
151	rare	25
605		100 %

extent of original forest, much of it the least disturbed, in the Organ Mountain Range. In addition, section V is the anticline region least propitious for pasture or agriculture because of the steepness

* Number of orchid species found in the various sections *

Section number	No. of Species Found	Percentage (%)
I	45	4
II	47	5
III	52	5
IV	98	9
V	158	14
VI	481	44
VII	48	5
VIII	163	14
	1092	**100%**

of the majority of the valleys. It is curious that Don João chose this inhospitable region in the 1820's for the first planned agricultural settlement for 2000 or so Swiss immigrants in the State of Rio.

The great concentration of orchid species is found between 800 and 1600 Metres (73%) which coincides with the greatest concentration of original forest on the scarp slopes of the Organ Mountain Range. Very little if any, original forest occurs below 600 Metres and any regrowth forest is also almost devoid of epiphytes with a few exceptions in sections II and VIII which we were unable to visit.

The anticline only shows epiphytic orchids on relict trees, oases and steep wooded ravines alongside permanent water courses.

* The altitudes at which the orchid species were found *

Altitude in M. asl.	Number of species	Percentage (%)
0-400	44	6
400-800	157	20
800-1200	351	44
1200-1600	232	29
1600-2000	11	1
	795	**100%**

You will see that originally the Organ Mountain Range was covered by densely humid wet forest on both the scarp and the anticline. It is thus not unreasonable to assume that the orchid content of both would have been similar. The current rainfall patterns are similar although some 40% less on the anticline overall which would indicate a similar within forest humidity pattern on both sides.

Rivers and streams would also have been permanently full (not like the flush toilets they resemble today). And the epiphytic coverage of bromeliads (as evidenced by the cover on relict trees and the

* The probable deficit of epiphytic orchids on the anticline *

	Non Pleurothallids	Pleurothallids	Total
Epiphytes	286	216	502
Less: epiphytic species also found on the anticline	(97)	(45)	(142)
Less: epiphytic species exclusive to the scarp	(35)	-	(35)
Deficit	154	171	325

ground cover along ravines and oases) would have created a suspended but shallow inland lake which would have acted as a humidity regulator throughout the dry season. Assuming the above we have estimated that some 325 epiphytic orchid species have become extinct or are very rare on the vast reaches of the anticline.

* The flowering periods of the orchids found *

Month	Epiphytes	%	Terrestrials and Lithophytes	%	Total	%
June	70	5	6	2	76	4
July	43	3	5	2	48	3
August	47	3	4	2	51	3
September	56	4	10	4	66	4
October	101	7	19	7	120	7
November	185	13	24	9	209	12
December	181	12	28	11	209	12
January	183	12	33	13	216	13
February	171	12	42	16	213	12
March	172	12	48	19	220	13
April	154	11	27	11	181	11
May	88	6	11	4	99	6
	1451	100	257	100	1708	100

The results shown on the above table confirm that which everybody knows, that the main flowering months are between November and April. However a study of the reproduction strategies used by the species that flower during winter which is colder, dryer and with 30% less daily light should prove interesting.

12. Descriptions of the Orchid Species Found

The genus
Phragmipedium Rolfe

Robert Rolfe worked for 40 years at the Kew Gardens herbarium in London England. He also founded and published The Orchid Review from 1893 until his death in 1921. He raised the genus *Phragmipedium*, from the Greek alluding to the slipper-shaped lip. A Central and South American genus of which six species are attributed to Brazil and only one to the Organ Mountain Range.

Phragmipedium vittatum
(Vell.) Rolfe | 4 (Plate 01)

3 cm

Etymology: Vittatum is Latin for longitudinally striped tube, probably referring to the stripe at the centre of the labellum.

Habitat: Found on a gentle granite west-facing 5°-10° slope, with a 5 cm. thick substrate of turf, which enjoys a year-round water seepage throughput, heavy in summer, light during the winter, and partly supported and shaded by coarse grasses but otherwise fully exposed to the elements at 900-1000 M.asl in section V; a similar habitat to that of *Epidendrum aquaticum*.

Occurrence: Rare, but when found, in significant colonies.

Plant: A large terrestrial with an unmistakable equitant fan of up to 10 light green, large stiff, leathery, conduplicate leaves most of similar size, up to 50 cm. long by 4.5 cm. These are curved like a bow, a narrow lanceolate and arise as new growths from a foreshortened rhizome. The root system is strong, profuse, adherent and penetrates downwards and around the plant through the turf substrate to the bedrock. Individual roots do not branch, are stiff and somewhat brittle, each up to 10 cm. long by 0.4 cm. in diameter.

Inflorescence and Flowers: A long, strong, sheathed, inflorescence emerges from the central base of the leaf mass in late October, which is over 1 cm in diameter at its base, tapering towards its apex and up to 1 metre tall. Single, successive well-sheathed flowers on 1 cm. stems, emerge in late November through to March. The dorsal sepal is oblanceolate, up to 3 cm. long by 1 cm. The laterals are fused, the same length but double the width. The petals are long, narrow, twisted and drooping, up to 5 cm. long by 0.5 cm. and are faintly hirsute at the margins. The lip is the typical lady's slipper or urn-shape, measuring 3 cm. in length by 1.7 cm, with the cup taking up half the length. The overall colour is green with faint wine tinges on the reverse of the petals.

Flowering Periods: As above and the flowers last for two weeks or so, thus the plant may show blooms for up to two months.

Pollination: Unknown. The plant reproduces vegetatively which may explain the large undisturbed colonies.

Cultivation: Note well the habitat conditions; the roots should be kept constantly moist, without getting saturated. Water quality is of prime importance and should lean towards the acid, pH 6.5-pH 7, with no calcium salts. The substrate should be forest floor black soil, sand, granite chips and chopped up tree fern in equal proportions. In practical terms plastic pots are best and the drainage layer of granite chips should take up one third of the pot. And remember only very weak fertilizer solutions should be used. Recommended is Peter's type NPK: 20-20-20 diluted to one quarter of the manufacturers recommendation, twice a year and one application of NPK: 10-35-15 to induce flowering. These are some of the recommendations by **Roberto Takase** in bulletin CAOB n° 42 December 2000 – his e-mail is taskerob@originet.com.br.

General: Probably benefits from grass fires – personal communication, **Lou Menezes**.

The genus
Habenaria Willd.

Ludwig Willdenow was the first botanist to analyse and describe the plant species from **Humbolt** and **Bonpland**'s huge collection made at the beginning of the 19th century, basically from Venezuela up to the Brazilian Amazon border. He died and **Kunth** spent the next 22 years completing the job. **Willdenow** raised the genus *Habenaria* referring to the long strap-like segments of the petals and lip in many species.

This genus is large, over 500 species, of pan tropical and pan subtropical terrestrial orchids; of these about 170 are found in Brazil and 36 attributed to the Organ Mountain Range. To date we have found 19. The majority of species found have been in artificial environments, such as roadside banks, particularly along the dirt roads of the interior anticline and generally at heights that cattle and horses can not reach. It is very difficult to find unbrowsed scrub and grassland, especially in the long dry season, when these animals will eat any thing even vaguely green. But it exists on some high mountain fields, or in the centre of steep, apparently bare rock faces where browsing animals don't usually roam; slopes where the bones below bear silent witness to the horses or cattle that tried to graze.

The second problem is identification. The best identification tools, without a doubt, are **F. C. Hoehne**'s *Flora Brasilica* vol XII coupled with **Pabst** and **Dungs** Alliance groupings and quick excursions into **Barbosa Rodrigues** *Iconographie des Orchidées du Brésil*.

Habenaria fastor
Warm. | *9* (Plate 02)

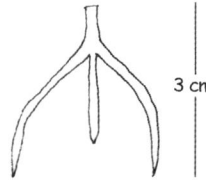

3 cm

Etymology: *Fastor*. We can't find a convincing identification for fastor. The best we can find is *fastosus*, Latin for proud or haughty which in a way fits this impressive inflorescence as it arises above the surrounding grasses.

Habitat: On roadsides and in swamps, always with a permanently wet base, from 1000-1200 M.asl, in section VI.

Occurrence: Common in its niche.

Plant: A large terrestrial, almost always living in colonies of up to seven - ten plants. It emerges in late December through January from a 3 cm. by 3 cm. white tuber. The round, 0.6 cm. diameter stem, which can reach up to 150 cm. or more, appears from within 4-7 light green, deeply centrally keeled and lengthwise ridged, lanceolate leaves, each 20 cm. long by 2.6 cm. These leaves totally sheath the stem up to the emergence of the next leaf. The root system is not extensive, composed of thinnish white, short roots interspersed with thicker, longer ones which form a tight bunch around, but not from, the tuber.

Inflorescence and Flowers: These are borne in a loose cone on the apical, 20-25 cm. of the stem, with up to 15 magnificent, typical Habenaria flowers, each with a trident lip and very long spur, 8.5 cm. by 0.2 cm. wide, which is rich in nectar. The sepals are a light green, pointed oval, 1.7 cm. long, by 0.8 cm. wide. The petals are a light greenish-yellow and sickle-shaped. The upper lobe is 1.8 cm. long by 0.2 cm. wide and fused to the dorsal sepal and the lower lobe is 2.4 cm. long by 0.1 cm. wide. The tri-lobed lip, with the side lobes 2.5 cm. long and 0.1 cm. wide and the mid-lobe, 2 cm. long and 0.1 cm. in width, is very distinctive.

Flowering Periods: January and February; the flowers last for up to a week.

Pollination: The significant nectary and its position would indicate large night flying moths; generally, over 70% of the flowers have fruit.

Cultivation: In large pots on a loose peat and coarse sand compost, watering well in its growing period with occasional misting thereafter, taking care that the substrate remains slightly damp.

General: The largest *Habenaria* in this region both in plant and flower size.

Habenaria trifida
H. B. K. | *27* (Plate 02)

Etymology: *Trifid*, referring to the three, almost equal lobes of the trident lip.

1.2 cm

Habitat: Steep, 80° inclined roadside banks with a heavy water throughput during summer and an almost dry autumn and winter period, exposed to full sun and air movement and supported by sparse, tall grasses around. In section VII at 1000 M.asl.

Occurrence: Occasional.

Plant: A medium-sized, somewhat delicate *Habenaria* with a 30 cm. high main stem emerging from a carrot-like tuber, 2 cm. long by 0.9 cm. diameter. The stem is round in section, light green with alternate bract-like leaves at 3 cm. intervals. These are 6 cm. long, 1 cm. wide, lanceolate and envelop the stem for half their length. The sparse roots are thin and short and arise from the tuber.

Inflorescence and Flowers: The few flowers are borne on the top 4 cm. of the stem. About nine, 1.6 cm. long, bracted, pure milk-white flowers are clustered towards the apex. The dorsal sepal is heart-shaped, and 0.5 cm. long by 0.3 cm. The laterals are an asymmetric lanceolate, 0.3 cm. by 0.2 cm. while the petals are bilobed, both lobes more or less the same size and shape, a sharper upper and a blunter lower lobe, 0.4 cm. long by 0.2 cm. The trilobed lip has a 0.4 cm. long centre lobe with 0.35 cm. side lobes and all are 0.1 cm. wide and slightly reflexed. The nectary is 2.8 cm. long.

Flowering Periods: December and the flowers last for a week or so.

Pollination: Unknown, but probably moths.

Cultivation: Refer to the habitat.

Habenaria angulosa
Barb. Rodr. | *28* (Plate 02)

6 mm

Etymology: Angulosa is Latin for angled or Greek *ankylos* - bent or crooked, referring to the angled symmetry of the flower parts.

Habitat: Found growing in full light amongst tall grasses, along roadsides at 1000-1200 M.asl in section VI.

Occurrence: Occasional.

Plant: A medium sized terrestrial herb, sentinel-like

in its erectness, growing up to 80 cm. tall, the stem arising from a central node in the root system. This is round, stiff, greenish white, 0.5 cm. diameter at the base and gradually tapering towards the apex and completely enveloped by the leaf bases. These leaves are linear lanceolate, plicate, alternate, up to 22 cm. long by 1.7 cm. at the base getting gradually smaller towards the apex where they appear as 1.5 cm. long flower bracts. The root system is not extensive, has no apparent tubers, and is composed of a number of thickish white, slightly hairy roots which penetrate the exposed argilic sub soil.

Inflorescence and Flowers: A loose cone of up to 30 greenish-yellow flowers appears on the stem's apical 15 cm. and are very similar in size, shape and colour to those of *H. montevidensis*. The heart-shaped green dorsal sepal is 0.4 cm. by 0.3 broad. The equally green lateral sepals are a slightly asymmetric lanceolate 0.5 cm. long by 0.2 cm. The yellow petals are a faintly heeled lanceolate, 0.4 cm. long by 0.25 cm. at the heel, and the trident lip is 0.55 cm. long to the apical lobe, by 0.4 cm. wide at the side lobes. The nectary is 0.8 cm. long.

Flowering Periods: March, April; the flowers last for up to 15 days.

Pollination: Night flying months with over 80% the flowers always fertilized.

Cultivation: As for *H. montevidensis*.

Habenaria parviflora
Lindl. | 29 (Plate 02)

Etymology: *Parvi-flora* is Latin for small, or puny flowers, referring to the flower size.

Habitat: A terrestrial from section VI at 1500 M.asl, among low scrub and grasses, enjoying high humidity, dappled light and a well-drained sandy substrate.

Occurrence: Occasional.

Plant: A small terrestrial herb similar to *H. montevidensis* in many aspects. The leaves are few and alternate at 4 cm. intervals from the base of the stem to around 25 cm. intervals higher up. They are dark green, narrowly lanceolate, deeply keeled and papery in texture and measure up to 9 cm. long by 1 cm. The root system is weak and composed of a few thick white roots interspersed with small, potato-like tubers.

Inflorescence and Flowers: The flowering stem arises from a node in the root system or from a tuber. It is erect and sturdy and about 30 cm. high and the top 5 cm. bears a cone-like inflorescence with 10-12 small green and yellow flowers. The sepals are dark green; the dorsal is heart-shaped, concave and 0.4 cm. long

by 0.25 cm., while the laterals are lanceolate, slightly asymmetric and 0.5 cm. long by 0.15 cm. The petals are a greenish yellow and a low-heeled lady's shoe-shape; the upper section is lanceolate and 0.35 cm. by 0.2 cm. while the tiny heel-like inferior section is barely visible. The trilobed lip is greenish yellow with the side lobes half the length of the stronger, spade-shaped central lobe. The lip is 0.45 cm. long by 0.3 cm. at its base.

Flowering Periods: May; the flowers last for ten days.

Pollination: Most flowers are pollinated, probably by night-flying moths.

Cultivation: In pots or frost-free garden beds using a sandy loam with good drainage. Ample water and feed is needed during the rapid growth and flowering. Avoid watering during winter but do not allow to dry out totally.

Habenaria hexaptera
Lindl. | 36

Etymology: *Hexa ptera* is Greek for six winged and probably refers to the fruit capsule.

Habitat: Found in grassy patches in clearings of regrowth forest at 900-1000 M.asl in section VI, enjoying full light and exposed to wind movement and a high humidity on a well drained sandy soil.

Occurrence: Occasional.

Plant: An up to 50 cm. tall, erect terrestrial with tightly packed, blue green, short erect lanceolate leaves emerging alternately and almost enveloping the stem until the base of the apical 15 cm. inflorescence. The root system is typically *Habenaria*; short thick white roots, some times with a small potato-like tuber.

Inflorescence and Flowers: This is one of the latest flowering *Habenaria*. The flowering spike emerges from a node in the centre of the root mass. The flowers emerge in a fairly tight cone on the apical 15 cm. and are a light green, opening rapidly but sequentially. The dorsal sepal is a concave, broad lanceolate 0.5 cm. long by 0.4 cm. The lateral sepals are a twisted, asymmetric, convex, lanceolate shape but are slightly longer and narrower and resemble the wings of a bird. The discrete petals are lanceolate, heeled, 0.4 cm. long, and 0.3 cm. wide at the heel. The lip is a long tongue-shape 0.8 cm. long by 0.15 cm. wide with two very short, ear-like basal side-lobes. The nectary is pronounced, 1.2 cm. long and yellowish, while the distinctive column head is very open mouthed.

Flowering Periods: April; the flowers last for a week or so.

Pollination: Probably nocturnal *Lepidoptera* and almost 100% of flowers are fertilized.

Cultivation: As for the section VI *Habenaria* species, always remembering to give good drainage.

Habenaria josephensis
Barb. Rodr. | *38* (Plate 02)

Etymology: Named after the Serra de S. José d'el Rey where **Barbosa Rodrigues** first found and described it.

Habitat: Found on the dryer forest floor, in a similar light situation to *Cyclopogon iguapensis* at 1200 M.asl and upwards, in section V and VI.

Occurrence: Occasional.

Plant: A medium-sized terrestrial. The stem is round and succulent, some 50 cm. tall, arising from the centre of a root mass. The leaves are alternate, deep green with a light silver sheen, lanceolate, keeled, 0 8 cm. long by 1.5 cm. wide at the base, gradually reducing in size towards the apical inflorescence where they become small flower bracts. The root system is fairly extensive, penetrating under decaying leaf matter and composed of swollen yellow roots.

Inflorescence and Flowers: These are held on a cone-like, apical inflorescence which is composed of some 20 whitish to light green flowers. The sepals are more or less equal in size. The dorsal sepal is a roundish triangle in shape while the laterals are slightly boomerang-like and 0.4 cm. long by 3.5 cm. wide. The petals are shaped like a high-heeled shoe, 0.4 cm. long, obtuse at the apex, with a 0.2 cm. heel. The trilobed lip has a 0.65 cm. long central lobe while the side lobes are minuscule at 0.1 cm. long.

Flowering Periods: In March and the flowers last for little over a week.

Pollination: Probably night flying moths.

Cultivation: On a leaf mould-rich compost in pots, watering well during the growing period, but only misting during the dormant stage and keeping out of direct sunlight.

Habenaria leptoceras
Hook. | *39* (Plate 02)

Etymology: Lepto is Greek meaning slender, and *ceras*, means a horn-like projection. This probably refers to the curious singular lip that appears like a slender horn-like appendage.

Habitat: Found in sporadic, small colonies, within tall grasses and scrub in high mountain fields above 1500

M.asl. in section VI.

Occurrence: Occasional.

Plant: A medium-sized terrestrial with an erect inflorescence arising from a central node in the tuber. The stem is thick, round, and up to 60 cm. in height and 0.8 cm. in diameter. It is almost enveloped, at 2 cm. intervals, by keeled, lanceolate almost erect, light green leaves, which measure 10 cm. long by 2 cm. wide at the base, and gradually diminish in size to 2.0 cm. long by 0.6 cm. wide just below the apical flower spike. The root system is quite extensive and apparently lightly tubered, and is composed of thick yellow succulent roots.

Inflorescence and Flowers: These appear on the top 15 cm. of the inflorescence as a spicate cone, formed of 30 pale, whitish-yellow flowers. The flower form is quite different to other *Habenaria* species in the region. The ovate dorsal sepal, 0.5 cm. by 0.4 cm, is partly fused with the petals and a high-heeled shoe shape, 0.5 cm. by 0.2 cm. with a short, 0.2 cm. lower lobe heel. The lateral sepals are also almost ovate, 0.7 cm.by 0.4 cm. The lip is only slightly trilobed. The sword-shaped central lobe is 0.8 cm. long by 0.1 cm. while the side-lobes are minuscule. The nectary is 1.2 cm. long by 0.1 cm. wide.

Flowering Periods: March and April; the flowers last for up to 10 days.

Pollination: Probably small night flying moths and 60% of flowers bear fruit.

Cultivation: As for *H. montevidensis*.

Habenaria montevidensis
Spreng. | *41* (Plate 02)

Etymology: One is tempted to assume that **Spreng.** found this orchid in Montevideo, Uruguay. However more prosaically, could be that it appeared - *videtur* - Latin on a mont - Latin growing on a mountain.

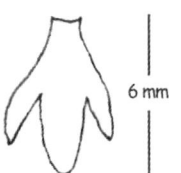

Habitat: Similar to the other *Habenaria* species at these altitudes; found on 45° roadside banks in full sun, also occasionally on the high mountain fields in a similar situation. In section VI at 1200-1000 M.asl.

Occurrence: Common.

Plant: A small terrestrial, in fact the smallest flowered Habenaria species in the region. The woody, erect stem arises in January from a central axis of the root system, and reaches some 30 cm. in height. The leaves are alternate, starting at the stem base, and completely envelop it up to the terminal inflorescence. They are light green, sharply lanceolate, deeply ridged and virtually erect. The basal leaf measures 9 cm. by 1.5 cm. and the remainder gradually reduce in size towards the apex. Roots are few, thickish and white, with several potato-like tuberoids attached.

Inflorescence and Flowers: These arise on cone-like apical inflorescences and 25 very small, sturdy, typical Habenaria flowers are produced. The distinguishing characteristic is the very thick centre lobe of the trident-shaped lip. The dorsal sepal, an obtuse ovoid, 0.3 cm. by 0.3 cm. is unusual while the petals have a proportionately minute lower lobe. The overall effect is a rich green interspersed with the yellow of the petals and lip.

Flowering Periods: Late February to April; the flowers last for up to three weeks.

Pollination: Probably small night flying moths; 90% of the flowers are normally pollinated.

Cultivation: On a compost made up of 50% coarse sand and 50% leaf mould, in pots, watering well in summer but only misting thereafter.

Habenaria crassipes
Schltr. | 50

Etymology: *Crassipes* is Latin for thick feet, referring to the plant's thick roots.

7 mm

Habitat: A terrestrial, found on roadside banks in section VI at around 1000 M.asl, amongst tall supporting grasses in shady conditions with a well-drained substrate and good air movement.

Occurrence: Occasional.

Plant: A medium to large terrestrial; the stem arises from a central node in the root system and is erect, reaching more than 1 metre high. It is fleshy, 0.4 cm. diameter and completely sheathed by the fused leaf bases. The leaves are lanceolate, alternate, plicate and a darkish, silvery-green. They diminish in size towards the stem apex and are 11 cm. long by 2 cm. at the stem base and 3 cm. by 0.5 cm. at the base of the inflorescence and they stand almost erect against the stem. The root system is composed of a few thickish white roots with evidence of small tubers along their length.

Inflorescence and Flowers: The flowers appear along the apical 20 cm. of the stem and up to 25 are held in a loose cone. The dorsal sepal is heart-shaped, 0.5 cm. long and wide, concave and deep green. The lateral sepals are also dark green but longer, narrower and an asymmetric lanceolate measuring 0.65 cm. by 0.3 cm. The bilobed petals are darkish yellow; the upper segment is lanceolate and measures 0.6 cm. by 0.3 cm. while the narrow anterior segment is 0.5 cm. by 0.1 cm. The

trident-shaped lip is 0.7 cm. long at the mid-lobe, 0.1 cm. wide and at the rounded tip. The side lobes are slightly shorter and 0.6 cm. long by 0.09 cm. The overall colour is deep yellow. A nectary measuring 2.2 cm. long by 0.15 cm. sits against the stem below the column.

Flowering Periods: Late February and early March and the flowers last for two weeks.

Pollination: Over 50% of the flowers set seed, and are probably moth-pollinated.

Cultivation: As *Habenaria parviflora*.

Habenaria minarum
Hoehne & Schltr. | 56

Etymology: *Minara* is Latinized Arabic for a tower or minaret, referring to the shape of the apical cone of flowers.

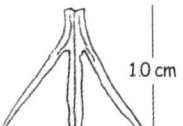

1.0 cm

Habitat: A terrestrial from section V, growing at 1000-1200 M.asl, on bare granite slopes among islands of vegetation with *Vellozia*, grasses, sedges and orchids such as *Epidendrum robustum*, *Oncidium batemannianum* and *Laelia cinnabarina*.

Occurrence: Common in its specific habitat. Interestingly, and unlike other high mountain *Habenaria* species, we have never found this plant on roadside cuttings at lower levels or in section VI.

Plant: A medium sized terrestrial herb up to 70 cm. high with a sequence of up to eight lanceolate, deeply keeled and softly plicate leaves which measure up to 7 cm. long by 1.4 cm. wide. The root system is composed of a number of thick, white and short roots, one or two of which may produce small tubers.

Inflorescence and Flowers: The stiff, round-sectioned, green inflorescence arises from a central node in the root system and reaches 70 cm. high. Between 35 and 40 flowers are arranged in groups of two to four at 1 cm intervals starting at 16 cm. from the stem base. The flowers are dull lime green; the dorsal sepal is an inverted, convex heart-shape measuring 0.7 cm. long by 0.6 cm. and has a serrated margin. The lateral sepals are an irregular lanceolate shape and 0.9 cm. long by 0.3 cm. at the widest part. The petals are greenish-yellow and shaped like a high-heeled stiletto shoe, the posterior lobe being longer at 0.8 cm. than the anterior at 0.7 cm. and half the width at 0.1 cm. The lateral lobes of the trilobed lip are 0.9 cm. by 0.1 while the central lobe is slightly shorter at 0.8 cm. by 0.1 cm.

Flowering Periods: The inflorescences appear after the spring rains in late November and early December and the flowers last for a week or so.

Pollination: The flowers are almost always pollinated

by, as **Dressler** suggests, night-flying moths.

Cultivation: In pots using a fibrous sandy loam, watering during spring and summer but only misting after dying down to prevent drying out totally.

General: To see this *Habenaria* species in the natural habitat is strenuous but worthwhile.

Habenaria achalensis
Krzl. | 63

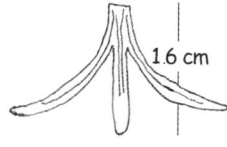

Etymology: Not known.

Habitat: We have found this species in regrowing forest at 1000 M.asl in section VI, where it enjoys shade, dappled sunlight, good drainage and air movement.

Occurrence: Occasional.

Plant: A medium sized solitary terrestrial, up to 50 cm. tall. The stem arises from a central node within the root system. The five to seven well-spaced leaves are broadly lanceolate, fleshy, alternate, plicate and dark green, measuring 14 cm. long by 4.5 cm. The leaf bases fold around the stem to where the previous leaf emerges. The root system has a few short, thick roots with no sign of tubers.

Inflorescence and Flowers: The apical 20-25 cm. of the stem bears up to 20 well-spaced, large flowers on a conical raceme. Each is light green with cream at the base of the petals and the lip. The dorsal sepal is concave, almost round and measures 0.55 cm. long and wide with a faintly pointed apex. The lateral sepals are a broad asymmetric lanceolate and measure 0.7 cm. by 0.55 cm. The two-segmented petals with a longer anterior segment are characteristic of the 16 plants in the **H. repens** alliance, sensu **Pabst**. The fine anterior segment is 0.9 cm. by 0.1 cm. and the posterior is an asymmetric narrow lanceolate shape, 0.7 cm. by 0.15 cm. wide. The trilobed, trident-shaped lip has equal sized lobes at 1.2 cm. length by 0.12 cm. These are green at their apices and cream at their bases. The nectary is 2.5 cm. long by 0.2 cm.

Flowering Periods: January and the flowers last for ten days.

Pollination: Pollination is probably by night flying moths.

Cultivation: As for **H. crassipes**.

Habenaria achnantha
Rchb. f. | 64 (Plate 02)

Etymology: Not known.

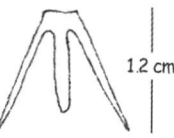

Habitat: A denizen of the high mountain fields, at over 1200 M.asl, occasionally straying to artificial fields at lower altitudes, the result of roadside-bank clearing or accidentally burnt mountain ridges. This is the first of many Habenaria species to flower in summer and is thus especially welcome. It is also the species that inhabits the most exposed terrestrial areas in section VI.

Occurrence: Occasional.

Plant: A small terrestrial, found in compact colonies of 10 or more plants. They emerge in early December from a nodule on a small tuberoid. The stem, which reaches 28 cm, bears 4 or 5 lanceolate and deeply keeled, shiny, dark green leaves. These are 8 cm. long by 1.4 cm. wide, alternate and at 3 cm. intervals. The root system is weak, composed of 3 to 5 thickish, though short, roots which spread for about 10 cm. under the stony strata around.

Inflorescence and Flowers: Nine or 10 flowers are borne alternately on the top 9 cm. of the inflorescence. The flowers are unmistakably *Habenaria*, with the trident lip, the long nectary and the curious sickle-shaped petal with its copied under-lobe which are uniformly narrow, pointed and yellow. The petals are 0.5 cm. long, the anterior segment being slightly longer at 0.6 cm. but thinner at 0.1 cm. thick. The lip side-lobes are linear, pointed, 0.9 cm. long by 0.1 cm. wide, while the central lobe is slightly shorter and marginally thicker. The sepals are a light green and quite conventional. The posterior petal lobes are partly fused with the cap-like dorsal sepal. The lateral sepals are a pointed wedge-shape, 0.7 cm. long by 0.3 cm. wide at the base.

Flowering Periods: December and the flowers last for up to 10 days.

Pollination: Probably small night flying moths.

Cultivation: In pots with a predominantly coarse sand base, mixed with about 20% leaf mould. Water as soon as the new growth shows but only mist after flowering.

Habenaria aff. marupaana
Schltr. | 70

Etymology: Marupâ is a district in Amapá from where **Schlechter** must have received the plant.

Habitat: Found in section VI at 900 M.asl., growing among tall grasses in young regrowth forest. We have found only one colony of about eight plants of this terrestrial on the swampy borders of a summer pond, enjoying good humidity, good dappled light and little air movement.

Occurrence: Occasional.

Plant: A medium sized terrestrial. The stem arises from a node in the root complex and reaches 60 cm. The alternate leaves arise at 5 cm. intervals and are light green, plicate and lanceolate. They measure up to 12 cm. long by 2.8 cm. broad and the basal 2 cm. surrounds the stem. The root system is composed of a few short, thick yellow roots with no sign of tubers.

Inflorescence and Flowers: The inflorescence appears as a cone of closely packed flowers on the stem's apical 15 cm. The dorsal sepal is light green, heart-shaped, concave and measures 0.6 cm. long by 0.4 cm. The lateral sepals are a similar colour and an asymmetric lanceolate shape, measuring 0.7 cm. long by 0.3 cm. wide. The two-segmented petals are yellow and the anterior segment is 1 cm. long by 0.1 cm. wide while the posterior segment is 0.55 cm. long by 0.15 cm. The trident-shaped lip is also yellow with the lateral lobes 1.1 cm. long by 0.1 cm. and the central lobe 0.65 cm. long by 0.14 cm.

Flowering Periods: Late March through to mid-April and the flowers last for up to ten days.

Pollination: Pollination is by small moths.

Cultivation: In a garden with the conditions described above. In pot culture, the compost should be rich in well-decayed litter, liberally watered in summer and just kept damp in winter.

Habenaria repens
Nutt. | 73 (Plate 02)

Etymology: Derived from the Latin for creeping which makes little sense.

Habitat: Found on almost 90° incline roadside banks, supported by tall grasses at 1200 M.asl in section VI with very wet summers, and long dry winters, growing on almost sterile subsoil, which at least has a nutrient throughput during the summer rains.

Occurrence: Occasional.

Plant: A very typical medium-sized *Habenaria*. It's stem arises from a node on the previous years tuber, which is like a small elongated potato, and holds a few small, short white roots. The leaves, which are alternate, are 7 cm. apart, partly enfold the stem and measure up to 14 cm. long at the base by 2 cm, diminishing in size towards the apex. They are a light green, semi-erect, mildly keeled, veined vertically, and lanceolate.

Inflorescence and Flowers: The inflorescence takes up around 27 cm. of the apical section of the 70 cm. stem and is a loose cone of up to 40 flowers. These bloom in sequence from the lowest. The concave dorsal sepal is a fat heart-shape, light green and 0.5 cm. long by 0.3 cm. long. The bilobed petals are shaped much like exaggerated stiletto-heeled shoes, 0.5 cm. long by 0.15 cm. while the heel measures, 0.6 cm. by 0.05 cm. The typical trilobed trident lip has a centre lobe 0.8 cm. long and 0.15 cm. broad, the side lobes are 0.6 cm. long and the hair-like petals and sepals are a yellowish white. The nectary is 1.2 cm. long by 0.2 cm. in diameter and a deep green.

Flowering Periods: During January; one of the first Habenarias to flower in early summer.

Pollination: Probably night flying moths.

Cultivation: As with any other *Habenaria* we have described. Look at the habitat description.

Habenaria riedelii
Cogn. | 74 (Plate 03)

Etymology: Dedicated to **Ludwig Riedel** from Germany who collected in Bahia and the Organ Mountain Range in the mid 1900s.

Habitat: Dark roadsides and forest clearings, requiring low light and high humidity but not swampy conditions. Found at 800 to 1100 M.asl in section VI.

Occurrence: Occasional, but generally found in colonies of up to eight plants.

Plant: A large terrestrial. From storage units of two or three small, 3 cm. by 2 cm, potato-like tuberoids, a tough and woody 120 cm. x 0.5 cm. diameter stem spurts. Some 14 dark green, lanceolate, deeply plicate leaves, 18 cm. long by 3.5 cm. wide, emerge alternately at 4 cm. intervals. The root system, composed of thick, white, non-branching roots, comes from a rhizome connecting the tuberoids. These roots are not long but permeate the rotting leaf detritus, and anchor themselves 1 or 2 cm. below the argilic topsoil.

Inflorescence and Flowers: These arise on the top 30 cm. of the stem as a spicate cone, with more than 50 curious and extremely hairy looking, but typical *Habenaria*-type flowers. The overall effect of the very thin, upward and inward curling petals and lip side-lobes, is of a spider's nest or 'old man's beard'. The sepals are a rich light green colour, more or less equal in size, a pointed oval, 0.9 cm. long by 0.4 cm. broad. The petals are yellow and bi-lobed; the lower portion measuring 2.2 cm. long by 0.1 cm. wide and the upper section, partially fused to the dorsal sepal and sickle shaped, is 0.7 cm. long by 0.1 cm. wide. The lip has the normal trident shape of the genus but only the central lobe, at 1 cm. long by 0.1 cm. wide, is pendent. The side lobes, 2 cm. long by 0.1 cm., curl upwards, helping with the spider's web effect. The nectary is 2.7 cm. long.

Flowering Periods: In February and March when the flowers last for up to a week.

Pollination: Probably pollinated by large night flying moths as the nectary length and flower colour would indicate. 60% of the flowers are normally pollinated.

Cultivation: As for *H. fastor*.

Habenaria rupicola
Barb. Rodr. | 75 (Plate 03)

Etymology: *Rupicola* is Latin for rock-loving, a situation in which, **Barbosa Rodrigues** states, 'this plant is pleased to grow'.

Habitat: Roadsides in clear sunlight, always on a 45° or steeper slope implying the need for good drainage, light and air movement. Found in huge colonies where these conditions are met, from 1000-1400 M.asl in sections IV & VI.

Occurrence: Common.

Plant: A medium-sized terrestrial up to 50 cm. tall growing from a node at the root system's centre. The storage organs are several potato-like tuberoids and thick yellowish white roots which penetrate just below the rock-like argilic substrate. The leaves are alternate, light green, deeply plicate and lanceolate, up to 12 cm. long by 2.5 cm. broad. They arise at short intervals from the tough erect, round stem, up to the apical inflorescence. The leaf petioles, where the deep leaf ridges originate, are fused to and sheath the stem, giving it a square cross section.

Inflorescence and Flowers: These arise on an apical, spicate, cone-shaped inflorescence up to 12 cm. long. The flowers show all the common *Habenaria* characteristics; the sepals are ovate and equal, 0.5 cm. long by 0.4 cm. wide and greenish-yellow. The upper lobe of the petals is partly fused to the dorsal sepal and measures 0.5 cm. long by 0.3 cm. broad at the thickish base. The lower lobe is slightly longer and thinner. The trident lip is fleshy and the central lobe is 0.9 cm by 0,12 cm; the sidelobes are slightly shorter; the petals and lip are a clear yellow.

Flowering Periods: February through April. The flowers last for up to three weeks.

Pollination: Probably small night flying moths. 90% of the flowers area pollinated.

Cultivation: As for *H. achnantha*.

Habenaria petalodes
Lindl. | 87 (Plate 03)

Etymology: *Petalodes* is Greek for petal-like, and probably refers to the unusual petals.

Habitat: A terrestrial from sections III, V and VI at 900-1000 M.asl, among tall grasses, often within young regrowth forest, growing on a thick layer of decayed vegetable matter with good light, high humidity and low wind movement, sometimes in large colonies.

Occurrence: Occasional.

Plant: A large to medium-sized terrestrial with alternate leaves emerging at 5 cm. intervals from a round sectioned, upright semi-succulent stem, 0.5 cm. in diameter. The leaves start at the stem base and occur for 80 cm; they are lanceolate and their bases envelop the stem for 3 cm. before unfolding. They are deeply keeled, dark green, slightly fleshy and lanceolate, smaller at the base and apex and, at mid-stem, they measure 16 cm. long by 3.8 cm. The root system is composed of a few thick, white roots with evidence of tubers.

Inflorescence and Flowers: The inflorescence forms a loose cone on the top 20 cm. of the one metre high stem. The flowers are quite atypical for a Habenaria as they are more elongate than compact and the petals and the lip have single lobes. The dorsal sepal is an almost round light green, concave heart-shape, and measures 0.9 cm long by 0.8 cm. The lateral sepals are a similar colour, an asymmetric lanceolate, 1.2 cm. long by 0.5 cm. The petals are an asymmetric square shape, yellowish-green and measure 1.0 cm. long by 0.6 cm. The lip measures 2.0 cm. long by 0.2 cm. wide and is convex and rectangular with two indistinct shoulders. The nectary measures 2.6 cm.

Flowering Periods: Late March and early April; the flowers last for up to two weeks.

Pollination: Probably pollinated by night-flying moths.

Cultivation: In pots using a rich organic compost, well drained and in good light conditions; well watered in summer and misted in winter to prevent drying out.

Habenaria rodeiensis
Barb. Rodr. | *106* (Plate 03)

Etymology: Named by **Barbosa Rodrigues** because it was found growing in argilic ravines near Rodeio in Rio de Janeiro.

Habitat: Found in cracks in the most inhospitable, bare, argilic, subsoil on 90° roadside banks, at 1000-1300 M.asl in section VI.

Occurrence: Occasional.

Plant: A medium to large terrestrial with an erect inflorescence arising from a central root node. The stem is up to 65 cm. in height, round and 0.3 cm. in diameter and completely enveloped by the fused sheathing leaf bases. The leaves are smooth, lanceolate and alternate, a light green in colour, 9 cm. long by 1.7 cm. wide and stand almost erect against the stem. Leaf sizes diminish gradually with height on the plant. The root system, composed of thickish, white, worm-like roots, is not extensive. No trace of tuberoids was found on the plants excavated.

Inflorescence and Flowers: Produced on the top 15 cm. of the stem on a spicate inflorescence with up to 14 flowers arranged in a loose cone. The dorsal sepal, 0.7 cm. by 0.6 cm. wide, is heart-shaped. The lateral sepals are somewhat boomerang-shaped and slightly longer: all are a rich green. The bilobed petals are unusual for Habenaria, in that the front lobe is ivory white, 0.7 cm. long by 0.4 cm. wide, a spear-head shape while the fine lower lobe is 0.8 cm. long. The trident-like lip has thin side lobes 0.8 cm. long; the slightly thicker centre lobe is 0.6 cm. The hidden nectary is 3.8 cm. long.

Pollination: Probably night flying moths. Fertilization is frequent.

Cultivation: In pots on a very coarse, well drained sandy soil. Water during growth and flowering, with only occasional misting thereafter.

Habenaria itatiayae
Schltr. | *161* (Plate 03)

Etymology: First found in Itatiaia, RJ, one presumes.

Habitat: Found on 45° inclined roadside banks amidst grasses and low scrub in large colonies, from 1000-1400 M.asl in section VI.

Occurrence: Common.

Plant: A medium-sized, robust terrestrial with a round and succulent stem which rises to about 75 cm. and is completely sheathed by the leaf petioles, giving the stem an apparent square cross-section. The leaves are plicate and lanceolate, 17 cm. long by 4 cm. broad, light to dark green, and alternate at 6 cm. intervals, diminishing in size towards the apex where they close up becoming flower bracts. The root system consists of six or seven thick but short yellow roots which form several small potato-like tuberoids at their extremities.

Inflorescence and Flowers: A very tightly clustered, thick, cone-like inflorescence with up to 60 yellowish-green flowers arises on the top 18 cm. of the stem. The dorsal sepal is heart-shaped, concave, yellow, and 0.4 cm. tall by 0.5 cm. wide. The acuminate laterals are 0.6 cm. long by 0.3 cm. wide and yellowish-green. The bilobed petals are a golden yellow with the curved upper lobe partly fused to the dorsal sepal and measure 0.45 cm. long by 0.1 cm. wide with the lower lobe slightly longer. The flower bracts are slightly longer than the flowers. The trident lip is also golden yellow and all three lobes are 0.7 cm. long by 0.1 cm. wide, but the central lobe is slightly thicker.

Flowering Periods: From February through to April and the individual flowers last for up to a week while the inflorescence may show flowers in bloom for up to a month.

Pollination: Probably by night flying moths. Pollination and fruit set is almost total.

Cultivation: As for *H. achnantha*.

The genus
Vanilla Sw.

Olof Swartz was Sweden's second most famous scientist, after **Linneaus**, in the late 18th and early 19th centuries, and named the genus from the Spanish word *Vanilla* meaning a long seed pod. The seeds of a plant or plants from this genus were used by the Aztecs in pre-colonial Mexico as a medicine or as a seasoning for a drink, made from the seeds of the Cacau tree, now known as chocolate.

This is a pan tropical, monopodial genus of which 31 species are attributed to Brazil, with a possible six to the Organ Mountain Range. We have certainly found at least three species but all are climbers over 10 metres in height and since the flowering periods are short. We have missed the flowers every year, except for one species, *V. edwallii*.

Vanilla edwallii
Hoehne | *206* (Plate 03)

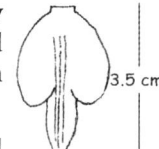

Etymology: Named in honour **Gustav Edwall**, a Swedish botanist who worked in São Paulo together with **Loefgren** and **Hoehne**.

Habitat: Found in both original and regrowth forest with equal frequency, in sections V and VI at 600-1000 M.asl, requiring average light, humidity and air movement.

Occurrence: Occasional.

Plant: A monopodial climber combining the characteristics of both epiphyte and terrestrial. The seed appears to germinate low down on a tree trunk or in leaf litter sending a root downwards. These roots are whitish-green, of uniform diameter, 0.3-0.4 cm. The plant finds a tree trunk and grows upwards producing broadly lanceolate leaves, 11-12 cm. long by 4-4.5 cm, at 3-4 cm. intervals. Each is fleshy, faintly keeled, pale green and long-lasting and emerges from the stem without an apparent petiole. Short, 2-3 cm. long single greenish roots emerge opposite the leaves and their function, we suspect, is more to anchor the plant than to feed it. The plant shows an almost limitless growth potential and we have seen plants of more than 20 metres.

Inflorescence and Flowers: Leafy inflorescences up to 17 cm. long, emerge from leaf axils starting at 75% of the plant's height and terminating a metre or so before the growing tip. A typical plant has four to six such inflorescences and each has 1 to 3 large, white and green waxy, showy, but short lived flowers. One flower is apical and all arise from a leaf axil with virtually no pedicel. Petals and sepals are almost equal sized, lanceolate and measure 4.0-4.5 cm. long by 1.0-1.4 cm. Their surface is flat on emergence but soon crinkles and curls backwards at the apex. The distinctive trilobed lip measures 3.5 cm. by 2.6 cm. across the side lobes and is a clear white with a creamy-yellow mid-lobe that is closely vertically lined with five to seven yellow ridges which run to the tip.

Flowering Periods: January and February and the flowers last only a few days.

Pollination: Probably pollinated by *Euglossine* bees.

Cultivation: It needs a botanic garden to house Vanillas as they often need to reach tree height before flowering. They grow well, starting in a pot of humus-rich compost then anchored onto a moss-pole or tree.

The genus
Epistephium H. B. K.

Humbolt - Bonpland - Kunth = H. B. K. Carl Kunth spent many years cataloguing and describing the species collected by **Humbolt** and **Bonpland** on their incredible journeys through northern South America in 1799 - 1804.

Pabst and Dungs attribute 12 species of this South American genus to Brazil and surprisingly three to the state of Rio de Janeiro of which *E. portellanum* is the only one to be found in the Organ Mountain Range.

Epistephium portellanum
Barb. Rodr. | *219*

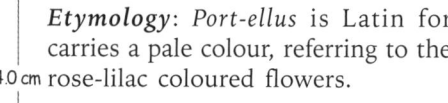

Etymology: *Port-ellus* is Latin for carries a pale colour, referring to the rose-lilac coloured flowers.

Habitat: A terrestrial found in very poor regrowth on ex-pasture in section VI at 1000-1100 M.asl. and growing on a north-facing dry slope in dappled light with good air movement and low humidity.

Occurrence: This plant, a relative of Vanilla, should not occur in section VI, which is essentially humid forest. However the effects of current deforestation are clearly opening up space for colonisers from the warmer and drier interior.

Plant: A large terrestrial which sprawls over undergrowth like *Epidendrum saxatile*, *Oncidium ramosum*, or, to a degree, *Gomesa glaziovii*. The annual new growth appears in February from a node on an underground rhizome. The stems measure up to 1.5 metres in length and are round, 0.4 cm. diameter, light green, stiff and woody. They bear up to 20 ovate, alternate leaves at between 5 cm. and 10 cm. intervals, which are smaller at the base and apex of the stem. At the mid-section they measure up to 15 cm. long by 8 cm, are thin but stiff and unorchid-like as they have six pronounced branching parallel veins arising from a thick, woody central vein. The leaves arise directly from the stem and are dark green and without a petiole. The root system is strongly developed and composed of thick, 0.6 cm. diameter white roots which arise from the base of each year's growth and penetrate the sandy, argilic soil substrate.

Inflorescence and Flowers: The flowers emerge on very short, 0.3 cm. pedicels from the stem's apex or a sub apical point, with one to three flowers on each of the two previous years' stems. Each flower is subtended by a small leaf-like bract. The sepals and petals are lanceolate, equal in shape and size, measuring 2.5 cm.

long by 0.7 cm, and are magenta. The two-lobed lip is purple and shaped like a ping pong bat with slightly crenate margins. There is a longitudinal callus, the basal half of which has a dense mass of backward pointing hairs.

Flowering Periods: February and March; the flowers last for a few days only, but as they open sequentially, any plant will have flowers in bloom for over a month.

Pollination: Pollinators are not known but possibly self-pollinated.

Cultivation: Requires year-round high light and a well-drained sandy loam with a dry winter period. The root system needs plenty of space.

The genus
Cleistes L. C. Rich.

Richard was a collector who visited French Guiana and Pará between 1781 and 1789 and named this genus *Cleistes*, from the Greek, referring to a closed flower as in fact many of the species' flowers open only at the tip. However the plants may well be cleistogamus as some species' ovaries seem to swell even before the flowers open. On the other hand, all flowers that we have examined show very attractive and elaborate lips, which if they are cleistogamus, i.e. fertilized within the unopened flower, does not make biological sense. In addition many species' flowers open on the same day, and only last for a day, throughout a region, which would strongly indicate cross-pollination.

These plants are almost impossible to transplant and further, we have proved quite unable to germinate seed in vitro. So it seems likely that these lovely plants will not be cultivated until a laboratory breakthrough is made.

Pabst and Dungs have attributed 45 species to Brazil and 19 to the Organ Mountain Range. All of the 11 species we have found were discovered 10 years ago, and try as we might, we have found no additional species in the interim.

Cleistes brasiliensis
(Barb. Rodr.) Schltr. | 232 (Plate 03)

Etymology: First found in Brazil one presumes.

Habitat: High mountain fields and scrub-covered ridges over 1200 M.asl in section VI.

Occurrence: Rare.

Plant: A small terrestrial with the stem, arising from the centre of the roots, round, thin, noded, light purple

in colour and very fragile. Four pointed oval, alternate, dark green, keeled leaves, 6 cm. long by 2 cm. wide, flow from the stem at 5 cm. intervals from the base. The roots are thick, yellowish-white and short, 0.5 cm. in diameter, with no apparent tuberoids.

Inflorescence and Flowers: These open in sequence. An apical and extremely attractive, pink flower opens widely. The sepals, pale pink, pointed and oval are 2.5 cm. long by 0.6 cm. wide and curl outwards. The petals are equal-sized and the same colour. The lip is striking and long and 2.4 cm. by 1 cm. at the flared basal lobe. It is trilobed and edged with a deep royal purple. The central callus from throat to base is ringed in white and is yellow with a tight series of Vs on the apical lobe.

Flowering Periods: February and the individual flower lasts for a day or so.

Pollination: **Dressler** suggests possibly bees by deceit as Cleistes flowers have no nectar, however, they are almost always fertilized.

Cultivation: Unknown.

Cleistes gracilis
(Barb. Rodr.) Schltr. | 233 (Plate 03)

Etymology: *Gracilis* is Latin -for thin or slender, referring to the stem.

Habitat: Found in high mountain fields, on ridges in elfin forest and scrub above 1300 M.asl in section VI.

Occurrence: Rare.

Plant: A small terrestrial and the stem, arising from a central node, is up to 44 cm. long, round, jointed, fragile and, for the first 15 cm, purple while the rest is green. It has up to 14 alternate, light green, lanceolate, keeled leaves, each 6.5 cm. long by 1.3 cm. wide and borne on the top 20 cm. of the stem. The roots are thick, yellow, relatively long and 0.6 cm. in diameter. There is no sign of tuberoids.

Inflorescence and Flowers: A single apical flower appears as a yellow tube. The sepals are 2.6 cm. long by 0.7 cm. wide. The ovate petals are slightly shorter but wider at 2.4 cm. x 0.75 cm. The lip is 2.3 cm. long, three lobed with a flared central lobe, white and lined deeply in royal purple. The long pronounced central callus is emerald green.

Flowering Period: February, when the flower lasts for a week.

Pollination: As for *C. brasiliensis*.

Cultivation: Unknown.

Cleistes ionoglossa
Hoehne and Schltr. | 235 (Plate 03)

Etymology: *Ionoglossus* which is Greek for violet tongued, referring to the colour of the lip.

Habitat: A terrestrial, found on roadside banks amongst tall grasses in full sunlight or in shade at between 1000-1200 M.asl in section VI, enjoying high humidity on a well drained substrate.

Occurrence: Rare.

Plant: A medium sized terrestrial herb. The stem, which emerges from a node in the root mass in February may reach 80 cm. in height and is round, succulent and light green. It is sparsely leaved. The first leaf, at 30 cm. from the stem base, is lanceolate, 2.5 cm. long by 1 cm. broad and light green. The second leaf, 30 cm. above, is slightly larger. The root system is typical of all *Cleistes* species with a few thick yellow roots which are short and show an occasional small tuber.

Inflorescence and Flowers: An untidy raceme of five or six flower buds is borne some 10 cm. above the second leaf. These open from the base in sequence at about 5 cm. intervals. The sepals are lanceolate, 4.5 cm. long by 1 cm. broad, violet towards their apices, ivory white at their bases. The petals are slightly shorter, at 4 cm. long, but broader at 1.5 cm, and have the same delicate violet colouring. The trilobed lip measures 4.5 cm. by 2 cm. broad and is violet at its apex tapering to ivory at its base. It shows a gross and long, longitudinal, yellow callus.

Flowering Periods: February and March. Individual flowers last for two days or so, but the plant may show flowers for up to three weeks.

Pollination: Possibly bees by deceit, as *Cleistes* flowers offer no nectar. In common with other *Cleistes* species, the flowers almost always bear fruit.

Cultivation: Unknown.

Cleistes itatiaiae
Pabst | 236 (Plate 04)

Etymology: First identified and named after a plant from Itatiaia, RJ, one presumes.

Habitat: Found in substantial colonies on very exposed roadside banks with little other cover and in full sunlight, at between 1100-1400 M.asl, and occasionally on the barer patches of high mountain fields in section VI.

Occurrence: Occasional.

Plant: A small to medium-sized terrestrial herb. The new growth stems from a nodule in the central root axis, first appearing above ground in late December. The root system is short with a few thickish, yellow roots, 3 or 4 of which swell up to form fairly substantial tuberoids, each 4 cm. long by 0.8 cm. broad. As the favoured substratum is rock-hard argilic pan, one is amazed at this plant's capacity to penetrate such a stratum, let alone form substantial tubers there. The stem is woody and round, 0.2 cm. in diameter and can reach up to 40 cm. in height. The leaves are fleshy, alternate, lanceolate, and silvery green, the first arising 20 cm. from the stem base. This basal leaf measures 6 cm. long by 1.3 cm. wide but they become progressively smaller towards the apex.

Inflorescence and Flowers: Several sequential flowers are produced on single short inflorescences arising from the joint of a leaf to the stem. Flowering is continuous and always from a new apex. The flowers are always fertilized and are generally only a half-open tube. The sepals are linear-lanceolate in shape, 4.5 cm. by 1.4 cm. The lip is trilobed, 4 cm. by 2 cm, wrinkled at the apex and ciliate. The overall colour is a pinkish blue.

Flowering Periods: Occasionally they appear in December and January, but usually concentrated in February and March.

Pollination: As for *C. brasiliensis*.

Cultivation: Unknown.

Cleistes lepida
(Rchb. f.) Schltr. | 237 (Plate 04)

Etymology: *Lepido* is Greek for scaly, presumably referring to the scaly callus.

Habitat: A terrestrial found growing on well-drained roadside banks amongst tall grasses in full sunlight at 1200-1300 M.asl in section IV.

Occurrence: Occasional.

Plant: A small terrestrial herb. A single stem arises from a node in the root mass and reaches a height of up to 40 cm. Between 3 and 6 leaves arise alternately at four to five cm. intervals. These are on false petioles up to 5 cm. long which also sheath the stem. The distinctly veined leaves are elliptic, a dark, dull green and measure up to 7 cm. long by 3 cm. broad. The root system is not extensive and composed of 2 or 3 thickish, yellowish roots sometimes with a small tuber.

Inflorescence and Flowers: Up to five pale pink flowers arise sequentially from a node in the axil of the topmost leaf bases. These are held on very short pedicels and, unlike many *Cleistes* species, open widely and are most attractive. The sepals are equal, 3.5 cm. long by 0.7

cm. broad and lanceolate. The petals are a similar shape but shorter and wider. The trilobed lip is distinctive, deeply veined in royal purple with the central lobe an almost blackish-purple and showing a gross waxy central yellow callus running the length of the lip. This measures 3.2 cm.long by 1 cm. broad.

Flowering Periods: January through to March. An individual flower lasts for two or three days only but the plant may show flowers for up to three weeks.

Pollination: *Cleistes* flowers in these regions are almost always pollinated. **Dressler** suggests that bees are attracted by deceit as *Cleistes* species offer no nectar.

Cultivation: Unknown; we have not yet discovered a technique to induce germination in vitro.

Cleistes sp. unnamed *237a* of the *C. lepida Alliance*

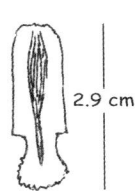

Etymology: *Lepido* is Greek meaning scaly. This could refer to the mealy or scaly callus.

Habitat: Terrestrial in a well-drained roadside bank environment at 1200-1400 M.asl in section VI, enjoying full sunlight, amongst tall grasses, and heavy summer rains during or just before its flowering period.

Occurrence: Rare.

Plant: A medium-sized terrestrial herb. A new growth arises from a node in the centre of the root system, first appearing above ground in January. The stem is woody, round, erect and can reach up to 50 cm. The leaves are fleshy, lanceolate, alternate and dark green, the first arising at 10 cm. from the stem base where they measure 6 cm. by 3 cm. broad but become progressively smaller towards the apex. The root system shows a few short thick yellowish roots some of which may have small tubers.

Inflorescence and Flowers: These are produced continuously during February and March. The new single flower emerges on a very short pedicel from a node in the axil of the next-to-apical leaf. The flowers appear as tubes and essentially only open enough to show the tip of the lip. The petals and sepals are ivory white with a faint pink tinge at their apices. The sepals are ± equal in size and shape, linear-lanceolate, 3.2 cm. long by 0.6 cm. broad. The petals are 2.9 cm. long by 0.7 cm. wide and are a broad lanceolate. The lip is trilobed, 2.9 cm. long by 0.8 cm. broad with a serrated, slightly ciliate, club-shaped apex which hangs down almost reflexed. The lip is ivory white with a purple tip to the apex and shows a mealy callus running its full length.

Flowering Periods: February and March. The individual flowers last for a few days only but the plant may show flowers for more than a month.

Pollination: Probably bees by deceit as *Cleistes* flowers offer no nectar (**Dressler**).

Cultivation: Unknown.

Cleistes pluriflora (Barb. Rodr.) Schltr. | *241* (Plate 04)

Etymology: *Pluriflorus* is from the Latin meaning several flowered.

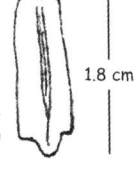

Habitat: Low regrowth scrub forest on high mountain ridges and fields, often in most inhospitable conditions, from 1200 M.asl upwards in section VI.

Occurrence: Occasional.

Plant: A small delicate terrestrial. The thin round, light purple stem is up to 47 cm. tall and arises from a root node. There are no apparent tuberoids but one or two of the 4 or 5, 0.3 cm. thick, white roots are swollen to 0.8 cm. in diameter. The 4 or 5 deeply-keeled, alternate, lanceolate, dark green leaves measure 5 cm. long by 1.7 cm. wide and arise on the top 30 cm. of the stem at 4 cm. intervals.

Inflorescence and Flowers: Single short, apical inflorescences arise with two or three dark pink flowers appearing sequentially at the new apex. They often do not open fully, but appear as pretty pink tubes slightly opened at the lip's apex. The sepals are pink, a pointed linear, 2.3 cm. long by 0.4 cm. wide. The pink, lanceolate petals are 2.1 cm. long by 0.5 cm. wide. The broadish ivory-coloured trilobed lip has an acuminate central lobe and is 1.8 cm. long by 0.8 cm. wide. It has the usual longitudinally running Cleistes callus which is yellow and the tip is tinged pink.

Flowering Periods: Late February and March and the flowers last for several days, but the plant may show flowers in bloom for up to a month.

Pollination: As for *C. brasiliensis*.

Cultivation: Unknown.

Cleistes metallina (Barb. Rodr.) Schltr. | 252 (Plate 04)

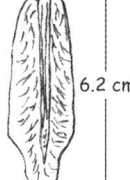

6.2 cm

Etymology: *Metallic* is Greek probably for like metal, and refers to the leaf texture.

Habitat: The habitat is similar to that of the other large Cleistes species and in fact they often grow together. Found from 800-1200 M.asl in sections, I, IV & VI.

Occurrence: Common.

Plant: This is another medium-sized terrestrial, very similar to *C. vinosa* and *C. calantha* and indistinguishable from either vegetatively.

Inflorescence and Flowers: Produced on a 70 cm. stem, typically in small roadside colonies, and arising continuously on single, short, apical and sub apical inflorescences. The flowers are the largest of the region's *Cleistes* species. The sepals are equal, tapered and a pointed linear, 10.1 cm. long by 1.4 cm. wide. The petals are broader, row-boat shaped and 9.8 cm. long by 2.3. cm wide. The sepals, a deep purple within, are a light ivory, tinged with pink without, while the petals are a deep purple. The real distinctive feature of *C. metallina* is in the lip, which is 6.2 cm. long by 3.4 cm. wide, extremely serrated at the apex and faintly trilobed. It is deep purple, almost black at the tip, phasing to white at the throat, and with a massive yellow, pencil-like callus, elevated at the base, running the length of the lip, and erupting into a myriad of cilia at the tip.

Flowering Periods: The flowers last for a day only but the plant may show flowers in bloom for over a month during March and April.

Pollination: We have seen bumble bees visiting the flowers.

Cultivation: Unknown.

Cleistes calantha Schltr. | 257 (Plate 04)

Etymology: *Calanthinus* is Latin for cup-shaped, probably referring to the lip.

Habitat: Occurs in small colonies on 45º roadside banks and in scrub forest from 1000-1400 M.asl. Almost always found in direct sunlight but protected by tall grasses and other herbs, in sections II, IV & VI.

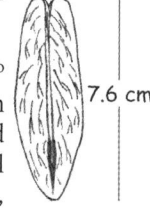

7.6 cm

Occurrence: Common.

Plant: A medium-sized terrestrial, very similar to *C. vinosa*. However the leaves differ slightly in shape and size and the plants tends to be larger. The stem, rising

up to 70 cm. is round, woody but brittle, and bears thick, wide, light green lanceolate leaves from near the base to the apex, at 10 cm. intervals. The larger leaves are 14 cm. long by 6 cm. wide and the smaller apical leaves are flower bracts. The stem arises from a nodule on a small tuberoid, in the centre of the root mass. The roots are yellow, few but thick and evidence was found of decayed tuberoids on the previous years root growth.

Inflorescence and Flowers: These arise continuously on single, short, apical and sub apical inflorescences. This has the second largest flowers of the regional *Cleistes* species. The overall colour is a rich purple with a thin, ridge-like callus on the starkly purple-veined labellum. The sepals are ± equal, pointed, thin and linear at 7 cm. long by 0.9 cm. wide. The petals are a wide pointed linear shape, 7.6 cm. long by 1.5 cm. wide. The lip has a single lobe, but is so serrated at the base that this is difficult to discern. It measures 7 cm. long by 2.5 cm. wide with the thin ridge-like callus almost reaching the tip.

Flowering Periods: March and April and the flowers last for a day only but the plant may show flowers for up to a month. All plants in a region will show their ephemeral flowers on the same days.

Pollination: We have seen the flowers being visited by bumble bees and the waxy look of the prominent calli might indicate visits from wax- or oil-gathering bees. But since the flowers are almost always bear fruit, and as they only offer deceit to pollinators they may be self-fertilized.

Cultivation: Unknown.

General: One of the most beautiful of the large orchids in the region.

Cleistes speciosa Gardner | 270 (Plate 04)

Etymology: *Speciosa* is Latin for showy, or splendid, which fits the flower.

Habitat: This is found in high altitude open regrowth scrub forest on or below mountain ridges at above 1200 M.asl in section I, V and VI. A good humid base of deep leaf detritus with support from the surrounding vegetation and dappled light satisfy this plant's needs.

8.2 cm

Occurrence: Occasional.

Plant: A large terrestrial plant, very similar to *C. metallina*, *C. calantha* and *C. vinosa* but distinguished by being slightly taller than the others and it prefers scrub forest at higher altitudes. The stem, which is round, stiff, jointed, woody and light green, emerges from a nodule in the centre of the root system in

February. By the flowering period, April, it has grown to just under 100 cm. Alternate, pointed, light grayish green, oval leaves emerge at intervals on the top 40 cm. of the stem. The apical leaves form flower bracts. The root system is composed of a few, thick, yellow roots.

Inflorescence and Flowers: These arise on short, single, sub apical inflorescences. Up to 5 flowers appear sequentially. These are large and showy, and overall, a rich dark purple. The sepals are lanceolate, narrow and pointed, 9 cm. long by 1.2 cm. wide. The petals are a similar shape but a little broader and shorter at 8 cm. in length and 1.8 cm. in width. The faintly trilobed lip, 8 cm. long by 2 cm. wide, has a blunt apex and a long thin callus. The flower opens well, with the petals and sepals curling backwards from their tips.

Flowering Periods: April and the flowers last a day only, but the plant may show flowers in bloom for a month.

Pollination: The same remarks as applied to *C. calantha*.

Cultivation: Unknown.

Cleistes vinosa
(Barb. Rodr.) Schltr. | *271* (Plate 04)

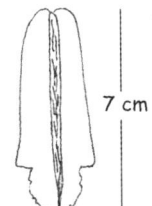

7 cm

Etymology: *Vinosus* is Latin for wine-coloured, purplish-red, referring to the flower.

Habitat: Found on 45° inclined roadside banks in section V with a cover of grasses and other herbs; in mountain fields and on scrub-covered ridges, at 1200 M.asl upwards and requiring full sunlight, a well drained substrate, in section VI.

Occurrence: Occasional.

Plant: A large single-stemmed terrestrial plant, sometimes found in colonies. The new growth comes from the centre of the root mass and emerges in December as a single erect to pendent, woody, round stem, 0.5 cm. in diameter which reaches up to 60 cm. in height. The alternate leaves, starting half way up the stem, are ovate, a rich green, powdered with silver and measure 12 cm. in length by 5.5 cm. in width. The root system consists of a few thick yellow roots, several of which develop into fairly substantial tuberoids which **Hoehne** states are rich in gluten.

Inflorescence and Flowers: These appear from a nodule at the base of the apical leaf on short single inflorescences. This process is continuous, and as a new apical leaf is produced so is another flower. The flowers are large, very showy and reminiscent of a *Laelia* species but very short-lived and delicate. The sepals are linear-subulate in shape, more or less equal, 7.2 cm. long by 1 cm. wide. Petals are 7.5 cm. long by 1.6

cm. wide and lanceolate. The lip is mildly trilobed and large at 7 cm. long by 2.5 cm. wide. It is serrate and ciliate at the tip and has a yellow, thick laminate callus. The overall colour of the flowers is a rich dark purple but the sepals are whitish ivory on the outsides.

Flowering Periods: In March and the flowers last for one or two days but the plant may show flowers in bloom for a month.

Pollination: See the remarks for *C. calantha*.

Cultivation: Unknown.

The genus
Psilochilus Barb. Rodr.

An Atlantic rain forest genus of two species, only one of which pertains to the Organ Mountain Range. **João Barbosa Rodrigues**, the lion of Brazilian botany, raised this genus named from the Greek and referring to the smooth lip. It has the same growth habit as *Erythrodes*.

Psilochilus modestus
Barb Rodr. | *281* (Plate 04)

Etymology: *Modestus* is Latin for humble in appearance, referring to the general aspect of the plant.

3.0 cm

Habitat: A terrestrial found in open scrub or regrowth forest but also in clearings in original forest from 700-1600 M.asl in section V and VI, either solitary or in small colonies where it enjoys dappled light and deep humid leaf detritus as a substrate.

Occurrence: Occasional, but it may be more common as when it is not in flower it is easily confused with small Gesneriaceae species or even a young *Tibouchina* plant.

Plant: A small to medium-sized creeping terrestrial plant. The leaves are ovate with a deep central vein and two side veins parallel to the margins as in many Melastomataceae. The leaves measure up to 6 cm. long by 3 cm. broad with a false petiole which sheaths the stem for over 4 cm. They diminish in size as they emerge alternately towards the apical inflorescence, becoming minute flower bracts at 0.3 cm. intervals where they are heart shaped, and 1 cm. long by 0.75 cm. The stem creeps under the leaf detritus and rises each year some ten to 30 cm. bearing the inflorescence. The roots are thick, short yellow and intensely hairy and arise at 3-4 cm. intervals on the horizontal stem. Each measures 6 cm. long by 0.5 cm. in diameter. A plant examined showed a 30 cm. long horizontal stem with seven such roots.

Inflorescence and Flowers: The inflorescence is an untidy raceme of up to five whitish green flowers with a purplish-tipped serrated trilobed lip, reminiscent of a *Maxillaria* flower, inasmuch as the lateral sepals curve inwards like ant's jaws. The dorsal sepal is a narrow lanceolate measuring 2.7 cm. by 0.5 cm. The lateral sepals are a lanceolate, tough boomerang shape, 2.2 cm. long by 0.6 cm, while the petals are also an asymmetric lanceolate, 2.0 cm. long by 0.4 cm. broad. Both sepals and petals are light green and translucent.

Flowering Periods: In December and the individual flowers last for four or five days while the plant may show flowers for a month.

Pollination: Either self pollinating or by small bees.

Cultivation: The plant transplants well if kept humid during the process, and due to its horizontal growth habit, probably grows best in a large pan, with a deep leaf detritus substrate which should be regularly misted and never allowed to dry out.

General: The epithet *modestus* does not do justice to the flowers which are quite beautiful.

The genus *Elleanthus* Presl

Presl, a natural history professor of Prague University, raised this generic name in 1827, which is attributed to **Helle**, daughter of **Athemas**, King of Thebes, who sent his two children away to safety on the back of a golden ram. Golden rams could fly in those pre-Jesus days. Unfortunately **Helle** fell off the ram just above the narrow water division between Europe and Asia at a spot later to be called the Hellespont. So our orchids and their genera get named.

The genus *Elleanthus* covers about 50 species from the Americas of which some seven are attributable to Brazil and three to the Organ Mountain Range. We have found all three and all are interesting in a botanical sense but of little value to the grower as their flowers are relatively insignificant, although **Hoehne** suggests they could form a group in a garden corner replacing bamboo; we, of course, are talking about a Brazilian garden.

Elleanthus brasiliensis
Rchb. f. | *283* (Plate 04)

Etymology: *Brasiliensis* means that of the 50 or so species in the genus, this is the Brazilian Elleanthus.

Habitat: This plant can be found most commonly at 400-800 M.asl but its range extends to 1200 M.asl at ground to mid-tree level in original forest where it requires dappled light, some air movement and high constant humidity within sections IV, V and VI.

Occurrence: Common.

Plant: A large epiphyte, terrestrial or lithophyte. Its stem and leaves are reminiscent of, at the same time a *Sobralia* species and a small bamboo. Canes, up to 1 metre in height emerge from nodules in the abbreviated rhizome. A typical plant would show up to 25 stems. The alternate leaves are dark green, a sharply pointed lanceolate shape, and 15 cm. long by 4 cm. wide, starting 20 cm. from the base. The root system is profuse and composed of thick whitish roots, extremely adherent and often totally embracing a thick bough.

Inflorescence and Flowers: Up to 25 small, roundish white and pink flowers appear forming a very attractive terminal rosette. The sepals are a pointed oblong, 0.8 cm. long by 0.3 cm. wide. The petals are club-shaped, 0.7 cm. long by 0.1 cm. wide. The serrated-edged, one-lobed lip is round, 1.1 cm. diameter.

Flowering Periods: In March and the flowers last for up to three weeks.

Pollination: The inflorescences secrete a large quantity of a clear sticky gel between the flowers similar to *Glomera* spp. in the Far East. This is probably a ruse to attract pollinators as we have observed a colony frequently visited by humming birds, butterflies and bees.

Cultivation: In large baskets with a fibrous, leaf mould compost, well drained and with 10% of coarse sand. Water well during spring and summer only keeping the substrate slightly moist thereafter.

General: There appear to be two distinct colour forms: pure white with two pink splotches at mid-lip and the second with the whole flower a uniform pinkish white.

Elleanthus crinipes
Rchb. f. | *286* (Plate 05)

Etymology: *Crinis* and *pes* are Latin for tufts long weak hair and feet respectively and probably refers to the hairy roots.

Habitat: Forming enormous colonies of 100-300 stems, *E. crinipes* creates its own environment, collecting all manner of detritus at it's base and penetrating this with its ramifying root system. Although grass-like, the leaves never wilt even in the hottest prolonged spells, suggesting that the base is ever moist. Found above 1000 M.asl in sections IV & VI, at mid-tree and on the ground, enjoying good light and wind movement and high constant humidity.

Occurrence: Common.

Plant: Found in large epiphytic or terrestrial colonies. The thin stems rise and arch gracefully. It lacks pseudobulbs but the stem bases are slightly swollen where they join the reduced rhizome. The stems rise 80 cm. and are rose-red pigmented, more so towards the base; dried leaf scales remain around the lower stem. The first leaves arise at 50 cm. from the base and are alternate, lanceolate, finely pointed, dark green and plicate, the bases sheathing the stem for 2 cm. They are 18 cm. long x 3 cm. wide. The roots are covered in hairy pith, are very tightly tangled and adherent to the substrate. They measure 0.5 cm. diameter. It is difficult to estimate root length but a 5 cm. seedling dissected from its substrate had 25 cm. of roots.

Inflorescence and Flowers: A terminal inflorescence with 25-30 small flowers is contained within one green and many pink pointed bracts, and forms a 5 cm. grouping at the apex of the stem. The sepals are ± equal, lanceolate, 1 cm. long by 0.3 cm. wide and the petals are linear lanceolate, 1 cm. long by 0.2 cm. The lip is cup-shaped and white on the inside but pink like the tepals on the outside. When flattened, the lip is vaguely bilobed, wedge-shaped, intensely serrated at its tip and with two basal, small, oblong, parallel and opposite calli.

Flowering Periods: February and March and the flowers last for up to three weeks.

Pollination: We have observed a colony frequently visited by humming birds, butterflies and bees probably attracted by the clear sticky gel secreted between the flowers.

Cultivation: As for *E. brasiliensis* but with slightly cooler conditions.

Elleanthus linifolius Presl. | *287* (Plate 05)

Etymology: *Linifolius* is Latin for flax-leaved, referring to the narrow, 4 mm linear leaves, typical of flax.

Habitat: On rock faces and at mid-tree level on original ridge forest scrub trees at 600-900 M.asl in section VI requiring good, but not direct, sunlight, fair wind movement and a constant high humidity.

Occurrence: Locally common.

Plant: A small intense tufted epiphyte found at high altitudes and very similar vegetatively to grasses and young bamboo. The stems, up to 20 cm high, arise very closely from nodes, tightly clustered on a thick, long wandering rhizome and are thin, round, with up to eight alternate and opposite, rich dark, grass-green,

long-lived leaves borne at 3 cm. intervals and whose bases totally envelop the stem. Each is a pointed linear 7 cm. by 0.4 cm. and keeled, with a hook at their acuminate apices. The root system, to a degree is very similar to the fleshy-rooted *Epidendrum* complex, is made up of up to 15 thick branching roots which arise from the clustered shoot bases and they penetrate, and adhere to the substrate.

Inflorescence and Flowers: Short, up to 1.0 cm. long, clustered flowers are held on the inflorescence as a tight raceme. Sepals and petals are equal, 0.4 cm by 0.1 cm. wide. The single lobed flared lip is 0.4 cm. wide by 0.3 cm. and serrated at the apex.

Flowering Periods: February and March and the flowers last for two weeks or so.

Pollination: Unknown.

Cultivation: Grow like a small *Epidendrum* on well-drained compost, feeding well during growth and never allowing to dry out.

The genus *Cranichis* Sw.

Olof Swartz, a famous Swedish botanist during the late 18th and early 19th centuries gave the genus the Greek name *Cranichis* referring to the helmet-like, non-resupinate, lip. It is composed of about 30 small terrestrial plants, natives of Argentina, Southern Brazil and Central America. The plant described is the only species that should be found in the Organ Mountain Range.

Cranichis candida (Barb. Rodr.) Cogn. | *310* (Plate 05)

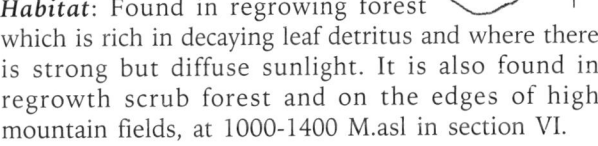

Etymology: *Candidus* is Latin for pure glossy white, referring to the flower colour, which ignores the green spots.

Habitat: Found in regrowing forest which is rich in decaying leaf detritus and where there is strong but diffuse sunlight. It is also found in regrowth scrub forest and on the edges of high mountain fields, at 1000-1400 M.asl in section VI.

Occurrence: Occasional.

Plant: A terrestrial plant which is often found in colonies of 50 or more. The inflorescence arises from a rosette of three or four shiny, oval, deep-green leaves. It is the group that draws ones attention, like a mass of ghostly sentinels. The effect is quite striking. The roots are thick and hairy, permeating the leaf litter and humus for 10-15 cm. around each plant. The thick roots perform any needed storage function and the new shoot arises from a short rhizome.

Inflorescence and Flowers: The 30 cm. tall, erect inflorescence arises from the centre of the rosette and is a 3 cm. diameter cone of 60 or more, minute, delicate white flowers. Individual flowers are small, about 3 mm. x 3 mm. and white. Their predominant feature is a shell-shaped, non-resupinate lip speckled with green spots.

Flowering Periods: June; the flowers last for a month.

Pollination: Maybe **Anthophorid** bees as the lip produces oil – **Dressler**. Seed is set on ca. 70% of flowers.

Cultivation: In pots on a rich decayed leaf mould, watering sparingly during the growth period and never allowing to dry out during the summer. Dappled light and low air movement are essential.

General: **Hoehne** attributes some seven species of the genus *Cranichis* to Brazil but one only to Rio de Janeiro State, *C. candida*. The remaining species are found in the dryer central Brazilian savannas. *C. candida* flowers in June when most other orchids are resting and so adds winter charm to tramping through the high mountain fields.

The genus *Prescottia* Lindl.

A South American terrestrial genus composed of 17 species attributed to Brazil of which 14 should be found in the Organ Mountain Range. **John Lindley** dedicated this genus to the English botanist and friend, **John D. Prescott**.

Prescottia glazioviana Cogn. | *315*

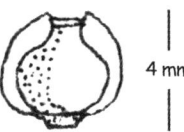

Etymology: Named in honour to **Glaziou**, a French botanist and landscape gardener whose projects are still extant in Rio de Janeiro, Petropolis and Nova Friburgo.

Habitat: A terrestrial found in open maturing regrowth forest at 900-1100 M.asl in section VI and requiring dappled light, low wind movement and a well-drained, slightly moist substrate of leaf litter.

Occurrence: Occasional.

Plant: A small terrestrial fitting well into the **Pabst** alliance grouping of *P. colorans* 'Lip internally glabrous, leaves elliptic, clearly petiolate'. Normally two elliptic leaves appear in March or April. These are opposite, light green, slightly fleshy and measure 11 cm. long

by 5 cm. broad, are deeply keeled and borne on an 8 cm. long petiole. They arise from a central node in the root system, itself composed of a few thick, white to yellow roots which penetrate for a short distance into the surrounding leaf litter.

Inflorescence and Flowers: The inflorescence arises on an up to 25 cm. long, round stiff erect stem, the apical 7 cm. of which has a loose cone of some 30 flowers that open rapidly and consecutively from the base. The sepals and petals are very small, no more than 2 mm. in length, and tend to backfold after a day, making them virtually invisible to the naked eye. Thus the small flower is quite dominated by the round, concave, cup-shaped non-resupinate lip, which measures 4 mm. wide and long and is a light yellow with a greenish tinge.

Flowering Periods: April and the flowers last for a day or so but the inflorescence may show open flowers for more than a week.

Pollination: The flowers are almost always fertilized, **Dressler** suggests possibly by very small oil-gathering **Anthophorid** bees.

Cultivation: In pots on a well-drained, loose, leaf mould, substrate which should be kept moist during summer and autumn.

Prescottia stachyodes Lindl. | *317* (Plate 05)

Etymology: *Stachyodes* is Greek for spike-like, possible referring to the pointed leaves or the spike-like inflorescence.

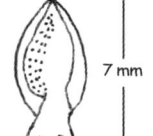

Habitat: Found mostly on high mountain ridges and fields at 1500 M.asl. This plant gets the prize of being the least attractive orchid in the regions of section V and VI, where it enjoys high or dappled light, a leaf mould substrate, summer rains and winter mists.

Occurrence: Occasional.

Plant: A medium-sized terrestrial with very thick roots which penetrate the substrate around. The foliage and inflorescence arise from the centre of a spider-like root system. A rhizome develops after flowering and forms a new spider-like complex some 10 cm. beyond the previous one. Four to five leaves are borne on the 17 cm. stems and each is a perfect light green pointed oval, 17 cm .long by 6 cm. wide, deeply keeled and with eight distinct parallel ribs.

Inflorescence and Flowers: These arise in a tight cone on the top 18 cm. of a 0.4 cm. thick, fleshy, 45 cm. erect spicate inflorescence, which springs from a leafy scale at the base of the root centre. The flowers are like a tiny green monk's hood which in fact is a superior lip; lateral minuscule sepals and larger petals are back-folded around the lip, and the seed pod is formed as the flowers open.

Flowering Periods: July and the individual flowers last for seven days or so but the inflorescence shows flowers blooming for a month or more.

Pollination: The flowers are almost 80% fertilized. Small flies, mosquitoes, and moths are attracted by the perfume from sundown and about three hours thereafter – **Rodrigo Singer**, Campinas University (Unicamp/Brasil), personal communication.

Cultivation: In pots or in a well drained shady slope in your garden, (if it is over 800 M.asl,) on a substrate of leaf mould, following the natural climatic conditions.

Prescottia montana
Barb. Rodr. | *320* (Plate 05)

Etymology: *Montanus* is Latin for pertaining to mountains or montane.

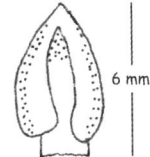

Habitat: A solitary plant, found in leaf litter on the dark, dryer forest floor, at over 1400 M.asl in section VI.

Occurrence: Occasional.

Plant: A small terrestrial, interesting because of its two beautiful leaves which measure 14 cm. long by 5 cm. wide, and are ovate, a very dark green, edged with white, with a thick, almost white central vein. They arise alternately on 6 cm. white, round petioles from the apex of a root-like rhizome. The second, slightly larger leaf, sheaths the flower stem. The short root system is composed of thick yellow roots, partly intertwined around the rhizome. Each years growth is separated by some 5 cm. of rhizome.

Inflorescence and Flowers: A spicate cone arises on the top 6 cm. of a 25 cm. long, green, erect, round and jointed inflorescence, with around 15 minute, green-looking flowers. The predominant feature is the disproportionately large, concave, green heart-shaped lip held above the flower and measuring 0.6 cm. by 0.35 cm. The petals and sepals are minute, a whitish green, and lanceolate. The lateral sepals are larger, 0.38 cm. long by 0.15 cm. wide. All parts curl back after a day or so leaving the superior lip standing alone.

Flowering Periods: March and April and the flowers last for only a few days, but the inflorescence blooms for up to three weeks.

Pollination: Unknown, but probably as with *P. stachyodes*.

Cultivation: As for *P. stachyodes*.

Prescottia nivalis
Barb. Rodr. | *321* (Plate 05)

Etymology: *Nivalis* is Latin for snowy, referring to the colouring of the flowers.

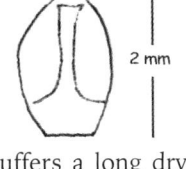

Habitat: We have found one extensive colony in cracks of a vertical rock face in section V at 800 M.asl where it suffers a long dry winter and is exposed to all the elements.

Occurrence: Occasional, but could be more common as it is very difficult to spot.

Plant: A very small terrestrial with a whorl of three or four small elliptical leaves which emerge from the centre of a root mass immediately after flowering. The leaves are pale green with a distinctive central vein and measure 4 cm. long by 1.8 cm. The root system is composed of up to eight fat, short, white, round hairy roots, each up to 5 cm. long by 0.5 cm. in diameter.

Inflorescence and Flowers: A thin, round, stiff, light green inflorescence, up to 15 cm. long emerges immediately after the first spring rains from the root mass centre and bears a tight cone of up to 40 tiny, clear milk-white flowers around the top 20%. These flowers open sequentially and rapidly from the base and are a very typical *Prescottia* type. They measure 0.2 cm. across and must rank with the smallest of all orchid flowers.

Flowering Periods: The timing varies depending on the arrival of the first spring rains but generally is in mid November. These flowers last for a few days but due to the sequential nature of their blooming, the inflorescence will show flowers for up to 10 days.

Pollination: Almost 100% fruit set, and probably fertilized by small flies or mosquitoes.

Cultivation: In small clay pots with a peat-based, coarse substrate, watering only during spring and summer.

Prescottia rodeiensis
Barb. Rodr. | *324* (Plate 05)

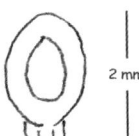

Etymology: Described by **Barbosa Rodrigues** as growing in the shady montane forests of Rodeio in Santa Catarina.

Habitat: Lithophytic and terrestrial on 30° - 40° inclined rock faces in and around cracks and fissures where there is a build up of detritus and a seepage of moisture. Exposed to all the elements with a hot wet short summer and a long dry winter following and found at 800-900 M.asl in section V.

Occurrence: Occasional.

Plant: A small to medium-sized plant with up to six leaves in a whorl and arising from the central root mass. The leaves are stiffish, dark green, slightly reflexed and keeled, elliptic, 9 cm. by 3.3 cm. wide, with a hint of a false petiole. The root system is composed of up to a dozen, thick yellowish, short hairy roots which only slightly exceed the plant's circumference.

Inflorescence and Flowers: A sectioned inflorescence, up to 30 cm. long by 0.3 cm. diameter arises from the centre of the root mass through the whorl of leaves. It is green, and has one or two additional shorter leaves at its base. The tiny flowers are held in a tight narrow cone on the apical 7 cm. of the stem. The flowers are minute; the sepals are a broad lanceolate ± equal, 0.2 cm. long by 0.1 cm. broad and the laterals are joined at their bases. The petals are narrow and linear, 0.2 cm. long by 0.1 cm. broad. All parts are a light green and reflex into a curl within a day or so of the flowers blooming, leaving exposed the light green, inverted, monks-hood-shaped labellum which is 0.2 cm. long by 0.1 cm. wide and deep green. This is so concave as to only leave a narrow hole or roundish slit to permit the entrance of the pollinator or the pollinator's proboscis.

Flowering Periods: Late May-June and the flowers last a day or so but, as they open sequentially from the base, the inflorescence may show flowers in bloom for up to three weeks.

Pollination: Probably tiny flies or moths: seed set is almost always over 50%.

Cultivation: As for *P. montana,* bearing in mind the climate in its natural habitat.

Prescottia epiphyta
Barb Rodr. | 329 (Plate 05)

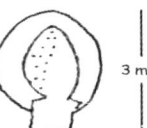

Etymology: *Epiphyta* is Greek for grows on another plant. This is the only *Prescottia* we have found as an epiphyte.

Habitat: Epiphytic or terrestrial; found along trails in both original and mature regrowth forest at low tree trunk level and requiring high humidity, dappled light, and little air movement in section VI at 1200 M.asl and upwards.

Occurrence: Occasional.

Plant: A small epiphytic and sometimes terrestrial plant. It differs from other *Prescottia* species in that its leaves are not decorative. The four or five leaves are in a whorl, arising from a central node in the root mass. Each is dark green, elliptic, deeply keeled and held on a very short false petiole. They vary in size from 4 cm. to 9 cm. in length by 1 cm. to 2 cm. broad. The roots are

few, thick, short white and densely hairy.

Inflorescence and Flowers: A central green, succulent and round stem, measuring up to 20 cm. emerges from the centre of the leaf whorl in late March. A tight cone of some 40 white flowers occupies the apical 6 cm. These are quite typical of all *Prescottia* flowers in shape. The cup-like superior lip dominates and is milk-white-almost round and measures 0.3 cm. by 0.25 cm. The sepals and petals are minute and reflex soon after opening.

Flowering Periods: During April and May and while individual flowers last for a day or two, the cone has open flowers for over a month.

Pollination: **Dressler** suggests small oil-gathering bees.

Cultivation: In pots, on a coarse fibrous substrate taking care that it is well drained but never dries out.

Prescottia lancifolia
Lindl. | 330

Etymology: *Lanci - folia* refers to the pointed leaf shape.

Habitat: Found in Section VI in well-established regrowth forest. A solitary terrestrial growing in a humid, decaying environment with low light and wind movement, high humidity and a substrate of leaf mould.

Occurrence: Occasional.

Plant: A small terrestrial more noticeable for its leaves than its flowers. The broadly lanceolate leaves, 10 cm. long by 3.5 cm. arise from a central node in the root mass on 6.0 cm. petioles and are pliable, floppy, centrally keeled, mildly veined and medium green, with a very mild pink tinge. The root system has a number of short, thick, spiranthoid-type yellow roots which are concentrated at the base of the new growth but also arise from the rhizome itself and are viable over several years.

Inflorescence and Flowers: A tall, sectioned and bracted inflorescence arises from a node at the growing end of a sheathed, stiff, horizontal and underground rhizome. It is up to 35 cm. tall with a cone of up to 40 light green flowers on the apical 11 cm. They open sequentially and are fused to the non-resupinate labellum which is a dark green hood.

Flowering Periods: Late March until April and the individual flowers last for a week or so while the inflorescence flowers for up to a month.

Pollination: This is frequent and probably effected by small flies.

Cultivation: Best grown in small pots on a loamy compost with added leaf mould.

Prescottia octopollinica
Barb. Rodr. | *331* (Plate 05)

Etymology: *Octo* is Latin for having eight and *pollinica* means pertaining to pollen.

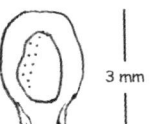

2 mm

Habitat: Found at around 1200 M.asl in section VI, in dryer regenerating forest on well-drained slopes and ridges where a leaf litter is forming and where the scrub forest is dense enough to percolate sunlight and retain some low humidity.

Occurrence: Common.

Plant: A medium-sized terrestrial plant with up to three leaves, each a dark green, faintly veined pointed oval, 12 cm. long by 4.5 cm. wide and arising alternately from the central inflorescence base. The root system is composed of a dozen or so short, thick, 0.5 cm. diameter, yellow roots which penetrate the substrate around to some 4 cm. below the surface.

Inflorescence and Flowers: These appear on an erect, and up to 80 cm. tall, spicate, thin, green, bracted stem. Up to 80 flowers are clustered in a cone around the top 28 cm. of the inflorescence. These are quite typical *Prescottia* models, green, enclosed and hooded by the large superior lip, and have a small opening permitting access to the pollen. Sepals and petals are back-folded around the protruding dark green lip.

Flowering Periods: September and October, the flowers last for several days and the inflorescence shows flowers blooming for up to three weeks.

Pollination: Unknown, but probably as with *P. stachyodes*. The plant sets seed regularly.

Cultivation: As for *P. stachyodes*.

Prescottia polyphylla
Porsch | *332* (Plate 06)

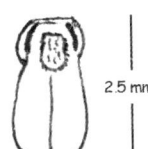

3 mm

Etymology: *Poly - phylla* is Greek for many leaved, referring to an exuberant plant.

Habitat: A lithophyte found at 600-800 M.asl in section V and growing best around moisture seepages from cracks and fissures on 30° - 40° rock faces, in extreme conditions of all the elements, together with a wet late spring and summer and a long dry autumn and winter.

Occurrence: Locally frequent.

Plant: A small to medium-sized epiphyte showing one to several stiff, lanceolate, slightly reflexed, keeled, dark green leaves measuring up to 12 cm. by 4 cm. wide. These arise in a whorl from a central node in the root mass and are held on 3 cm. false petioles. The root system is composed of up to a dozen thick, white, shortish, hairy roots that fix themselves to the rock and the peat-like substrate but do not extend much further than the plants circumference.

Inflorescence and Flowers: An up to 45 cm. long, 0.4 cm. in diameter, stiff green, sheathed and noded inflorescence arises in mid-April from the centre of the root mass in the leaf whorl. The very small flowers are held in a long, tight, narrow cone covering 14 cm. of the stem's apex. The flowers are minute. The dominant feature is the "monks cap" which is the inverted lip and it is emerald green and 3 mm. long. The other significant feature is an acuminate, lanceolate bract which protrudes to the front of each flower. The sepals are a narrow lanceolate while the narrow tiny petals are linear. All reflex in a curl soon after the flower opens and all are a greenish white. The lip's interior is intensely covered by very short, almost wart-like hairs.

Flowering Periods: May and June and as the flowers open sequentially, the plant will be in bloom for up to a month.

Pollination: Probably small flies; seed set is normally up to 50%.

Cultivation: In pots on a coarse sand and fiber substrate following the habitat conditions.

The genus
Wullschlaegelia Rchb. f.

Heinrich Reichenbach dedicated this small American genus to *Wullschlaegel*, a keen collector in Jamaica.

Wullschlaegelia aphylla
Rchb. f. | *336*

2.5 mm

Etymology: *Aphylla* is Latin for without leaves, which these plants lack.

Habitat: A saprophyte found in small colonies in deep leaf litter in both original and regrowth forest between 500 and 600 M.asl, in section VI. It requires a high and constant humidity and low light.

Occurrence: Occasional to rare but because of its modest form it may be more common.

Plant: A small leafless terrestrial saprophyte, almost invisible amongst the leaf litter on the mature forest's floor, only becoming faintly visible when the milk-white seed pods are formed. The root system is abnormal for an orchid and composed of a dozen or so, thin brown, unbranched roots up to 6 cm. long and arising from a central point.

Inflorescence and Flowers: A thin, round fleshy inflorescence emerges from the centre of the root system. It is pink to white and up to 30 cm. high, bearing 25-30 minute white flowers in a tight cone on the apical 9 cm. The petals are fused and when separated, are elliptic, about 1.5 mm. long and 0.5 mm. broad. The lip is spade-shaped and slightly larger with a minute hairy, ball-like callus at the throat. The interesting feature is the fruit which is 8 mm. long, 4 mm. wide and ovoid. All flowers are usually fertilized.

Flowering Periods: In January and the flowers last for only a day or two.

Pollination: Unknown: the fruits ripen within a month.

Cultivation: Unknown; but follow the habitat conditions.

General: This plant is a saprophyte, living on decaying vegetable matter through the medium of symbiotic fungi which inhabit the roots and surrounding humus.

The genus *Sauroglossum* Lindl.

Lindley raised the small genus of one, possibly two, Brazilian species. These are large terrestrials. The name comes from the Greek compound word meaning Lizard's tongue and refers to the shiny large leaves.

Sauroglossum nitidum
(Vell.) Schltr. | *353* (Plate 06)

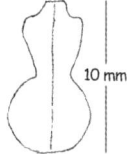

Etymology: *Nitidus* is Latin for shining or polished and refers to the leaves.

Habitat: A terrestrial, found in dryer regenerating forest where the sparse topsoil is very well-drained and the brush and first colonizing tree cover allow diffuse sunlight and a level of constant, though low, humidity. Found in such an environment at around the 900-1200 M.asl in all sections.

Occurrence: Common.

Plant: A medium to large-sized terrestrial plant. The leaves are lanceolate, mid-green, up to 30 cm. long by 5 cm. wide and deeply keeled. As many as 10 arise from

the centre of a spiders legs' format of 15 to 20 very thick, 1 cm. diameter, tuberous roots, stretching up to 30 cm. from the plant's centre, and completely enveloped in fungal mycelium. The plant retains its leaves year round.

Inflorescence and Flowers: The inflorescence is a cone on the apical 20 cm. of the 60 cm. tall, 0.5 cm. diameter, green, bracted stem and shows up to 70 small green and white flowers. The petals are fused to the dorsal sepal. Together, they are light green, spathulate, concave and measure 0.7 cm. long by 0.3 cm. wide. The dorsal sepal is hairy on its upper side and at its fused margins, whereas the petals are entire. The lateral sepals are concave and spathulate measuring 0.7 cm. by 0.2 cm. broad and curve upward and inwards to touch the column at its apex rather like a high-board diver before the winning plunge. The lip is a blunt spoon-shape, single lobed and flared at its apex, measuring 1.0 cm. by 0.4 cm.

Flowering Periods: The flowers last for up to a week. However the inflorescence may show flowers in bloom for a month during September & October.

Pollination: *Sphingid* spp. Moths – **Rodrigo Singer** ,Unicamp, personal communication. The flowers are normally over 50% fertilized.

Cultivation: In large pots on a well-drained rich leaf mould compost, watering in the spring and misting thereafter. In flowering habits, this is a temperamental plant, and just the correct amount of dappled light is the key.

Sauroglossum sp.
(Possibly a new species) *353a* (Plate 06)

Habitat: A large terrestrial plant found amongst rocks and boulders in high mountain fields from 1900-2300 M.asl in section VI.

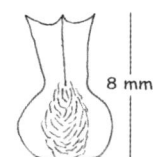

Occurrence: Occasional.

Plant: The foliage is held in a whorl arising from a central root node. The leaves are elliptic, 25 cm. by 8 cm, light green, shiny and somewhat fleshy and have distinct petioles 2.5 cm. long. The root system is intense and composed of many thick, yellowish-white, hairy, shortish roots.

Inflorescence and Flowers: A thick, light green, inflorescence up to 130 cm. tall arises from the centre of the leaf whorl. Its height depends on the height of the grasses and scrub around, which it tends to surmount. Up to 60 light green and white flowers appear as a loose cone on the apical 30 cm. of the inflorescence. The dorsal sepal is a broad lanceolate, 1 cm. by 0.3 cm, light green and finely hairy and is shaded by a 3.5 cm. long, 1.0 cm. broad, lanceolate, green bract. The lateral sepals fall below the lip and are also a light green, hairy, narrow lanceolate 0.8 cm. by 0.4 cm. The petals are linear lanceolate, green and fused with the dorsal sepal, 0.8 cm. by 0.2 cm. The lip is entire, perhaps slightly bilobed, narrowing towards the throat and broadly flared and stiffly reflexed at the apex. It measures 0.8 cm. by 0.4 cm. at the apex.

Flowering Periods: December and January after the spring rains. The flowers last for a week or so but the inflorescence may show flowers for up to three weeks.

Pollination: Probably hawk-moths.

Cultivation: Is surprisingly easy, in a large pot, with a peat enriched substrate; kept wet and humid during spring and early summer, but during the autumn and winter dormant periods simply prevent the plant from drying out completely. It needs high light and reasonable air movement.

General: The inflorescence of this plant is much taller and thicker in diameter than *S. nitidum*. The flower is shaded by a broad lanceolate leaf-like bract which is lacking in *S. nitidum*, and in addition, its flowering period is two months later and the habitat is ± 500-900 metres higher.

The genus
Cyclopogon Presl

K.B. Presl, a natural history professor of Prague University, named this genus in 1827 from the Greek *cyclo* and *pogon*, meaning surrounded with a beard, and referring to the hirsute sepals, viewed from the front. This is a ± 70 species genus from Florida to Brazil, within the sub tribe Spiranthinae which is constantly under revision. At the time **Pabst and Dungs** published in 1973, it was made up of some 40 species attributed mainly to eastern Brazil of which the majority should be found in the Organ Mountain Range. All species, except possibly one, are small to medium sized terrestrials with small, usually white, not very decorative flowers. Some, however, have very attractive leaves and would be classed as Jewel Orchids.

The majority are found in deeply shaded, humid, original, or maturing regrowth forest.

Cyclopogon argyrifolius
Barb. Rodr. | *358* (Plate 06)

Etymology: *Argyri-folius* is a Latin compound word meaning silvery-leaved, referring to the decorative silvery streaks on the leaves.

Habitat: Found as a terrestrial growing in full open light, exposed to all climatic conditions, on a peaty substrate on otherwise bare granite rock faces, with neighbours such as **Oncidium batemannianum**, **Cyrtopodium glutiniferum**, **Laelia cinnabarina**, **Pseudolaelia corcovadensis** and three species of **Vellozia**, in section V at 900-1100 M.asl.

Occurrence: Occasional.

Plant: A small to medium-sized terrestrial distinguished by its four or five, lanceolate, to elliptic, leaves, which are dark green, save for three distinct parallel silver veins, and measure up to 10 cm. by 5 cm. broad. They arise in a whorl, on short false petioles, from a central node in a root mass and are stiff and somewhat coriaceous. The root system is composed of a number of thick white, though short, sausage-like roots, which do not exceed the plant's circumference.

Inflorescence and Flowers: The thickish heavily bracted stem arises from the centre of the leaf whorl for some 35 cm. and bears a bracted, 12 cm. inflorescence carrying 20-25 small flowers. The concave dorsal sepal is lanceolate, 1 cm. long by 0.2 cm. broad. The lateral sepals are awl-shaped and slightly longer. All sepals have pilous margins. The petals are club-shaped, a narrow spathulate, 0.5 cm. long and white. The lip is a waisted rectangle measuring 6 mm. long by 3 mm. wide at the flared apex and is white.

Flowering Periods: The flowers last for two weeks.

Pollination: Probably bumble bees (**Dressler**).

Cultivation: In pots with a peat-based substrate, well drained but never waterlogged and never dry.

Cyclopogon longibracteatus
(Barb. Rodr.) Schltr. | *360*

Etymology: *Longus-bracteatus* is Latin for long bracts, and appropriately refers to those protecting the flowers.

Habitat: Found at 300-700 M.asl in humid original forest in section IV and VI, growing as a low-tree epiphyte or terrestrial. It enjoys low light and air movement and a constant high humidity.

Occurrence: Occasional.

Plant: Up to five light green, oblanceolate leaves appear

in early spring. These have 4-5 cm. long false petioles and emerge as a tight whorl from the stem base. They are finely veined, somewhat fleshy and are 8.0 cm. long by 2.5 cm. The root system is made up of a few thick, 0.5 cm. diameter, short, white and hairy roots, each up to 8 cm. long.

Inflorescence and Flowers: The central stem emerges from the base of the leaf whorl and reaches 14 cm. It bears a cone of nearly 35 small, whitish, clustered flowers on the apical half. Each flower is protected by an infolded superior, light green, narrowly lanceolate, acuminate bract which is as long or longer than the flower. The dorsal sepal and the two thin petals are fused to their apices and, as a group, measure 0.7 cm. long by 0.2 cm. The lateral sepals are milk-white, tinged with green at their bases, forward thrusting and parallel to the lip, a blunt lanceolate shape measuring 0.8 cm. long by 0.2 cm. The lip is white, faintly tinged with green at the margins and measures 0.9 cm. long by 0.25 cm, and 0.3 cm. by 0.45 cm. wide at the base, centre and tip respectively. The tip is vaguely bilobed and serrated at the margins.

Flowering Periods: September, when the flowers last for ten days.

Pollination: They are probably pollinated by bumble bees or male *Euglossine* bees.

Cultivation: In pots on a well-drained compost containing rotting wood and leaf mould and observing the habitat details.

Cyclopogon calophyllus
Barb. Rodr. | 363

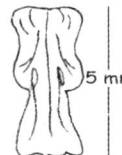

Etymology: *Calo* and *phyllus* are Greek for beautiful leaves.

Habitat: A terrestrial found in very humid stream-side situations, growing on leaf litter in low light at 900 M.asl in section V in regrowth forest.

Occurrence: Rare.

Plant: A small terrestrial, notable, in fact unmistakable, by its broadly silver striped, dark green leaves. Up to four of these emerge from a central node in the root mass and are shortly petiolate, a pointed oval with a venation exactly like that of the **Tibuchina** genus. They measure up to 9 cm. long by 4 cm. broad, are slightly fleshy and floppy with light green undersides. The root system is composed of up to eight, white, thickish though short roots which do not stretch beyond the leaf circle.

Inflorescence and Flowers: A thin, segmented inflorescence up to 20 cm. high arises from the centre of the leaf whorl bearing a few insignificant flowers in

a loose cone at its apex. The lip is typical of the genus; long and narrow with a flared apex, an infurled centre and base, 0.5 cm. long by 0.2 cm. broad at the apex. The smaller petals are fused to the dorsal sepal, while the slightly longer free lateral sepals project forward accompanying the lip.

Flowering Periods: November and December when the flowers last for a week or so.

Pollination: Unknown.

Cultivation: Grow on a loose substrate of decaying leaves, always kept damp, with low light and air movement.

Cyclopogon elegans
Hoehne | 365 (Plate 06)

Etymology: *Elegans* is Latin for elegant, referring to the attractive leaves.

Habitat: On loose leaf litter in humid conditions on the mature forest floor, growing in low light in section I at 1200 M.asl.

Occurrence: Occasional, but in groups.

Plant: A small terrestrial plant with up to three strikingly pretty leaves arranged in a whorl with two opposite, subtending leaves, each on a 3 cm. petiole. The leaves are a broad oval with slightly pointed apices, 6 cm. long by 3.5 cm. and have a Melastomataceae-like pattern of silver lines. There are up to six roots which are white, thick and hairy and 0.6 cm. diameter by 6 cm. long.

Inflorescence and Flowers: This data is taken from the *Iconografia de Orchidaceae do Brasil* by **F. C. Hoehne**, Plate T45, as we have not yet flowered this plant. A slim, light green and sparsely bracted inflorescence, up to 25 cm. tall arises form the centre of the leaf whorl, bearing a loose cone of seven to nine, greenish-white, elongated flowers. The sepals are more or less equal, lanceolate, 1 cm. long by 0.3 cm, milk-white at their apices and greenish at their bases. The petals are half the sepal length and width, club-shaped and similarly coloured. The lip is 1.1 cm. long with a widely flared apex and a narrow waist and throat, and white with a greenish centre.

Flowering Periods: Probably March and April.

Pollination: Unknown.

Cultivation: In pots on a loose, moist leaf mould in diffuse light and intermediate conditions.

Cyclopogon variegatus
Barb. Rodr. | *367* (Plate 06)

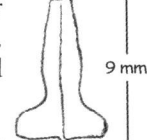

Etymology: *Variegatus* is Latin for variegated, meaning various leaf colours, which may also arise from infection and disease.

Habitat: Rarely terrestrial but more often found on sound or rotting tree trunks up to two metres above the forest floor. It is a solitary plant, unlike our other *Cyclopogon* species which form terrestrial colonies. Found at around 1400 M.asl in section VI, in deep, open original forest and always with its roots thickly moss-covered.

Occurrence: Common.

Plant: A low tree-trunk epiphyte with three to five basal, dark green, oblanceolate, petiolate leaves which arise from the centre of the root system. The stem cross-section has the curious shape of a railway line. There is no pseudobulb, but the root system is extensive and composed of thick adherent and penetrating roots which spread under moss and into rotting wood.

Inflorescence and Flowers: These appear on a stiff, fleshy, erect, spicate inflorescence up to 25 cm. tall. Up to 20 flowers form a loose cone on the top 9 cm. of the inflorescence. The first impression of the flower is of the wide, white, back-curled, two-lobed lip, 0.5 cm. wide x 1.1 cm. long. The two brown, pointed petals are fused with the green, ciliate dorsal sepal. The former are 0.8 cm. in length by 0.1 cm. wide, while the latter are the same length but three times wider. The green, narrow, pointed, ciliate lateral sepals embrace the lip up to the downward curve. The lip itself measures 0.9 cm. long by 0.17 cm. wide.

Flowering Periods: In October and November and the flowers last for two weeks or so.

Pollination: Unknown; the flowers, however, are almost always fertilized.

Cultivation: In pots on a substrate of rotting or dead wood mixed with coarse sand, grown in shady and damp conditions but never allow to be water logged. This plant retains its leaves all year round.

Cyclopogon congestus
(Vell.) Hoehne | *371*
Now: *Beadlea congesta*
(Vell.) Garay | (Plate 06)

Etymology: Congestus is Latin for crowded together, referring to the close knit flower cluster.

Habitat: A terrestrial growing on a thick

leaf litter substrate in transitional regrowth forest in section V at 600 M.asl and requiring a short wet period from October through March with a long dry period thereafter.

Occurrence: Occasional.

Plant: A small to medium-sized terrestrial with a whorl of 4 - 6, oblanceolate emerald green leaves, each held on a 3 cm. false petiole. They are keeled, somewhat lax and fleshy, on average 12 cm. long by 2.5 cm. and they emerge from a central node in the root system. The roots are short, white, thick, non-branching and numerous, spreading to a circumference of some 7 cm. from the central point.

Inflorescence and Flowers: One or two round, bracted, fleshy inflorescences, up to 23 cm. tall, arise from a central point in the root mass. They are curved towards the apex, and bear a tight raceme of some 30 light green and white flowers which open sequentially from the base. The dorsal sepal, which is light green, is tenuously fused with the greenish-white petals. As a unit they measure 0.7 cm. long by 0.3 cm. and are lanceolate. The lateral sepals are a light whitish-green, a clumsy lanceolate 0.8 cm. by 0.3 cm. broad, projected alongside the jutting lip. This is white, very vaguely bilobed at the flared apex, and measures 0.9 cm. long by 0.4 cm. at the apical lobes.

Flowering Periods: July, when the inflorescence shows flowers in bloom for up to a month.

Pollination: Unknown; but always up to 90% of flowers bear fruit.

Cultivation: On a substrate of loose leaf mould in pots large enough to retain moisture for some time, watering well during the growing period but just misting after flowering.

Cyclopogon iguapensis
Schltr. | *381* (Plate 06)

Etymology: Refers to the city of Iguape, SP.

Habitat: Found in deep, dripping, damp shade under tall original forest, generally in sunless gullies and valleys between 1000 and 1400 M.asl enjoying low light, low wind movement and constant damp in section VI.

Occurrence: Common.

Plant: A terrestrial from the deep forest. There are five or six sequential leaves at 1 cm. intervals starting at ground level. These are dark green, pointed and oblanceolate, borne on a thick 8 cm. long petiole and up to 20 cm. long by 8 cm. at the widest point. They are deeply keeled and tend to flop down to the ground at their tips. The roots are thick, yellow and when young, hairy. The root system is extensive under the leaf litter and organic topsoil. No storage organ has been observed.

Inflorescence and Flowers: On an erect, 70 cm., jointed, round, stem which is 0.75 cm. thick at the base and 0.4 cm. thick where the flowers start, at 60 cm. from the base on a racemose inflorescence. Between 25-30 small white and green flowers appear, giving a cone-shaped apex to the inflorescence. The lip is broad, white and 1.1 cm. long. The distinctive feature is the white-tipped dorsal sepal, fused with the petals for 9/10 of their length, while the 1 cm. long, lanceolate lateral sepals are light green and appear almost as bracts.

Flowering Periods: July up to September and the flowers last for a few days but the inflorescence may show flowers blooming for up to three weeks.

Pollination: Probably bumble bees (**Dressler**).

Cultivation: In pots with a peat or leaf mould substrate, always kept damp but not water logged and grown in shade. This plant retains its leaves year round.

Cyclopogon candidus (Krzl.) Pabst | 388 (Plate 06)

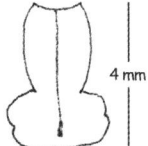

Etymology: *Candidus* is Latin for glossy white, which refers to the flowers. However they do have green stripes which **Pabst** overlooked when naming it.

4 mm

Habitat: A gypsy orchid found in regrowth and high mountain scrub from 1200-1600 M.asl. It moves to a convenient environment on an annual or biennial basis. It requires diffused light, a leafy substrate and a low but constant humidity. Its propensity to prolific seeding obviously facilitates its desire to migrate constantly around section VI.

Occurrence: Common.

Plant: A small terrestrial, which emerges in a ghostly sentinel way, as the 25 cm. inflorescences arise from the substrate before the leaves appear. There are two or three dark green leaves, deeply keeled and narrow pointed ovals on significant petioles and measuring 4 cm. long by 1.3 cm. wide. There is a fat, translucent white tuber, from which flow a few thin roots.

Inflorescence and Flowers: Four or five alternate small flowers appear on the spicate, 25 cm, round, sparsely-bracted inflorescence and are concentrated on the top 4 cm. The overall effect is white, streaked with thin green stripes which are particularly significant on the down-curved, wide lip.

Flowering Periods: September and October and the flowers last for about a week.

Pollination: Unknown.

Cultivation: Unknown.

The genus *Hapalorchis Schltr.*

This has been taken out from *Cyclopogon* for reasons we cannot quite understand. However, the genus was originally raised by **John Lindley** and now contains two species in the Organ Mountain Range.

The genus name is a Greek compound composed of *hapal* meaning soft and *orchis* meaning orchid which could be interpreted as meaning a delicate orchid.

Hapalorchis lineatus (Lindl) Schltr. | 389

Etymology: *Lineatus* is Latin for many fine lines, referring to the distinct leaf pattern.

9 mm

Habitat: A terrestrial found in deep leaf litter in original humid forest at 1000-1200 M.asl. in section V and VI receiving low light, high humidity and low air movement.

Occurrence: Rare.

Plant: A surprisingly delicate clumped terrestrial; the two basal leaves are a broad lanceolate, mildly indented at their margins, lightish green and somewhat flimsy, up to 6 cm. long by 3 cm. broad with a strong central vein and quite visible minor veins, They are held on a 7 cm. wine coloured petiole and arise from a central node in the root system. The roots are banana-shaped, thickish, white and hairy, swollen for the major part but thinner at both ends.

Inflorescence and Flowers: A tall, thin, erect noded, inflorescence up to 25 cm long, arises from the centre of the root mass in September. The few flowers form a loose cone on the apical 6 cm. The dorsal sepal is a broadish shouldered lanceolate, 1 cm. long by 0.4 cm. The lateral sepals are a narrow linear-lanceolate 1.1 cm. by 0.3 cm. wide while the petals are smaller, baseball-bat shaped and 0.6 cm. long by 0.2 cm. wide. The lip is stout, thick-necked and bilobed, 1 cm. long by 0.7 cm. wide at the apical lobes and reflexed. The overall colour is a whitish, greenish-wine.

Flowering Periods: In September and October and the flowers last for a week or so.

Pollination: Unknown; but generally successful.

Cultivation: In pots on a loose, leaf litter substrate always kept damp, and grown in low light and air movement with high constant humidity.

Hapalorchis micranthus (Barb. Rodr.) Hoehne | 390

Etymology: *Micranthus* is Greek for small-flowered, and presumably refers to the small number of flowers shown by this plant or their size.

Habitat: A terrestrial growing on thick, damp, leaf mould in shade under shrubs at 1500 M.asl in section VI, enjoying mists all year round a very wet summer and a cool, much drier winter.

Occurrence: Occasional.

Plant: A small delicate plant very similar to *Cyclopogon candida*. One basic difference being that the one or two leaves appear before the flowers open. The leaves are carried on short white petioles up to 1 cm. long, and arise from a node at the end of a short white potato-like tuber 1 cm. long by 0.4 cm. thick. The leaves are a broad lanceolate 2 cm. by 0.8 cm. broad, keeled, silky and light green. One or two very short, hair-like roots emerge from the tuber.

Inflorescence and Flowers: The inflorescence emerges from the leaf petiole base at the apex of the tuber and is up to 12 cm. long, thin, delicate and white, and bears one to four sequential flowers, small, white and tinged with green on very short pedicels. The dorsal sepal is creamy white, fused with the petals for 75% of its length, 0.6 cm. by 0.25 cm. and lanceolate. The lateral sepals are the same size and shape but milk-white and greenish at their base. The petals are an asymmetric lanceolate, 0.65 cm. long by 0.1 cm. wide, white with a green central vein. The vaguely bilobed lip is butter-yellow at its base for half its length, while the rounded crenate apex is flared and white, with three short green, longitudinal lines, 0.7 cm. long by 0.3 cm.

Flowering Periods: September and the flowers last for up to a week.

Pollination: Unknown.

Cultivation: On a substrate of loose, well-aerated leaf mould in shade and never allowed to dry out.

The genus Sarcoglottis Presl

Raised by **Karl Presl**, a professor of natural history at Prague university in the 1840s who studied the collections made by **Haenke** in the Spanish South Americas. *Sarcoglottis* is a Greek compound word meaning a fleshy tongue, which indicates the texture and aspect of the lip. A terrestrial genus of 50 or so species found throughout South America with 29 species attributed to Brazil by **Pabst** and **Dungs** and five to the Organ Mountain Range.

Sarcoglottis grandiflora (Hook.) Kl. | 403

Etymology: *Grandi-flora* is Latin for large-flowered.

Habitat: A terrestrial from section VI found at 800 M.asl. in deep, very moist original forest on a mass of leaf mould and requiring modest air movement, low light and high humidity.

Occurrence: Rare.

Plant: A large terrestrial plant which, at first sight looks like *Sauroglossum nitidum* or a *Cyclopogon*. The leaves are a broad oblong-elliptic shape, fleshy and bright green, measuring up to 20 cm. long by 10 cm. They arise in a tight whorl of four to five from the central nucleus of the root system. There are up to a dozen of these thick, white equal-sized roots, each up to 15 cm. long and 1 cm. thick at their base.

Inflorescence and Flowers: The thick, stiff stem arises from the centre of the leaf whorl and is densely covered by fine white, short hairs which give it a halo effect when viewed from a few metres away. It is up to 45 cm. tall and bears a narrow cone of up to 30 light brown, strongly bracted flowers which surround the 25 cm. apical section. The dorsal sepal is lanceolate, 2.2 cm. long and 0.5 cm. broad. The petals are narrowly spathulate and partly fused with the dorsal sepal, measuring 1.8 cm. long by 0.45 cm. The lateral sepals are asymmetrically lanceolate, like a boomerang and are 2.2 cm. long by 0.4 cm. The long lip is constructed in two parts: the upper basal two thirds is spathulate and 2.5 cm. long by 0.7 cm, while the almost round apical one third is flared and measures 0.7 cm. long by 0.8 cm.

Flowering Periods: November, and the flowers last for ten days.

Pollination: **Dressler** suggests the flowers are pollinated by bumble bees or *Euglossine* bees.

Cultivation: In large pots with a fibrous, well-drained compost, needing plenty of water in spring and summer and simply misting to prevent drying out in autumn and winter.

The genus *Eurystyles* Wawra.

Heinrich Wawra was an Austrian who accompanied several collecting expeditions to Brazil in the 1830s. He named the genus using the Greek *eury* meaning broad and Latin *stilus* a base, referring to the extension of the ovary, bearing the stigma. All four species from this small Brazilian genus can be found not 200 metres from our house, at 1450 M.asl, in Macaé de Cima. Additionally they can be found in all sections of the Organ Mountain Range, in original or regrowth forest where there is a modicum of constant humidity. **Pabst and Dungs** state that this genus is a good example of a terrestrial evolving into an epiphyte. It could well be.

Eurystyles actinosophila (Barb. Rodr.) Schltr. | 421 (Plate 07)

Etymology: *Actino* is Greek for ray, star-like, or radiating from a centre, and phyll is also Greek for relating to leaves, clearly describing the plant.

Habitat: Like most *Eurystyles*, this one grows in colonies on low tree trunks or on saplings, in low light and little air movement but with significant humidity. Found in original forest or in regrowth between 1300-1600 M.asl, in section VI and VIII.

Occurrence: Common.

Plant: A small epiphyte with a rosette of 6-8 light green spathulate leaves, so shiny on both sides that they appear to be permanently moist. These glabrous leaves, faintly keeled, measure 3.5 cm. long by 1.5 cm wide. The rather weak root system is composed of 5 or 6 grey-green, thick, generally moss-covered roots which do not reach beyond the plants circumference.

Inflorescence and Flowers: These appear slowly in a composite arrangement, carried on a 2.5 cm. pendulous, hairy stalk arising from a central node. The inflorescence contains up to 15 flowers, each separated by an almost equilateral triangular-shaped, serrated-edged bract. The small white flowers appear as open-mouthed tubes, always facing downwards. The ± linear dorsal sepal and the partly fused petals are roughly equal in size and shape, though the petals are more obtuse at the apex and measure 0.4 cm. long x 0.15

cm. wide. The lateral sepals are twice the size and subulate. The 0.4 cm. long lip is faintly trilobed, appearing like a shafted spear head.

Flowering Periods: In February and March and the flowers last for up to three weeks.

Pollination: Unknown.

Cultivation: Grow on plaques of soft tree fern, kept moist always, in dappled shade.

Eurystyles cotyledon Wawra | 422 (Plate 07)

Etymology: *Kotule* is Greek meaning cup or hollow of equal, fleshy, free, but tightly touching petioles to the leaves.

Habitat: Always found within a few feet of the ground yet never on it, this species grows on tree trunks both in regenerating and original forest from 1000-1500 M.asl in all sections. The plant looks healthy even during prolonged periods of very hot, dry weather.

Occurrence: Common.

Plant: This interesting and unusual little plant is more reminiscent of a Saxifrage than an orchid. The leaves, which form a rosette, are a translucent grey-green oval shape, tapering to a point and with a distinct white stalk. They are shiny and appear slimy or wet-looking and measure 4 cm. x 1 cm. and a lens shows the edges to be finely ciliate. The colony of 12 plants we studied were closely bunched and bore 12 flowering spikes, slightly longer than the leaves. There are twelve leaves in each rosette declining in size towards the centre. The roots cover a similar area to the plant, 1.5 cm. diameter, and are grey but yellowing towards the older parts, and appear densely entwined in a fine powdery substrate, looking like nest of worms.

Inflorescence and Flowers: A composite-like inflorescence arises on a fleshy, round, 0.2 cm. thick, pendulous, green stem. There are up to 30 flowers, each separated by a short, lanceolate hairy bract. The dorsal sepal and the petals are similar in shape, partly fused, with a greenish tinge at the obtuse apex; club-shaped and 0.35 cm. long by 0.15 cm. broad. The sepals are twice as long as wide, an extended oval shape and white. The single lobed lip is club-shaped, 0.6 cm. long by 0.25 cm. wide, white with pronounced shoulders and a greenish tinged, ciliate apex.

Flowering Periods: February and March; the flowers last for up to three weeks.

Pollination: Unknown.

Cultivation: As for *E. actinosophila*.

Eurystyles cogniauxii (Krzl.) Pabst | 423 (Plate 07)

Etymology: Named in honour of **Cogniaux** a famous Belgian botanist in the 19th century.

Habitat: We have found this curious little ex-terrestrial, now a low-tree epiphyte, from 1000-1500 M.asl in all sections growing in deep shade and high humidity from almost ground level up to 4 metres, on humid, moss-covered trunks or saplings.

Occurrence: Common.

Plant: Very similar in form to all our other *Eurystyles* species and inhabiting a similar environment. It is a small, low tree-trunk epiphyte often forming quite extensive colonies. The leaves are a darkish green and oblanceolate in shape, slightly petiolate, 3.4 cm. long by 1.2 cm. wide, with a central keel and arranged in a rosette. The root system is neither extensive nor adherent and the roots are thick and white to green in colour.

Inflorescence and Flowers: This is the first *Eurystyles* to flower in spring, on a 6 cm, fine, bracted, pendulous inflorescence with a large terminal 3 leaved bract enclosing the flowers. There are generally three, sometimes four, flowers, whose clear distinguishing characteristic is the white overall effect with thick claret lines on the petals. The petals are partly fused with the dorsal sepal and are milk-white, 1 cm. long x 0.2 cm. wide and blunt at the apex. The dorsal sepal is pointed, the apical half is a milk-white and the basal half almost emerald green. The lateral sepals have a similar colour arrangement but are slightly more pointed and curvaceous, 1.2 cm. long x 0.4 cm. wide. The lip is almost rectangular, 1 cm. long by 0.4 cm. wide and is distinguished by two large emerald blotches at the half way point.

Flowering Periods: October and November and the flowers last for three weeks.

Pollination: Unknown.

Cultivation: As *E. actinosophila*.

Eurystyles lorenzii (Cogn.) Schltr. | 424

Etymology: Not known.

Habitat: Similar to that of other *Eurystyles* species.

Occurrence: Common.

Plant: A small epiphyte, quite different in general appearance from the other species of this genus. It sports 2-4 light green shiny, leaves, each an extended pointed oval, 2.5 cm. long by 0.8 cm. wide, surrounding the emergent inflorescence. The root system is weak, consisting of a few short, thickish, greyish-white roots.

Inflorescence and Flowers: One or two white flowers appear on a pendant, 2 cm. long, densely hairy, 0.1 cm. diameter inflorescence, arising from a central root system node. The sepals are lightly ciliate, triangular in shape, + equal, 1.0 cm. long by 0.3 cm. wide at base. The petals are sword-shaped and slightly smaller. The lip is a distinct spear head, 1 cm. long by 0.25 cm. wide at the base.

Flowering Periods: February and March and the flowers last for three weeks.

Pollination: Unknown.

Cultivation: As for *E. actinosophila*.

The genus Pelexia Poit. ex L. C. Rich

Richard, a Frenchman, collected in French Guiana and Pará between 1781 and 1789. He named the genus *Pelexia* probably because the dorsal sepal of species in this genus resembles a *pelex*, Greek for a helmet. A South American genus of 50 species of which 38 are attributed to Brazil and six to the Organ Mountain Range, generally at high altitudes.

Pelexia laminata Schltr. | 427 (Plate 07)

Etymology: *Lamina* is Latin for blade-like, referring to the sepal shape.

Habitat: A terrestrial from the high mountain fields at 1800-2200 M.asl, growing among tall grasses, scanty shrubs in a well drained but almost peat-like substrate composed of dead grasses. The climate is heavy rains in spring and summer, cold mists in autumn and winter with high light and constant winds. Found in section IV and V.

Occurrence: Solitary and occasional.

Plant: A medium-sized terrestrial plant with the leaves appearing before the stem, which they surround. The stem emerges from a node in the central root system. The basal leaves are 15 cm. long by 2.5 cm. wide with a pink central vein. They are opposite, strongly veined and deeply keeled, stiff, dark green, lanceolate and sheathed by the initial, infant leaves which are pink. The root system is vigorous and composed of up to 15 thick, short 0.5 cm. long, white roots which appear to be viable for several years.

Inflorescence and Flowers: The stem reaches 40 cm. with up to four erect linear-lanceolate leaves which fold around the 0.8 cm. diameter stem. Up to 50 flowers are held in a tight cone on the apical 6 cm. and are overall a pale whitish-green. The dorsal sepal, which has a light green pilose cap and is partly fused to the petals, is lanceolate, white to greenish and measures 1.4 cm. long by 0.5 cm. The lateral sepals are a slightly asymmetric lanceolate shape measuring 1.3 cm. by 0.4 cm. and white but with a deep central green vein. The partly fused petals are lanceolate, acuminate and 1.2 cm. long by 0.25 cm. The lip is faintly bilobed, longitudinally green-veined on a cream background and measures 1.8 cm. long by 0.8 cm. at the shoulders. The column has a significant nectary.

Flowering Periods: March and the flowers last for a week.

Pollination: Is probably by bumble bees.

Cultivation: Grow in a peat-rich, well-drained compost with added coarse sand.

Pelexia itatiayae Schltr. | 434 (Plate 07)

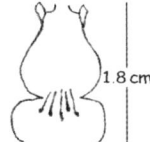

Etymology: Presumably first found in the Itatiaia mountains, RJ.

Habitat: A terrestrial found in original forest and at the edges of clearings in section VI at 1300-1500 M.asl and requiring a substrate of leaf litter, dappled light, high constant humidity and little wind movement.

Occurrence: Occasional.

Plant: A small to medium-sized terrestrial plant always hidden in the under forest. The fleshy leaves, normally four or five, appear as a rosette and are dark green on the upper side but reddish underneath, with a well defined central vein. They are oblanceolate and measure 10 cm. long by 5 cm. wide and are held on a petiole which is pink and up to 14 cm. long with an oval cross-section. The root system is extensive, composed of a number of thick, hairy roots which emerge from a central node.

Inflorescence and Flowers: The inflorescence emerges in October from the centre of the leaf whorl and may reach up to 65 cm. in height. The stem is hairy, round and three sectioned, each section is surrounded for up to 4 cm. by a short bract which is pink and 1.5 cm. wide and lanceolate. The flowers appear in a tight cone of about 20 on the apical 15 cm. of the stem. The individual flowers are attractive and the dominant feature is the trilobed lip, whose central apical lobe reflexes and shows two orange calli on a milk white background at its base. The lip is 2 cm. long by 1 cm. broad. The almost hidden sepals and petals are milk white, lanceolate with hairy margins and measure 1.8 cm. long by 0.5 cm. broad. The petals, fused to the dorsal sepals are narrower.

Flowering Periods: The flowers, which appear in late November, early December last for four to seven days, while the inflorescence may show flowers for up to three weeks.

Pollination: Possibly bumble bees (**Dressler**).

Cultivation: Unknown.

General: **Cyclopogon**, **Sarcoglottis** and **Pelexia** are genera of the Spiranthinae sub-tribe and it is the Devil's Own Job to separate them if the eye is not perfectly attuned. Add to this the difficulty of dissecting the flowers and one has all the ingredients necessary for misidentification.

Pelexia hypnophila (Barb. Rodr.) Schltr. | 453

Etymology: From the Greek *hypno* for sleep and *phila* for loving.

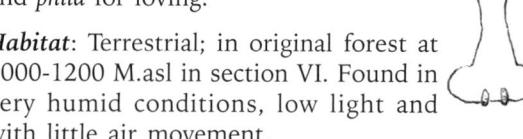

Habitat: Terrestrial; in original forest at 1000-1200 M.asl in section VI. Found in very humid conditions, low light and with little air movement.

Occurrence: Rare.

Plant: A small to medium-sized terrestrial. Three or four fleshy leaves are produced in a loose whorl from a central node in the root system. These are elliptic, measure up to 17 cm. by 5 cm. broad, a darkish green, with a pronounced central vein and are held on 5 cm. long white petioles. The root system is composed of a mass of thick white hairy roots, up to 10 cm. long, that emerge from the stem base. These are narrow at the junction with the stem, soon swelling into what in fact are elongated tubers.

Inflorescence and Flowers: The stem arises in September from the centre of the leaf whorl, up to a height of 40 cm. It is round, stiff, green, sectioned, lightly bracted and shows a loose cone of flowers on the top 10 cm. These are attractive and are dominated by the relatively large, white, reflexed apical lobes of the lip. The dorsal sepal is a concave lanceolate shape, measuring 0.7 cm. long by 0.4 cm. broad while the petals, fused to this sepal, are narrowly club-shaped and measure 0.75 cm. by 0.2 cm. broad. The lateral sepals are linear to club-shaped and measure 1.6 cm. by 0.3 cm. broad and are a light yellow. The lip is 1.6 cm. long, trilobed and thinly waisted with a large, milk-white, reflexed, ovate apical lobe, measuring 0.5 cm. broad by 0.4 cm. deep.

Flowering Periods: October and the flowers last for a

week or so although the inflorescence may show flowers for up to three weeks.

Pollination: **Dressler** suggests possibly bumble bees.

Cultivation: Unknown. However due to the delicacy of the root connections with the stem and the nature of the roots, transplanting would be precarious.

Pelexia novafriburguensis (Rchb. f.) Garay | 453a (Plate 07)

Etymology: Presumably the type was found in Nova Friburgo.

Habitat: Found growing loosely on leaf mould on original high mountain forest floor at 1500 M.asl in section VI. It requires leaf litter in partial shade with high constant humidity and some wind movement.

Occurrence: Occasional.

Plant: A medium sized terrestrial, similar to *P. itatiayae*, but with distinctive long wine coloured leaf petioles. There are several leaves, a deep green and veined like most Melostomataceae species. The petiole is up to 15 cm. long and centrally hollowed while the leaf is oblanceolate, up to 13 cm. long by 5.5 cm, and arises from a central root mass composed of many thick white short, banana-like, non-branching roots which do not stretch beyond the plant's circumference.

Inflorescence and Flowers: The 40 cm. high, light pink fleshy, inflorescence arises from a central node in the root mass, bearing a loose cone of up to 12 flowers which take up the top 10 cm. The only distinctive flower characteristic is the broad, reflexed apical lobe which is pure white and unwrinkled. The green lateral sepals are an elongated narrow lanceolate, while the dorsal sepal is spathulate, green and fused with the narrow petals. All tepals are hairy on the outside.

Flowering Periods: Late September and early October and the flowers last for two weeks or so.

Pollination: Unknown.

Cultivation: In a large pot with a loose, leaf mould substrate mixed with 20% coarse sand. Some air movement and a constant high humidity are essential, in good but shady light.

The Genus *Cogniauxiocharis* Hoehne

A terrestrial genus of two species from Brazil named after **Cogniaux** and the Greek word *charis* meaning graceful.

Cogniauxiocharis glazioviana (Cogn.) Hoehne | 466

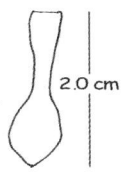

Etymology: Named in honour of **Glaziou**, a Belgian botanist and landscape gardener who lived in Brazil during the late 19th. Century and was responsible for many parks and gardens still extant in Rio de Janeiro, Petropolis and Nova Friburgo.

Habitat: A terrestrial from deep original forest with high humidity, low light and air movement, growing on a bed of leaf litter around 800 M. asl in Section II.

Occurrence: Occasional.

Plant: A terrestrial orchid very similar to *Eltroplectrus roseo-alba* and nearly as attractive. It has a whorl of four or five spathulate leaves on 4.0 cm. false petioles which are 20 cm. long, 5.0 cm wide, dark green, floppy and shiny. The root system is typically Spiranthinae with short, gross, hairy and numerous yellow, finger-like roots.

Inflorescence and Flowers: The inflorescence arises from a node in the centre of the root mass with up to 10 bracted and alternate, attractive white flowers which open in sequence. The dorsal sepal is a narrow lanceolate, pilose and 1.3 cm. by 0.3 cm. wide. The laterals are also pilose, asymmetric lanceolates and 2.0 cm. long by 0.6 cm. The lip is elongate, spathulate and single lobed, 2.0 cm. long by 0.6 cm. at the broadest part. There is no perceptible callus.

Flowering Periods: January, when the flowers last for a week or so, but the sequential flowering lasts for a month.

Pollination: Unknown.

Cultivation: Grow in pots on a substrate of leaf mould and loam, well-drained in deep humid shade, watering in summer and merely misting in other seasons

The genus *Lankesterella* O. Ames.

Oakes-Ames, founder of the **Oakes-Ames** herbarium in the USA, gave this genus name to a plant found in Costa Rica and dedicated to one **Lankester**, a collector and grower who lived in Costa Rica. **R. Schlechter** had previously separated the species out from the genus *Stenorrhynchos* into a new genus *Cladobium*. However, this generic name had previously been used, so they reverted to the name given by **Oakes-Ames**. Such are the travails of small plants in the hearts and minds of taxonomists! This is a small Brazilian genus of eight species, all of which are confined to the humid Atlantic rain forest. Six of these species should be found in the Organ Mountain Range.

Lankesterella ceracifolia (Barb. Rodr.) Mansf. | 468 (Plate 07)

Etymology: *Ceraceus* is Latin for waxy and *folia* is Latin meaning leaf, which describes the apparent waxiness of the leaves of the plant.

Habitat: Found in sections V and VI on low, moss-covered tree trunks in almost any epiphytic situation in original forest, at 1200-1500 M.asl with adequate humidity and light.

Occurrence: Common.

Plant: A small epiphyte, very similar in morphology to *Eurystyles* and growing as an irregular rosette of light green, keeled, shiny, ovate leaves, each 2 cm. long by 1 cm. wide and originating from a central node in the root system. This is neither extensive nor adherent and is composed of thick, white, short roots always under moss.

Inflorescence and Flowers: Borne on a 2 cm, fleshy, round and very hairy-stemmed inflorescence, with one or two white, projecting flowers. The dorsal sepal is triangular, white, 1.5 cm. long by 0.6 cm. wide at the base. The lateral sepals curve outwards and are linear lanceolate, 2.1 cm. long by 0.3 cm. wide. The petals are club-shaped and measure 1.5 cm. long by 0.3 cm. wide close to the pointed apex. The lip is distinctive, 10 mm. long by 7 mm. wide, tri-lobed, almost violin shaped, with two clear, green-veined vertically running calli.

Flowering Periods: February and March when the flowers last for 10 days.

Pollination: **Dressler** suggests probably bumble bees.

Cultivation: On plaques of tree fern, in dappled shade, always keeping them damp.

Lankesterella gnomus (Krzl.) Hoehne | 470 (Plate 07)

Etymology: *Gnomus* is Old French for gnome, probably meaning a very small ugly plant.

Habitat: Low moss-covered tree trunks in original forest, at around 1300 M.asl in section VI and higher, requiring a high humidity, medium light and low air movement as its basic requirements.

Occurrence: Common.

Plant: A tiny epiphyte with five or six small, light green, pointed, ovate leaves, each 2 cm long by 1 cm. wide on short 2 cm. stems originating from a central root base; a typical terrestrial turned epiphyte. Roots are relatively thick, brown, always underneath moss, in no way adherent, and the system is not extensive.

Inflorescence and Flowers: One to several flowers arise on a thin, very hairy, racemose inflorescence up to 4 cm. long. The flowers are large in proportion to the plant and are an overall white. The sepals are more or less equal, 1.5 cm. by 0.5 cm. while the petals are slightly smaller and incurved around the dorsal sepal. The lip is white with a greenish centre, large, almost 2 cm. long by 0.7 cm. wide and both sepals and petals are thrust forward round this long lip.

Flowering Periods: March to May; and the flowers last for two weeks.

Pollination: Probably bumble bees.

Cultivation: As for *L. ceracifolia*.

Lankesterella aff. *longicollis* (Cogn.) Hoehne | 471 (Plate 07)

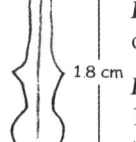

Etymology: *Longi collis* are Latin for long hill or in this case long column.

Habitat: Original high montane forest at 1500 M.asl in section IV, on low mossy tree trunks in reasonable light and air movement in a permanently humid environment.

Occurrence: Locally frequent.

Plant: A small, low tree trunk epiphyte. A whorl of up to 12 light green leaves arises closely from a short white stem which in turn emerges from a node in the centre of the root system. These leaves are a pointed spathulate oval measuring up to 2.5 cm. long by 1.3 cm. broad, slightly keeled and shiny. The root system is composed of up to 10 hairy thickish yellowish roots, running beneath the moss but reaching only slightly further than the plant's circumference.

Inflorescence and Flowers: A short, 2.5 cm., hairy inflorescence arises from the central root whorl and holds one, or occasionally two, surprisingly large, white flowers, tinged with green. The dorsal sepal is fused with the two petals. The former is more or less broadly lanceolate, 1.9 cm. by 0.4 cm. wide while the latter are a narrow lanceolate 2 cm. long by 0.35 cm. The lateral sepals are awl-shaped 2.5 cm. long by 0.5 cm. The lip is elongated, flared at the apex, though single lobed, measuring 1.8 cm long by 0.5 cm at the apex. All segments are milk white for the apical half but thickly veined in green at their bases.

Flowering Periods: Early February and the flowers last for about a week.

Pollination: **Dressler** suggest probably pollinated by bees.

Cultivation: On bark or tree fern plaques observing the environmental data above.

The genus
Mesadenella Pabst & Garay

Raised by **Pabst** and **Garay**, both lions of orchid taxonomy, one amateur and Brazilian the other American and professional who transferred this genus from *Cyclopogon* for good reasons that we are not aware of. No more can we fathom from our Greek and Latin botanical dictionaries, how they came across the name *Mesadenella*. This is a Brazilian genus of two or three terrestrial orchids, according to **Pabst & Dungs**, which were hauled out from the emasculated genus *Cyclopogon* and only one of which is attributed to the Organ Mountain Range.

Mesadenella esmeraldae
Lindl. & Rchb f. | *473*
Now: *M. cuspitata*
Lindl. & Garay
Synonym: *Cyclopogon albopunctatus*
Barb. Rodr. (Plate 08)

Etymology: A plant, all of whose synonyms are beautifully descriptive, *Esmeraldae* is one which comes from old Latin *smaraldus* meaning emerald-coloured referring to the leaf colour 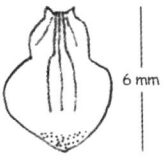 while *cuspitata* refers to spittal on the leaves and *albopuncta* refers to white spots on the leaves.

Habitat: Found in original forest in deep shade on a bed of decaying vegetable detritus at 400 to 600 M.asl in sections IV, V and VI, and requiring high constant humidity, low wind movement and shade.

Occurrence: Occasional.

Plant: A medium sized terrestrial, easily recognised when not in flower by its foliage. The leaves, which emerge in a whorl are held on short false petioles and measure up to 15 cm. by 5 cm, a broad lanceolate, light green and covered in small irregular white blotches. They are flimsy, slightly fleshy and keeled. The root system is vigorous, composed of numerous thick, white, almost tuberous, short roots which are faintly hairy.

Inflorescence and Flowers: A many noded and bracted, stiff, erect, round inflorescence emerges from a node in the centre of the root system, up to 30 cm. tall and showing a narrow cone of up to 40 small white and green bracted flowers on its apical half. The dorsal sepal, which is fused with the petals, is a broad lanceolate 0.4 cm. by 0.2 cm. wide. The lateral sepals are an asymmetric lanceolate, 0.7 cm. long by 0.2 cm. All three are faintly hairy at their margins. The petals are slightly smaller but the same shape as the dorsal sepal. The lip is single lobed, a broad spathulate and somewhat reflexed for its apical half, 0.6 cm. long by 0.4 cm.

Flowering Periods: March; the flowers last for 10 days or so.

Pollination: Unknown.

Cultivation: In a large pot on an organic, fibrous, well-aerated substrate; watering well throughout the year but never water logged. Keep in the shade with some air movement and a high atmospheric humidity.

General: It is probable that the white leaf markings, strongly reminiscent of bird guano, make the leaves less appetizing for grazing animals.

The genus
Stenorrhynchus L. C. Rich.

Richard, a collector who visited French Guiana and Pará, between 1781 and 1789, raised this terrestrial genus, based on the Greek compound word for a narrow beak and refers to the almost tubular flower parts. This is an American genus of about 60 species of which **Pabst&Dungs** have attributed 18 species to Brazil and four should be found in the Organ Mountain Range.

Stenorrhynchus lanceolatus (Aubl.) L. C. Rich. | 476 (Plate 08)

Etymology: Lanceolatus is Latin meaning like the point of a spear or lance, which well describes this plant's tepals.

2.0 cm

Habitat: It is found in all sections; a roadside bank species at around 800-1200 M.asl and extremely common in such exposed bare or grassy situations. An invader, we suspect, carried in on mule hooves and later, car or lorry tyres. Like *Eulophia alta*, neither species is found far from such locations and also like this latter species, it has a geographical range from North America to Argentina.

Occurrence: Common.

Plant: A medium-sized terrestrial plant with no bulb or corm but thick yellow roots which take up the storage function. These roots branch out spider-like and penetrate almost any type of soil for 15 cm. around the plant, some 8-10 cm. below the surface. Leaves appear after flowering and in the wild are almost never noticed. They are deep green, sword shaped and multiveined, arising from the inflorescence base on a 10 cm. thick, green stem up to 18 cm. long by 4 cm. wide.

Inflorescence and Flowers: Appearing on a thick inflorescence up to 70 cm. high which varies in colour from light green at the base merging into a light pink towards the apical flower cone. The flowers, upwards of 40, are a pure salmon flesh-pink tube approximately 2 cm. long by 0.5 cm. wide. They are thickly clustered around the apex. The predominant features are an apparent wide pointed dorsal sepal which in fact is a fusion between the small dorsal sepal and the two petals which dominates a pointed but narrower down-folded lip.

Flowering Periods: Late August to early October and the individual flowers last for up to a week while the inflorescence shows flowers in bloom for three weeks.

Pollination: 80% of the flowers are usually pollinated and humming birds are the Pollinators – **Rodrigo Singer** ,Unicamp, personal communication. Subsequent personal observation together with **Singer** has confirmed at least four species of humming bird as pollinators.

Cultivation: Grow in pots, on a rich, peat-like compost with a high percentage of coarse sand, watering well in spring and while the foliage remains, but only keeping the compost damp thereafter It requires high light and good drainage, with warmer growing temperatures in summer as the growth cycle ends, then keeping cooler to simulate the habitat and to encourage flowering during the following year.

General: This plant received a gold medal at the 1996 World Orchid Congress in Rio de Janeiro. The life cycle is very swift and from flowering to seed shedding can be as little as three weeks and coincides with the spring rains.

The genus *Eltroplectris* Raf.

This genus is composed of nine species of which four possibly could be found in the Organ Mountain Range. We have found only one, which by all accounts is the most beautiful of them all. The genus was named by **Constantine Rafinesque** a Turkish botanist, who became a professor of natural history in the United States. *Eltroplectris* is a Greek compound word of which we can find no trace of meaning.

Eltroplectris roseo-alba (Rchb. f.) Garay & Sweet | 501 (Plate 09)

1.9 cm

Etymology: Roseo alba is Latin for pink and white, referring to the flower colour.

Habitat: A terrestrial found at 600 M.asl, in sections V and VII in original forest remnants. It enjoys rains from October to March and a dry April through to September and grows on a substrate of deep leaf mould in fresh but shady conditions.

Occurrence: Rare but always found in colonies.

Plant: A very attractive medium-sized terrestrial. Up to five spathulate, delicately fleshy, pale green leaves, sometimes speckled with light silver streaks, emerge in a tight whorl from a central root node. The leaves have keels, almost no petiole and measure up to 20 cm. long by 10 cm. broad and die back after flowering. The root system consists of up to six thick, hairy roots up to 15 cm. long and 1 cm. diameter.

Inflorescence and Flowers: A fleshy, intensely bracted inflorescence, up to 36 cm. high and 1 cm. diameter emerges from the centre of the leaf whorl. There may be a second, shorter inflorescence. As many as 15 sequentially opening flowers are borne on the apical 18 cm. appearing alternately but nearly opposite at 0.5 cm. intervals and forming a loose cone. Each has a 1 cm. pedicel and a bract for its entire length; the flowers are milk-white and streaked with dark pink. The dorsal sepal is 2.5 cm. long by 0.5 cm, lanceolate and fused to the petals with only its acuminate tip showing. The lateral sepals are a slightly asymmetric lanceolate, 2.8 cm. long by 0.7 cm. broad. The partly fused petals are

1.9 cm. long by 0.7 cm. and a distorted lanceolate shape. The lip is distinctive, tri-lobed, with a heart-shaped apical lobe and doubly serrated margins on the basal half. The basal side-lobes are strongly veined in rose and overall, the lip measures 1.9 cm. long by 1.2 cm. broad.

Flowering Periods: April when the flowers last for one week.

Pollination: The species appears to be self-fertilising although **Dressler** suggests bumble bees.

Cultivation: In pots with a well-drained fibrous compost with some coarse sand and decayed vegetable matter. Needs good air movement and high summer humidity.

The genus *Erythrodes* Blume.

Karl Blume was a German botanist who studied plants in Java including *Erythrodes*, a pantropical genus which surprisingly has more species in the Americas than in Asia. **Pabst and Dungs** attribute 33 species to Brazil and 10 of these to the Organ Mountain Range. *Erythrodes* is a Greek compound meaning a red stem which many of the Asian species have, but not the species we have found. Known as one of the Jewel orchids because of its beautiful leaves, the flowers are usually insignificant, except for *E. commelinoides* whose flowers form a striking orange cone.

Erythrodes arietina (Rchb. f. & Warm.) Ames | *511*

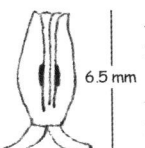

Etymology: *Ari tina* is Latin for like a small moth, referring to the flowers.

Habitat: A terrestrial found in original forest and old regrowth in sections IV and VI or on boulders, always near water and in constant, extremely humid conditions, with filtered light and little air movement.

Occurrence: Common in its very specific habitat.

Plant: A small terrestrial plant from the deep dark forest floor. A new growth emerges from a nodule on the previous year's stem, creeps under dead leaves for some 10 cm. until it curves upward, becoming erect for a further 20 cm. or so. Up to seven broadly lanceolate, dark green, floppy, mildly plicate, leaves are borne alternately in a whorl at 4 cm. intervals. Each is about 12 cm. long by 5 cm. wide and the narrow bases envelop the stem for 1.5 cm. The stem narrows towards the

apex from where emerge two narrow but opposite leaves. A small lighter green lanceolate leaf measuring 4 cm. long by 1.3 cm. subtends the apical inflorescence. The root system is weak and composed of some six, 10 cm. long, thin white roots which are enveloped by a fibrous hairy mass. The roots on previous years' growths appear active.

Inflorescence and Flowers: Up to 60 small flowers form a tight cone on the apical 17 cm. of a 30 cm. round-sectioned noded, inflorescence which is covered by fine, white, short hairs. The flowers are almost sessile and the outsides of the flower parts are verrucous and covered by many liquid-filled short hairs, each topped by a minute transparent ovoid-shaped sphere. The dorsal sepal is 0.4 cm. long by 0.1 cm. and a laterally compressed ovate, while the lateral sepals are 0.3 cm. long by 0.1 cm. wide and are slightly sickle-shaped. The petals are a similar shape but slightly smaller. The distinctive lip is 6.5 mm. long with a green basal triangle and an apical white half and is 0.15 cm. wide. The apical tip has two protruding hooks measuring 0.3 cm. long. There is a 0.4 cm. long, green nectary at the base of the lip.

Flowering Periods: November through to January and the flowers, which open more or less simultaneously, last only a few days.

Pollination: Is probably by small moths.

Cultivation: Unknown.

Erythrodes commelinoides (Barb. Rodr.) Ames | *528* (Plate 08)

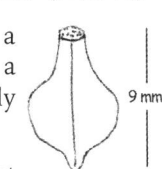

Etymology: *Commelin* and *oides* make a Greek compound word meaning like a plant from the Commelinaceae family (Spiderworts).

Habitat: A terrestrial, found in the most humid and dark under forest with low light and wind movement, growing on a reasonably drained substrate of well decayed leaf litter in section VI at 1000-1500 M.asl.

Occurrence: Occasional.

Plant: A medium-sized terrestrial from the base of the humid original forest. It is a creeping plant similar to *Psilochilus modestus* inasmuch as each years growth is an extension of the previous years stem creeping under the leaf litter. This stem is a rose-tinted green, fleshy, round, 1 cm. in diameter, but narrowing towards the apex. The leaves arise alternately from the stem at 2 - 4 cm. intervals and are dark green, elliptic and measure 10 cm. by 3.5 cm. broad diminishing in size towards the apex. They are deeply keeled and held on a short white false petiole which also envelopes the stem.

Thickish, white roots emerge from the stem while it is under the substrate, opposite and at short intervals.

Inflorescence and Flowers: Up to 30 very attractive orange flowers are borne in a tight cone on the apical 8-10 cm. of the stem. The sepals are lanceolate, 1 cm. long by 0.3 cm. while the petals, fused to the dorsal sepal are somewhat acuminately club-shaped, 0.9 cm. long by 0.3 cm. at their apices, narrowing for their basal halves. The lip is vaguely trilobed and in fact resembles a pointed spade 9 mm. long by 0.5 cm. broad and is attached at its base to a significant nectary, projecting backwards below the ovary and measuring 1 cm. long and 0.3 cm. wide at the nectar holding pocket.

Flowering Periods: In March and the individual flowers last for a few days but the inflorescence shows open flowers for up to a month.

Pollination: Possibly butterflies – **Dressler**, or night-flying moths; fertilization is always over 80%.

Cultivation: Unknown, but it must be extremely difficult to reproduce the natural conditions.

Erythrodes nobilis (Rchb. f) Pabst | 530 (Plate 08)

Etymology: From the Latin *noble* and among the meanings is well-known and tall.

Habitat: Found in small colonies at original forest edges at 1200 M.asl in section VI growing on leaf litter, and enjoying high constant humidity, low light and wind movement.

Occurrence: Occasional.

Plant: A whorled rosette of dark green leaves, broadly veined in silver, are this plant's major feature. These, up to 7 in number, are a broad lanceolate, 2 - 6 cm. long by 1 - 2 cm. broad, and slightly fleshy. They emerge as new growths from the apex of a long, thick rhizome which runs just beneath the leaf litter. The roots are few, extremely hairy, short and unbranched. The thick rhizome also performs a storage function.

Inflorescence and Flowers: The 15 - 20 cm. inflorescence arises from the leaf whorl centre, round, reddish, 3-noded, covered with short hairs and bearing a cone of up to 15 yellowish to white flowers on the apical 5 cm. The sepals are ± equal, a slightly acuminate, lanceolate, at 0.55 cm. by 0.2 cm. broad, a pale white with a small light pink blotch at the centre. The petals are a curious asymmetrical spathulate, vaguely trilobed 0.5 cm. by 0.3 cm. broad and dentate at the lower lobe margin. The lip is trilobed, a butter yellow, 0.5 cm. long by 0.4 cm. with the margins of all three lobes intensely dentate at their apices. The petal shape shows quite clearly

how the lip, third petal, evolved. Needless to say all this can only be really appreciated through a microscope or a strong hand lens. A long, 0.4 cm. nectary extends from the base of the lip.

Flowering Periods: January and the flowers last for a week or so.

Pollination: Possibly butterflies or moths.

Cultivation: **Hoehne** describes this process beautifully: 'without a doubt in *Erythrodes* we have plants of the most delicate kind, perfect to enhance a humid-cool green house under dappled shade. Vegetable detritus, a humid atmosphere and little wind movement are the perfect conditions for growing healthy and vigorous plants. In these conditions all species develop admirably and can be multiplied by the simple division of the stem'.

Erythrodes bidentifera (Schltr.) Garay | 532 (Plate 08)

Etymology: *Bi - denti - fera* is a Latin compound word meaning bearing-two-teeth, referring to the tooth-like projections at the lip base.

Habitat: As with the other *Erythrodes* species, this plant inhabits the original forest floor in deep shade and intensely humid conditions, sometimes in small colonies but more often solitary, at between 650-1400 M.asl in sections V, VI, where such conditions exist.

Occurrence: Rare.

Plant: A small terrestrial with the creeping habit shared with other members of the genus. The leaves are the distinctive feature of this species. An average plant shows up to four heart-shaped leaves, 4 cm. long by 1.5 cm. wide, tightly clustered around the stem base. These are bordered with the darkest of greens and have a central section, around the mid-vein of light silvery green. The new stem emerges from a node at the up curve of the previous stem's base and reaches a height of some 10 cm, after creeping beneath decaying leaf detritus for up to 15 cm. The roots are short, up to 3 cm. long, thick and hairy, and emerge at short intervals from internodes on the stem and appear viable for several years.

Inflorescence and Flowers: A thin, stiff, pink, hairy inflorescence of up to 20 cm. tall emerges from the stem's apex in November. A tight cone of up to 10 small flowers is held on the apical 6 cm. The dorsal sepal is lanceolate, 0.5 cm. long by 0.2 cm. broad, and translucent white. The lateral sepals are the same shape but slightly larger and show a distinct down-facing brown "V" at their centres. The petals are fused to the dorsal sepal and are lanceolate and measure 0.3 cm. by 0.15 cm. broad, speckled with brown on a white background and toothed at their outer margin. The lip is trilobed, translucent yellow, measuring 0.4 cm. long by 0.25 cm. wide, and is intensely indented at all its margins, and the two side lobes have a pointed tooth at their apices, hence the name.

Flowering Periods: December and the flowers last for a few days, while the inflorescence may show open flowers for 10 days or so.

Pollination: **Dressler** suggests possibly night flying moths.

Cultivation: Unknown.

The genus
Liparis L. C. Rich.

Richard was a collector who visited French Guiana and Pará between 1781 and 1789. *Liparos* is the Greek for greasy or shiny referring to the leaves of many species. It is a pandemic genus of more than 240 species and interestingly, one species is found in Britain and is only one of two pseudobulbous orchids found in that country. **Pabst&Dungs** attribute three species to Brazil but only one to the Organ Mountain Range.

Liparis nervosa
(Thunb.) Lindl. | *540* (Plate 08)

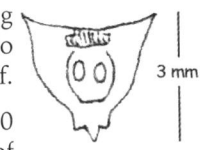

Etymology: *Nervosa* is Latin for veins, referring to the strong veining of the leaves.

Habitat: A terrestrial, found in young regrowth forest which has been consistently burnt over the years, in sections I, III, V and VI at 800-1100 M.asl, enjoying good though dappled light, a well-drained soil with leaf detritus, heavy rains during spring and summer and a dry autumn and winter.

Occurrence: Common in its specific habitat.

Plant: A pseudobulbous terrestrial which is in fact an invader to deforested areas in these regions, coming from the much drier and frequently burnt-over interior scrub forest. The pseudobulbs are a thick conical fusiform shape, quite reminiscent of those of various *Catasetum* species, partially sheathed and measure up to 10 cm. long by 1.5 - 2.0 cm. diameter. Initially these are upright but after a year or so they collapse and become immersed by leaf litter. The new growth emerges on a very short rhizome from the base of the previous year's pseudobulb and shows up to five elliptic, plicate, dark green leaves which are strongly veined and

measure up to 20 cm. long by 5 cm. The root system is not extensive and is composed of a number of thinnish white roots which penetrate the vegetable detritus around.

Inflorescence and Flowers: A loose cone of some 20-30 flowers is carried on the 15 cm. long apex of the 30 cm. apical inflorescence, triangular in cross section. It is light green, thin and erect. The flowers are at 1.5 cm. to 2 cm. intervals and are an overall red to purple. Both petals and sepals are reflexed; the petals strongly so. The dorsal sepal is lanceolate and measures 0.7 cm. long by 0.25 cm. broad. The lateral sepals are the same shape but broader, while the petals are club-shaped and measure 0.8 cm. by 0.2 cm. at the club-like head. The lip is the distinguishing feature, projected forwards, bi-lobed, with the apical side lobes reflexed backwards and measuring 0.6 cm. long by 0.55 cm. wide and deep purple. A 1 cm. nectary extends behind the lip base.

Flowering Periods: Late October through November, but we have also found this plant in flower in March and April. The flowers last for a week or so while the inflorescence may show open flowers for over a month.

Pollination: Unknown; but the long nectary would indicate butterflies or moths.

Cultivation: In pots with a well-drained substrate made up of 50% organic material and 50% coarse sand and some ashes or charcoal.

The genus
Malaxis Sw.

This is a tropical and temperate genus of some 300 species mainly concentrated in Asia. **Pabst& Dungs** attribute nine species to Brazil of which four should be found in the Organ Mountain Range. **Olof Swartz** was a Swedish botanist who studied under **Linneaus**'s son. He travelled to North America and the Caribbean between 1783 - 1790. He used a Greek word, *Malaxis* meaning soft and referring to the soft leaves.

Malaxis excavata
(Lindl.) Ktze. | *544* (Plate 08)

Etymology: *Excavata* is Latin meaning hollowed out in a curve, referring to the hollowed out petiole to the leaf.

Habitat: Found at around 1500 M.asl and above in the leaf litter of dense scrub forest or original forest, requiring little light and high humidity, in section VI.

Occurrence: Occasional.

Plant: A small to medium-sized terrestrial, generally found in scattered colonies in deep, damp, dark, forest, in leaf litter. A small underground pointed ovoid pseudobulb is produced annually from a very short rhizome. The new pseudobulb, about 2.5 cm. tall, extends to two apical, opposite dark green and deeply keeled velvety leaves and is a squat pointed oval, up to 8 cm. long by 3.5 cm. wide. The leaves are held on a hollow laterally-compressed succulent stem, 0.4 cm. wide and up to 4 cm. tall and are retained for a year or so.

Inflorescence and Flowers: Up to 60 very small inverted flowers, each on a thin white 1 cm. stalk are borne on a sub-umbellate inflorescence up to 10 cm. tall, which crowns the stem and each opens from the base. The deep overall effect is a circle about 3 cm. in diameter looking rather like a bicycle wheel, with the tyre of light orange and the spokes a translucent white. The sepals are equal, ± 0.3 cm. long by 0.1 cm. wide. The petals are acute, very thin ± 0.2 cm. long and the faintly trilobed inverted lip is triangular and about 0.3 cm. long by 0.2 cm. wide at the base.

Flowering Periods: November and December and the flowers last for two weeks or so.

Pollination: Not known but possibly small flies.

Cultivation: Grow in pots, on a well drained, loose, leaf litter compost, watering well up to flowering and thereafter keeping damp, in deep shade.

Malaxis parthonii
C. Morr. | *547* (Plate 08)

Etymology: From the Greek *parthenos*, a virgin, this probably refers to forming seed without fertilization or partheogenesis.

Habitat: On deep leaf litter in intensely humid conditions with low light and wind movement at 900-1000 M.asl in regrowth forest in section VI.

Occurrence: Rare.

Plant: A small terrestrial orchid more delicate than its robust cousin M. excavata. The new shoot arises from an underground rhizome and ultimately develops a slender, sheathed pseudobulb. Generally there are two opposite, fleshy, dark green, keeled, elliptic leaves, on long false petioles which sheath the stem, each 11 cm. long by 7 cm. wide. Thickish white to brown roots emerge from the new shoot's base and spread through the leaf litter around.

Inflorescence and Flowers: The apical inflorescence rises to some 12 - 15 cm. above the leaves, thin, light green and sectioned like a star. The flowers are held at the

apex in a sub-umbellate grouping. Some 30 tiny resupinate flowers are held in a circle on 0.9 cm. pedicels looking like the spokes of a bicycle wheel. The sepals are ± equal in size and shape 0.27 cm. long by 0.15 cm, light green and a bluntish broad lanceolate. The petals are linear and very narrow, 0.2 cm. long. The lip is vaguely trilobed, triangular, relatively thick with a double crater-like space towards its throat, only to be enjoyed under a microscope!, and is 0.21 cm. long by 0.18 cm. wide.

Flowering Periods: January; the flowers, opening sequentially, last for up to three weeks.

Pollination: Unknown; possibly small flies, or self.

Cultivation: As for **M. excavata**.

The genus
Polystachya Hooker

Hooker, of Kew Gardens, named this pantropical genus of over 150 species (12 to Brazil), using the greek composite word *poly* meaning many and *stachys* a wheat spike, describing well the flower spikes of the three species attributed to the Organ Mountain Range.

Polystachya caespitosa
Barb. Rodr. | *553* (Plate 08)

Etymology: *Caespitosus* is Latin and means growing in dense tufts, which describes the plant well.

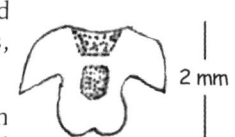

Habitat: Found at mid-tree in original or regrowth forest in all sections. It is wide ranging, from 400-1100 M.asl. with good air movement, reasonable humidity and light, but happier at lower altitudes.

Occurrence: Common.

Plant: This is a small epiphyte with a line-up of quite distinctive, clustered pseudobulbs which are ovoid, ringed and partly sheathed. There are two or three apical leaves and one is long-lived, lanceolate, keeled, light green in colour and measures 6.5 cm. by 0.9 cm. The roots are thick, 0.3 cm. in diameter, white and the system is extensive.

Inflorescence and Flowers: Arising on a thick apical 5 cm. racemose inflorescence, bearing a dozen or so yellow, heather-bell-like flowers appearing as a loose cone. The dorsal sepal is ovate, obtuse and 0.3 cm. long by 0.2 cm. wide. The laterals are slightly larger and acute. The petals are 0.2 cm. long by 0.1 cm. wide, linear and tridentate at apex. The lip is distinctly trilobed with a pronounced central callus.

Flowering Periods: January and the flowers last for a week while the inflorescence may show flowers blooming for three weeks.

Pollination: Unknown but possibly small bees.

Cultivation: On bark plaques, requiring dappled light, some air movement and reasonable constant humidity.

Polystachya estrellensis
Rchb. f. | *554* (Plate 10)

Etymology: *Stella* and *ensis* are Latin for star-like, referring to the star-like aspect of the flower.

Habitat: Epiphytic in regenerating and original forest from 600-1200 M.asl, at mid to upper-tree level in all sections, requiring good dappled light and wind movement together with reasonable humidity.

Occurrence: Common.

Plant: A small to medium-sized pseudobulbous epiphyte and the sheathed pseudobulbs are little more than slight swellings of the stem base only distinctly appearing after two or three years when the leaves die off. They are an extended cone-shape and arise closely packed in a line from a thin rhizome, and are 2 cm. long by 1.5 cm. in diameter. Up to six leaves arise alternately from the base to half way up the stem, overlapping and sheathing the stem with their false petioles up to 6 cm long. The leaves measure up to 25 cm. by 2 cm, and are slightly leathery, deeply keeled, dark green and linear lanceolate. The root system is robust and strongly adherent, consisting of numerous thin white roots which arise from each new stem base and are viable for many years.

Inflorescence and Flowers: An erect, apical, stiff, inflorescence up to 35 cm long, appears in January or February. The flowers are borne in a loose apical cone or sometimes in paniculate form with smaller sub-inflorescences, holding fewer flowers, and arising from nodes on the upper half of the stem. The flowers are butter yellow and bell-like. The sepals are heart-shaped and measure 0.5 cm by 0.2 cm. The lip is trilobed, 0.3 cm. long by 0.3 cm. wide at the side-lobes and the apical lobe is 0.2 cm wide.

Flowering Periods: February to April and the individual flowers last for two weeks or so while the inflorescence may show open flowers for over a month.

Pollination: Most flowers are fertilized, **Dressler** suggests probably by small bees that gather the pseudo-pollen from the lip surface.

Cultivation: Grow in pots on a substrate of tree fern fibre and coarse sand, well drained, with plenty of water

during spring and summer, but just misting during the rest of the year.

Polystachya micrantha
Schltr. | *559*

Etymology: *Micrantha* is Greek for small flowered which refers to the relatively small flowers of this plant.

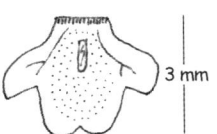

Habitat: In regrowth forest at 400-900 M.asl in sections V and VI at mid-tree level enjoying good light and air movement and reasonable year round humidity.

Occurrence: Common.

Plant: A medium-sized epiphyte, very similar to the other *Polystachya* species described. The pseudobulbs arise tightly on a thickish green, sheathed, rhizome, and are ovoid, internoded, partly sheathed and vary greatly in size. There are two or three long-lived, light green, lanceolate, keeled leaves which may be up to 8 cm. by 1.0 cm. wide. The roots are typically thick, white, occasionally branching and make an extensive system.

Inflorescence and Flowers: A thick green, apical inflorescence holds a number of racemes of small, bell-like, yellow flowers. The dorsal sepal is heart-shaped 0.3 cm. long by 0.2 cm., while the lateral sepals are linear, 0.3 cm. by 0.1 cm. The lip is broad and three-lobed, almost like a clover leaf, 3 mm. long by 3 mm., with a short nose-like callus at its base and a central lobe which is 90% covered by very faint warting. The tepals are all convex.

Flowering Periods: January and the flowers last for up to a month.

Pollination: Unknown, but seed is frequently set.

Cultivation: As for *P. caespitosa*.

The genus
Galeandra Lindl.

John Lindley raised this new world genus *Galeandra* from the Latin for a helmet, referring to the helmet-like anther cap. This is a small terrestrial genus mainly from the tropical Americas of which **Pabst and Dungs** attribute about 22 species to Brazil and one or possibly two to the Organ Mountain Range. Our one representative **G. beyrichii** is sporadic but has an enormous range, from Argentina through to the United States.

Galeandra beyrichii
Rchb. f. | *577* (Plate 10)

Etymology: Named in honour of **Beyrich** a distinguished German botanist.

2.0 cm

Habitat: Over the years we have found this plant on roadside banks, in 20 year old regenerating forest, in bracken-covered steep granite slopes between 900-1200 M.asl in sections V and VI. A very widespread species. **Dunsterville** found it in the cloud forest at 1300 M.asl above Caracas, Venezuela and he also describes the plant as being rare.

Occurrence: Rare.

Plant: A large terrestrial herb with a penchant for roadside habitats. It produces underground dorso-ventrally compressed pseudobulbs 5.5 cm. long by 2 cm. wide on a thickish rhizome. There are no leaves apparent before flowering. The root system is fairly dense and composed of thick white non-branching roots arising from the pseudobulb base, which spread to no great distance around the plant, in the loose decaying leaf detritus.

Inflorescence and Flowers: Borne on a long, apical up to 100 cm., tall and stout, 0.5 cm. diameter, woody and surprisingly, paniculate inflorescence. The 15-40 flowers which populate the apical 40 cm. of the inflorescence are an overall light greenish white. Petals and sepals are more or less equal, lanceolate, 2.3 cm. long by 0.6 cm. wide and light green. The faintly trilobed lip, which folds around the column, is relatively huge, 2.8 cm. wide by 2.0 cm. long, and longitudinally striped with thick green lines, on a creamy white background, which is intensely ciliate at the apical margins.

Flowering Periods: November through to February and the flowers last for up to three weeks while the inflorescence may show flowers in bloom for up to two months.

Pollination: **Dressler** suggests *Euglossine* bees, and these bees are to be found at these altitudes during its flowering periods. Up to 50% of the flowers are usually pollinated.

Cultivation: In pots on a peat compost with a 30% addition of coarse sand. It requires watering during growth but only keeping the substrate damp during dormancy.

The genus
Bulbophyllum Thou.

Aubert Thouars, a French botanist, so named this genus in 1822 from *bulbo* Greek for a bulb and *phyllum* Greek for a leaf, hence a leafed bulb or a leaf that is thick. One of the largest genera of pantropical orchids with over 1000 species. These are mainly concentrated in South East Asia but also found in Australia, New Zealand and Africa. Brazil has over 53 species, according to **Pabst&Dungs**, only 20 of which are found in the Organ Mountain Range. *Bulbophyllum* species may be a superb example of an early evolutionary development in South East Asia, which managed a migratory movement before the continents separated. Be that as it may, unfortunately, the Brazilian, at least the Organ Mountain Range species, are somewhat dull compared to their African and Asian counterparts. We have found 11 of the 20 species and probably overlooked most of the remainder as all are similar vegetatively, with roundish or squared pseudobulbs, well separated and borne on a stout woody rhizome. Their lips are all balanced and rock with the slightest wind movement. The only single-flowered plant is the pretty **B. napelii** found covering tree top branches on high mountain ridges.

Bulbophyllum atropurpureum
Barb. Rodr. | *585* (Plate 10)

7 mm

Etymology: *Atropurpureus* is Latin and means dark purple, referring to the flower colour.

Habitat: Colonies several metres in diameter can be found around mountain ridges at 1000-1300 M.asl in original forest on the middle branches of stunted and wind-pruned trees. It requires good light, strong air movement and frequent mists brought on the up draughts which are so characteristic of this habitat in section VI.

Occurrence: Occasional.

Plant: This is a small mid-tree epiphyte with conical pseudobulbs that are four-sided, measuring 2.2 cm. high and 1.7 cm. broad, and arising at 1 cm. intervals on a stout rhizome. They are light green and the older pseudobulbs frequently reactivate so that the plant appears to creep in clusters rather than having a line of bulbs. There is a single, stiff, light green, apical leaf, oblanceolate and long-lived, with a firm central vein. The root system is very strong with up to 10 fine, wiry roots per bulb, penetrating beneath the bark through any decomposing material.

Inflorescence and Flowers: A stiff, fine, purple

inflorescence arises in June and curves down wards at a right angle just before the first flower emerges, some 18 cm. from the base of the pseudobulb. There are eight alternate flowers at 0.5 cm. intervals, a deep claret colour with a greenish tinge. The yellow column has two tooth-like projections and this, combined with the nodding deep purple lip and the deep claret interior of the large dorsal and lateral sepals, makes the flower easy to recognize.

Flowering Periods: June and July; the flowers open sequentially and last for a week or so, while the inflorescence may show flowers in bloom for over a month.

Pollination: Unknown, however it could be flies or wasps.

Cultivation: In wide pots or flat pans on a rich compact compost that must be well drained and always damp.

Bulbophyllum campos-portoi
Brade | *586* (Plate 10)

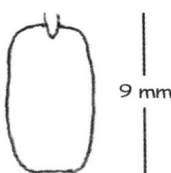

Etymology: See *Stelis campos-portoi - 1020*.

Habitat: This is a tree-top species, occurring in significant colonies, also surviving at lower levels if light and wind conditions are adequate. It is found at 1000 to 1500 M.asl, throughout sections VI, V and II in both old regrowth and original forest. By insinuating its root system beneath moss it retains moisture during dry spells. Good light and wind movement are its other needs.

Occurrence: Common.

Plant: A small clustered epiphyte. The pseudobulbs are light green-brown and smooth, ovoid when young but aging gives them four ridges, wrinkles and a yellowish colour. They are 2 cm. high by 1.5 cm. wide and are strung out on a thick grey, branching rhizome, 1.5 cm. apart. There is a single dark green linear lanceolate, deeply keeled leaf to each pseudobulb, each 7.5 cm. long x 2 cm. broad and long-lived. Roots are greyish, 6-8 to a pseudobulb, and up to 15 cm. in length, adherent and branching, at once attached to and penetrating the substrate. This gives almost 9 metres of roots to a hundred bulb colony, generally underneath moss.

Inflorescence and Flowers: Between 5 and 8 flowers are produced on the top 8 cm. of a tough, thin, spicate, basal inflorescence up to 30 cm. long. This always curves abruptly downwards to allow the alternate flowers to be presented correctly. The large and dominant heart-shaped dorsal sepal, 1 cm. in height and 1 cm. across, is olive yellow, thinly speckled with

claret. The pointed lateral sepals, curved backwards for half their length, are the same colour pattern. The petals are minute, while the balanced lip, looking like an ox tongue, has a deep purple throat and the remainder is thickly dotted with claret.

Flowering Periods: From July up to November and the flowers last for up to two weeks but as they open in sequence the inflorescence may show flowers in bloom for over a month.

Pollination: Unknown, but could be flies or wasps, and the flowers are frequently pollinated.

Cultivation: In wide pots or in shallow pans, as for *B. atropurpureum*.

General: We have also found a rare alba variety.

Bulbophyllum aff. campos-portoi
586a (Possible new species) (Plate 10)

Etymology: As yet unnamed.

Habitat: Section VI at 600 M.asl in scrub forest on rocky hillocks with good light, serious wind movement and high humidity.

9 mm

Occurrence: Sporadic in its niche.

Plant: An epiphyte so very similar to all *Bulbophyllum* species in the Organ Mountain Range, and very much so to *B. campos-portoi*. However the plant comes from altitudes far below the regions favoured by *B. campos-portoi* and the flowering period is quite different. The pseudobulbs arise on a tough woody rhizome at 5 cm. intervals. These are khaki, 3 cm. tall, ± 4 cornered, cone-like and 1.5 cm. wide at the base. They have one stiff leathery, deeply keeled, long-lived, linear lanceolate dark green leaf. The root system is strong and composed of a dozen or so wiry roots arising directly from the base of each new pseudobulb.

Inflorescence and Flowers: The basal inflorescence is thin, wiry up to 23 cm. long and holds up to 7 flowers pendent from the apex. These are a dark green and speckled with claret. The dorsal sepal is a conduplicate, heart-shaped concave, 1 cm. by 0.8 cm. broad. The lateral sepals are an asymmetric, very acuminate lanceolate 1.4 cm long. by 0.8 cm. The petals are minute. The labellum is ox-tongue-shaped, 0.9 cm, by 0.5 cm, and a deep purple. The column head shows two 0.6 cm. long cream teeth at each side and this is perhaps the most distinctive facet of the flower.

Flowering Periods: February and the flowers last for 10 days.

Pollination: Unknown.

Cultivation: As for all Brazilian *Bulbophyllum* species.

Bulbophyllum cantagallense (Barb. Rodr.) Cogn. | 587 (Plate 10)

Etymology: Presumably named because the first plant examined came from the municipality of Cantagalo-RJ.

Habitat: Found at low to mid-tree level in ridge or original scrub forest, requiring good light and wind movement, heavy summer rains and it is tolerant of a dry winter period at altitudes of 500-700 M.asl in sections V and VI.

Occurrence: Occasional.

Plant: One of the smaller *Bulbophyllum* species. A small creeping, epiphyte, sometimes lithophyte, with cone-like pseudobulbs 2.7 cm. tall by 0.7 cm. diameter at the base, green tinted with light yellow and tending to ridge vertically with age. They arise at up to 1.6 cm. intervals on a woody 0.15 cm. diameter, sheathed rhizome and hold one apical, stiff, leathery, keeled, long lasting, pointed oval leaf, up to 5 cm. long by 0.8 cm.

Inflorescence and Flowers: The inflorescence is basal, thin, wiry and horizontal, arising from the most recent pseudobulbs. At 10 cm. it curves abruptly downwards to the vertical for the apical 7 cm, whereon are held up to six, alternate, delicate flowers. These open sequentially and the overall effect is of firm dark vertical parallel stripes on a greyish white background. The dorsal sepal is heart-shaped, 0.45 cm. long by 0.4 cm. broad with three firm lines coming together just below the rounded apex. The lateral sepals are larger, a broad asymmetric acuminate, lanceolate, and 0.8 cm. long by 0.5 cm. broad at their bases and showing the same colour configuration as the dorsal sepal. The petals are microscopic, ± round and dark. The lip is a grooved, round-ended rectangle, vaguely trilobed showing a curious hooked small spur at it's under base. The balanced lip is 0.5 cm. long by 0.3 cm. wide, back folded at its apex, and is a dirty white with a dark wine central groove.

Flowering Periods: May and the flowers last for up to three weeks.

Pollination: Unknown, but could be flies or wasps.

Cultivation: As for other *Bulbophyllum* species described.

Bulbophyllum luederwaldtii Hoehne & Schltr. | 592 (Plate 10)

Etymology: Named after **Herman Luederwaldt** (1865-1934) who collected in São Paulo in 1912.

Habitat: Found in original forest from

1300 to 1600 M.asl on mountain ridges bur also in the forest canopy. The roots are always moss-covered. Plenty of light and wind and reasonable moisture are this plant's needs, in sections II, IV and VI.

Occurrence: Occasional.

Plant: A creeping epiphyte. It's pseudobulbs are conical, 4-ridged, squat, 1.7 cm. tall by 1.3 cm. wide and borne at regular 1.5 cm. intervals on a tough, 0.2 cm. wide light-grey, branching rhizome. The leaves are stiff, thin, oblanceolate, apical and single, a light reddish-green, deeply keeled and 6 cm. long by 1.3 cm. wide. The roots are thin, wiry, and branching, and form an adherent and extensive system.

Inflorescence and Flowers: Appearing on a long, 23 cm. wiry, spicate, basal inflorescence, divided into 2 cm. sections, which sets out horizontally, the apical 8 cm. dropping at 90° towards the ground giving up to 20 closely packed alternate flowers opening sequentially from the base. The flowers are smallish. The dorsal sepal is 0.6 cm. long by 0.5 cm. wide, a greenish light orange, deeply speckled with claret. The lateral sepals are the same colour, 0.9 cm. long by 0.3 cm. wide, with tapering apices that curl backwards and are fused for half their length. The petals are microscopic. The lip is fleshy, 0.5 cm. long by 0.3 cm. wide and is striking due to the two basal horn-like protuberances, and the apical half being a royal purple colour.

Flowering Periods: November and December and the individual flowers last for 10 days or so while the inflorescence may show flowers in bloom for over a month.

Pollination: Unknown, but could be flies or wasps and the flowers are frequently fertilized.

Cultivation: As for **B. atropurpureum**.

Bulbophyllum paranaense Schltr. | 594 (Plate 10)

Etymology: Presumably first found in Paraná.

Habitat: Found from 1500 M.asl upwards, in dwarf regrowth forest along mountain ridges in section II, IV and VI.

Occurrence: Common in its niche.

Plant: A small clustered epiphyte. Most *Bulbophyllum* species in the region are vegetatively almost identical and this one is no exception. Squat, four-cornered pseudobulbs, 1.7 cm. tall by 1.5 cm. wide, arise at short intervals on a thick branching rhizome. They are light brown, probably due to exposure to intense sunlight. The single, apical, almost linear leaf, 5.4 cm. long by 1.5 cm. wide, is light green, stiff, deeply keeled and

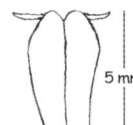

long lived. The root system, composed of greyish white, thin and long roots, is intensive, extremely adherent and invariably under coarse moss.

Inflorescence and Flowers: These are borne on typical Bulbophyllum inflorescences which are erect, 14 cm. long up to the right-angled down-turned, 5 cm. long, apical flower-bearing stretch. Up to 12 bilaterally compressed, cleistogamous flowers are borne alternately. The pinkish red buds, which look so promising, do not open. If dissected, the heart-shaped dorsal sepal measures 0.6 cm. tall by 0.4 cm. wide; the laterals look like extracted rose thorns, 0.5 cm. long by 0.4 cm. wide. The petals are minute while the faintly trilobed, shoe-horn shaped lip measures 0.7 cm. by 0.2 cm.

Flowering Periods: March.

Pollination: Self.

Cultivation: As for other **Bulbophyllum** species described.

Bulbophyllum ricaldonei
J. E. Leite | 596

Etymology: Named in honour of **Ricaldone**.

Habitat: Found in high montane original dwarf forest at 1600 M.asl in section VI requiring nightly humidity, high air movement and good light.

Occurrence: Rare.

Plant: Typical of most of the high mountain Bulbophyllum scrub epiphytes. The squat, four cornered pseudobulbs are held on a tough woody rhizome at up to 2 cm. intervals, sheathed when young, 1.5 cm. tall by 1.5 cm .in diameter, a light green to brown. There is a single apical leaf, light green, linear lanceolate up to 6 cm. long by 1.5 cm, stiff, keeled and long lasting. The root system consists of up to eight thin, wiry, white, green tipped roots, arising from the base of new growths and remaining viable for many years.

Inflorescence and Flowers: The inflorescence is basal and arises from the most recent pseudobulbs. It is semi-pendulous, stiff, measures 10 cm. before turning down to show a pendent inflorescence of up to six flowers. The flower is confused. The dorsal sepal is almost round, 1 cm. broad by 0.75 cm. high. The lateral sepals are partially fused and boomerang-shaped with acuminate apices. The petals are minute, triangular and measure 0.4 cm. by 0.5 cm. The lip looks like a gross and wrinkled tongue, ± oval and 0.4 cm. by 0.25 cm, with two horns at its shoulders.

Flowering Periods: February and March and the flowers

last for a week or so. However the inflorescence may show flowers for a longer period.

Pollination: **Dressler** suggests probably nectar or carrion visiting flies.

Cultivation: As for *B. cribbianum*.

Bulbophyllum cribbianum
Toscano | 620

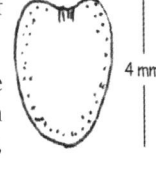

Synonym: B. micropetalum **Barb. Rodr.**

Etymology: Dedicated to **P. Cribb** curator of the Kew Gardens herbarium.

Habitat: Epiphytic, growing at 1000 M.asl, in section VI and found on mid-tree trunks and tree-top branches enjoying high rainfall and humidity in summer and mists in winter, with good air movement and high light, in original forest.

Occurrence: Occasional.

Plant: A small to medium-sized creeping and branching epiphyte with pseudobulbs arising at 1 cm. intervals on a thick, tough, woody and partly sheathed rhizome. The pseudobulbs are round and squat, 1.4 cm. tall and 1.5 cm. diameter, sheathed when young, smooth and light green. There is a single light green, linear lanceolate apical leaf up to 6 cm. long by 1.5 cm. wide and long lived. The root system is vigorous and composed of up to eight short, stiff and wiry white, green-tipped roots which emerge from the base of the new growth and remain active for many years.

Inflorescence and Flowers: A stiff, thin 11 cm. stem arises from the bases of the most recent pseudobulbs, and a raceme of flowers is borne on the distal 7 cm. or so. The flowers open singly or two at a time at most and the inflorescence bends at right angles to the stem. The heart-shaped dorsal sepal is 0.9 cm. long by 0.8 cm. wide and has a silver-white background, deeply longitudinally veined with purple. The lateral sepals are fused, the same colour for 70% of their length and are a broad asymmetric lanceolate shape, measuring 1.0 cm. by 0.8 cm. wide at the base. The petals are almost invisible: hence, appropriately, micropetalum, the name given by **Barbosa Rodrigues**, now a synonym. The lip is oval in silhouette with an interior smaller oval section and has two small ear-like protuberances on its shoulders. The overall measurements are 8 mm. long by 6 mm and the colour is a dirty, dark purple.

Flowering Periods: Late March and April and although the flowers last for about a week, the inflorescence lasts for over a month.

Pollination: Probably pollinated by nectar-collecting or carrion visiting flies.

Cultivation: On bark plaques observing the natural conditions.

Bulbophyllum mirandaianum Hoehne | *621* (Plate 10)

Etymology: Presumably, first found in the district of Miranda.

Habitat: Low mountain, semi-vertical rock slopes in sub-scrub forest around 500 M. asl in Section VI, requiring high light, good wind movement and high year-round humidity.

Occurrence: Occasional in its rare environmental niche.

Plant: A small creeping epiphyte, somewhat similar to *B. napellii* in size. The pseudobulbs are yellowish-green, semi-conical and four-sided and borne on a thick, woody branching rhizome at 1.0 cm. intervals. Pseudobulbs are 1.5 cm. high by 0.5 cm. broad at their bases with single apical, linear-lanceolate leaves which are 1.3 cm. long by 0.5 cm, deeply keeled and leathery.

Inflorescence and Flowers: The thin, wiry inflorescences are up to 20 cm. long and arise from the most recent pseudobulb bases with flowers borne on the apical hooked 6.0 cm. They open sequentially and are very pretty and delicate with deep purple striped petals and sepals on a muddy green-white background. The dorsal sepal is heart-shaped, 0.4 cm. long and broad, while the half-fused laterals are similar but with marked tails, 0.6 cm. long by 0.5 cm. The petals are minute, and the oval, finely-balanced lip has a short central line of tiny balls like a microscopic string of pearls, reaching from the throat for half its length.

Flowering Periods: The flowers last for a week or so during June but as they open sequentially, the plants may bloom for a month.

Pollination: Unknown.

Cultivation: As with all other Organ Mountain *Bulbophyllum* species, this should be grown on bark or tree fern plaques, with high humidity during spring, summer and autumn and with a semi-dormant period in winter.

Bulbophyllum regnellii Rchb. f. | *622*

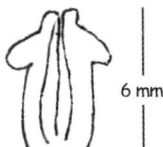

Etymology: Named after **Anders F. Regnell** (1807-84), a Swede who collected in Minas Gerais.

Habitat: On wind-blown dwarf regrowth scrub at 1500 M.asl on mountain ridges from almost ground level up to three metres, usually at the spot where a branch joins the trunk or subdivides and always bedded deep in moss. Searing sunlight and desiccating winds show this plant to be a very tough survivor in section VI.

Occurrence: Common in its niche.

Plant: A small epiphyte, rather more of a clusterer than other *Bulbophyllum* species in the region, with pseudobulbs emerging very closely together on a tough woody rhizome. The pseudobulbs are squat, hardly angled and almost round, 1.5 cm. high by 1.5 cm. broad and light brown in colour. They bear a single, long-lived, oblong, deeply keeled, dark green, stiff apical leaf, 4 cm. long by 1.3 cm. broad. The root system is composed of fine, long, branching and extremely adherent roots in some profusion. The plant is always found packed tightly in a bed of coarse moss.

Inflorescence and Flowers: The inflorescences are long, stiff, spicate, and basal and arise from the most recent pseudobulbs. They are round, jointed, partly sheathed, generally angled upwards for the first 14 cm. when they bend downwards to the vertical, presenting all the flowers at precisely the same angle as those of any neighbouring colony. The flowers are alternate, small, bilaterally compressed and partly sheathed. The dorsal sepal is heart-shaped while the lateral sepals are an asymmetric, acuminate heart-shape, 0.5 cm. long by 0.5 cm. broad. The petals are a round, pointed rectangle and measure 0.4 cm. by 0.2 cm.

Flowering Periods: February and March when the flowers last for 10 days or so although the inflorescence may show flowers in bloom for over a month.

Pollination: Unknown.

Cultivation: As for other *Bulbophyllum* species described.

Bulbophyllum napellii Lindl. | *634* (Plate 11)

Etymology: Named in honour of Napell.

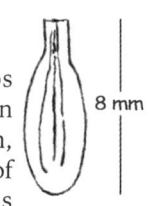

Habitat: Found on some mountain tops and ridges at 1200-1600 M.asl often covering entire branches of wind-blown, stunted trees; also on canopy branches of original forest trees. Mists and moss ensure its high humidity requirements, the other basics being good light and a lot of wind movement, in section VI.

Occurrence: Common and locally abundant in its niche.

Plant: This is a small epiphytic clusterer. The pseudobulbs are a light green, elongated cone, 1.5 cm. tall x 0.8 cm. wide, strung at 1.5 cm. intervals on a tough brownish, branching rhizome. There is a single apical, long-lived leaf per bulb, dark green, broad, leathery, linear and deeply keeled, 2.5 cm. long by 1 cm. broad. The root system is not extensive. Six or 7 thin and wiry roots arise from a new bulb base, each up to 5 cm. in length, grey, adherent, always under moss or within bark cracks.

Inflorescence and Flowers: These appear on thin, wiry, 4 cm. basal inflorescences, each bearing a single flower. The dorsal sepal is a pointed spade shape, 0.5 cm. long and wide. The lateral sepals are similar but merge to a narrow point and curve backwards at the merging. The petals are smaller, each a pointed rectangle, 0.3 cm. long x 0.2 cm. wide. All have an opaque greenish tinge with fine red stripes. The lip is tongue-like and also red striped, and when extended, measures 8 mm. long by 0.4 cm. wide.

Flowering Periods: January and February and the flowers last for 10 days or so.

Cultivation: As for other **Bulbophyllum** species but requires high light and strong air movement more than most.

Pollination: Unknown.

General: The only single flowered **Bulbophyllum** in these regions although occasionally we have found an inflorescence with two flowers.

The genus
Amblostoma Scheidw.

Scheidweilder named this genus in 1838 using the Greek composite word: *amblo* meaning blunt-obtuse while *stoma* means mouth, possibly referring to the very flat column head. There are about nine species from tropical South and Central America with one from the Organ Mountain Range.

Amblostoma tridactylum (Lindl.) Rchb f | *635*

Etymology: *Tri-dactlylum* is Greek for three-fingered, referring to the distinctive trilobed lip.

Habitat: Found around 1000 M.asl along river banks in sections VI and VII, requiring reasonable light and wind movement with high constant humidity at low to mid-tree level in regrowth forest.

Occurrence: Rare.

Plant: This can initially be confused with an **Epidendrum** or **Tetragamestus** species when not in flower. However closer inspection shows a distinctive pseudobulb which eliminates the **Epidendrum** and no sequential pseudobulbs on the same stem which eliminates the **Tetragamestus**. The pseudobulbs are narrowly cylindrical, up to 20 cm. long and 0.7 cm. in diameter at their central most swollen point. They arise closely from a thinnish wandering rhizome. The leaves are alternate, sub apical to apical, light green, mildly keeled and flexible. The root system is extensive, composed of thin white sometimes branching roots, which extend for some distance beyond the plant's circumference.

Inflorescence and Flowers: The inflorescence is apical, pendant and holds one to three racemes of small yellow flowers, up to 40 of which may appear closely on each branch at 0.3 cm. intervals. The flowers are quite unmistakable. The lip, being the most prominent factor is aggressively trilobed-, 0.5 cm. long at all lobes and appears as a trident, with a nose like callus running down the centre of each lobe for over half their length. The sepals are a broad lanceolate 0.4 cm. long by 0.2 cm. wide, while the petals are spathulate 0.4 cm. long by 0.12 cm. wide. The lip is totally fused to the column and forms a tube. All tepals are convex and incurved.

Flowering Periods: December and the flowers last for three weeks.

Pollination: Unknown.

Cultivation: As for the reed stemmed **Epidendrum** species.

General: This is another of **George Gardner**'s collections from the Organ Mountain Range and described by **John Lindley** as *Epidendrum tridactyle* in 1838. **Reichenbach** transferred it to *Amblostoma* in 1864.

The genus
Encyclia Hooker

William Hooker was the director of Kew Gardens, London, England in the mid part of the 19th century when, in 1838, he split out this genus from *Epidendrum*. Quite recently two sections, *Osmophyta* and *Encyclia* have been separated by **Dressler** on the grounds that Osmophyta members have triangular fruit in section, and fusiform, sometimes laterally compressed, pseudobulbs while the Encyclia group have round fruit in section, and the pseudobulbs are round to conical. To further complicate matters, *E. pygmaea* was transferred into the *Osmophyta* sub genus from the genus *Hormidium* by **Pabst and Dungs**.

Encyclia comes from the Greek word *Enkyklein* meaning circular, referring to the way that the lip's lateral lobes encircle the column. *Osmophyta* is a Greek compound word meaning scented plant, rather a weak name for this group as few species have any perceptible scent.

The true Encyclias are quite distinguishable, and all seem to conform to the generic name with their lips encircling the column. **Carl Withner**, in volume VI of *The Cattleyas and their Relatives*, lists some 80 species attributable to South America, of which around 50% can be found in Brazil and only six in the Organ Mountain Range; of these we have discovered four. Of the *Osmophyta* section, **Pabst & Dungs** have attributed 24 species to Brazil of which 10 should be found in the Organ Mountain Range and of these we have discovered six, five of which are common and are concentrated in section VI and one, *E. fragrans*, found in sections V and VIII, is rare (however we also found it in Pará).

While species in the section *Osmophyta* are easily distinguishable from each other, the true *Encyclia* species' flowers are extremely difficult to differentiate; so much so that there are literally hundreds of synonyms for plants in this group. Many new species have been discovered since 1973 (**Pabst and Dungs** analysis) but mostly in the Brazilian hinterlands, and few or none in the Organ Mountain Range.

Encyclia bracteata
(Barb. Rodr.) Schltr. | *648* (Plate 11)

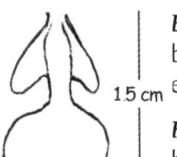

Etymology: *Bracteata* is so-called because of the marked bracts which envelop the new pseudobulbs.

Habitat: On shrubs and moss-covered boulders at about 800 M.asl, in section VIII where it enjoys a reasonably high humidity, good light and air movement; a wet summer and a dry but misty winter.

Occurrence: Rare but locally frequent.

Plant: A small epiphyte and sometimes lithophyte with very tightly clustered egg-shaped pseudobulbs, light to dark green and 2.5 cm. high by 1.5 cm. diameter and bearing a single, long-lived apical leaf. This is deeply grooved, dark green, linear lanceolate, somewhat pendent, held on a 2 cm. petiole and measuring up to 25 cm. long by 0.5 cm. The root system is composed of a few thickish white, short roots that burrow effectively into the substrate.

Inflorescence and Flowers: An inflorescence of up to 12 cm. emerges from the pseudobulb apex, bearing one apical and an occasional, second lateral flower on 1.5 cm. pedicels about 1 cm. apart. The lanceolate sepals are more or less equal, 2 cm. long by 0.6 cm. The petals are the same shape but slightly smaller. The lip is trilobed; the apical lobe is almost round and 1 cm. wide by 0.8 cm. deep. The lateral, forward-pointing side lobes start at the throat, measure 0.5 cm. by 0.2 cm. wide, are a blunt lanceolate and envelop the column. The entire lip is 1.5 cm. long, the apical lobe being pink to magenta while the remaining flower parts are a light green, speckled with brown.

Flowering Periods: Early December and the flowers last for about ten days.

Pollination: Unknown, but probably bees.

Cultivation: On tree fern or cork bark plaques watering and feeding well during summer but only misting when not in active growth to keep the pseudobulbs from shrivelling.

General: We were shown this plant by a collector in section VIII of the Organ Mountain Range who assured us of its local provenance. However the literature we have limits it to Espírito Santo and Minas Gerais.

Encyclia flabellifera
Hoehne & Schltr. | *655* (Plate 11)

Etymology: *Flabellifera* is Latin for like a fan, presumably referring to the fan-like lip.

Habitat: Epiphytic in original mountain and scrub forest at 900-1200 M.asl at mid-tree level, in section VI enjoying good, though dappled light, good air movement and a high constant humidity. These plants appear to be very choosy as to habitat. On one mountain ridge they appear in swarms while on an adjacent ridge at the same altitude and in apparently the same conditions no plants appear.

Occurrence: Common in a specific habitat but absent from large areas of original forest which have apparently ideal conditions.

Plant: A medium to large epiphyte with conical-ovate pseudobulbs, somewhat squatter than those of *E. oncidioides* or *E. odoratissima*. These arise closely and alternately on a thick woody rhizome. The pseudobulbs vary in size greatly between plants, from 7.5 cm. tall and 4 cm. diameter at the base to 2.5 cm. tall by 2 cm. A large plant we examined showed 25 pseudobulbs, all apparently active. There are two or three apical leaves which are long lived, strap-shaped and measure up to 40 cm. long by 2.5 cm. broad, dark green, and deeply keeled. The root system is prodigious and immeasurable, composed of long, thickish branching roots, 0.2 cm. in diameter, grey and which run up and down mid-tree branches and seem to be viable for many years.

Inflorescence and Flowers: A stiff, erect, slightly pendulous, paniculate inflorescence up to 55 cm. long emerges in October. Around 35 waxy, yellow brown, stiff flowers are shown, that produce a strong and agreeable perfume which can be detected at 20 metres distance on a still morning. The flowers are almost 4 cm. wide. Sepals and petals are more or less equal in size and shape, 2 cm. long by 0.7 cm. broad, a rounded lanceolate to spathulate and appear rather like a four-bladed propeller. The trilobed lip is projected frontward and downwards, showing a central lobe which is almost round, with margins both crisp and dentate, a whitish yellow speckled in purple. The lateral lobes are oblong, 1 cm. long by 0.4 cm. and enfold the column while the lip length is 1.4 cm.

Flowering Periods: In November and December and the flowers last for three weeks.

Pollination: Probably bee pollinated – **Dressler**. The very strong scent coupled with the fact that when pollination does occur, it occurs intensely, would indicate *Euglossine* bees.

Cultivation: In pots with a well drained fibrous substrate, taking care that the plants are never over watered, or on tree fern plaques. In nature they grow well on live bark but thrive when the roots have access to rotting wood or accumulated detritus where branches bifurcate.

General: **Antonio Toscano do Brito** recently discovered that this species is synonymous with *Encyclia ionosma* (Lind.) **Schltr**, the violet-scented *Encyclia*.

Encyclia odoratissima
(Lindl.) Schltr. | *663* (Plate 11)

Etymology: *Odoratus* is Latin for having a sweet scent, and the superlative *-issima* means a very sweet smell.

Habitat: Rarely found above 1200 M.asl. It appreciates a reasonably high humidity, filtered light and good air movement at the mid-tree level, in both original and maturing regrowth forest in section VI.

Occurrence: Occasional.

Plant: A medium to large tufted epiphyte. The pseudobulbs are a round cone-shape, alternating on a thick fibrous dark green rhizome, and measure 4 cm. tall by 3 cm. wide, wrinkling with age. There are two to three apical leaves, linear-lanceolate in shape, dark green, leathery, deeply keeled and retained for several years. The root system is extensive and adherent and composed of thick white, sometimes branching roots, generally insinuating into cracks in bark and usually under moss.

Inflorescence and Flowers: Around 30 greenish flowers are produced which open more or less simultaneously in a paniculate terminal inflorescence which occupies half of the 60 cm. tall, apical, erect, stiffish, raceme. The petals and sepals are a light olive green. All are spathulate, similar in length, 1.5 cm. by 0.7 cm, although the petals are somewhat more club-shaped and wider. The lip is white, with a very faint mauvish tinge at its base, two up-curling wings and a down-folded labellum. The distinctive feature is a strong but pleasant perfume.

Flowering Periods: In November and the flowers last for up to three weeks.

Pollination: As with *E. oncidioides* and *E. flabellifera* pollination is frequent, often up to 90%, suggesting a common pollinator for all three structurally similar species.

Cultivation: Best grown in pots on a fibrous well-drained and aerated compost, watered well in summer, only misted thereafter and kept in dappled light with good air movement.

Encyclia oncidioides
(Lindl.) Schltr. | *664* (Plate 11)

Etymology: *Oncidium*-like, though this is not very convincing.

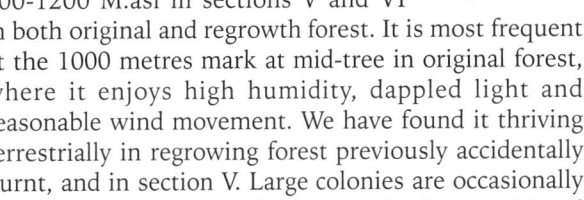

Habitat: This species is found from 700-1200 M.asl in sections V and VI in both original and regrowth forest. It is most frequent at the 1000 metres mark at mid-tree in original forest, where it enjoys high humidity, dappled light and reasonable wind movement. We have found it thriving terrestrially in regrowing forest previously accidentally burnt, and in section V. Large colonies are occasionally found on relict original forest trees in the transitional zone.

Occurrence: Common.

Plant: A clustered epiphyte and sometimes terrestrial, similar to *E. odoratissima* although on average the plant is some 30% smaller. The pseudobulbs are conical, dark green, 3-4 cm. tall by 2-3 cm. wide and alternate from a stiff rhizome. The young bulbs are smooth and sheathed and the old bulbs are wrinkled. The two apical leaves are deeply keeled, dark green, linear-lanceolate, up to 15 cm. long by 1.8 cm. wide, and long lived. The root system is profuse and composed of thick white penetrating and adherent roots which are active for many years.

Inflorescence and Flowers: These are borne on a stiff, 40 cm. long, apical stem bearing up to 30 flowers in a paniculate terminal inflorescence which occupies half of the stem. Sepals and petals are 1.6 cm. long by 0.5 cm. wide, somewhat spathulate and a brownish, yellow green. The lower lip is rounded and protrudes and the wings are yellowish and long.

Flowering Periods: From June through to September; the plants at lower altitudes are the first to bloom in June while those at the higher altitude limit flower much later and the flowers last for up to three weeks.

Pollination: As for *E. odoratissima*.

Cultivation: As for *E. odoratissima*.

General: The only obvious difference between this and *E. odoratissima* is the latter's strong scent. An interesting point: **Richard Harrison** was the first to flower this plant in England in 1822.

Encyclia pygmaea (Hook.) Dressler | *674* (Plate 11)

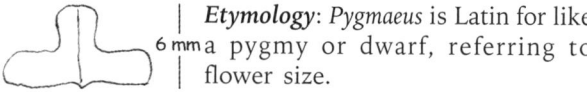 *Etymology*: *Pygmaeus* is Latin for like a pygmy or dwarf, referring to flower size.

Habitat: Found in the high tree zone in original forest around rivers and streams. It appreciates good light, high air movement and humidity at around 1000 M.asl in section VI.

Occurrence: Common.

Plant: A small creeping epiphyte forming large colonies at mid- to tree top level. The pseudobulbs are light green and fusiform, 2.5 cm. by 1 cm. wide, with a light brown sheath, and arise at 1 cm. intervals from a long, thick, branching rhizome. The root system is profuse and adherent, composed of thick, white, sometime branching roots forming a mat around or along a bough. The two apical leaves are light green, stiff, leathery, deeply keeled and lanceolate, 3 cm. long by 0.8 cm. wide and long lived.

Inflorescence and Flowers: One to six inverted flowers arise on a very short fasciculate inflorescence. The flowers are quite typical of the *E. calamaria* alliance but much smaller, as the name suggests. The petals, sharply acute, 0.6 cm. long by 0.2 cm. wide, are a yellowish cream at the base with 5 vertical, pinkish brown stripes. The petals are narrow, pointed, 0.6 cm. long by 0.1 cm. wide, and cream in colour. The lip, when spread, is white and faintly tri-lobed, and has the lateral wings folded around the column and a small purple streak at the apical point.

Flowering Periods: December through to February and the flowers last for up to three weeks.

Pollination: The flowers are frequently pollinated even when the plant is moved to much higher altitudes, which suggests a common and widespread pollinator.

Cultivation: Due to its vigourous growth and creeping habit this plant is best set on fairly large tree fern or bark plaques with good dappled light, air movement and a year around constant humidity.

General: This has been moved from the genus *Hormidium* to *Encyclia* by **Pabst & Dungs**.

Encyclia calamaria (Lindl.) Pabst | *679* (Plate 11)

Etymology: *Calamaria* is Latin for squid, referring to this plant's capacity to raise itself up on its roots thus collecting leaf litter and other detritus.

Habitat: This is a mid-tree plant found in original and occasionally in old regrowth forest from 900-1400 M.asl, forming large colonies on thick branches or around the trunk. Its habit of raising itself above the surrounding stratum on new root growth, like a squid standing on its tentacles, has resulted in its name and enables the plant to collect dead leaves in its foliage, contributing both to the humidity in dry periods and ultimately to its nutrition when the collected leaves decay. We have also found it growing as a terrestrial in accidentally burnt regrowth scrub in sections V and VI where the squid-like character is even more pronounced.

Occurrence: Common.

Plant: A small epiphytic clusterer, with light green, fusiform pseudobulbs, 5 cm. long by 1.2 cm. wide, arising at 1 cm. intervals from a tough, green, branching rhizome. There are two or rarely, three, apical, lanceolate leaves which are erect, opposite, deeply keeled and dark green, 8.5 cm. tall by 1 cm. wide, and long lived. The root system is extensive, arising ± uniformly from the rhizome base and the roots are thick, penetrating and adherent, tending to lift the plant somewhat above its chosen platform.

Inflorescence and Flowers: Four to six flowers are produced on a thick green racemose, 6 cm. apical inflorescence. The flowers at first sight are a small version of *E. inversa*, but the lip is single lobed, spade shaped, 1.0 cm. by 0.6 cm. broad, with a nose-like central callus for half its length. The 1.5 cm. long sepals are finely pointed and the petals are a narrow lanceolate, 1.1 cm. by 0.2 cm. broad. The overall colour is white, slightly greenish with a few purple streaks on the lip.

Flowering Periods: April and May and the flowers last for three weeks or so.

Pollination: As for *E. pygmaea*.

Cultivation: As for *E. pygmaea*.

Encyclia kautskyi Pabst | *681* (Plate 11)

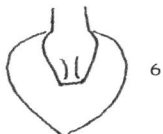

Etymology: Named in honour of **Roberto Kautsky**; a famous botanist from Espirito Santo.

Habitat: Large colonies of half a square metre can be found on thick horizontal mid-tree branches in original forest, although we have also found it growing terrestrially in young regrowing forest that was accidentally burnt. Good light and air movement and moderate humidity levels cover its needs at 900-1200 M.asl in sections V and VI.

Occurrence: Common in the dryer slopes but becoming sparse as we reach the wetter Atlantic facing scarp slopes.

Plant: This is an epiphyte with pseudobulbs that alternate, 2 cm. apart, on a thick, tough, green rhizome. Branches occur and form an intense mat with thick horizontal leads. The pseudobulbs are a compressed ovoid, dark green and smooth, measuring 3 cm. tall x 1 cm. The pair of apical leaves measuring 2 cm. x 1.2 cm. is dark green, spiny and stiff. The thick roots are white and form an extensive system which seems not very adherent.

Inflorescence and Flowers: A thick, apical inflorescence, 2 - 4 cm. high is produced in mid-summer and holds up to six flowers which are typical of the *E. inversa* group in shape, small, and an overall yellowish brown. The petals and sepals are a sharply pointed lanceolate, 0.75 cm. long by 0.25 cm. with the petals slightly smaller. The projecting lip is heart-shaped, 0.7 cm. long by 0.4 cm. and presented uppermost with the whole flower apparently upside down.

Flowering Periods: In December and January and the flowers last for three weeks.

Pollination: As for *E. pygmaea*.

Cultivation: As for *E. pygmaea*.

General: This plant is easily confused with **Sophronitis brevipedunculata** or **Maxillaria cerifera** when not in flower.

Encyclia inversa (Lindl.) Pabst | *687* (Plate 12)

Etymology: *Inversa* is Latin for turned upside down, referring to the position the flowers is held on the inflorescence, non-resupinate.

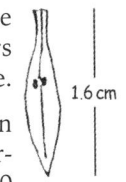

Habitat: Common throughout the region in sections VI, V and II in the mid- to upper-tree zone in original forest, above 1000-1450 M.asl. It forms extensive colonies and is a leaf litter gatherer. Filtered light and a high humidity, usually supplemented by neighbouring bromeliads on hot dry days, and fair air movement make up this plants ideal conditions.

Occurrence: Common.

Plant: A medium-sized epiphyte. The pseudobulbs are dark green, elongated and oval, 7 cm. tall and 2 cm. broad and carried at 2 cm. intervals on a thick, sometimes branched rhizome. There are two dark green, erect, long-lived, apical leaves for each bulb, each broadly lanceolate and measuring 13 cm. long by 2.5 cm. and deeply keeled. The root system arises from the underside of the rhizome in loose groups of 2-4 per cm, and the thick, white roots attain 15-25 cm. in length, and are adherent, surrounding the host branch, firmly anchoring the plant.

Inflorescence and Flowers: These appear on an erect and fleshy inflorescence of 10-12 cm. with small light green sheathing bracts. There are 7 to 10 flowers, 3.5 cm. high x 4 cm. across and upside down compared with most orchids, non-resupinate. The sepals are 2 cm. by 0.5 cm. broad, an acuminate lanceolate, white, with wine coloured stripes for 2 mm. at the base. The petals are an acuminate elliptic measuring 1.8 cm. by 0.5 cm. broad. The pointed lip 1.6 cm. long, by 0.4 cm. broad, is folded at the shoulders with marked wine striping, fading to the tip and with three small calli. The column is also wine-striped.

Flowering Periods: January and February but sometimes in other months and the flowers last for two weeks.

Pollination: As for E. pygmaea. Pollination is frequent at all altitudes and often as many as 70% of the flowers are fertilized, and soon show three winged capsules, which presupposes a common and widespread, pollinator.

Cultivation: In large pots well filled, with a fibrous, well aerated and well drained compost in dappled but good light and air movement. Water well in spring and summer, thereafter only misting.

Encyclia suzanensis (Hoehne) Pabst | 694 (Plate 12)

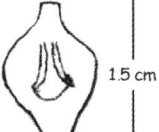

Etymology: Presumably referring to Suzana a town near São Paulo from where the specimen examined came.

Habitat: A mid-tree epiphyte from section VI of the Organ Mountain Range growing at 900-1000 M.asl. It needs dappled light, good air movement and a constant high humidity.

Occurrence: Rare.

Plant: A large epiphyte, tall and ungainly with compressed fusiform pseudobulbs tending towards a long ovate shape. They measure 18 cm. high and up to 3.5 cm. wide and arise alternately from a thick straightish rhizome at short intervals. The two to three alternate and opposite apical leaves are lanceolate, keeled and flexible and measure up to 20 cm. long by 4.0 cm. and are long-lived. The root system is vigorous and active for many years and is composed of numerous thick, white and very adherent, long, branching roots. The plant is identical with *Encyclia vespa*, itself variable in size, profile, range and even habitat.

Inflorescence and Flowers: A 30 cm. apical inflorescence bears a cone of 25 flowers on the apical 20 cm. The flowers are resupinate and waxy; the sepals are equal in size and a pointed spathulate shape measuring 1.8 cm. long by 0.8 cm. wide. The petals are the same shape but 1.5 cm. long by 0.8 cm. and blunter at the tips. All parts have a light cream background heavily speckled with light chocolate brown. The lip measures 1.5 cm. long by 0.7 cm. and is vaguely trilobed with a thick, central longitudinal callus. It is creamy-white and a blunt arrowhead shape.

Flowering Periods: April and May and the flowers last for up to three weeks.

Pollination: Probably pollinated by bees.

Cultivation: In pots with a sandy, fibrous bark-based mixture or on bark or tree fern rafts. A hardy plant requiring natural ambient conditions as above, good watering in spring and summer and misting while dormant.

General: *Encyclia vespa* flowers from September to December at altitudes up to 1400 M.asl, while *E. suzanensis* appears in April and May. Also, the lip of the latter is longer and slightly larger. We are not convinced that this plant is a separate species but rather a variant of *E. vespa*.

Encyclia vespa (Vell.) Dressler | 695 (Plate 12)

Etymology: *Vespa* is Latin for wasp, presumably referring to the flower's wasp-like colouring and appearance.

Habitat: Filtered light, good humidity and mild air movement are this plant's basic needs. It spreads rapidly as a terrestrial in accidentally burnt regrowth forest and is only excluded by extreme shade. At 1200 M.asl we have found it forming dense terrestrial colonies in such situations on south facing slopes, with *Elleanthus crinipes*, *Zygopetalum pedicellatum* and some larger *Epidendrum* species. As an epiphyte, it occurs in similar mid-tree situations to *E. inversa* but such colonies tend to be smaller in dimension to the terrestrials. These are found at 1200-1500 M.asl in original forest or in maturing regrowth in sections V and VI.

Occurrence: Common.

Plant: A large epiphyte and a successful terrestrial found on fallen logs or well-drained leaf litter. It is tall and slightly ungainly, but a large stand in flower is most attractive. The pseudobulbs are a compressed fusiform and green, 24 cm. high x 2.5 cm. wide, covered with papery scales on the lower parts. They arise from a thick, 0.5 cm. straight rhizome at 1 cm. intervals. There are 2 to 4 apical leaves, 24 cm. long x 4.5 cm. wide, deep green, a pointed oblong and retained for 4 years on healthy colonies. Roots are produced from the base of the rhizome at 4 to 6 roots per cm. of rhizome length. They are thick, 0.4 cm. in diameter, and average 25 cm. in length. A five bulb colony we examined had a total length of over 10 metres of adherent roots.

Inflorescence and Flowers: A 30 cm. tall apical inflorescence has up to 40 waxy non-resupinate flowers. Each flower is 2 cm. high x 2.8 cm. wide and the petals and sepals are yellow-green and claret, spotted with green. The sepals are a broad spathulate, 1.1 cm. by 0.6 cm. broad. The sepals are more club-shaped and 1.2 cm. by 0.5 cm. broad. The lip is fused up to half the column length and has a distinct maroon tip. It is spade-shaped, 1.2 cm. long by 0.4 cm. broad.

Flowering Periods: September to December and the flowers last for up to three weeks. The plant is very altitude conscious, the lower level plants flower in September to October while those at the upper limits bloom in November and December.

Pollination: **Dressler** suggests bees or wasps. While the three winged capsules are not so common on this plant, five per cent of the flowers on average are pollinated. This is consistent at all altitudes, which suggests a common and widespread pollinator.

Cultivation: In large pots on a coarse fibrous compost.

Water well in spring and summer and thereafter just keep the compost damp, in dappled light with some air movement. This is a tough plant, difficult to kill even by beginners.

Encyclia fragrans (Sw.) Lemée | 696 (Plate 12)

Etymology: *Fragrans* is Latin for fragrant, referring to the flower's pleasant scent.

Habitat: An epiphyte, in sections V and VII at 500-800 M.asl in transitional forest at mid-tree level, requiring good light and air movement, a long dry winter with high summer moisture.

Occurrence: Occasional.

Plant: It is unusual for Organ Mountain Encyclias. The pseudobulbs arise at 1.8 cm. intervals from a thick noded rhizome. They are up to 6 cm. long, 1.3 cm. wide, partly sheathed, bilaterally compressed and 0.7 cm. thick, light green and elliptical. The single apical leaf is leathery, keeled, long-lasting, infolded at its base, light green, elliptical and up to 16 cm. long by 4 cm. The roots emerge at the pseudobulb base and also sparsely from the rhizome. They are on average about 6 cm. long, tend to branch at their tips and are thickish and somewhat stiff.

Inflorescence and Flowers: The short thick apical inflorescence, up to 6 cm long, holds a cone of up to six attractive flowers and arises from the new growth. The non-resupinate flower is dominated by the lip which is fused to the column base. It is singly lobed, a broad heart-shape, 1.5 cm. long by 1.4 cm. broadly acuminate at the apex thickly veined in deep purple on a creamy-white background. The sepals and petals are ± equal in size and a narrow to broadish acuminate, lanceolate, cream-white with a tinge of green and measure up to 2.8 cm. by 0.5 cm. broad. The orange anther cap contains two pairs of pollinia and the column is blotched purple.

Flowering Periods: In November and the flowers last for upwards of two weeks.

Pollination: Probably carpenter bees – **Rodrigo Singer**, Campinas University (Unicamp/Brasil), personal communication.

Cultivation: In pots on a coarse fibrous mixture, watering well in summer but only keeping moist during winter.

General: This was the first South American orchid to be flowered at Kew Gardens in 1782, collected in the Caribbean and described by **Olof Swartz** in 1788 as *Epidendrum fragrans*.

The genus *Epidendrum L.*

This was the first new world genus described by **Linnaeus,** the father of modern systematics and taxonomy, in 1753. *Epidendrum* is a Greek compound meaning growing on trees. It is a new world genus found from North Carolina through the Caribbean and in all Central and South American countries with the possible exception of Chile. About 110 species are attributed to Brazil by **Pabst and Dungs** and of these, 44 to the Organ Mountain Range. We have found 36 species, in almost all sections.

The range in plant size is very wide, from the huge pendent plants of *E. paranaense* in the humid forests on the scarp and the very large lithophyte, *E. robustum*, on the anticline, to the small fleshy plants of *E. latilabre* and the tiny flowered *E. armeniacum*.

The variation of conditions in which the varied species are found has resulted in many adaptations; lithophytic or terrestrial on the dryer anticline, epiphytic on the wet scarp - the climate has forced these pseudobulbless plants, to 'dance according to the music'.

All inflorescences are apical, sometimes single, many are paniculate and almost always attractive. The root systems are generally extensive and strongly developed and cultivation is relatively simple, with the exception of *E. ecostatum* which rejects any substrate that is not a live tree.

Epidendrum armeniacum Lindl. | 699 (Plate 12)

Etymology: *Armeniacus* is Latin for apricot coloured, referring to the overall flower colour.

Habitat: Typically found on low tree-trunks or lianas in dark, dripping, dank forest and by the sides of streams and rivers. The plant's requirements are for a very high humidity, low light and only gentle air movement and these conditions should be constant all year. It is found in original forest in section V and VI from 600-1300 M.asl, wherever such conditions exist.

Occurrence: Common.

Plant: This is a medium-sized epiphyte with dark green leaves, each shaped as a long, narrow, pointed oval, 15 cm. long and 2 cm. wide; three or more alternating on a thick, green, jointed and sheathed 12 cm. high stem. This arises from a convoluted, thickish rhizome. The root system is profuse, with thick greenish-white roots spreading for some distance in all directions around the plant.

Inflorescence and Flowers: A 17 cm, stiff, terminal, somewhat pendent, inflorescence holds a tight cone of up to 180 orange flowers, of typical *Epidendrum* shape. The dorsal sepal is an acuminate heart-shape 0.26 cm. by 0.35 cm broad and the lateral sepals are the same size but slightly asymmetrical. The minute petals are a slightly bent linear, 0.25 cm, long by 0.1 cm. broad, while the complicated trilobed lip is 0.5 cm. long, undulated at the side lobes and infolded, with an acuminate central lobe.

Flowering Periods: November and December and the flowers last for two to three weeks.

Pollination: Normally up to 30% of the small flowers are pollinated at all altitudes, suggesting a widespread and common pollinator; the flower size and plant habitat suggest small flies but **Dressler** speculates on *Lepidoptera*.

Cultivation: It grows well in pots on a coarse fibrous substrate, kept in the shade, always humid but never waterlogged.

Epidendrum addae
Pabst | *702* (Plate 12)

Etymology: Not known.

Habitat: It is found on low, thick horizontal branches in stunted, original, elfin forest 20 metres or so below mountain ridges, at 1100-1300 M.asl. It requires diffuse light, good humidity, accumulated vegetable detritus and good air movement within sections V and VI.

Occurrence: Locally frequent, but absent from many apparently ideal areas.

Plant: A medium-sized epiphyte with a jointed stem rising to 15 cm. from an inconspicuous rhizome. The stem is thin at the base but broadens and flattens towards the inflorescence. The first leaves are at 10 cm. from the base, sheathing the stem and are thin, floppy, very dark green with purple undersides. There are up to four, measuring 10 cm. x 2 cm, pointed and oblanceolate with a distinct keel. The root system is extensive, very adherent, and composed of thick, grey, slightly branching roots generally found along cracks in rough bark and under moss.

Inflorescence and Flowers: Two or three very attractive flowers are borne on a short, bracted, terminal, pendent inflorescence. The sepals are oval and concave, pinkish and 0.9 cm. long by 0.5 cm. broad. The petals are spathulate, pink, incurved and measure 0.7 cm. by 0.3 cm wide. The dominant four-lobed convex lip is a darker pink, a mild ellipse, saving the insets at the margins

marking the lobes, and measures 1.2 cm. by 1.4 cm. wide. Two short white, very apparent calli are shown at its throat. An alba form is not uncommon.

Flowering Periods: This plant can be found in flower throughout the year with a climax in November to January and the flowers last for three weeks.

Pollination: We have rarely found these flowers with capsules even though the major plant colony found runs spasmodically for four kilometres along a mountain ridge, which indicates that they may be at the end of the pollinators range.

Cultivation: In pots, on a compost which is a mix of coarse fibre, small wood chips and leaf detritus, in dappled light, fair wind movement, keeping damp all year round but never waterlogged. This is a difficult plant to grow well.

Epidendrum chlorinum
Barb. Rodr. | *711* (Plate 12)

Etymology: *Chlorinum* is Latin for yellow green, referring to the flowers' overall colour.

Habitat: While this plant can be found at all levels in the original forest from 1000-1600 M. asl, it really thrives as a terrestrial on high mountain ridges in low regrowth scrub amongst coarse grasses, growing in full sunlight and exposed to all the elements, in section VI.

Occurrence: Common.

Plant: An epiphyte found at most levels of the canopy but also sometimes growing as a terrestrial on mountain ridges. When not in flower it is discovered by it's dense root mass since the drooping habit and deep green foliage camouflages it against the background of the smaller twigs. It produces a reed stem of 45 cm, the lower 18 cm. or so of which are covered with papery bracts from the older leaf bases. The leaves sheathing the stem are deep green, a narrow elliptic shape, 12 cm. long x 1.5 cm. wide and usually 8 in number. The previous years shoot retains its leaves and becomes lichen-encrusted. The roots are about 4 mm. in diameter and fleshy, radiating out from the shoot base like earthworms. On one plant 13 roots were counted with an average length of 34 cm.

Inflorescence and Flowers: A terminal inflorescence is produced carrying 7-12 translucent, pendent, tightly packed, scented, lime-green flowers. The sepals are a blunt, concave oval and measure 1 cm. long by 0.75 cm. broad. The petals are more spathulate, also concave and somewhat smaller. The lip is 1 cm. long, single lobed, spade-shaped and 1 cm. wide at the shoulders, with two small distinct calli at the base.

Flowering Periods: November to March and the flowers last for up to three weeks.

Pollination: Half of the flowers are generally pollinated in all areas where this plant is found, once again indicating a common and widespread pollinator. **Dressler** suggests *Lepidoptera* and we have observed small butterflies pollinating this plant.

Cultivation: In large pots on a peat-like, though loose compost mixed with 25% of coarse sand; watering year around but never allowing water logging.

Epidendrum aff. geniculatum
Barb. Rodr. | 714 (Plate 12)

Etymology: *Geniculatum* is Latin for bent like a knee which may refer to the shape of the lip.

Habitat: Found at mid height on a relict tree at 1000 M.asl in section V, enjoying dappled light, a wet humid summer and a long dry winter.

Occurrence: Rare.

Plant: A small to medium-sized epiphyte, lithophyte or terrestrial, found growing in the same manner as *E. chlorinum*. The stem, which arises from the base of the previous years growth on a very short rhizome, is up to 30 cm. tall, oval in cross-section and thin. The leaves emerge alternately at up to 2 cm. intervals and are dark green, linear lanceolate, stiff and mildly keeled, measuring up to 9 cm. long by 1.5 cm. Their bases envelop the stem for about 2 cm. and the leaves are short lived. The root system is strong, composed of numerous, short white, green-tipped, branching thickish roots.

Inflorescence and Flowers: A short slightly pendulous, terminal inflorescence holds up to 4 small green to yellow, sessile flowers. Two to three sequential, very slightly bracted racemes may appear. The sepals are a broad lanceolate 0.8 cm. long by 0.45 cm. with a somewhat obtuse or blunt apex. The abruptly forward-pointing petals are a narrow elliptic shape, 0.6 cm. long by 0.2 cm. The lip is single lobed, almost round, 0.8 cm. long, including a 0.3 cm. narrow neck, fused to the column, by 0.65 cm wide, showing two parallel ridges stretching to almost 1/3 of the circular lobe, while the lobe's centre is perceptively uplifted to a quasi callus.

Flowering Periods: August and the flowers last for two weeks of so, but the inflorescence may show blooming racemes for up to a month.

Pollination: Unknown.

Cultivation: In smallish pots on a well-drained coarse tree fern based substrate.

General: We have labelled this plant *aff.* because the drawing of the lip in *Barbosa Rodrigues's Iconographie* and **Pabst and Dungs'** line drawing, both show similar characteristics but with markedly different emphases.

Epidendrum henschenii
Barb. Rodr. | 715

Etymology: Named after **Salomon Henschen**, born in 1847.

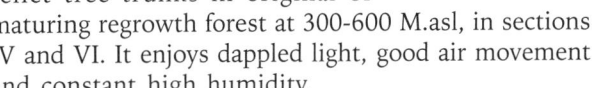

Habitat: An epiphyte growing on relict tree trunks in original or maturing regrowth forest at 300-600 M.asl, in sections IV and VI. It enjoys dappled light, good air movement and constant high humidity.

Occurrence: Common.

Plant: A large clumped, reed-stemmed epiphyte. A typical plant has as many as 20 leafy canes and half as many leafless, living canes, each up to 60 cm. high and 0.5 cm. diameter. The dark green leaves are lanceolate and paper-like in texture, conduplicate and mildly keeled and arise alternately at roughly 4 cm. intervals from the stem. Each cane has up to 14 long-lived leaves which measure 14 cm. long by 2.5 cm. and the leaf bases fold around the stem for 4 cm. New growths arise from nodes within the mass of thick worm-like green to white roots, each 0.3 cm. in diameter. This root system is extensive, extremely adherent and long which is to be expected when supporting such a large heavy plant on a vertical substrate. Five to six-year-old roots appear viable both as anchors and feeders.

Inflorescence and Flowers: Short, racemose, bractless inflorescences arise from the cane's apex bearing up to five closely packed alternate flowers which may also appear on older, leafless canes, but not on new growths. The sepals, pinkish on the outside and greenish-white within, are acuminate-ovate and similar in size at 1.2 cm. long by 0.45 cm. wide. The petals are narrowly spathulate, a perfect baseball-bat shape, white and measure 1.05 cm. long by 0.2 cm. The lip is distinctly bilobed, much wider than long and measures 0.6 cm. long by 1 cm. wide enveloping the column.

Flowering Periods: December and January and May and June and the flowers last for ten days.

Pollination: Possibly pollinated by bees.

Cultivation: We suggest growing this in a basket using a well-drained and aerated fibrous compost which is well-watered during spring and summer but only misted during autumn and winter.

Epidendrum aff. hololeucum
Barb. Rodr. | 716 (Plate 13)

Etymology: *Holo-leucum* is Greek for completely white, referring to the overall flower colour.

Habitat: An epiphyte from section IV and VI growing at 200-600 M.asl on small branches and trunks, in very humid under forest with low light and low air movement.

Occurrence: Common.

Plant: A variable, small to medium-sized tufted epiphyte resembling **Epidendrum mantiqueiranum**. The stems arise from the centre of the root mass and reach up to 30 cm. high. They are dark green, sheathed when young and up to 0.5 cm. diameter. There are three to five alternate, lanceolate dark green leaves, at 3.0 cm. intervals on the top 10-15 cm. of stem and they measure 15 cm. long by 2.5 cm. wide. These are papery and lightly keeled and their bases enfold the stem for 3 cm. to the point of the previous leaf's emergence. The root system is strong, and the worm-like mass of 0.2 cm. diameter, whitish-green and very adherent roots anchor the plant tightly to its substrate.

Inflorescence and Flowers: A short terminal raceme bearing one to several small milk-white flowers appears at the apex of both old leafless and younger, leafy canes. The sepals are lanceolate and measure 0.5 cm. by 0.2 cm. The petals are a similar shape but narrower and measure 0.95 cm. long by 0.2 cm. wide. The lip is trilobed, somewhat oval and 0.9 cm. long by 0.6 cm. with a three-lobed callus which is finer and longer than that of *E. mantiqueiranum*.

Flowering Periods: April and May and the flowers last for two weeks.

Pollination: **Dressler** suggests that pollination is by small bees; curiously, however, we have found that the flowers don't always open and could thus be self-pollinated.

Cultivation: In baskets with humus-rich, well-drained compost taking note of the habitat data above.

Epidendrum mantiqueiranum
Pôrto & Brade | 718 (Plate 13)

Etymology: Presumably the plant examined came from the Mantiqueira Mountain Range.

Habitat: Epiphytic in humid original under forest at 700-800 M.asl, in section VI. It grows on low tree trunks but it may also grow lithophytically or even as a terrestrial, enjoying high humidity, dappled light and low air movement.

Occurrence: Common.

Plant: A medium-sized, tidy reed-stemmed species, quite catholic in its habits and habitats. The pendent canes can reach up to one metre on a mature plant, and are round to oval in section and up to 0.5 cm. diameter. They arise from a node within the root mass and more rarely, from a node on a cane. Six to eight alternate, lanceolate leaves are borne on the apical 20 cm. of a mature cane. These are stiff, leathery, faintly keeled and measure up to 10 cm. long by 2 cm. wide. The colour varies from light green to dark wine depending on the plant's exposure to direct sunlight. The root system is robust and composed of 0.3 cm. diameter, very adherent and branched roots, sometimes lime green tinged. If torn from the substrate, the roots cease to function and the plant remains dormant until new roots grow. Old undisturbed roots remain functional for many years.

Inflorescence and Flowers: A short raceme with up to five small cream flowers is produced from the stem apex. The sepals are lanceolate, more or less equal in size and shape and measure 0.8 cm. long by 0.6 cm. wide. The petals are linear-lanceolate and measure 0.8 cm. long by 0.3 cm. The single lobed lip is a fat oval, 0.5 cm. long by 0.7 cm. wide with a distinct three-pronged callus running for most of its length.

Flowering Periods: April through to July when the flowers last for two weeks.

Pollination: **Dressler** suggests that pollination is by small bees.

Cultivation: This plant seems to like hanging baskets with a well-aerated fibrous compost. It will also grow well on tree fern or bark slabs. The habitat details point to the correct culture.

Epidendrum robustum
Cogn. | 725 (Plate 13)

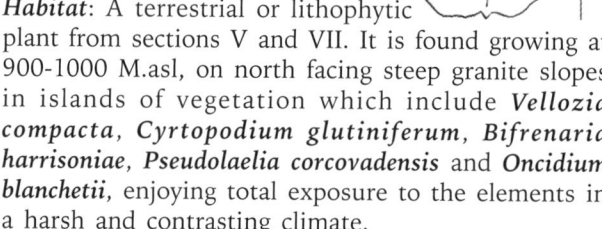

Etymology: *Robustum* is Latin indicating the robust nature of the plant.

Habitat: A terrestrial or lithophytic plant from sections V and VII. It is found growing at 900-1000 M.asl, on north facing steep granite slopes in islands of vegetation which include *Vellozia compacta*, *Cyrtopodium glutiniferum*, *Bifrenaria harrisoniae*, *Pseudolaelia corcovadensis* and *Oncidium blanchetii*, enjoying total exposure to the elements in a harsh and contrasting climate.

Occurrence: Where it occurs it is quite abundant but it is absent from many apparently suitable habitats.

Plant: A large terrestrial and lithophytic, cane-stemmed *Epidendrum* which grows in colonies of up to 30 or so plants on otherwise bare 30° to 40° granite slopes. The canes can reach 1.5 m. high and 3 cm. diameter and are sturdy, woody and stiff. They arise from a node at the centre of the root system. The leaves are alternate at 2 cm. intervals in an absolutely regular form and the 3 cm. long bases envelop and sheath the stems. The leaves are broad ovals, extremely stiff but leathery, mildly keeled in the central vein and measure 14 cm. long, including the stem sheath, 6 cm. wide and tend to infold on hot sunny days. A stem may hold 70 of these dark green leaves in ideal conditions. Fires and extended droughts mean that ideal conditions rarely occur in such regions and the stems are often leafless. The plants form 'keikis' which fall off in high wind and hail storms and these are swept down the rock faces to form new colonies where they can attach. The root system is enormous, composed of a plethora of 0.3 cm. thick, branching roots which glue themselves to the granite rocks in all directions for over a metre and accumulate vegetable matter and mineral fragments brought down by the heavy summer rains.

Inflorescence and Flowers: The inflorescences are apical or sub apical, closely packed panicles, each bearing up to 15 flowers. The largest number of inflorescences we have found on a single plant was five, each bearing 50 flowers. More commonly one finds two with about 25 greenish, wine coloured flowers on a 6 cm. inflorescence. The sepals are equal in size and oval and measure 2.5 cm. long by 1.2 cm. wide, while the petals are oblanceolate and measure 2.2 cm. long by 0.8 cm. The lip is almost round when flattened, vaguely bilobed and 1.5 cm. long by 1.8 cm. broad with two parallel white egg-like calli at the throat. The lip's centre is white, surrounded by a greenish wine shade. The flowers are frequently fertilized when they tend to take on an orange sheen.

Flowering Periods: October through to December and the flowers last for over three weeks.

Pollination: **Dressler** suggests that pollination is by butterflies or moths.

Cultivation: In large pots with a well-drained compact compost, fed lavishly and watered well during growth.

General: We have found plants in section VII with dark emerald green flowers.

Epidendrum ecostatum
Pabst | *726* (Plate 13)

Etymology: *Exo-status* is a Greek-Latin compound meaning projecting outwards, referring to the way the flowers are held on the inflorescence.

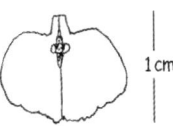

Habitat: A low tree epiphyte found in both original forest and maturing regrowth at 1200-1300 M.asl in section VI. It enjoys good dappled light and air movement and reasonable humidity.

Occurrence: Rare.

Plant: A small to medium-sized epiphyte which produces several woody stems from the root base, each stem being round, scaly and stiff. A new shoot is produced each year from an internode near the apex of the previous years growth, and these are alternate, occurring at 3 cm. intervals, each a vertical replica of the normal horizontal growth pattern of epiphytes. Each shoot is 7 cm. long and holds two or three apical and opposite, dark green to purple, elliptic, keeled, pliable, leaves which last for a year and measure 3 cm. by 1.5 cm. broad. The roots are profuse, long, thick and adherent.

Inflorescence and Flowers: A short, 2 cm. apical, lightly pendulous, inflorescence holds up to 15 tightly packed light green to purplish flowers. The dorsal sepal is vaguely spathulate, concave 1.3 cm. by 0.4 cm. broad. The lateral sepals tend to an asymmetric, concave lanceolate and are slightly longer and broader, while the petals are the most narrow of pointed spathulates, 1.2 cm. by 0.15 broad. The lip is single lobed, 1 cm. long by 1.6 cm. broad, an oval shape with 2 small, but clear, calli at its throat.

Flowering Periods: In January and the flowers last for 10 days or so.

Pollination: The flowers are rarely pollinated and the plant itself is rare to occasional. There are some conclusions to be drawn from this concerning the possible pollinator.

Cultivation: Extremely difficult as the plant appears to prefer only live bark as a substrate, and detests being transplanted. However, one could experiment with pots on a thick fibrous substrate, well fed and watered during the year through, but never waterlogged, and kept under dappled light. Success would be well rewarded as the flower clusters are very attractive.

Epidendrum janeirense
Pôrto & Brade | *727* (Plate 13)

Etymology: Presumably referring to the fact that the specimen described was from Rio de Janeiro State.

Habitat: An epiphyte found at low to mid-tree in original and regrowth forest, requiring low to medium light, wind movement and fair constant humidity, in section VI at 900-1200 M.asl, generally on saplings.

Occurrence: Common.

Plant: An untidy epiphyte with a woody branching stem, 0.5 cm. in diameter, sectioned and up to one metre long. New shoots are produced alternately at 15 cm. intervals and measure 8 cm, generally showing two apical, long lived leaves. These are lanceolate, 9 cm. long by 2 cm, light green and papery in texture. The main stem emerges from a node in the centre of the root mass. The roots are numerous, 0.5 cm. thick, white, long, and worm-like, brittle, very adherent and sometimes branching. A mature plant may show upwards of 25 roots, totalling over four metres in length.

Inflorescence and Flowers: The inflorescence is apical and pendulous consisting of up to 10 short pedicelate flowers held on 2 cm. stems. These flowers are similar to those of *E. ecostatum*, a yellowish-green tinged with purple, however the growth habit and flowering periods allow for no confusion in identification. The dorsal sepal is lanceolate, 1.5 cm. long by 0.6 cm. The lateral sepals are broader, an asymmetric lanceolate, measuring 2 cm. long by 1 cm, while the petals are smaller, club-shaped, 1.4 cm. long by 0.5 cm. All are an underlying light green shaded in purple. The three-lobed lip is almost oval, 2 cm. broad by 1.2 cm. deep and is a muddy purple.

Flowering Periods: In June and July and the flowers may last for up to three weeks.

Pollination: Unknown.

Cultivation: On tree fern plaques, repeating the natural conditions as far as possible.

Epidendrum obergii
Hawkes | *728*

Etymology: Named in honour of Oberg.

Habitat: Epiphytic at low tree or scrub level in reasonably open forest requiring good dappled light and air movement with a high constant humidity. The plants we have found were at 800 M.asl, over streams or rivers in transitional forest in section V.

Occurrence: Occasional.

Plant: One of the several woody, branching species from the *E. proligerum* alliance, the stem arising from a node in the centre of the root mass. After 10 cm. it branches and continues to do so alternately at 5-10 cm. intervals. The leaves are opposite and held sub-apically. Normally two leaves are shown on new growths and these are a broad lanceolate, papery, light green and 10 cm. by 4 cm. The root system is extensive, composed of many thick, grey-white roots which run up and down the substrate for up to a metre and are extremely adherent.

Inflorescence and Flowers: The inflorescence is apical and holds up to eight light green flowers in a slightly pendulous tight raceme. The fruits are much larger than those of others in the *E. proligerum* alliance and this is perhaps one of the most distinguishing features. The dorsal sepal, 1.5 cm. long by 0.5 cm. broad is lanceolate and light green. The lateral sepals are similar in size and shape but slightly asymmetric. The petals are linear, 1.6 cm. by 0.2 cm. broad. The lip is almost round, 1.4 cm. broad by 1.5 cm. and single lobed.

Flowering Periods: August to October and the flowers last for up to three weeks.

Pollination: **Dressler** suggests probably by butterflies or moths.

Cultivation: Difficult; the species in this alliance tend to enjoy live tree bark as their substrate and further, the roots seem to require to be firmly glued to the substrate to be functional.

Epidendrum ochrochlorum
Barb. Rodr. | *729* (Plate 13)

Etymology: *Ocraceus-chloro* is a Latin - Greek compound word meaning yellowish brown to green flowers.

Habitat: Epiphytic on low scrub forest, on tree trunks on wind-swept mountain ridges at around 1500 M.asl and over in section VI.

Occurrence: Occasional.

Plant: A woody-stemmed medium-sized epiphyte. The stem, up to 30 cm. tall, arises from a central node in a root mass, and the annual growth appears as a new branch from the base, or from an internode on the previous year's stem. The leaves are alternate, lanceolate, dark green and keeled, each 10 cm. long by 1.5 cm. broad and extending to the apex of the new growth. Generally, leaves on the old growth fall each year. The root system is extensive, composed of numerous thick, greyish, long roots always in bark cracks or under coarse moss.

Inflorescence and Flowers: The inflorescence is racemose or paniculate, short apical and pendulous, with 7-12 light greenish-yellow, large lipped flowers. The dorsal sepal is lanceolate, 1.4 cm. long by 0.4 cm. broad. The laterals are 1.5 cm. long by 0.5 cm. wide and also lanceolate, but with a bulge on the lower half. The petals are light yellow, pointed and narrow, at 1.5 cm. long by 0.15 cm. wide, and the back-folded, trilobed lip is a sulphurous green, 1 cm. deep by 1.45 cm. wide.

Flowering Periods: Irregular as we have found this plant in flower both in March and July. The flowers last for three weeks.

Pollination: **Dressler** suggests *Lepidoptera*.

Cultivation: In pots on a compost of thick fibrous material mixed with 25% of coarse sand, keeping moist through out the year in good dappled light.

Epidendrum parahybunense
Barb. Rodr. | *730* (Plate 13)

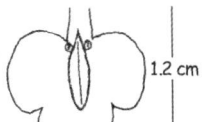

Etymology: The author of this orchid's name, **Barbosa Rodrigues**, also named n° 711 and 729 which are very similar, and he must have had difficulty in finding an appropriate descriptive name, so resorted to giving it the name of the River Parahybuna on whose banks it was found.

Habitat: On mountain ridges in section VI, at over 1500 M.asl in dwarf forest, on thin tree trunks, with its entire root system moss-covered.

Occurrence: Occasional.

Plant: An epiphyte of medium size. The stem is thick and branching, up to 30 cm. long, very woody and quite similar to that of *E. ecostatum*, extending annually almost in the same manner as a pseudobulbous plant. It is jointed, a mauvish colour and arises originally from a central point in the root system. The two apical leaves are dark green, deeply keeled, lanceolate, 4.5 cm. long and 1 cm. wide and only remain on the current years growth. The root system is composed of thick, white, long and adherent roots, in abundance.

Inflorescence and Flowers: A short apical, fasciculate inflorescence arises bearing up to 7 light Bordeaux wine-tinged, green flowers. The sepals are equal, an extended oblanceolate, 1.3 cm. long by 0.5 cm. wide. The petals are narrow at the base, but expand towards the pointed apex, and measure 1.2 cm. by 0.2 cm. The lip, a pinkish light green, is faintly tri-lobed and 6 mm. long by 1.3 cm.

Flowering Periods: February, March and July and the flowers last for three weeks.

Pollination: As for *E. ochrochlorum*.

Cultivation: As for *E. ochrochlorum*.

General: An attractive plant, similar to *E. ecostatum* and *E. proligerum*, but even more tolerant of harsh conditions.

Epidendrum proligerum
Barb. Rodr. | *731* (Plate 13)

Etymology: From the Latin *prolix* meaning drawn-out and referring to the extended stem.

Habitat: Found in the low- to mid-tree zone on trunks in deep shade in original forest and on trunks in scrub forest, and branches along mountain ridges, often a consort of **Sophronitis**. It is occasionally a terrestrial on these open ridges, from 1000 to 1600 M.asl in section VI.

Occurrence: Common.

Plant: An epiphytic *Epidendrum* with a thick, green, woody, branching stem up to 30 cm. tall. The two or three leaves are apical, dark green, leathery, deeply ridged, a pointed linear shape and 12 cm. long by 1.3 cm. wide and retained for one year. The roots arise from a central point and are thick and white, turning green with exposure, and run up and down the supporting branches under moss and within bark cracks. The system is extensive.

Inflorescence and Flowers: 4-6 large-lipped, yellowish-green pendent flowers which turn yellow with age are borne on a short, umbellate, terminal inflorescence. The sepals are ± equal, 1.2 cm. long by 0.4 cm. wide and lanceolate. The petals are the same length but club-shaped and narrower. The lip is very broad, yellow, discretely tri-lobed and 2 cm. wide by 1.2 cm. deep.

Flowering Periods: February to July and the flowers last for up to three weeks.

Pollination: **Dressler** suggests *Lepidoptera*.

Cultivation: As for *E. ochrochlorum*.

General: Quite similar to *E. chlorinum* but by no means as robust or as widespread and the latter's heart-shaped lip and non-branching stem remove any doubt.

Epidendrum aff. caldense
Barb. Rodr. | 732

Etymology: Presumably referring to Poços de Caldas in Minas whence the specimen examined came.

1.1 cm

Habitat: Found in elfin forest two to four metres from the forest floor at 900-1000 M.asl, in section VI, exposed to speckled light, good wind movement and a constant high humidity.

Occurrence: Rare.

Plant: Perhaps the most delicate of all the epiphytic reed-stemmed species of **Epidendrum** that we have found. A small epiphyte rarely with more than two stems and these emerge from the centre of a worm-like mass of 0.2 cm. diameter roots, and are stiff, rarely more than 15 cm. long, 0.15 cm. thick and sheathed by the leaf bases. Three or four alternate, narrow, linear-lanceolate dark green leaves are borne at 2 cm. intervals up to the tip measuring 8 cm. long by 0.4 cm. We understand that at much lower altitudes these plants are more robust.

Inflorescence and Flowers: A short racemose inflorescence appears at the stem apex bearing four closely held, mostly pink flowers. The sepals are more or less equal in shape and size, a broad lanceolate measuring 1 cm. long by 0.3 cm. The petals are a narrow lanceolate and measure 1 cm. long by 0.15 cm. The lip is distinctly tri-lobed and almost square with side-lobes which are larger than the squarish central lobe which has a small dot at the tip. The lip is white tinged with pink,

Flowering Periods: Late February until early March.

Pollination: Probably pollinated by bees.

Cultivation: In pots or bark rafts giving plenty of water and feed during growth and misting thereafter.

Epidendrum filicaule
(Sw.) Lindl. | 733

5 mm

Etymology: *Fili-caule* is a Latin compound word meaning slender-stem, and refers to the plant's fine stem.

Habitat: Found in section V on the transition line between coastal rain forest and seasonal forest at 1000 M.asl., growing epiphytically on relict original forest trees. Since these are rare and intact blocks of original forest rarer still in this section, this plant is an unusual find.

Occurrence: Rare.

Plant: A small epiphyte producing a new shoot from the base of a previous growth. This stem is totally sheathed by the leaf bases and is a narrow ellipse in cross-section, 0.3 cm. long by 0.1 cm. wide, slightly thicker at the base and the apex and rarely exceeding 10 cm. in height. The leaves arise alternately at 2 cm. intervals and are short-lived, linear-lanceolate, stiff, deeply keeled, light green and measure up to 8 cm. in length by 1 cm. wide. The root system is composed of a few thick, green to white roots.

Inflorescence and Flowers: The short apical inflorescence holds a raceme of up to 12 tightly-packed, small, pendent and attractive, yellowish green flowers. The sepals are a broad lanceolate, light green faintly smudged with claret and 0.7 cm. long by 0.38 cm. wide. The petals are linear-lanceolate, 0.6 cm. long by 0.15 cm. and yellowish green. The yellowish green lip is almost round, faintly pointed at its apex and measures 0.5 cm. by 0.5 cm.

Flowering Periods: June and July and the inflorescence has flowers in bloom for over a month.

Pollination: Probably butterflies or moths.

Cultivation: In small pots in a compost of fibrous material which must be kept slightly damp but never waterlogged, and in dappled light.

Epidendrum aff. tenue
Lindl. | 734 (Plate 13)

Etymology: *Tenuis* is Latin for fine or slender, aptly describing the stems of this plant.

Habitat: Found at mid-tree level in large colonies on relict trees at 500-600 M.asl, in section VIII and requiring medium shade and wind movement, a wet summer and a dry winter.

Occurrence: Occasional.

Plant: A medium-sized, bushy, untidy epiphyte with many fine branched stems vaguely reminding one of **E. filicaule** or **E. saximontanum**. The stems have a total length of 25 cm. with several alternate branches of up to 10 cm. and the terminal 5 cm. of each branch bears up to five keeled, slightly stiff to papery, light green, alternate, linear-lanceolate leaves, each 7 cm. long by 0.2 cm. The root system, as with other frail-looking bush-like **Epidendrum**, is remarkably profuse and composed of a multitude of long, branching, thickish, white medium-sized roots, many of which are aerial and the remainder penetrating into the loose substrate of rotting wood.

Inflorescence and Flowers: A short, 2 cm. inflorescence arises from the apex of each new branch, bearing a short raceme of one to four pale yellow-green flowers.

The linear dorsal sepal is 0.8 cm. long by 0.2 cm. and back folded, while the lateral sepals, similarly back folded are the same size but with distinct claws at the tips. The petals are clavate and very thin, 0.6 cm. long by 0.1 cm. at the widest point. The lip is composed of four distinct lobes. The basal shoulder lobes resemble a butterfly's forewings while the distal lobes look like the minor wings. The callus has one single central line running to the apex and two, shorter lateral calli which turn outwards from the throat.

Flowering Periods: November and December and the flowers last for about a week.

Pollination: Pollinators not known.

Cultivation: A coarse pine bark compost with some old tree fern will allow the roots to penetrate and spread. Water well in summer and just mist in winter.

Epidendrum setiferum Lindl. | *737* (Plate 14)

9 mm

Etymology: *Setiferum* is Latin for bearing bristles.

Habitat: Like *Laelia cinnabarina* this plant grows in high mountain fields above 1000 M.asl in sections V and VI together with groups of tough grasses in poor soil which has had most of the organic matter washed from the surface and then been baked by the sun. The plant survives by its huge root network which anchors it firmly and covers a wide area for support and nutrition. Like a lot of other terrestrials in the area, growth is only stimulated after consistent spring and summer rains, the plants then growing and coming to flower within a month.

Occurrence: Rare.

Plant: A tufted terrestrial with thin stems up to 50 cm. tall. The alternate leaves start in two opposite rows 20 cm. from the base. The stem is woody-looking and broadens slightly towards the apex. There are 12-24 dark green leaves with a red tinge to the undersides, 9 cm. long and 1.2 cm. broad, tapering to a finely angled point with an apparent midrib. The roots radiate out in a circle from the plant and spread in a layer under 5 cm. of arid topsoil. This soil is in reality a mineral subsoil on the surface after the constant exposure to spring and summer rains. The little organic matter that remains is in the 3-5 cm. deep layer where these roots penetrate. This horizontal network is white, brittle and immense for a small plant.

Inflorescence and Flowers: An inflorescence of 10 cm. emerges at the apex of the previous years shoot. This inflorescence is two-branched, each with up to 10 closely packed slightly pendent white, pink-streaked flowers. The sepals are similar, lanceolate, 0.8 cm. long

by 0.3 cm. wide, (laterals are slightly broader). The petals are thinner and a deeper red, 0.7 cm. x 0.2 cm., and club-shaped. The single-lobed lip is rounded, a yellow green with a claret tinge and a slight point to the tip, 0.9 cm. wide by 0.75 cm. deep.

Flowering Periods: December and January and the flowers last for two weeks.

Pollination: **Dressler** suggests *Lepidoptera*.

Cultivation: In medium-sized pots on a coarse fibre and sand mixture keeping moist the year through, watering well in spring when growth starts.

Epidendrum purpureum Barb. Rodr. | *748*

Etymology: *Purpureus* is Latin for dull red with a slight dash of blue, referring to the flower's colour.

1.5 cm

Habitat: Found in clearings on mountain ridges at 1200-1400 M.asl in section VI amongst tall grasses and shrubs, enjoying high light and wind movement and constant high humidity.

Occurrence: Occasional.

Plant: A terrestrial cane-stemmed *Epidendrum*, similar to *E. denticulatum* and *E. xanthinum* but larger than the former and with a different flower structure and colour than the latter. The plant's several stems are thickish, arising from nodes in the root structure and reach up to 1.7 metres tall. The deep green alternate leaves start at about 50 cm. from the stem base and terminate just below the apical inflorescence. They are elliptic, fleshy and keeled, enfolding the stem for 1 cm. and measure 6 cm. long by 3 cm broad and appear at 2 cm. intervals. The roots are thickish, white and short when aerial, however, at the base, the roots spread and branch under the decaying leaf substrate for half a metre around the plant.

Inflorescence and Flowers: A 10 cm. terminal inflorescence is held at the stem's apex in a typical fasciculate loose spray of 15-20 flowers which are an unvarying magenta with a creamy white lip centre. The sepals and petals are more or less equal, lanceolate, 1.3 cm. long by 0.75 cm. broad. All four margins of the four lobed lip are deeply serrate. Its distinctive feature is the creamy yellow nose-like callus and its surroundings, and the deep cleavage between the apical lobes.

Flowering Periods: In December and the flowers last for up to three weeks.

Pollination: Probably by diurnal or nocturnal *Lepidoptera* according to **Dressler**.

Cultivation: If you have a garden above 800 M.asl, treat as you would any of the cane-stemmed terrestrial *Epidendrum* species, planting on a substrate of decayed leaf matter on a well-drained slope in dappled light, or in large pots on a similar well-drained substrate.

Epidendrum ansiferum
Rchb. f. | 751

Etymology: Unknown.

Habitat: Original scrub montane forest, amongst tall grasses at 1580 M.asl in section VI, enjoying dappled light, high winds and constant humidity.

Occurrence: Sporadic in its niche.

Plant: A sturdy, reed-stemmed *Epidendrum*, up to one metre tall, similar to *E. purpureum* or *E. xanthinum*. A terrestrial which experiences a rigorous climate and has an enormous root system, composed of many long, thickish white roots which deeply penetrate the turf-like substrate around. The leaves are closely alternate on the single thick stem and are leathery, deeply keeled, a very dark green, oblanceolate and measure 9 cm. by 3 cm. and are long lived.

Inflorescence and Flowers: The inflorescences are apical on several sections of a branching stem, each 4 cm. long and holding a tight umbelliferous rosette of up to 20 smallish flowers. The tepals are ± equal in size and shape broadly lanceolate, 0.9 cm. by 0.3 cm. wide and a pinkish colour. The lip is indistinctly four lobed, almost round, with much indented margins some 0.9 cm. across by 1 cm. deep.

Flowering Periods: February and the flowers last for up to 14 days.

Pollination: Probably small butterflies.

Cultivation: As for all reed-stemmed *Epidendrum* spp.

Epidendrum denticulatum
Barb. Rodr. | 753 (Plate 15)

Etymology: *Denticulatus* is Latin for with very small teeth, referring to the intense small indentations on the lip.

Habitat: Grows in dry scrub, on mountain ridges and tops, and on dry road banks above 800 M.asl in all sections. We have found one massive colony containing all the colour forms, growing on a flat rock, 5 metres square, with over 250 flowering canes which, at an average of 50 flowers a cane, gave over 12,500 flowers.

Occurrence: Common.

Plant: A very variable reed-stemmed terrestrial, essentially from the high mountain fields, ridges and tops and their artificial and temporary parallel habitats: the roadsides and clearings. It grows in full sunlight. There is an enormous temptation amongst the "splitters" as opposed to the "lumpers" to describe several species instead of this one, *E. denticulatum*. The stems range in height to over 1 metre by 0.4 cm., growing from a central base or from one of the nodes of an old flowering shoot. There are 10-16 dark green alternate leaves, a thick leathery pointed linear shape, 8.5 cm. long by 2 cm. wide, on the top 20 cm. of the stem. The roots are often aerial, short, thick and white, and the anchor root system is extensive, penetrating leaf detritus and topsoil for up to a metre around the plant.

Inflorescence and Flowers: A rosette of between 10-100 flowers appears on an apical, fasciculate inflorescence. The petals and sepals are an equal pointed oval, 1 cm. long by 0.4 cm. wide. The lip is four-lobed showing two opposite triangular wings at the base, while the apical lobes are a flared, divided wide triangle. It is 1 cm long by 0.6 cm wide and all margins are distinctly serrated. There are 5 distinctive colour forms; pink, orange, cream, white and yellow, the commonest being pink.

Flowering Periods: June through to March with an emphasis on the later months; the flowers last for up to a month.

Pollination: We have observed butterflies visiting the flowers; pollination is common, most plants show up to 30% of flowers with fruit, particularly during the summer and autumn.

Cultivation: In large pots on a coarse organic compost, in full light, watering well during the summer, just keeping damp thereafter. It can also be grown in gardens in full sunlight in a well drained, preferably sloping, flower bed.

Epidendrum elongatum
Jacq. | 755

Etymology: *Elongatum* is Latin for stretched out, referring to protruding column which the flower shows.

Habitat: Found on roadside banks, in grass tufts on almost bare rock faces, exposed to all the elements at 800-900 M.asl in section V, enjoying a long dry winter and a short wet summer.

Occurrence: Common.

Plant: A somewhat shortish reed-stem *Epidendrum*, more compact and tidy than its companions in the **E. denticulatum** alliance, and forming small groups of up to 10 stems. We have found no colour variations to date. The stem rises up to 40 cm. from a node at the base of a mature stem. It is stiff, woody, round and generally wine-coloured. The closely alternate leaves are green to wine arising at 10 cm. from the stem base and are long lasting, leathery, lanceolate, almost elliptic, 6 cm. long by 2 cm. The root system arises from the stem base and is composed of many thick, non-branching white roots, up to 20 cm long, which penetrate the peaty substrate and rock cracks and are extremely adherent.

Inflorescence and Flowers: The inflorescence is apical, fasciculate and short, holding a rosette of up to 15 small light purple to magenta flowers. The sepals are ± equal 0.7 cm. by 0.3 cm. wide. The petals are 0.9 cm. by 0.4 cm. wide, slightly acuminate, spathulate, and the same colour as the petals. The lip is vaguely four lobed with heavily dentate margins to all lobes. The basal side lobes are almost round, 0.8 cm. long by 0.9 cm. wide, while the vaguely bilobed apical lobe is 0.6 cm. wide by 0.4 cm. long. The narrow throat is fused to a relatively long column, 0.5 cm long, and shows two nose-like orange calli.

Flowering Periods: January to March and the flowers last for up to a month.

Pollination: Seed set is frequent but the pollinator is unknown.

Cultivation: As for *E. denticulatum*.

Epidendrum xanthinum
Lindl. | *761* (Plate 15)

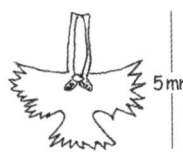

Etymology: *Xantho* is Greek for yellow, referring to the flower colour.

Habitat: Found in all sections, from 1000-1400 M.asl. This species will not tolerate the extreme desiccation and aridity that E. denticulatum accepts but grows well in slightly richer, moister soils which have *Lycopodium*, bracken and shrubs providing leaf litter and partial shade.

Occurrence: Common.

Plant: Although similar to **E. denticulatum** in flowering habit, this species also shows marked differences in habitat and growth form being more robust and requiring more fertile sites. It is a tall and sturdy terrestrial plant over 1 metre high without the inflorescence. The new shoots, arising from the base of the old are succulent and pink, like asparagus spears. Up to 40 alternate leaves are produced on a 120 cm.

tall, thick stem. Each succulent and lanceolate leaf measures 10 x 4 cm., is dark green, deeply keeled and sheaths the stem. Leaves are retained after flowering for at least a year. Aerial roots are not seen as in E. denticulatum, nor are side-shoots, typical of that species. The roots are white, 3 mm. thick and penetrate the surface leaf mould layer, to the extent of 4 metres of root to a single shoot.

Inflorescence and Flowers: A 50 cm. apical inflorescence bears a tight fasciculate cluster of bright yellow flowers which turn orange with age. About 50 flowers are produced, each 2 cm. high x 1.7 cm. across. Sepals and petals are similar in size and shape, broadly lanceolate, measuring 1 cm. long by 0.5 cm. wide and the trilobed lip is 0.6 cm. wide by 0.5 cm. deep. The colour is consistent except for the base of the three-lobed lip, which has a deeper yellow, heart-shaped callus.

Flowering Periods: November through February and the flowers last for up to a month.

Pollination: As for *E. denticulatum* and normally up to 30% of the flowers bear seed pods.

Cultivation: As for *E. denticulatum*, but the pots should be larger, the compost richer and the plant should receive dappled light.

General: Another of the high mountain scrub, reed-stem *Epidendrum* species that most people are familiar with. However, the richer growth and more attractive and intensive flowering single this species out.

Epidendrum difforme
Jacq. | *766*

Etymology: *Difformis* is Latin for differently formed. This plant was named by Jacquin at the end of the 18th century when very few American orchids were known 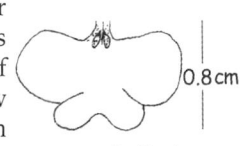 and the fleshy nature of this plant seemed distinct.

Habitat: Found in sections II, IV and VI at 200-600 M.asl, in humid regrowth or original forest. It requires medium light and air movement and fairly constant humidity.

Occurrence: Common.

Plant: An erect, clumped epiphyte, or sometimes, lithophyte. In appropriate conditions, a well-established plant will have up to 30 stems, each 30 cm. long, oval in cross-section and 0.5 cm. by 0.3 cm. Stems are fleshy, light green and thickly sheathed by leaf bases. The leaves are arranged alternately at irregular intervals and the new leaf bases enfold the stem as far as the previous leaf's lamina. The leaves vary in size, are light green, leathery to fleshy, keeled, oblanceolate and measure up to 7 cm. long and 1.8 cm. wide and have a slightly emarginate apex. The root system is strong and composed of thick, branching, white and very adherent roots which embrace the substrate far beyond the plant's circumference.

Inflorescence and Flowers: One to many flowers are borne on apical inflorescences. A mature plant may bear 15 inflorescences and over 50 light green to yellow flowers. The lanceolate sepals are a similar shape and size and measure 1.4 cm. by 0.4 cm., but the laterals are slightly asymmetric. The much thinner petals are linear-lanceolate and 1.3 cm. by 0.2 cm. The almost oval lip measures 1.4 cm. wide by 0.8 cm. and is vaguely trilobed when flattened out, with two small basal calli.

Flowering Periods: March, and the flowers last for about a week.

Pollination: Pollination may be by butterflies or moths according to **Dressler**.

Cultivation: In pots with a fibrous, well-aerated substrate giving much water until autumn and then only misting.

General: A wide ranging species which occurs from Florida and throughout tropical South America.

Epidendrum latilabre
Lindl. | *767* (Plate 15)

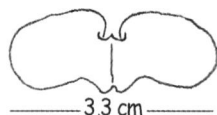

Etymology: *Latilabris* is a Latin compound word for broad lipped, referring to the flower's very broad lip.

Habitat: Epiphytic and sometimes lithophytic, this plant is generally found in colonies in the lower to mid-tree area in original forest where it is exposed to high humidity, filtered light and little air movement between 600 to 1000 M.asl in sections V and VI.

Occurrence: Occasional but locally abundant.

Plant: Easily distinguished from most other *Epidendrum* species and indeed most other orchids, by the succulent appearance of the rounded fleshy leaves. These are up to 10 cm. long by up to 3 cm. broad and arise alternately at 2 cm. intervals. Shoots arise from the thick, ridged, 0.7 cm. diameter rhizome, reach about 10 cm. in height and are sheathed by the thick leathery leaf bases. About 12 roots are produced per new growth and these reach about 6-8 cm. in length. They are 2 mm. in diameter, branched and adhere to the substrate.

Inflorescence and Flowers: These are yellow with a greenish tinge. The sepals are ± equal in size and shape, 2 cm. long by 0.7 cm. broad and lanceolate, while the petals are obtuse at the apex, linear and 1.9 cm. long by 0.3 cm. wide. The lateral sepals are hidden behind

the broad lip which is 3.3 cm. wide x 0.8 cm. with two high calli and is indistinctly two lobed with a hint of two smaller lobes towards the centre.

Flowering Periods: During December through to March and the flowers last for three weeks.

Pollination: Most flowers are pollinated, **Dressler** suggests *Lepidoptera*.

Cultivation: Best grown in pots on a coarse fibrous compost, always moist, never waterlogged and in dappled light.

Epidendrum aquaticum
Lindl. | *774* (Plate 15)

Etymology: *Aquaticum* is Latin for of water which aptly describes where the plant grows.

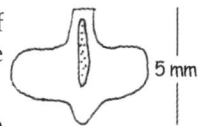

Habitat: In section V at 700-800 M.asl., found beside swamps, on cliff sides always by water seepage, growing in full sunlight, open to air movement with water running over its roots.

Occurrence: Locally very common in these unusual habitats.

Plant: A medium sized terrestrial or lithophyte; this many caned *Epidendrum* is similar to *E. dendrobioides* but only a third of its size. The canes arise in groups of two or three from the base of previous years' canes and are connected at 5 cm. intervals by a thick underground rhizome which is in turn a fallen cane. These measure up to 40 cm, 0.4 cm. in diameter and in their second year show one or two short apical branches. The leaves arise alternately at 3 cm. intervals and are a narrow lanceolate, 7 cm. by 0.8 cm. broad, light green, very faintly keeled and coriaceous. Their bases envelop the stem for 4 cm. The root system is profuse and composed of thick, white, 0.2 cm. diameter roots which spread for some distance around the plant.

Inflorescence and Flowers: The flowers arise on a short branched, apical inflorescence with one to 7 small alternate flowers on each of three short sub inflorescences. The dorsal sepal is broadly lanceolate, 0.5 cm. by 0.2 cm. broad. The laterals are an asymmetric lanceolate 0.7 cm. long by 0.3 cm. The petals are linear, 0.6 cm. by 0.1 cm. broad. The lip is essentially conduplicate, trilobed, with two side-lobes which together form an extended oval, 0.6 cm. broad, while the small obtuse apical lobe measures 0.5 cm. long from the throat. The overall colour is a light yellow.

Flowering Periods: November to March and the flowers last for two weeks.

Pollination: **Dressler** suggest frequently seen with fruit and probably pollinated by butterflies or moths.

Cultivation: Difficult. Perhaps on a wall on a peaty substrate with water permanently passing through, in high light and windy.

Epidendrum dendrobioides
Thunb. | 775

Etymology: *Dendrobioides* is a Greek compound word for tree-like, and in fact, the plant's growth pattern is like that of a small tree.

Habitat: 900-1000 M.asl in section V, in amongst tall grasses and sedges in swampy areas or near streams, enjoying full sunlight, high humidity on peat-like soil with water running through and a fair wind movement.

Occurrence: Occasional.

Plant: A tall, reed-stem, terrestrial *Epidendrum*. The canes can reach 1.30 metres high and arise mainly from nodules at the base of the previous years growth and in the second year the stems branch from 2/3 of their height. The canes are round, regular and up to 0.7 cm. diameter for their total length. The leaves are a narrow lanceolate, up to 9 cm. long by 1.8 cm. broad and arise alternately at 5 cm. intervals. They are stiff, leathery to succulent, distinctly keeled and long lasting, their bases enveloping the stem for 4 cm. The root system is extensive and composed of thick round 0.2 cm. diameter, long branching roots which arise irregularly from the first 10 cm. of the stem base and these penetrate the peaty substrate around. In addition aerial roots emerge from internodes on the stem, slightly above and opposite where the leaf emerges, for a further 40 cm. from the stem base. These roots are white, green tipped, long and penetrate the thickets of surrounding grasses only reaching the substrate should the cane fall.

Inflorescence and Flowers: Two or three spicate racemes of 7-12 alternate flowers, 0.5 cm. apart and up to 7 cm. long are held on a short 2 cm. round, fleshy, apical inflorescence. The flowers are yellow to orange and, if pollinated, remain until the capsule dehisces. The dorsal sepal is a broad lanceolate 0.9 cm. long by 0.4 cm. The laterals are an asymmetric lanceolate 1.0 cm. by 0.5 cm. wide. The petals are a narrow lanceolate 0.8 cm. long by 0.2 cm. broad. The trilobed lip is composed of two almost round side-lobes and a bluntly pointed apical lobe and overall it measures 1.0 cm. wide by 8 mm. deep. The side-lobes fold around the column immediately after pollination.

Flowering Periods: March to May and the flowers last for two weeks or so but the inflorescence may show flowers in bloom for up to six weeks.

Pollination: Infrequent and **Dressler** suggests by

butterflies or moths.

Cultivation: In large well-drained pots, on a tight peaty substrate. The canes, which will require support, need to grow in high light and air movement with the root area never allowed to dry out.

Epidendrum nutans
Sw. | 778

Etymology: *Nutans* is Latin for nodding presumably referring to the flowers which are constantly nodding in the slightest breeze.

Habitat: Terrestrial, lithophytic or epiphytic at low tree level in humid, original forest, requiring low air movement, low light and high humidity. Found at 400-800 M.asl, in sections I, IV and VI.

Occurrence: Occasional.

Plant: A large ungainly epiphyte with a prodigious root system composed of numberless thickish, white roots which emerge from the base of the new growth. They are long, branching, 0.2 cm. diameter, turning greenish with age and very adherent in order to support such a large plant. The plants we have examined had between five and 20 leafy canes up to 60 cm. high and 0.3 cm. or more diameter. The leaves are dark green, lanceolate, measuring 14 cm. by 3.5 cm. and are conduplicate, papery in texture, slightly keeled and long lasting, arising alternately at 3-4 cm. intervals starting at 10 cm. from the base and folding around the stem for 3-4 cm.

Inflorescence and Flowers: A 30 cm long apical inflorescence arises in summer, which may be racemose or racemose-paniculate and can bear up to 50 flowers. The sepals and petals are lanceolate and yellow-green. The sepals are 1.2 cm. long by 0.5 cm. and the lateral sepals are narrower but longer than the dorsal. The petals are lanceolate but much narrower at 1.1 cm. long by 0.2 cm. The lip has four ± equal but indistinct lobes, and its overall square shape measures 1.1 cm. deep by 0.7 cm. with two calli at the base and a narrow horse-shoe shaped, nose-like protuberance reaching to the tip. The lip is mostly white, finely speckled with green.

Flowering Periods: This depends on altitude. At 1000 M.asl, the flowers appear in February to March but at lower altitudes they flower in January or even December. The flowers last for ten days or more.

Pollination: **Dressler** suggests that pollination is by diurnal or nocturnal *Lepidoptera*.

Cultivation: In heavy pots using a very fibrous and well-aerated compost, fertilizing heavily and watering well during growth.

| 215 |

Epidendrum paniculatum
Ruiz and Pavón | *779* (Plate 15)

Etymology: *Panicula* is Latin for much-branched referring to the inflorescence.

Habitat: A terrestrial growing in high light in deep pockets of leaves and leaf mould at 400-500 M.asl in section I and V, under transitional forest and enjoying a wet summer and a long dry autumn and winter.

Occurrence: As occasional as its environment.

Plant: A large variable terrestrial with stems up to 1.5 metres, round and over 1 cm. diameter. The leaves are long-lived, arising at 4 cm. intervals alternately from the base to the apex. These average 20 cm. long, 4 cm. of which enfold the internodes, and 5 cm broad. They are broadly lanceolate, light greyish-green, flimsy and many-veined, but not plicate, and mildly keeled. The root system is massive, composed of very many thickish white long and branching roots which mainly emerge from the stem base.

Inflorescence and Flowers: A paniculate inflorescence arises at the apex. It can hold over one hundred light green flowers on the several racemes. The sepals are ± equal in size and shape, lanceolate, 1 cm. by 0.4 cm. broad while the petals are a club shape 0.8 cm. by 0.2 cm. wide. The lip is slightly four lobed with the two basal lobes much larger and dominating and it is fused to the column apex, and overall white. A ridged callus runs down the centre with a triangle of three small round warts at its mid point.

Flowering Periods: It can flower in any summer month but principally in November and December. The flowers may last for up to three weeks.

Pollination: **Dressler** suggests either day or night flying *Lepidoptera*.

Cultivation: Simple if you live in a tropical country; find a well drained rocky niche on a slope in your garden, fill it with leaf mould and leaves as deep as possible in semi-shade in a sheltered area and nature will look after the rest. So massive can the plants become it is difficult to conceive of them in baskets or pots.

General: First described by **Ruiz&Pavón** in 1798 from Peru; this plant is found throughout the tropical Americas.

Epidendrum saxatile
Lindl. | *783* (Plate 15)

Etymology: *Saxtilis* is Latin for found among rocks where it is occasionally seen..

Habitat: **Hoehne** describes this plant as an epiphyte on *Vellozia compacta* in the Organ Mountain Range. However, it is found on mountain ridges and high mountain fields above 1000 M.asl, also on roadside banks. As a terrestrial it prefers partial shade and the company of dwarf shrubs to support its sprawling growth habit in most sections of these mountains.

Occurrence: Occasional.

Plant: We have only found it growing terrestrially, and *E. saxatile* is well camouflaged as an epiphyte or terrestrial since the long horizontal woody stems blend in well with the surrounding twigs. These stems, similar to the leafless reeds of *E. denticulatum*, but more polished, sprawl through the undergrowth. They are about 25 cm. long and the leaves are borne on vertical shoots arising from the nodes on the horizontal stems. Of the 5-6 leaves, the top and bottom are smaller while the largest measure 10 cm. x 2.4 cm. New shoots appear on old parts of the rhizome so there may be 2 or 3 shoots to a node. There are many roots, clustered at the nodes, aerial and thick and not reaching the ground even though a short distance from it. The oldest nodes have a profusion of roots stretching for metres under the leaf detritus.

Inflorescence and Flowers: The flowers are especially beautiful. Large quantities are produced in succession on an apical inflorescence up to 10 cm. long. Groups of these pretty cream flowers are found nodding in the slight up draughts. The sepals are ± equal in size and shape, lanceolate to spathulate, 0.9 cm. long by 0.3 cm. broad while the petals are linear with an obtuse apex, measuring 0.8 cm. by 0.1 cm. broad. The lip as a whole measures 1.3 cm. wide by 1 cm. deep with two short central nose-like basal calli. The relatively large basal lobes are almost round with entire margins while the apical lobes appear as two overlapping fans with densely serrate margins.

Flowering Periods: November through to March when the flowers last for two weeks or so although the inflorescence may show flowers in bloom for several months.

Pollination: Rarely pollinated and as it is a consorciant of many other reed stem *Epidendrum* species, which are always pollinated, this seems to indicate a specific pollinator which may be absent from our search areas.

Cultivation: If you have small shrubs in your greenhouse, this plant could be grown in a pot of rich compost in such a way as to sprawl through the shrubs.

Epidendrum paranaense
Barb. Rodr. | *787* (Plate 15)

Etymology: The specimen of this plant that **Barbosa Rodrigues** examined came from the Serra de Uitupava in Paraná.

Habitat: Lurks in the deep, dank, humid, sunless gullies and in the dense, mid- to lower-tree riverside vegetation beloved by *Pabstia jugosa* and *Dichaea pendula*, in original forest. Also on ridges between mountains in similarly humid conditions at 1000 to 1600 M.asl in sections V and VI.

Occurrence: Common.

Plant: A large, sprawling, branching epiphyte with thick, pendulous but upward curving stems often stretching for several metres in a light-finding exercise, before producing an inflorescence. The stem is thick, fleshy, branching and pendulous. The new growth is always slightly up curled, carrying alternate, dark green, deeply keeled, leathery lanceolate leaves, 15 cm. long by 3.5 cm. wide. At the base, the root system is adherent and profuse and composed of medium thick greyish roots.

Inflorescence and Flowers: Produced on a short, thick, apical, fan-shaped inflorescence looking rather like a hand of cards. There are up to 9 ivory-white, fleshy flowers, each one surrounded by a stiff sheathing bract. The petals and sepals are about equal, 1.3 cm. long by 0.5 cm. broad, lanceolate and ivory white. The lip is the same colour and a pointed heart-shape, 1.5 cm. long by 1 cm. wide, showing two small egg-like calli at its throat.

Flowering Periods: From March to May and the flowers last for two weeks.

Pollination: Almost always, all the flowers are pollinated at all altitudes which indicates a common and widespread pollinator, probably a moth.

Cultivation: Due to this plant's growth habit, it would be best grown on a shaded trunk, over water. We have had total success on a live *Dicksonia sellowiana* tree fern, in deep shade and high year-round humidity and the upturning stems have attained 1.5 metres and have just flowered with a flower fan on almost every extremity.

General: This is another curiosity that in fact reaches downwards in the search for light.

Epidendrum pium
Rchb. f. & Warm. | *789* (Plate 15)

Etymology: Derived from the Latin pejorative *pium* meaning lesser, referring to the small stature of this plant.

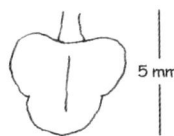

Habitat: Found in section V in mature transitional regrowth forest on thick mid-tree horizontal branches with good light and air movement, experiencing a long dry winter and a wet humid summer, between 600-800 M.asl.

Occurrence: Rare.

Plant: A small, tufted, pseudobulbless epiphyte which can form large colonies in suitable conditions. The stems arise originally from a central node in the root mass and subsequently from the stem bases, reaching up to 12 cm. They tend to be erect and ramified and the long-lasting leaves are dark green, leathery, linear-lanceolate, and deeply keeled. There are up to 7, arising alternately at 1 cm. intervals and measuring up to 5 cm. including a 1 cm. false petiole that sheathes the stem. Up to 5 ramicauls may appear on a new stem. The root system is prolific, composed of thin brittle white roots which may branch and involve themselves in the substrate. A fair sized colony will collect detritus within its mass of stems.

Inflorescence and Flowers: Up to three, alternate, emerald green flowers are held on a short 3 cm. terminal on each branched stem inflorescence. These flowers are strongly sheathed up to half their length. The sepals are a broad lanceolate, 0.6 cm. by 0.3 cm. broad, while the petals are a narrow linear, 0.6 cm. by 0.12 cm. broad. The lip is vaguely trilobed, shaped like a club in a pack of playing cards, 0.5 cm. by 0.5 cm.

Flowering Periods: February and the flowers last for up to three weeks.

Pollination: Probably butterflies or moths.

Cultivation: Grow on tree fern or bark plaques, leaving space for future growth, in good light but not direct sunlight, well watered during summer but only slight misting in winter.

General: An interesting comment by **G. F. J. Pabst** stating that **F. Dungs** found it at 1200 M.asl near N. Friburgo in moist forest, but previously only recorded from São Paulo and Minas Gerais – *Orchid Review*, 1971.

Epidendrum ramosum
Jacq. | 790 (Plate 16)

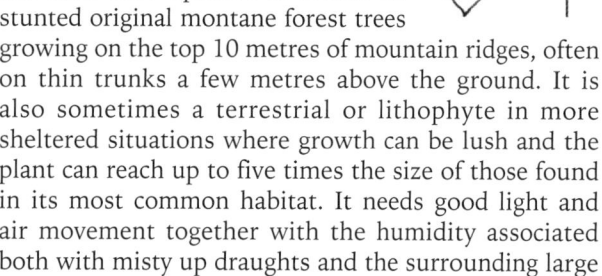

Etymology: *Ramosus* is Latin for much branched, referring to the plant's growth habit.

Habitat: This species is found on stunted original montane forest trees growing on the top 10 metres of mountain ridges, often on thin trunks a few metres above the ground. It is also sometimes a terrestrial or lithophyte in more sheltered situations where growth can be lush and the plant can reach up to five times the size of those found in its most common habitat. It needs good light and air movement together with the humidity associated both with misty up draughts and the surrounding large colonies of water-storing bromeliads, at 1200-1600 M.asl. in section VI.

Occurrence: Common.

Plant: This is a small to medium-sized plant. The leaves are linear, dark green, thin and stiff with a spine, and measure 5 cm. long x 0.7 cm. Each branching woody stem has many leaves. The roots flow from the thick bunched rhizome, and are profuse, thick and white, spreading 15-20 cm. below the colony, completely embracing the thin supporting trunks or anchoring in cracks of rock faces and boulders.

Inflorescence and Flowers: The 2 cm. long apical inflorescences each bear a single light-green to yellow flower. The sepals are ± equal in size, lanceolate 0.6 cm. long by 0.2 cm. broad. The petals are 0.5 cm. long by 0.1 cm. broad and linear. The distinctly veined yellow, spade-shaped lip is 0.5 cm. long by 0.3 cm. wide at the shoulders with two tiny ball-like calli at its base.

Flowering Periods: January and the flowers last for three weeks.

Pollination: As for *E. rodriguesii*.

Cultivation: In pots on a fibrous compost; watering in summer but just keeping damp thereafter.

General: Another tough, high mountain *Epidendrum* species which is able to survive testing extremes of climate.

Epidendrum rodriguesii
Cogn. | 791 (Plate 16)

Etymology: Named in honour of **Barbosa Rodrigues**, the most famous Brazilian orchid botanist of all time.

Habitat: This hardy species is typically found on stunted shrubs and wind-pruned trees on mountain tops and their connecting ridges, exposed to all the elements. Generally it is found with a moss clump around its roots and as often as not, surrounded by water-storing bromeliads, which alleviate the taxing low humidity on dry days. We have also found this plant in the canopy of high montane forest from 1400-1600 M.asl in all such areas in the Organ Mountain Range.

Occurrence: Common.

Plant: This is a small, tough but fleshy epiphyte which produces many-leaved, branching stems from a stout rhizome. Mature stems can measure 7 cm. and are sheathed by alternate leaves folding over a pronounced spine. There are up to 7 leaves on a stem and they measure 3 cm. long by 1 cm. The root system is profuse and composed of a multitude of varicose, white roots running down the support and often totally embracing a thin, stunted tree trunk.

Inflorescence and Flowers: A short raceme is produced bearing three or four small greenish-yellow, half-opened, typically *Epidendrum*-shaped flowers, each protected by a stout, opaque, snail-shell like, yellowish sheath. The sepals are ± equal, 0.75 cm. long by 0.3 cm. broad and lanceolate. The petals are linear, 0.75 cm. long by 0.1 cm. broad and the lip is a perfect heart-shape, 0.5 cm. long by 0.3 cm. broad at the shoulders.

Flowering Periods: In April and May and the flowers last for a week or so.

Pollination: Most flowers are pollinated and **Dressler** suggests *Lepidoptera*.

Cultivation: On tree fern or bark plaques, always kept humid with perhaps more water in spring and summer and in full light.

General: One of the toughest of high mountain epiphytes which appears to thrive in harsh conditions.

Epidendrum saximontanum
Pabst | 792 (Plate 16)

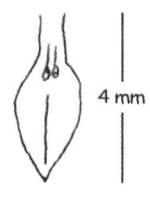

Etymology: *Saxi-montanum* is Latin for found on rocky mountains.

Habitat: Found in transitional forest at 500 M.asl in section V in good light and wind movement, enjoying a short wet summer and a long dry winter.

Occurrence: Occasional.

Plant: A delicate version of *E. ramosum* having all the same characteristics and forming quite a fair sized, pendulous clump with many ramifications. The many leaves are stiff, light green, linear-lanceolate, keeled, reflexed, up to 6 cm. long by 0.4 cm. wide and long-lived, with an emarginate apex. They are held alternately around the apex of the many branches. New growth arises from internodes on the thin woody stems. The root system is extensive, composed of many long white branching roots that firmly anchor the plant to its substrate.

Inflorescence and Flowers: Up to three flowers are held on short apical inflorescences, each partly sheathed around the ovary and borne on very short pedicels. The sepals are a broadish lanceolate 0.4 cm. long by 0.1 cm. wide. The lip is narrow, single lobed, a pointed spade-shape, 4 mm. by 0.2 cm. wide. The overall colour is a pale lime.

Flowering Periods: April and May and the flowers last for 2 weeks or so but the plant may show flowers for up to a month.

Pollination: Unknown, but frequent.

Cultivation: As for *E. ramosum*.

Epidendrum vesicatum
Lindl. | 797 (Plate 16)

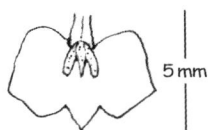

Etymology: *Vesica atum is* Latin for provided with small bladders, referring to the cauliflower like, compact flower grouping.

Habitat: An epiphyte found at low to mid-tree trunk levels in original forest on mountain ridges at 400-1000 M.asl, in section VI. It enjoys dappled light, good air movement and constant humidity.

Occurrence: Occasional and local.

Plant: A curious pendulous epiphyte, classified with a group of three other equitant species; equitant meaning that the leaf bases overlap and partly envelop the succeeding leaf, Iris-like. The few plants we have examined had 12 alternate leaves arising from a thin, pendent stem at 2 cm. intervals. The leaves are 4 cm. long at the stem base, increasing in size towards the inflorescence where they may measure 11 cm. They are light green, paper-like, lanceolate and sharply conduplicate. The smallest is 2 cm. wide and the longest 3.5 cm. The stem arises from a central point in a weakly developed root system; subsequent growths arising from the stem bases. The roots are few, short, 0.1 cm. diameter, brittle and white.

Inflorescence and Flowers: A short, apical inflorescence arises in November, partly hidden by the topmost infolded leaves. Up to 15 small white flowers are borne on a raceme in a tight umbel, resembling a small cauliflower. The sepals are a broad lanceolate, the same size and shape, 1 cm. long by 0.15 cm. wide. The petals are smaller and slightly club-shaped, 1 cm. long by 0.7 cm. wide. The lip is bilobed and almost square when flattened, measuring 1 cm. long by 0.7 cm. and has two small parallel nose-like calli at its base.

Flowering Periods: November and the flowers last for ten days.

Pollination: Probably pollinated by butterflies or moths.

Cultivation: Because of the plant's pendulous habit, it is best grown on fibre or bark plaques, watering well in summer and autumn but only misting in winter to prevent drying out.

Epidendrum infaustum
Rchb. f. | 798

Etymology: A Latin adjective signifying unhappy, sad.

Habitat: Found in mature transitional regrowth forest at 800 M.asl at mid-tree level on gross horizontal branches, enjoying a long dry winter, but with tolerable humidity and a wet, four month-long summer, in good light and wind movement in section V.

Occurrence: Rare.

Plant: A medium to small-sized cane *Epidendrum*. Normally the plant shows three or four leafy, erect canes, up to 40 cm. in length, oval in section and the two most recent bear inflorescences and are sheathed by leaf bases, some 0.5 cm. wide by 0.3 cm. The canes arise from an imperceptible rhizome or from nodes in the root mass. The leaves are a narrow ellipse, dark green, faintly keeled, leathery, long-lasting, up to 12 cm. long by 2 cm. wide with a sheathing petiole up to 4 cm. long and they arise alternately at 4 cm. intervals. The root system is vigorous, composed of thick, sometime branching, grey-white roots, up to 0.25 cm long, almost always visible and running in all directions on the substrate similar to *E. addae*.

Inflorescence and Flowers: Several inflorescences, one apical and others from sub apical internodes, arise either sequentially or simultaneously on a short 0.5 cm. - 1 cm. stem. Each is a raceme of up to seven small attractive flowers. The sepals are a broad lanceolate 0.9 cm. by 0.4-0.5 cm. broad and are a cream colour, very faintly tinged in pink. The petals are a very narrow club-shape, creamy white and 0.9 cm. long by 0.15 cm. at the apical 1/3. The lip is the same colour, very distinctly trilobed. The side lobes are somewhat roundish, while the centre lobe is acuminate towards the apex and 0.5 cm. long by 0.9 cm. broad and with an almost imperceptible callus at the throat.

Flowering Periods: February, and the flowers last for a week or so.

Pollination: Unknown; but up to 70% of flowers bear fruit.

Cultivation: As for *E. chlorinium*.

The genus
Lanium Lindl.

John Lindley raised this genus taking it out of the *Epidendrum/Encyclia* grouping to which it is closely allied. The name comes from the Latin *lana* meaning wool-like and referring to the woolly aspect of the flowers. Only one species is found in the Organ Mountain Range, while the other three are from either the north or the interior.

Lanium avicula
Benth. | 802 (Plate 16)

Etymology: One of a minority of our orchids adequately named; *lanium* means woolly and *avicola*, bird-like, is most descriptive; good for Benth!

6 mm

Habitat: We have found isolated colonies at the mid-tree zone in original forest fragments at 900-1000 M.asl in section VI. **Pabst & Dungs** examined herbarium specimens taken from these pluvial forests, while **Hoehne** seems to think that it enjoys a drier climate. We have also found it in profusion in an isolated forest pocket in section V. **George Gardner** introduced this plant to European culture in 1841 from the Organ Mountain Range.

Occurrence: Occasional.

Plant: A small creeping epiphyte which, when not in flower is very easily confused with *Encyclia pygmaea*, *E. calamaria* or *E. kautskyi*. The pseudobulbs are a pallid green, fusiform, 3 cm. long by 0.9 cm. and arise at 1 cm. intervals alternately from a thickish, woody, branching, sheathed rhizome. There are two opposite apical leaves which are deeply keeled, leathery, long-lived, a dark green, lanceolate and measure 2.5 cm. long by 0.8 cm.

Inflorescence and Flowers: An erect terminal inflorescence is produced from the most recent pseudobulbs, 8 cm. long with up to 10 flowers in a very loose cone. These flowers are wide open and a yellow green, delicate and delightful, though small. The lanceolate sepals are more or less equal in size, 0.7 cm. long by 0.25 cm, while the narrower petals are 0.6 cm. long by 0.1 cm. The lip, in the shape of a rather acuminate heart, is a clear green, 0.6 cm. long by 0.4 cm. The inflorescence stem and the reverse side of the sepals are lightly hairy.

Flowering Periods: In March and April and the flowers last for up to three weeks although the inflorescence may show open flowers for over a month.

Pollination: Unknown.

Cultivation: This plant seems happiest with its root bases exposed, and due to its climbing and creeping habit, tree fern plaques are recommended as a substrate. These should be sufficient for many years growth as the plant detests transplanting.

The genus
Cattleya. Lindl.

The genus *Cattleya* was named by **John Lindley** in 1828 and dedicated to **William Cattley** of Barnet, London. **Cattley** was a patron of horticulture, a collector of rare plants and at that time had perhaps the best collection of orchids in England. During the rest of the century Cattleyas became the principal orchids in all European collections "and at the present is still the Queen of the orchid cultivators' world" (**Withner**).

Cattleya species are distinguished from *Laelia* by having four pollinia instead of eight, and from *Epidendrum*, because the flowers are much larger and further that the flowers' lip is not fused with the column. These are really artificial separations and don't stand up to a chromosome/gene/DNA attack. However just like the Euro will take a generation to absorb the change so will it take a generation of orchidophiles to absorb these changes - or not.

The genus is composed of some 50 species and a large number of natural intergeneric hybrids. Its geographic range is from Central America, through Peru, Colombia, Venezuela to Brazil. There appears to be a sub-group of Bifoliate Cattleyas that have developed in the Atlantic rain forest, of which 7 species are attributed by **Pabst&Dungs**, to the Organ Mountain Range. Some are so rare, due to over collecting or destruction of their natural habitat, which we hardly bothered to look for them. These are, or rather were, *C. porphyroglossa*, *C. loddigesii* and *C. labiata*. Others were found by pure chance; *C. bicolor* and *C. guttata*, while *C. dormaniana* and *C. velutina* took years of frustrating search before we were rewarded. Also *C. harrisoniana* is the only one that still can readily be found around the foot of the scarp slope of the Organ Mountain Range. True also that *C. forbesii* and *C. intermedia* can still be found on off-shore islands and also in existing swamps near to the shore line but these are outside our brief.

It is curious, but notable, that *C. dormaniana*, the rarest of them all, will probably be the last *Cattleya* survivor in this Organ Mountain Range, due only to its present almost inaccessible redoubts.

Cattleya bicolor
Lindl. | 809 (Plate 16)

Etymology: *Bicolor* refers to the fact that the flower shows two colours.

Habitat: As the area where this plant should occur in the Organ Mountain Range is being totally devastated, we frankly did not expect to find it wild in this region. **Hoehne** makes many references to this plant in his wanderings from Resende, in the state of Rio, down the Paraíba River up the Paraíbuna to Juiz de Fora and beyond; also in the regions around Belo Horizonte and down to the borders of Espirito Santo and Rio de Janeiro, circling the Organ Mountain Range. He associates it with *Laelia perrinii, Lanium avicula, Miltonia clowesii* and *M. candida, Brassavola tuberculata, Bifrenaria harrisoniae, Sophronitella violacea, Sophronites cernua et al.* So we hunted out these associates on the anticline slopes of the Organ Mountain Range and finding them in remnants and in gallery forests along rivers, we extended our search to spot this bifoliate *Cattleya*. Persistence was rewarded. Friends showed us a colony on a dangerous rock face in a gallery forest and on an ancient tree in section I. A far cry from **Hoehne**'s descriptions of clumps showing 100 canes, with inflorescences holding up to 20 flowers but *C. bicolor* all the same. It requires high, but not direct light, good wind movement, heavy but short summer rains and a long dry winter ,in a somewhat humid ambient, at 500-1000 M.asl in section I. In boletim CAOB N° 45 September 2001, **Joaquim Barreto** describes an 'oasis' in Minas Gerais where he came across a small humid piece of forest in the middle of an arid region where amazingly he found a large number of *C. bicolor* plants in this niche, and this, we guess, in how many orchid species are now found in the devastated interior.

Occurrence: Rare.

Plant: A typical bifoliate *Cattleya* which may form huge clumps, either as an epiphyte or lithophyte. The pseudobulbs, or canes, can be up to 75 cm. long and up to 1 cm. in diameter, sectioned, and sheathed for several years. There are two apical, opposite, stiff, leathery, mid-green leaves, keeled and elliptic, up to 15 cm. long by 6 cm. and long-lasting. The root system is prolific with thickish white branching roots covered by lichen or on moss-covered rocks and branches helping to accumulate a substrate of detritus around the plant.

Inflorescence and Flowers: The apical, sheathed inflorescence arises in January, pauses to rest for a while and the three to five flowers then emerge on an extended 14 cm. inflorescence. These vary in colour; the tepals are brownish, orange to green or copper, and sometimes spotted. All are a fattish lanceolate while the dorsal sepal is slightly asymmetric and measures 4 cm. long by 2 cm. The lip is uncomplicated, slightly bi-lobed, 4 cm. long by 2.5 cm. at the apical lobes. It is an overall light magenta with a yellow "V" at the throat.

Flowering Periods: February and March when the flowers last for two weeks, however, they droop rapidly if pollinated.

Pollination: **Dressler** suggests probably bee pollinated.

Cultivation: As for *C. guttata*, remembering the long dry winter.

Cattleya velutina
Rchb. f. | 810 (Plate 16)

Etymology: *Velutina* is Latin for densely covered by fine short erect hairs, referring to the velvety throat of the lip.

Habitat: An epiphyte or occasional lithophyte growing at 700-1000 M.asl, in humid conditions and enjoying good but dappled light and good air movement. Found on eastern scarp slopes.

Occurrence: Rare, partly because it has been over collected, but it was probably always an occasional plant.

Plant: A sturdy epiphyte or lithophyte with thickened stems, which reach 40 cm. high and almost 1 cm. in diameter, and arise at 1 cm. intervals from a wandering rhizome. A typical plant will have about six canes. Two opposite leaves are borne at the apex of the new stem and are a pointed oval, leathery, stiff and long lived. They measure 10 cm. long by 3 cm. and are dark green and deeply keeled. The root system is reasonably strong and composed of long white and sometimes branching roots which arise from the base of the new growth and envelop tree branches under moss and lichens. They are very adherent.

Inflorescence and Flowers: Each recent cane produces an apical sheathed inflorescence rising about 8 cm. above the leaves. There are up to four flowers with petals and sepals of similar size and lanceolate shape, measuring 5 cm. long by 1.6 cm. wide, dark orange, blotched with dark brown. The lip is an almost perfect violin shape when flattened out but on the intact flower the lateral lobes are folded around the column and the central lobe is significantly infurled and with crisped margins. The lip measures 4.5 cm. long by 3 cm. wide at the apical lobe, and is basically whitish, heavily lined with brown veins. The throat has a covering of minute soft hairs.

Flowering Periods: February and the flowers last for three weeks.

Pollination: Dressler suggests that the plants are bee pollinated but we suspect humming birds to be involved.

Cultivation: In pots using a normal bark-based compost, allowing good drainage. Water and feed well during spring and summer up to and after flowering and keep slightly moist while resting. The plant needs good but dappled light and, remember, this is a cooler growing *Cattleya* and is a difficult plant.

Cattleya dormaniana
Rchb. f. | *815* (Plate 16)

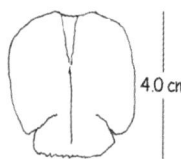

Etymology: Named for **Charles Dorman** of England who first flowered it in 1880. Interestingly, it seems that **Dorman** was the father-in-law of Sir **Ernest Shackleton** of Antartic Fame.

Habitat: An epiphyte or occasional lithophyte growing in original scrub forest on the eastern scarp slopes at 700-900 M.asl. It needs dappled light and good air movement and all the plants we have found were growing on or around a tree of the Clusiaceae: probably *Clusia organensis*.

Occurrence: Locally frequent, but it has a tendency to inhabit inaccessible locations, in very specific environments.

Plant: A somewhat weak-looking bifoliate *Cattleya*. A typical plant has 8-12 thin, sheathed canes which measure up to 30 cm. long by up to 1 cm. wide. Each cane carries two apical but slightly alternate opposite leaves which are a broad lanceolate shape and measure 18 cm. by 4.5 cm. The dark green leaves have rounded keels and are long-lived, and their alternate nature distinguishes this species from **C. velutina** when not in flower. The root system is quite strong and composed of many whitish adherent roots, almost always moss or lichen-covered and insinuated within bark cracks.

Inflorescence and Flowers: A sheathed, apical inflorescence about 8 cm. long appears in late March on the new growth, bearing one or two flowers which are quite striking although not very large. The tepals are more or less equal in size and shape, a bluntish lanceolate measuring 4-4.5 cm. long by 1.2-1.5 cm. wide. The lip is trilobed and when flattened, almost round and measures 4 cm. long by 5 cm. across. The apical lobe is flared and projected and the side-lobes envelop the column showing their lighter purple undersides. The tepals are a uniform dark chocolate brown and form a delightful backdrop to the vivid purple lip.

Flowering Periods: April and the flowers last for two weeks.

Pollination: We have observed the flowers being visited frequently and pollinated by the species of humming bird - **Phaethornis eurymome** but **Dressler** suggests that bees are responsible.

Cultivation: In pots with standard well-drained compost. Water well in the summer up until flowering, and thereafter, keep moist. It requires good dappled sunlight and is a cooler-growing *Cattleya* species.

General: The largest colony we have found after a four year search, bore 40 canes mostly still with leaves, and had 18 inflorescences of which three were twin flowered. A truly awesome sight when one considers that many authorities had considered the plant to be extinct in nature.

Cattleya guttata
Lindl. | *816* (Plate 17)

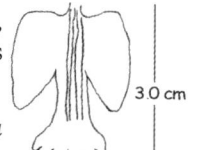

Etymology: *Guttata* means spotted, referring to the densely spotted tepals of this plant.

Habitat: We have found *Cattleya guttata* from Parati through to the Restinga reserve in Carapebus, always imagining it to be a coastal, sea-level plant. However, reading a Boletim of the Guido Pabst Herbarium for August 1994 (vol. V, nº 3) **Lucio Leoni** found *Cattleya guttata* at 550 M.asl, in Carangola, Minas Gerais, illustrated as an epiphyte on tortured scrub in a rocky area above the Carangola River. So, when we researched certain river gullies on the anticline of the Organ mountain Range we always kept **Lucio Leoni**'s article in mind. Finally, after many years, we found the habitat on a river bank, a tributary of the Paraíba River, with a remnant of almost original forest above a 150 Metre long series of rapids at 300 M.asl, on section VII. The climate is hot and heavily humid all year long. The plants were associated with masses of **Oncidium baueri** and many Pleurothallidinae.

Occurrence: As rare as its sparse habitat on the anticline.

Plant: A stout bifoliate epiphyte that can grow up to 1.5 metres. The pseudobulb's stem has several internodes and when young, is sheathed. The apical oval leaves are opposite and thick, fleshy and keeled, measuring 12 cm. long by 7 cm. The root system is prodigious and composed of numerous, 0.3 cm. thick white branching roots which may spread for several metres around the plant.

Inflorescence and Flowers: The inflorescence is apical. Up to five flowers on a 10 cm. inflorescence burst through the sheath where they have been developing for some time. The sepals and petals are equal sized, 6 cm. long by 2 cm. and green with chocolate spots. The rather squat, trilobed lip is dominated by the two side-lobes which envelop the column, and are white, bluntly lanceolate and measure 3 cm. long by 1 cm. The deep purple-magenta apical lobe is smaller and truncated.

Flowering Periods: January through to March and the flowers last for three weeks or so.

Pollination: Possibly humming birds, but probably bumble bees or carpenter bees.

Cultivation: *Cattleya* as a group, enjoy good light. As a guide, yellow leaves mean too high light and dark green, too little. Watch for roots emerging from the new growth and then start to water and feed. On completion of growth, ease the watering but keep the growths turgid by spraying.

Cattleya harrisoniana
Batem. ex. Lindl. | *822* (Plate 17)

Etymology: Collected by the **Harrison** brothers in the Organ Mountain Range and thus dedicated to them.

4.3 cm

Habitat: Epiphytic in section IV and VI from 0-400 M.asl in adequate habitat, very rarely up to 1000 M.asl. Typically this is a swamp orchid, growing in large colonies on stunted and twisted scrub-like trees in waterlogged areas, enjoying almost sauna-like conditions during summer days, cool nights and somewhat dryer conditions during winter, with high light and some air movement.

Occurrence: Once common in most such habitats at low altitudes at the base of the Organ Mountain Range, but now only occasional to rare since most such areas have been drained for agriculture and the remaining specimens collected out for roadside sale.

Plant: A medium to large sized epiphyte, found in large colonies. It has two leaves, apical and opposite which emerge from a thin pseudobulb some 40 cm. tall. The leaves are almost ovate, dark green, fleshy, deeply keeled and long lived, measuring 12 cm. long by 6 cm. broad. The pseudobulbs are held on a thick wandering and branching rhizome. The root system is composed of thick, very adherent, long and branching roots which are viable for many years.

Inflorescence and Flowers: An apical inflorescence emerges in November and December at low altitudes but in February and March for ecotypes at around 1000 M.asl. It is thick and fleshy, up to 14 cm. long and holding up to four, pink to lilac coloured, large flowers. The sepals are similar in size and shape, a broad lanceolate and measure 6.0 cm. long by 1.9 cm. The petals are more elliptical, but the same size. The lip, as with all, *Cattleya* species, envelops the column with its side lobes, while the central lobe presses firmly upwards against the anther, making the pollinator work hard for its reward. When flattened, the trilobed lip measures 4.3 cm. long by 4 cm. wide at the shoulders. Its margins are very crisped. Inside it is a yellow to cream with three longitudinal magenta parallel veins immediately below the anther cap. The flared apex is magenta.

Flowering Periods: November through January at low altitudes; March to early April higher up. The flowers last for up to three weeks.

Pollination: *Euglossine* bees: an observation by **Daniel Barroso do Amaral**. We also suspect that humming birds may participate.

Cultivation: In pots on a rich fibrous compost requiring good drainage, high light, constant reasonable humidity and some air movement. Water while growing but only mist when dormant.

The genus
Laelia Lindl.

Raised by **John Lindley** in 1831 and dedicated to one of the Roman vestal virgins – a group of girls, virginal teenagers, who sat around, stayed chaste and kept the goddess *Vesta*'s stove permanently burning; *Vesta* was the goddess of the hearth. *Laelia* species, without doubt and together with the Cattleyas are the most beautiful of all orchids, and have been considered so for nearly 200 years.

The Laelias are most curious in their distribution; there are Mexican Laelias and the genus then hops over to Brazil where it splits up into three or four distinct sections.

The *Cattleyodes* section includes *L. crispa, L. fidelensis, L. perrinii* and *L. virens*. We also have *L. pumila* in the *Hadrolaelia* section and *L. cinnabarina* in the section *Parviflorae*. We have found all the above except *L. fidelensis*. This is a local species and is far commoner in cultivation than in the wild, so the cost of looking for it would not really add to this work. Of the remainder, *L. perrinii* is truly at risk from habitat loss and collectors as is *L. pumila*. Only *L. virens* and *L. cinnabarina*, due to their preferred habitats, are safe, while *L. crispa*, currently common in its habitat of original forest, will continue as long as the forest does. It has recently been suggested, after extensive DNA analysis, and to the fury of amateur growers, that all the Brazilian Laelias should be transferred to the genus *Sophronitis*.

Laelia crispa
(Ldl.) Rchb. f. | *863* (Plate 17)

Etymology: *Crispa* is Latin for crisped, irregularly waved and twisted, kinky, curled, which adequately describes the shapes of the parts of the flower.

Habitat: Although once common throughout sections II, IV, VI and VIII, at 400-1000 M.asl *L. crispa* has been almost extinguished, together with its natural habitat, and is now only found in vestiges of original forest. However it is still found growing naturally and in profusion in the forests in the north-eastern scarp sector in the thick upper forks of tall trees where it appears like snow when in flower and where good light, strong wind movement and the inherent humidity of original forest ensure its annual flowering.

Occurrence: Common in its habitat; almost completely absent from regrowth forest.

Plant: A large epiphyte also found growing lithophytically. The pseudobulbs are a laterally compressed fusiform, light green and smooth when young, to light yellow and wrinkled with age, 30 cm. tall by 5 cm. wide and arising at 2 to 3 cm. intervals on a 1.8 cm. thick, sturdy, radially ridged rhizome. There is a single, occasionally 2, apical leaf, shaped as a pointed oval, darkish green, stiff and leathery and with a deep central keel. It measures 25-30 cm. long and 8 cm. broad and is retained for several years. The root system is prodigious; 0.6 cm. thick, white, green-tipped roots coming from the base of a new pseudobulb, simply race up and down a tree trunk in the rough bark cracks and continue growing for many years. Two years root growth on a ten-pseudobulb re-established plant showed a full 40 metres of roots.

Inflorescence and Flowers: A fleshy 25 cm. racemose, apical, sheathed inflorescence is produced in February and March bearing 5-9 large flowers. Overall, these are white with a convoluted lip in stunning clerical purple, crisply tinged with white. The sepals and petals are up to 8 cm. long, rolled into a cylinder but are crisped and flared for the apical 2 cm., while the flattened trilobed lip is 5.8 cm. long by 5 cm. broad and densely convoluted and serrated at the apical margins and spade-shaped.

Flowering Periods: February and March and the flowers last for 10 days.

Pollination: We have observed humming birds and also *Euelma* (*Euglossine*) bees, pollinating the flowers.

Cultivation: In large pots on a thick fibrous well-drained compost, remembering that the root system is viable for many years, as are the back bulbs, so judicious rhizome cutting may give you a plant with many inflorescences. As with other allied species, bright light

is necessary for successful flowering.

General: We have also found a carmine lipped colour form.

Laelia perrinii
Lindl. | *867* (Plate 17)

Etymology: Perrinii refers to **Mr. Perrin**, the Liverpool **Harrison**'s very successful gardener, after whom **Lindley** first named it *Cattleya perrinii* in 1838, transferring it to *Laelia* in 1842.

Habitat: Somewhat catholic in its tastes, this plant has been found growing on rock outcrops in section V and as a mid tree epiphyte in section VI at 600 and 400 M.asl respectively. In the first instance it suffers a 5 month dry winter season while in the second, around 3000 mm. of year around rain and mist.

Occurrence: Once common but now occasional due to habitat elimination and over collecting.

Plant: A variable epiphyte/lithophyte growing on rocky outcrops in high light where it shows a tendency to be red pigmented both in leaves and pseudobulbs, whereas in its epiphytic habitat all are dark green and almost twice the size. The pseudobulbs, which arise tightly from a tough woody and round, 0.5 cm. diameter rhizome are up to 20 cm. tall by 2.5 cm. broad, a narrow elliptical, generally well ridged. They hold one, very rarely two, linear-lanceolate apical leaves, deeply keeled, stiff, leathery, up to 30 cm. long by 5 cm. broad. The root system is extensive and strong, composed of brittle, thick, white, long, branching, very adherent roots. The plant can form, and is most happy, in thick clumps.

Inflorescence and Flowers: Up to three flowers are held on a thick, round, fleshy apical inflorescence which reaches up to half the length of the leaf. This conveniently bends downwards to permit the flower's beauty to be shown. The sepals are ± equal, linear-lanceolate, up to 7 cm. long by 1.3 cm. broad. The petals are broader and more elliptical 7 cm. by 2.2 cm. broad. The tepals are a pale whitish pink. The lip is faintly trilobed and measures 4.0 cm. by 3.8 cm. at the middle, the lobes funnelling the column. The side and apical lobes are irregularly serrated, a deep magenta except for the throat which is a clear milk white.

Flowering Periods: March; the flowers last for two weeks.

Pollination: Due to the lip colouring we suspect humming birds. However, large *Euglossine* bees are also suggested.

Cultivation: In large clay pots on a substrate of rich fibrous compost; water in the two months up to flowering and thereafter keep well drained but in good humidity. However it is a plant which must be watched as few growers seem to be really successful.

General: The plant collected in the Organ Mountain Range near Nova Friburgo in the 1830's was sent to **William Harrison** in Liverpool.

Laelia virens
Lindl. | *870* (Plate 17)

Etymology: *Virens* is Latin for green, which, applied to the flower is not quite fair as the flowers are an attractive light yellow.

Habitat: This is a mid-tree species found frequently in original forest and in vestiges at 1200-1400 M.asl in section VI. It is also found in scrub forest along mountain ridges. A shade and humidity loving plant, requiring moderate wind movement.

Occurrence: Occasional.

Plant: This is a large epiphyte with laterally compressed, fusiform pseudobulbs, dark green and smooth, arising from a thick green rhizome at 2 cm. intervals and measuring 10 cm. tall by 3 cm. wide. There is a single apical, dark green, leathery, deeply keeled leaf, a pointed oval, 16 cm. long and 5 cm. broad and retained for many years. The root system is composed of thick white roots which penetrate bark cracks for some distance and are generally moss-covered.

Inflorescence and Flowers: Up to 5 half-opened, light yellow flowers are carried alternately on a thick, dark green, 10 cm. long apical and sheathed inflorescence. The sepals and petals are of similar size, pointed ovals, 2.8 cm. long by 1.5 cm. wide, invariably incurled around a whitish, down-curled, trilobed lip, 2.2 cm. long by 1.3 cm. wide.

Flowering Periods: From April to June and the flowers last for two weeks.

Pollination: The flowers are invariably pollinated. We suspect that this is by self-pollination as no actual visitors have been observed.

Cultivation: As for *Laelia crispa*.

General: It is the only epiphytic *Laelia* found at high altitudes in these regions and its flowers are disappointing. Introduced to European culture in 1837 by **George Gardner** from the Organ Mountain Range, but understandably it has never caught on.

Laelia pumila
(Hooker) Rchb. f. | *876* (Plate 17)

Etymology: *Pumila* is Latin for close growing, short which describes the plant's habit.

Habitat: We have found this plant in small quantities in forest fragments in section V. It is also rumoured to exist in sections VI and VIII but to date we have found no trace. It inhabits 30 year old secondary forest at mid-tree level in humid conditions, enjoying good light and air movement at 800 M.asl.

Occurrence: Rare.

Plant: A small pseudobulbous epiphyte. These pseudobulbs are round to oval, up to 5 cm. tall by 0.5 cm and emerge from a tough, woody, green, round, rhizome at short intervals. Each has a single apical leathery, light green, long, keeled and tongue-like leaf retained for many years, measuring 10 cm. long by 2.3 cm. The root system is composed of thick white, long, sometimes branching roots which are extremely adherent.

Inflorescence and Flowers: A single flower is borne on a short apical inflorescence. It is large and showy and is held just above the leaves. The sepals are ± equal in size, shape and colour, elliptic to narrow lanceolate, 4 cm. long by 1 cm. broad. The petals are oval and tend to reflex, 4.3 cm. by 2 cm. broad. All tepals are a rose-magenta paling towards their bases. The lip is loosely tubular, the basal half infolding and vaguely trilobed, is 4.3 cm. by 3.5 cm. at the shoulders, and the apical lobe and side-lobes are a deep magenta while the throat is white. A faint three-lined callus runs from the throat to the apical lobe.

Flowering Periods: March and April and the flowers last for up to three weeks.

Pollination: Frequent; the colour and the guide-line callus would indicate humming birds. However large *Euglossine* bees may also be involved.

Cultivation: In clay pots on a rich fibrous compost. The colonies we have found suffer a five month long, very dry winter. However they were near waterfalls and rapids and so must be enjoying a fairly high constant humidity year round.

Laelia cinnabarina
Batem. | *895* (Plate 17)

Etymology: *Cinnabarinus* is Latin for vermilion, which is the overall colour of the flower.

Habitat: A true species of the high

mountain fields wherever scrub is low or absent, but also a colonizer of cleared roadside banks above 900 M.asl. It thrives in bright sunlight in exposed conditions at high altitudes where mists and strong winds prevail, in all sections of the Organ Mountain Range.

Occurrence: Occasional.

Plant: Found as a very variable clustered terrestrial with pseudobulbs ranging in size from 6 cm. to 25 cm. tall and from 1 cm. to 3 cm. wide at the base. They are light green and cylindrical although tapering somewhat towards the apex. There is a single apical leaf, an almost perfect pointed oblong, dark green and deeply keeled, 12 to 20 cm. long, 2 to 3 cm. wide and retained for several years. About 10 roots arise from the new pseudobulb base and these are white, thick and occasionally branch at the tip. They spread in all directions under the surface for an average of 20 cm., giving a mature plant of 6 pseudobulbs a total root length of 12 metres. The roots are tenaciously adherent in poor soil, securing the plant against fierce sun and torrential runoff water.

Inflorescence and Flowers: An apical, sheathed racemose inflorescence up to 40 cm. long bears 8 alternate flowers some 1.5 cm. apart, from the top 12 cm. of stem. These are long-lived and a stunningly bright cinnabar-orange. The petals and sepals are long and narrow and almost identical, 3 cm. long and 0.6 cm. wide. The lip is curiously trilobed and the two long basal lobes furl around the column while the long stemmed apical lobe is oval, serrated at its margins, projected forwards and downwards and strongly veined with a touch of bright orange at the tip. It measures 1.8 cm. long and 1.5 cm. wide at the side lobes, when flattened.

Flowering Periods: October through December when the flowers last for three weeks or so.

Pollination: **Dressler** suggests bees, though we imagine that the flowers' colouring and venation could induce humming birds. Seed set is frequent.

Cultivation: We have heard that growers have difficulty with this plant. This should not be so when one considers the appalling conditions found in its natural habitat. It should be grown in a fair sized pot on a well-drained fibre and sand compost, always kept damp and in high light. As an alternative grow as a rupicolous species simply in a clay pan filled with granite chippings which will require careful attention when feeding and watering.

The genus *Schomburgkia Lindl.*

John Lindley named this genus in honour of **Sir Richard** and **Robert Schomburgk**, two Prussians who worked for the British government, collecting plants principally in British Guiana in the 1840's. They also collected for **Charles Darwin**. While **Pabst and Dungs** only attribute one species to Brazil, albeit the type species, we have a feeling that other species have been and will be discovered in western and north western Brazil. **Carl Withner** discusses, in his book *The Cattleyas and their Relatives* Vol III, 22 species from northern South America, the Caribbean and through to Mexico.

Schomburgkia crispa Lindl. | *931* (Plate 18)

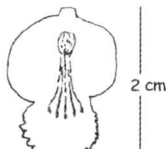

Etymology: *Crispa* is Latin describing the crisped and convoluted form of the flowers.

Habitat: Found in the dryer anticlines of sections I, III and V at 100-600 M.asl, generally in swampy situations or along riversides. Because of the almost complete deafforestation of this area, this plant is now mainly found on ancient Mango trees and Paineiras, *Mangifera indica* and *Chorisia speciosa* respectively, which were used as shade trees around the great houses during the coffee era in the 19th century. The trees often remained, well after the great houses had crumbled and fallen. Very occasionally found in large quantities in an "oasis" of original forest in these barren areas.

Occurrence: Occasional.

Plant: A large epiphyte with fusiform pseudobulbs arising alternately at 5 cm. intervals from a thick, branching, creeping rhizome. The pseudobulbs are sheathed when young and are 9-15 cm. tall by 4 cm. broad and greenish-yellow, becoming deeply ridged with age. Each bears two apical leaves which are lanceolate, light green, keeled, tough and leathery and measure 25 cm. long by 6 cm. The root system is strongly developed and composed of numerous thick, long, white, branching roots which stick strongly.

Inflorescence and Flowers: A woody, erect, tightly sheathed, apical stem arises from the most recent pseudobulbs and may measure as much as 1 metre in height. The flowers appear as a dense terminal raceme, umbel-like, on the top 5 cm. of the stem with up to 20 bracted flowers blooming simultaneously. The sepals and petals are more or less equal in size, light yellow-brown and 2 cm. long by 0.7 cm. wide, and linear with deeply crisped margins. The lip is trilobed, up to 2 cm. long and 1.2 cm. wide at the basal side-lobes but 0.7 cm. wide at the lower central lobe. It is orange-yellow, sometimes lightly speckled in faint purple and densely veined with raised longitudinal ridges in the middle with folded and crisped margins.

Flowering Periods: May and June and the flowers last about ten days on the anticline, but flowering in May at 100 M.asl on the scarp.

Pollination: Probably bee pollinated; we have observed a *Euglossine*, *Eulaema* sp. visiting.

Cultivation: In large pots using a rich but well-drained compost. Allow room for five years growth since the species resents being repotted. Water well during growth in summer and mist for the remainder. It requires good light and air movement and a medium but constant humidity. If grown too cool the flowers hardly open.

The genus *Pseudolaelia* Pôrto & Brade

This genus was raised in the 1930s by **Pôrto & Brade** which they called the false *Laelia*. All kinds of polemic have arisen concerning the origin of this species - an ancient mix between an *Encyclia* and a *Laelia* etc. -. In any event it is a most curious genus due to its association with host species, principally *Vellozia* and *Cacti*. **Campos Pôrto** was the grandson of **Barbosa Rodrigues** and a director of the Rio de Janeiro Botanic Gardens. **Alexander Brade** was a German who worked as a botanist together with **Pôrto** in the same institution.

Pseudolaelia corcovadensis Pôrto & Brade | *934* (Plate 18)

Etymology: Named in 1935 from plants discovered on Corcovado mountain in Rio de Janeiro.

1.2 cm

Habitat: Found growing only on *Vellozia* bushes, surrounded by short grasses and sedges in fissures on sheer high mountain granite slopes in sections II, III and V at 700-1000 M.asl. where it experiences moderately wet summers and dry winters.

Occurrence: Where it is found it may be locally common, but not all *Vellozia* colonies carry this orchid.

Plant: A curious, almost bushy, epiphyte which insinuates itself through a grove of *Vellozia* shrubs. The orchid seed appears to germinate on the trunk or lower branches between 10 and 30 cm. from the ground and from there the plant sprawls upwards and sideways for several metres on a thick woody rhizome. New growths and the resulting pseudobulbs arise at 4-12 cm. intervals. The pseudobulbs are conical and measure up to 8 cm tall by 3 cm. wide, initially sheathed and pale green but later turning to claret. Up to eight close, alternate, apical light green leaves are produced, which closely resemble those of the *Vellozia* host in size and in colour. They measure up to 15 cm. long by 1.75 cm. wide and are a narrow lanceolate. However, where the orchid leaves are smooth and conduplicate, the *Vellozia* leaves are intensely plicate. *Vellozia* leaves are probably indigestible and the apparent mimicry of the orchid may well be a protective device against grazing animals during the dry season. The root system is difficult to detect as it disappears within the multitude of old leaf bracts of the *Vellozia* stems. Roots arise principally from the pseudobulb base but also from the internodes on the rhizome. They are thin, wiry, sparse and penetrate their substrate with apparent ease.

Inflorescence and Flowers: A stiff, round-sectioned, thin, racemose inflorescence, up to one metre tall, arises from the pseudobulb apex bearing one to 14 flowers consecutively on the top 25 cm. of the inflorescence. The flowers are overall a deep magenta and highly visible as the inflorescence tends to rise well above the *Vellozia* grove. Only one or two flowers are open at the same time. The petals and sepals are the same size and shape; a broad blunt lanceolate measuring 1.7 cm. long by 0.7 cm. wide. The lip is trilobed with the narrow wing-like side-lobes jutting out horizontally from the base and not enveloping the column like *Encyclia*. The lip measures 0.75 cm. by 0.3 cm., while the mid-lobe, shaped like a spade, tapers towards the throat and has a number of yellow lines running half way. At its widest, the mid-lobe measures 1.2 cm. long by 1 cm.

Flowering Periods: April to August, at the end of the wet season and the beginning of the next. While individual flowers last only a few days, the inflorescence bears flowers for over a month.

Pollination: Is probably by bees but we have found small beetles in and around the flowers which are also visited by humming birds.

Cultivation: We have only seen this plant growing in **Vellozia** 'gardens' at the above mentioned altitudes.

The genus
Brassavola Rob. Brown

The genus was raised by **R. Brown**, and the name was given in honour of **Dr. Brassavola**, an eminent Venetian botanist and professor at Ferrara. It is a new world genus of which some eight species are attributed to Brazil and two to the state of Rio de Janeiro, both of which are included here.

Brassavola flagellaris
Barb. Rodr. | *942* (Plate 18)

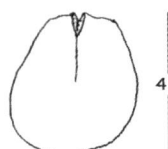

Etymology: *Flagellaris* is Latin for whip-like, referring to the long whip-like leaves.

Habitat: The largest grouping we have found were on large bushes of *Vellozia compacta* in section V within an extensive rocky, sometimes bare, granite slope at 600-800 M.asl, enjoying a short wet spring and summer from November through to March and a long dry autumn and winter.

Occurrence: Locally common.

Plant: A large grass-like, pendulous epiphyte or lithophyte. The leaves are narrowly terete, light green, up to 45 cm. long by 0.4 cm., flexible, long lasting and pendulous. These are held on a stem up to 15 cm. long, round, sheathed when young, 0.3 cm. diameter which in turn arises from a thick 0.5 cm. diameter, sectioned, dark green, branching rhizome at short intervals. The root system is strong, extensive and fiercely adherent, composed of long, strong, white, branched roots which integrate themselves within the bracts of the *Vellozia* lower branches and trunk.

Inflorescence and Flowers: A fairly prolific flowering plant and the flowers emerge in a loose panicle of up to six on a short stiff inflorescence which arises from the junction of the stem and the leaf on one year old growths on an 8 cm. pedicel which is mostly the ovary. A mature plant with much branching growth may show up to 30-40 flowers. The sepals are awl-shaped, up to 4.9 cm. long by 0.7 cm. broad. The petals are the same shape, up to 4.5 cm. long by 0.4 cm. The overall colour is a faintly lime-white with occasional very faint claret specks. The lip is a clear white fat oval, 4.0 cm. long by 3.0 cm. with a round, spreading light lime-green blotch around the throat to the centre.

Flowering Periods: In cultivation we have seen flowers during July but in nature, September and October. The flowers last for up to three weeks.

Pollination: Fruits are frequent, **Dressler** suggests that the flowers are probably pollinated by hawk moths.

Cultivation: As for *B. tuberculata*.

Brassavola tuberculata
Hooker | *945* (Plate 18)

Etymology: *Tuberculata* is Latin for covered by wart-like projections.

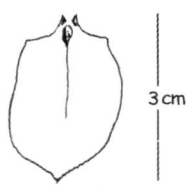

Habitat: An epiphyte or lithophyte, often in large colonies, growing from sea level up to 800 M.asl, in sections I, III and V. Curiously it grows in the coastal mountains of Rio de Janeiro city and state but it skips the scarp slopes of the Organ Mountain Range and we have only found it on the dryer anticline.

Occurrence: Common.

Plant: A medium to large epiphyte or lithophyte with terete, pendent leaves which measure 12 cm. long by 0.6 cm. and have a deep groove on the 'upper' side. The leaves have a 5 cm. long, partly sheathed, round petiole, 0.5 cm. diameter at the leaf junction and 0.2 cm. at the rhizome. Leaf and stem colour vary widely, depending on their exposure to sunlight, from pale green to deep wine and all colours in between. The leaves arise from a stiff, woody and noded rhizome at intervals of 1 cm. The root system is composed of thick, white, rarely branching but adherent roots which in large compact colonies, tend to be aerial for up to 12 cm. before penetrating the substrate or gluing to bark or rock. This growth habit, similar to **Maxillaria rigida** or **Encyclia calamaria** allows the plant to collect falling detritus.

Inflorescence and Flowers: One, sometimes two, white to yellowish flowers emerge from the most recent leaf bases on 4 cm. pedicels. The ovary is up to 8 cm. long. The sepals and petals are almost equal, subulate and measure 3.2 cm. long by 0.8 cm. and are acuminate. They are creamy-yellow and faintly tinged pink at the tips of the petals only. The lip is almost round, vaguely bilobed, 3 cm. long and wide, white with a lemon-yellow central blotch and a narrow, straight greenish central vein for half its length. The flowers have a pleasant perfume which is stronger at night.

Flowering Periods: June and the flowers last for ten days.

Pollination: Is probably by Sphingid moths, according to **Dressler**, which would fit with the night scent and white colouration.

Cultivation: In baskets with a well-drained compost, requiring high light and wind movement and more water in summer than in winter.

The genus
Isabelia Barb. Rodr.

A two species genus dedicated by **Barbosa Rodrigues** to **Princess Isabel**, daughter of **Pedro II**, Emperor of Brazil during the middle of the 19th century. Both species are small but very beautiful and more or less confined to the Atlantic rain forest.

Isabelia virginalis
Barb. Rodr. | 949 (Plate 18)

Etymology: This plant has a curious pedigree; introduced to European culture by **George Gardner** in 1837 from the Organ Mountain Range, it apparently remained officially nameless until **Barbosa Rodrigues** adopted it in 1877 and named it after our virgin princess, Isabelia.

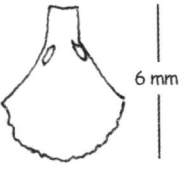

6 mm

Habitat: Found in remnants of original forest at 1000 M.asl and in original forest at 800 M.asl in sections IV, V and VI. A mid-tree to tree-top epiphyte enjoying high light and wind movement with a high constant humidity.

Occurrence: Occasional, but when found it may be in large colonies, but only in original forests or on relict trees.

Plant: A small creeping epiphyte that forms large mat-like colonies from the mid to upper-tree level. It is one of those orchid species that can be immediately recognized when not in flower. The pseudobulbs emerge very closely and alternately from a thin but woody, branching rhizome. Both are sheathed by the most intricate and regularly spun hessian-like veil, almost like an open meshed jute sack. The pseudobulbs are ovoid, 0.75 cm. high, 0.5 cm. in diameter and turn reddish with age. A single long-lived, needle-like, 4 cm. long leaf emerges from the pseudobulb apex. The root system is intense, adherent and penetrating, composed of numerous thin white roots which emerge from the base of new shoots and remain viable for many years.

Inflorescence and Flowers: The flowers are apical, generally single and borne on very short pedicels. The sepals are lanceolate, white on a faintly rose-coloured base and measure 0.7 cm. long by 0.35 cm. wide. The petals are shorter and narrower but with the same colouring. The bilobed lip is almost spade-shaped, 0.6 cm. long by 0.5 cm. at the flared base. It is milk-white.

Flowering Periods: May; the flowers last for a week or so.

Pollination: Unknown.

Cultivation: On tree fern or bark plaques of large size as the plant is capable of rapid expansion through its branching rhizomes and it detests replanting. A curiosity; we have found numbers of plants on branches fallen from relict original forest trees. Yet no matter what we do, in the same valley, at the same altitude and suffering the same climate these plants, well established, will not flower for us.

The genus
Sophronitis Rchb. f.

The genus name *Sophronitis* is derived from the diminutive of the Greek word *sophronia* which means chaste, modest or small; about the least appropriate name possible for this dramatic genus.

We have lived amongst *Sophronitis* species for nearly 30 years in a very wide valley whose enclosing mountain ridges and slopes of high montane original forest hold hundreds of thousands *Sophronitis* plants between 1200-1700 M.asl. We are thus reasonably sure that every possible variation of *Sophronitis coccinea* has been seen herein repeatedly; large plants with tiny flowers, tiny plants with huge flowers and many colour forms. So here we disagree with **Jack Fowlie** who invented new geographical species at the drop of a hat, mostly published in *The Orchid Digest*. We can agree with the following species and varieties:

1.	*Sophronitis cernua*	
2.		Var. *pterocarpa*
3.	*Sophronitis coccinea*	
4.		Var. *pygmaea*
5.		Var. *rosea*
6.		Var. *flava*
7.		Var. *mantiqueirae*
8.		Var. *acuensis*
9.	*Sophronitis brevipedunculata*	
10.	*Sophronitis wittigiana*	
11.		Var. *brevipedunculata*

It is quite possible that all the varieties will develop into distinct species over the next thousand years or so, but we truly believe that **Jack Fowlie** jumped the gun. All except 1, 2 & 10 can be found in Macaé de Cima.

Sophronitis brevipedunculata (Cogn.) Fowlie. | 950 (Plate 18)

Etymology: *Brevi* is Latin for short and *pedunculata* for flower stem, referring to the flower's very short stem.

Habitat: This plant seems more tolerant of lower altitudes than *S. coccinea* and can be found on regrowth scrub as low as 1000 M.asl. However the larger colonies have been found at around 1200 M.asl in original high montane ridge forest, growing on narrow tree trunks at up to four metres from ground level, enjoying constant mists, dappled light and good air movement in section VI.

Occurrence: Common.

Plant: A small variable epiphyte with almost round pseudobulbs 0.6 cm. in diameter and a dark greyish-green colour, arising closely and alternately from a short, thick branching rhizome. There is a single apical leaf of similar colour which is a fleshy oval and faintly keeled, measuring 1.6 cm. x 1.5 cm. The leaves are long-lived and a mature colony will have up to 20 leafed pseudobulbs, resembling a succulent. The roots are thick and white and always find cracks under moss. The root system is not extensive.

Inflorescence and Flowers: True to its name, the apical single flower's peduncle is very short. In fact this flower is held mainly on its 2.5 cm. long ovary. The dorsal sepal is a narrow elliptic up to 3 cm. long by 1 cm. broad. The lateral sepals are slightly shorter and narrower and more lanceolate. The trilobed lip is composed of two lateral and one apical almost equilateral triangles. The overall colour is scarlet.

Flowering Periods: August through to October and occasionally in April; the flowers last for up to three weeks.

Pollination: By humming birds (personal observations) however we have found plants pollinated in a bird, but not insect, proof, orchid house which presumes that bees may also contribute.

Cultivation: Laboratory-raised plants grow much better in captivity than those plucked from the wild, which is good news indeed. This plant is so commonly cultivated, that some very good tips are given by various expert growers. The consensus seems to be: good light, high constant humidity and good air movement on a base of sphagnum or any other medium that holds moisture well; in pots or on tree fern slabs, always remembering that water-logging is fatal and that these plants love being overcrowded and resent division and repotting. Also remember that young and transplanted plants develop roots faster in shade.

General: Claimed by **Jack Fowlie** to be restricted to *Vellozia* bushes in the state of Minas Gerais but clearly it is not.

Sophronitis cernua Lindl. | 951 (Plate 19)

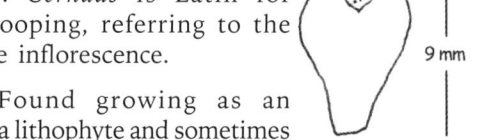

Etymology: *Cernuus* is Latin for slightly drooping, referring to the habit of the inflorescence.

Habitat: Found growing as an epiphyte or a lithophyte and sometimes forming large masses on horizontal branches or boulders in sections I, III and IV at 500-700 M.asl. It requires high light, good wind movement and reasonable humidity. This is another plant like *Brassavola tuberculata*, n° 945, which may be found on the coastal mountains but is absent from the hydrophilic forests of the Organ Mountain Range, appearing again on the western anticline.

Occurrence: Occasional in its specific environment which has been devastated and what is left is often so accessible that the species have been over collected.

Plant: A small creeping epiphyte or lithophyte and a mature grouping will have many round to ovate pseudobulbs, each measuring up to 2 cm. long by 1.3 cm. thick. These are closely and compactly held on a thickish, branching rhizome and have a single apical leaf which, in ideal conditions, is long-lived. The leaves are ovate, light green, thick and fleshy, darkening with age, becoming keeled and leathery, and measure up to 2.8 cm. long by 0.7 cm. broad. The root system is weak and composed of a few thick, white unbranched roots which are almost always found beneath lichens or mosses.

Inflorescence and Flowers: The short, erect, inflorescences are up to 5 cm. long and bear loose groupings of up to five flowers. The sepals are more or less equal, a broad ovate and are normally cinnabar-red and measure 1.25 cm. long by 0.5 cm. The petals are the same colour and shape but broader. The triangular lip measures 1 cm. long by 0.7 cm. at the shoulder base and is yellow from the base to the centre where it merges from orange to red at the tip. The column has purple tips on its wings.

Flowering Periods: Flowers from May to July and the flowers last for about three weeks.

Pollination: Our own observations suggest that humming birds may be the pollinators.

Cultivation: *Sophronitis* spp. enjoy being overcrowded and detest being repotted. Grow in smallish pots which should suffice for five years growth, adding some chopped moss to the compost, or on mossy tree fern or bark plaques. Water from spring through to autumn, misting thereafter but never let the substrate dry out.

Sophronitis coccinea
(Lindl.) Rchb. f. | 952 (Plate 19)

Etymology: *Coccineus* is Latin for deep red - from scarlet to carmine, referring to the flowers' colour.

Habitat: Mid-tree to tree-top sites in original forest are the preferred locations, always over 1300 M.asl in all sections. It may also be in large colonies in scrub forest on mountain tops and ridges. It requires good light, strong air movement and high, stable humidity. This humidity is caused by constant mists and, on high mountain ridges, by evaporation from the intense bromeliad cover, their rosettes holding significant amounts of water.

Occurrence: Common.

Plant: A small variable epiphyte, occasionally found growing terrestrially or on rocks. The short, dark green, fusiform pseudobulbs are borne on a thick branching rhizome at intervals of 2 cm. and measure 2.5 cm. x 0.8 cm. A typical plant may have up to 10, often suffused with claret where exposed to high light. The leaves are the same colour, and a single apical leaf is borne on each pseudobulb. The leaves are lanceolate, leathery and deeply keeled measuring 5.5 cm. x 2.1 cm. and are long-lived. The root system, arising from the rhizome at the pseudobulb base, is adherent but not extensive, with 4 or 5 thick, white roots of average length 10 cm, burrowing into cracks in bark or covered with moss.

Inflorescence and Flowers: A mature colony will have 4 or 5 single flowers from the new bulb apex, on a 2 cm. long, fleshy pedicel. The size varies but the colour is always an aggressive scarlet. The sepals are a pointed oval, 2.4 cm. long x 1.3 cm. The petals are heart-shaped and 3 cm. x 2.7 cm. wide. The lip is broad and abruptly pointed. We have quite often found inflorescences bearing two flowers and only once a three-flowered inflorescence.

Flowering Periods: August through to October is the principal flowering period with a lesser flush in March and April. However the same plant very rarely flowers in both periods; the flowers last for three weeks or more even after pollination. This may be a system whereby the colony continues to attract the pollinators for the later blooming flowers by its nectar and colour mass.

Pollination: By humming birds: personal observations.

Cultivation: As for *S. brevipedunculata*.

Sophronitis coccinea
(Lindl.) Rchb. f. var. *pygmea*, | 952a
var. *rosea(b)*, var. *flava(c)*, orange(d)
(Plates 19-20)

Etymology: *Pygmeaus, roseus*, and *flavus* are Latin and refer respectively to the flowers small size, and rose, pale yellow and orange to the flower's colouring.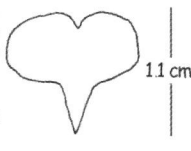

Habitat: **Var.** *pygmea* is often found at 1300 M.asl in section VI in dwarf forest on mountain ridges, generally above a terrestrial bromeliad garden with **Quesnelia lateralis** dominant. The colour variants are found together with the predominant deep red colour form.

Plant: **Var.** *pygmea*; as the name suggests is a small variant of **S. coccinea**. The pseudobulbs, borne on a branching rhizome are a very short ovoid, tightly bunched, light to dark green and 1 cm. tall by 0.4 cm. wide. Leaves are apical, single and long lasting, lanceolate, flat, and slightly keeled, emerald green when young, and 3 cm. long by 1 cm. wide. The root system is surprisingly extensive, a dozen or so relatively thick white, greenish roots stretching up and down branches for 16 cm. or more, under moss. The plants of the three colour variants are exactly like those of the predominant deep red colour form.

Inflorescence and Flowers: **Var.** *pygmea* shows a single apical flower from a new pseudobulb on a 3 cm. inflorescence of which over 3/4 is the ovary. The sepals are about equal, 1.2 cm. long by 0.5 cm. wide and a pointed oval. The petals are obovate and 1.5 cm. long by 1 cm. wide. The trilobed lip is broad, almost triangular and abruptly pointed at the tip. The overall colour is scarlet with an orange centre to the lip base. The yellow colour form is very rare and has only been found once in section VI some years ago. The pink and orange colour forms are also in section VI and one or two turn up most years.

Flowering Periods: In February and March or September and October and the flowers last for three weeks or so.

Pollination: Probably humming birds.

Cultivation: As for *S. brevipedunculata*.

General: We had resisted accepting **var.** *pygmea* as a subspecies, considering it as a normal *S. coccinea* growing under exceptionally harsh conditions. However in February 1993 we came upon swarms of these delightful elves in flower in less harsh conditions and so have been converted in as far as we consider it as a distinct variety.

Sophronitis coccinea
var. acuensis | 952e (Plate 20)

Etymology: Named *acuensis* since it is found around the Peak of Açu Mountain near Petrópolis Rio de Janeiro State.

Habitat: Grows at around 2000 M.asl in section I on scrub trees in wooded gulleys. This mountain receives constant mists at all seasons which keep the plant's lichen and moss-covered roots constantly moist.

Occurrence: Local and rare.

Plant: We have not been able to examine a large population so can only comment on one or two plants. These show thin, almost fusiform, very dark green pseudobulbs 2.0 cm. tall by 0.5 cm, each with one apical, dark green, faintly keeled leaf, a pointed oval, leathery, long-lived, with dark wine undersides and measuring 4 cm long by 0.5 cm. The root system is similar to that of all *Sophronitis* spp; a few thickish white roots almost always below moss and lichens.

Inflorescence and Flowers: A single flower bud, borne on a very short, 0.2 cm. peduncle emerges from within the new infolded leaf. The flower is typically *Sophronitis* type, slightly rounder in all its parts than other species or varieties. The sepals are oval with a blunt tip 2 cm. long by 0.9 cm. The petals are almost round 1.8 cm. by 1.6 cm. All tepals are vivid scarlet. The trilobed lip is less angular and more rounded than other *Sophronitis* spp, and 1.7 cm. long by 1.2 cm. wide at the side-lobes with 3 parallel longitudinal, strong red veins on a butter yellow background reaching from the throat to ¾ of the lip length. The remainder is the same overall scarlet as the tepals.

Flowering Periods: November and December and the flowers last for up to 3 weeks or so.

Pollination: Probably humming birds.

Cultivation: As for other *Sophronitis*, remember that these plants resent being repotted. Good light and humidity with constant air movement are the key elements.

Sophronitis cernua var. pterocarpa
Lindl. | 955

Etymology: *Pterocarpa* is Latin for winged-fruit, a distinct characteristic of the ovary and developing seed pod.

Habitat: Found in section V at 500-600 M.asl as an epiphyte on vertical trunks or horizontal branches requiring good light, reasonable wind movement

enjoying a short, wet five month spring and summer period with the remaining months very dry.

Occurrence: Occasional.

Plant: A small clustered creeping epiphyte that holds its pseudobulbs and leaves closely adpressed to the substrate. The pseudobulbs tend to be oval, heavily bracted 2 cm. long by 1.3 cm, greyish green and also superficially covered with a hessian-type fibre-like network, reminiscent of *Isabelia virginalis*. The leaves are single and apical, fleshy, oval, a greyish green, keeled and long lived. The root system is essentially weak, composed of a few thick and white non-branching roots.

Inflorescence and Flowers: Three to five flowers arise closely on a very short racemose inflorescence, held on even shorter pedicels. The ovary however is long at 1.5 cm. and it is that which separates this flower from *S. cernua* inasmuch as its ovary is often significantly winged which is not the case of the latter. The flowers are small, star-like, with sepals and petals more or less equal, a broad lanceolate, 1 cm. long by 0.6 cm. broad. The lip is similar in shape, hardly modified from the petals but slightly smaller and with a vague smooth swollen neck that is a darkish yellow. The tepals and apical part of the lip are orange.

Flowering Periods: May and June and the flowers last for up to three weeks.

Pollination: Frequent and probably by humming birds.

Cultivation: As for *S. cernua*.

General: Withner states that the original plant was discovered by **George Gardner** in the Organ Mountain Range. This must have been in 1837 or 1840; curious then that **Lindley** only named it in 1863.

Sophronitis wittigiana
Barb. Rodr. | 956a
var. brevipedunculata (Plate 20)

Etymology: Named in honour of **Emile Wittig** who found it in Espirito Santo and sent it to London in 1880.

Habitat: A scrub epiphyte found occasionally in small groups on wind-beaten tree trunks on original forest ridges in section VI from 1000-1500 M.asl. It needs high light and strong air movement and a constant high humidity.

Occurrence: Occasional.

Plant: Similar in form to a number of *S. coccinea* variations but the leaves are significantly deeply keeled and coriaceous to succulent. A typical plant may have half a dozen leaves which are oblanceolate rather than ovate. The pseudobulbs are not so elongated as on many *S. coccinea* plants nor as round as in *S. brevipedunculata*. The root system is poorly developed, as in most *Sophronitis* species, and is composed of a few relatively thick, white and sometimes branched roots always under moss and lichens.

Inflorescence and Flowers: One or two flowers appear on an apical inflorescence from the new growth on 1-2 cm. peduncles. The flowers vary in size but may reach 7 cm. across. The sepals are equal in shape, lanceolate and 3 cm. long by 1.5 cm. wide. The petals are almost round and up to 4 cm. long by 3.9 cm. wide. The trilobed lip has a slightly more pointed apical lobe than *S. coccinea* and measures 3 cm. long by 2.5 cm. wide at the side lobes. The overall colour leans towards magenta. We have found paired inflorescences and also a tiny flowered variety, possibly **var.** *pygmaea*.

Flowering Periods: April and May; the flowers last for three weeks.

Pollination: We have observed humming birds visiting the flowers.

Cultivation: Like all *Sophronitis* species this likes to be crowded in small pots with a well-drained compost. Because of its reluctance to produce new roots, the plants should not be divided on repotting and, after potting on should be kept in the shade until new roots form. Water well in spring and summer, misting only in winter enough to stop the compost from drying out.

The genus
Sophronitella Schltr.

Rudolf Schlechter called this genus a small *Sophronitis*, but as such it is not really very convincing, being about the same size as most other *Sophronitis* species. In his 53 years, **Schlechter** was arguably the most productive of taxonomists, naming 31 Brazilian genera and 226 Brazilian species.

Sophronitella violacea
(Lindl.) Schltr. | *957* (Plate 20)

Etymology: *Violaceus* is Latin for violet, referring to the colour of the flower.

1.6 cm

Habitat: Found from 600-900 M.asl in section VI at mid to high-tree in original forest often in large colonies, enjoying dappled but high light, good wind movement and the inherent humidity that montane rain forest offers.

Occurrence: Occasional, but when found is normally in large groups.

Plant: A small epiphyte with dark green and generally sheathed pseudobulbs, each a narrow ovoid 2.5 cm. long by 0.7 cm. wide, tightly clustered and on a wandering rhizome. There is a single, thin, deeply keeled, dark green, stiffish, linear-lanceolate and apical leaf 7-8 cm. long by 0.5 cm. wide which is long lived. The root system is prolific with the roots arising from the rhizome, and concentrated at the pseudobulb base. They are thin, wiry, strong and the system is adherent and extensive.

Inflorescence and Flowers: A one or two-flowered inflorescence, about 5 cm. long, emerges from the apex of new growth. The sepals and petals are more or less equal, pointed, lanceolate and 2 cm. long by 0.5 cm. wide. The lip is almost oval, narrowing at the throat, with an apical point, and 1.6 cm. long by 0.7 cm. wide. The overall colour is a violet-purple.

Flowering Periods: August and September and the flowers last for two weeks.

Pollination: Possibly humming birds. Most plants carry seed pods both in the wild and in cultivation, suggesting that self-fertilization is frequent.

Cultivation: Similar to *Sophronitis* species with somewhat less light and humidity. Another plant that resents repotting, so leave space for 5 years growth; also grows well on plaques of cork bark. The species also seems to resent over-watering. Experience has shown that plants should by sprayed before mid-day in the green house to allow the plants to dry before nightfall.

General: **George Gardner** introduced this plant to European culture from the Organ Mountain Range in 1837. It is a single species genus only found in the eastern Brazilian mountain cloud forests.

The genus
Leptotes Lindl.

John Lindley raised this genus using the Greek word *leptotes* meaning slender, thin, or delicate, and referring to the terete leaves. A small South American genus with three species of which two should be found in the Organ Mountain Range.

Leptotes bicolor
Lindl. | 963 (Plate 20)

Etymology: *Bicolor* is Latin for two-coloured, referring to the two-coloured flower.

Habitat: From 600-1000 M.asl, in original forest in sections V and VI, a low to mid-tree epiphyte, enjoying low light and wind movement and high constant humidity.

Occurrence: Occasional and sporadic.

Plant: A small pendulous or pendent epiphyte and sometime lithophyte. The dark green leaves are cylindrical and slightly curved with a deep central groove, resembling rats' tails and measuring 10 cm. long x 0.8 cm. wide. Older leaves attract lichen growth. Leaves arise from a thick succulent 1.5 cm. diameter stem or reduced pseudobulbs, themselves arising from a thick, wandering rhizome. The root system, while adherent, is not extensive, and the thin, branching, white roots fan out laterally from the base of new growths for 6 to 7 cm. around the plant.

Inflorescence and Flowers: The short, thin inflorescences may be single flowered or bear up to four delicate medium-sized flowers. These are unmistakable, with incurved white petals and sepals surrounding a relatively large pink to purple lip. The dorsal sepal is awl-shaped, 3 cm. long by 4 cm. The lateral sepals are similar in shape though marginally smaller, while the petals are the same size and shape though narrower. The lip is trilobed, elongated, the basal third composed of two opposite half-moon shaped lobes, while the apical lobe is a long pointed spearhead.

Flowering Periods: September and October and the flowers last for 10 days to two weeks.

Pollination: Unknown but could be humming birds.

Cultivation: In pots on a fibrous compost or mounted on cork bark. Always keep moist, watering well during growth and the flowering period but never water logging. This is a plant that also resents repotting.

General: Terete leaves indicate that this plant evolved when the Atlantic Rain Forest underwent a dryer era.

Leptotes tenuis
Rchb. f. | 964 (Plate 20)

Etymology: *Tenuis* is Latin for thin, fine or slender, referring to the shape of the leaves.

Habitat: To us this plant is an enigma. We have found two colonies only on relict original forest trees at 1000 M.asl in section VI,

and try as we might, we have not found it in similar original forest locations all over.

Occurrence: Rare.

Plant: A small tufted epiphyte found in large colonies. The pseudobulb is a short fleshy light green cylinder, 1 cm. long by 0.2 cm. wide, arising from a tough, branching rhizome. The apical leaf is dark green, fleshy, infolded, practically terete and 4 cm. long by 0.25 cm. wide, long lived and clearly has a storage function. The roots are white, tough, adherent and penetrating, growing from the pseudobulb base and spreading around for ± 8 cm. from the plant's centre.

Inflorescence and Flowers: A single flower appears on a thin, wiry, 2.5 cm. apical inflorescence arising from the new pseudobulbs and there may be many flowers in a colony. The flower is a gem, dominated by 3 equal sepals each 0.9 cm. long by 0.4 cm., pointed and lanceolate, deep yellow and mildly veined with pink. The petals are smaller, a pointed ovate shape 0.9 cm. long by 0.3 cm. wide and are a clear creamy yellow. The broad four-lobed lip is distinguished by a deep claret-veined, arrowhead-shaped centre while the very undulated central lobes are a thick cream colour, 0.9 cm. long by 0.8 cm. wide.

Flowering Periods: October and November and the flowers last for up to three weeks.

Pollination: As for *L. bicolor*.

Cultivation: As for *L. bicolor*.

The genus
Loefgrenianthus Hoehne

Fredric Hoehne who, after **Barbosa Rodrigues**, contributed more to Brazilian orchidology than any other, named this one-species genus in honour of **Albert Loefgren**, a Swedish botanist who studied the Minas Gerais flora and later headed the Botanical section in the Rio de Janeiro Botanic Garden.

Loefgrenianthus blanche-amesii
Hoehne | 966 (Plate 21)

Etymology: The genus was created by **Hoehne** in honour of the Swede who found the plant in 1896. The specific name refers to **Blanche Ames**, wife of Professor **Oakes Ames** who, together with **Loefgren**, rediscovered it in 1915.

Habitat: Found in the cool, humid original forests in section VI at 1000-1300 M.asl, and requiring shade, high humidity and good air movement.

Occurrence: Very rare.

Plant: A small, creeping, low-tree epiphyte, insignificant when not in flower when it could easily be confused with a creeping **Pleurothallis** sp. The plant creeps on a long, thin rhizome with three nodes between leaves, over the surfaces of tree trunks and has leaves borne at 1 cm. intervals, and each is up to 3 cm. long, lanceolate, keeled, 0.4 cm. wide and long-lived. Each leaf is borne on a minute pseudobulb 0.2 cm. high by 0.1 cm. which is surrounded by fine filamentous bracts, as is the rhizome. The root system is quite strong and worm-like, with pinkish, unbranched roots which penetrate the mosses and moist substrate from the bases of the older pseudobulbs.

Inflorescence and Flowers: The single-flowered inflorescences arise from the apices of two or three of the most recent pseudobulbs, on 1 cm. pedicels. The flowers are all white with some colouring on the lip. The sepals and petals are more or less equal, broadly lanceolate, 1 cm. long by 0.4 cm. and the petals are slighter. The lip has two curious small ears on the apical sides of the round basal lobe and a five fingered central callus. It measures 0.7 cm. by 0.3 cm. at the apical lobe. The tip is egg-yolk yellow, striped with claret and is drawn upwards in a cup-shape towards the column and is fringed with a number of fine hairs. The anther cap is a striking purple.

Flowering Periods: December; the flowers last for a week or so.

Pollination: Unknown, but the predominant white colour indicates night-flying Lepidoptera.

Cultivation: Anchor the plant in a fine bark compost or on a small piece of tree fern and keep cool and moist all year round.

General: The plants we found on fallen branches could almost be called twig epiphytes which contrasts with all the literature we have read. Remember that this is a tiny plant and a very small flower so only very close examination of fallen branches will reveal it.

The genus
Tetragamestus Rchb. f.

Heirich Reichenbach raised this genus alluding to the squared-off stigma. It is a one species South American genus.

Tetragamestus modestus Rchb. f. | *980* (Plate 21)

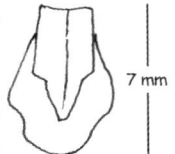

7 mm

Etymology: *Modestus* is Latin and refers to the modest, small size of the flower.

Habitat: **T. modestus** grows on the high mountain ridges, generally epiphytically, on low horizontal thick branches where it receives good light, air movement and a high humidity. It is also found successfully colonizing regrowing forest as a terrestrial in deep humus, from 1000-1300 M.asl in sections II, IV and VI.

Occurrence: Common.

Plant: There are many stems each up to 75 cm. long which can form a large epiphytic clump. The pseudobulbs are up to 25 cm. long x 0.6 cm. and are light green, narrow and cylindrical. Quite often they are formed continuously or in tandem with as many as six stems arising from the apex of the previous pseudobulb. The first pseudobulbs arise from a convoluted, tough rhizome. The leaves are a pointed linear shape and apical with a distinct central spine, light green and measuring 8 cm. x 0.8 cm. The root system is profuse, arising from the base of the anchoring pseudobulbs and composed of thin wiry roots which penetrate bark and detritus and are often moss-covered.

Inflorescence and Flowers: Up to three small, light greenish-pink single, sheathed and bracted flowers are produced on very short peduncles from the base of the apical leaves and from the apices of older pseudobulbs. The sepals are a broad, concave oval, the dorsal 0.5 cm. by 0.3 cm. while the laterals are 0.7 cm. long by 0.5 cm. The petals are bluntly lanceolate, 0.5 cm. by 0.2 cm. broad. The distinctive feature, apart from the unusual colour, is the rather curious reflexed, broad, wedge-shaped, pink-throated lip which measures 0.7 cm. by 0.5 cm. broad. The column has broad wings.

Flowering Periods: March and April and the flowers last for 10 days or so.

Pollination: Unknown.

Cultivation: On tree fern plaques or in pots on a coarse fibrous moisture retaining compost in dappled light, requiring misting the year round and light air movement.

The genus
Ponera Lindl.

John Lindley, of Kew Gardens, London, England fame, raised this one-species genus with the Greek word, which we can not find in any dictionary. However, **Bechtel, Cribb** *et. al.* say that *Ponera* could be an allusion to the small unattractive flowers (although we find them extremely attractive) or the thin and weak appearance of the plant (although the plants we have found have been very robust). So nobody is perfect.

Ponera striata
Lindl. | *981* (Plate 21)

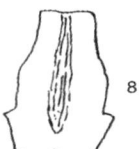

8 mm

Etymology: *Striata* is Latin for lined or streaked, referring to the striped flower parts.

Habitat: Found at 500-700 M.asl in sections III and V on rocky outcrops amongst large bromeliad colonies, enjoying a short wet summer and a dry rest of the year, growing in the company of *Schomburgkia crispa*, *Brassavola tuberculata* and *Laelia perrinii*.

Occurrence: Locally frequent.

Plant: A tall clumped lithophyte, sometimes epiphyte, often found in large groups and showing many long thin, noded, stiffish, round stems up to 1 metre tall. The long-lived leaves are a light green, narrow linear-lanceolate with a distinct keel and up to 14 cm. long by 0.8 cm. They arise alternately and opposite at 3 cm. intervals and sheath the stem for 3 cm. The root system is prodigious and composed of countless thick white roots, up to 0.75 cm. in diameter which arise from a shortish, thick, woody, branching rhizome, forming a massive worm-like ball which anchors the plant so firmly to the substrate that only by cutting the system can the plant be removed. These roots clearly constitute a storage system.

Inflorescence and Flowers: A small tight cluster of light pink flowers arise from the apex of new growth, according to **Hoehne**. But the plants in flower we have found, only show flowers on the internodes of leafless stems. The ± equal sepals are a very broad lanceolate, 0.5 cm. long by 0.5 cm., while the petals are a narrower lanceolate - almost elliptic- and 0.6 cm. by 0.3 cm wide. The lip is almost rectangular, vaguely bilobed with faintly indented margins 0.8 cm. long by 0.3 cm. wide.

Flowering Periods: April to May and the flowers last for two weeks or so.

Pollination: Unknown, but probably the same as for

reed-stem *Epidendrum* species.

Cultivation: In large pots on a tight fibrous substrate, watering well through summer but only lightly misting for the remainder of the year.

General: One of the few wild orchids we have found to be plagued by scale insects.

The genus
Isochilus R. Br.

Brown used the Greek compound *iso* meaning like or equal and *chilus*, lipped, or rather, with lip equal in length to other parts of the flower. This is a small genus with a single species in Brazil but ranging from Argentina through to Mexico and all parts between, including the Caribbean.

Isochilus linearis
R. Br. | *987* (Plate 21)

Etymology: *Linearis* is Latin for linear, referring to the leaf shape.

Habitat: This is emphatically a mid-tree species in original forest around the 1200-1400 M.asl mark in section II, V, VI and VIII where it forms large colonies strung up and down from the parent plant. It requires good light and air movement with significant ambient humidity. It really comes into its own when the tree dies but before the bark falls.

1 cm

Occurrence: Common.

Plant: A tufted epiphyte found in large colonies and with the same habitat requirements as *Elleanthus crinipes*. New growths arise from a very short rhizome and the shoot stem is wiry, thin and up to 40 cm. long, with short, dark green and deeply keeled, alternate, linear leaves, each 4 cm. long by 0.2 cm. wide and found for some 75% of the shoot length at 1 cm. intervals. The root system is extensive, composed of some surprisingly thick, storage roots 0.5 cm, in diameter, and thinner, grey to green roots which run along bark cracks but rapidly extend when they penetrate dead or rotting wood.

Inflorescence and Flowers: One to seven flowers are produced, alternating on a terminal inflorescence from September to November. These are a pinkish-mauve, somewhat tubular and surprisingly reminiscent of a heather flower. The sepals are a pointed sword-shape, 0.7 cm. long by 0.3 cm. wide, while the petals are slightly smaller and an extended pointed oval. The lip is narrow, pointed and over 1 cm. long.

Flowering Periods: September to November and the

flowers last for two weeks.

Pollination: Unknown; but probably small flies.

Cultivation: Grow on thick tree fern slabs or in pots, allowing space for the large root mass, or on a coarse fibrous compost which must never dry out or become waterlogged.

The genus
Hexadesmia Brogn.

A genus of two species in Brazil characterised by long leaves and found growing in tufts of dozens of pseudobulbs. Named by **Brogniart** (1801-76) for the six pollinia, from the Greek *Hexa*, six and *desmia*, parts.

Hexadesmia sessilis
Rchb. f. | *989*

4 mm

Etymology: *Sessilis* refers to the short pedicel.

Habitat: Found at 200 M. asl in Section I at the edge of very old, 200 year, regrowth forest and requiring good light, reasonable wind movement and humidity at mid-tree on trunks and thick side branches.

Occurrence: Rare.

Plant: A curious and unusual plant not dissimilar to *Tetragamestus modestus* and *Amblostoma tridactylum*. One of its synonyms is *Leoa monophylla* **Barb. Rodr.**, the generic name referring to the resemblance of the frontal view of the flower to a lioness. The narrow, spindle-shaped, dark green pseudobulbs measure 10 cm. long by 0.4 cm. and are borne on a 4.0 cm. round, sectioned, stiff and 0.4 cm. diameter stem. There is a single apical and one subtending leaf and the apical leaf is long, linear, lanceolate, keeled, dark green and up to 24 cm. long by 0.4 cm and long-lived. The root system has many short, thin, adherent and branching roots which extend beyond the plant's circumference.

Inflorescence and Flowers: Up to six small yellow-green alternate and sequential flowers arise on a short inflorescence from the leaf base. The dorsal sepal is a broad, almost round, acuminate lanceolate, 0.25 cm. long by 0.2 cm. The laterals are similar but slightly asymmetric and larger at 0,3 cm. by 0.23 cm. wide. The violin-shaped lip is distinctly bilobed with a deep inset at the apex and is 0.4 cm. long by 0.2 cm.

Flowering Periods: January until April and the flowers last for a week or so.

Pollination: Unknown.

Cultivation: Grow on bark or tree fern plaques, watering in summer but just misting in winter.

The genus
Bletia Ruiz & Pavón

This is a New World genus of approximately 50 species whose centre of distribution is Mexico. First described by **Ruiz & Pavón**, two Spanish botanists, who in fact were looking for medicinal plants in the central Americas and Mexico during the 1790's. They named it in honour of **Don Luis Blet** a fellow Spanish botanist. Only one species has been found in Brazil and almost only in the central and western parts, so this find on the Organ Range anticline represents a significant discovery.

Bletia catenulata
Ruiz & Pavón | *1540* (Plate 21)

Etymology: *Catenulatus* is Latin for chain-like or resembling a little chain, which probably refers to the chain-like flower sequence.

2.1 cm

Habitat: The colony which was discovered was intermixed with a large colony of over 50 plants of **Phragmipedium vittatum**, on a 10° granite slope facing west, with an 8 cm. deep humus layer which enjoys year-round water seepage, and found at 900-1000 M.asl in section V. The plants were protected by tall rough grasses but otherwise open to the elements.

Occurrence: Rare.

Plant: A tall terrestrial with nearly round pseudobulbs, very reminiscent of *Grobya amherstiae*, and measuring 5 cm. high by 6 cm., dark green and 3-4 cm. under the turf. The roots are few, thick and short but cling tenaciously to the substrate. There are 3 or 4 apical leaves, long, thin, plicate, linear-lanceolate and 60 cm. long by 2.5 cm. and supported by the tall surrounding grasses.

Inflorescence and Flowers: A tall basal inflorescence arises in October, stiff, round, erect and 100 cm. by 0.4 cm. The apical 20 cm. holds 3 groupings of flower buds which open sequentially from the basal group of 6 through the middle with 6 and the apical group with 8. The dorsal sepal is bluntish spathulate-lanceolate, 2 cm. long by 0.7 cm. The lateral sepals are similar but slightly asymmetrical and narrower. The petals are broadly oval, slightly pointed at both apex and base, 2 cm. long by 1.4 cm. The four-lobed lip is dominated by two large opposite, faintly indented side-lobes with a short bilobed apex, 2 cm. long, and 2 cm. wide at the side lobes. The centre has a number of elevated veins running fan-like towards the side-lobe margins. The overall colour is magenta with raised orange venation at the throat.

Flowering Periods: The flowers open in sequence from late October through to April.

Pollination: Unknown; the colour possibly attracts humming birds.

Cultivation: Unknown.

General: The association of this with *Phragmipedium vittatum*, also reported by **Fowlie**, could be significant.

The genus
Catasetum L. C. Rich

Richard named this genus using a Greek compound word meaning downward-bristle probably referring to the appendages, sometimes called antennae which extend downward from the column towards the lip cavity of the male flower and when touched, trigger the explosive release of the pollen on to the back, or thorax of the bee pollinator. This is a tropical American genus, with only a few of the 200 or so species attributed to the Organ Mountain Range. We have definitively identified three species mainly on the dryer and warmer anticline. However, the large numbers of plants without flowers, found even on otherwise epiphyte-free trees, in section I, III, V and VII of the anticline, suggests that there are other species to be found, probably invaders from Minas Gerais. However, we don't take even back bulbs from a plant in a dry hot climate at 600 M.asl, to watch it die at 1500 M.asl in Macaé de Cima's cool wet clime. So other visits in other years will either confirm or deny our theory.

What is remarkable about the genus and its three allied genera, is the fact that you can find plants with male flowers only, with female flowers only or one with both. This seems to be unique in the orchid world, though not so in the plant kingdom. Many logical explanations for this have been made but they all seem to be inadequate.

Catasetum aff. macrocarpum
L. C. Rich. | *1557* (Plate 21)

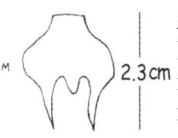

Etymology: *Macro* is Greek for long, large or great and *carpus* is Greek for fruit. These put together mean large fruited, which it is.

Habitat: Found principally in the transitional zone in section I, III and V, at 600-900 M.asl at mid-tree level

on old trees and occasionally on mature Eucalyptus. However, also but rarely, found in section VI in regrowth forest at 900 to 1000 M.asl which suggests a capacity to invade dryer deforested areas.

Occurrence: Occasional.

Plant: A large, clustered epiphyte similar in growth habit and form to most of the genus. The pseudobulbs are conical, fusiform, 15 cm. long by 5 cm. wide, light green and sheathed diaphanously. The root system is curious, profuse and composed both of very thick roots penetrating rotting wood, and thin surface roots that point upwards like a crew-cut, probably to catch dew or to trap detritus. There are 4 or 5 apical leaves, oblanceolate, light green and longitudinally veined, 20 cm. long by 5 cm. wide and rarely retained for more than a year.

Inflorescence and Flowers: Five to seven flowers arise on a short, thick, fleshy, erect 25 cm. inflorescence, bearing both male and female flowers. The male flowers are at the tip, probably as a response to higher light. In dryer regions we have found plants with all male or all female flowers. Female flowers have ± equal sepals, 2 cm. long by 1.1 cm. wide, a pointed ovoid. The petals are pointed though rounder, 2.5 cm. long by 1.4 cm. wide. The basic colour is light green, heavily speckled with dark burgundy bottle dregs. The upside-down lip looks like a crude helmet. The male flower, which is 'right side up', has more or less equal oblanceolate sepals and petals 4.0 cm. long by 1.0 cm. The lip is a gross trident shape 2.3 cm. long by 2.4 cm. wide. The hanging triggered column seems poised to eject pollen at any stage to the underlying female flower.

Flowering Periods: January and February and the flowers last for two weeks.

Pollination: *Euglossine* bees. The fruit capsules are very large.

Cultivation: Everybody has their own pet method of growing *Catasetum* species; the most successful we have heard of is from a friend in Brasília who ignores the five month recommended dry-off period, plies the plants with moisture year round in pots containing a compost of charcoal and fibrous material mixed with some horse dung and dry castor oil seed crushings. To judge by his photos, his success is total. A more readily obtained compost would be oak or beech leaf mould with added sand or perlag.

Catasetum cernuum
(Lindl.) Rchb. f. | *1567* (Plate 21)

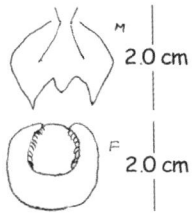

Etymology: *Cernuum* is Latin for slightly drooping, referring to the inflorescence.

Habitat: Prefers dryer transitional forest but also occasionally in pluvial forest at 300-800 M.asl in sections V and VI requiring high light, good wind movement, a long dry season and a somewhat shorter wet one.

Occurrence: Locally common.

Plant: A very variable epiphyte. Depending on its climate and substrate, a mature flowering plant may show many thick, internoded, diaphanously-sheathed cigar-shaped pseudobulbs, 10-20 cm. long 2-5 cm. in diameter. The leaves are plicate, elliptic, up to 20 cm. long by 4 cm., apical, light green and up to 6 in number. They appear on new growths at the beginning of the rainy season and in normal circumstances die off during the dry period that follows. The root system is intriguing, composed of medium to thick white branching roots which anchor the plant to its substrate and penetrate the gradually decomposing wood which is its normal or preferred base. In addition the plant may produce a dense maze of vertical upward growing roots, rather like a crew hair cut, whose function may by to trap dew or detritus.

Inflorescence and Flowers: A basal inflorescence emerges from the new growth. This is fleshy, round, a light green and up to 30 cm. long, holding up to 20 alternate and/or opposite flowers in a semi-pendent raceme. These flowers may be all male or all female or mixed. Our experience seems to show that in stressful conditions the flower sexes are mixed where as in ideal situations the racemes tend to hold flowers of one sex only. The male flower has lanceolate sepals and petals which are ± equal in shape size and colour; 3 cm. long by 1 cm. and with a dark green background, heavily blotched in coffee-brown. The lip is a gross three-lobed trident 2 cm. across at the shoulders, the side-lobes gracefully acuminate, 2 cm. long while the centre lobe is 1.4 cm. long and blunt at the apex. The female flower shows elliptic petals and sepals, broader and shorter than those of the male and a cup shaped lip, all are green, lightly blotched in coffee-brown.

Flowering Periods: October to December; depending on the arrival of the first heavy rains. The flowers last for several weeks.

Pollination: Euglossine bees and seeding is frequent.

Cultivation: In pots on a rich compost. If you water in winter make sure that air movement is strong and light high.

Catasetum hookeri
Lindl. | *1577* (Plate 21)

Etymology: *Hookeri* means it was dedicated to **Hooker** one of the most dynamic of Scotland's 19th century botanists.

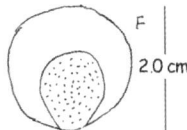

Habitat: Mainly on dead or rotting branches and trunks between 600 and 700 M.asl in transitional forest in section V, enjoying a short wet summer and the remainder of the year very dry but with a reasonable constant humidity due to year-round seepage from the mountains above. A slightly shady situation with reasonable air movement.

Occurrence: Rare.

Plant: A medium to large, variable epiphyte with a predisposition to grow well only on dead wood which by definition limits its life. The medium to large pseudobulbs may grow up to 40 cm. long by 5 cm. diameter in ideal conditions. They arise from either internodes on mature pseudobulbs or on a very short rhizome from the base of the most recent mature pseudobulb and are a fat fusiform, partly sheathed and hold up to 3 subtending and two apical leaves which tend to die off in a very dry winter. The leaves are plicate, up to 60 cm. long by 9 cm. broad. The subtending leaves are shorter and all are a narrow elliptic with a significant short false petiole. The root system is prodigious and composed of thick white branching roots which proliferate in the preferred, decaying substrate.

Inflorescence and Flowers: We have only found plants with male flowers. These are held on a thickish, greenish basal, umbellate apical grouping up to 40 cm long. There are up to 12 flowers, a butter-yellow, completely round, resupinate with a round gap to allow the entrance of pollinators. The dorsal sepal is ovate, 20 cm. long by 13 mm. with 15 faint parallel veins. The laterals are a pointed ovate 25 mm. long by 15 mm. with more venation. The petals are similar, slightly asymmetric with a short stalk and a blunt apex. The lip is broadly trilobed, 3 cm. wide and 2.0 cm. long, with visible venation on the side-lobes. The antennae are parallel, running closely and then bent sideways, towards the lateral lobes.

Flowering Periods: April, lasting up to 10 days unless the pollen is taken when death is ? instantaneous.

Pollination: *Euglossine* bees.

Cultivation: In pots on a loose compost with added leaf mould and rotten wood fibre.

General: The books suggest that the flower parts are heavily blotched in red-brown, so we may have come across a concolor variety.

The genus *Cycnoches* Lindl.

So named by **Lindley** from the Greek *kyknos*, a swan and *anchen*, a neck because the column is long and curved like a swan's neck, principally on masculine flowers, but apparently much shorter on female flowers, which we have not seen. There are about four species of this very variable, Central and South American genus attributable to Brazil and only one to the Organ Mountain Range where it is sporadic to rare.

Cycnoches pentadactylon Lindl. | *1614* (Plate 22)

History and Etymology: Discovered by **William Lobb**, a collector for the firm of **Messrs Veitch** of England, in 1841 in the province of Rio de Janeiro and described in 1843 by **John Lindley**. As far as we know it is the only *Cycnoches* species recorded for Rio de Janeiro state. *Pentadactylum* is Greek a compound word meaning digitally divided into 5 finger like lobes, referring to the male flower.

Habitat: Found in section V in young regrowth forest at 950 M.asl at two metres height on a thickish 30 cm. diameter, dying horizontal tree trunk on a substrate of very rough bark enjoying dappled light, good wind movement and low humidity.

Occurrence: Rare, but it may be more frequent as when it is not in flower it is indistinguishable from some *Catasetum* species.

Plant: A large epiphyte with oblong-fusiform pseudobulbs which may measure up to 15 cm. long by 3 cm. in diameter, many noded and partly sheathed. The several leaves are apical and lanceolate, somewhat plicate and measure up to 20 cm. by 4 cm broad. New growth may arise from an internode on the pseudobulb or from a new shoot on a tough rhizome at the pseudobulb base. The root system is extensive and composed of numerous, thickish, white, long roots.

Inflorescence and Flowers: The male inflorescence arises from a nodule near the apex of the pseudobulb and is pendent with many large flowers. The sepals and petals are yellowish-green to whitish, banded with red brown and all reflexed. The sepals are a narrow lanceolate, 3.5 cm. long by 1 cm. broad while the petals are oblong, the same length but broader. The lip is superior on a 3 cm. extension and is roundish, fleshy and acuminately trilobed with several horns at its base. The curved column is long and narrow, up to 3.5 cm. The female inflorescence is erect with few flowers. The sepals and petals are similar to those of the male flower but slightly

smaller and whiter with a reddish maroon base. The column is thick, slightly club-shaped, 2-3 cm. long and upturned. The superior lip is fleshy, an extended ovate with two calli at the base.

Flowering Periods: December to February and the flowers last for three weeks.

Pollination: A complex process achieved by male *Euglossine* bees, according to **Dressler**.

Cultivation: As for *Catasetum* species, in pots on a well-drained compost of fern fibre and sand with added leaf mould or rotted dung, carefully watering during the growing period and virtually drying out until growth starts in spring.

The genus *Oeceoclades* Lindl.

This genus is, like *Eulophia*, basically an African genus, only one species appearing in South America where it appears in all countries on the Atlantic side of the Andes and up to Florida.

Oeceoclades maculata Lindl. | *1615*

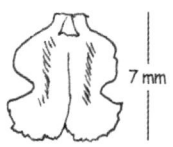

7 mm

Etymology: *Maculata* is Latin for blotched, referring to the mottled leaves.

Habitat: A very adaptable terrestrial plant found throughout the Organ Mountain Range at sea level and on the dryer interior anticline but rarely in the wet scarp forests. It is found in dappled light situations amongst scrub on sandy soils from 0-600 M.asl but also growing on rocks by the sea.

Occurrence: Common outside the humid original forest.

Plant: A small pseudobulbous terrestrial plant with a wide range of habitats and found throughout the Americas and Africa. The broadly lanceolate leaf arises from a very short rhizome and is subtended by four alternate, short-lived sheaths. A small, 2 cm. tall x 1 cm. wide, almost rectangular pseudobulb, bilaterally compressed, forms at the leaf base and soon becomes wrinkled. The single apical leaf is up to 20 cm. long and 5 cm. wide and is dark green, heavily streaked with much darker green flashes. The beautifully mottled leaves are reminiscent of some *Paphiopedilum* species. There is a strong root system composed of four to ten, 0.5 cm. diameter, unbranched, 10 cm. long roots which are white, densely calloused and verrucous. Older roots have a crumbly texture.

Inflorescence and Flowers: The basal inflorescence reaches up to 40 cm. high and bears a raceme of up to 15 flowers. Overall, these are a pinkish-orange. The dorsal sepal is linear-lanceolate and 1 cm. long x 0.25 cm. wide. The lateral sepals are a similar size but asymmetric. The petals are a similar size but slightly broader. The four-lobed lip is violin-shaped with the opposing basal lobes sloping downwards at 30° and joined to the broad, flared apical lobes by a thick neck. The lip is 0.7 cm. long and 0.7 cm. across at the broadest part. There is a bilobed tear-shaped callus at the base of the lip and a short underlying spur.

Flowering Periods: February and March and unpollinated flowers last for about six weeks.

Pollination: The pollinators are not known but may be small bees; fruit formation is frequent.

Cultivation: In pots and growing in a well-drained compost enriched with peaty loam and enjoying good light, reasonable constant humidity and air movement.

The genus
Eulophia R. Br.

Robert Brown was a Scottish taxonomist, one of the best in the 19th century, and he named this genus using the Greek words *eu* and *lophos* meaning true crest referring to the furrowed crested lip. This is a large genus of over 200 species which are predominantly terrestrials and found mainly in Africa. *E. alta* appears to be the only species in Brazil, but it is also found in all Latin America up to Costa Rica. There is also an *alba* form (bulletin CAOB n° 23. September 1998 New Brazilian Orchids). How did it escape from Africa?

Eulophia alta
(L.) Fawc. & Rendle. | *1616* (Plate 22)

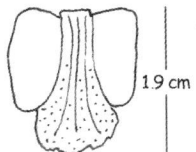

19 cm

Etymology: *Alta* is Latin for elevated or tall, referring to the very tall flower spike.

Habitat: This is almost always a roadside plant, in our experience, throughout sections I and VI between 900-1200 M.asl amongst low shrubs and tall grasses, enjoying shaded light and a fair constant humidity.

Occurrence: Common throughout the Americas.

Plant: A large terrestrial species with subterranean tuber-like corms the size of a small potato. These measure about 4 cm. by 3 cm. and are formed each year at the base of the shoot on a stout, 1.3 cm. thick, white, round rhizome. They are found about 3 cm. apart and about 9 cm. below the soil surface. Up to five alternate leaves are borne on a 15 cm., stout central shoot arising from the old corm's apex. The leaves are light green, plicate, measuring 50 cm. x 6 cm. and are sharply pointed at the apex, narrowing abruptly towards the stem, and can arise from any point on the stem.

Inflorescence and Flowers: A stiff, erect racemose inflorescence up to 100 cm long arises from the corm base bearing 25 flowers which form a cone around the top 30 cm. of the inflorescence; the flowers are unmistakable, with light-green, pointed oval sepals, 2 cm. long x 0.5 cm. wide. The petals are shorter, broader and purple and all five jut forward above the deeply heeled, partly reflexed, centrally hairy purple lip.

Flowering Periods: In February and March.

Pollination: This is almost invariably over 80% and probably effected by small bees.

Cultivation: In large pots on a compost composed of 50% coarse fibrous material and 50% coarse sand. Water well after new growths appear and through flowering, thereafter keeping the compost slightly damp.

The genus
Cyrtopodium R. Br.

Robert Brown was a very important Scottish taxonomist in the early 18 hundreds. *Kyrtos* is Greek, meaning upturned and *podium* is foot referring to the upturned column foot. **Pabst&Dungs** discuss some 26 Brazilian species of which only two pertain to the Organ Mountain Range. For a short but excellent biography of **Rob Brown** (1773-1885) see Genus *Cyrtopodium*, Brazilian Species pp. 25-28 by **Lou Menezes**.

Cyrtopodium glutiniferum
Raddi | *1618* (Plate 22)

Etymology: **Lou Menezes** states that the correct name for this species is *Cyrtopodium glutiniferum* **Raddi**, described in 1832 and so named because the sticky, glutinous sap was used to glue leather boots by cobblers.

3.2 cm

Habitat: Found in sections I, III, V and VI in high mountain fields, the margins of steep bare rock faces up to 1000 M.asl, generally amongst tall grasses and shrubs and various bromeliad species of the genus *Vriesia*, enjoying total exposure to the elements, a peat-based substrate, summer rains, a dry winter with mists and frequent brush and grass fires during this period.

Occurrence: Common.

Plant: A large clustered pseudobulbous terrestrial or lithophyte and sometimes epiphyte. The pseudobulbs are shaped like an enormous fat cigar measuring over 50 cm. in length, 6 cm. in diameter, many noded, partly sheathed and a clear green. The leaves are light green and linear-lanceolate, plicate and long-lasting, measuring 40 cm. by 5 cm. broad and up to 12 emerge alternately from the apex of a young pseudobulb. The root system is intense and extensive with long white branching roots, very adherent to rocks or hard soil. The root system and the plant itself rapidly accumulate detritus, creating a peaty soil around.

Inflorescence and Flowers: The basal inflorescence arises from new growths in August. It is paniculate and may reach up to a metre in height, stiff, round and erect, with a panicle of up to 100 flowers on its apical third. These flowers resemble vaguely those of its neighbouring oncidiums, *O. batemannianum* or *O. blanchetii*, but much larger. The sepals and petals are of similar size and shape, broadly spathulate to oval, 2 cm. long by 1.5 cm. broad and a clear yellow. The lip is trilobed; the upper two opposite lobes, separated by a faint callus, are almost oval, each measuring 2 cm. long by 2 cm. broad. The apical lobe is also more or less oval and 1.2 cm. deep by 2 cm. wide and a clear yellow.

Flowering Periods: September and October and the flowers last for three weeks or so.

Pollination: Probably *Euglossine* bees.

Cultivation: Grow in large well-drained clay pots on a loam-based compost with added coarse grit, with copious water and feed and strong sunlight during growth. It will also grow well in tropical gardens above 600 metres altitude on a similar substrate or on boulders on a very well-drained slope.

General: *Cyrtopodium andersonii* comes from northern South America and the Carribean and **Lou Menezes** correctly speculated that there was no way it could have hopped the intervening 3000 miles to the Organ Mountain Range.

Cyrtopodium gigas
(Vell.) Hoehne | *1624*

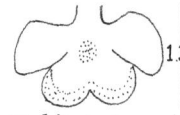

Etymology: *Gigas* is Greek for giant or very large which aptly describes the plant's size.

Habitat: An epiphyte, also growing terrestrially around and upon scrub trees and swamps at 20-200 M.asl in section V and VI, accepting high humidity and light with some air movement.

Occurrence: Occasional; originally common but now sparse due to over collection and habitat elimination.

Plant: A large robust epiphyte, sometime terrestrial, that thrives in the scrub forests around swamps and almost always found together with **Cattleya harrisoniana** and **Oncidium flexuosum**. The pseudobulbs are long, gross and cigar-shaped, up to 60 cm. high by 4 cm., heavily sheathed and noded. The leaves are ultimately apical, a tuft of 8-10 dark green, linear-lanceolate and plicate, measuring up to 30 cm. by 3 cm. wide, remaining for several years provided the humidity remains high. The root system is very extensive and prolific, composed of a mass of thickish white, long, branching roots which lock this large and heavy plant to its substrate.

Inflorescence and Flowers: A basal inflorescence appears together with the new growth. This is round, sheathed, erect and rises higher than the leaf tuft. It holds a broad panicle of many large flowers which are tiger-striped in yellow and dark brown. The sepals and petals are ± equal, almost round and 1.5 cm. long by 1.2 cm. broad, yellow but speckled with brown. The lip is four-lobed, two back-lobes reaching up to embrace the column, both are a bright yellow. The two apical lobes are also yellow but with brown serrated margins. The throat shows a short club-shaped warty callus.

Flowering Periods: October/November and the flowers last for three weeks..

Pollination: Infrequent; **Dressler** suggests probably by *Euglossine* bees.

Cultivation: For those with a similar climate, grow in a very well-drained sloping bed with a substrate of coarse sand and fibrous compost. The plant is tolerant and will grow well up to 1000 M.asl if the above conditions are observed. In greenhouse conditions, grow in very large pots, with a similar well-drained substrate in high light and humidity with some air movement.

The genus
Govenia Lindl. ex Lodd.

John Lindley raised this genus in honour of **J. R. Gowen**, an English gardener and horticulturist. It is terrestrial and only one species is attributed to Brazil but it ranges from Argentina right through to Venezuela, the West Indies and Costa Rica.

Govenia gardneri
Hook. | 1646

Etymology: Dedicated to **George Gardner**, Scottish Botanist who collected in Brazil from 1836-41.

Habitat: The plant has a predilection for moist ground where it appears and flowers with regularity. It seems that terrestrial orchids such as Govenia, have the capacity to remain underground in less suitable positions so missing a year's flowering. We have found this plant on riverbanks at 1000 M.asl on mountain ridges at 1500 M.asl and in between, in section VI.

Occurrence: Occasional but usually in colonies.

Plant: A medium-sized terrestrial perennial with a subterranean tuber, potato- or even artichoke-like. The previous years tuber is a translucent yellow, 3 cm. in diameter and irregularly round. The current years growth arises on a square-sectioned stem from the previous tuber, rapidly swelling at the base to produce the next. The sturdy stem produces two large roundly ovoid, pointed plicate leaves – 28 cm. by 10 cm, which are like a folded fan and may be opened out by pulling when young. The leaf bases and the flower spike are contained in a cylindrical red plastic-looking sheath which may contain water. The roots, which may penetrate the bark of fallen logs, arise from the stem base and penetrate the leaf litter for 10 cm. around. There is a total of more than 1 metre of roots and each one is about 0.15 cm. in diameter and hairy.

Inflorescence and Flowers: The inflorescence is tall, to 50 cm., with an apical arrangement of 20-25 separate, creamy-white flowers opening in sequence; the flowers are 3 cm. wide x 3.5 cm. high. The sepals and petals are free and finely freckled with vinous markings. The dorsal sepal is a somewhat irregular narrow elliptic, 1.2 cm. long by 0.4 cm. broad. The laterals are an asymmetric banana-shaped narrow oval, 1 cm. long by 0.5 cm. wide, while the petals are a broad asymmetric lanceolate 1.0 cm. long by 0.7 cm. broad. The lip is similar in length to the petals, vaguely trilobed and 0.6 cm. broad at the shoulders and with three maroon dots at the apex.

Flowering Periods: December to February and the individual flowers last for a short period although the inflorescence may show flowers blooming for up to a month.

Pollination: Not known, probably by bees and over 50% of the flowers are normally pollinated.

Cultivation: In pots on a rich leaf mould compost in good dappled light and always kept damp but not waterlogged.

The genus
Grobya Lindl.

John Lindley dedicated this genus to **Earl Grey** of Groby a munificent patron of horticulture, in the early 1800s. There are three species to this Brazilian genus, all located in the Atlantic Rain Forest with two in the Organ Mountain Range, both of which we have found.

Grobya amherstiae
Lindl. | 1647 (Plate 22)

Etymology: Named in honour of **Lady Amherst** who was the first to flower this plant in England, and was the British consul's wife in Rio de Janeiro.

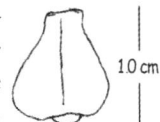

Habitat: Found throughout all sections of the Organ Mountain Range, mainly in regrowth forest but also in original forest on dying trees or branches at the low to mid-tree level. Although seedlings begin to grow on the bark of healthy trees, good colonies are only found on a). living trees adjacent to dead boughs, or b). dead parts of living trees, and c). dead trees or on long dead logs. The decline of these colonies occurs as the decomposing substrate erodes.

Occurrence: Common.

Plant: A medium-sized epiphyte or sometimes terrestrial known locally as the onion orchid for the shape of the bulb which is round and 3-4 cm. in diameter. It carries up to 6 long, thin, leaves measuring 40 cm. x 1.8 cm. which are grass-like at first glance, and deep green. These are not retained on exposed colonies beyond the second year. The roots are thick, branching and may roam for great lengths under the bark of a dead or dying tree. Measured on a old decaying stump, 3 metres of root were found on the leading mature pseudobulbs, penetrating the peaty soil and invading twigs and even old empty pseudobulbs.

Inflorescence and Flowers: A good specimen in flower is most attractive. Two basal spikes are produced per pseudobulb with up to 16 flowers on each spike. Inflorescences produced from the previous year's pseudobulbs are rarely erect, sometimes pendent and usually drooping. The sepals are narrow at the base and thick but broadening, convoluted and curling outwards, thin, tapering, yellowy-pink and measuring 1.8 cm. by 0.7 cm. wide. The laterals are fused for 10% at their bases. The petals are wide, 2.0 cm. long by 1.5 cm. and are a curiously translucent yellow with greenish-yellow veins, bearing rows of irregular wine-coloured rings which may coalesce. They have a heady scent of slightly stale honey. The labellum is yellow with a red tip and deep orange at the throat and it is finely balanced to tip potential pollinators against the wine-marbelled white column

Flowering Periods: February and March; the flowers last for three weeks.

Pollination: Infrequent but during the summer when it happens, usually a hot dry one, pollination is extensive. This coincides with the arrival of far travelling *Euglossine* bee groups and these may well be the pollinators.

Cultivation: In pots on a thick fibrous compost mixed with an amount of dead wood or bark chips in good but dappled light always keeping the substrate damp.

Grobya galeata
Lindl. | 1649 (Plate 23)

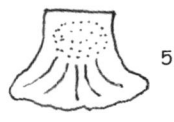

5 mm

Etymology: *Galeata* is Latin for helmet-like and refers to the gathered dorsal sepal and petals which give a helmet-like appearance.

Habitat: Found on rotting or decaying tree trunks or branches at 1200 M.asl, in humid original or maturing forests in section I, and requiring shade and some air movement.

Occurrence: Rare.

Plant: A medium-sized epiphyte with light to dark green, ringed, globose to attenuated, ovate and lightly sheathed pseudobulbs, which can be 6 cm. high by 2.5 cm. diameter at the widest part, and are closely grouped on a short rhizome. There are up to nine apical strap-shaped, grass-like, linear-lanceolate leaves which are keeled, dark green, up to 40 cm. long by 1.5 cm. and last for several years. The many roots are white, thickish, worm-like and penetrate the decaying substrate around. The species is significantly more robust than *G. amherstiae.*

Inflorescence and Flowers: An inflorescence emerges from the bases of the newly matured pseudobulbs. This is up to 12 cm. long, semi-pendent and bears up to 10 spaced flowers which are smaller than those of G. amherstiae. The dorsal sepal, which is fused to the two petals, is a blunt, lanceolate, club-shape, 1.4 cm. long by 0.5 cm. The lateral sepals are also partly fused at the apex and the base, asymmetrically lanceolate, 1.0 cm. long by 0.3 cm. The lip is complex, five-lobed with two dominating side-lobes which are an obtuse asymmetric lanceolate and 0.5 cm. by 0.25 cm. The flared apical lobe is short, 0.4 cm. wide and ridged. The neck is dominated by a complex warty callus with two short wings inclined towards the short narrow throat. The overall colour is brownish-green rather

than the pale honey of *G. amherstiae.*

Flowering Periods: March and April.

Pollination: *Euglossine* bees, suggested by **Dressler.**

Cultivation: Similar to *G. amherstiae.*

The genus
Promenaea Lindl.

One more for **John Lindley** of Kew Gardens, London, England. This is a small Brazilian genus of 15 species of which three should be found in the Organ Mountain Range. **Lindley** named the genus by dedicating it to a priestess of *Dodona*, no doubt of outstanding beauty, *Promencia.*

Promenaea xanthina
Lindl. | 1662 (Plate 23)

Etymology: *Xantha* is Greek for yellow, referring to the bright yellow flowers.

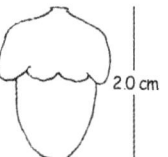

2.0 cm

Habitat: Found up to mid-tree level in both original and regrowth forest between 1200-1500 M.asl in section VI. It is demanding in habitat, preferring shady, vertical trunks with a throughput of water but not sodden with detritus. It grows best on a vertical, damp and mossy trunk in indirect light. We have found large colonies growing under the nest-epiphyte Miltonia cuneata and also directly underneath bromeliads, enjoying a dampness in dry periods.

Occurrence: Common.

Plant: The Brazilian vulgar name of "silken leaves" is perhaps an exaggeration, although the texture and pale green colour identifies the genus at once when not in flower. The pseudobulbs are closely-packed angled ovals measuring 2 cm. x 1.5 cm. New growths have four leaves, two apical and two which develop into bulb sheaths. Pseudobulbs older than two years can retain a single leaf. In stagnant air the leaves soon become yellow and water-soaked and are lost. The roots are fine but not fibrous, grey and about 2 mm. diameter, becoming brown at about 18 cm. from the tips. They can reach 50 cm. in any direction from the plant and frequently girdle the host tree. A 10 bulb colony examined had 34 roots from the new growths averaging 15 cm. in length and giving a total of 5.1 metres of root.

Inflorescence and Flowers: The inflorescences are about 6 cm. long and arise from the sides of recently matured bulbs. Each bears one or two lemon-yellow flowers which measure 4 cm. across. Sepals and petals are ±

equal, a broad lanceolate, 2 cm. long by 0.1 cm. wide. The trilobed lip has two up-pointed opposite almost rectangled basal side-lobes. The reflexed, half oval, apical lobe has a ribbed throat. The lip measures 2 cm. by 1.4 cm. broad. The lip base and lobes have maroon markings similar to those on the column base.

Flowering Periods: November and through January and the flowers last for up to two weeks.

Pollination: Most probably by *Euglossine* bees. Pollination is more frequent in long hot summers coinciding with the visits of these far ranging insects. After pollination the flowers rapidly turn white, which on a single flowered plant can cause confusion with a not uncommon *alba* form.

Cultivation: In pots on a coarse fibrous compost which allows for good drainage, or on bark or tree fern plaques always allowing water to flow through the roots but never stagnating, and in dappled light with good air movement, to avoid leaf spotting, and reasonable humidity.

Promenaea stapelioides (Link & Otto) Lindl. | *1664* (Plate 22)

Etymology: So-called because of the resemblance of the heavily-barred maroon flowers to those of the genus *Stapelia*.

Habitat: In humid original forest, around streams and river banks above swampy ground on moss-covered boulders, low to mid-tree trunks from 400-800 M.asl in sections IV, VI and VIII, enjoying high humidity, low light and air movement.

Occurrence: In such environments it may be common.

Plant: Like *P. xanthina* in all its aspects but perhaps the leaves are a little longer and narrower but they have the same silky texture and light green colour.

Inflorescence and Flowers: Arising from the bulb base on a 3 cm. pendulous inflorescence, bearing a single dark claret and white flower. The sepals are a pointed oval, 2.4 cm. long by 1 cm. wide, with a greenish-white base, heavily barred with lateral claret stripes. The petals are rounder at 2 cm. long by 1.2 cm. broad and are heavily speckled by short claret lateral bars. The trilobed lip is practically black with the lateral lobes showing touches of a greenish white base, 1.8 cm. wide by 2 cm. long.

Flowering Periods: In February and March and the flowers last for up to a month.

Pollination: Probably Euglossine bees; seed set is frequent.

Cultivation: As for *P. xanthina*, but in more shade,

somewhat less air movement and in warmer conditions.

General: These two plants, in spite of their similarities in habitat and growth patterns, very rarely overlap in altitude range.

The genus *Houlletia* Brongn.

The genus was raised by **Adolphe Brongniart**, a French botanist, who was the director of the Jardin des Plants in Paris in the mid 1800's. He dedicated this genus to **Houllet** who found the species described below near Rio de Janeiro. **Houllet** later became head gardener of the Jardin des Plantes. **Pabst** and **Dungs** state that this is a small American genus of seven species, two of which seem to be exclusively from Venezuela, only two from Brazil and one of these from the Organ Mountain Range.

Houlletia brocklehurstiana Lindl. | *1671* (Plate 22)

Etymology: Named in honour of **Lady Brocklehurst** who flowered it in England in 1841.

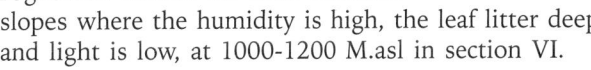

Habitat: A terrestrial in original or regrowth forest on well-drained slopes where the humidity is high, the leaf litter deep and light is low, at 1000-1200 M.asl in section VI.

Occurrence: This species is an enigma; try as we might over a 10 year period at all altitudes between 600-1500 M.asl we have found no other colonies in apparently ideal situations so we classify this species as rare through over colletion.

Plant: A large terrestrial species with clustered pseudobulbs 10 cm. x 4 cm. high, regularly angular, and six-sided. The leaves are large, dark green, single, plicate. pointed ovals, 50 cm. x 16 cm., on a 53 cm. petiole, and attached to the circular apex of the pseudobulbs. The roots, springing from the pseudobulbs' base, are covered by moss and heavy leaf mould, and reach 45 cm. from the colony. They are 0.4 cm. in diameter, white with yellow tips running through the litter, and fungal threads can be seen reaching out from them into the substrata.

Inflorescence and Flowers: The inflorescence, produced from the pseudobulbs base, is 65 cm. long, red-brown with bracts at each of five nodes. Each bract is 2.5 cm. long and pointed. The arching inflorescence bears up to nine flowers, which are heavily scented of cloves. Each flower is 8 cm. high and 8.5 cm. wide. The sepals measure 2.5 cm. by 4 cm. and the petals are 2 cm. by

3.8 cm., and waisted and tapered towards the base. The lip is shaped like a chair, white with deep burgundy spots, and has two curved horns rising from the side and recurved towards the labellum tip; a very spectacular orchid.

Flowering Periods: February through April and the flowers last for three weeks or so.

Pollination: *Euglossine* bees; according to **Dressler**. However, in ten years we have not seen fertilized flowers in the three known colonies.

Cultivation: In large pots on a rich compost of loose humus leaving space for substantial root growth, keeping damp throughout the year, never allowing water accumulation, in dappled light with low wind movement.

The genus *Stanhopea* Hooker

Sir **William Hooker** dedicated this genus to the **Rt. Hon. Philip Stanhope**, president of the London Medico-Botanical Society. A South to Central American genus of about 25 species of which three should be found in the Organ Mountain Range. The difficulty is to find appropriate forest at reasonably low altitudes. Quite the most dramatic genus in the Americas due to the size of the flowers, their heady scent and their complexity.

Stanhopea guttulata Lindl. | *1678* (Plate 24)

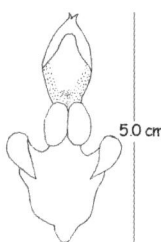

Etymology: *Guttulatus* is Latin for sprinkled with dots, referring to the speckled flower parts.

Habitat: Found at 700-1000 M.asl in very humid original forest on rocks covered by rotting wood and vegetable detritus, growing in low light, low wind movement and very high humidity from nearby water, in section III, V and VI. Found in large groups together with equally large groups of *Xylobium* and could thus be confused.

Occurrence: Occasional, but when found, in large colonies.

Plant: A large clumping lithophyte or terrestrial on steep rock ledges or banks. A typical mature plant will have up to 15, leafed pseudobulbs clustered on a short branching rhizome. The conical, somewhat wrinkled, pseudobulbs are dark green, sheathed for several years, 5 cm. tall by 3 cm. Each carries a dark green, plicate, though slightly fleshy, apical leaf, held on an 8 cm.

deeply grooved stem, 0.4 cm wide. They are elliptic and spathulate with five firm equidistant veins running parallel with the margins and measure 44 cm. long by 10 cm. The root system is profuse, arising from the pseudobulb bases and viable for many years. The roots are short, branched, white, 0.15 cm. in diameter and reach to the plant's circumference, penetrating the decomposing, relatively loose, substrate in all directions.

Inflorescence and Flowers: Single pendent sheathed 10 cm. inflorescences emerge from the bases of the newest pseudobulbs. These bear one or two pendent flowers which are almost indescribable; so better to just look at the photograph or otherwise imagine a large winged *Pterodactyl* coming down to land. The flower is about 10 cm. tall and nearly the same width, a yellowish pink, lightly speckled in wine on all its parts. The overpowering scent is of nail varnish; amyl acetate.

Flowering Periods: Early November and the flowers last for up to 3 days. However, sequential flowering is the norm, so healthy plants may show blooms for a month or so.

Pollination: Male *Euglossine* bees.

Cultivation: As the inflorescence only grows from the base of the pseudobulb, and always descends, it is necessary for the plants to be grown in an open compost, in slatted baskets to allow the flower spikes to emerge from the bottom; dappled light is recommended. These plants will enjoy both cool and intermediate temperature conditions. Copious amounts of good quality water throughout the year is vital.

The genus *Gongora* Ruiz & Pavón

Ruiz&Pavón named this species in honour of the bishop of Córdoba and Viceroy of Colombia, Don **Caballero Y Gongora**. This is a small South and Central American genus of about 25 species of which only one is found in south-eastern Brazil from Pernambuco to Santa Catarina, on the foot hills of the Atlantic rain forest. The other five Brazilian species are confined to the northeast and the Amazon regions.

Gongora bufonia Lindl. | *1685* (Plate 24)

Etymology: *Bufo* is Latin for toad, presumably referring to the flower shape.

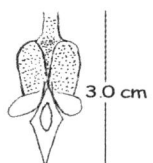

Habitat: An epiphyte or lithophyte in section IV at 200-300 M.asl, found in 40

year-old regrowth forest in hot, humid conditions, growing on large boulders and low tree trunks in filtered light, low air movement and a constant high humidity.

Occurrence: Occasional.

Plant: A medium to large epiphyte with distinctive shiny, fleshy light-green pseudobulbs. These are a ridged octagonal narrow cone, and are up to 9 cm. high and 2.5 cm. diameter arising alternately from a thin 0.5 cm. diameter rhizome at 1.5 cm. intervals. There are two apical long-lived leaves, 43 cm. long by 9 cm. wide, broadly lanceolate, plicate and with a distinct central vein, two lesser veins and held on a 5 cm. false petiole. The root system is reasonably strong and composed of short, white branching roots, mostly from the base of the new growth and spreading for about 20-25 cm. around the plant.

Inflorescence and Flowers: A pendent inflorescence up to 30 cm. long is produced from the base of the most recent pseudobulb. We have found plants with several such inflorescences. These are pink and thin but tough, with up to 30 coral-pink alternate flowers borne on thin pedicels. The dorsal sepal is broadly lanceolate to ovate, reflexed and measures 2 cm. long by 1 cm. The lateral sepals are broader, a slightly asymmetric lanceolate and measure 2.9 cm. long by 1.6 cm. wide, reflexed with sharply acuminate tips. The basal halves of the petals are fused to the column and are incurved for 30% of their length. When flattened they are a linear-lanceolate boomerang-shape measuring 1.5 cm. long by 0.2 cm. wide. The fleshy lip is so complex as to be almost impossible to describe. From the side it is composed of two base-to-base isosceles triangles. The basal lobes are stiff, waxy, infolded upwards and partly fused at their bases. The two apical lobes are totally fused and have a distinct chin at their bases. The basal and apical lobes are joined by a thick, 0.4 cm. long by 0.2 cm. wide neck and are respectively, 1 cm. long by 0.5 cm. wide and 1.3 cm. long by 0.5 cm. wide. The column is slightly downward curved and relatively long and slim, measuring 2.5 cm. long by 0.2 cm. wide, keeled on its upper surface and bears two pollinia. The overall flower colour is coral pink, speckled with darker brown spots. The flowers have an indistinct scent.

Flowering Periods: December, and the flowers open simultaneously and can last for three weeks if not pollinated.

Pollination: Pollination is by *Euglossine* bees, which can destroy the flowers while still in bud, by trying to reach the perfumed oil therein.

Cultivation: Grow as *Stanhopea* or *Cirrhaea* in hanging baskets with a well-drained and aerated substrate ensuring dryer conditions in winter but well-watered and fed in summer.

The genus *Cirrhaea* Lindl.

Named by **John Lindley**, from the Latin *cirrus* and referring to the rostellum which is extended in the form of a small tendril or curl of hair. One more small, six-species genus which appears to be confined to the Atlantic rain forest and four of these to the Organ Mountain Range, all of which we have found. These are mainly confined to the Atlantic-facing original forest, the scarp slopes although occasionally we have come across *C. dependens* in humid stream-side situations on the anticline.

Cirrhaea dependens Rchb. f. | *1699* (Plate 24)

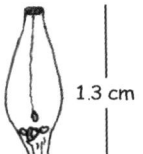

1.3 cm

Etymology: *Dependens* is Latin for suspended or hanging down, referring to the pendant nature of the inflorescence.

Habitat: Deep shade, high humidity and almost no air movement, or the flowers abort, are the necessities of this plant on low to mid-tree trunks or on lianas. It is almost always found in original forest, at 600 to 1200 M.asl in sections IV, V VI and VII, but the greatest concentration at 700 M.asl.

Occurrence: Common in sections IV and VI rare in sections V and VII.

Plant: An epiphytic clusterer forming large colonies. The pseudobulbs are roughly 5-sided but with intermediate ridges, light green, conical and 4 cm. tall x 2 cm wide. The leaves are long-lived, apical and single, papery, light green, a compressed ovate 30-40 cm. long, 6-8 cm. wide with 5 or 6 very pronounced longitudinal veins and with a stiff yellow-green stalk, up to 10 cm. long. The root system is fine but not extensive, reminiscent of *Miltonia* but also with some thickish white penetrating roots.

Inflorescence and Flowers: A long pendulous claret-coloured, racemose inflorescence arises from the base of a recently matured pseudobulb and bears up to 15 non-resupinate large flowers of extreme beauty. The overall colour effect is of deep claret, though colouring can be very varied. The sepals are ± equal, 2.5 cm. long by 0.9 cm. broad, pointed, a light red-wine colour, striped or speckled with darker claret; the petals are the same colour and length, but are thin, 0.2 cm. wide. The lip is a curious, narrow, pointed spade-shape while a long slender and curved column hangs above. On their pendulous inflorescence the flowers appear like upturned umbrellas. We have found 2 colour varieties: var. *cornata* **Hoehne** and var. *tigrina* **Porsch**.

Flowering Periods: In February and March and the flowers last for a week or so but many consecutive inflorescences in a colony will show flowers in bloom for over a month.

Pollination: *Euglossine* bees make a spectacular sight when the group arrives and in 15 minutes 90% of the flowers are pollinated – personal observation.

Cultivation: In baskets to allow the pendulous inflorescences their full glory, and on a rich compost always damp but never waterlogged, in deep shade with little air movement.

General: This species is commoner than we imagined in its specific environment. The inflorescence, so dramatic in a greenhouse, blends with the shade in the forest and only an educated eye will spot it.

Cirrhaea loddigesii
Lindl. | 1700

Etymology: Named in honour of the *House of Loddiges* one of the major orchid nurseries in the 19th century in England.

Habitat: An epiphyte or sometimes lithophyte, growing in dark, humid original forest on low tree trunks or moss-covered boulders at 700-800 M.asl, in section VI. It enjoys sparse light, constant high humidity and low air movement.

Occurrence: Occasional.

Plant: Indistinguishable from *Cirrhaea dependens* when not in flower and even with difficulty when in flower because of the variety of colour forms of *C. dependens*. A clustered epiphyte, forming medium-sized groups of up to six active pseudobulbs which are conical, ridged and light to dark green, up to 3 cm. high by 2 cm. wide. The single, apical long-lived leaves are plicate and lanceolate and measure up to 30 cm. long by 6 cm. wide, dark green and borne on an 8 cm. petiole. The prolific root system is composed of adhesive roots but is not extensive.

Inflorescence and Flowers: One or two pendent, 20 cm. long, fragile racemes emerge from the bases of the recently matured pseudobulbs. These bear up to 15 alternate and curiously-shaped non-resupinate flowers. The dorsal sepal is undulate and measures 2.3 cm. long by 0.8 cm. wide. The lateral sepals are 1.5 cm. long by 0.8 cm. and are oblong and slightly acuminate. The petals are much narrower, a curved linear shape and measure 1.5 cm. by 0.25 cm. The overall colour of the tepals is pink with a few purple blotches at their tips. The curious lip is trilobed and resembles and arrow or a diving bird. The side lobes are 1.1 cm. long and 0.15 cm. wide and flow backwards from the top third of the mid-lobe. The lobe itself is 1.5 cm. long and 0.15 cm. wide and the lip is dark purple.

Flowering Periods: November; the flowers last for a week.

Pollination: We have observed pollination by *Euglossine* bees.

Cultivation: In hanging baskets in a rich fibrous and well-aerated compost observing the natural habitat conditions of the plant.

General: The flowering period separates this from **C. dependens** which flowers as regular as clockwork in February and March while **C. loddigesii** flowers in November.

Cirrhaea longiracemosa
Hoehne | 1702 (Plate 24)

Etymology: *Longi-racemosa* is Latin for long racemose inflorescence.

Habitat: At 400-500 M.asl at the edges of original forest in section VI, at tree 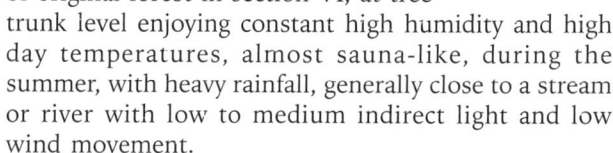 trunk level enjoying constant high humidity and high day temperatures, almost sauna-like, during the summer, with heavy rainfall, generally close to a stream or river with low to medium indirect light and low wind movement.

Occurrence: Rare.

Plant: A large clustered epiphyte or lithophyte which may form significant clumps. The pseudobulbs are closely packed, arising from a 1 cm. diameter thick, woody, branching rhizome. These pseudobulbs are four-cornered, slightly laterally compressed, a clear light green, reminiscent of **Gongora bufonia**, and up to 8 cm. tall by 4 cm. wide. They have a single apical, long-lived, stiff, light green, papery leaf with 4 very distinct nerves running parallel to the margins, and a deep central keel. These are a broad lanceolate, 35 cm. long by 9.5 cm. and emerge on a short, 3 cm. long, false petioles. The root system is profuse, composed of numerous thin, branching, very adherent white roots.

Inflorescence and Flowers: One mature large clump composed of 30 pseudobulbs, bore 12 thin wine-coloured pendent basal inflorescences arising from both old and new pseudobulbs. The average length of the inflorescence was over 40 cm., whilst the average number of flowers was 45. These flowers are non-resupinate, a faintly greenish yellow, emerge both opposite and alternately at short but irregular intervals and on 4 cm. thin, wine-coloured pedicels forming an attractive, loose, pendent cone. The pedicels are horizontal for 2 cm. and vertical for the apical flower-holding part which is in fact the ovary. The dorsal sepal is a broad lanceolate, 1.7 cm. long by 0.4 cm. The lateral sepals are the same shape but slightly larger but with a tendency to fold. These are a pale yellow with very faint wine speckles. The butter-yellow petals are a narrow, lanceolate, 1.7 cm. long by 0.4 cm. The lip and column form a butter-yellow half-circle with the lip uppermost. The lip is almost anchor shape with two horn-like side-lobes and a distinct speckled claret-coloured pouch for the apical lobe.

Flowering Periods: In December and the flowers last for 10 days or so, but the plant may show blooming inflorescences for much longer.

Pollination: By *Euglossine* bees.

Cultivation: Grow as *C. dependens* remembering well the habitat description.

Cirrhaea saccata Lindl. | 1703 (Plate 24)

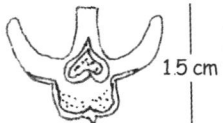

15 cm

Etymology: *Saccatus* is Latin for pouch or bag-shape, referring to the mid-lobe of the lip.

Habitat: Found in deep original forest like other *Cirrhaea* species, growing in low light, high humidity and with little air movement at around 400 M.asl, in section VIII.

Occurrence: Rare.

Plant: Many tightly clustered, ridged, 7 cm. high pseudobulbs are borne on a very short rhizome. There is a single apical leaf on a 3 cm. false petiole, broadly ovate and measuring up to 30 cm. long by 9 cm. wide and with 5 marked veins. The roots are 2 mm. thick, branched and penetrate and bind the leafy substrate.

Inflorescence and Flowers: The inflorescence is basal, pendulous and wine-coloured, up to 25 cm. long and bearing up to 20 green to yellow-green non-resupinate flowers. The sepals are a similar shape, concave and thick. The dorsal measures 2.3 cm. long by 0.9 cm. while the slightly smaller laterals are 1.9 cm. by 0.9 cm. and a slightly asymmetric oval. The petals are baseball-bat shaped and are 2.6 cm. long by 0.35 cm. at the widest part. The complex lip is shaped like a horseshoe, bearing a small hood at the tip and suspended by a strap from the column base. The flower has the scent of menthol-mint toothpaste.

Flowering Periods: December and the flowers last for 15 days or so.

Pollination: *Euglossine* bees.

Cultivation: Best grown in a basket with a substrate of tough humus, peat and leaf litter, in the sort of climate described above and allowing space for the developing inflorescences. It is useful, when the inflorescences emerge, to stick a flexible label beside them to prevent them from growing into the compost.

The genus *Xylobium* Lindl.

John Lindley named the genus using the Greek compound word meaning wood-loving or living on wood probably alluding to the epiphytic preference for some of the species. Four species are attributed to Brazil and all of these also to Central America. The number of synonyms for the two species attributed to the Organ Mountain Range shows the variability in colour and even shape of the flowers.

Xylobium variegatum (Ruiz & Pavón) Mansf. | 1708 (Plate 24)

Etymology: *Variegatum* is Latin for diversified, having many forms, referring to the many colour varieties of this plant's flowers.

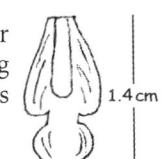

1.4 cm

Habitat: Epiphytic and sometimes lithophytic or terrestrial between 200-900 M.asl, in sections II, IV and VI. It requires dappled light and low wind movement and mostly grows at mid-tree level in very humid original forest.

Occurrence: Common.

Plant: A bulky epiphyte or terrestrial with pseudobulbs arising alternately and closely on a thick woody rhizome, often forming large clumps. The pseudobulbs are ovoid, smooth and dark green and measure up to 8 cm. high by 2 cm. diameter at the base. The two apical long-lived leaves are oblong-elliptic and measure 40 cm. long by 6 cm. They are plicate, strongly ribbed, shortly petiolate and dark green. The root system is strong and composed of many short, medium-thick white roots which penetrate the decaying substrate.

Inflorescence and Flowers: Short stout fleshy inflorescences arise from the bases of the most recent pseudobulbs and barely exceed their height. Each bears a raceme of up to ten whitish-pink tightly clustered flowers. The dorsal sepal is lanceolate and measures 1.3 cm. long by 0.5 cm. wide while the lateral sepals are 2 cm. long by 0.7 cm. and an asymmetric lanceolate. The lip is distinctly trilobed, 1.4 cm. long by 0.8 cm. The central lobe is narrow, projected and deep purple while the side-lobes are whitish-pink. A nose-like central callus reaches down for half the lip length. The fruits of this plant are very distinctive, triangular in cross-section with three longitudinal straps between the angles from which, when the fruit is ripe, the seed will escape.

Flowering Periods: The main flowering periods are

November and December but flowers have been found during most months of the year.

Pollination: Our observations show pollination by small, nectar-collecting, stingless bees.

Cultivation: In pots or baskets with a well-drained compost.

The genus *Bifrenaria* Lindl.

John Lindley raised this genus, including the genus *Stenocoryne*. The name *Bifrenaria* is derived from the Latin *bi* and *fraenum* meaning two straps referring to the two caudicles attaching the pollen. *Stenocoryne* is a Greek compound word meaning a delicate plant. There are some 25 species attributed to South America, the majority to Brazil and while **Pabst&Dungs** list 14 to the Organ Mountain Range.

Samantha Koehler (Unicamp) in her recent master's dissertation – A Synopsis and Cladistic Analysis of *Bifrenaria* Lindley "sensu lata"; Orchidaceae, of which she kindly sent us a copy, shows that:

1. The plants from the original *Bifrenaria* alliance are all very similar morphologically and also variable within each species;

2. That the flowers of *B. harrisoniae* are intensely variable as to colour, giving rise to a number of species which are in fact synonyms;

3. That the plants of the group (Ex. *Stenocoryne*) show many species which have been confused with each other.

In other words she has mostly straightened out a serious mess. The tentative conclusion considering the paucity of material for some species is:

Organ Mountain Range

Bifrenaria	Ex. *Stenocoryne*
B. tetragona	*B. stefanae* (new species)
B. mellicolor	*B. vitellina*
B. calcarata	*B. aureo-fulva*
B. atropurpurea	*B. clavigera*
B. inodora	*B. charlesworthii*
B. thyrianthina	
B. harrisoniae	*B. racemosa* ? = *B. charlesworthii*
B. melanopoda?	

Even with this wealth of material we are still somewhat confused.

Bifrenaria atropurpurea (Lodd.) Lindl. | *1709* (Plate 24)

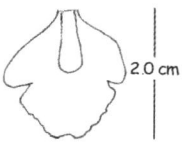

Etymology: *Atro* is Latin for dark and *purpureo* Greek for the shellfish that yields the purple dye , and referring to the dark purple flowers of this plant.

Habitat: A mid-tree species found in original forest at 1000-1500 M.asl in sections V and VI.

Occurrence: Common.

Plant: A medium-sized variable epiphyte with light to dark-green pseudobulbs, each a squat 4-sided cone-shape, 7 cm. long by 4 cm. wide and arranged in tight clusters on a short, thick branching rhizome. The colony we are describing has over 30 pseudobulbs and three new shoots, within a 25 cm. sided square. The single apical leaf on each pseudobulb is a broad pointed oval, with a central stem and 7-9 parallel ribs, paper-like, dark green and stiff, 25 cm. long by 7 cm. wide, and long-lived. The root system is vigorous, with 8 to 12 thickish branched roots to each pseudobulb, anchoring the plant firmly to its preferred vertical support.

Inflorescence and Flowers: A thick erect 7 cm. fleshy inflorescence arises from the pseudobulb base. There are up to 5 flowers, ranging in colour from deep claret to an olive green with only a hint of purple. The sepals are a pointed oval, 2.5 cm. long by 1.2 cm. wide; the petals are smaller but the same shape. All are concave and incurve around the trilobed lip giving the appearance of a Crocus.

Flowering Periods: In October through December and occasionally until March and the flowers last for up to 20 days

Pollination: Unknown. Seed set is normally up to 20%.

Cultivation: In pots on a humus-rich compost, well-drained with strong light during the dormant period to induce flowering, and shade during growth.

General: This one was introduced into European culture by **F. Warre** from the Organ Mountain Range in 1828.

Bifrenaria aurea Barb. Rodr. | *1710* (Plate 24)

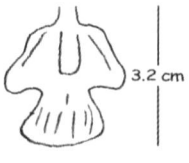

Etymology: *Aurea* is Latin for golden-yellow describing the flower colour. **Samantha Koehler** (Unicamp, 2001) follows **Castro & Campacci**, in reducing *B. aurea* to a synonym of *B. harrisoniae*.

Habitat: Found in section V at 1000 M.asl in

transitional forest, at mid-tree level in dappled light, good wind movement with high humidity during spring and summer followed by five months of drought.

Occurrence: **Barbosa Rodrigues** described this species as very rare. The appropriate environment has almost been eliminated in the Organ Mountain Range, but **R. Singer** informs us that this plant is not uncommon in the São Paulo coastal mountains at 1000 M.asl.

Plant: When not in flower it closely resembles *B. atropurpurea* and *B. harrisoniae*. Perhaps the leaves are rounder and shorter, while the pseudobulbs are squatter and more triangular than those of the above species, but as all are variable, only the environment is diagnostic. The pseudobulbs are a dark green cone up to 7 cm. long by 4 cm. wide and arise closely on a short thick rhizome. The leaves are single and apical on a very short petiole, plicate, a broad ellipse, up to 15 cm. long by 5 cm. broad, dark green and long-lived. The root system, composed of thickish, short white roots is extensive and adherent.

Inflorescence and Flowers: Three or four flowers are held on short fleshy, green, single, basal inflorescences from the most recent pseudobulbs. These are an overall yellow-gold. The sepals are a broad lanceolate, ± equal at 3 cm. by 2.4 cm. broad. The petals are similar in shape but smaller. All are slightly concave and the trilobed lip is 3.2 cm. long by 2.5 cm. The side-lobes are infolded towards the column while the apical lobe is reflexed. The overall colour is a golden-yellow with orange to magenta veining. It has a split nose-like callus running from the throat to a third of its length and both this and the apical lobe are slightly hairy.

Flowering Periods: Late September early October; the flowers last for up to 20 days.

Pollination: Infrequent; probably bumble bees.

Cultivation: As for *B. atropurpurea*.

General: Probably a colour variety of *B. harrisoniae*.

Bifrenaria calcarata
Barb. Rodr.   *1711* (Plate 24)

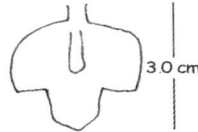

Etymology: *Calcarata* is Latin for spur, referring to the nectar-producing appendage created by the fusion of the lateral sepals at their bases.

Habitat: Original forest at mid-tree level above 1500 M.asl in section VI requiring good light and air movement with a reasonable constant humidity.

Occurrence: Rare.

Plant: Very similar to others in the *B. harrisoniae* alliance though perhaps slightly smaller. The

pseudobulbs are tightly clustered on a thick wandering rhizome and often produce two new shoots. The pseudobulbs are conical, squat and distinctly four-cornered, light green to yellow, and up to 3.5 cm. by 2 cm. broad. The single apical leaves are stiff, mildly plicate, elliptic, dark green, long-lived and 15 cm. long by 6 cm. The root system is extensive, composed of sometimes branching, thick, long white and brittle roots.

Inflorescence and Flowers: A short basal one or two-flowered inflorescence grows slightly higher than the pseudobulbs. The large flowers are almost 5 cm. across. The dorsal sepal is elliptic, up to 3 cm. tall by 1.6 cm. The lateral sepals are a broadish lanceolate, fused and elongated at their bases into a spur or false nectary, 3 cm. long by 1.7 cm. wide. The petals are smaller and irregular, spathulate-lanceolate, 2.4 cm. by 1.5 cm. broad. The lip is 2.8 cm. long by 2.0 cm., trilobed with a nose-like callus reaching from its throat to a third of its length. The overall colour is a greenish-pinkish-white with all tepals showing purplish margins, while the lip is more densely coloured towards the apices of all three lobes. The central lobe is hairy from the throat to the apex.

Flowering Periods: August and the flowers last for several weeks.

Pollination: Unknown.

Cultivation: As for *Bifrenaria atropurpurea*.

Bifrenaria harrisoniae
(Hook.) Rchb. f. | *1712* (Plate 25)

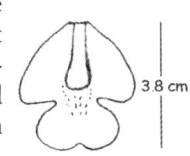

Etymology: Named after the **Harrisons**, a family of traders that operated in Rio de Janeiro from 1820-1860, and who significantly helped the botanists who worked the Organ Mountain Range.

Habitat: Found irregularly in sections II, IV and VI as a lithophyte or epiphyte at 1000 M.asl. Its true habitat is in the transitional zone between these sections and sections I, III and V, on almost vertical rock faces or on 45° bare granite slopes beneath clumps of *Vellozia* between 700-1000 M.asl. Almost collected out save on the most inaccessible rock faces, this magnificent plant deserves a better fate.

Occurrence: Occasional, rapidly becoming rare.

Plant: A medium-sized lithophyte, and sometimes an epiphyte. The pseudobulbs are a 4-sided cone, generally wrinkled, dark green to bright yellow, 6 cm. long by 2.5 cm. wide and borne in tight clusters on a short, thick branching rhizome. There is a single dark green apical leaf per bulb, 20 cm. long by 7 cm. wide, a broad pointed oval in shape, paper-like with a central keel and with parallel ribs, and long-lived. The roots are thick, white and numerous, penetrating and involving the strata around. The colony we are describing is lithophytic and smaller than epiphytic plants in all aspects.

Inflorescence and Flowers: Up to four large whitish-pink flowers arise on a short fleshy racemose inflorescence. The overall colouring and size is variable and they emit a faint agreeable perfume. The sepals are more or less equal, 3.8 cm. long by 2.2 cm. wide, a slightly concave broad oval, while the petals are slightly smaller and more ovate. The large trilobed lip is yellow, hairy, intensely veined with reddish purple with a gross orange callus for over a third of its length. The lip measures 3.8 cm. by 3 cm. wide.

Flowering Periods: In August to October through November; the flowers last for 10 days or so.

Pollination: Unknown; probably *Euglossine* bees but more probably bumble bees. We observed one such taking out the pollen on its upper thorax and wandering about for four days, very irritated, trying to get rid of it.

Cultivation: In pots on a compressed peat-based compost, watered during growth in spring and summer just damping thereafter, always in high light.

Bifrenaria inodora
Lindl. | *1713* (Plate 25)

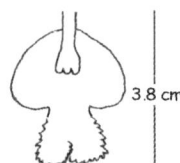

Etymology: *Inodora* is Latin for without scent, named by **Lindley** in 1839 for a plant sent from Rio de Janeiro. In 1882 **Barbosa Rodrigues** named a similar plant *B. fragrans* only to be told later that in fact it was B. inodora, which must have annoyed him greatly.

Habitat: On relict trees at 800-1000 M.asl in section VI and VIII. We did not expect to find it in these regions. **Hoehne** describes large clumps found in Santa Catarina and Minas Gerais; the first at 200 M.asl near the sea and the second deep into the mountains. Our first examples were found by a friend swimming at the junction of the rivers near Lumiar, RJ, where he came across a large branch simply covered with these plants, towed the branch to the shore and saved them. They showed their gratitude; two plants flowered three months later.

Occurrence: Occasional.

Plant: Identical to all the large species of the **B. harrisoniae** alliance and thus very variable. Our examples showed, four-cornered, squat, somewhat stressed, yellow bulbs, 5 cm. tall by 3.9 cm. wide, arising

alternately and closely on a thick rhizome. Each has an apical long-lived leaf which is stiff, leaning to plicate, lightly keeled, held on a 1 cm. long false petiole, elliptic and 20 cm. long by 7 cm. The root system, emerging from the base of new growth, is composed of a number of thick white roots which penetrate the substrate successfully.

Inflorescence and Flowers: A short inflorescence holding a raceme of up to four flowers arises from the base of the most recent pseudobulb. The sepals and petals are an olive-green both inside and out. The dorsal sepal, 3.2 cm. long by 2 cm., is a broad lanceolate. The lateral sepals are slightly asymmetrical but the same shape, fused at their bases into a 'chin' and 3.8 cm. by 1.8 cm. wide. The three-lobed lip is magenta, pilose, 3.8 cm. long by 2.9 cm. wide at the side-lobes. A nose-like callus protrudes from the throat to half the lip's length.

Flowering Periods: October and November; the flowers last for two weeks or so.

Pollination: Unknown.

Cultivation: As for **B. atropurpurea**.

Bifrenaria mellicolor
Rchb. f. | *1715* (Plate 26)

Etymology: *Melleus* and *color* is a Latin compound for honey-coloured referring to the colour of the flower parts.

Habitat: A mid-tree epiphyte of medium size found at 1200 M.asl in section VII in original forest enjoying a long wet spring and summer, a misty autumn and winter, with dappled light and the constant high humidity associated with above mountain river situations.

Occurrence: Occasional.

Plant: This has a concentrated, untidy pseudobulb grouping on a short woody, wandering rhizome. The pseudobulbs are conical, olive green, up to 7 cm. tall by 3 cm. Each has a broad lanceolate, long-lived apical leaf, stiffly plicate, many veined and up to 10 cm. long by 3.1 cm., and carried on a short, false petiole. The root system is like most in the **B. harrisoniae** alliance, composed of thick white, long branching and adherent roots, which greatly exceed the plant's circumference.

Inflorescence and Flowers: The plants examined were stressed and only showed single flowers on erect 3 cm., basal, inflorescences. The dorsal sepal is lanceolate, broad, concave, and acuminate, 2.5 cm. long by 1.5 cm. The laterals are ± equal but slightly larger. All three are a light brown with a greenish central base and apex. The petals are a slightly asymmetric, broad, lanceolate, 2.2 cm. by 1 cm. broad, and a whitish pink. The trilobed lip is 2.5 cm. by 2.0 cm. The side-lobes rise almost infolding the column and are slightly emarginate with few hairs at the centre and a short nose-like callus protruding at the throat. The apical lobe parts are a light wine.

Flowering Periods: November and December and the flowers last for a week or so.

Pollination: Unknown.

Cultivation: As for *B. atropurpurea*.

General: Very like *B. calcarata* but the spur is smaller.

Bifrenaria tetragona
(Lindl.) Schltr. | *1717* (Plate 25)

Etymology: *Tetra-gona* is Greek for four-angled, referring to the distinctive pseudobulbs.

Habitat: An epiphyte from 500-1000 M.asl, in section VI, found at mid-tree level in original forest where it receives good filtered light, slight air movement and a constant high humidity.

Occurrence: Rare in section VI but when not in flower it could easily be confused both with *B. atropurpurea* and *B. harrisoniae* whose habitats overlap.

Plant: A large epiphyte with greenish-yellow pseudobulbs up to 9 cm. high by 3 cm. and an elongated four-sided cone. These arise alternately and tightly from a short, 0.5 cm. thick rhizome. The single, apical, dark green plicate leaves are lanceolate and long-lived, with a 4 cm. false petiole and are up to 35 cm. long by 7 cm. A typical colony would have four leaves and seven pseudobulbs. The root system is strong and composed of thick, white penetrating roots.

Inflorescence and Flowers: A 0.4 cm. thick, green raceme, 3 cm. long, arises from the base of the previous year's pseudobulb with up to five alternate flowers which open simultaneously. These are borne on thick round green pedicels, sheathed at their bases and measuring 3 cm. long, including the ovary, and 0.3 cm. diameter. The dorsal sepal is a broad lanceolate, 3.5 cm. long by 2.2 cm. with an acuminate apex, slightly concave and light green, streaked with light brown longitudinal lines. The lateral sepals are fused to the column foot, have a deep chin and measure 3.5 cm. long by 3.0 cm. wide at their bases. They are a very broad lanceolate, acuminate and green, streaked with purple and very dark purple at their bases. The petals are broad, lanceolate and concave measuring 2.6 cm. long by 2.0 cm. and are greenish-yellow, with occasional deep purple blotches and white markings on their undersides. The trilobed lip is the darkest purple, almost black, with some white markings at the central apex and the erect side-lobes. When flattened, the lip

measures 2.5 cm. long by 2.7 cm. across the side-lobes.

Flowering Periods: Late January through to early February. The flowers last for three weeks and have a distinctive and delightful perfume.

Pollination: Probably pollinated by bumble bees.

Cultivation: As for *B. atropurpurea*. It grows well on bark plaques or in baskets with well aerated compost. It needs high light and little water in winter to promote flowering but this plant is still difficult to flower.

Bifrenaria thyrianthina
(Lodd.) Rchb. f. | *1718*

Etymology: A reference to the Tyrian Dye or purple, collected by the Phoenicians from gastropod shellfish.

Habitat: Lithophytic and sometimes epiphytic in section V and VI at 1000 M.asl in full or dappled light, good wind movement and humidity during spring and summer followed by a five month very dry period.

Occurrence: Rare in the Organ Mountain Range, however according to **Hoehne**, common in the mountains of Minas Gerais.

Plant: A medium to large plant, differing slightly from *B. harrisoniae* and *B. atropurpurea* as the pseudobulbs are taller and narrower and the single apical, long-lived leaf has a distinctive false petiole 3 cm. long, more reminiscent of *B. tetragona*. The pseudobulbs are 8 cm. tall, four-cornered and 3 cm. wide forming a dark green, sharply tapered four-sided cone, while the leaf is plicate, a narrow ellipse up to 25 cm. long by 7 cm. wide. Mature plants may form large clumps.

Inflorescence and Flowers: Short, 9 cm., partly sheathed, somewhat stiff inflorescences emerge from the bases of the most recent pseudobulbs, each bearing one to four large waxy flowers of uncommon beauty. The sepals are a strong lilac colour; the petals are white to cream while the lip is a creamy white, deeply and intensely veined in lilac or magenta with indented apical margins. The dorsal sepal is a broad, obtuse-apexed oval, 4 cm. by 2.3 cm. wide. The lateral sepals are an asymmetric, slightly pointed oval and measure 4 cm. by 2 cm. long with tails which accompany the nectar-free column spur. The petals are a pointed spade-shape, and measure 3.8 cm. long by 2.3 cm. and all are slightly concave. The lip is distinctly trilobed. The side-lobes are erect and parallel while the apical lobe protrudes, reflexes and is almost round. The lip centre is hairy and the gross, orange nose-like callus runs longitudinally from the lip base to half its length and is split at its apex. The lip measures 3.4 cm. by 3.3 cm. wide.

Flowering Periods: Late September to early October and the flowers last for up to 20 days.

Pollination: Infrequent but possibly bumble bees.

Cultivation: As for **B. atropurpurea**.

General: Probably a colour variation of **B. harrisoniae**.

Bifrenaria melanopoda
Kl. | 1722 (Plate 25)

Etymology: *Melan-podus* is a Greek compound word, very dark-based, referring to the purple base of the lip.

Habitat: Found in either original or regrowth forest at low to mid-tree, or even hanging on vines, at between 1000-1400 M.asl in section VI and requiring dappled light, high humidity and fair air movement.

Occurrence: Occasional.

Plant: It is very similar to **B. wendlandiana** in all aspects. A small tightly-clustered epiphyte, with ovoid pseudobulbs measuring 3.5 cm. long by 2 cm. broad which do not show the characteristic **Bifrenaria** squared-off corners but are bilaterally compressed. A single long-lasting elliptic apical leaf, dark green, stiff, paper-like and distinctly veined measured up to 9 cm. long by 4 cm. The roots are numerous, thick, adherent, white and penetrate the substrate around for quite some distance.

Inflorescence and Flowers: A thin, wiry, erect, greenish-claret, basal inflorescence rises to a few centimetres above the leaf apex. Up to seven creamish, translucent flowers are borne, emerging as a racemose cone on the apical 5 cm. of the flower stem. The sepals are more or less equal, an elongated heart-shape, 1.5 cm. long by 0.8 cm. broad with a faint brownish trace on the cream base. The petals are smaller, elliptic, 1.2 cm. by 0.6 cm. The lip is single-lobed, white, faintly streaked with purple veins at its apex but strongly purple at its base and shoulders and shaped like a broad spade, measuring 1.4 cm. long by 1.45 cm. broad. Its most distinctive feature is a central orange callus reminiscent of two fat pollinia, lightly covered with hair.

Flowering Periods: In June and July and the flowers last for up to three weeks.

Pollination: Probably pollinated by bees, but its flowering period excludes **Euglossine** bees as during this cold, short-day period, they do not venture up to these altitudes.

Cultivation: In our experience it is only happy on live bark; however, it should be tried on a tree fern plaque with significant feeding. Its creeping nature probably precludes pots.

Bifrenaria racemosa
Hooker | 1723

Etymology: So called for its racemose inflorescence.

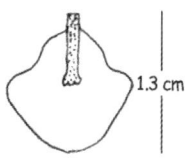

Habitat: Found in original forest and high montane scrub forest at low to mid-tree level in section VI at 1200-1600 M.asl, enjoying constant humidity, good wind movement and good light and often forming large groups on decaying wood.

Occurrence: Common.

Plant: A small to medium-sized epiphyte, similar in all ways to **B. vitellina**. The squat pseudobulbs are four-cornered, closely packed and 4 cm. tall by 0.3 cm. The single apical leaf is long-lived, stiffly plicate, lightly green, lanceolate to narrowly elliptic and up to 30 cm. long by 4.5 cm. and held on a false petiole of up to 5 cm. The profuse root system is composed of short white, sometimes branching roots which profoundly penetrate decaying matter.

Inflorescence and Flowers: A basal inflorescence up to 25 cm. long arises from the most recent pseudobulbs. This holds a loose raceme of up to 8 flowers on its apical 9 cm. The dorsal sepal is almost oval, 1.5 cm. long by 0.7 cm. wide. The lateral sepals are blunt, fat, lanceolate and somewhat asymmetric, 1.7 cm. long by 0.7 cm. The petals are elliptic, 1.3 cm. long by 0.6 cm, while the lip is very vaguely trilobed, spade-shaped and 1.3 cm. long and wide. It has a nose-like callus running from the throat for half its length and the centre is deeply lined in magenta. The overall colour is a creamy white.

Flowering Periods: March and April and the flowers last for nearly 3 weeks.

Pollination: Infrequent and not known.

Cultivation: As for **B. vitellina**.

Bifrenaria aureo-fulva
(Hooker) Lindl. | 1724 (Plate 25)

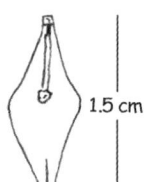

Etymology: *Aureo* and *fulva* are Latin for golden-yellow and yellowish brown respectively and refer to the flower's colour.

Habitat: Found in original montane forest in section VI at 1000-1300 M.asl, at trunk to mid-tree level enjoying high constant humidity, good air movement and reasonable light.

Occurrence: Rare, but could be commoner as it is very similar to **B. vitellina** and **B. racemosa** when not in flower.

Plant: A small to medium-sized epiphyte. The squat pseudobulbs are four-cornered, generally dark green and held closely and tightly on a thick wandering rhizome and measure 4-6 cm. by 4 cm. broad. There is a single apical long-lasting leaf, dark green, stiffly plicate, lanceolate, 20 cm. by 4 cm. broad with a false petiole of up to 4 cm. The root system is intense, composed of a profusion of thickish short, grey to white, green-tipped roots which tend to collect and invade leaf litter.

Inflorescence and Flowers: A thin, stiff, round, basal inflorescence emerges from the most recent pseudobulb. This is up to 20 cm. long and has a loose untidy raceme of up to 10 orange to golden flowers towards the apex. These do not open fully, have a distinct chin and are presented as horizontal tubes. The sepals are ± equal 2.3-2.7 cm. long by 0.5 cm., orange-gold and fading towards the bases where a number of very fine claret lines spread from the centre towards the apices. The petals are similar in colour configuration and shape but smaller, 2.1 cm. long by 0.4 cm. The lip is vaguely trilobed with the similar but more golden pattern, lanceolate and 1.5 cm. long, 1 cm. broad and showing a nose-like callus for 1/3 of its length from the throat.

Flowering Periods: February and the flowers last for two weeks or so.

Pollination: Infrequent; could be bees.

Cultivation: Grow in pots on a well-drained coarse fibrous compost in light shade always well misted but never waterlogged.

Bifrenaria wendlandiana (Krzl.) Cogn. | *1726* (Plate 25)

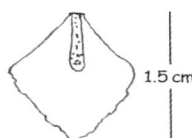

Etymology: Two generations of the **Wendland** family were botanists in Herrenhausen, Germany. **Herman Wendland** (1825-1903) collected in the Americas in 1857.

Habitat: Found at 1200-1500 M.asl on low branches and almost to ground level in the stunted elfin forest on mountain ridges and in regrowth. It enjoys leaf mould detritus around the roots and grows well on decaying sections of trunk in dappled light with reasonable humidity and air movement in section VI.

Occurrence: Common.

Plant: An epiphyte with tightly-clustered squat, cone-shaped pseudobulbs which are four-ridged, light green, and measure 2.5 cm. x 2.0 cm., arising from a tough, thick rhizome. There is a single apical, narrow pointed oval, papery, matt-green leaf, 14 cm. long x 2 cm. wide with a firm central, and parallel minor veins. It may be

retained for three years. The root system is extensive and consists of thickish, white branching roots which penetrate surrounding substrate and are very adherent. **Samantha Koehler** follows **Castro** and **Campacci** (2000) reducing this plant to be a synonym of B. clavigera.

Inflorescence and Flowers: Three or four flowers arise alternately on a thin, wiry, wine-coloured inflorescence about 7 cm. long, from the base of newly matured pseudobulbs; the flowers are extraordinarily beautiful with petals and sepals of approximately equal size, pointed ovals, measuring 1.1 cm. x 0.7 cm. and are greenish-mauve on the outside and creamy, greenish-yellow inside. The lip is large, faintly trilobed and almost square with a nose-like callus running from its throat for half its length with intense serrations on its apical margins and measures 1.5 cm. x 1.5 cm. It is white, deeply veined with purple, particularly at the throat.

Flowering Periods: September to December and the flowers last for up to three weeks.

Pollination: Rarely pollinated and the agents are unknown.

Cultivation: In medium-sized pots on a coarse fibrous mix always kept damp and in dappled light with good air movement.

Bifrenaria vitellina Lindl. | *1729* (Plate 25)

Etymology: *Vitellinus* is Latin for a dull yellow just turning to red, referring to the lip.

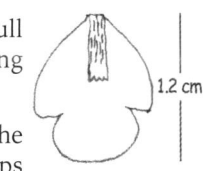

Habitat: The preferred ambient of the plant is within the high mountain tops and ridges. It shows a penchant for good air movement, humidity and good light. We have also found this plant at mid-tree in original montane forest from 1200-1600 M.asl in section VI.

Occurrence: Common.

Plant: An epiphytic species from the higher altitudes where it occurs at most levels on the contorted and wind-blown trees. The pseudobulbs are rhomboid, closely packed and angular, measuring 4 cm. high x 3 cm. across. They are found on living and dead wood where they form dense colonies, often dome-shaped with the youngest growth on top. The leaves, sheathed in young shoots, are single and apical and grow to 40 cm. long x 5.5 cm. wide, deep green and evenly, finely ribbed. They are retained on the previous three years' growth. The roots are thick, branching and fleshy, a white ochre and penetrate into thick bark or into the crumbling substrate, forming a dense matrix around dead or living wood. A 22 pseudobulb clump had over 20 metres of roots.

Inflorescence and Flowers: The inflorescence is 22 cm. long, and bears 8-10 flowers. These are partly open and tube-like. The dorsal sepal and the petals are ± equal in size and shape, a broad lanceolate, 2 cm. long by 0.8 cm., while the lateral sepals are an asymmetric lanceolate 2.3 cm. by 0.9 cm. broad and cream with a reddish tinge. The trilobed lip is a broad triangle, 1.2 cm. by 1.2 cm. at the side-lobes, while the apical margins are serrated, light yellow on the sides, and dark orange in the centre.

Flowering Periods: December and January through to April and the flowers last for 2 weeks.

Pollination: This is frequent, up to 30%; agents unknown.

Cultivation: In pots on a coarse fibrous compost in dappled light good air movement keeping damp all year round.

Bifrenaria stefanae
V. P. Castro. | *1729a* (Plate 27)

Etymology: Not known.

Habitat: As for *B. vitellina*.

Occurrence: As for *B. vitellina*.

Plant: As for *B. vitellina*.

Inflorescence and Flowers: The plant is slightly smaller than *B. vitellina*. The petals are whiter and the lip has more carmine and in addition the sepals run parallel to the column whereas in *B. vitellina* the sepals are more splayed.

Flowering Periods: This plant flowers a month before *B. vitellina*.

Pollination: Unknown.

Cultivation: As for *B. vitellina*.

The genus
Pabstia Garay

Garay named the genus in honour to **Pabst**. Previously it was in the long standing genus *Colax*, however **Garay** discovered that *Colax* had already been used for a different genus, so he invented this one. **Garay** a Hungarian, and naturalized USA citizen is one of the most knowledgeable taxonomists and works in the Oakes-Ames orchid herbarium at Harvard University. The five species genus, of which two are questionable, are attributable to Brazil and to the Organ Mountain Range. Try as we might we have found only one with various colour variations.

Pabstia jugosa
(Lindl.) Garay | *1736* (Plate 27)

Etymology: *Jugus* is Latin for paired, presumably referring to the paired pollinia.

2.0 cm

Habitat: Found in the second layer of original forest, on 12-20 cm. wide boughs or on trunks of large forest trees, rarely above 3 metres from the ground, very rarely on the ground and then only on steep slopes showing that some air movement around the roots and good drainage are important. It also inhabits sunless moss-covered rock faces and rock walls of mountain streams in deep forest. In such a dripping damp environment, this plant requires little light, and the habitat variation is minimal, in sections II, IV, VI and VIII at 900-1500 M.asl.

Occurrence: Common in its niche.

Plant: A medium-sized epiphyte with clustered pseudobulbs, each a green translucent ovoid, 6 cm. long, and 2.7 cm. wide and viable for many years. Clusters with over 20 pseudobulbs are common, generally bearing 20 leaves. On a new shoot, these are long, with five ribs, up to 45 cm. in length and 7 cm. broad at the widest part, a dark green and Roman broad-sword shape. There are four leaves to a new pseudobulb; two apical and two basal. The apical leaves last over three years and are thin, paper-like, supple and all but the new leaves soon have a high percentage of their surface covered by lichens and mosses. The roots branch and are almost always moss-covered and green, not especially adherent, and short. They average four per pseudobulb, each about 13 cm. long giving about 10 metres of roots to a 20 pseudobulb plant.

Inflorescence and Flowers: An erect inflorescence, 20-25 cm. tall arises from young growths at the new pseudobulb base. Each has between one and six flowers. The most flowers to one plant that we have found is

18, from 6 inflorescences, and these inflorescence-bearing shoots will form the new pseudobulbs. The sepals are ± equal, a broad, vaguely pointed oval, 2.9 cm. by 1 cm. wide and white; the petals are the same size, a slightly pointed spathulate, while the four-lobed lip is almost rectangular, 2 cm. long by 0.9 cm. wide. The petals and lip are speckled and stained with purple.

Flowering Periods: From July through to December and the flowers last for around three weeks. However, late blooming plants often abort flowers due to over soaking by the spring rains.

Pollination: This is frequent, up to 20% and seedlings are invariably found within 30 metres of a mature plant; pollinator unknown.

Cultivation: Follow faithfully the habitat description, if you can. If you cannot, don't waste your time.

General: This is one of very few epiphytic orchids which will grow in this habitat, where it is common, but it is never found outside this environment; the flowers appear as striking jewels in the gloom.

The genus
Zygopetalum Hooker

Hooker raised this genus and used the Greek compound word *Zygo-petalum* meaning joined petals or petals yoked together which doesn't mean too much, nor is it particularly distinctive. This is a small Brazilian genus of ten species, two of which are epiphytic, the remainder terrestrial. The Organ Mountain Range is host to eight species according to **Pabst&Dungs**. The majority, six of these species are essentially from the high mountain fields, ridge, scrub montane forest and high artificial situations such as constantly cleared roadside banks. Some of these species are in the throes of natural hybridisation and introgression which accounts for the extraordinary variability of flower colour and size. The species that are terrestrial and are exposed to full sunlight do not lose their flowers after fertilization. On the contrary the tepals and lip have chloroplasts which become active to help in the photosynthesis process. Just like their deep forest-living cousins *Neogardneria murrayana*, *Promenaea xanthina* and *Pabstia jugosa*. Thus we assume that the high mountain field species used to be forest dwellers, aeons ago, but did not lose their curious capacity when they migrated to a full light situation.

An interesting point is that it seems that New Zealanders are this genus's greatest current fans.

Zygopetalum brachypetalum Lindl. | *1741* (Plate 27)

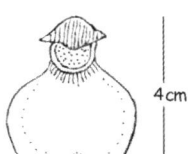

Etymology: *Brachy* is Greek for short and petalum is Latin for petals, referring to the shorter petals shown by this species compared to others.

Habitat: From high mountain fields and mountain ridges, interspersed with, but not as common as, *Z. mackayi*. It appreciates strong light with some shade, a humid rooting base and light winds. It is rarely found below 1000 M.asl, but present in all sections.

Occurrence: Common.

Plant: It is very similar to *Z. mackayi*; a large terrestrial with up to 15 tightly clustered pseudobulbs. These are a yellowish-green, arising from a stout rhizome, conical, 6 cm. long by 4 cm. diameter. The leaves are apical, up to eight, light green, lanceolate with a central keel and eight less pronounced parallel ribs, and up to 40 cm. long by 4 cm. wide. Three leaves remain after the bulb is fully formed. The root system is prolific, consisting of thick, white, brittle, burying roots which penetrate deeply into the detritus around.

Inflorescence and Flowers: A thick, fleshy, racemose inflorescence up to 60 cm. long, higher than the leaf tip, arises from the base of the new pseudobulb. Six or seven large showy flowers are carried on the apical 20 cm. with sepals and petals of similar size, each a pointed oval, 3.5 cm. long by 1.5 cm. Overall, they are an uninterrupted reddish brown. The lip is roundish, large, flared and bilobed, 4.0 cm. long by 3.9 cm. and white with dense purple veining.

Flowering Periods: From April through to August and the flowers last for up to a month.

Pollination: **Dressler** suggests *Euglossine* bees; however we have used scented baits at 1400 M.asl during flowering periods but no such bees appeared in late autumn early winter and we surmised that this altitude was outside their range at this season. However pollination is frequent, over 20%, like *Z. mackayi* and this plant blooms at the same time and place and as both species' flowers are very varied, though similar, we feel that many plants must be natural hybrids. **Rodrigo Singer M.Sc.** Unicamp suggests carpenter bees through deceit and also self-pollination – personal communication.

Cultivation: In large pots on a leaf mould base mixed with 20% coarse sand. Keep the substrate damp but never waterlogged and in good light and ventilation.

Zygopetalum crinitum
Lodd. | *1742 and 1742a* (Plate 27)

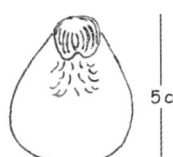

Etymology: *Crinitus* is Latin for having tufts of long weak hairs, referring to the lines of hairs on the lip.

Habitat: Normally found terrestrially in dense leaf litter on exposed mountain ridges or in regenerating forest, but it also explores the mid-tree region, in detritus-collecting large forks in original forest from 1000-1500 M.asl in section VI. It requires shade, high humidity and slight air movement.

Occurrence: Common.

Plant: A medium to large, generally solitary, terrestrial and sometimes epiphytic plant. The pseudobulbs have brown leaf bases remaining at the apex and as side sheaths. The rhizome is thick and short. The leaves, usually five, two apical and 2 + 1 subtending the pseudobulb, are retained until the third year when they are shed. Apical leaves are 25-30 cm. long and 4-4.5 cm. at the widest point, pale green with a midrib and parallel fainter ribs. They are broad and papery, tapering to a broad tip and narrowing at the attachment to the pseudobulb. The three sheathing leaves are successively shorter. The roots are 0.5 cm. thick, white when young, and brittle, stretching for 50 cm. in all directions from the colony but, because of frequent branching the total length is substantially increased. About 30 roots on the most recent two bulbs give a total of over 22 metres to a 10 bulb colony.

Inflorescence and Flowers: The inflorescence is always shorter than the leaf height. There are up to three, thick, light green, 25 cm. racemose inflorescences arising within the new growths' subtending leaves from the new pseudobulb base. They have 6 or 7 large, showy, sweetly-scented flowers. The petals and sepals are convoluted, roughly equal in size, 4 cm. long by 2 cm. wide. The lip is almost round, 5 cm. long by 4.5 cm. broad, with a crisped undulate margin, pure white but closely veined in red/purple, and has the distinguishing feature of lines of fine, similarly coloured, 1 mm. high hairs, concentrated toward the throat. There is a horseshoe-shaped callus with a deep central cleft.

Flowering Periods: The main emphasis is between December and January but plants in flower can be found in most months and the flowers last for two weeks.

Pollination: Unknown but size and scent would indicate large bees or night flying *Lepidoptera*.

Cultivation: In large pots on a rich leaf mould compost, always damp but never waterlogged, grown in shade with some air movement.

General: A magnificent plant, arriving in European cultivation in 1829 but now not so frequently grown, except oddly, in New Zealand.

Zygopetalum intermedium
Lodd. | *1744* (Plate 27)

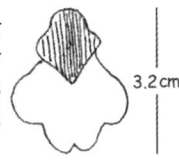

Etymology: *Intermedius* is Latin for between two things, possibly suggesting that this plant falls between two other *Zygopetalum* species.

Habitat: Found above 1200 M.asl in section VI on ridges and mountain tops, as a terrestrial, enjoying shade, humid conditions and some air movement.

Occurrence: Common.

Plant: This is another of the high mountain field *Zygopetalum* species that are vegetatively so similar one to the other. It is generally solitary, like **Z. crinitum**, rather than in huge groups like **Z. mackayi**. It is a large terrestrial clusterer, with up to 20 conical yellow-green pseudobulbs, 6 cm. long by 4 cm. broad. The new shoot has up to 5 leaves of which generally 3 remain at the apex after the bulb is formed. Leaves are sword-shaped, plicate with a strong keel, dark green and up to 50 cm. long by 5 cm. wide. The root system is profuse, consisting of masses of 3 mm. thick, white, brittle long roots penetrating the decaying leaf detritus around for a considerable distance.

Inflorescence and Flowers: A long thick, fleshy, racemose inflorescence up to 40 cm. high is produced in April and through to June. It arises from the incipient bulb and has up to 6 flowers at the height of, or just higher, than the leaf tips; the flowers have similar sepals and petals, 3 cm. long by 2 cm. wide and are a basic light green with claret coloured streaks and blotches. The lip is round and flared, 3.2 cm. long and 3 cm. wide, white with purple veining and just a hint of purple hairs at the throat. The callus is horseshoe-shaped with no central cleft.

Flowering Periods: April through to June and the flowers last for up to a month.

Pollination: Unknown, but the size and scent would indicate large bees or night-flying *Lepidoptera*.

Cultivation: In large pots on a rich leaf mould compost, always damp but never waterlogged, in shade with some air movement.

Zygopetalum mackayi
Hooker | *1745* (Plate 27)

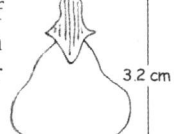

Etymology: **Mackay** was the curator of the Dublin – Ireland, Botanic Garden and **Hooker** named this plant after him.

Habitat: The colony described, like many others, was growing on exposed, cleared ground with little leaf litter and the root network is spread into the soil horizontally. This is a plant of the high mountain fields and ridges from all sections at 900-1600 M.asl. where it can rapidly populate cleared roadside banks or regenerating burnt forest, if conditions are similar to its original habitat.

Occurrence: Common.

Plant: A large tufted terrestrial plant found in large swarms of plants with up to 25 pseudobulbs. Older pseudobulbs are half buried, 8 cm. x 6 cm. diameter and rounded egg-shaped. All arise tightly from a thick branching rhizome. There are up to seven leaves per shoot. The three apical leaves are long-lived and large, plicate and up to 50 cm. x 6 cm. while the basal group are short-lived and smaller and all are an olive green.

Inflorescence and Flowers: A thick green inflorescence, arises from the outermost leaf but one from the young growing shoot and bears 8 to 10 flowers. It measures 75 cm. long, higher than the leaves, 53 cm. of which is flowerless. The sepals are similar in shape, 3.5 cm. x 1.5 cm. wide, and are green with maroon dots and streaks which coalesce towards the tip. The petals are thinner, 3 cm. x 1 cm. The lip is broadly flared, hairless and 3.2 cm. high by 2.4 cm. wide with 2 large lobes and mauve-blue lines following the lip contours. The flowers smell of pepper. The callus is horseshoe-shaped, stepped at the edge and has no cleft. The colour variations of the lip range from almost pure white to a deep reddish mauve.

Flowering Periods: From April to August the flowers last for up to a month.

Pollination: As for *Z. brachypetalum*.

Cultivation: As for *Z. brachypetalum*.

Zygopetalum maxillare
Lodd. | *1746* (Plate 27)

Etymology: Presumably **Lodd.** thought that the flowers were similar to those of the genus *Maxillaria*. *Maxilla* is Latin for a jaw-bone, and relates to the incurved lateral sepals, rather like an ant's mouth.

Habitat: Found from 600-1500 M.asl through out the Organ Mountain Range, but only on two species of tree fern, principally **Dicksonia sellowiana** and requiring low light, reasonable humidity and low air movement. It seems that its seeds will only germinate on these two ferns.

Occurrence: Almost as common as its tree fern host.

Plant: This is a quite distinctive plant differing from all other *Zygopetalum* species by its dimorphic growth habit and its intimate association with two tree-fern species. Pseudobulbs are formed at up to 30 cm. intervals on an 0.8 cm. thick, green, ringed rhizome during the first growth stage, which generally starts at mid-trunk on the tree fern as the plant rushes upwards to the fern's crown. Here it becomes a clusterer, all around the crown. The pseudobulbs are light green, 9 cm. long, 3 cm. wide and a compressed ovate. The three apical leaves, 45 cm. long by 6 cm. wide, and two slightly shorter basal leaves are light green, lanceolate with a distinct central keel and 10 parallel veins, and retained for several years. The thick white roots, arising from the rhizome, dig into the tree-fern.

Inflorescence and Flowers: There are up to 7 alternate flowers on a fleshy, light green, 30 cm. inflorescence produced from the pseudobulb base. A mature colony may have over 80 flowers on 15 inflorescences around the tree-fern crown; the flowers are absolutely distinctive because of the paramount effect of the deep purple, horseshoe-shaped lip, callus and throat. The sepals and petals are similar in shape and size, a broad lanceolate 2.8 cm. by 1.2 cm. broad, green, speckled and blotched with light brown, while the lip is purple and white, broad, flared and almost round measuring 2.6 cm. by 2.5 cm.

Flowering Periods: February and March but occasionally extending through May and the flowers last for three weeks.

Pollination: Is frequent, and up to 20% of all flowers bear fruit.

Cultivation: In pots on a rough fibrous medium. Flask-grown plants have no problems in adapting to the growers preferred medium. Grow in shade and keep constantly damp, but never overly so and with light air movement.

Zygopetalum pedicellatum (Thunb.) Garay | 1748 (Plate 27)

Etymology: *Pedicellatum* is Latin for stalk, referring to the plant's long stem.

Habitat: A fairly catholic terrestrial, found from 1200-1500 M.asl in section VI on mountain ridges, at forest edges or in regenerating forest where there is plenty of light and the ground cover retains humidity.

Occurrence: Common.

Plant: The unusual growth form, essentially monopodial, separates this from other species of *Zygopetalum*. Growth is apical and continuous, producing a long rhizome which lies under the leaf litter, curling gently up through shrubs. Live rhizome lengths of over 2 metres have been found and these may branch giving green side shoots which in turn develop their own rhizome, which is 0.6 cm. diameter and thickly sheathed with old leaf bases. The youngest 50 cm. of continuous growth bears green leaves, the largest measuring 30 cm. x 3.5 cm. with a deep central keel and 8 finer veins. The thick roots, 0.5 cm. diameter, are channelled backwards along the rhizome and are contained within the old papery leaf bases for over 1 metre of rhizome length. About 4 white roots are produced per node and about 4 metres of root per metre of rhizome which is fixed very firmly in the substrate.

Inflorescence and Flowers: The inflorescences are 25-40 cm. long, and arise from leaf nodes 5-7 cm. from the apex of the green shoot and 4-5 flowers are produced on each inflorescence with a distinct peppery smell. Each flower measures 4.5 cm. across and in height. The sepals and petals are green with deep maroon blotches, 1.8 cm. long by 1.0 cm. broad. The lip is 2.5 cm. long, vaguely four-lobed, triangular and white with fine radiating purple lines and a distinct sharp basal callus.

Flowering Periods: January through March and the flowers last for three weeks.

Pollination: **Dressler** suggests *Euglossine* bees and the summer flowering period would indicate their presence. Pollination is up to 20%.

Cultivation: In large pots on a substrate of rough leaf mould in good dappled light with some air movement, allowing space and support to accommodate its growth habit.

General: It is curious that a distinctive plant has so many synonyms - *Z. mosenianum* Barb. Rodr., *Z. caulescens* Rolfe and *Cymbidium pedicellatum* Thunb.

The genus Neogardneria Schltr.

It is almost infuriating that **George Gardner**, who travelled though Brazil in the 1830's, wrote a book on his travels, which is a classic, sent thousands of plants to **Hooker** in Kew Gardens for classification, yet as a prize had only one orchid genus named in his honour and that is *Neogardneria*. Probably one of the most literate and scientific of all the collectors, he spent considerable time in the region around Teresópolis, which is the emotional heart of the Organ Mountain Range.

A one-species, Organ Range genus which reminds us of him.

Neogardneria murrayana (Gardn. ex Hook.) Schltr. | 1751 (Plate 28)

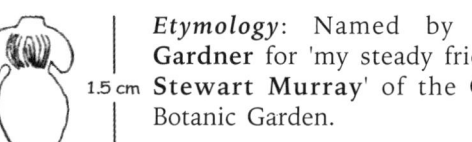

Etymology: Named by **George Gardner** for 'my steady friend, **Mr. Stewart Murray**' of the Glasgow Botanic Garden.

Habitat: A denizen of the deep forest, and of mountain ridges from 1200-1600 M.asl., the plant likes neither much wind, light nor excessive humidity. Unlike *Pabstia jugosa*, it adapts to other environments with some success. The trunks of old trees on dry or well-drained ridges are favourite spots in section VI.

Occurrence: Occasional.

Plant: Easily confused with *Pabstia jugosa* or a small *Zygopetalum crinitum* when not in flower, this species is a long-lived epiphyte, rarely found more than 3 metres above ground level, commonly with clusters of up to 18 pseudobulbs. These are green, ovoid, 8 cm. in height and 4 cm. in width. It has a strong system of white, thin, medium-adherent roots and there are about 5 to a pseudobulb, short and averaging 20 cm. in length. A 15 pseudobulb plant showed 10 metres of roots. The leaves are long, green, thin, paper-like and shiny with five parallel ribs, a roman sword-shape from 25 to 46 cm. long and 3 to 6 cm wide. There are four or five long-lived leaves to a new pseudobulb, two apical and two basal.

Inflorescence and Flowers: Two to six-flowered spikes, 15 cm. high, arise from the young shoot base. Flower colour ranges from medium green to light yellow with red/purple freckles on the labellum and there is a faint smell of apple. Sepals and petals are a broad lanceolate, ± equal and measure 2 cm. by 1 cm. broad. The lip is trilobed. The two narrow, small, sickle-like side-lobes surround a veined, circular callus at the throat while the central apical lobe is a broad lanceolate measuring 1.5 cm. by 0.7 cm broad.

Flowering Periods: December through to March and the flowers last for three weeks.

Pollination: **Dressler** suggests *Euglossine* bees. Fruit set is frequent, over 30%.

Cultivation: In a large pot on a coarse fibrous compost, in shade with some air movement, always kept damp but never excessively so.

General: Another of **George Gardner**'s contributions to European cultivation, found in the Organ Mountain Range in the 1830's.

The genus
Warrea Lindl.

John Lindley raised this genus in honour of **Frederick Warre** who collected this terrestrial plant in Brazil in 1829. It is a genus of seven species ranging from Guatemala to Brazil and Chile.

Warrea warreana
(Lodd. ex Lindl.) | *1776*
C. Schweinf (Plate 28)

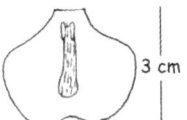

Etymology: Named after **F. Warre**, a collector who sent plants to **John Lindley** from Brazil in 1829.

Habitat: In swampy areas beneath original forest, usually associated with a sub forest of giant bamboo – *Dendrocalamus giganteus*, in sections IV, V and VI., at 450-1000 M.asl, mostly at 700-800 M.asl.

Occurrence: Occasional.

Plant: A large terrestrial, which we suspect is associated with the mycorrhizal fungi around the roots of *Dendrocalamus giganteus*. There are 3 or 4 leaves, 60 cm. x 10 cm., lanceolate, intensely plicate on a long, 10 cm., false petiole arising on an incipient, long, narrow, three-noded, pseudobulb. This is held closely on a thick underground rhizome. The root system is strong and extensive and composed of thick, white, succulent roots penetrating the decaying leaf substrata for 50 cm. around the plant.

Inflorescence and Flowers: A 1 metre tall, round, erect, thick, noded and spicate inflorescence arises from a basal bract of the new pseudobulb at ground level, with 10 to 12 large-sized white flowers with a wide, orange-throated, purple lip. One of its many synonyms is *W. tricolor*. The dorsal sepal is ovoid, 2.7 cm. long by 1.6 cm. wide. The laterals are half as big again. The petals

are similar in shape and size to the dorsal sepal. The bi-lobed lip is almost round, purple, with an orange throat and a three-ridged callus, 3 cm. by 3 cm. broad.

Flowering Periods: In February and March and the flowers last for a week or so but the inflorescence may have flowers in bloom for over a month.

Pollination: Infrequent. **Dressler** suggests *Euglossine* bees. The flowering period and the altitude would permit this.

Cultivation: In large pots on a compost of leaf mould intermixed with coarse fibrous material and grown in shade with light air movement, always damp but never waterlogged.

The genus
Huntleya Lindl.

This is now represented by a single Brazilian species. In fact while in Brazil a whole stack of varieties has been reduced to a single species, in Venezuela, Colombia and in Central America, at least six and possibly ten distinct species have been described. If you put your money on **Jack Fowlie**, Orchid Digest Vol., 3. May-June 1974, there are even more. The genus name is dedicated to the **Rev J. T. Huntley** an enthusiastic grower of orchids in England during the 19th century.

Huntleya meleagris
Lindl. | *1782* (Plate 28)

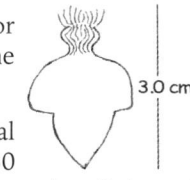

Etymology: *Mel* and *gris* are Latin for honey and pearl-grey, referring to the flower's colour.

Habitat: An epiphyte found in original under forest in section VI at 450-550 M.asl, with intensely humid conditions, low light and low air movement.

Occurrence: It is possibly more common in its specific environment than we think. At first sight it is also difficult to distinguish from the mass of bromeliads found in the same conditions.

Plant: An exuberant, pseudobulbless, medium-sized, fan-like epiphyte. There are up to six leaves on each growth arranged like a fan and measuring up to 40 cm. long by 3.5 cm., long lasting, light green, linear-lanceolate, paper-like but conduplicate. The new growths arise at intervals of up to 6 cm. from the 0.75 cm. thick, light green rhizome. A mature plant may have ten such fan-like growths. The root system is intense and composed of dozens of greenish-white, 0.2 cm. diameter thickish roots which originate from the base of the new growth and are active for many years.

They are extremely adherent and hug the substrate for some distance around.

Inflorescence and Flowers: A single flowered basal inflorescence arises in January from the base of the most recent growth. Each is up to 12 cm. long and bears an extraordinary single, large and attractive flower. The sepals are an elliptic-lanceolate and measure 3.0 cm. long by 2.0 cm. The sepals are slightly smaller and measure 2.5 cm. long by 1.5 cm., but are essentially the same shape. All tepals are acuminate and an irregular barred light brown on a butter-yellow base. The curious lip is trilobed; the apical central lobe is heart-shaped, cream-coloured towards the throat and orange-brown on the apical third. There is a tufted callus at the throat and the lip measures 3.0 cm. long by 2.0 cm. at the apical lobe.

Flowering Periods: November through to March and the flowers last for about a week.

Pollination: Probably pollinated by *Euglossine* bees.

Cultivation: We have not grown this species, but in natural conditions, the plant is curious. Its coiled habit resembles *Zygopetalum maxillare*, thriving about 2 metres above the forest floor on thin under forest saplings. We suspect that its nutrients are won from leaves and twigs that fall into the exuberant fans of foliage.

The genus *Cochleanthes* Raf.

A small Central and South American genus composed of nine species, one of which is attributed to the Organ Mountain Range. **Constantine Rafinesque**, a Turkish botanist, named it from the Greek words *kochlias* and *anthos* - flower shaped like a snail shell.

Cochleanthes candida
(Lindl.) Schultes & Garay | *1784*
(Plate 28)

Etymology: *Candida* is Latin for pure white which refers to the colour of the tepals, but ignores the lovely blue blotch on the lip.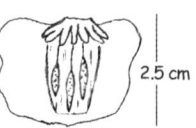

2.5 cm

Habitat: Epiphytic and lithophytic in under forest or older regrowth forest on saplings up to five metres above the forest floor. Found at 500-600 M.asl, in sections V and VII enjoying low levels of dappled light, low wind movement and good humidity.

Occurrence: Locally frequent.

Plant: Vegetatively indistinguishable from *Huntleya*

meleagris, lacking pseudobulbs and with a fan-like group of leaves, generally five, which emerge from a thick, woody and wandering rhizome at 3 cm. intervals. The leaves are long-lived, a narrow lanceolate, with a prominent central keel, light green and with a 6 cm. false petiole, measuring up to 35 cm. long by 2.8 cm. A mature plant can be large and clustered with up to 30 such fans. The root system is prodigious and composed of countless tough, thick, short, white roots which wander through the plant while some fix it tightly to the substrate. The plant looks like a bromeliad, and like them, it collects detritus within the leaf fans thus gaining nutrients

Inflorescence and Flowers: Single flowers, borne on 2 cm. diameter round-sectioned inflorescences up to 10 cm. long are produced from the bases of the outer leaf bracts on the new growths. As the new growths are precocious, up to ten flowers can be seen on a mature plant. The sepals and petals are more or less equal, broadly lanceolate and measure 2.6 cm. long by 3 cm. wide and are milk-white. The single-lobed lip is a broad rounded spade-shape, 2.6 cm. long by 3 cm. broad at the shoulders. A blue swathe, 0.7 cm. wide runs from the base to the tip while a five veined blue claw-like landing pad sits up from the throat. The rest of the lip is milk-white while the slightly blue-veined shoulders embrace the column. The flower has a distinctive sweet scent.

Flowering Periods: The flowers last for a few days during April but the plant can produce flowers for three weeks or so.

Pollination: **Dressler** suggests that pollination is by *Euglossine* bees.

Cultivation: According to **Hoehne**, this plant is easy to cultivate in baskets or pots in a shady, cool environment with a substrate of chopped tree fern fibre and added decaying leaves. In its natural habitat it receives heavy rain in spring and summer until autumn when it flowers. The winter climate is dry, cool and misty.

The genus
Maxillaria Ruiz & Pavón

These were two Spanish botanists who in the 1790s went looking for medicinal plants in the Central Americas and Mexico. They named this genus *Maxillaria* since the lateral sepals curve around, resembling the maxillae or jaws of a soldier ant.

This new world genus is composed of over 300 species stretching through the American tropics to Argentina. Like the Laelias there is very little in common between the Mexican species, those from north of the Amazon, with those in Southern Brazil, principally in the Atlantic rain forest zone of influence.

Pabst&Dungs attribute some 140 species to Brazil, of which ± 40 pertain to the Atlantic rain forest and of these, 40 can be legitimately assigned to the Organ Mountain Range. We have discovered 43 distinct species of which at least three are possibly new. Obviously there are more to find, invaders from dryer climes.

Maxillaria desvauxiana
Rchb. f. | *1787*

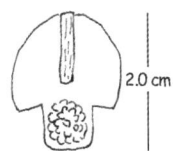

Etymology: Named after **Étienne-Émile Desvaux**, (1830-54).

Habitat: Epiphytic between 300-600 M.asl, in original forest, growing at mid-tree level and enjoying constant high humidity with medium wind movement and light, in section IV.

Occurrence: Occasional in the foothills of the Organ Mountain Range but it is also found from Colombia to the southern Brazilian state of Santa Catarina but not in the hot lowland forests of the Orinoco and Amazon systems.

Plant: A large, distinctive epiphyte with ovate, laterally compressed pseudobulbs borne in a tight cluster and which arise from a short, thick, wandering rhizome. They measure 3.5 cm. tall by 2.4 cm. wide and are dark green and generally unwrinkled. There is a single apical, long-lasting leaf which is borne on a stiff, dark green, oval-sectioned 15 cm. petiole. The leaf is oval or broadly lanceolate and measures up to 34 cm. long by 5 cm. and is a stiffish papery texture, firmly keeled and a shiny dark green. A mature plant may bear up to 12 such leaves. The root system is extensive, although not as strong as one might imagine for such a top-heavy plant, and is composed of a mass of thin, white to brown, branching roots which do not extend much beyond the plant's circumference.

Inflorescence and Flowers: The single flowers are about the height of the pseudobulbs and are borne on 2 cm. long, heavily-sheathed inflorescences from the pseudobulb bases. Each pseudobulb bears one or two salmon-pink flowers. The sepals are more or less equal in size and shape, a broad lanceolate, 2.4 cm. long and 1 cm. wide and lying parallel to the column. The lip is distinctive, trilobed and measures 2 cm. long by 1.5 cm. wide. The central lobe is covered by dark purple shiny warts for most of its length and breadth with a typical *Maxillaria* nose-like callus for half of the basal section.

Flowering Periods: March and the flowers last for a week or so.

Pollination: Probably pollinated by stingless bees.

Cultivation: This plant looks good in a hanging basket in a well drained compost with added peat or leaf mould for the fine roots to spread into. Allow to rest in a bright place after the pseudobulbs are mature to ripen them prior to flowering.

Maxillaria leucaimata
Barb. Rodr. | *1791* (Plate 28)

Etymology: *Leuco* is Greek for pale white, referring to the colour of the flower.

Habitat: This plant is found only in extremely humid conditions, normally hanging on creepers over a river or 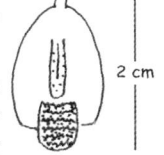 stream, around a waterfall, in diffuse light but with reasonable air movement and between 1000-1200 M.asl in original forest in section VI.

Occurrence: Common in its niche.

Plant: An exuberant, clustered epiphyte with small pseudobulbs and long leaves. The pseudobulbs are a flattened oval, light green, 3 cm. long by 2 cm. broad, and tightly clustered on a fine stiff rhizome. The leaves are dark green, deeply keeled, long and pointed. The single erect to pendulous apical leaf is 35 cm. long by 3 cm. and there are two small sheathing leaves. The roots are profuse, reminiscent of *Miltonia cuneata*, forming a veritable bird's-nest of thin, wiry but not very adherent roots.

Inflorescence and Flowers: Many flowers are produced on single fleshy, round, sheathed, 10 cm. inflorescences arising from the bulb base. The overall effect is quite dramatic in the dark under-forest and resembles a white star surrounding a splotch of purple. The sepals are broad at the base, narrowing and pointed at the apex, 3 cm. long by 0.8 cm. broad. The dorsal sepal is slightly smaller. The petals are 2.4 cm. long by

0.5 cm. wide and are more acutely pointed. All are a translucent milky white and the lateral sepals curving down and backwards, strongly reflect light. The lip is slightly trilobed and oval, 2 cm. long by 0.8 cm. wide. The side-lobes are a deep purple and the central lobe is white, rounded at the serrated apex and covered by a yellowish waxy farina.

Flowering Periods: In November and December and the flowers last for up to three weeks.

Pollination: This is frequent, up to 40%. **Dressler** suggests wax-gathering bees.

Cultivation: In baskets on a coarse leaf mould compost in dappled light, high humidity and some air movement; always damp, but never waterlogged.

Maxillaria modesta
Brade | *1792* (Plate 28)

Etymology: *Modestus* refers to the modest, almost shy aspect of the flower.

1.3 cm

Habitat: Found on low to mid-tree, on moss-covered trunks, with intense humidity, low light and air movement. It is found from 1000-1500 M.asl in such a niche. The plant is sporadic in original forest and also in old regrowth but forms significant colonies where conditions are right in section VI.

Occurrence: Occasional.

Plant: A medium-sized, mid-tree epiphyte, with small pseudobulbs, each a compressed oval, 1.8 cm. wide x 1.5 cm. tall and light green. Initially there are two leaves, one apical and one basal, both petiolate, lanceolate and 15 cm. long by 2 cm. wide, dark green and with a deep central vein. The basal leaf dies in its second year. The root system is weak and composed of very thin, wiry, mildly adherent roots which spread up and down a tree trunk, under moss for some 15 cm. in each direction.

Inflorescence and Flowers: The flowers are sparse, only one or two, borne on sheathed basal inflorescences, up to 9 cm. long. The predominant characteristic is a very typical *Maxillaria*-type incurved flower, an overall light pink, running to ivory at the lip base. The sepals are pointed, acute, 3 cm. long by 0.8 cm. wide, and the lip is trilobed with a distinctive waxy nose-like callus running for half its length.

Flowering Periods: September through March and the flowers last for up to three weeks.

Pollination: Seed set is frequent and often hundreds of small plants can be found on the trunk below the parent plant. **Dressler** suggests wax-gathering bees.

Cultivation: Mounted or in small pots on a coarse fibrous compost in shade, low wind movement and always damp.

Maxillaria caparaoensis
Brade | *1798* (Plate 28)

Etymology: Caparão is a mountain on the borders of Minas Gerais and Espírito Santo, where presumably **Brade**'s plant came from.

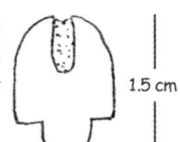

1.5 cm

Habitat: An epiphyte found growing at 800-1000 M.asl, in section VI on low trunks in scrub forest. The plants are reasonably lit but with no direct sunlight, nor excessive humidity.

Occurrence: Rare in section VI but probably more common elsewhere.

Plant: A distinctive, medium-sized epiphyte resembling **M. modesta** and **M. rufescens** but twice the size of both. The leaves are sessile, single and apical, long-lived and measure up to 35 cm. long by 4 cm., an elongated lanceolate, deeply keeled and leathery. The pseudobulbs are tightly clustered on a short rhizome, laterally compressed, oval, normally unwrinkled and measure 4 cm. high by 3 cm and are sheathed by hessian-like bracts. The root system is strong and composed of a myriad of thin, white, branching roots which rapidly penetrate the substrate and are extremely adherent making the plant difficult to move and transplant.

Inflorescence and Flowers: One or two single 10 cm., sheathed inflorescences, 0.3 cm diameter, arise from the new pseudobulb base and open into attractive pink and ivory flowers. The sepals are more or less equal, fleshy and an elongated linear shape with a blunt apex and measuring 2.7 cm. long by 0.7 cm wide. The petals are similar but narrower, 2.3 cm. long by 0.5 cm. and are mildly acuminate at the tips. The lip is proportionately smaller, trilobed, 1.6 cm. long by 0.7 cm. wide and with the typical Maxillaria nose-like longitudinal callus extending from the lip base to over half the length of the central lobe.

Flowering Periods: April and May and the flowers last for a week or so.

Pollination: We have observed small black stingless bees removing pollinia from newly opened flowers, which they hold on the upper abdomen.

Cultivation: In pots or hanging baskets using a firm, fibrous and well-aerated substrate and giving the mature pseudobulbs good light to induce flowering.

Maxillaria cleistogama
Brieger & Bicalho | *1799* (Plate 28)

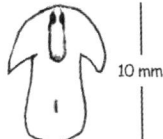

Etymology: *Cleistogama* is Latinised Greek for fertilized within the unopened flower, and referring to the facts that: 1). This flower does not open; and 2). It is always fertilized.

Habitat: An epiphyte from sections IV and VI, found in humid conditions on low tree trunks in original forest at altitudes from 500-1300 M.asl.

Occurrence: Occasional throughout its range. However, as this plant is indistinguishable from **M. rufescens** when not in flower, it may be more common than we imagine.

Plant: A small to medium-sized clustered epiphyte. The pseudobulbs are aggregated into a tight group and each is a bilaterally compressed, ridged, bright green, ovoid, 4 cm. long by 2 cm. wide. They arise from a thick, branching rhizome and have a single, apical, long-lived leaf which is lanceolate, stiff, dark green, deeply keeled and measures 15 cm. long by 3 cm. wide. The root system is profuse, strong and composed of long, branching, adherent roots of medium size and initially wine-coloured.

Inflorescence and Flowers: The buds appear in January and February and occasionally in other months. They are single and borne on short, round, green, partly-sheathed pendent inflorescences up to 2 cm. long and arising from the base of the previous year's pseudobulb. The reddish-brown bud does not open, and within a few days the ovary begins to swell, the bud remaining with the fruit which matures in about six months. When a 'mature' bud is dissected, it contains a perfect *Maxillaria*-type flower; in fact a half-sized version of **M. rufescens**. The dorsal sepal is ovate, obtuse at the tip and measures 1.3 cm. by 0.6 cm. The lateral sepals and petals are similar in shape but slightly smaller and the petals are narrower and vaguely asymmetrical. The trilobed lip, shaped like an upended anchor, measures 1 cm. long by 0.8 cm. wide at the side-lobes and the central lobe has a typical *Maxillaria* nose-like waxy callus for half its length The overall flower colour is russet-brown with the lip speckled in dark wine.

Flowering Periods: January and February although sporadic flowering occurs through the year.

Pollination: Self-pollinated.

Cultivation: Grows well on fibrous compost or rough bark plaques with dappled light, constant high humidity and medium air movement.

Maxillaria monantha
Barb. Rodr. | *1801*

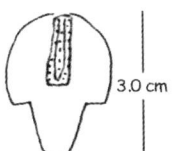

Etymology: *Monantha* is Latin for one flower, referring to the paucity of single flowers shown by this plant.

Habitat: Found at 700-800 M.asl in section VIII in the mid to upper branches on 30-40 metre tall original forest trees, requiring good dappled light with underlying humidity and serious air movement. Obviously, this plant is only found when a forest giant falls.

Occurrence: Probably locally frequent in its specific original habitat.

Plant: A small, tightly clustered, pseudobulbous epiphyte. The single leaves are apical, sessile long-lived and held on a short 3 cm. long false petiole, linear lanceolate, leathery, mildly keeled, and up to 20 cm. long by 2 cm. wide and dark green. They arise from the apex of distinctly ridged, ovate, slightly bilaterally-compressed, dark green pseudobulbs which measure 3 cm. long by 2.5 cm. and are on a very short thickish rhizome. The root system is profuse and composed of long wiry branching roots, brick-red when young, turning to a dirty white as they mature. They both penetrate and adhere to a healthy substrate and are viable for many years, arising from the pseudobulb base.

Inflorescence and Flowers: One, occasionally two, single, strongly scaped inflorescences emerge from the most recently matured pseudobulb base. The single flower is an overall light yellow. The dorsal sepal is a broad lanceolate up to 3 cm. long by 0.5 cm. The laterals are similar but narrower. The petals are the same length but club-shaped while the three-lobed lip is again the same length but 1.4 cm. broad at the side-lobes and shows the typical *Maxillaria* central nose for slightly more than half its length, with some faint purple blotches.

Flowering Periods: January, and the flowers last for two weeks or so. Seed pods are frequent.

Pollination: Unknown.

Cultivation: On bark plaques, well-drained, with high humidity year around and always with good indirect light and significant air movement.

Maxillaria osmantha
aff. H. Barbosa | *1802*

Etymology: *Os* is Latin for scented and *antha* for flower, referring to the plant having scented flowers.

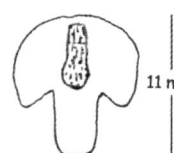

Habitat: Found in section VI in original and regrowth forest at low tree level in good light, high humidity and low wind movement, between 900-1200 M.asl.

Occurrence: Occasional.

Plant: A small epiphyte, sometimes terrestrial, very similar to **M. modesta**, for which it is easily mistaken when not in flower. The pseudobulbs are tightly clustered on a thickish branching rhizome and are compressed ovals, measuring up to 3 cm. by 2.5 cm. with single apical leaves on very short false petioles. They are lanceolate, light green, keeled, slightly stiff and up to 20 cm. by 3 cm. wide and long-lived. The root system is fairly profuse and composed of short white, thin, sometimes branching roots which penetrate and adhere to the substrate.

Inflorescence and Flowers: One to three basal inflorescences arise from the most recent pseudobulbs. These are sheathed, fleshy and may rise up to half the leaf height and show a relatively large single, apical flower. Overall, this is creamy and slightly orange. The sepals are the same size and shape, a broad lanceolate, 2.5 cm. long by 1.4 cm. The petals are spathulate and narrow, 2.3 cm. by 0.5 cm. wide. The trilobed lip is 2.7 cm. long by 1.5 cm. wide at the side-lobes. A nose-like callus runs from the throat for half the lip's length.

Flowering Periods: In September and the flowers last for two weeks or so.

Pollination: **Dressler** suggests probably wax-gathering bees.

Cultivation: As for **M. modesta**.

Maxillaria rufescens
Lindl. | *1803* (Plate 29)

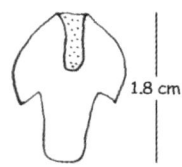

Etymology: *Rufus* is Latin for reddish, referring to the flower's overall colour.

Habitat: Low tree-trunk sites in very humid situations in original forest. Sporadic, at 1000-1500 M.asl in sections VI and VIII.

Occurrence: Common.

Plant: A medium-sized epiphyte with pseudobulbs aggregated in a tight cluster. Each is ovoid, bilaterally compressed, ridged, and 4 cm. long by 2 cm wide.,

arising from a thick branching rhizome. There is a single apical and long-lived leaf, briefly petiolate, lanceolate, dark green and deeply keeled, 15 cm. long by 3 cm. wide. The root system is profuse and strong, composed of medium thickness, wine-coloured, long, branching, adherent roots.

Inflorescence and Flowers: The flowers are single and fleshy on short reddish-brown inflorescences, produced from the most recent pseudobulb base and raising the flower a centimetre or so above the pseudobulb apex. The sepals are a slightly pointed oblong ± equal at 2.8 cm. long by 1.2 cm. wide. The petals are a pointed oval and slightly smaller at 2.1 cm. long by 0.9 cm. wide, and all are a rich pinkish-brown. The trilobed lip is square at the base with two pointed side-lobes, 1.7 cm. long by 1.3 cm. wide, dark brown, speckled with an even darker brown.

Flowering Periods: In January and February and the flowers last for 2-3 weeks.

Pollination: Frequently up to 30% of flowers bear fruit. **Dressler** suggests wax-gathering bees.

Cultivation: As for **M. modesta**.

Maxillaria
(*rufescens* Alliance) | *1803a*
A possible new species (Plate 28)

Etymology: Open.

Habitat: Found in section VI at 900-1000 M.asl, in original, elfin forest at low to mid-tree level where it receives filtered light, good air movement and a fair constant humidity provided by night mists and a ground cover of various water-holding bromeliad species.

Occurrence: Rare.

Plant: A medium-sized to small epiphyte, closely related to **M. rufescens** and **M. cleistogama**. The tightly-clustered pseudobulbs arise alternately from a thick, wandering rhizome and are oval, bilaterally-compressed, faintly ridged and a light greyish-green. They measure 4.5 cm. high by 2.0 cm. and bear a single apical, long-lived, deeply keeled stiff and leathery, lanceolate leaf which is dark green and measures 18 cm. long by 3 cm. The root system is profuse and composed of innumerable, long thin and branching roots that emerge from the base of the new growth. The roots are brick-red where they emerge and turn green as they penetrate the substrate and branch.

Inflorescence and Flowers: One to four, single-flowered semi-pendent inflorescences arise from the bases of new pseudobulbs. Each is borne on a 2 cm., fleshy, sheathed and green pedicel which differentiates this species, superficially, from others in the Alliance. The dorsal sepal is a bluntish-lanceolate measuring 1.4 cm. long by 0.6 cm. The lateral sepals are shorter and asymmetric, measuring 1.25 cm. long by 0.65 cm., while the petals are virtually linear and 1.2 cm. long by 0.4 cm. The lip is trilobed and measures 1.1 cm. long by 0.9 cm. at the side-lobes and has a faintly lined nose-like callus reaching half way down the central lobe. It is greenish-yellow with a few irregular claret blotches close to the margins.

Flowering Periods: The flowers last for just a few days in January.

Pollination: Pollinators are not known.

Cultivation: As for **M. rufescens**.

General: We are not able to identify this plant from the available literature. It differs slightly in sepals and petal size and shape and in the lip silhouette, shape and size from **M. rufescens** and **M. cleistogama**. In addition its basically green colour form sets it apart. It may be a new species or an ecotype of either of the above mentioned species.

Maxillaria consanguinea
Kl. | *1808* (Plate 29)

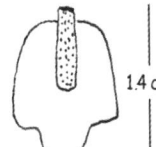

Etymology: *Consanguinea* is Latin for with a blood-red colour which more or less describes the flower colour.

Habitat: Epiphytic or lithophytic at 800 M.asl in sections IV, VI and VII at mid-tree level in exposed conditions in original forest at 1000-1200 M.asl. We have also found it as a massive lithophytic colony on huge boulders in degenerated coffee plantations, enjoying high light, heavy rainfall in summer, good wind movement and a dry winter.

Occurrence: Occasional.

Plant: Similar morphologically to other species in the **M. picta** Alliance – *sensu* **Pabst**. The pseudobulbs are normally more wrinkled, yellow and squatter than its allies, probably because this plant prefers more exposed conditions. These pseudobulbs are tightly packed together and measure 3.5 cm. tall by 3 cm. broad, elongated ovoids that arise alternately from a thick branching rhizome. There are two apical leaves that look exactly like an erect pair of rabbit ears, lanceolate, coriaceous, deeply keeled, long-lived, dark green and measure 17 cm. by 2.5 cm. The root system is strong, composed of dark brick-red roots that emerge from the base of the new growth and remain viable for many years. They branch under the substrate turning white and forming an extensive adherent network.

Inflorescence and Flowers: Depending on the plant's condition, from one to eight single inflorescences emerge from the base of each of the most recently matured pseudobulbs. The flower stem, including the ovary, is 8.5 cm. long, fleshy and partly sheathed. The overall flower colour is a clear reddish-brown showing yellow at the bases of the sepals and petals. The sepals are equal in size and shape, oblanceolate and measure 2.5 cm. by 1 cm. The lanceolate petals are 2.3 cm. long by 0.6 cm. The trilobed lip, when flattened, is 1,4 cm. long by 1.4 cm. at the side-lobes' shoulders. The apical lobe, slightly crisped at the margins, is almost round and white and the side-lobes are a yellowish-white with bold purple streaks. The usual nose-like callus runs from the base to half the lip's length.

Flowering Periods: January and February and the flowers last for around two weeks.

Pollination: Probably bees.

Cultivation: In pots on a rich fibrous substrate mixed with coarse sand, repeating as far as possible the natural conditions described above.

Maxillaria phoenicanthera
Barb. Rodr. | *1810* (Plate 29)

Etymology: *Phoeniceus* is Greek for bright red, scarlet, probably referring to the splotches and speckles on the tepals of the flowers.

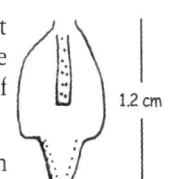

Habitat: Found on the high mountain ridges at 1000-1200 M.asl in section VI on thick horizontal branches in medium shade and air movement with high humidity.

Occurrence: Rare.

Plant: A large mid-tree epiphyte which may easily be confused for **M. ubatubana** or **M. picta** when not in flower. The pseudobulbs are large, khaki to light green and conical although slightly compressed laterally. They measure 6 cm. tall x 3 cm. and have 8 or so well-defined, vertically running ridges. There are two long-lived, dark green, apical leaves, 24 cm. x 2 cm. which are folded around a distinct spine. Pseudobulbs arise alternately at 1 cm. intervals on a long, sometimes branching, thick rhizome. The roots are profuse, white with a reddish tinge, and are of medium thickness, adherent and penetrating.

Inflorescence and Flowers: Up to 12 delicate flowers arise from the base of the mature pseudobulbs. Each is borne on a 9 cm. long, succulent, sheathed inflorescence and is a typical **Maxillaria** flower form, with sepals and petals a uniform yellow, speckled with red spots like the column and lip. The sepals are ± equal in size, lanceolate, 2.5 cm. by 0.6 cm. broad. The petals are linear-lanceolate, 2 cm. long by 0.3 cm. while the trilobed lip is 1.3 cm. long by 0.6 cm. showing a waxy nose-like callus for half its length.

Flowering Periods: Late May and early June when the flowers last for 3 weeks.

Pollination: Infrequent and the pollinator is unknown.

Cultivation: Grow in large pots on a rich fibrous compost, in light shade with good air movement and always kept damp. This is a robust plant and can be neglected.

Maxillaria picta
Hooker | *1811* (Plate 29)

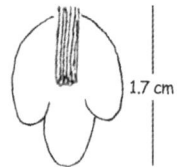

Etymology: *Pictus* is Latin for coloured, painted, referring to the colour of the splotches on the tepals.

Habitat: Mid to upper-tree in original forest from 1100-1500 M.asl in sections II, IV, VI and VIII.

Occurrence: Probably common but the similarity to *M. ubatubana* confuses the issue.

Plant: A medium-sized variable epiphyte, similar to *M. ubatubana* with which it is constantly confused. The tightly-clustered pseudobulbs are slightly laterally compressed cones, dark green to yellow, ridged and often wrinkled, 4 cm. tall and 3 cm. wide, and arising from a thick, tough and branched rhizome. The two apical leaves are erect, dark green, deeply keeled, leathery, rabbit's-ear-like and long-lived. The root system is strong. Red wiry roots spring from the latest pseudobulb, becoming white and branching as they penetrate the surrounding substratum extensively.

Inflorescence and Flowers: Up to six flowers grow from the newly matured pseudobulb base on pale green, 8 cm. tall, inflorescences. The flowers are smaller than *M. ubatubana* but maintain the same characteristics; a trilobed lip measuring 1.7 cm. by 1.2 cm. broad with a waxy nose-like callus running for a third of its length, white, flecked with purple. The sepals are fleshy and roundly pointed, 2.5 cm. long by 1 cm. wide and yellow, faintly specked with purple. The petals are smaller but the same colour.

Flowering Periods: May and June and the flowers last for over two weeks.

Pollination: Frequently over 20% of the flowers bear pods. **R. G. Singer** and **A. A. Cocucci** state the flowers lack nectar or any other reward but the strong honey-like fragrance appears to attract stingless bees, *Trigona spinipes*, for the first 5 days or so after the flowers open and these perform pollination for free (*Lindleyana* 14 (1): 47-56 1999).

Cultivation: In large pots on a coarse fibrous compost in good light and air movement, watering well during growth but only misting thereafter; this is a very easy plant to grow and ideal for beginners.

Maxillaria porphyrostele
Rchb. f. | *1813* (Plate 29)

Etymology: *Porphyro-stele* is Greek for the purple column shown by this flower.

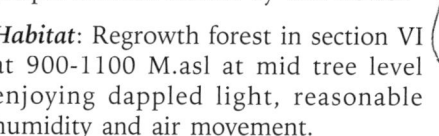
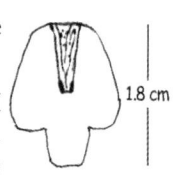

Habitat: Regrowth forest in section VI at 900-1100 M.asl at mid tree level enjoying dappled light, reasonable humidity and air movement.

Occurrence: Rare; though due to its similarity to other plants in the **M. picta** Alliance, *sensu* **Pabst**, it may be less rare than we suppose.

Plant: A medium-sized clustered epiphyte. The pseudobulbs are ovoid, somewhat laterally compressed and emerge alternately at 1 cm. intervals from a thick, sheathed, branching rhizome. They are 4 cm. tall by 2 cm. broad, khaki, and become grotesquely wrinkled under stressful conditions. The pseudobulbs show two apical, long-lived lanceolate, deeply-keeled, dark green leaves which measure 7-10 cm. long by 2 cm. broad. The root system is vigorous and adherent and in common with other allies in this grouping, roots are brick-red at their source, only whitening when they penetrate the substrate.

Inflorescence and Flowers: One or two inflorescences emerge from both sides of the recently mature pseudobulbs. These are single flowered, heavily sheathed, fleshy and measure up to 4 cm. long as far as the ovary. While still in bud the colour is an attractive magenta. The flowers open wider than normal for a *Maxillaria*, and viewed from the front they are a yellowish-cream with the margins tinged in pink while the exterior of both petals and sepals retain the bud's pink colour. The sepals are lanceolate, 2.5 cm. long by 0.4 cm. The lip is large, 1.8 cm. long, clearly trilobed, white and speckled with dark pink spots and lines. A distinct nose-like callus runs for half its central length.

Flowering Periods: In July and the flowers last for up to 15 days.

Pollination: Probably bees but probably not *Euglossine* as the flowering period is too cold for their visits, at this altitude.

Cultivation: As for other species in the Alliance mentioned above.

Maxillaria rupestris
Barb. Rodr. | *1814* (Plate 29)

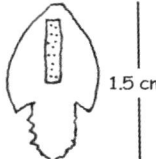

1.5 cm

Etymology: *Rupestris* is Latin for rock-growing which has no reference to this plant's growth habits in these regions. **Barbosa Rodrigues**, who described it, found the plants on trees on the summit of the Serra do Pedra Branca in Minas, so it may mean mountain dwelling.

Habitat: This species is found at mid-tree in original forest. Shade, high ambient humidity and reasonable air movement are its basic needs, at 900-1200 M.asl in section VI.

Occurrence: Occasional.

Plant: A medium-sized epiphyte, very typical of the **M. picta** alliance and so common in these high mountain regions. The pseudobulbs are a stout cone, 3 cm. tall by 2 cm. broad, light green with 6 distinct ridges arising immediately from a short tough branching rhizome. There are 2 apical, lanceolate leaves, dark green and leathery with a marked keel, 18 cm. tall by 2.3 cm. wide. Roots from the new pseudobulb base are red, medium thick and wiry until they penetrate the substrate around, when they whiten and branch. It is a competent root system; spreading, penetrating and adherent.

Inflorescence and Flowers: Two to four single-flowered, sheathed inflorescences of up to 10 cm. arise from the pseudobulb base. The overall flower colour is brownish-orange. The dorsal sepal is a broad lanceolate, 20 cm. long by 0.7 cm. broad. The lateral sepals are longer but narrower, 2.5 cm. by 0.5 cm. broad. The petals are shorter and narrower still, linear-lanceolate 2.0 cm. long by 0.3 cm. All are incurved around a broadish, but pointed, lighter coloured, trilobed lip, 1.5 cm. long by 0.9 cm. broad at the side-lobes with the typical nose-like waxy callus for one third of its length.

Flowering Periods: During October and November; the flowers last for three weeks.

Pollination: Infrequent; **Dressler** suggests wax-gathering bees.

Cultivation: In pots on a coarse fibrous substrate, in medium light and air movement, always kept damp but never waterlogged.

Maxillaria ubatubana
Hoehne | *1815* (Plate 29)

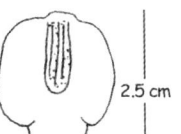

2.5 cm

Etymology: Presumably first found at Ubatuba, São Paulo; for us this is a variety of **M. picta**.

Habitat: The plant is very catholic, and does well from high tree-top sites to ground level, forming large colonies in all situations and flowering most freely with adequate light. It is a very tough plant and can survive light frosts but not excessive damp or shade. It inhabits practically every epiphyte-accepting tree from 900-1550 M.asl in sections II, IV and VI, but requires good drainage on the ground.

Occurrence: Common.

Plant: A large epiphyte with tightly clustered ovoid pseudobulbs on a thick branching rhizome. These are green, yellower in exposed conditions, many-ridged, and wrinkled. The size varies tremendously depending on the habitat. Tree-top mature bulbs are 4 cm. in length, 3 cm. wide but these measurements could be trebled for a plant living on a rotted fallen tree and the surrounding humus where it really thrives, and forms vast colonies. There are two apical lanceolate leaves which are erect when the plant is small, but droop when longer. They are dark green, leathery, shiny, narrow, centrally spined, and long-lived; the size varies, 20 cm. long x 2 cm. wide at tree top level but 70 cm. long x 5 cm. on the ground. The roots arise from the new pseudobulb base and are a red brown, thinnish, prolific and very adherent but not long. They thicken, turn white and branch freely when they penetrate the growing medium. The average length is 12 cm. at tree top: 25 cm. on the ground. There are approximately 15 roots to a pseudobulb, giving nearly 2 metres of roots in a tree-top specimen and nearly 4 metres per pseudobulb on a thriving plant established on the ground.

Inflorescence and Flowers: One to five inflorescences, up to 15 cm. long arise from the base of a new pseudobulb; the flowers are individual, large, fleshy, 7 cm. wide x 6 cm. long, and showy. The sepals are similar in size and shape, lanceolate and 3.3 cm. long by 1.0 cm. The petals are the same shape, 3 cm. by 0.5 cm. broad. The trilobed lip is 2.5 cm. long by 1.0 cm., yellow and white, striped with dark claret and showing the typical nose-like callus for one third of its length. The tepals are predominantly light yellow with claret speckles on the backs and the petals are incurved. The column and the anther are a deep claret.

Flowering Periods: In August and September, the flowers last for two weeks.

Pollination: Frequent and up to 20% of flowers bear fruit. See also n° *1811*, **M. picta**; probably the same pollinator.

Cultivation: Perhaps the least demanding and easiest of all the mountain-loving orchids to grow. Best in large pots on a coarse fibrous substrate in good light and wind movement, in reasonable humidity.

Maxillaria gracilis
Lodd. | *1817* (Plate 29)

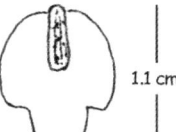

Etymology: *Gracilis* is Latin for thin or slender, describing both the plant as a whole and its leaves.

1.1 cm

Habitat: An epiphyte in original forest at low to mid-tree level in section VI, enjoying dappled light, high humidity and good air movement at 1000-1400 M.asl.

Occurrence: Occasional.

Plant: A small distinctive, clustered epiphyte. The pseudobulbs arise in a tight line from a thin, sometimes branching, rhizome. These are a narrow cone-shape, dark green and measure up to 3.5 cm. long by 1.5 cm. broad with a strong tendency to wrinkle. There are two apical leaves, long and narrow, linear-lanceolate, dark green, and long-lived, measuring up to 23 cm. long by 0.8 cm. wide. These are this plant's distinctive feature. The root system is strong, composed of numbers of stiff light brown roots which whiten and branch on penetrating the substrate. They emerge from the new growth base, are adherent and remain viable for years.

Inflorescence and Flowers: Single inflorescences arise from the bases of recently mature pseudobulbs. Each is 3 cm. long by 0.2 cm. in diameter, fleshy and thickly sheathed and holds a single flower. These are a butter-yellow, streaked with dark red. The sepals are more or less equal in size, a broad lanceolate, 1.5 cm. long by 0.5 cm. broad. The petals are a linear-lanceolate shape and measure 1.3 cm. by 0.3 cm. broad. The trilobed, thick trident-shaped lip is 1.2 cm. long by 0.9 cm. broad at the side-lobes. The distinctive feature of the lip is that the side-lobe margins are outlined by a thin, almost black line. The lip is heavily veined and is irregularly spotted with small dark purple blotches. It shows the typical *Maxillaria* nose-like callus for half its length.

Flowering Periods: In January and the flowers last for two weeks or so.

Pollination: Probably bees.

Cultivation: Grow in pots, on a thick fibrous compost with tree fern fibre containing charcoal and 10% of coarse sand, always well-drained and watering during spring and summer but misting thereafter.

Maxillaria kautskyi
Pabst | *1818*

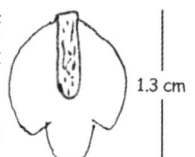

Etymology: Named in honour of **Robert Kautsky** a famous orchidist from Espírito Santo state.

1.3 cm

Habitat: Original forest at mid-tree level in section VI, at 1200 M.asl enjoying medium light, good air movement and constant humidity.

Occurrence: Rare.

Plant: On first appearance it looks like a miniature **M. picta**, however the pseudobulb shape and the linear-lanceolate leaves give rise to the first doubts. The pseudobulbs are tightly clustered on a thick wandering rhizome and are light green, many ridged and almost conical measuring 3.5 cm. by 2.2 cm. wide. There are two apical long-lived leaves; smooth, keeled, slightly stiff, linear-lanceolate and 20 cm. long by 1.7 cm. The root system is composed of numerous brick-red, brittle branching roots arising from the new growth base, viable for many years and becoming greyish-white when they penetrate the substrate.

Inflorescence and Flowers: One to three single, sheathed and fleshy inflorescences arise from the bases of the most recent pseudobulbs. These are taller than the pseudobulb and hold single yellow flowers, slightly tinged with claret blotches near their extremities. The sepals are ± equal, a broad lanceolate 1.7 cm, by 0.55 cm, wide. The petals are narrower, 1.4 cm. long by 0.3 cm and flank the column. The lip is trilobed and altogether a broad oval when flattened, 1.3 cm. long by 0.9 cm., and a nose-like callus runs down from its throat to half its length.

Flowering Periods: The flowers last for 10 days or so.

Pollination: Unknown but due to the flower's similarity to **M. picta**. the comments by **R. G. Singer** and **A.A.Cucucci** concerning the pollination of that species may be relevant.

Cultivation: As for **M. picta**.

Maxillaria chrysantha
Barb. Rodr. *1819* (Plate 30)

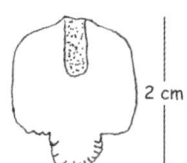

Etymology: *Chrysantha* is Latin for golden flowered, referring to the flower colour.

2 cm

Habitat: Sections I, III and V in transitional, mature regrowth forest at mid to upper tree level, also on rock faces, tolerating full sunlight, but also quite reasonable shade with good air movement, a long dry winter and a wet humid summer at 600-800 M.asl.

Occurrence: Locally common.

Plant: A large variable epiphyte or lithophyte, often found in large sprawling colonies both on large trees and on exposed rock faces. What differentiates this plant from others so similar is the distance between the pseudobulbs. These are held on a thick woody rhizome at up to 5 cm. intervals and may stretch for over two metres. The pseudobulbs are broad, faintly laterally compressed, ovoid, green to yellow depending on the light intensity, and measure up to 7 cm. tall by 4 cm. There are two long-lived apical leaves which are deeply keeled, stiff, dark green, linear-lanceolate and up to 25 cm. long by 3 cm. The root system is extensive, arising principally from the pseudobulb bases, composed of numerous 0.2 cm. thick, brick red, roots, sometimes branching and some of which may be permanently aerial.

Inflorescence and Flowers: The flowers arise from the most recent pseudobulb base on single, sheathed inflorescences up to 12 cm. long and may be few or many depending on the substrate. The sepals are ± equal, a broad, blunt lanceolate 2.5 cm. by 0.1 cm. wide. The narrower petals are lanceolate, 2.2 cm. by 0.6 cm. wide; all are orange, faintly speckled in claret. The trilobed lib is typical with the familiar nose-like callus projecting for 50% of the throat to the apex. Its measurements are 2.0 cm. by 1.3 cm. wide at the side-lobes.

Flowering Periods: **Barbosa Rodrigues** reports **var. macrobulbosa** flowering in January as this does, and the flowers last for up to three weeks. Other varieties flower in September.

Pollination: Probably the same pollinator as for *M. picta*, but fruits are infrequently found

Cultivation: As for *M marginata*.

Maxillaria marginata
Fenzl. | *1820* (Plate 30)

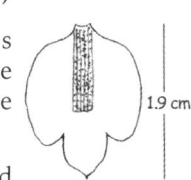

Etymology: *Marginata* is Latin, referring to the coloured margins of the tepals.

Habitat: A mid-tree epiphyte requiring high humidity, dappled light and good air movement, found in original forest at 500-800 M.asl, in section VI.

Occurrence: Common in it's niche.

Plant: A very sturdy mid-tree epiphyte with large, ridgcd pseudobulbs, each a four-cornered cone, dark green and measuring 5 cm. high and 3 cm. wide and bearing a pair of very long-lived, opposite, apical, linear-lanceolate leaves. These are mildly coriaceous, faintly keeled and measure up to 19 cm. long by 2 cm. wide. The pseudobulbs arise from a 1.2 cm. thick, woody, sheathed and branching rhizome at intervals up to 2.5 cm. The root system is prolific and composed of many penetrating and branching roots. 0.25 cm. thick, brittle, and a brick colour.

Inflorescence and Flowers: Up to six, single-flowered, sheathed, 8 cm. inflorescences arise from the bases of the new growths. These flowers are surprisingly small for a large plant and yet are very attractive. The sepals are linear-lanceolate and 2.2 cm. long by 0.5 cm. wide, a creamy light yellow with their margins having a dark wine-coloured unbroken line. The petals are the same colour as the sepals, and a narrow lanceolate measuring 1.8 cm. by 0.25 cm. broad. The very distinctive trilobed lip measures 1.6 cm. long by 0.95 cm. wide at the side-lobes while the pronounced central lobe extends for 0.7 cm. and is 0.3 cm. broad. A typical nose-like waxy central callus extends longitudinally from the lip base for 0.7 cm. The background colour is creamy-white but heavily blotched with dark wine at the margins of the central lobe and the apices of the side-lobes.

Flowering Periods: These vary from December to February and the flowers last for a week or so.

Pollination: Probably by stingless bees.

Cultivation: Because of the long distance between the pseudobulbs, it is best grown on bark plaques of sufficient thickness and moss cover to retain humidity.

Maxillaria piresiana
Hoehne | *1821* (Plate 30)

Etymology: After hours of fruitless research we can only assume that the plant examined came from somewhere called **Pires**.

Habitat: Epiphytic from section VI and found at 800 M.asl, at mid-tree level where it receives medium intensity light and air movement and constant reasonable humidity.

Occurrence: Rare, but when not in flower it may be confused with the more common **M. marginata** or **M. chrysantha** and thus it may be under recorded.

Plant: A large epiphyte, similar vegetatively to the species of the **M. picta** and **M. marginata** alliances but with somewhat squatter pseudobulbs, well separated from each other. These are almost conical and measure 7 cm. high by 5 cm. wide and are eight-winged and dark yellow and emerge from a stout 0.5 cm. diameter rhizome alternately at intervals of 1.5 cm. to 2.0 cm. Each bears two apical leaves which are linear-lanceolate, dark green, deeply keeled, leathery and long-lived and measure up to 25 cm. long by 3.5 cm. wide and one is slightly shorter and broader than the other. The intense root system is composed of many brown, branching roots which arise from the base of the new growth and whiten after entering the substrate.

Inflorescence and Flowers: Up to ten single-flowered inflorescences emerge from the bases of the most recent pseudobulbs. These are 11 cm. long, thick, fleshy, green and partly sheathed. The dorsal sepal is brownish on a yellow base and measures 2.2 cm. by 1.1 cm. The lateral sepals are the same size and colour but a slightly asymmetrical lanceolate. The petals are smaller and narrower, essentially yellow and measure 1.95 cm. by 0.6 cm. and a classic lanceolate. Both sepals and petals are speckled with purple blotches on their outer apical halves. The stout, three-lobed lip measures 1.9 cm. long by 1.6 cm. wide and has a number of finely speckled purple veins on a yellow background, running longitudinally to the side-lobes. The central lobe is blunt and crenate at the tip and a typical Maxillaria nose-like central callus terminates at exactly half way from the lip base to the tip.

Flowering Periods: January, and the flowers which open simultaneously, last for about two weeks.

Pollination: Probably pollinated by stingless bees.

Cultivation: Culture is as for species in the **M. picta** Alliance which enjoy being mounted on bark plaques or in baskets using a well-drained compost with added fibre for the dense root system. This is an easy plant to grow and a rewarding intermediate house species.

Maxillaria bradei
Schltr. | *1825* (Plate 30)

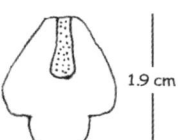

Etymology: Bradei is named after the German **Alexander Brade** (1881-1971), a famous systematic botanist who worked in the National Museum and the Rio de Janeiro Botanic Garden.

Habitat: The plant is found epiphytically as a low to mid-tree species in deep humid forest and also found terrestrially in burnt, regrowing forest. The plant is very sensitive to environment and lives in a very narrow band of deep shade, high humidity and little air movement at 1000-1200 M.asl in section VI.

Occurrence: Occasional.

Plant: A small to medium-sized and very variable epiphyte. The pseudobulbs are small, 3 cm. long by 2 cm. wide, green, flattened laterally, sheathed by current or dead leaves, and borne on a short, thin rhizome. There are 4 leaves but only one grows to any length; 24 cm. long x 2 cm. wide. This will remain on the

pseudobulb for 3-4 years and is a dull matt green, almost velvety in texture and its rounded shape distinguishes this plant when not in flower. The roots are thin, fibrous and white, rather like hessian; with only a few on young growths but older bulbs have many roots which appear to be functional. A pseudobulb may have 20 or more roots of 15 cm. length but because of branching this gives about 50 cm. in total, and thus an average 20 bulb colony may have over 10 metres of tightly-meshed roots, rather like a doormat.

Inflorescence and Flowers: Single flowers are borne on 10 cm. basal stalks and 5 or more may grow from each bulb in succession. The sepals are an acuminate lanceolate, 4 cm. by 1 cm. broad at their bases. The petals are the same shape but slightly smaller. The trilobed lip is 2 cm. long by 1.2 cm. at the side-lobes. They are honey-scented in the afternoon. We have found colonies with flowers ranging from pure white through pink to yellow.

Flowering Periods: November through March and the flowers last for two weeks but the plant may show flowers in bloom for very much longer.

Pollination: Is frequent with up to 30% of flowers setting seed. **Dressler** suggests wax-gathering bees to be the vector. Seedlings are found in numbers beneath the plant.

Cultivation: Best grown in shallow pots on a rich fibrous well-drained compost in shade, with low air movement and humidity throughout the year.

Maxillaria ochroleuca
Lodd. ex Lindl. | *1827* (Plate 30)

Etymology: *Ochr-leuca* is Greek for pale ochre colour referring to the flower.

Habitat: Epiphytic, found at 900-1500 M.asl, in sections II, IV and VI in original forest at mid-tree level together 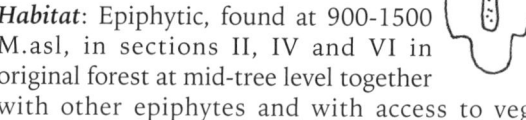 with other epiphytes and with access to vegetable detritus.

Occurrence: Common throughout the region in its niche. It is also a successful survivor of forest fires and is found growing as a terrestrial in regrowth that has not been used for agriculture.

Plant: This plant is identical to **M. rodriguesii** and as both species vary according to the substrate, they are impossible to tell apart when not in flower. See **M. rodriguesii**.

Inflorescence and Flowers: As with **M. rodriguesii**, numerous sweetly scented flowers arise basally from between the lateral pseudobulb sheaths on 10 cm, erect and sheathed inflorescences. The differences between these and those of *M. rodriguesii* are that these are 20% smaller, the lip is narrower, generally orange and bears no hairs.

Flowering Periods: December and January and the delicate flowers last for up to ten days, weather permitting.

Pollination: Probably pollinated by stingless bees.

Cultivation: Requires dappled light, constant and reasonable air movement and humidity and grows well in hanging baskets or in pots on standard compost. It develops best when overcrowded.

Maxillaria rodriguesii
Cogn. | 1829

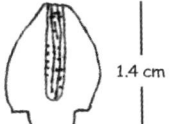

1.4 cm

Etymology: Named in honour to **Barbosa Rodrigues**, Brazil's most famous orchid botanist.

Habitat: Epiphytic plants found at 900-1500 M.asl, in sections II, IV and VI at mid-tree level in original forest together with other epiphytes and with access to vegetable detritus.

Occurrence: Occasional, but may be commoner that we imagine as it is so similar vegetatively to *M. ochroleuca*.

Plant: A medium-sized epiphyte with a clump of tightly clustered, distinctly bilaterally compressed ovoid pseudobulbs which are 6-7 cm. high and 4-5 cm. wide with a single apical leaf and three basal leaves. The apical leaves are deeply keeled and measure 30 cm. long by 3.5 cm. wide and are linear-lanceolate tapering to a crenate apex. They are deep green but become paler towards the false petiole attachment to the pseudobulb. These and one subtending leaf are retained for some years. Roots are plentiful; a fine fibrous network penetrating any leaf litter, branching frequently and reaching up to 50 cm. around the plant. This branching, net-like structure makes length assessment difficult and is reminiscent of the nest-epiphyte system found in *Miltonia* species.

Inflorescence and Flowers: Numerous single, sweetly-scented flowers arise continuously from between the pseudobulb sheaths, on 15 cm. erect, sheathed and basal inflorescences. The flowers are 8 cm. across and white. The sepals are equal size, subulate, and measure 4.2 cm. long by 0.5 cm. wide at the bases. The petals are shorter and narrower, also subulate and measure 3.4 cm. long by 0.4 cm. broad. The trilobed lip is 1.4 cm. long by 0.7 cm. wide at the side-lobes, faintly hairy and lemon-yellow. A veined, nose-like central longitudinal callus projects from the base to the start of the central lobe. The curved column is 0.8 cm. long.

Flowering Periods: During December and January and

the delicate flowers survive according to the weather.

Pollination: Probably by stingless bees.

Cultivation: This plant requires dappled light, constant reasonable humidity and air movement and grows well in a hanging basket or pot. Like many *Maxillaria* spp, it suffers when not overcrowded.

General: In his original drawing, **Barbosa Rodrigues** notes 'I think only a form of *M. ochroleuca* Lodd. - bulb larger'. **Barbosa Rodrigues** named the species *M. longipetala* but **Cogniaux** had already named it *M. rodriguesii*. Interestingly, *M. ochroleuca* was also found in three separate geographical locations in Venezuela, from the Gran Sabana to the Andes where variability appears to be confined to the length of the pedicel and the plants with larger flower stems gained the varietal name of *M. ochroleuca* var. *longipes*.

Maxillaria brasiliensis
Brieger & Bicalho | 1833 (Plate 30)

Etymology: Presumably first found in Brazil.

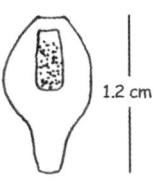

1.2 cm

Habitat: Original forest at mid-tree level from 800-1300 M.asl. in section VI, requiring dappled light, low air movement and high constant humidity.

Occurrence: Common.

Plant: A medium-sized epiphyte, almost a copy of *M. discolor* and since both plants are variable depending on substrate and environmental conditions, they are impossible to tell apart when not in flower. The new growth emerges from a very short thick rhizome. The pseudobulbs are oval, light green and very compressed laterally. There are one apical and three subtending leaves. The subtending leaves sheath and hide the pseudobulb for several years. The leaves are a broad linear, deeply keeled, dark green up to 20 cm. by 3.5 cm. wide, stiff, leathery and long-lived. The root system is composed of numerous, thickish, brick-red, branching roots which very firmly anchor the plant to its substrate, but do not extend much beyond its circumference.

Inflorescence and Flowers: One to several single inflorescences up to 6 cm. long, arise from the most recent pseudobulb base and force their way through the sheathing of the subtending leaves. The flower is an overall butter-yellow. The sepals are ± equal in size and shape; lanceolate, 1.6 cm. long by 0.7 cm. wide. The petals are the same shape but much narrower and slightly shorter. The lip is vaguely trilobed with a typical nose-like callus running from its throat for half its length. It is a clear butter-yellow, 1.2 cm. long by 0.9

| 273 |

cm. and does not show the wax-like substance on its apical lobe which is one of the features which distinguishes it from *M. discolor*.

Flowering Periods: November and March and the flowers last for two weeks.

Pollination: Almost 100% but pollinator unknown.

Cultivation: As for *M. discolor*.

Maxillaria aff. discolor (Lodd. Ex Lindl.) Rchb. f. | *1834*
(Plate 30)

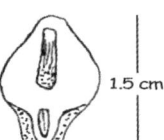

Etymology: *Dis-color* is Latin for to change in colour, referring to the change in colour as the flower ages.

Habitat: An epiphyte from section VI, found at 800-1300 M.asl, in humid original forest at mid-tree level and requiring dappled light, low air movement and high humidity.

Occurrence: Common in its niche.

Plant: A medium to large epiphyte with an almost equitant leaf arrangement. New growths arise at very short, 0.5 cm. intervals from a 0.5 cm. diameter rhizome with five leaves, one apical and four subtending the pseudobulb in a fan-like arrangement only permitting the pseudobulb to appear when the smaller leaves start to wither after three or four years. The leaves are a broad linear shape with an emarginate rounded apex, deeply keeled, leathery, dark green and measure up to 27 cm. long including a 6 cm. long Iris-like sheath around the pseudobulb, hence the synonym *iridifolia* and 4.5 cm. wide. The pseudobulbs are strongly compressed laterally and up to 6 cm. long by 2.5 cm. wide, almost rectangular and pale green. The root system is vigorous and a five-year-old plant has hundreds of light brick-coloured branching, brittle roots of 0.15 cm. diameter which do not extend much beyond the circumference of the plant.

Inflorescence and Flowers: One to several single, 5 cm. long fleshy inflorescences arise from the pseudobulb base and force their way consecutively up through the tight subtending leaf sheaths. The single waxy flowers appear just above the pseudobulb with the flower and ovary showing. The sepals and petals are a pale yellowish-orange tinged with faint green at their tips. The sepals are about equal in size, 1.5 cm. long by 0.8 cm. wide and a bluntish lanceolate. The petals are shorter and narrower, 1.5 cm. by 0.4 cm. and also lanceolate. The trilobed lip is darker orange, almost oval and strongly speckled with wine-coloured dots and measures 1.45 cm. long by 0.85 cm. A typical nose-like callus stretches from the base to the middle of the lip. The apical lower lobe centre has a waxy substance

arranged as an elongated callus.

Flowering Periods: November and December and the flowers last for two weeks.

Pollination: **Dressler** suggests that pollination is by stingless bees. However, the flowers have no perceptible scent, are invariably fertilised and we have evidence of self-pollination.

Cultivation: An elegant plant which is easy to grow on bark plaques, in pots or baskets in standard compost.

Maxillaria cogniauxiana Hoehne | *1844* (Plate 31)

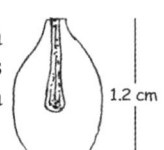

Etymology: Named after **Cogniaux**, a famous Belgian botanist, who sometimes worked in tandem with **Barbosa Rodrigues**.

Habitat: Found in mid to upper tree in original forest from 1200-1450 M.asl in sections VI and VIII.

Occurrence: Common.

Plant: A small epiphyte, growing in drooping tufts and similar to *M. acicularis* to which it is closely related. The pseudobulbs are cylindrical, light green and slightly ridged, 2 cm. tall and 1 cm. wide arising at 0.5 cm. intervals from a short vertical brown sheathed rhizome. Only anchor bulbs appear to have roots. However, as with *M. acicularis*, roots from the new bulbs are channelled back along the rhizome beneath the sheaths. The two apical leaves are dark green, narrow, linear-lanceolate, tending to terete, and 3 cm. long and 1.5 cm. wide, deeply keeled and long-lived. The anchor root system consisting of fine white roots is very adherent but not extensive.

Inflorescence and Flowers: During April and May, a single flower arises from the new bulb base on a 2 cm. sheathed stem. The flower is overall claret and the predominant feature is the broad curled-back, spade-like lip with a dark wet-looking centre and the typical nose-like waxy callus running for half its length. It measures 1.3 cm. by 0.75 cm. broad. The sepals are broad and pointed, roughly equal, at 1 cm. long by 0.5 cm. wide with an orange base and deeply speckled with claret. The smaller petals are the same colour.

Flowering Periods: During April and May; the flowers last for up to three weeks.

Pollination: Is infrequent and **Dressler** suggests wax-collecting bees or that the shiny, wet-looking lip attracts any large insect during the predominantly dry months of April and May.

Cultivation: On tree fern or bark plaques in dappled

light, good air movement and always damp.

General: One more example of a species that evolved during one of the periods when the Atlantic Rain Forest underwent a dryer era.

Maxillaria madida
Lindl. | *1846*

Etymology: *Mador* is Latin for wetness, referring to the wetlooking lower half of the lip.

Habitat: An epiphyte from section VI found at 1000-1400 M.asl, in original forest at mid to upper-tree level where it receives dappled light, constant good humidity and air movement.

Occurrence: Occasional to common.

Plant: A small tufted, somewhat pendulous epiphyte found in large colonies. The pseudobulbs are an elongate ovate measuring 2 cm. long by 0.4 cm. and arise at 0.5 cm. intervals alternately from a thick, densely sheathed and branched rhizome. The pseudobulbs are dark green and smooth, wrinkling with age, turning greyish and bear two apical, stiff, opposite, long-lived, linear-lanceolate leaves which are infolded, deeply keeled, dark green and measure 3.5 cm. long by 0.5 cm. The root system is curious. The first growth forms the basic feeder and anchor roots. Roots from newer growths are channelled back along the rhizome and beneath the sheathing until they reach the substrate.

Inflorescence and Flowers: Thick, short and well-sheathed inflorescences arise from the bases of the most recent pseudobulbs bearing single flowers that reach to the height of the leaf. The flowers vary in colour but the overall tone is claret sometimes fading to a claret-speckled butter-yellow on the sepals, petals and lip tips. The dorsal sepal measures 1.5 cm. by 0.8 cm. and is a broad lanceolate while the lateral sepals are longer, 1.9 cm. by 0.8 cm. and an asymmetric lanceolate. The petals are also asymmetric lanceolate and measure 1.2 cm. by 0.4 cm. The lip shows the discrete difference between this species and **M. cogniauxiana**, being faintly bilobed whereas the latter's is single-lobed. The lip measures 1.7 cm. by 1.0 cm. and has the typical waxy, nose-like, vertical callus on its basal half.

Flowering Periods: April; the flowers last for two to three weeks.

Pollination: Probably by stingless bees.

Cultivation: As for **M. neuwiedii**.

Maxillaria neuwiedii
Rchb. f. | *1847*

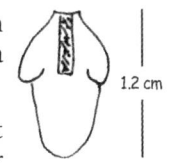

Etymology: Named after **Maximilian Wied-Neuwied**, (1782-1867), a Prussian collector.

Habitat: Found on original forest relict trees as a mid-tree pendent epiphyte or lithophyte in section V at 400-600 M.asl enjoying good but dappled light, good air movement, experiencing a long dry winter with a short wet summer.

Occurrence: Occasional.

Plant: A medium-sized, branching, pendulous plant which can form colonies up to one metre long. The rhizomes are anchored tightly at their bases to the rock or bark substrate by a plethora of fine white roots. Pseudobulbs arise at short intervals on the main rhizome and on branches. These are oval, bilaterally compressed, light green and measure 1.5 cm. by 0.75 cm wide. There are both basal, short-lived and apical, long-lived leaves which are lanceolate, keeled, light green, papery but stiff and measuring up to 5 cm. long by 1 cm. The rhizomes are thickish, 0.3 cm., round, heavily sheathed and underneath this sheathing runs a root system which, when reaching the substrate, penetrates and spreads.

Inflorescence and Flowers: One, sometimes two, single short-stemmed flowers arise from the most recent, mature pseudobulb bases. As branching is frequent, quite a number of flowers may be shown on a mature plant. These are yellow to orange with a deep purple centre. The sepals are more or less equal, lanceolate, up to 1.6 cm. long by 0.6 cm. The petals are lanceolate to baseball-bat in shape, 1.3 cm. by 0.4 cm. broad. The lip is spade-shaped, very vaguely bilobed with a nose-shaped, deep purple, central callus, 1.3 cm. long by 0.3 cm. deep, reflexed at the apex and appearing moist.

Flowering Periods: In September and October and the flowers last for 10 days or so.

Pollination: Unknown and infrequent.

Cultivation: On a rough bark substrate held at 70°-80° so that the plant can rest on the substrate and root from the branching rhizome. Allow for a long dry dormant period and water well before and after flowering.

Maxillaria ferdinandiana
Barb. Rodr. | *1849* (Plate 31)

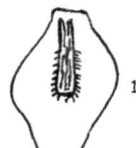

1.2 cm

Etymology: Not known.

Habitat: We have found this plant as an epiphyte in intensely humid conditions in section VI and as a lithophyte in transitional dry forest in section V both at 500 M.asl, and requiring good indirect light, tolerating wildly variable humidity and with some air movement.

Occurrence: Occasional

Plant: A pendulous epiphyte or lithophyte sometimes found in large clumps. The pseudobulbs emerge more or less alternately on long branching rhizomes which may reach up to 50 cm. and are elliptic, 2.5 cm. long by 0.7 cm., strongly bilaterally compressed, generally ridged and a light green. There are single apical, deeply keeled, dark green, leathery, stiff, long-lived linear-lanceolate leaves, measuring up to 12 cm. by 0.7 cm. broad. The root system is extensive. It runs back along the rhizome, underneath the sheathing, from the pseudobulb bases to the substrate where the roots branch, penetrate and anchor the plant. Should the pendulous rhizomes be in contact with a substrate then the roots depart the rhizome sheaths and at that point penetrate the substrate. In this instance, the plant may form a very large grouping. The roots in themselves are interesting, thin white, intensely warted with short hairs.

Inflorescence and Flowers: Single flowers are borne on short inflorescences, heavily sheathed, from both sides of the most recent pseudobulb bases. The sepals are broadly elliptical, 1.25 cm. by 0.5 cm. broad, the laterals being slightly broader and vaguely asymmetric. The petals are also elliptic but smaller. The overall colour is a whitish yellow speckled with small wine spots. The lip is typical of most *Maxillaria* spp, vaguely trilobed with side-lobes reaching upwards towards the column and a nose-like central callus, reaching down from the throat to one third of the lip's length. It measures 1.25 cm. by 0.75 cm. broad at the shoulders. The margins are finely and intensely serrate. The anther cap to the column is distinctive and acuminate with a purple tip. The lip has a shiny purple apical lobe; the remainder is a dull white with intense purple specks.

Flowering Periods: November/December; the flowers last for two weeks.

Pollination: Unknown and infrequent.

Cultivation: Grow on a tree fern or on long lasting large bark plaques with good, indirect light, reasonable air movement and humidity to suit.

Maxillaria minuta
Cogn. | *1851* (Plate 31)

7 mm

Etymology: *Minuta* is Latin for very small, minute, referring to the flower size.

Habitat: Found in transitional and gallery forest in section V at 600-700 M.asl, in large colonies as a lithophyte or a low to mid-trunk epiphyte on mature trees, enjoying year round good humidity, medium light and wind movement.

Occurrence: Occasional, but when found it is usually in large colonies.

Plant: A small, tight, pendulous, many branched epiphytic, sometimes lithophytic plant in the *M. pumila* alliance. The basal pseudobulbs are strongly anchored to the substrate and the plant forms these branches prodigiously downwards on thick short, rhizomes between the thickly sheathed pseudobulbs. These are a narrow rugby football in shape, ridged, dark green, 1 cm. long by 0.3 cm. and very long-lived. There are single, apical, very long-lived leaves, flat to terete, dark green, fleshy though stiff and up to 3.2 cm. long by 0.2 cm. broad. The root system is similar to all in this alliance; a firm anchoring system from the basal growth, while roots from new, hanging growth run back beneath the rhizome sheaths. If and when they reach the substrate they spread out to help with the anchoring and feeding systems. These roots are fine, white and branching but do not spread far on or under the substrate.

Inflorescence and Flowers: One or two small flowers arise from the bases of new growths on very short, heavily sheathed, peduncules. Due to rapid branching, there may be many flowers. The dorsal sepal is very broad, lanceolate and 0.7 cm. long by 0.4 cm. The lateral sepals are broader and slightly longer but essentially the same shape, while the petals are elliptic to lanceolate and slightly smaller than the dorsal sepal. The lip is vaguely bilobed, a slightly indented rectangle with the familiar nose-like callus reaching half way up the lip.

Flowering Periods: March, April and October when the flowers last for 3 days or so.

Pollination: Is frequent, probably by wax-gathering bees.

Cultivation: As for *M. pumila* or *M. plebeja*.

Maxillaria plebeja
Rchb.f. | *1855* (Plate 31)

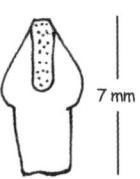

7 mm

Etymology: Probably from the Latin *plebs* meaning common - which it is.

Habitat: Found in large matted colonies at 700-1000 M.asl in section VI, at mid-tree level in original forest, enjoying dappled light, good wind movement and humidity.

Occurrence: Common.

Plant: A tufted epiphyte and a smaller replica of **M. cogniauxiana**, for which it could well be confused when not in flower. The pseudobulb is a ridged dark green cylinder measuring ± 1 cm. in length and 0.2 cm. width. There is a single light green apical, thick and very stiff, lightly keeled linear-lanceolate leaf which is very long lived. The pseudobulbs arise at very short intervals from a thickish branching rhizome and the roots are channelled back along this rhizome so only the anchoring pseudobulbs appear to have roots.

Inflorescence and Flowers: A short, heavily sheathed inflorescence grows from the most recent pseudobulb base. The dorsal sepal is lanceolate, 0.7 cm. long by 0.2 cm. wide and the lateral sepals are slightly larger. The petals are smaller, thinner but obtuse at the apex and the down-curling lip is bilobed, and club-shaped measuring 0.8 cm. by 0.3 cm. The overall colour of the flower runs from light brownish-green to a central lip ridge of dark claret.

Flowering Periods: During January and February and the flowers last for three weeks.

Pollination: Is frequent, up to 15% and **Dressler** suggests wax-gathering bees.

Cultivation: As for **M. cogniauxiana**.

Maxillaria pumila
Hook. | *1856* (Plate 31)

Etymology: *Pumilus* is Latin for dwarf, close growing or short, referring to the 8mm plant's habit.

Habitat: Found at mid-tree and lower upper branches in original forest at 600-900 M.asl often in large colonies, enjoying dappled light, good wind movement and humidity in sections V and VI.

Occurrence: Occasional but when found, in large numbers.

Plant: The twin of **M. plebeja** and indeed included within the same alliance whose covering description says; all plants usually pendant fixed only by their basal root-bearing pseudobulbs, the subsequent pseudobulbs, apparently rootless. Pseudobulbs, unifoliate, leaves flat, sometimes terete or looking terete. The small pseudobulbs are a ridged egg-shape, precisely like a rugby football, 1.3 cm. long by 0.3 cm. wide. The single apical leaf is dark green, fleshy and flat, long-lived, 3.8 cm. long by 0.4 cm. wide. The roots of all but the anchor pseudobulb are not apparent but

are channelled back around the short stiff rhizome under light brown paper-like sheaths. The root system is extensive and adherent, composed of many long thin, white roots.

Inflorescence and Flowers: Single, relatively large flowers are produced from the base of the most recent pseudobulb on short, sheathed inflorescences. As colonies are large, many flowers may appear at once. The lanceolate dorsal sepal is light orange, 0.75 cm. long by 0.3 cm. wide and the lateral sepals are longer and wider. The petals are narrower and slightly shorter. The lip is distinctly four-lobed, a bulging rectangle, 0.8 cm. long by 0.5 cm. wide and orange with a wide wet-looking, dark claret, callus.

Flowering Periods: February and March and the flowers last for up to three weeks.

Pollination: as for **M. cogniauxiana**.

Cultivation: as for **M. cogniauxiana**.

General: The difference between this plant and **M. plebeja** is principally in the lip shape. The latter's central lobe is flat at the base whereas the former's has a deep cleft. This is one more example of a species that evolved during one of the periods when the Atlantic Rain Forest underwent a dryer era.

Maxillaria spannagelii
Hoehne | *1857*

Etymology: Not known.

Habitat: Found in original forest at 900-1000 M.asl in section VI at mid- to upper tree level, often forming large colonies on thick horizontal branches, requiring good light and air movement with a high consistent humidity.

Occurrence: Occasional.

Plant: A small creeping and clustered epiphyte very similar to **M. plebeja** and **M. pumila** and may be confused with either when not in flower. The pseudobulbs are held closely on a thickish sheathed rhizome which may branch and are up to 0.7 cm. long by 0.4 cm. and ovate, laterally compressed, dark green with a tendency to wrinkling. There are single apical dark green, thick, leathery long-lived leaves which are linear-lanceolate and 3 cm. long by 0.4 cm. wide. The root system runs back down the rhizome from the base of the new growth until the roots meet the substrate where they penetrate and branch.

Inflorescence and Flowers: One or two short, sheathed inflorescences arise from the bases of the most recent pseudobulbs with small single flowers to a height of a third of the leaf. Overall they are purple, lightly and faintly blotched with cream. The sepals are ± equal, a broad lanceolate 0.8 cm. long by 0.5 cm. The petals are the same shape but narrower, 0.75 cm. by 0.2 cm. wide. The lip is long, narrow and faintly four-lobed, 0.8 cm. long by 0.5 cm. wide at the basal lobes. The typical, somewhat bulbous, nose-like callus extends from the throat to half the lip length.

Flowering Periods: November and December and the flowers last for up to three weeks.

Pollination: Unknown, but flowers are quite frequently pollinated.

Cultivation: As for **M. cogniauxiana**, but also does well in pot culture on a coarse fibrous substrate in good indirect light with good air movement, reasonable consistent humidity but never over watered.

Maxillaria aff. paulistana
Hoehne | 1858a (Plate 31)

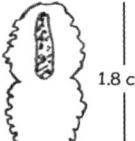

Etymology: Possibly a new species.

Habitat: A lithophyte growing at 600 M.asl in section V, in low light with high humidity in summer, but drying out seriously during the autumn and winter months, and with very little air movement.

Occurrence: Rare.

Plant: A medium-sized lithophyte which also could be an epiphyte in the **M. paulistana** alliance. The single leaves are dark green, terete to plane, linear and fleshy, up to 16 cm. long by 0.5 cm. and held apically on a short 2 cm. tall by 0.4 cm. cylindrical pseudobulb. These are borne closely spaced on a thick sheathed rhizome. The root system is basal, spreading, attached to the substrate, quite similar to **M. acicularis** in the sense that roots from new growths run back along the rhizome until they reach the substrate.

Inflorescence and Flowers: Single short inflorescences emerge from the bases of the most recent pseudobulbs. The dorsal sepal is lanceolate, 1.6 cm. by 0.5 cm. The laterals are similar but somewhat asymmetric 1.7 cm. by 0.7 cm wide. The petals are narrower, 1.2 cm. by 0.3 cm. All are a dirty orange with reddish-brown tips and bases. The vaguely four-lobed lip is 1.8 cm. by 0.9 cm. wide with a divided nose-like callus running from the throat down to half the lip length.

Flowering Periods: April and the flowers last for up to two weeks.

Pollination: Unknown.

Cultivation: As for others in the **M. acicularis** alliance.

Maxillaria acicularis
Herb. | 1859 (Plate 31)

Etymology: *Acicularis* is Latin for narrow, stiff, pointed like a needle, and refers to the leaves.

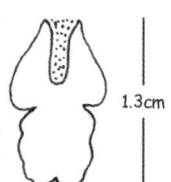

Habitat: Found in the mid to tree-top zones in original forest, from 900-1500 M.asl in section VI. The dense network of needle-like leaves trap detritus which is usually damp within. The plant soon dies back if it falls to the ground suggesting that it cannot tolerate sustained wetness around the leaves, but enjoys good light and air movement.

Occurrence: Common.

Plant: An epiphytic plant growing in dense tufts and looking like clustered pine needles. The dark green cylindrical pseudobulbs are 3 cm. long and 0.5 cm. diameter and separated by a short branching rhizome. There is a pair of dark green apical, needle-shaped leaves about 13 cm. long by 1 mm. wide, flattened on the facing sides giving a 'D' shape section. The 6 most recent pseudobulbs retain their leaves but because the growth is not strictly annual these do not represent 6 years growth. As the colony grows out from its support, the leading 6-8 pseudobulbs are apparently without roots. However, on closer examination the small connecting rhizome is found to be surrounded with sheaths which make a tube. Roots from the youngest bulbs are channelled backwards in parallel for 5-8 cm into the substrate where they branch and penetrate bark and dead tissue. The thin fibrous roots are thus protected from desiccation. On a colony of 12 bulbs, rooting is not very extensive with about 50 cm. in total length.

Inflorescence and Flowers: The single flowers arise from the base of the most recent pseudobulbs and are 3 cm. long x 3 cm. across, sessile and held on short sheathed inflorescences and unscented. They are yellow, with delicate freckling on the sepals, deepening to a shiny dark smudge on the labellum. The sepals are a broad lanceolate, 1.3 cm. by 0.4 cm. broad. The petals are a faintly pointed, spathulate 1.2 cm. long by 0.4 cm. The faintly trilobed lip is almost rectangular, 1.3 cm. by 0.6 cm. broad with a faint nose-like callus at its throat.

Flowering Periods: December and January; the flowers last for three weeks.

Pollination: Infrequent; **Dressler** suggests wax-gathering bees.

Cultivation: This attractive and unusual species is worth growing for its ease of cultivation and regular flowering. It has unusually large blooms for such a small plant. Grow as for **M. cogniauxiana**.

General: One more example of a species that evolved during one of the periods when the Atlantic Coastal Forest underwent a dryer era.

Maxillaria subulata
Lindl. | *1863*

Etymology: *Subulate* or awl-shaped refers to the grooved leaves.

Habitat: A mid- to high tree species, sometimes in large colonies, found in humid original forest at around 1200 M.asl in section VI.

Occurrence: Occasional.

Plant: A small pendent epiphyte; small but stouter than the rest of the **M. madida, M. pumila** and **M. subulata** alliances with the exception of **M. aff. paulistana** The root system seems to stem from the original pseudobulb rhizome, but in fact runs back from each of the closely packed, sequential pseudobulbs, under sheaths surrounding the thick woody rhizome, until they hit the substrate where they penetrate and branch. These roots are long, thin, profuse and branching. The closely held pseudobulbs are dark green, spindle-shaped, nearly 2.0 cm. tall by 0.5 cm. in diameter with two equal, apical, almost terete, dark green leaves up to 6.0 cm. long by 0.2 cm., deeply grooved and long-lived.

Inflorescence and Flowers: The flowers appear in November on short basal inflorescences. We have only examined dead and dying flowers on pollinated ovaries so consequently can not furnish an adequate description.

Flowering Periods: November.

Pollination: Unknown, but the colony we found was over 50% fertilized.

Cultivation: As for all in the above mentioned alliances but always remembering a high level of humidity and good light.

Maxillaria subulata
Alliance | *1863a* (Plate 31)

Etymology: Possibly a new species.

Habitat: An epiphyte from section VI found at mid-tree level in original elfin forest at 900-1100 M.asl, where it enjoys dappled light, good humidity and air movement.

Occurrence: Rare. We have found this plant on only one mountain ridge.

Plant: A medium-sized pendent epiphyte which can form large colonies. The pseudobulbs are ovate to cylindrical and measure 3 cm. long by 1 cm. diameter and arise closely and alternately from a densely sheathed, thick rhizome and a string of 20 pseudobulbs may measure 15 cm. long. There are two dark green, needle-like, long-lived leaves at the apex of each pseudobulb which measure 15 cm. long by 0.15 cm. in section. The root system is like others in this Alliance with roots from the new growths running back along the rhizome and under the sheaths until they reach the substrate where they branch and penetrate cracks in the bark and detritus. These roots are fine, white somewhat brittle and extensive.

Inflorescence and Flowers: Pairs of flowers on short pedicels appear from opposite sides of the most recent pseudobulbs and are densely sheathed for a third of their length by fine brownish tissue. The flowers are attractive, cup-shaped and deep dark wine colour, particularly so at the base of the protruding lip. The sepals are almost equal, lanceolate and measure 1.5 cm. by 0.5 cm. The petals are smaller, lanceolate and measure 1.2 cm. by 0.5 cm. The lip is vaguely trilobed, a broad spade-shape with two rounded shoulders at the base and a nose-like callus dropping from the central base to one third of the lip's length. The overall colour is an even darker red wine while a yellow streak runs for half the length of the callus.

Flowering Periods: Late March until early May when the flowers can last for up to 18 days.

Pollination: Probably pollinated by stingless bees.

Cultivation: As this plant is distinctly pendent, it is best grown on a thick bark plaque with its base firmly anchored to the substrate and applying the natural conditions where possible.

General: We are unable to identify this plant from available literature and our visiting experts have taken specimen flowers and subsequently lost them. Either this is a new species or a tetraploid form of **Maxillaria acicularis**, flowering in April and May where the latter flowers in December and January.

Maxillaria vernicosa
Barb. Rodr. | *1864*

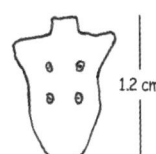
1.2 cm

Etymology: *Vernicosa* is Latin for varnished, referring to the shiny, varnished sheen of the lip.

Habitat: A tree-top epiphyte from original elfin forest at 900-1100 M.asl, on mountain ridges in section VI where it receives high light and humidity and the strong winds typical of cloud forest.

Occurrence: Occasional, although it can be confused with a stunted **M. acicularis** when not in flower.

Plant: A small pendent epiphyte often found in large clusters. The small pseudobulbs are an elongated egg-shape and measure 1 cm. high by 0.4 cm. They are smooth and dark green but become wrinkled and brownish with age and emerge closely and alternately, almost in clumps, from a thick, heavily scaled rhizome. There are two apical, needle-like leaves which are almost terete, dark green, long-lasting and measure 2 cm. long by 0.15 cm. The pendulous plant, hanging in thick strings of up to 10 cm. is firmly anchored to the substrate, firstly, by a plethora of thin roots which emerge from the basal pseudobulb and secondly by additional roots emerging from the bases of the new growths which are ducted to the substrate under the sheaths and scales that surround the rhizome.

Inflorescence and Flowers: Short, 1 cm. long single-flowered inflorescences arise from the bases of the most recent pseudobulbs with pale purple flowers. The sepals are broadly lanceolate and 1.3 cm. long by 0.5 cm. The petals are the same shape but narrower and measure 1.15 cm. by 0.47 cm. The lip is vaguely trilobed, almost ovate and measures 1.3 cm. by 0.7 cm. with dark purple, low, wart-like protuberances at the tip and a faintly yellow, vertical, nose-like 0.4 cm. callus at the base.

Flowering Periods: May and June and the flowers last two weeks.

Pollination: Probably pollinated by stingless bees.

Cultivation: Best on bark plaques and hung in full daylight with due attention to the humidity and air movement requirements. As the plant is pendent and with a root structure which runs from the new growths along the rhizome to the base, care must be taken to fix the base firmly to the substrate.

Maxillaria valenzuelana
(A. Rich.) Nash | 1867 (Plate 32)

Etymology: Named after **Eloy Valenzuela** (1755-1833) who collected in Colombia.

1.1 cm

Habitat: Found in section VI at 700-800 M.asl on moss-covered trunks at mid-tree level in original humid forest and enjoying medium light, high humidity and low air movement.

Occurrence: Occasional.

Plant: A small to medium-sized pendulous epiphyte which resembles a small Iris. There is no apparent pseudobulb or stem. The leaves are a narrow lanceolate, light green and vary in size on mature plants from 10 cm. to 27 cm. long and from 1.0 cm. to 1.8 cm wide. The leaves are folded inwards and fused for 20% of their length from the tip and the new leaf emerges from this sheath. The fused apical section is thickish

and succulent and presumably has a storage function. The root system is extensive, branching and adherent and composed of many brown stiffish roots.

Inflorescence and Flowers: A 3 cm. inflorescence arises from the base of the most recent leaf and appears just above the leaf sheath, bearing a single flower which is generally borne on the plants shaded underside. The yellow sepals are almost equal and lanceolate, measuring 0.9 cm. long by 0.6 cm. The petals are the same shape, narrower and the same butter-yellow. The lip is a vaguely trilobed spade-shape, yellow and speckled with blood-red blotches and with a waxy nose-like callus running the length of the central lobe. It measures 1.0 cm. long by 0.7 cm.

Flowering Periods: February and March and an unpollinated flower may last for two weeks.

Pollination: Unknown.

Cultivation: We recommend growing on bark plaques ensuring that the base is firmly fixed to the substrate, and attending to the habitat conditions.

General: This species was originally named *M. iridifolia* (Batem.). **Hoehne** moved it to his genus *Marsuparia Hoehne* in a section Iridiphyta, plants with apparently no stem or at least a stem hidden by equitant sheaths holding gomiform leaves, perpendicularly rather than transversally flattened. **Pabst** reinstated it in *Maxillaria* as *M. valenzuelana*.

Maxillaria johannis
Pabst | 1870

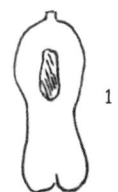

Etymology: Not known.

1 cm

Habitat: Grows as a lithophyte on humus-covered granite rocks in full sun in large colonies with over 30 stems, in sections V, VI and VIII within both transitional and original scrub forest at between 500-700 M.asl. It requires high humidity during spring, summer and fall, with a dry winter and moderate air movement.

Occurrence: Rare to occasional.

Plant: A medium to large-sized plant forming large clumps and colonies. A 1 cm. thick rhizome produces an apical ovoid pseudobulb which is bi-laterally compressed, light green, 10 cm. tall by 3.5 cm. wide and 1.5 cm. thick. The rhizome extends upwards producing a 30 cm. leafy stem densely sheathed with overlapping bracts formed by old dried leaf bases. It holds 10-12 almost opposite, light green, linear-lanceolate, keeled leaves up to 15 cm. long by 2 cm. wide at the broadest part. The roots are 0.3 cm. in diameter, branched and dense, arising from all parts of the old rhizome.

Inflorescence and Flowers: Single, butter-yellow, half-opened flowers arise on a short 2 cm. inflorescence from the 6th to the 3rd apical leaf axils. The sepals are a blunt tipped ovate, 1.4 cm. long by 0.5 cm. and the laterals form a short spur at their bases. The petals are smaller, 1 cm. long by 0.4 cm wide. The lip is a slightly indented rectangle, 1 cm. long by 0.5 cm. and has a single drop-shaped central callus running to half its length and flushed red. There are also red speckles at the column base.

Flowering Periods: November; the flowers last for two weeks or so.

Pollination: Unknown and infrequent.

Cultivation: We have experimented with back bulbs attached both to tree fern plaques and planted in pots on a coarse mixture of 90% humus and 10% sand. The back bulbs have reacted well, flowering in two years in both situations.

Maxillaria cerifera
Barb. Rodr. | *1871* (Plate 32)

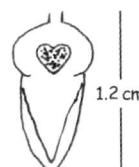

Etymology: *Cera-fera* is Latin for wax-carrying, referring to the waxy substance shown on the V of the lip.

Habitat: A mid- to tree-top species throughout section VI at 900-1500 M.asl in original forest, thriving in good light and capable of surviving long periods without significant moisture in winter, when it flowers profusely. It is very resilient but does not enjoy deep shade or high humidity. The plants die quickly when a host tree falls.

Occurrence: Common to abundant.

Plant: A small epiphytic clusterer, whose pseudobulbs arise at short intervals from a thick branching rhizome. They are ovate, green, ridged and small, 1.5 cm. long x 1 cm. wide. The plant forms colonies which cover 3 or 4 square metres of 12 cm. thick tree-top branches. There are two apical leathery, erect rabbit-ear-shaped leaves which are deeply keeled and measure 3 cm. long x 0.8 cm. wide. In large colonies 80% of the pseudobulbs have leaves. The roots are thin, fibrous, light grey, very adherent and 6-8 per pseudobulb each about 3 cm. long. A colony of 100 pseudobulbs, taking up 50 cm. of a 6 cm. diameter decaying tree top branch, has about 20 metres of roots, of which 30% are aerial and the remainder penetrate the bark and ultimately, rotten wood.

Inflorescence and Flowers: The flowers are single on stiff inflorescences rising up to 8 cm. from the pseudobulb base with one or two per pseudobulb. The flower is small but showy, 2.5 cm. long x 2.8 cm. wide, creamy yellow, with a vivid white V on the lip. The sepals are equal and a broad lanceolate, 1.3 cm. long by 0.5 cm. broad. The petals are narrowly elliptic, 1 cm. long by 0.3 cm. The lip is a blunt arrow-head shape 1.2 cm. long by 0.5 cm. broad at the shoulders, with a small heart-shaped callus at the throat. The above-mentioned, typical colony had 68 flowers all rising above the leaves.

Flowering Periods: December, and May through to July and the flowers last for three weeks.

Pollination: Seed set is frequent. We have noticed small bees collecting the chalk-like substance on the lip margin, but have not actually observed pollination.

Cultivation: Although a so called 'botanical', its profuse flowering habit with all the flowers standing proud of the leaves and its highly concentrated growth, can make a very attractive show. Due to its sprawling habit it should be grown on large bark or tree fern plaques or in wide baskets on a coarse fibrous compost, in good light and air movement with regular misting.

Maxillaria notylioglossa
Rchb. f. | *1871a* (Plate 32)

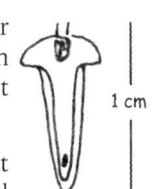

Etymology: *Notylia-glossa* is Greek for dorsal hump and tongue which presumably refers to the somewhat curious lip of this flower.

Habitat: An epiphyte from section VI at 1000-1500 M.asl, growing in original forest, from mid-tree to tree-top, on trunks and thick branches, enjoying good light and air movement and reasonable humidity.

Occurrence: Occasional, but because of its similarity to M. cerifera in all aspects, it may be commoner than we imagine.

Plant: A small, creeping or sprawling epiphyte often forming large groups. The almost ovate pseudobulbs are slightly laterally compressed, light green when young, turning brown with age and measure up to 1.6 cm. high by 1.0 cm. wide. Each bears two apical, opposite oblanceolate stiff, deeply-keeled dark green and long-lived leaves, with emarginate apices, up to 6 cm. long by 0.8 cm. The pseudobulbs emerge from a 0.3 cm. thick, tough, branched and tightly scaled rhizome at about 1.5 cm. intervals. The root system is not extensive and is composed of stiff, short, thin, light brown roots which penetrate the bark substrate very effectively. The roots arise along the rhizome length, but principally at the pseudobulb bases. As the plant has a capacity for multiple branching, it can form a dense mat on a tree trunk or totally envelop a thick upper branch.

Inflorescence and Flowers: Usually two single-flowered 6 cm. long inflorescences, tightly enveloped with many sheaths, arise from opposite sides of the new pseudobulb and appear just above the leaves. The sepals are lanceolate and measure 1.7 cm. by 0.4 cm. The petals are much narrower and 1.5 cm. long by 0.2 cm. The simple lip is a perfect arrow-head shape and measures 0.9 cm. long by 0.5 cm. wide at the base and like *M. cerifera*, it has a V-shaped line of waxy deposit next to the apical half of the margin and a small oblong callus at the base. The flower is a yellowish pale green.

Flowering Periods: The flowers last for ten days.

Pollination: As the flower grows older, the waxy white lip deposit reduces, or vanishes altogether, indicating visits by wax-collecting stingless bees.

Cultivation: As for *Maxillaria cerifera*.

Maxillaria jenischiana C. Schweinf. | *1874* (Plate 32)

Etymology: Not known.

Habitat: It is found on tree stumps along the high mountain ridges from 1000-1300 M.asl in original forest in section VI. We have also found it on the topmost forks of branches of giant forest trees in the river valleys. It needs good light and air movement together with reasonable humidity,

Occurrence: Occasional.

Plant: An epiphyte of medium size with light green pseudobulbs each an elongated, compressed oval emerging alternately from a long, thick and mostly aerial rhizome at 1.5 cm. intervals. The pseudobulbs are 2.5 cm. tall by 1 cm. wide and are generally smooth unless the plant is stressed. The rhizome, which may carry up to eight pseudobulbs over a 17 cm. length, is sheathed. The pseudobulbs are also partly sheathed. A typical colony will have up to 10 such active aerial rhizomes. The leaves are single or paired, apical, linear, dark green, 9 cm. long by 1 cm. wide, long-lived and deeply keeled. The root system is profuse with an enormous number of thin long, branching white roots partly arising from the base of the first bulb. On the rhizome, roots from new growths run back along and under the thickly bracted and sheathed rhizome and may reach the substrate. However as the plant is a leaf and detritus-gatherer, these roots also serve in the nutrient-gathering process.

Inflorescence and Flowers: One to three creamy-white flowers borne on individual 2 cm. inflorescences arise

from the latest pseudobulb base. The sepals and petals are ± equal in size and shape, lanceolate and measure 1.5 cm. long by 0.4 cm. The lip is three-lobed with two short opposite basal side-lobes and a rounded tear-shaped apical lobe and measures 1 cm. long by 0.7 cm broad at the side-lobes. A ten pseudobulb colony can make quite a display with 30 or so of these medium-sized and typical Maxillaria flowers.

Flowering Periods: Between September and December and occasionally in other months; the flowers last for 10 days or so and are frequently destroyed by insects.

Pollination: Frequent, up to 20%, and probably by wax-gathering bees.

Cultivation: Due to its pendulous nature, this plant is best grown on tree fern plaques, in good light and air movement, misted throughout the year.

Maxillaria loefgrenii (Cogn.) Pabst | *1876* (Plate 32)

Etymology: Named in honour of **A. Loefgren**, a Swedish botanist who was director of the Rio de Janeiro Botanic Garden.

Habitat: Grows at 1000-1200 M.asl in section VI in tall original forest at mid-tree level, in deep shade, amongst other epiphytic plants. It needs high humidity, limited air movement and appreciates detritus around its anchorage.

Occurrence: Locally common.

Plant: A curious epiphyte, drooping through the mid-tree often for several metres, anchored to a thick branch or trunk. The pseudobulbs are spherical, laterally compressed, 3 cm. high by 2 cm. wide, green, rapidly becoming grey-brown, and arise at 10-15 cm. intervals on a 0.6 cm. thick scaly rhizome. Five leaves are produced, four of which sheath the bulb and the single apical leaf, which is retained for two years, measures 27 cm. x 5.5 cm. On the young shoot, the leaves are all turned laterally towards the light giving the impression that they spring from the ground. They are deep green and distinctly keeled. The roots are very sparse and arise from the current pseudobulbs, some turning back and growing along the rhizome penetrating the scales. In the older parts of the colony, there is extensive rooting from the anchor pseudobulbs forming a mat in the peat and moss.

Inflorescence and Flowers: The flowers are small and produced from between the remaining leaf sheaths of recently matured pseudobulbs. The sepals and petals are a deep grey-brown, up to 0.9 cm. in length, while the cream labellum has two distinct lobes and is folded back on itself.

Flowering Periods: In November and December and the flowers last for a week or so.

Pollination: Frequent, but unknown.

Cultivation: As for *M. rigida*.

Maxillaria rigida
Barb. Rodr. | *1879* (Plate 32)

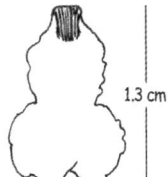

1.3 cm

Etymology: *Rigidus* is Latin for stiff, rigid, unbendable, probably referring to the texture of the flower parts.

Habitat: Grows on 60° angle rock faces forming large colonies but also as an epiphyte at mid-tree level in original forest and in scrub forest on mountain ridges from 1000-1300 M.asl in section VI where it enjoys high light and wind movement and constant mists.

Occurrence: Locally frequent.

Plant: A large pseudo-lithophytic plant which may form vast colonies, up to 10 metres square. Also an epiphyte. The pseudobulbs are ovoid, 2.5 cm. tall by 1.5 cm. wide, light green and smooth, borne on a 1 metre long, very tough, 0.5 cm. diameter, branching rhizome, at 4 cm. intervals. The leaves are lanceolate, light green, leathery, infolded and lightly keeled. There is a single apical and up to four basal leaves, 15 cm. long by 1.2 cm. wide. The roots are stiff, reddish brown, long and branching from the rhizome but concentrated at the pseudobulb base. The roots remain aerial until the weight of the growing rhizome puts them in contact with a substrate. The anchor roots adhere fiercely to rock and sedges and over the years the colony accumulates a very substantial quantity of detritus around its base.

Inflorescence and Flowers: One or more single flowers arise from the basal leaf axils on short pedicels. The sepals are pointed, 1.5 cm. long by 0.5 cm. broad and the petals are narrower, 1.5 cm. long by 0.4 cm. broad. The trilobed lip is flared at the apex with a cinched waist and is 1.3 cm. long by 1 cm. wide at the apex. The flower's colour is a light mauve, slightly tinged with green.

Flowering Periods: In January and the flowers last for 10 days or so.

Pollination: This is frequent but the pollinators are unknown.

Cultivation: A difficult candidate due to the large space between pseudobulbs and its sprawling habit. However, it could be grown on large tree fern plaques against a wall, in good light and air movement, constantly misting.

The genus
Ornithidium Salsb. ex. R. Br.

Richard Salsbury, an English naturalist in the late 18th early 19th centuries used a Greek compound word birdlike, or little bird referring to the small flowers of this genus. This seems to have been subsumed into *Maxillaria*. However as the flowers are so different from other *Maxillaria* species, we have left it in its one species genus as **Pabst & Dungs** listed it.

Ornithidium parviflorum
(P. & E.) Rchb. f. | *1880* (Plate 32)

Now: *Maxillaria parviflora*.

Etymology: *Parvi-florum* is Latin for small flower.

4 mm

Habitat: Found in humid original forest in section VI at 550 M.asl, enjoying reasonable light and wind movement and a high constant humidity.

Occurrence: Rare.

Plant: A untidy sprawling, pendent medium-sized epiphyte which morphologically fits into the *Maxillaria pumila* alliance. It is very similar to *M. ferdinandina* in that the roots of the unanchored pseudobulbs run back along the rhizome to the base or re-anchor on the substrate beforehand. The pseudobulbs are a broad linear, laterally compressed and light green, 4.5 cm. long by 1 cm., and arise at varying intervals from a thickly-sheathed, wandering and ever branching rhizome. There is a single apical, sheathed linear-lanceolate, dark green, long-lived and keeled leathery leaf, up to 12 cm. long by 1.5 cm. The root system, composed of many fine white roots is extensive both within the rhizome, aerial and basal.

Inflorescence and Flowers: A sequence of tiny whitish flowers emerge singly from the bases of the most recent pseudobulbs. These by contrast, are most un-*Maxillaria*-like and form a small cup. The sepals are ± the same size and shape; a broad convex lanceolate 0.4 cm. by 0.2 cm. The petals are smaller but much the same shape. The lip is curious as it appears trilobed but cup-shaped with no callus and is 0.35 cm. by 0.25 cm. broad.

Flowering Periods: March and the flowers last for a few days but the sequence may continue for a month or so.

Pollination: Unknown.

Cultivation: As for *Maxillaria ferdinandiana*.

The genus
Scuticaria Lindl.

Lindley raised this five species, essentially Brazilian and Amazonian genus, and called it by the Latin word *scutica* meaning whip like and referring to the long, pendent, whip like leaves. These leaves are apparently a relic of evolutionary change when South America passed through a long dry period and such leaves evolved to limit water loss. If this is so, we already know one of the survivors of the next dry millennium.

1. *Scuticaria hadwenii* (Lindl.) Hooker | *1881*
2. var. *strictifolia* | *1881a* (Plates 32-33)

Etymology: Isaac Hadwen was a collector from Liverpool, England.

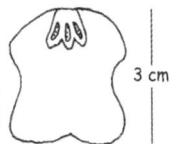

Habitat: Substantial colonies are most often found in original elfin forest at high altitude and at low level in the canopy between 1400-1600 M.asl in sections IV and VI. More rarely it is found at a lower altitude higher in the canopy. The widely stretching pigmented roots are unusual. It is often found together with *Sophronitis coccinea* and a sure sign that the forest is original.

Occurrence: Occasional.

Plant: A mid- to high-tree plant unless in high altitude elfin forest over 1500 metres, where it may appear at eye-level, but not lower. The rhizome is very short and the pseudobulbs are much reduced and sheathed. The colony appears as a clustered hanging bunch of green rats' tails. Unless you are observant, you may pass these by as the divided hanging leaves of a palm or strands of cactus or fern. Rat-tail or terete leaves are not uncommon among orchids and in *S. hadwenii* they may be pendulous and long or erect and short - the latter are referred to as *S. strictifolia* by **Hoehne**. **Pabst & Dungs** suggest that the erect forms are a response to light and dryness, the shade of the underside of branches inducing greater growth. Certainly the *strictifolia* forms occur in exposed positions, and mixed colonies occur. No differences in flower colour or features have been observed between the two forms. The leaves are pendulous, to 65 cm. in length, 0.5 cm. in diameter, circular in cross-section with a small groove along one side. The *S. strictifolia* leaf is 24 cm. x 0.4 cm (Plate 33). The roots are sparse, thick, 0.4 cm. diameter, strongly adherent and mahogany-red but stretching over a wide area. A 21-leaf colony had 25 roots averaging 60 cm. each, branching and giving the colony over 16 metres of root. Globular swellings are often seen on the roots and are probably galls.

Inflorescence and Flowers: The flowers measure 6 cm. square. They are large and showy and held on 7 cm. pedicels of which the major part is the ovary. The sepals and the petals are a greenish-yellow with maroon blotchings and are ± equal in shape and size, a broad lanceolate, 3.5 cm. by 1.5 cm. broad.

Flowering Periods: In October through to January. The strictifolia form blooms later in February and March and the flowers last for two weeks or so.

Pollination: Is not frequent; **Dressler** suggests *Euglossine* bees which are likewise infrequent at these altitudes.

Cultivation: On tree fern or bark plaques or even better on live tree fern, in dappled light with good wind movement and humidity. It resents transplanting so the longer the plant remains on its original substrate the more flowers you will get.

General: An example of a plant which evolved during one of periods when the Atlantic Rain Forest underwent a dryer time. Terete leaves are typical of plants that evolve in dry areas. They limit the water loss which broad-leaved plants cannot control on hot, dry, sunny days.

The genus
Dichaea Lindl.

Lindley named this genus using Greek *dichus* for two-fold because the leaves run in two close symmetrical rows on either side of the stem.

This is a Central and South American genus, of which **Pabst & Dungs** attribute some 22 species to Brazil and seven to the Organ Mountain Range. Within this grouping there are two distinctive lines; the completely pendulous which inhabit intensely humid environments, don't lose their leaves in dry spells and have fruit with soft spines. And those more tolerant of less humid conditions which, except for one species, **D. anchorifera** with spiny fruit, are semi-pendent, lose their basal leaves in dry periods and have smooth-skinned fruit. These latter species are colonizers of regrowth forest where there is a hint of reasonably constant humidity.

Dichaea muricata (Sw.) Lindl. | *1892* (Plate 33)

Etymology: *Muricata* is Latin for rough with short hard points like the shellfish *Murex*, probably referring to stiff hairs on the fruit.

Habitat: Found in section VI at 1000-1500 M.asl, in extremely humid original montane forest, almost always around or above a river or stream and growing on low trunks or small under forest branches. The conditions are intense humidity but low light and little air movement.

Occurrence: Common within its specific habitat.

Plant: We have found continuously growing stems of up to a metre in length which sometimes branch and are completely sheathed by the permanent and closely alternate leaf bases. These are 3 cm. long by 0.6 cm. and are lanceolate, deeply keeled, semi-succulent and deep green. The basal halves envelop the stem and are themselves partly enveloped by the preceding leaf. Up to six stems grow from a thin rhizome. The plant is anchored by a weak root system composed of short, white roots and usually covered by moss.

Inflorescence and Flowers: In ideal conditions, up to 20 semi-open light yellow to cream flowers emerge on short pedicels from the bases of the leaf bracts in a random manner, towards the apical half of the stem. The concave, broad lanceolate petals and sepals are similar in size, 1 cm. long by 0.5 cm. The lip is a squat anchor-shape and measures 0.7 cm. long by 0.9 cm. wide.

Flowering Periods: Late November until December and lasting for a week at 1000 M.asl. The distinctive fruit is covered by stiff, white hairs.

Pollination: **Dressler** suggests that *Euglossine* bees are the pollinating culprits. We question this as such bees are rare in this habitat but the short-lived flowers are almost always pollinated.

Cultivation: As the plant is entirely pendulous and its root system must not dry out, we suggest growing it on a moss-covered bark or in a basket on a moist, well-aerated substrate with the natural conditions.

General: This plant, and all its parts except the flower, is twice the size of **D. pendula**, and the flower, while similar in size, is a pale yellow to cream as opposed to the speckled white tepals and purple lip in **D. pendula**. Also, we have not found **D. pendula** above 900 M.asl, nor **D. muricata** below 1000 M.asl.

Dichaea pendula
(Aubl.) Cogn. | *1893* (Plate 33)

Etymology: *Pendula* is Latin for hanging down, referring to the plant's growth habit.

Habitat: Epiphytic and sometimes lithophytic in conditions of extreme high humidity, low light and low air movement. Found growing on tree trunks, lianas and under forest trees around streams in original forest at 300-600 M.asl, in sections IV and VI.

Occurrence: Common in its very specific habitat.

Plant: This is a pocket-size edition of **Dichaea muricata** and while it shares all the same characteristics, all segments of the plant are significantly smaller and tighter. It has multiple pendulous stems up to 35 cm. long. The leaves which are closely alternate, are lanceolate, deeply keeled and 1.7 cm. long by 0.4 cm broad. The basal half enfolds the stem, itself partly enfolded by the previous leaf. A typical stem has over 50 identical leaves and in high humidity they show no disposition to wither and fall. The root system is weak and composed of short, thin, white roots which precariously fix the plant to its substrate. Additional roots emerge at the leaf bases, running along the stems underneath the sheathing leaves like **Zygopetalum pedicellatum** and **Maxillaria acicularis**.

Inflorescence and Flowers: This plant does not flower freely and one rarely finds more than half a dozen flowers on the typical six to eight stemmed plant. The flowers have short pedicels and emerge from the leaf bracts of the distal half of the stem. The flowers are extremely pretty. The concave sepals and petals are similar size, broadly lanceolate and measure 1.0 cm. long by 0.6 cm. wide, light yellow, speckled with wine-coloured blotches. The typical Dichaea anchor-shaped lip is squat, wider than long and measures 0.7 cm. long by 1.1 cm. wide and is a beautiful dark purple colour.

Flowering Periods: Late November to early December and the flowers last only for a few days.

Pollination: Pollination is by *Euglossine* bees.

Cultivation: As for **D. muricata** but requiring a higher winter temperature.

Dichaea aff. bryophila
Rchb. f. | *1898*

Etymology: *Bryo-phila* is Greek for moss-loving, referring to this plants propensity for humid places.

Habitat: An epiphyte, from sections IV and VI between 250 and 600 M.asl, found in dark, almost windless, intensely humid original forest on low tree trunks and lianas, and always close to a water course.

Occurrence: Common in its niche.

Plant: A semi-pendulous epiphyte; a good specimen can have up to 15 stems, each up to 30 cm. long which emerge from a contorted rhizome where the root system also appears. The stems are a pointed oval in cross-section and are tightly sheathed by the basal 1-2 cm. of the leaf bases of both current and fallen leaves. The leaves are alternate at 1 cm. intervals and appear shortly above the stem bases. They are linear-lanceolate and measure up to 8 cm. long by 0.5 cm. and are dark green. In dry periods during winter or in dryer habitats, the older leaves die off. The root system is confused and composed of a mass of thin, white and brittle roots which sometimes branch and arise from all the stem bases and sometimes from stem internodes. These axillary roots sometimes project at right angles to the stem or a short distance but more frequently run back down the stem, underneath the sheathing to join the basal root mass.

Inflorescence and Flowers: Flowers arise on 1 cm. pedicels from the leaf bases and as many as 40 flowers may be produced on an eight-stemmed plant. All parts of the flower are milky white, intensely speckled with deep purple. The concave, lanceolate dorsal sepal is 0.8 cm. by 0.2 cm. and the similarly shaped lateral sepals are 0.9 cm. by 0.3 cm. The petals are more densely speckled and a narrower lanceolate. The lip measures 0.7 cm. long by 0.65 cm. at the anchor points and the fruit is egg-shaped and smooth surfaced.

Flowering Periods: December and the flowers last only for a few days.

Pollination: **Dressler** suggests *Euglossine* bees are the pollinators.

Cultivation: It is difficult to simulate the conditions that this species enjoys in nature. However, and as the name suggests, bryophila = moss-loving, our experience has shown that it grows well on a moss-covered tree fern or bark plaque in shade, with constant high humidity and mild air movement.

General: The flowers are similar to *D. cogniauxiana* but are larger with a more pointed lip tip and distinctive hairy ear-like appendages to the column. The plant is also larger and the leaves more freely spaced than *D. cogniauxiana*.

Dichaea cogniauxiana
Schltr. | *1899* (Plate 33)

Etymology: Named in honour of **Alfred Cogniaux**, a famous Belgian botanist.

Habitat: Principally found on the slopes below mountain ridges and on tree trunks in scrub forest. However it is also found throughout original forest and in regrowth at mid-tree level. It requires medium humidity and light and some air movement, at between 600-1400 M.asl in all sections.

Occurrence: Common.

Plant: Although it has some of the characteristics of *D. muricata*, this species neither requires the intense

humidity needed by the former and further is a pioneer plant in regrowth forest. In particularly dry periods, old leaves fall. It is monopodial with a continuously growing stem of 0.3 cm. diameter bearing closely alternate and opposite leaves along its length. These are 3 cm. x 0.4 cm., narrow, lanceolate, dark green and deeply keeled. An average of 15 leaves is retained on the growing stems which branch and form intermediate nodules, roots emerge which penetrate the substrate after about 7 cm. of aerial growth. The anchoring roots at the colony base are gross and tangled, adherent and always covered by moss and detritus.

Inflorescence and Flowers: Short 1 cm. stems with single flowers arise from the leaf sheaths of the previous years growth; the flowers are delicate and very beautiful, a translucent white, tinged with pink and dotted with wine coloured spots. The petals and sepals are of similar size, 0.7 cm. by 0.3 cm. and a broad lanceolate. The lip is anchor-shaped, white and spotted, 0.6 cm. long by 0.6 cm. broad at the anchor points. In appropriate conditions, a mature colony can have 20 or so of these delicate flowers.

Flowering Periods: During October and November and the flowers last for a week or so.

Pollination: **Dressler** suggests male *Euglossine* bees. However, these are rare at 1400 M.asl so there must be another pollinator. Pollination is frequent, with up to 50% of the flowers with fruit.

Cultivation: On tree fern plaques, in good dappled light and air movement, always kept slightly damp and with high humidity levels maintained throughout the year.

Dichaea mosenii
Cogn. | *1909* (Plate 33)

Etymology: Named after **Hjalmar Mosén**, a Swede who collected in Brazil from 1873-1876.

Habitat: Found in section VI at 400 M.asl in original humid forest at low tree trunk level requiring very high humidity, low light and air movement.

Occurrence: Rare.

Plant: An unusual *Dichaea* in this section since the growth habit is quite different from all other local species. A well-developed plant shows up to 10 reed-like sheathed stems which are erect, a flattened oval in section and up to 30 cm. long and with continuous growth, monopodial. Alternate leaves arise from the base section to the apex, at 2 cm. intervals. The leaves are linear-lanceolate, keeled, light green and measure 7 cm. by 0.4 cm. broad. Older leaves tend to fall during long dry spells. The root system is strong, composed of thinnish, pearl-white roots which arise in quantities from the stem bases.

Inflorescence and Flowers: The flowers arise on very short single 1 cm. inflorescences from the leaf axils on the previous years stem's growth. Two to 5 flowers appear which are an overall yellow through the sepals and petals, while the broad lip is cup-like and creamy white. The sepals are ± equal, lanceolate and 0.6 cm. long by 0.3 cm. broad. The lip is vaguely reminiscent of a blunt anchor, almost round with a broad throat and a rounded apex, the opposite mid-sides showing the two faint anchor points. It measures 0.5 cm. by 0.55 cm. broad.

Flowering Periods: November and the flowers last for only a few days; however, as they tend to open sequentially a pretty show can be seen for up to a week.

Pollination: Infrequent and probably by *Euglossine* bees, according to **Dressler**.

Cultivation: On tree fern plaques, in shade, high humidity and low air movement.

Dichaea anchorifera
Cogn. | *1911*

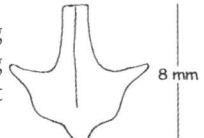

Etymology: *Ankyra* is Greek, meaning anchor and *fera* is Latin meaning bearing, referring to the very distinct labellum.

Habitat: Found in original forest on tree trunks, vines and tree ferns in deep shade, with low air movement and high humidity at 600-800 M.asl in section VIII.

Occurrence: Locally frequent in its niche.

Plant: A small to medium-sized semi-pendulous epiphyte whose distinctive characteristics are a hirsute fruit, the most anchor-shaped lip of all the *Dichaea* species in Brazil and that the leaves seem not to drop off during dry spells. The plant is monopodial with continuously growing flattish oval stems, up to 30 cm. long by 0.3 cm. wide which arise from a central tangled mass. Roots emerge from the internodes, which may penetrate and adhere when they contact the substrate. The leaves are closely alternate and opposite, starting at the base infolding the stem for about 1.0 cm., linear-lanceolate, keeled, papery, light green, long-lived and up to 3.5 cm. long by 0.5 cm broad.

Inflorescence and Flowers: Short 1 cm. single-flowered inflorescences emerge from the leaf sheaths of the previous years growth. The flowers are very similar to *D. cogniauxiana*, but slightly smaller and more bell-like, and pale pink speckled with light brown. The sepals are a narrow, sharply pointed lanceolate 0.8 cm. long by 0.2 cm. at the base. The petals are narrower, baseball-bat in silhouette, 0.5 cm. long by 0.15 cm. The lip is

an almost classic anchor-shape, 0.8 cm. long, narrow at the waist and 0.4 cm. wide at the anchor tips.

Flowering Periods: January and the flowers last for a few days.

Pollination: Pollination is frequent and **Dressler** suggests male *Euglossine* bees.

Cultivation: As for *D. cogniauxiana* with a little less light but more constant humidity.

General: **Pabst & Dungs** do not show a line drawing of the flower, however, they place it in the *D. brachypoda* alliance which is the only group outside of the *D. pendula* alliance showing an ovary with soft spines.

The genus
Gomesa R. Br.

Robert Brown named this genus in honour to **Dr. Bernardino Gomes**, a surgeon in the Portuguese navy who was also a botanist and described and collected many Brazilian plants during the early 19th century in Rio de Janeiro. His statue still stands in the Lisbon Botanical Gardens. Way back when we wrote Orchids of the High Mountain Atlantic Rain Forest in Southeastern Brazil we were very sceptical of the number of *Gomesa* species recorded, imagining many of them to be "variations on a theme", in our sainted innocence.

The genus seems to be exclusively Brazilian and further, the centre of dissemination is the Organ Mountain Range. Of the 13 species discovered 11 should be found in this area and we have found 8.

The cultivation of these species is very easy in a temperate Brazilian mountain climate and in European horticulture. All are easy to germinate in vitro and are perfectly happy when de-flasked, potted, and placed on the window sill above the sink in the northern hemisphere's taxing winters.

Some of these species are the first colonizers of regrowth forest on the eastern scarp slopes similar to their *Warmingia* and *Notylia* cousins on the anticline.

Gomesa crispa
(Lindl.) | *1914*
Kl. & Rchb. f. (Plate 33)

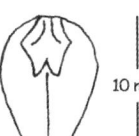

Etymology: *Crispus* is Latin for crisp, irregularly waved and twisted, referring to the tepal margins.

Habitat: It is found in all areas in the Organ Mountain Range at 500-1500 M.asl. It is common at ground level in deep forest, in regenerating forest on mossy stumps and in scrub. High and constant humidity and little air movement are the conditions where the best plants are found.

Occurrence: Common.

Plant: A medium-sized epiphyte or terrestrial. The plant is very variable in size but the pattern is always recognizable. Clusters of 20-30 pseudobulbs are not unusual, sometimes hanging from a tangled root mass. The pseudobulbs are 8 cm. high x 3.5 cm. and are light green, a flattened oval, smooth with fine white lining at the base, becoming ridged at the apex and deeper ridges and yellowing develop with age. The rhizome is short and indistinct. Two apical, lanceolate leaves measure 20 cm. x 3.5 cm., and there are 2-3 small subtending leaves at the pseudobulb base. The leaves are dark green, lighter green on more exposed plants, and are retained on the previous 4 or 5 pseudobulbs. The roots are thin, 1-2 mm in diameter, almost wiry and 20 or more produced per bulb. For a 10 bulb colony growing epiphytically, 22 metres were measured running vertically up and down the supporting tree. On terrestrial specimens, a thick network is produced, much contorted and giving a texture like a wire pan scrubber.

Inflorescence and Flowers: Two to five inflorescences, up to 30 cm. long, are produced from the base of the previous mature pseudobulb, which arch and open to give 30-40 greenish-yellow flowers each in the form of a figure of a baggy-panted clown. The sepals are ± equal, between lanceolate and spathulate, 1.2 cm. long by 0.5 cm. broad. The laterals are fused at the base only. The petals are the same shape but slightly smaller. The lip is reflexed to 90°, crisped at the margins 0.8 cm. by 0.4 cm. broad, with two calli at the throat. The dorsal sepal and petals are curved around the column area. The lateral sepals are expanded slightly to give the baggy-pants effect.

Flowering Periods: May through to July and the flowers last for three weeks or so, becoming yellower with age.

Pollination: This is sometimes frequent; the flowers have a very strong agreeable scent. **Dressler** suggests bees.

Cultivation: An easy orchid to grow as it can accept neglect and punishment, however if it is well grown it is quite a sight. Grow in pots on a rich fibrous compost, in dappled light, some wind movement and a reasonable constant humidity and it should flower in the northern hemisphere around Christmas.

Gomesa barkeri
(Hooker) Regel | *1917* (Plate 33)

Etymology: Not known.

Habitat: Epiphytic in the low tree zone in original forest at 1300-1400 M.asl in section VI enjoying low light, high humidity and moderate air movement.

Occurrence: Occasional, but it may be more common than we think as when not in flower is identical to other *Gomesa* species.

Plant: A small clustered epiphyte similar to all the other Gomesa species which in themselves are variable. Perhaps the pseudobulbs are narrower and more wrinkled than other species. These are oval, light green, laterally compressed are 4-5 cm. tall, 1.5 cm. broad and 0.5 cm. in section. There are two long-lived apical leaves which are dark green, stiff, coriaceous and lanceolate, measuring 11 cm. long by 1.7 cm. wide. One leaf is always slightly longer than the other. The root system is not as well developed as in *G. recurva* or *G. crispa*, but is more adherent and without aerial roots.

Inflorescence and Flowers: The flowers are in a raceme of up to 20 on a 10 cm. basal inflorescence. They are delicate, reminiscent of *G. glaziovii* in form and colour and also have entire margins. The sepals and petals are more or less equal, lanceolate and measure 1.3 cm. long by 0.2 cm. broad. The lateral sepals are partly fused and the ovate lip has a central elevated callus with two short pin-like parallel protuberances at its apex. It is 1 cm. long by 0.25 cm. wide.

Flowering Periods: In November and the flowers last for up to three weeks.

Pollination: **Dressler** suggests small bees.

Cultivation: As for other *Gomesa* species, grown on a fibrous substrate in pots, watering during the summer but only misting thereafter.

Gomesa fischeri
Regel | *1919* (Plate 33)

Etymology: Not known.

Habitat: An epiphyte found between 800-1400 M.asl, in section II, IV and VI at low levels, in both original and regrowth forest with low light and air movement but high humidity.

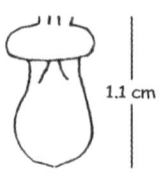

Occurrence: Occasional.

Plant: A very variable low-tree epiphyte or even terrestrial in deep leaf litter with typical *Gomesa crispa* alliance morphological characteristics. The pseudobulbs are closely packed and arise from a 0.4 cm. diameter rhizome at 0.5 cm. intervals and are ovoid, bilaterally compressed and up to 5 cm. long by 1.5 cm. wide, bearing an apical pair of unequal, linear-lanceolate leaves, each up to 30 cm. long by 3 cm wide, mildly conduplicate, dark green and long-lived. The root system is partly aerial with up to ten white, unbranched roots up to 30 cm. long, which arise from the pseudobulb bases and are longer when attached to or penetrating a substrate.

Inflorescence and Flowers: Up to 50 greenish flowers arise on a 25 cm. long pendent cone-shaped inflorescence. The flowers are densely packed, apple-scented and open simultaneously. The dorsal sepal is linear-lanceolate and 0.7 cm. long by 0.2 cm. broad. The lateral sepals are fused for 70% of their length and together measure 1.0 cm. long by 0.5 cm. The petals are the same shape as the sepals and 0.8 cm. long by 0.15 cm. The simple, shield-shaped lip measures 1.0 cm. long by 0.4 cm. wide with a double toothed basal callus which shows a hint of a red rim to the stigmatic cavity.

Flowering Periods: December and the flowers last for two weeks.

Pollination: Probably pollinated by bees.

Cultivation: Treat as the more common *G. crispa* and grow in a pot or basket with general compost, watering and feeding well during active growth.

Gomesa laxiflora
(Lindl.) Kl. & Rchb. f. | 1921
(Plate 34)

Etymology: *Laxi-flora* is Latin for loose-flowers, referring to the free and vaguely irregular arrangement of the flowers.

Habitat: An epiphyte from section VI at 900-1200 M.asl, found in regrowth forest at low tree trunk level, and requiring reasonable humidity, dappled light and low air movement.

Occurrence: Occasional but it may be commoner than we think due to its similarity to other *Gomesa* species when not in flower.

Plant: A small, delicate epiphyte with tightly clustered pseudobulbs which arise from a short thin rhizome and are ovate, light translucent-green, strongly bilaterally compressed and 2.5 cm. high by 1.5 cm. wide at the broadest point. The two unequal apical, long-lived leaves are broadly elliptical, lightly keeled, stiffish and pale green, measuring 6 cm. long by 2 cm. wide. The root system is extensive and adherent with fewer aerial roots than other *Gomesa* species and these roots

are thin, brittle and white to green and a typical plant has ten main roots, some branching and, in total, over 2 metres of root length.

Inflorescence and Flowers: A short, 10 cm. long, pendent, basal inflorescence, with up to ten small white to pale green alternate flowers, is produced in spring. The sepals and petals are equal in size and shape, measuring 0.8 cm. long by 0.3 cm. wide and are elliptical to lanceolate with the laterals fused for 66% of their length. The wedge-shaped lip measures 0.8 cm. by 0.4 cm. and is curved backwards with two distinct short, tram-like calli. There is a faint red margin around the stigmatic cavity.

Flowering Periods: October and the flowers last for two weeks.

Pollination: Probably pollinated by small bees.

Cultivation: As easy as other members of the **G. recurva** alliance.

Gomesa recurva
R. Br. | 1923 (Plate 34)

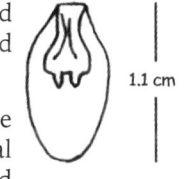

1.1 cm

Etymology: *Recurvus* is Latin for curved backwards, referring to the backward curving lip.

Habitat: Found in the low to mid-tree in open forest and often a terrestrial on roadside banks. It requires good light, air movement and not excessive humidity and is found from 1000-1400 M.asl throughout all sections of the Organ Mountain Range.

Occurrence: Common.

Plant: A tightly clustered epiphyte or terrestrial with light green to yellow pseudobulbs depending on exposure to light, each 7 cm. long by 3.5 cm. wide and a flattened pointed oval on a thickish branching rhizome. The two erect lanceolate leaves are dark green, pointed and deeply keeled, 15 cm. long by 3.5 cm. wide and long-lived. The root system, composed of a profusion of thin, long, white roots, is penetrating and adherent but rarely aerial as in the similar G. crispa.

Inflorescence and Flowers: The flowers are alternate, up to 20 in number and produced on a pendulous, basal, green, spicate inflorescence up to 20 cm. often with 4 or 5 inflorescences to a plant. Flowers arise from the whole length of the inflorescence and are most commonly yellow to orange. The petals and sepals are of similar size, 1 cm. long by 0.13 cm. wide, pointed at the apex with the lateral sepals partly fused. The apical half of the lip is curved downwards and backwards and is 1.0 cm. long and 0.45 cm. wide with twin, central, ridge-like calli at the throat.

Flowering Periods: November through to February and

the flowers last for three weeks or so.

Pollination: As for *Gomesa crispa*.

Cultivation: As for *Gomesa crispa*.

General: Differs from **G. crispa** in the following ways:

1. Flowers in midsummer, **G. crispa** flowers in June.

2. The plant is smaller and more compact than **G. crispa**.

3. Petals and sepals are smooth edged and not all over crinkled as with **G. crispa**.

4. **G. crispa** prefers deep under forest with a high humidity whereas **G. recurva** is more catholic in its choice of ambient.

5. **G. recurva** has lateral sepals which are always partly fused whereas **G. crispa**'s may be separate.

Gomesa sessilis
Barb. Rodr. | 1924 (Plate 34)

Etymology: *Sessilis* is Latin for stalk less or apparently so, referring to the flowers attachment to the inflorescence.

Habitat: Humid forest at low tree trunk level requiring shade, some wind movement and reasonable humidity in section V at 800 M.asl.

Occurrence: Occasional but may be more frequent as when it is not in flower it could be confused with any other *Gomesa* species in the *G. recurva* alliance.

Plant: A typical *Gomesa* with pseudobulbs which arise at 1 cm. intervals and are light green, bilaterally compressed, oval, 8 cm. long by 1.5 cm. wide and 1 cm. broad. The two apical leathery leaves are opposite, light green, keeled, lanceolate, long-lived and supple, 20 cm. long by 2 cm. wide. The untidy root system partly adheres to the substrate but also has a number of aerial roots. These are long, white, sometimes branching and numerous, arising from both the pseudobulb base and the thick connecting rhizome.

Inflorescence and Flowers: The plant we are describing had a 55 cm. long, basal, pendent, light green inflorescence with some 40 whitish green flowers borne alternately at 1 cm. intervals on its apical 45 cm. While these flowers showed the typical recurved lip of this alliance, the major difference was in their size. The dorsal sepal, broadly lanceolate, is 1.5 cm. long by 0.3 cm. wide. The lateral sepals, fused for 40% of their length, are the same shape but slightly narrower, 1.8 cm. long by 0.29 cm. The petals are similar in shape and size. The lip shows two significant vertical parallel "ears" on either side of the throat. The shape is a broad

obtuse lanceolate measuring 1.4 cm. by 0.4 cm. broad. And, of course, the pedicel is minute.

Flowering Periods: January and the flowers last for two weeks or so.

Pollination: By bees.

Cultivation: As for other species in this alliance.

Gomesa glaziovii
Cogn. | 1935 (Plate 34)

Etymology: Named in honour of **Glaziou**, a French botanist and landscape gardener whose work is still extant in Rio de Janeiro, Petrópolis and Nova Friburgo.

Habitat: This plant is frequently found on mountain ridges at 1200-1500 M.asl in section VI, in colonies that start growth on the ground but straddle up nearby shrubs reaching nearly 3 metres in height. The roots are only slightly adherent and easily removed suggesting that they are not the main agent of this climbing. The colonies are so interwoven with other true climbers that the support seems to come from other faster-growing species which trap the rhizome onto the tree.

Occurrence: Common.

Plant: This species is a straggler and with its long rhizomes can attain heights of 2 to 3 metres or more although it usually starts at ground level. The rhizome branches frequently and on the oldest parts it may decay so reactivating the pseudobulbs below the break. The pseudobulbs are a green translucent, flattened oval, 3.0 cm. x 1.5 cm. wide through which vertical veins can be seen. There is a single apical leaf, a regular pointed oval of 11.0 cm. x 2.0 cm. Three smaller leaves subtend the pseudobulb and 8 or so smaller leaves surround the previous years rhizome. These will form the papery scales. Roots are produced from the first to third scales behind the pseudobulb and each produces a single root up to 24 cm. long with a white velamen and a green tip. These are aerial and are much reduced in the second year.

Inflorescence and Flowers: Paired basal spicate inflorescences are produced, about 10 cm. long and bear 10-15 pale green flowers, similar to **G. crispa** but smaller and more delicate and have a faint citrus smell. The sepals and petals are equal in size and shape, lanceolate, 1.1 cm. long by 0.5 cm. broad, but the lateral sepals are fused for 2/3 of their length. The lip is single-lobed, reflexed for the apical half with a twin-nosed callus which is almost a round platform at its throat. Instead of the marked crimson spot at the top of the lip in most species, there is a yellow spot.

Flowering Periods: January and the flowers last for three weeks or so.

Pollination: This is infrequent and **Dressler** suggests bees.

Cultivation: Tricky, we have planted them on tree fern plaques with some success, however perhaps pots on a rich fibrous compost with a stick of cut tree fern to climb in shade, good air movement and a constant high humidity.

The genus
Binotia Rolfe.

Raised by **Rolfe**, the name refers to the founder of the famous orchid nursery *Orquidário Binot*. It is a single species genus, so far only found in the Atlantic rain forest of the states of Rio de Janeiro and São Paulo.

Binotia brasiliensis
(Rolfe) Rolfe | *1936* (Plate 34)

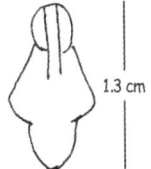

1.3 cm

Etymology: Requires no interpretation.

Habitat: Found in section VI at 400-600 M.asl, on low tree trunks and shrubs in fairly open maturing regrowth forest, enjoying dappled light, some wind and all year round, fairly high humidity.

Occurrence: Rare; but may be commoner as when it is not in flower, it is easily confused with *Gomesa* or *Rodrigueziella*, both of which are earlier synonyms.

Plant: A small epiphyte with pale green oval, bilaterally-compressed pseudobulbs which are 2.5 cm. long by 2.0 cm. wide and arise closely and alternately from a thickish sheathed and woody rhizome. There are two apical, lanceolate, papery green leaves which are 15 cm. long by 1.5 cm. and faintly keeled. The roots are long, thin, light green, fragile, branched and adherent.

Inflorescence and Flowers: A wine-coloured, bracted, semi-pendulous inflorescence, up to 25 cm. long, arises from the base of the most recent pseudobulb. Eight or so flowers are borne alternately at 1 cm. intervals on the apical 9.0 cm. Smaller inflorescences may be produced, flowering sequentially. The sepals and petals are about equal, lanceolate and 1.3 cm. long by 0.4 cm. wide. The lip is vaguely trilobed, a blunt lanceolate, 1.2 cm. long by 0.6 cm. wide at the side-lobes and with a central hairy section. The overall colour is greenish-white with some brown blotches on the sepals and petals.

Flowering Periods: December and January and the flowers last for ten days or so.

Pollination: Unknown.

Cultivation: On tree fern or cork plaques in good dappled light, with some air movement and good year round humidity.

General: The plant is sensitive and its correct growing niche must be found.

The genus
Rodrigueziella O. Ktze.

Established in 1891 by **Otto Kuntze** from Germany, who was a collector in Itatiaia, to replace the old name *Theodorea*. and dedicated to our very own **Barbosa Rodrigues**. But he got mixed up with the last letter and substituted a z for an s. In any event it is an interesting Brazilian genus of five species, of which all are attributed to the Organ Mountain Range.

Rodrigueziella gomezoides
(Barb. Rodr.) Pabst | *1938* (Plate 34)

Etymology: *Gomezoides* means that the plant resembles a *Gomesa*.

10 mm

Habitat: R. *gomezoides* is found throughout region VI at 1000-1500 M.asl, generally in low scrub and along ridges but also in young regrowing forest close to the ground and on the upper branches of trees in original forest.

Occurrence: Occasional.

Plant: A small epiphytic clusterer with smooth, laterally compressed ovoid pseudobulbs. These are a light translucent green, borne alternately and closely packed on a tough, fibrous and branching rhizome and measure 3 cm. x 1.5 cm. There is a pair of apical, light green lanceolate leaves, 12 cm. long x 1.6 cm. which are retained for several years. The roots are profuse, white and adherent and attain a length of 20 cm., branching under rough bark and moss.

Inflorescence and Flowers: Each new pseudobulb will produce a pair of 12 cm. pendulous racemose inflorescences, with up to 14 flowers each. A mature colony may have up to 8 inflorescences with 70 or so flowers. The incurved pointed sepals and petals are overall a light yellow with a brown base. The lip, curving to the back, is a cream colour with orange marks on the throat. The sepals and the petals are ± equal, 1.8 cm. by 0.3 cm. broad, lanceolate, while the vaguely four-lobed lip is 1 cm. by 0.4 cm. wide at the heart-shaped lower lobe. The callus is almost separate and triangular with two nose-like, small protuberances which take up the entire throat.

Flowering Periods: In January and February and the flowers last for over three weeks.

Pollination: Seed pods are frequent, up to 50% take and **Dressler** suggests bees.

Cultivation: As for *Gomesa crispa*.

Rodrigueziella handroi (Hoehne) Pabst | *1939* (Plate 34)

Etymology: Not known.

Habitat: *R. handroi* is found together with *Gomesa* spp. in the scrub forest on the sides of high mountain ridges on twigs and thin branches a few metres above the ground, at 1300-1550 M.asl in section VI. Its basic requirements are high humidity, low light and low wind movement.

Occurrence: Occasional.

Plant: This long-lived and medium-sized epiphyte has pseudobulbs which are ovoid and laterally compressed, light green, closely alternating on a short tough rhizome and measuring 6 cm. x 2.4 cm. The two apical leaves are pointed, narrow and deeply keeled ovals measuring 9 cm. x 2.7 cm. and are light green and stiff and remain for two years on the pseudobulb. The root system is strong and adherent with medium thick roots arising from the pseudobulb base and penetrating the surrounding substrate with ease, anchoring the plant firmly to the host.

Inflorescence and Flowers: A pair of racemose inflorescences is generally produced from the bases of the most recent pseudobulbs. Each is about 18 cm. long with up to 12 flowers, with slightly incurved sepals of yellowish green. Sepals and petals are of similar size, about 0.7 cm. x 0.3 cm., all with three bold, claret coloured lines running their length. The lip is yellow and has a typical *Gomesa* reflexed form.

Flowering Periods: July though to October and the flowers last for three weeks.

Pollination: This is frequent and **Dressler** suggests bees.

Cultivation: As for *Gomesa crispa*.

Rodrigueziella jucunda (Rchb. f.) Garay | *1940*

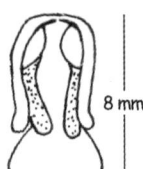

Etymology: Not known.

Habitat: Found in section VI at 1100 M.asl in regrowth forest or scrub, at mid to low tree level requiring shady light, high constant humidity and some wind movement.

Occurrence: Occasional.

Plant: A small clustered epiphyte vegetatively like a *Gomesa*. The pseudobulbs are light green, a strongly bilaterally compressed oval, 2.5 cm. long by 0.5 cm. Two apical, long-lasting, dark green, lanceolate, slightly keeled, stiffish leaves are held on short false petioles and are up to 10 cm. by 1.8 cm. wide. The root system consists of a few long, thin to thickish, green to white roots which run up and down the thin branches forming the substrate.

Inflorescence and Flowers: A 10-15 cm basal inflorescence arises from the most recent pseudobulb, initially subtended by a short basal leaf. Four to six alternate flowers are held on the apical 6 cm. at 0.5 cm. intervals. The sepals are ± equal, a very broad lanceolate, 0.6 cm. long by 0.32 cm. The petals are almost the same shape and size but slightly acuminate and wider. The lip is very distinctive, consisting of a round apical lobe 0.3 cm. by 0.3 cm., and a complex almost rectangular basal lobe showing erect sail-like, parallel protuberances at the throat, extending in diminished form to the distal lobe. The overall size is 0.7 cm. by 0.3 cm. wide. The flower's base colour is a light yellow, with all parts showing faint reddish parallel lines.

Flowering Periods: February and the flowers last for up to two weeks.

Pollination: Unknown.

Cultivation: As for *Gomesa* species.

Rodrigueziella petropolitana Pabst | *1941*

Etymology: Presumably the first specimen examined came from Petrópolis, RJ.

Habitat: A single colony of half a dozen plants was found in scrub forest at 1600 M.asl in section VI,. enjoying good air movement, high light and mountain-top high humidity.

Occurrence: Rare.

Plant: Very similar to *R. gomezoides* and possibly a variation of it. A small clustered epiphyte with pseudobulbs tightly held on a short branching rhizome. These are light green, ovate, laterally compressed and up to 3.5 cm. long by 1.5 cm. wide. There are two apical and two subtending opposite leaves on the new growths. The apical leaves are long-lived, while the basal pair last for a year only. All leaves are a narrow, keeled lanceolate, up to 8 cm. long by 1.2 cm. wide. The root system is strong, *Gomesa*-like and composed of numerous long, white, sometimes branching roots, some of which are aerial while the remainder adhere to the substrate.

Inflorescence and Flowers: One or two opposite inflorescences emerge from the bases of the most recent pseudobulbs. These are thin, stiff, up to 8 cm. tall and hold a raceme of up to 12 small, orange to yellow flowers. The sepals are lanceolate, 1.3 cm. by 0.25 cm. wide. The petals are smaller and slightly broader but the same shape and 1.1 cm. long by 0.28 cm. The lip is vaguely trilobed, 1.3 cm. long by 0.45 cm., acuminate, with a pale orange, narrow, longitudinal callus running from the throat to half the lip's length.

Flowering Periods: January and the flowers last for two weeks.

Pollination: Unknown.

Cultivation: As for *Gomesa* species.

The genus *Oncidium* Sw.

Olof Swartz, a famous Swedish botanist during the late 18th and early 19th centuries gave the genus name using the Greek word *onkos* meaning mass or pad and referring to the swollen calli on the lips of almost all species. *Oncidium* is a New World genus composed of about 400 species of which 105 are attributed to Brazil. Of these some 46 are attributed to the Organ Mountain Range. We have found 38 species and a number of natural hybrids, the most significant of which are between *O. marshallianum* and *O. crispum* and *O. forbesii* and *O. gardneri*. Our favourites, without a doubt, are the glorious yellow *O. marshallianum* and the smaller white and pink *O. truncatum* and *O. waluewa*. However, the Oncidiums have something for everybody, large and small, terrestrial or epiphyte.

Oncidium pumilum
Lindl. | *1949* (Plate 34)

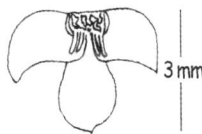

Etymology: *Pumilus* is Latin for dwarf, close-growing, short and in this instance probably refers to the very close growing bunched small flowers.

Habitat: Though we have found isolated plants at 1000 M.asl in scrub regrowth in section VI, they are probably garden escapes. We have also found *Coelogyne fimbriata* and *Dendrobium nobile* in similar circumstances. Its true home seems to be in the drier transitional forest in section V. **Bechtel Cribb & Launart** state that it was introduced to European culture in 1825 from Botafogo, Rio de Janeiro, now perhaps the most densely inhabited piece of urban real estate in the world, with blast furnace heat in summer.

Occurrence: Occasional.

Plant: A small epiphyte, with numerous, alternate minute, squat ovate, dark green, sheathed pseudobulbs, 0.4 cm. tall by 0.3 cm. wide, arising from a thick branching rhizome. The leaves, in a splayed-out arrangement, are single, apical, long-lived, oval, dark greyish green, thick and keeled, 8 cm. long by 4 cm. wide. The root system is strong with 6 or 7 white, relatively thick, non-branching, adherent roots up to 20 cm. long, which arise from each pseudobulb and remain active for many years.

Inflorescence and Flowers: A stout basal, erect, paniculate, inflorescence of some 10 cm., bears up to 40 small yellow and brown flowers. These are unmistakable, due to the curious lip; wider at 0.6 cm., than long, 0.3 cm., trilobed, anchor-shaped, yellow, streaked by dark magenta on the upper edge, and a thick inverted 'V' shaped central callus. The sepals and petals are the same size, all incurved, yellow, streaked horizontally with brown, roundish to rectangular and 0.3 cm. long by between 0.15 cm. and 0.2 cm. wide.

Flowering Periods: In March and the flowers last for three weeks.

Pollination: Unknown.

Cultivation: In pots on a coarse fibrous compost in good light with some air movement and good humidity during summer and autumn, misting thereafter.

Oncidium divaricatum
Lindl. | *1951* (Plate 35)

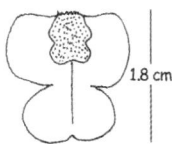

Etymology: *Divaricatus* is Latin for spreading apart at a wide angle probably referring to the paniculate inflorescence.

Habitat: An epiphyte found at mid-tree level in original forest, in section VI at 800-1200 M.asl, enjoying dappled light, good wind movement and a constant medium humidity.

Occurrence: Occasional.

Plant: This **Oncidium** belongs to the popularly named Mule-ear group. In more systematic terms it belongs to section **Pulvinatum**, *O. pulvinatum* Alliance, *sensu* **Pabst**, which contains this plant and three others. A compact, tufted epiphyte with closely packed, discus-like pseudobulbs. These measure 2 cm. tall by 2.5 cm. broad, 1 cm. in section and are light green. They have a single elliptic, deeply keeled stiff, coriaceous, long-lived, light green apical leaf measuring 20 cm. by 5 cm.

The root system has an abundance of long, thick, very adherent roots, white along their length and green at their growing tips. These emerge from the base of the new growth, itself arising from a thick, tough, wandering rhizome.

Inflorescence and Flowers: A stiff, erect, 25 cm. long, paniculate inflorescence emerges from the base of the most recently matured pseudobulb with up to 25 small yellow and brown flowers. The sepals are more or less equal, spathulate, 1 cm. long by 0.6 cm. broad, mustard-yellow for the apical half and light brown at their bases. The petals are a somewhat squarer spathulate but otherwise the same measurements and colouring as the sepals. The lip has four lobes, two upper and two lower, each is almost round and 1.2 cm. in diameter, lemon-yellow, splotched with reddish brown.. The overall size is 1.8 cm. by 1.5 cm. broad. The powder-puff-like callus at its throat is likewise four-lobed, 0.7 cm. long by 0.7 cm. broad.

Flowering Periods: February and May and the flowers last for three weeks.

Pollination: Probably **Anthophorid** bees, according to **Dressler**, but rarely found with fruit.

Cultivation: In pots, on a coarse, fibrous, well-drained substrate requiring water during spring and summer and regular misting thereafter.

Oncidium pulvinatum
Lindl. | 1952

Etymology: *Pulvinatus* is Latin for cushion-shaped, or strongly convex, referring to the cushion-like callus at the lip's throat.

Habitat: The species prefers horizontal branches in the mid-tree area where the leaf network collects fallen detritus. Found in original forest at 1000-1200 M.asl in section VI, enjoying dappled light, good air movement and high humidity.

Occurrence: Occasional.

Plant: This is a mid-tree epiphyte which resembles O. sphegiferum vegetatively although it is more of a clusterer. The pseudobulbs are elliptic and tightly compressed, green and smooth and measure 3.5 cm. long x 2 cm. There are one or two stiff, row-boat shaped apical leaves which measure 25 cm. x 6 cm. and are light green with a distinct keel and retained for several years. The root system is extensive and penetrating.

Inflorescence and Flowers: A mature plant will sport several semi-erect paniculate inflorescences. These can be 50 cm. long and bear 50 or so flowers each. The flowers are similar in form and colouring to those of

O. sphegiferum although they are smaller. The sepals are spathulate, 1.1 cm. long by 0.75 cm. while the petals are the same size but the spoon-like apical section is almost square. The four lobes to the lip are almost equal, each a semicircle. The overall length is 1.2 cm., breadth is 1 cm. There is an undivided ± round power-puff-like callus at its throat.

Flowering Periods: In April and the flowers last for up to a month.

Pollination: Infrequent. It is possible that the powder-puff-like callus acts as a deceit mechanism, simulating an insect when the flower nods in a breeze, and thus attracts bees, suggests **Dressler**.

Cultivation: In pots on a rich compost in dappled light, good air movement and reasonable constant humidity.

Oncidium robustissimum
Rchb. f. | 1953

Etymology: *Robust issimum* is a Latin superlative for most robust referring to the plant's habit.

Habitat: Found growing epiphytically within enormous numbers of bromeliads, on relict original forest trees at 100-200 M.asl, in open pasture land at the foot of section IV. It experiences full sunlight, high humidity and wind movement and has access to the detritus surrounding the bromeliad garden.

Occurrence: Although once common, it is now occasional to rare, due to the lack of forest in its original habitat.

Plant: A large, clustered epiphyte forming colonies with up to 15 active pseudobulbs. These are almost round, strongly laterally compressed, pale green, and measure 8 cm. long and 6 cm. wide and closely packed on a thick, woody rhizome. There is a single apical long-lived leaf which is short, tough to leathery, deeply keeled and pale green, measuring up to 15 cm. long by 6 cm. wide. The root system is extensive and penetrating and very difficult to measure as the plant is intertwined with many other epiphytes. The roots are long, white and thickish and sometimes branched.

Inflorescence and Flowers: A tough stem, semi-erect, round and 45 cm. long by 0.25 cm. wide emerges from the bases of the most recent pseudobulbs. This produces a wide paniculate inflorescence 25 cm. square with up to 30 flowers. The sepals are more or less equal in size, obovate and measure 1.0 cm. by 0.5 cm. The petals are irregular ovals measuring 0.9 cm. long by 0.6 cm. and are a mustard-yellow with faint brown markings. The four-lobed lip forms a divided square, 1.1 cm. long and broad. The apical lobes are pale yellow and the basal lobes are lightly brown-speckled with crenate margins on the underside. The callus is a small roundish powder-puff at the throat of the basal lobes.

Flowering Periods: Late November to mid-December and the flowers open simultaneously lasting for two weeks.

Pollination: Probably pollinated by *Anthophorid* bees.

Cultivation: This is a warm growing *Oncidium* which would receive heavy rain during the spring, summer and autumn but only mist during winter. It grows well on live bark but also in a pot with compost and decaying matter.

Oncidium sphegiferum Lindl. | *1954* (Plate 35)

Etymology: Not known.

Habitat: A mid-tree species, always found in mixed epiphytic company. It is sporadic and virtually confined to original forest where it is exposed to good light, high humidity and regular air movement, at around 1200 M.asl in section VI.

Occurrence: Occasional.

Plant: This is one of the larger mule-eared *Oncidium* species in the region. The pseudobulbs are elliptic, bilaterally compressed, smooth and a clear light-green. They measure 5 cm. long x 4 cm. and are clustered on a tough rope of rhizome. There is a single apical leaf measuring 21 cm. long x 10 cm. which is retained for four or five years. Leaves are light green, stiff and leathery and have a defined keel. The root system is extensive and penetrating but difficult to measure as the plant is generally mixed with other dense epiphytic growth.

Inflorescence and Flowers: A two metre long pendulous, paniculate inflorescence arises from the base of the most recent mature pseudobulb. As it grows it insinuates and supports itself through the surrounding foliage, eventually opening over 100 flowers. The sepals are spathulate, 1.3 cm. by 0.75 cm. broad, while the petals are the same size, but the spoon-like apical segment is almost square. The lip has four half-circle lobes with a two segmented powder-puff-like callus at its throat and is 1.75 cm. long by 1.5 cm. broad. The overall colour is mustard to clear yellow with some brown.

Flowering Periods: During December and the flowers last for up to a month.

Pollination: Infrequent; as for *O. pulvinatum*.

Cultivation: As for *O. pulvinatum*.

General: *O sphegiferum* is the epiphytic equivalent of *O. blanchetii* and *O. hydrophilum*, inasmuch as its long inflorescence with its myriad of flowers requires significant support from the surrounding epiphytic growth. There is no record of European cultivation which is a shame as the longevity of the flowers and the dramatic inflorescence merit the attention of growers.

Oncidium harrisonianum Lindl. | *1955* (Plate 35)

Etymology: Introduced to Europe in 1828 from the Organ Mountain Range by **William Harrison**, an avid collector of orchids in Rio de Janeiro State.

Habitat: Grows on moss-covered trunks at mid-tree level, generally near streams. The plants respond to strongish but diffuse light, high humidity and low air movement, constant through the year. It is found in original forest in section VI from 600-1300 M.asl, with a preponderance at the lower altitudes.

Occurrence: Common.

Plant: This is a small, variable mule-ear *Oncidium*. The pseudobulbs are elliptical, bilaterally compressed, light green and measure 1.7 cm. x 1.5 cm. wide. There is a single canoe-shaped leaf which is deeply keeled and 6 cm. long by 1.5 cm. wide. The leaves are greenish grey, stiff and leathery and survive for many years. The rooting system is not extensive but is adherent and penetrating and often moss-covered.

Inflorescence and Flowers: A 30 cm. long semi-erect, paniculate inflorescence is produced from the base of the most recent pseudobulb. Up to 40 half-open small and long-lived flowers are borne on several branches. The predominant effect is of a close, nodding reddish-brown group with yellow blotches and the mustard colour of the petal and sepal tips glowing discreetly. The dorsal sepal is a roundish oval, 0.6 cm. long by 0.4 cm., the lateral sepals and petals are similar in shape and size, rectangular, 0.6 cm. by 0.4 cm. broad. The four-lobed lip is 0.9 cm. long by 0.7 cm. broad at the flared apical lobes. The distinctive feature is an ivory 5-horned callus at the throat.

Flowering Periods: November and December but also in March and April and the flowers last for three weeks.

Pollination: Seed pods are frequent.

Cultivation: On fern or bark plaques observing the habitat descriptions.

General: We have also found a variety which is a pure butter yellow; var. *flava* or var. *concolor* (Plate 35).

Oncidium cornigerum
Lindl. | 1960 (Plate 35)

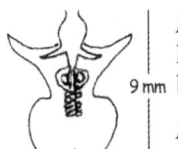

Etymology: *Cornis* is Latin for horn, horned and refers to the horns at the base of the callus.

Habitat: This is a denizen of the deep original forest, often found together with *Promenaea xanthina* and *Masdevallia infracta* on low tree trunks and lianas. It requires a highly humid and diffusely lit site, and specimens found outside such areas are invariably struggling and reach only 10% of their potential. Found at 600-1000 M.asl in sections IV, V and VI.

Occurrence: Occasional.

Plant: This is a medium-sized very variable epiphyte. The dark green pseudobulbs are cigar-shaped, tightly clustered and with fine longitudinal wrinkles. They measure up to 15 cm. x 3 cm. The dark green apical, floppy leaves have a pronounced midrib and measure 23 cm. x 4 cm. The extensive fine, white root system is often found under moss, and a 19 bulb colony had over 70 metres of roots.

Inflorescence and Flowers: Arching inflorescences of up to 75 cm. are produced, bearing up to 80 medium-sized, half-open, cupped flowers. The main feature is the yellow lip surrounded by a yellow barred fringe of chestnut petals and sepals. This is the only *Oncidium* species that we have observed with both an apical and basal inflorescence on the same pseudobulb. The dorsal sepal is a broad spathulate, 0.7 cm. long by 0.5 cm wide. The lateral sepals are fused for 95% and together are 0.7 cm. by 0.5 cm. broad. The petals are a broad lanceolate, 0.8 cm by 0.4 cm wide, while the four-lobed lip has two opposite basal horn-like side-lobes surrounding a complex red, gnarled callus. The apical lobes form a flared circle.

Flowering Periods: October and November but also in March and April and the flowers last for three weeks.

Pollination: Infrequent; **Dressler** suggests probably by oil-gathering bees.

Cultivation: In pots on a rich compost in deep shade and high constant humidity with some air movement.

General: Introduced to European cultivation in the 1880's from the Organ Mountain Range.

Oncidium aff. cruciatum
Rchb. f. | 1961 (Plate 35)

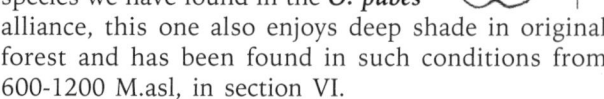

Etymology: *Cruciatum* is Latin for cross-wise, referring both to the shape of the flower and the callus.

Habitat: As with all the **Oncidium** species we have found in the **O. pubes** alliance, this one also enjoys deep shade in original forest and has been found in such conditions from 600-1200 M.asl, in section VI.

Occurrence: Occasional.

Plant: The dark green pseudobulbs are cigar-shaped and vary in size from 10-15 cm. long by 1.5-3.0 cm. diameter and are slightly laterally compressed. They bear two opposite, dark green, leathery and short-lived leaves which are a broad blunt lanceolate and measure 15 cm. long by 5 cm. This is probably the plant's most distinguishing feature. The root system is extensive and composed of numerous thin white and branching roots. This species resents transplanting and the roots soon lose their viability when taken from their original substrate.

Inflorescence and Flowers: A panicle of flowers is borne on a long, round, stiff inflorescence up to 30 cm. long, bearing up to 20 flowers. The overall colour is brown tinged with yellow. The petals and sepals are concave; the dorsal sepal is spathulate, obovate and truncate as are the petals, and all measure 1.2 cm. long by 0.9 cm. The lateral sepals are fused almost to their tips and the narrow spathulate pair together measure 1.2 cm. by 0.9 cm. The lip is extremely complex, trilobed and measures 1.3 cm. long by 1 cm. It is convex, with a broad, warty callus running for most of its length and the column has two distinct wings.

Flowering Periods: October and November; the flowers last for three weeks.

Pollination: Fruits are rarely seen and the pollinating organisms not known.

Cultivation: Treat as **O. cornigerum**.

Oncidium kautskyi
Pabst. | 1963

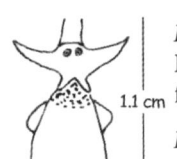

Etymology: Dedicated to **Roberto Kautsky**, a well known orchidophile from Espírito Santo.

Habitat: Found at 1400 M.asl in original forest in section VI, growing in shade, high humidity and low air movement.

Occurrence: Rare.

Plant: A mid-tree-trunk or under-scrub epiphyte. The pseudobulbs are cylindrical and tightly clustered, somewhat grosser than those of *O. truncatum* which it closely resembles, and are up to 7 cm. long by 2 cm. and dark green. They hold a pair of erect lanceolate apical leaves, up to 9 cm. long by 2.5 cm. and dark green. The roots are thickish, grey-white and extend greatly beyond the plants circumference.

Inflorescence and Flowers: A basal, pendent and branching inflorescence up to 20 cm. long, bears up to 10 flowers. The dorsal sepal, when flattened, is a squared-off spathulate, 1.1 cm. long by 0.6 cm. The lateral sepals are fused for 98% of their length and again, when flattened, together measure 0.8 cm. by 0.7 cm. wide with the same apical squaring off. The petals are spathulate, slightly divided at their apices and 1.1 cm. by 0.65 cm. wide. The lip, like *O. truncatum*, is 6-lobed with essentially the same configuration. The differences are in the composition; *O. kautskyi*'s lip is narrower in all its parts and its callosity is less complex. Overall it measures 1.1 cm. long by 0.6 cm. The petals and sepals are a very pale green towards their bases while their apices are a translucent milk-white. The lip is white with orange.

Flowering Periods: December and January and the flowers last for two weeks.

Pollination: Unknown.

Cultivation: As for *O. truncatum*

Oncidium lietzei
Regel | 1965 (Plate 35)

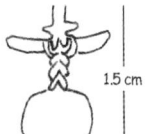

Etymology: This was discovered by **H. W. Lietz** in the 1880's and introduced to Europe at that time.

Habitat: A low-tree epiphyte generally growing on moss-covered branches and lianas. It requires high humidity, little air movement and only diffuse light. It is an orchid of the deep humid forest and has been found at all levels between 800-1100 M.asl in sections V and VI.

Occurrence: Occasional.

Plant: A medium-sized epiphyte with cylindrical pseudobulbs formed in a tight cluster, easily confused with O. cornigerum but is distinguished because it is smaller and flowers only in April whereas O. cornigerum flowers mainly in November. The pseudobulbs are up to 10 cm. long and 1-2 cm. wide. The two apical leaves are long-lasting, up to 15 cm. long and 4 cm. wide. The root system is not very extensive and the roots are buried under moss.

Inflorescence and Flowers: Flowers are produced on an arching 30 cm. long, paniculate inflorescence with up to 20 small, incurved flowers forming an attractive brownish dancing spray with the tip of the yellow lip winking. The dorsal sepal is broadly spathulate, 1.0 cm. by 0.7 cm. wide. The laterals are much smaller, fused for 99% of their length and together they are 0.7 cm. long by 0.4 cm. wide. The petals are narrowly spathulate and 1.0 cm. long by 0.5 cm. wide. The lip is complex, four-lobed with two opposite horn-like basal lobes and the apical lobes are round.

Flowering Periods: April and the flowers last for over three weeks.

Pollination: As for *O. cornigerum*.

Cultivation: As for *O. cornigerum*.

Oncidium truncatum
Pabst | 1969 (Plate 36)

Etymology: *Truncatus* is Latin for ending very abruptly, as if cut straight across, and refers to the lip's apex.

Habitat: A low tree-trunk epiphyte found in original forest in section VI at 1000-1400 M.asl. It requires deep shade, no direct sunlight and a high average humidity. We have never found this plant in regrowing forest.

Occurrence: Rare.

Plant: A small plant that is long-lived in a stable environment. The pseudobulbs are cylindrical and tightly clustered measuring 7 cm. x 1 cm., dark green and tapered towards the apex. There is a pair of apical leaves, 8 cm. long x 2.5 cm., erect, with a spine and are retained for several years. The root system, generally hidden under moss, is profuse and composed of fine white roots which run for distances up and down the support, branching towards their tips.

Inflorescence and Flowers: A spectacular pendulous inflorescence measuring up to 40 cm. is produced from the base of the most recent mature pseudobulb. As many as 50 delicate pink and white flowers flow out from the branched stem. The dorsal sepal is a very broad spathulate as are the similar sized petals which measure 0.9 cm. by 0.5 cm. broad. The lateral sepals are fused for two thirds of their length. Individually they are elliptic and 0.7 cm. by 0.3 cm. broad. The lip could be considered six-lobed. There are two basal opposite, small triangular horns followed at mid-lobe by two larger opposite triangular lobes which surround a warty callus. The apical lobe is slightly divided and quite square. The overall length is 0.9 cm. by 0.6 cm. wide at the mid-lobes.

Flowering Periods: January and February and the flowers

last for two weeks.

Pollination: Unknown and sadly, we have never found seed pods.

Cultivation: In pots or hanging baskets, due to the pendulous inflorescence, and on a rich fibrous compost, always kept damp but not overly so, and in shade with some air movement.

General: *O. truncatum* is not in cultivation although it should be. Its beauty, long-lived flowers and colour combination are all assets.

Oncidium fimbriatum
Lindl. | 1973

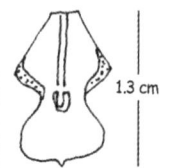

Etymology: *Fimbriatum* is Latin for having a fringe of hairs, referring to the hairs around the apical labellum lobe's base.

Habitat: Found in original forest at mid-tree level in section VI at 1000-1200 M.asl, requiring dappled light, high humidity and some air movement.

Occurrence: Occasional.

Plant: A medium-sized though variable epiphyte with cylindrical dark green pseudobulbs which are sheathed for the first year and up to 15 cm. long by 3 cm. broad. Two equal and opposite leaves emerge from the apex which are dark green, keeled, somewhat floppy, elliptic, lanceolate, long-lived and up to 25 cm. long by 4 cm. The root system, composed of long, white sometimes branching and adherent roots, is extensive and remains viable for many years.

Inflorescence and Flowers: The inflorescence is basal, long and pendulous and very similar to that of **O. sphegiferum**, up to 1.80 metres in length with up to 20 panicles each with 20 or so closely packed alternate flowers on the apical half. The dorsal sepal is a broad spathulate, 1 cm. long by 0.7 cm. The lateral sepals are lanceolate, 1 cm. long by 0.4 cm. wide and fused for 0.2 cm. at their bases. The petals are vaguely trilobed and further, mildly dentate at their apices and their overall shape is a broad and crude spathulate 1.5 cm. long by 0.6 cm. The lip almost defies description; an apical lobe almost round, 0.7 cm. deep by 0.8 cm. wide with a neck that holds two opposite linear wings, each 0.4 cm. long and a central marbled callosity which reaches from the throat to the start of the lower apical lobe.

Flowering Periods: December and March and the flowers last for three weeks or so.

Pollination: Infrequent and unknown.

Cultivation: As for **O cornigerum**.

Oncidium waluewa
Rolfe | 1976 (Plate 36)

Etymology: Not known.

Habitat: An epiphyte found in cool, scrubby original ridge forest at 1500 M.asl in section IV, enjoying high humidity, shade and some wind movement. The plant described was surrounded by healthy specimens of **Sophronitis coccinea** enjoying a long warm wet summer and a cool misty winter.

Occurrence: Rare.

Plant: A small to medium-sized epiphyte with tightly clustered pseudobulbs on a very short rhizome. These are smooth to faintly ridged, dark green, almost cylindrical but slightly laterally compressed and 3-4 cm. tall by 1.2-1.5 cm. wide. Each has a single dark green apical leaf on a short 1 cm false petiole, which is shiny, keeled, lanceolate, leathery, and measures 10-11 cm. long by 2-2.5 cm. wide and lasts for 2-3 years. The root system is extensive and composed of many short, white, 0.1 cm. thick, sometimes branching roots usually under mosses and lichens.

Inflorescence and Flowers: The pendent inflorescence, up to 10 cm. long arises from the base of the most recent pseudobulb. This is a light ivory colour, 0.15 cm. in diameter with up to 15 flowers held alternately at very short intervals, starting from the very base of the inflorescence. These are also ivory, with the petals showing an irregular number of close, wine coloured, horizontal stripes. The dorsal sepal is intensely concave above the column head, spathulate, 1.1 cm. long by 0.5 cm. The lateral sepals are fused to their apices and together measure 1 cm. by 0.5 cm. broad. The petals are a table-tennis-bat shape, 0.9 cm. long by 0.7 cm. wide showing close, fine, horizontal wine lines at their bases and apices, on a light ivory background. The lip consists of a bilobed, round apex joined by a thin bridge to a bilobed base. The basic colour is light ivory heavily blotched with wine; the length is 1.3 cm. by 0.45 cm. at the lobes. The column head is distinguished by two long drooping wings on each side of the stigma.

Flowering Periods: Late March and April and the flowers last for up to three weeks.

Pollination: Unknown.

Cultivation: As for **Sophronitis** species but remember the pendulous flowering habit which looks best on an ever humid plaque.

Oncidium trulliferum
Lindl. | 1977 (Plate 36)

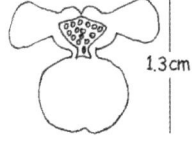

Etymology: *Trulliformus* is Latin for shaped like a bricklayers trowel, broadest below the middle with two equal straight sides meeting at the apex and two others shorter straight sides meeting at the base which describes perfectly the pseudobulbs.

Habitat: An epiphyte from sections III and VI found at the mid-tree level in original forest between 600 and 700 M.asl where it enjoys high humidity, reasonable air movement and dappled light.

Occurrence: Rare.

Plant: An epiphyte with distinctively elongated pseudobulbs which appear cylindrical, but on closer inspection, are strongly bilaterally compressed. They arise tightly from a thick rhizome and measure up to 20 cm. long by 3.5 cm. There are two or three lanceolate, apical leaves which are long-lived, deeply keeled, stiff and yellowish-green and measure up to 25 cm. long by 5 cm. The root system is not particularly strong and composed of a small number of thickish, white adherent roots which neither spread widely nor penetrate the substrate with any enthusiasm.

Inflorescence and Flowers: A stiff, tight paniculate inflorescence up to 60 cm. long, arises from the base of the most recent pseudobulb and can bear up to 150 smallish light yellow to mustard-coloured flowers of significant beauty. Perhaps it is the tightness of the panicle which gives the effect of a huge yellow mustard-blob. The dorsal sepal is a table-tennis-bat shape and measures 0.5 cm. long by 0.45 cm. wide. The lateral sepals are similar in size and shape but slightly asymmetrical. All have a base of yellow, strongly tiger-striped in brown. The petals are similar in shape and colour but slightly larger at 0.7 cm. long by 0.5 cm. wide. The four-lobed lip is complex and practically indescribable. It measures 1.3 cm. long and is mostly yellow.

Flowering Periods: March and April and the flowers last for three weeks.

Pollination: Pollination is probably by wasps; seed set seems infrequent.

Cultivation: Grow on bark plaques or in baskets and pots with a thick layer of well aerated compost. The plant must be well anchored because of its potential size and the natural habitat conditions should be reproduced where possible.

Oncidium hians
Lindl. | 1983 (Plate 36)

Etymology: Not known.

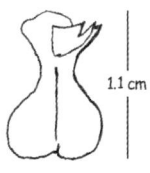

Habitat: Epiphytic in original forest at low to mid-tree-trunk level in section VI, enjoying dappled light, high humidity and reasonable air movement at 1000-1200 M.asl.

Occurrence: Occasional, but when found, it is normally in large colonies.

Plant: A small, tightly clustered, and charming epiphyte showing a marked similarity to **O. harrisonianum** and included in the mule-ear group. The pseudobulbs are discus-shaped, 1 cm. long by 1 cm. and 0.7 cm. in section, bilaterally-compressed, light green, emerging closely and alternately on a thick branching rhizome. Each pseudobulb has one, apical long-lasting, elliptic, deeply keeled, stiff, coriaceous, greyish-green leaf which measures up to 8 cm. long by 2.0 cm. broad. The root system is extensive and adherent, composed of thickish white to green roots which run up and down a tree trunk within bark cracks and generally under moss. A typical 15 pseudobulb plant has some 3 metres of active roots.

Inflorescence and Flowers: A 15 cm., erect, stiff, sectioned, claret-coloured, basal inflorescence emerges in October with a panicle of some 10 flowers whose main characteristic is an apparently narrow and long, bilobed yellow lip. This is in fact slightly back-folded at the central margins before the flared apex. It measures 1.1 cm. long and 0.4 cm. broad at the apex. The callus is extremely pronounced but simple and composed of two pairs of closely parallel, up-curved, short horn-like protuberances about 4 mm. long. The sepals are yellowish brown, more or less equal in size, elliptic and measure 0.5 cm. long by 0.3 cm. broad. The petals are the same size and colour but slightly squarish. All are slightly concave.

Flowering Periods: October and November and the flowers last for three weeks.

Pollination: **Dressler** suggests probably small **Anthophorid** bees.

Cultivation: Best grown on tree fern plaques or in pots on a compost formed of tree fern fibre, charcoal and 10% of coarse sand, requiring watering during spring and summer and misting thereafter.

General: This is one of few **Oncidium** species found in significant colonies. We have also found an all yellow variety; var. *flava*.

Oncidium kraenzlinianum
Cogn. | 1984 (Plate 36)

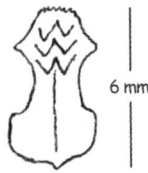

6 mm

Etymology: Named after *Fritz Kraenzlin*, a taxonomist held in some contempt by **Rudolf Schlechter**.

Habitat: Found at midtree at 1000 M.asl in remnants of original forest in section VII, a habitat similar to that of *Oncidium hians*.

Occurrence: Very rare.

Plant: A small delicate **Oncidium** with narrow, discus-like pseudobulbs, 1 cm. tall by 0.8 cm. and light green, gathered tightly on a short woody rhizome. The single apical leaf is light green, long-lived, floppy, up to 8 cm. long by 0.6 cm., keeled and linear-lanceolate. The root system, which spreads from the pseudobulb base, extends beyond the plants circumference and is composed of up to 8 ultimately branching roots, white at first, turning brown with age, adherent and penetrative.

Inflorescence and Flowers: A short, unbranched inflorescence, up to 15 cm. bears only three separated flowers on short pedicels. We doubt that this is typical, however only one ailing plant was found mixed with a colony of **Oncidium harrisonianum** – the plant has subsequently been revived in the Rio de Janeiro Botanic Garden orchidarium. The dorsal sepal is vaguely spathulate to lanceolate, and measures 0.5 cm. long by 0.35 cm. The lateral sepals are a very slightly asymmetric lanceolate, 0.5 cm. long by 0.25 cm. The petals are semi-spathulate, 0.5 cm. long by 0.3 cm. The overall colour is yellow, speckled with reddish brown. The lip´s apical lobes are round or obovate. There are two small side-lobes at the base, whitish below but running to brown and yellow. The apical area is fringed and there are purple-brown raised veins running towards the apex. The size is 0.6 cm. long by 0.4 cm. at the apical lobes. The two pollinia are held in a curious open-ended capsule like a column head and almost directly above the unusual short but elevated venation at the base of the lip.

Flowering Periods: March.

Pollination: Unknown.

Cultivation: Grow on tree fern plaques in shady humid conditions.

General: The plant was described in Martius's: *Flora Brasiliensis*, Vol. III Part 6 1904-1906. However this description was preceded by one in the June 18, 1896 *Gardeners Chronicle* under the synonym **Cyrtochilum micranthum. Krzl. N.** SP. The plant was introduced from Brazil by Messrs. **Sander** and **Co.** and it flowered in June 1896. However not one of the imported plants had a provenance, beyond simply Brazil.

Oncidium hookeri
Rolfe (Plate 36) | 1985
Oncidium loefgrenii
Cogn. | 1986
Oncidium raniferum
Lindl. | 1988

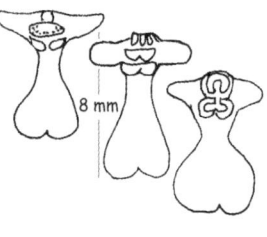

8 mm

Etymology: Named in honour of **Hooker** and **Loefgren**, respectively English and Swedish, and famous 19th century botanists. While *raniferum* means frog-like in Latin referring to the frog-like callus.

Habitat: These are all found from the topmost twigs of a tall forest tree, through the rotting stumps in regenerating forest, to leaf litter at 900 to 1500 M.asl in sections V, VI.

Occurrence: Common.

Plant: All three are very variable, widespread epiphytes with aerial-type roots which can also colonize rotting wood and the loose-leaf upper soil of scrub forest. The pseudobulbs are light green, conical, slightly compressed and deeply furrowed by purple tinged grooves. They measure up to 5 cm. long x 1.3 cm. There are two apical, ribbed light green leaves up to 15 cm. x 1.5 cm. which remain on the plant for three or four years. The root system is immense and in the case of a specimen that had taken over a rotting stump, we found over 30 metres of root arising from 15 sound pseudobulbs.

Inflorescence and Flowers: An erect basal paniculate inflorescence up to 40 cm long, bears up to 400 flowers. Their principal feature is a disproportionately large, claret-coloured callus shaped like a small frog and surrounded by tiny yellow sepals and petals. The sepals and petals are ± equal, broadly lanceolate, 0.3 cm. long by 0.1 cm. The four-lobed lip has two opposite small yellow horn-like lobes on either side of a complex warty wine-red callus, while the flared apical lobes are butter-yellow, skirt shaped and measure 0.6 cm. by 0.45 cm. broad.

Flowering Periods: From November to March and the flowers last for up to a month.

Pollination: Frequent; the flower shape and distinctive wine-red callus may deceive small bees to compete, or the flowers may possess oil glands to attract oil-gathering bees. Whatever the device, it is efficient.

Cultivation: One would imagine that plants with such catholic and wide-ranging natural habitats would be easy to grow; not so in our experience. Best in pots on a loose fibrous compost in good light with some air movement, never over watered but always kept slightly damp.

General: European cultivation dates from **George Gardner**'s collection in the Organ Mountain Range in 1838. Only with a hand lens can one differentiate between the three species; *O. loefgrenii*, having a grosser callus and a clearer bilobed lip while *O. raniferum* has the most frog-like callus of the three.

Oncidium concolor
Hook. | *1989* (Plate 36)

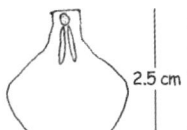

Etymology: *Concolor* is Latin for of one colour throughout, referring to this yellow flower.

Habitat: Found sporadically at mid-tree level in original forest where it appreciates shade, moderate air movement and a high, stable humidity in sections VI and VII at 1200-1400 M.asl.

Occurrence: Rare but locally common.

Plant: This is a medium-sized epiphyte. The pseudobulbs are a smooth deep-green and a compressed oval measuring 6 cm. long x 2 cm. wide. Plants are long-lived with up to twelve bulbs on a tough, short rhizome. The older pseudobulbs often reactivate giving a colony that may flower frequently. Two apical, opposite, lanceolate leaves are retained for some years and are light green and leathery measuring 9 cm. x 2 cm. The root system is prolific with eight strongly adherent roots per pseudobulb. A mature plant can have 28 metres of fine grey roots.

Inflorescence and Flowers: Up to nine flowers are borne on a 30 cm. arching inflorescence. The dorsal sepal is elliptic, 1.5 cm. long by 0.7 cm. broad. The petals are similar but the lateral sepals are fused for a third of their length, asymmetric lanceolates, 1.8 cm. by 0.5 cm wide. The bilobed lip is a blunt spade-shape 2.5 cm. long by 2.0 cm. wide and the entire flower is a brilliant sulphur yellow. The lip is disproportionately large and instead of the typical warty callus of an *Oncidium* there are the twin tracks of a *Miltonia* or a *Rodriguezia* reaching down into the throat.

Flowering Periods: November and December and the flowers last for three weeks.

Pollination: Infrequent.

Cultivation: A difficult plant to grow well. It also resents transplanting. Its extensive root system takes a long time to establish or re-establish. Grow in a large pot on a rich fibrous compost, in shade with some air movement and always kept slightly damp.

General: Introduced into Europe by **George Gardner**

in 1838 from collections made in the Organ Mountain Range in Rio State. In many aspects this species is more like a *Miltonia* than an *Oncidium*, especially considering the fine root system and the lip.

Oncidium cogniauxianum
Schltr. | *1992* (Plate 37)

Etymology: Dedicated to *Alfred Cogniaux* a famous Belgian 19th century botanist.

Habitat: Given that a principal requirement of this golden gem is a high humidity around the roots, the plant goes on to break all other rules. It is found on the topmost moss-covered twigs of trees in original forest, bleached by full sun, lashed by high winds and exposed to huge variations in temperature. But we have also found it flowering quite happily in the stable and gloomy environment of the under forest so beloved by *O. truncatum* and where no ray of sunlight penetrates, in section VI at 1000-1500 M.asl.

Occurrence: Common.

Plant: A small twig epiphyte that can form large colonies under favourable conditions. Tree-top specimens have light yellowish-green pseudobulbs which are ovoid, compressed and 2 cm. tall in 6 cm. clusters. There are one or two light green opposite apical leaves 10 cm. long by 1 cm. wide. The root system is extensive, fast growing and adherent.

Inflorescence and Flowers: Up to seven surprisingly large flowers arise from the base of the most recent pseudobulb on a 10 cm. racemose inflorescence which may appear twice each year. They are predominantly yellow with brown barring. Back bulbs frequently reactivate and a small colony may have several inflorescences. The dorsal sepal is elliptic, 0.7 cm. long by 0.4 cm. broad, while the laterals, fused for 5% of their length, are a narrow lanceolate, 1 cm. long by 0.4 cm. wide. The petals are a broadish, blunt lanceolate and measure 0.7 cm. by 0.25 cm. broad. The four-lobed lip has two opposite horn-like basal lobes, fringed at their lower margins and surrounding a complex warty callus. The widely flared apical lobes are together 0.7 cm. long by 1 cm. broad.

Flowering Periods: April through June and the flowers last for up to three weeks.

Pollination: Pollinators are unknown and seed set is rare.

Cultivation: In small pots on a coarse fibrous compost, in good light and air movement and always kept slightly damp.

General: The plant is very precocious and may flower

at three months from deflasking! On the other hand like most twig epiphytes it is not long-lived.

Oncidium longipes
Lindl. | 1994 (Plate 37)

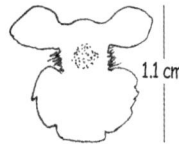

Etymology: *Longi* is Latin for long and *pes* is Latin for foot, probably referring to the length of the column-foot, or it could refer to the length of the peduncle.

Habitat: It is generally found among a collection of other epiphytes making it hard to spot, in the mid to upper-tree zone, in original forest at 800-1400 M.asl in sections V and VI. Here it enjoys a high humidity and dappled light with reasonable air movement. Perhaps it is the lack of these conditions which prevents it colonizing regrowing forest.

Occurrence: Occasional.

Plant: A small epiphyte found in the mid to upper tree zone mainly in original forest. The pseudobulbs are clustered, elongate-ovoid, compressed and often wrinkled, deep green and measure 3 cm. x 1 cm. and arise from a thick rhizome. There is a single apical leaf which is dark green and canoe-shaped, measuring 8 cm. x 1.5 cm. and has a distinct keel. The root system is extensive, white, fine, adherent and penetrating.

Inflorescence and Flowers: One to five flowers are produced on a thin erect racemose inflorescence up to 17 cm. long. The flower is large for a small plant and the main features are a large and hairy callus and wings within a deep yellow lip against a background of fawn and yellow barred sepals and petals. The dorsal sepal is lanceolate, 1.3 cm. long by 0.45 cm. broad while the laterals are fused for nearly half their length, lanceolate, 2 cm. long by 0.4 cm.. The petals are spathulate, 1.3 cm. by 0.6 cm. broad. The four-lobed lip has two roundish opposite basal lobes and a wide, faintly divided apical pair, 2 cm. broad by 1.1 cm. which are joined by a broad, fringed neck enclosing a complex noduled callus.

Flowering Periods: March to May when the flowers last for up to a month. At lower altitudes it may flower in December/January.

Pollination: This is very sporadic but when it happens, it pollinates en masse which indicates a group of far-ranging bees are responsible. The seed pods are curious, long, triangular in section, and need 12 months to mature.

Cultivation: As for **O. cogniauxianum** except use larger pots as this plant sprawls. Our most successful planting is on live tree fern where a single shoot fixed 10 years ago now has hundreds of flowers.

Oncidium uniflorum
Booth | 1995a (Plate 37)

Etymology: *Uniflorum* is Latin for one flower on an inflorescence.

Habitat: An epiphyte from section VI found between 1100 and 1400 M.asl, in original forest on thickish, mid- to upper branches and tree trunks and often in large colonies growing in high light with strong wind movement and a constant underlying humidity.

Occurrence: Common.

Plant: A small, creeping epiphyte with ovate elongate pseudobulbs measuring 2 cm. high by 0.5 cm., slightly laterally compressed and arising at 1 cm. intervals from a thick 0.4 cm. diameter, sheathed, branching and woody rhizome. Each pseudobulb bears a single, long-lived, lanceolate apical leaf, light green, deeply keeled and measuring 4.5 cm. long by 1 cm. The root system is composed of thin, wiry, tough and shortish roots which are adherent but not extensive.

Inflorescence and Flowers: A short basal inflorescence arises from the new pseudobulb. The spathulate dorsal sepal is 1 cm. long by 0.7 cm. while the oblanceolate lateral sepals are larger, partly fused at their bases and measure 1.3 cm. long by 0.4 cm. The petals are a ping-pong-paddle-shape, 1 cm. long by 0.7 cm. wide and all are a pale khaki with faintly yellow margins. The lip has four distinct lobes; the basal, opposite pair are smaller and almost round while the apical pair are much larger but the same shape. Overall, the lip measures 1.3 cm. long by 1.5 cm. wide and is golden yellow. The neck, joining the upper and lower lobes, is a broad triangle, intensely serrated at the margins and has a mass of short, yellow, finger-like calli on a reddish brown background.

Flowering Periods: Late December and January and the flowers last for three weeks.

Pollination: They are visited by Euglossine bees – personal observation.

Cultivation: In pots with a well-drained compost. Like a lot of **Oncidium** species, it thrives on living bark but with access to rotting wood.

General: While the flowers are smaller than those of *O. longipes* and are mostly single, an amateur would be forgiven for calling this plant **O. longipes** growing on a less than favourable substrate.

Oncidium barbatum
Lindl. | *1996* (Plate 37)

Etymology: *Barbatum* is Latin for bearded referring to the indentations on the sides of the lip.

1.1 cm

Habitat: An opportunistic epiphyte from sections VI and VIII, frequently found in abandoned orchards and coffee plantations around 600-700 M.asl, where it enjoys good light and air movement and reasonable humidity.

Occurrence: Sporadic.

Plant: A small, compact epiphyte with ovate, light green, laterally compressed and 'squared-off' pseudobulbs which measure 4 cm. long by 2.5 cm. wide and arise closely together from a thick woody rhizome and normally bear one apical and one smaller subtending leaf. The apical leaf is coriaceous, keeled and oblanceolate, dark green and measures up to 11 cm. long by 2.7 cm. The root system is vigorous and adherent and composed of a hessian-like mass of fine, long, greyish roots more reminiscent of a *Miltonia* than an *Oncidium*.

Inflorescence and Flowers: A stiff, woody and erect inflorescence, up to 40 cm. long arises from the base of the most recent pseudobulb with up to 30 flowers on the apical panicle. The dorsal sepal and petals are spathulate, crisped at the margins and measure 1.3 cm. long by 0.7 cm. The lateral sepals are a broad lanceolate and are fused for a third of their length. The overall tepal colour is yellow, lightly blotched with pale brown. The lip is deep yellow, distinctly trilobed and all three lobes are equal in size, almost round and measure 0.5 cm. diameter. The central structure has wart-like calli and is doubly serrate at the margins. The apical lobe and the calli vary from plant to plant.

Flowering Periods: April and May and the flowers last for three weeks.

Pollination: They are probably pollinated by bees.

Cultivation: Grow in pots, on bark or tree fern slabs. The compost should be well-drained and aerated and the plant requires dappled sunlight and the natural conditions described.

Oncidium aff. ciliatum
Lindl. | *1998* (Plate 37)

1.3 cm

Etymology: *Ciliatum* is Latin referring to the hairy fringe around the lip's central lobe.

Habitat: Found at mid-tree level on relict trees at 700 M.asl in section VI and VIII enjoying

high light and wind movement and good constant humidity.

Occurrence: Occasional.

Plant: A small to medium-sized creeping epiphyte. The pseudobulbs are elongate ovate, light green, deeply ridged, measuring up to 8 cm. long by 1.9 cm. wide and arising at up to 2.5 cm. intervals from a thick round, 0.5 cm. diameter, woody creeping and branching rhizome which is densely scaled. There are two apical, short-lived leaves, linear lanceolate, floppy, light green and deeply keeled which measure 22 cm. by 1.5 cm. wide.

Inflorescence and Flowers: Single short basal inflorescences emerge from new growths, measuring up to 5 cm. long, including the ovary, which is the only part appearing, the remainder being tightly swathed by short, light green, alternate sheaths. The tepals are all similarly shaped. The dorsal sepal is a blunt lanceolate measuring 1.4 cm. long by 0.4 cm. wide, while the laterals are fused for half their length and measure 2 cm. long by 0.4 cm. The petals are slightly broader and measure 1.5 cm. long by 0.5 cm. broad. All segments are slightly crisped at their margins. The trilobed lip is divided into upper and lower sections. The apical half is almost round, bright yellow and measures 0.8 cm. tall by 1 cm. wide with lightly crisped margins. The basal section has two more or less oblong opposite wings, each bright yellow and measuring 0.5 cm. long by 0.3 cm. wide. At their centre, an irregularly ridged round group of calli appears, flanked by faint reddish brown speckles. Two faint small tooth-like appendages appear at the lower base of each side-lobe. Sepals and petals are faintly barred with brown on a predominately bright yellow base.

Flowering Periods: April and the flowers last for up to three weeks.

Pollination: Probably Euglossine bees, according to **Dressler**.

Cultivation: As for *O. longipes* and *O. uniflorum*.

Oncidium crispum
Lodd. | *2006* (Plate 37)

Etymology: *Crispum* is Latin for crisped, irregularly waved and twisted, kinky curled which adequately describes the shapes of the flower parts.

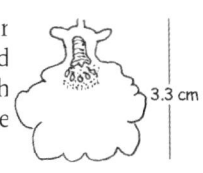

3.3 cm

Habitat: A mid- to high-tree epiphyte found from 800-1400 M.asl in all sections, in both original and regrowing forest on horizontal branches in good light, with significant air movement and wide ranges of temperature and humidity. It responds well to rotting wood and is one of few orchids that has no problem in colonizing regrowth.

Occurrence: Common.

Plant: A large plant which may be very long-lived. On prime specimens, we have found as many as 20 pseudobulbs, often with secondary colonies arising from reactivated back-bulbs. The pseudobulbs vary in size in response to nutrient availability and arise from alternate sides of a stout rhizome at 2 cm. intervals. They are dark-green to brownish depending on light. The paired apical leaves are lanceolate, deep green with a red tinge and measure 30 cm. x 5 cm. The root system is large; each pseudobulb can have 8 to 10 thick, white and branching roots often penetrating dead wood. A typical ten-year-old colony would have upwards of 50 metres of adherent roots.

Inflorescence and Flowers: Up to 100 large and showy flowers are produced on a 60-100 cm. stiff, branching paniculate inflorescence arising from the mature bulb base. They are reddish brown, but with the hook-nosed callus and the waist of the labellum a sulphur yellow; the flowers are variable and based on two main colour forms; firstly, that mentioned above and secondly an olive-brown type. The dorsal sepal is spathulate, 3 cm. long by 1 cm. broad, the laterals, fused for 30% of their length, are lanceolate and 3 cm. long by 1 cm. wide. The petals are large and almost round, 3.2 cm. long by 3.0 cm. broad, while the almost round lip is 3.3 cm. long and wide. All segments have crisped margins.

Flowering Periods: December and January and the flowers last for three weeks.

Pollination: Infrequent seed set probably pollinated by oil-gathering bees.

Cultivation: In large pots on a loose coarse fibrous compost taking care of the immense quantity of roots that will be produced; these only appear to be viable when they glue to a substrate, but will remain viable for many years. Grow with good light and air movement and keep damp all year round.

General: Introduced to Europe by **Loddiges** from Nova Friburgo in 1832 where it has been popular ever since. There is also a well known variety *O. crispum* var. *olivacea*, which is frequently found in section VI.

Oncidium marshallianum
Rchb. f. | *2008* (Plate 37)
Natural hybrid (Plate 38)

Etymology: **William Marshall** was a Scots collector in the 1850's.

Habitat: Found between 900-1500 M.asl in section VI growing as a mid

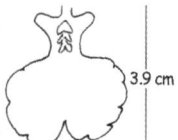

3.9 cm

to high-tree epiphyte in original forest, although we also find it in regenerating forest on the ground and in low scrub. It is most often found on medium to thick vertical branches with high air movement, good light and a wide range of temperature and humidity.

Occurrence: Common where it has not been collected out.

Plant: A large, long-lived plant with a strong tendency to reactivate back bulbs, resulting in colonies of many inflorescences. The vegetative characters are similar to *O. crispum* but differ as the pseudobulbs are smoother and greener and the leaves lack the reddish tinge. There are two or three floppy, 'rabbit-ear'-like apical leaves measuring 30 cm. x 6.5 cm. and these are retained for two years. The root system is extensive, tending to grow up and down trunks with 8-10 adherent and fast growing roots per bulb. An 8 year old plant had 75 metres of root. The plant really thrives when its roots come into contact with, and penetrate decomposing wood.

Inflorescence and Flowers: A stiff, hollow 75 cm. paniculate inflorescence arises from a mature bulb base branching dramatically and 60-70 large yellow flowers adorn this fan-like spray which can span 80 cm. width. So vivid is the yellow mass that it can easily be seen across a 500 metre valley with the naked eye and even by moon light! The dorsal sepal is elliptic, 1.5 cm. long by 1.0 cm. wide. The laterals are fused for 10% of their length are an asymmetric lanceolate, 2 cm. long by 0.8 cm. broad. The petals are violin-shaped, 2 cm. by 1.5 cm. wide. The lip is dominated by an almost round, divided apical lobe, 4 cm. long by 3.5 cm. broad. At its throat two opposite, triangular small wings appear surrounding a vaguely 5-nosed callus.

Flowering Periods: Mid-October to mid-November when the flowers last for three weeks or so.

Pollination: Infrequent and **Dressler** suggests oil-gathering bees.

Cultivation: As for *O. crispum*.

General: Quite the most superb *Oncidium* in the region but its flamboyance is also its death warrant because, during the flowering season, it is hunted mercilessly by amateur collectors. It is curious that it was only introduced to Europe in 1865 whereas its less exciting cousin, *O. crispum*, preceded it by 30 years. Occasionally a plant is found with flowers even more attractive than either *O. marshallianum* or *O. crispum* and this is most probably a natural hybrid between the two, as sometimes their flowering periods overlap (Plate 38).

Oncidium praetextum
Rchb.f. | 2009

Etymology: *Praetextum* is Latin for bordered, fringed and presumably refers to the brown colour margins to the labellum.

Habitat: High montane forest at over 1500 M.asl in section IV receiving heavy rains from September to April, with heavy mists and constant cloud for the rest of the year.

Occurrence: Occasional, and probably restricted.

Plant: Exactly the same as *O. forbesii* and *O. gardneri* to the point that one can only fantasise that this plant is a natural hybrid between the two with a chance that sports of *O. marshallianum* and *O. crispum* could have sneaked in. The pseudobulbs are somewhat rounded, laterally compressed, ovoid, variable in shape and size and measure up to 6 cm. tall by 5 cm. wide. These arise closely and alternately from a thick woody rhizome. The paired apical strap-like, dark green, narrowly lanceolate, keeled, floppy leaves measure up to 40 cm by 2.0 cm. and live for two or three years. The root system is extensive, similar to others in the *O. forbesii* group and are long, branching, white and adherent.

Inflorescence and Flowers: The flowers arise on a nodding, basal panicle, up to 70 cm. long, which may hold up to 30 flowers. The dorsal sepal is ping-pong-bat-shaped, yellow to light brown, 1.6 cm. long by 1.3 cm. The laterals are fused for 50% of their length, a broad lanceolate, horizontally striped in yellow and orange-brown, 1.7 cm. long by 0.7 cm. The petals are broadly paddle-shaped, irregular at the margins, a light orange-brown, 2.3 cm. long by 2.1 cm. while the narrow-necked lip is faintly bilobed, 3 cm. long by 2.7 cm., with a moon-like yellow centre, an orange brown margin, and a twin, hook-nosed pair of calli.

Flowering Periods: Late January and February when the flowers last for three weeks or so.

Pollination: Is infrequent but the pollinators are probably as for *O. enderianum*.

Cultivation: As for *O. forbesii*.

Oncidium curtum
Lindl. | 2010

Etymology: *Curtus* is Latin for short, referring to the short space taken up by the callus.

Habitat: An epiphyte found at mid-tree level in original or regrowth forest between 1000 and 1400 M.asl, in section VI, experiencing good air movement, dappled light and constant reasonable humidity.

Occurrence: Occasional.

Plant: An epiphyte which resembles all the species in the *O. forbesii* Alliance. As these plants are naturally variable and we further suspect that natural hybridisation accounts for some of the variation as it is nearly impossible to differentiate between them when not in flower. The typical plant described is over 15 years old with laterally compressed, almost oval pseudobulbs which measure 6 cm. high by 5 cm. wide, and arise closely and alternately from a 1.2 cm. diameter, woody and sheathed rhizome. There is one or occasionally two, apical, coriaceous, light green leaves which are lanceolate, deeply-keeled and measure up to 25 cm. long by 5.5 cm. wide. The root system is vigorous and composed of many long white branching roots, 0.3 cm. diameter which are exposed, very adherent and active for many years.

Inflorescence and Flowers: An erect, paniculate inflorescence, which is basal, stiff and round and up to 45 cm. long by 0.4 cm. diameter, bears up to 30 flowers on the apical 15 cm. The overall flower colour is an orange-brown base with the petals and lip bordered with an irregular bright yellow margin, while the lip also has a bright yellow collar at the throat. The dorsal sepal is a slightly concave oval and measures 1.7 cm. long by 1.2 cm. The partly fused lateral sepals are 2.0 cm. long by 1.0 cm. and are oval. All the sepals are faintly streaked yellow horizontally. The petals are an irregularly-edged pong-pong bat-shape, 2.3 cm. long by 2.0 cm. The flared-margined lip is two lobed and almost oval, measuring 2.0 cm. long by 2.4 cm. wide with a distinctive two-nosed callus facing the apex and repeated, slightly smaller, facing the throat and appearing as a warty cross.

Flowering Periods: January and the flowers last for two weeks.

Pollination: Unknown.

Cultivation: In pots on a well-drained compost or on cork bark of some size since root production can be prolific. This plant thrives on living bark but with access to decaying wood or vegetable detritus.

General: The feature that distinguishes this species from *Oncidium forbesii* is the wings to the lip base which are more pronounced in *O. curtum*.

Oncidium enderanum
Hort. | 2011 (Plate 38)

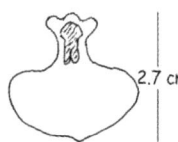

Etymology: Not known.

Habitat: *O. enderanum* rarely penetrates the mountains above 1000 metres altitude. It thrives in bright sunlight and is at its most spectacular when colonizing a decomposing tree trunk or branch and is often found in regrowing forest on these media, growing from 600-1000 M.asl in section VI.

Occurrence: Occasional.

Plant: A large, long-lived and variable epiphyte, and because of its propensity to reactivate back bulbs, can form colonies with over 50 pseudobulbs and 8 inflorescences bearing up to 500 flowers. Isolated small plants have 8 pseudobulbs, not unlike *O. forbesii*, with a 10 flowered inflorescence. The pseudobulbs are compressed, ridged, ovoid, 8 cm. long by 5 cm. wide, red-brown and spaced 2 cm. apart on a tough 1 cm. branching rhizome. There are one or two apical, olive-brown leaves measuring 20 cm. x 3 cm. and turning reddish in high light. The roots are profuse, thick, white and adherent and twine up and down a branch which, if decomposing, benefits the plant greatly.

Inflorescence and Flowers: A 70 cm. paniculate inflorescence arises from the base of the most recent pseudobulbs. It is arching but stiff and much branched, bearing up to 80 flowers which, in a group, give the visual effect of splotches of gold surrounded by dark brown circles. The predominant features of these flowers are the pronounced winged callus, the gold waist and neck of the lip and the wing-like lateral sepals. The dorsal sepal is a roundish spathulate 2.2 cm. by 1.2 cm. broad and the laterals, fused for 1/3 of their length, are a broad lanceolate 2.6 cm. long by 1.0 cm. The petals are slightly bilobed, asymmetric, spathulate, 2.7 cm. long by 1.8 cm. The trilobed lip is 2.5 cm. broad at the side-lobes and 2.7 cm. long and has a complex noduled callus at its winged throat.

Flowering Periods: March and April and the flowers last for three weeks.

Pollination: Infrequent; **Dressler** suggests oil-gathering bees.

Cultivation: As for *O crispum* but with more light.

General: This plant has been confused with *O. gravesianum* for 80 years but was recently the subject of a polemic when it was established that *O. gravesianum* flowers in October, lives only in far off Pernambuco, whereas this species conforms exactly to the type in the Kew Gardens' herbarium and flowers in March and April.

Oncidium forbesii
Hooker | 2012 (Plate 38)

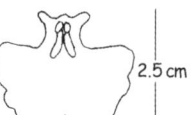

Etymology: Named after **H. O. Forbes**, the orchid grower at Woburn in the 19th century.

Habitat: Thickly spread throughout the upper levels of original forest, this species can quickly colonize regrowing woodland, where it is also found as a terrestrial. It ranges from below 800 metres up to 1600 metres altitude in all forested sections with a strong concentration in sections II, IV and VI.

Occurrence: Common.

Plant: There are two distinct varieties of this epiphyte in forests, one flowering in February-April which tends to be a prolific tree top to mid-tree species. The other, less common, is a mid- to lower-tree-trunk inhabitant flowering in November. The plants are very variable, with small mature specimens having six or so small pseudobulbs and a raceme of flowers of 15 cm. Large specimens can have pseudobulbs twice the size and a 70 cm. nodding inflorescence of 30 flowers. The pseudobulbs are ovoid and compressed, green, tinged brown and vary from 3 cm. x 2 cm. to 7 cm. x 3.5 cm. There is a pair of strap-like apical leaves, dark, leathery and measuring 30 cm. x 4 cm. The roots are white, thick and stretch for long distances from the plant. This feature is useful for finding specimens on a fallen tree; simply follow the roots for a metre or two and you will come to the plant.

Inflorescence and Flowers: These are a reddish-brown with an irregular margin of gold and are variable in both shape and size. The February-April type tends to be rounder with sepals and petals quite concave whereas the November flowering variety has narrower, flatter sepals and lip. The dorsal sepal is elliptic, from 1.4-2.0 cm. long and from 1-1.4 cm. broad. The laterals are elliptic to lanceolate, up to 30% fused and 1.5-2.3 cm. long by 0.5-0.9 cm. broad. The petals are a broad irregular spathulate 1.5-2.5 cm. long by 1.2-2.3 cm. broad. The lip is a rounded oval on its side, vaguely four-lobed, 2.2-3.2 cm. broad by 1.9-2.5 cm. deep, with a distinctive two-nosed callus at the thickish winged neck.

Flowering Periods: November and February-April and the flowers last for a month or so.

Pollination: As for *O. crispum*.

Cultivation: As for *O. crispum*.

General: Introduced to Europe from Nova Friburgo in 1837 and is still a popular plant. It is an easy and rewarding orchid for the beginner, with long-lived, beautiful flowers.

Oncidium gardneri
Lindl. | 2013 (Plate 38)

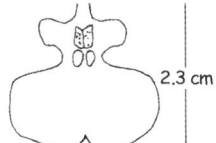

2.3 cm

Etymology: Named in honour of **George Gardner** who collected it in the Organ Mountain Range in 1838 where he teamed up with the **Harrison Brothers** and along with **George Forrest**, was probably Scotland's most famous naturalist in that century.

Habitat: A high mountain forest, mid-tree trunk orchid which grows above the range of *O. crispum* but is quite neighbourly with *O. forbesii*, at 1400-1600 M.asl in original forest in section VI, enjoying high humidity, good dappled light and good air movement.

Occurrence: Occasional.

Plant: This species is generally found growing as a mid to upper-tree trunk epiphyte. The dark green, smooth pseudobulbs are more rounded than those of *O. forbesii*, variable in shape, ovoid, compressed and measure about 6 cm. long x 5 cm. They arise at 1 cm. intervals on a tough rhizome. The single or paired apical leaves are strap-like, sharply pointed, dark shiny green and are 42 cm. by 3.5 cm., with a central ridge and are retained for two years. The root system is extensive and adherent but the roots are much finer and shorter than *O. forbesii*, helping to distinguish between them.

Inflorescence and Flowers: These arise on a 50-70 cm., nodding, paniculate inflorescence with up to 40 flowers. The dorsal sepal is a very broad spathulate 1.4 cm. by 1.1 cm. wide. The laterals, fused for up to 30% of their length, are an irregular ellipse and 1.5 cm. long. The petals are a very broad, irregular spathulate, 2 cm. long by 1.4 cm., while the bilobed lip is a crisp margined oval on its side, 2.7 cm. wide by 2.3 cm. deep, held on a thick winged neck that encloses a small bunch of grape-like calli. Their general effect is of numbers of large, striking, bright yellow blobs surrounded by fringes of chestnut brown. The ratio of yellow to brown in the flowers of different clones varies considerably. The flowers are sweetly scented in the morning. They are smaller than *O. crispum* flowers but about the same size and shape as the February-April *O. forbesii*.

Flowering Periods: November and December and the flowers last for three weeks or so.

Pollination: As for *O. crispum* and as infrequent.

Cultivation: As for *O. crispum*.

General: Because of the overlapping habitats and flowering periods with both *O. marshallianum* and *O. forbesii* it may well be an ancient natural hybrid between these two species. The synthetic hybrid does, in fact, resemble *O. gardneri*. The great variety of its flowers can also be attributed to continuing back-crossing, introgression, between itself and *O. forbesii*.

Oncidium flexuosum
Sims | 2024 (Plate 38)

Etymology: *Flexuosus* is Latin for bent, zigzag, alternately in opposite directions and refers to the way the flowers are arranged on the inflorescence.

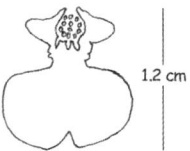

1.2 cm

Habitat: An epiphyte found principally at low altitudes and in scrub forests in sections II, IV, VI and VIII, generally in swampy areas and permanently inundated slopes where large tree cover is missing. It is often found with **Cattleya harrisoniana**, which may also be found sporadically as high as 1000 M.asl, in regrowth forest. It enjoys high light and humidity and good air movement.

Occurrence: Common.

Plant: An epiphyte on scrub trees, often forming large colonies. The pseudobulbs arise alternately at 2 cm. intervals from a thick woody rhizome and are ovate, laterally compressed, light green and measure up to 8 cm. high and 3 cm. wide. There is a pair of strap-like apical leaves, linear-lanceolate, flexible, coriaceous and conduplicate which measure 25 cm. long by 2.5 cm. wide. The root system is fairly prolific and composed of numerous thin white roots many of which are aerial and appear to be reaching for a substrate.

Inflorescence and Flowers: A paniculate inflorescence with up to 200 flowers is borne on a stiffly erect, metre-long stem. The flowers are smallish and golden-yellow (hence the local name 'little golden drops'). The dorsal sepal is a concave pointed oval, 0.4 cm. long by 0.5 cm. wide, and yellow with two thick horizontal light-brown bars. The partly fused lateral sepals are similar. The petals are also similar, slightly larger and with irregular margins. The lip is by far the largest flower segment and measures 1.5 cm. across by 1.2 cm. long. The two lower lobes are divided by a central cleft and make up a broken round-cornered square which is a deep golden-yellow. The callus on the upper lobe is fleshy, cristate and yellow surrounded by light brown.

Flowering Periods: December and January with the flowers lasting nearly a month.

Pollination: Probably pollinated by small bees.

Cultivation: This is a very adaptable plant which accepts varied conditions and situations. It may be grown in pots in a well-drained compost but is perhaps best on large tree fern or bark plaques. Locally, it is grown in gardens on living wood at chest height.

General: Because of the numerous, beautiful and long-lived flowers, this is universally used in the cut flower trade. Curiously, the major producer for the world market of this Brazilian flower and its hybrids, is Thailand.

Oncidium batemannianum
Parm. | 2031 (Plate 38)

Etymology: **James Bateman** collected mainly in Guatemala and Mexico.

Habitat: A terrestrial found in high mountain fields, the margins of steep bare rock faces, roadside banks above 1100 M.asl, and in previously burnt regenerating young forest, in sections I, III, V and VI, amongst tall grasses and various bromeliads of the *Vriesia* genus, enjoying total light, a peat-based substrate and good air movement, summer rains and winter mists.

Occurrence: Common.

Plant: A large, exuberant, pseudobulbous terrestrial, fast growing and long flowering, easily confused with *O. blanchetii* and *O. hydrophilum*. The pseudobulbs are tightly clustered, a yellowish green, ridged and erect, ovoid and slightly bilaterally compressed. They come in all sizes, depending on the quality of the substrate, and can measure up to 9 cm. tall by 6 cm. wide and arise on a thick branching rhizome. There are three long, narrow, pliable, linear-lanceolate, light green, deeply keeled apical leaves which measure up to 40 cm. long by 3 cm. broad and are long-lived. There are also two shorter, opposite subtending leaves. The root system is robust and extensive, composed of thick white brittle roots which penetrate the substrate way beyond the plant's circumference.

Inflorescence and Flowers: A long, green, round stiff and thick inflorescence emerges in October from the most recent pseudobulb base. In open ground exposed to full light it will reach a metre or so tall, but in partial shade it may double this length. There is a panicle of few to many flowers on the apical fifth of the stem. These are a brilliant sulphur-yellow, freckled with reddish brown. The dorsal sepal is oval, concave, light brown and measures 0.8 cm. long by 0.45 cm. while the laterals are the same colour but an asymmetric elliptic and measure 1.2 cm. by 0.4 cm. broad. The lip is four-lobed, the two basal sections are small, separated by the calli and jut out like two bat ears. The complex callus section which is wedge-shaped, measures 0.7 cm. long by 0.6 cm. broad at its base and consists of some half dozen clawed, red, small protuberances. The apical lobes are each almost round, flared, a brilliant sulphur-yellow and together measure 2.5 cm. across by 1.4 cm. deep. This feature is the most obvious difference with *O. blanchetii* whose lip is half the size.

Flowering Periods: September through to February and the individual flowers last for up to a month, while the inflorescence produces flowers for up to four months.

Pollination: Bees. On slopes where large colonies of this plant thrive, many flowers are pollinated.

Cultivation: In large pots on a peaty substrate with a 20% coarse sand. It must have a good throughput of water during the summer but remain only damp during the five months surrounding winter.

Oncidium blanchetii
Rchb. f. | 2032 (Plate 38)

Etymology: Introduced to Europe in the 1840's by **J. S. Blanchet**, a prolific collector in this period, mainly in Bahia.

Habitat: This is a terrestrial found in sections I, III and VI in high mountain fields at 1000-1600 M.asl but also found in isolated colonies in section VI in regrowing forest after burning, and above 1200 metres altitude.

Occurrence: Common.

Plant: The pseudobulbs are found in tight clusters; smooth, erect ovoid-oblong, to 8 cm. x 3.5 cm. and varying in colour from yellow to light green. There are two leaves per bulb which are linear, sharply pointed, deep green and measuring 50 cm. x 3.5 cm. and are retained for many years. The roots are thick and white and permeate the leaf litter and humus layers for over a metre.

Inflorescence and Flowers: These are predominantly yellow with brown bars on both the sepals and petals. The large warty callus and the large sulphur-yellow trilobed labellum are distinctive features. The long paniculate inflorescence arises from within the bract of the most recent pseudobulb. The dorsal sepal is elliptic, 0.75 cm. by 0.3 cm. broad. The laterals are an asymmetric lanceolate 0.75 cm. long by 0.2 cm. The petals are an irregular spathulate 0.8 cm. by 0.5 cm. broad. The apical lobe of the lip is an irregular oval 1.3 cm. wide, 1.0 cm. deep. The callus is highly complex and like a bunch of grapes flanked by two short wings. In savannah, the inflorescence will only be 30 cm. long but can reach 2 metres in dense scrub with up to 50 flowers.

Flowering Periods: October through to April and the flowers last for up to three weeks but the inflorescence will flower for three months.

Pollination: Frequent if the plant is in colonies but isolated plants rarely have fruit. Probably pollinated by oil-collecting bees.

Cultivation: In large pots on a peat like substrate which is very well drained, in high light, dry in winter but well watered during spring and summer.

Oncidium hydrophilum
Barb. Rodr. | 2033a (Plate 39)

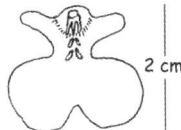

Etymology: *Hydro* and *philus* are Greek for water and loving. **Barbosa Rodrigues** really emphasized that this plant loves marshy places.

Habitat: It is a terrestrial confined to mountainous marshy scrub-covered areas. It is as sporadic as its micro-environment in section VI at 1300 M.asl.

Occurrence: Rare.

Plant: Perhaps because of its liking for swampy ground, the tightly-clustered, smooth, ovoid, dark-green pseudobulbs can become massive, measuring up to 10 cm. x 6.5 cm. They bear a huge root system composed of thick branching white roots which penetrate the sodden topsoil for metres around. There are two leaves per bulb; apical, erect and with a distinct spine, measuring 50 cm. x 3.8 cm., deep green and retained for many years.

Inflorescence and Flowers: These are borne towards the end of a two metre inflorescence which branches widely, pushing its way through bracken and the scrub that will support it. Each can bear over 100 flowers but the flower stalk remains alive for up to two years, pushing out new branches at any time. These flowers are larger than *O. blanchetii* but essentially similar with a more dramatic labellum and more pronounced calli. The dorsal sepal is oval, 0.6 cm. by 0.4 cm. broad. The laterals are lopsided lanceolates 1.0 cm. long by 0.4 cm. The petals are an irregular square 0.8 cm. by 0.75 cm. wide. The bilobed lip is composed of two apical, almost circular lobes which together are 2 cm. wide by 1.7 cm. deep. The complex multi-horned callus is flanked by two wings.

Flowering Periods: Between November and June and the flowers last for three weeks or so while the inflorescence may flower for up to five months.

Pollination: As for *O. batemannianum*.

Cultivation: As for *O. batemannianum* but with a shady site and wetter.

General: Very rarely found in cultivation but of obvious interest because of its robust growth, almost permanent flowering and great beauty. It is also a transient colonizer of swampy areas in regrowing forest.

Oncidium ramosum
Lindl. | 2034 (Plate 39)

Etymology: *Ramosus* is Latin for bearing branches, much branched which fits this plant perfectly.

Habitat: We have found this plant in quantity in scrub mountain forest from 800-1600 M.asl on five mountain ridges in section VI, and in addition, in regrowth scrub in previously burnt but cultivated forest. It enjoys a well-drained substrate, in high light and subject to the mists and rains of its environment.

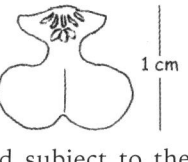

Occurrence: Common.

Plant: A terrestrial with a most uncharacteristic growth habit for an **Oncidium**. The pseudobulbs arise at 15 cm. intervals on a round, green, thick, though brittle, woody, rhizome up to 250 cm. long, sheathed for its full length and coursing through low scrub. The pseudobulbs are light green, a lengthy ovoid, bilaterally compressed and 3.5 cm. long by 1.5 cm. There are up to four fan-like leaves, generally only found on the new pseudobulbs which are apical, stiff, light green and keeled, linear-lanceolate and 7 cm. long by 1.8 cm. Roots are not profuse and only those on back bulbs penetrate the deep leaf litter, both anchoring the plant and searching for nutrition. The airborne pseudobulbs produce one or two thick, white, short roots which only make contact when the plant falls and they can enter the substrate. On an excavated plant we were surprised to find only a single root per pseudobulb which branched prodigiously under leaf litter.

Inflorescence and Flowers: A 28 cm. long, stiff, erect, paniculate inflorescence arises from the base of the developing pseudobulb during April. There are up to 40 pale yellow flowers, lightly and faintly striped with claret on both petals and sepals. The dorsal sepal is lanceolate, 0.6 cm. by 0.3 cm. The partly fused laterals are an asymmetric lanceolate 0.6 cm. by 0.2 cm. broad. The petals are almost square, 0.6 cm. by 0.45 cm. broad. The distinctly bilobed labellum has two ± round apical lobes, which are 1.2 cm. wide by 0.75 cm. deep. The complex callus at the throat, between two short triangular wings, is like a bunch of grapes.

Flowering Periods: April and the flowers last for two weeks or so.

Pollination: Infrequent and unknown.

Cultivation: This plant can be tamed to shorten its growth significantly in high light. Try it in pots on a rich fibrous compost in high light, good air movement and constant misting.

General: We are worried about the identification of this plant because experts from Espirito Santo and Rio de Janeiro, who have given it another name, always lose the flowers we send them. It is inconceivable to us that a plant so flamboyant and common in the Organ Mountain Range, walked over by hundreds of botanists for two centuries, could suddenly turn up as a new species in Espírito Santo though, God help us, we are not taxonomists.

Oncidium baueri
Lindl. | 2045

Etymology: Named after **Franz Bauer** (1758-1840), the famous orchid painter.

Habitat: Found in huge colonies encircling the trunks of trees in humid areas over streams and rivers in section VII at around 300 M.asl.

Occurrence: Rare on the interior anticline, since most of the original habitats have now been diminished or destroyed.

Plant: This species, although not a large plant, has perhaps the longest inflorescence, certainly of Oncidiums. The pseudobulbs are bilaterally very compressed, ovate and measure 6 cm. high by 2.5 cm., the older ones yellowish and the new ones green and arise at 2 cm. intervals from a tough 0.6 cm. diameter rhizome. The new growth has two or three lanceolate leaves, each with a marked midrib and measuring 21 cm. by 2.5 cm. wide. The roots are fine, white and wiry and some are clearly aerial.

Inflorescence and Flowers: The inflorescence can be up to 3 metres long and our specimen had 30 branches, each with 6-8 flowers totalling over 250, each 1.5 cm. by 1.5 cm. The dorsal sepal is 1 cm. long by 0.4 cm., semi-clavate and pointed, yellow, barred with chestnut. The lateral sepals and petals are similar, a pointed spathulate and measure 1 cm . by 0.3 cm. The lip is 1.2 cm. wide by 1.5 cm., flared and bright yellow with an ochre waist, two lobed and has two distinct circular wings like **O. barbatum**. The callus is finger-like and the column has short, yellow and chestnut barred wings with two downward and outward pointed teeth.

Flowering Periods: October and November and while each flower lasts for a week or so, the whole inflorescence may last for months.

Pollination: Infrequent in nature and **Dressler** suggests that it occurs by deceit through bees.

Cultivation: From observations of the habitat it is probably best grown in baskets with a high proportion of rotting leaf mould and tree fern chunks, kept well watered and fed throughout the growing period.

The genus
Miltonia Lindl.

John Lindley raised this genus in honour of **Viscount Milton**, an important orchid fancier and grower of Wentworth House in Yorkshire during the early 19th century. A small, but very important, South American genus of which nine species are attributed to Brazil and seven of these to the Organ Mountain Range. We have found five and most in quantity and all in original humid forest. The sixth, **M. spectabilis**, which eluded us for years, was found by our friends from the Rio de Janeiro Botanic Garden in the Tinguá State Park in the summer of 2001.

Miltonia flavescens
Lindl. | 2051 (Plate 40)

Etymology: *Flavescens* is Latin for pale yellow, referring to the flower colour.

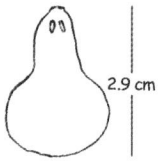

Habitat: Epiphytic from section IV, growing between 200 and 400 M.asl, in the mid- to upper-tree in original forest, usually above a river or stream and almost always in large colonies. A huge colony covering most of the mid to top tree branches of a large tree is a sight to take your breath away.

Occurrence: Locally common in its specific habitat.

Plant: A medium-sized epiphyte growing at its best in closely-packed, large colonies. The pseudobulbs are an extended ovate, bilaterally-compressed, smooth and light green and measuring up to 10 cm. long by 2.5 cm. There is a pair of linear-lanceolate apical leaves, pale green and measuring up to 30 cm. long by 1.5 cm. and long-lasting. New growths have two or three opposite subtending leaves which are the same shape and short-lived. The root system is prodigious and composed of a hessian-like mass of thin, long branching roots which envelop the substrate for far beyond the plant's circumference.

Inflorescence and Flowers: One or two 50 cm. opposite, racemose inflorescences arise from the bases of the new growths. These are stiff, slightly pendulous and bear up to 12 front-facing alternate flowers on short pedicels. The sepals and petals are linear-lanceolate, almost equal in shape and size and measure 5 cm. long by 0.5 cm. The single-lobed lip is spade-shaped and measures 3 cm. long by 0.5 cm. and has an undulate, crisped margin. The lip base has two short tram-line-like vertical calli and longitudinal veins appear as faint purple.

Flowering Periods: Late October and November and the flowers last for 10 days.

Pollination: **Dressler** suggests that pollination is by bees but we have also watched humming birds visiting the flowers. Fruits are frequently found.

Cultivation: As for *M. clowesii*.

General: *M. flavescens, M. cuneata* and *M. clowesii* are indistinguishable when not in flower, however, *M. cuneata* will not be found below 1000 M.asl, and the other two don't appear above 600 M.asl, in these mountains. This species was introduced into European cultivation by the **Harrison** brothers in 1832.

Miltonia spectabilis
Lindl. | *2052* (Plate 40)

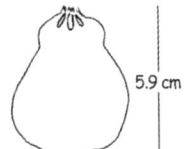

Etymology: *Spectatus* is Latin for esteemed, referring to the desirability of the flowers.

Habitat: Very old regrowth forest around 300 M. asl in Section II, requiring good light, wind movement and reasonable humidity..

Occurrence: Locally frequent.

Plant: Similar to all other *Miltonia* species described but as it is much more exposed to light. the pseudobulbs and leaves take on a yellowish colour. The pseudobulbs are bilaterally compressed ovals, narrowing towards the apex and up to 8 cm. tall, 3.5 cm. wide and 2 cm. thick at the base, held at 1.0 cm. intervals on a thick, branching rhizome. The apical leaves are stiff, linear-lanceolate, 20 cm. long by 2.5 cm., deeply keeled and long-lived.

Inflorescence and Flowers: Single, large and colourful flowers are borne on bracted, stiff basal inflorescences up to 20 cm. long. The tepals are lanceolate, ± equal, 4.0 cm. long by 1.5 cm. broad and faintly pinkish-white. The large lip is single-lobed with two small shoulders, pinkish-white with a tiny, bilobed yellow callus at its throat. The basal half, including the shoulders has a magenta blotch dripping downweards in veins towards the tip. It is up to 6.0 cm. long by 4.5 cm. wide.

Flowering Periods: February through to April and the flowers last up to three weeks.

Pollination: Unknown.

Cultivation: Grow on tree fern plaques or bark with a fibrous base, watering well in summer but allowing for a dryer autumn, winter and spring.

Miltonia candida
Lindl. | *2053* (Plate 40)

Etymology: *Candida* is Latin for pure white probably referring to the colour of the lip.

Habitat: Epiphytic at the edges of original forest and growing at 600-700 M.asl, on lianas about two to four metres from the forest floor in section VI. It experiences dappled light, good air movement and high humidity.

Occurrence: Occasional.

Plant: So similar are the three *Miltonia* species that in the '*Manual of Cultivated Orchid Species*' only *M. clowesii* is described to cover all three. The plant is a large, strong, clustered epiphyte found in large colonies. The pseudobulbs are ovate and up to 10 cm. long by 2.5 cm., tapering towards the apex, slightly compressed and arising at 1 cm. intervals from a 1 cm. thick, tough, woody, scaly and sheathed rhizome. The pair of apical leaves are linear-lanceolate, long-lived, light green, deeply keeled and up to 30 cm. long by 2.5 cm. Newer growths have two or three similarly shaped smaller subtending leaves on either side of the pseudobulb base. The intense root system is composed of a myriad of thin white tenacious roots which form a hessian-like mass completely enveloping the substrate. The plant gathers leaves and detritus and the larger the colony, the greater the compost heap collected.

Inflorescence and Flowers: One or two basal inflorescences arise from either side of the most recent pseudobulb. These are erect and woody and reach 50 cm. long, the apical 20 cm. bearing a raceme of up to eight 5 cm. diameter alternate flowers. The sepals and petals are more or less equal, 4 cm. long by 1.5 cm, pointed ovals, and red-brown with a few yellow bars. The lip, when flattened out, is almost round, crisped at the margins and measures 3 cm. by 3 cm., milk-white when opening, turning yellow with age and with a light purple sheen at its base which also shows five slightly raised short veins.

Flowering Periods: The flowers appear in late April and early May and can last for three weeks.

Pollination: Probably pollinated by oil-gathering bees.

Cultivation: Similar to the other species in this group, *M. cuneata* and *M. clowesii*, always remembering that all three species enjoy being overcrowded.

Miltonia clowesii
Lindl. | 2054 (Plate 40)

Etymology: Named in dedication to the **Rev. Clowes** in Liverpool England, an avid orchid collector and grower in the 1830's and 40's.

3.4 cm

Habitat: Found in sections IV and VI at 200-500 M.asl, in original forest at the mid- to upper-tree levels. Usually found above watercourses, the species enjoys high humidity and air movement with dappled light.

Occurrence: Common in its fairly specific environment.

Plant: A medium-sized epiphyte normally found in large colonies covering the mid to upper branches of original forest trees. The pseudobulbs are narrowly oblong to ovate, laterally compressed and light green, measuring up to 10 cm. long by 2.5 cm. Each bears a pair of opposite, apical leaves which are darkish green, linear-lanceolate, deeply keeled, long-lived and a stiff papery texture, measuring up to 30 cm. long by 2.5 cm. Newer growths show two or three similarly-shaped subtending leaves, smaller and short-lived. New shoots arise on a short, 1 cm. diameter, woody, wandering rhizome at intervals of 2 cm. The root system is intense and composed of an uncountable mass of thin, white, hessian-like threads which envelop the substrate completely. Like the other species in this group, this is a detritus-collector forming its own compost heap.

Inflorescence and Flowers: One or two erect, basal inflorescences arise from opposite sides of the most recent pseudobulbs. These are stiff racemes, 30-60 cm. long, with up to 10 alternate flowers on the apical 20 cm. The sepals and petals are similar in size, lanceolate and measure 3.2 cm. long by 1 cm. and are dark yellow with a series of transverse, irregular light brown bars. The lip is violin-shaped with a white apical lobe, while the neck and base are purple-tinged with white at the shoulders and with five or six raised short unequal ridges at the throat - probably guidelines for the pollinator. The lip measures 3.4 cm. long by 2 cm. wide at the apical lobe.

Flowering Periods: March and early April with the flowers lasting two weeks.

Pollination: **Dressler** suggests that bees are the pollinators but we have also observed humming birds visiting these plants.

Cultivation: Brazilian *Miltonia* species from this Alliance enjoy overcrowding in whichever medium they grow. In pots the compost should be fibrous and well aerated. It requires generous watering and feeding in summer but only misting in winter.

General: **George Gardner** introduced this species in 1835 after discovering it in the Organ Mountain Range. The local name is 'little dancing girl' from the effect of the inflorescence nodding in the breeze.

Miltonia cuneata
Lindl. | 2055 (Plate 40)

Etymology: *Cuneatus* is Latin for wedge-shaped which probably alludes to the lip shape.

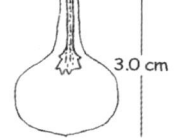

3.0 cm

Habitat: Found on the top 1/3 of old trees only in original forest. Wild Guava is a favourite host. It tends to form massive colonies even on thin branches from 1000-1500 M.asl in section VI. The many major colonies we have found were on very large trees directly over water courses, guaranteeing year-round intensive humidity but, at the same time, plenty of light and wind.

Occurrence: Common.

Plant: A large epiphytic clusterer on a thick branching rhizome. The pseudobulbs are ovate, narrow and green and form large groups and colonies. Mature pseudobulbs are 9 cm. tall x 3 cm. wide, characteristic for Miltonia but bigger, greener and more abundant. The two apical leaves are dark green, long, lanceolate, semi-erect with a distinct central spine. Size, texture, colour and tone vary considerably with the environment. In a humid, mid-tree position they will be 30 cm. long x 4 cm. wide and dark green, whereas in dryer situations, only half that size, stiffer and a paler green. The roots are thin, white, branching, adherent and almost countless. A mature pseudobulb will have up to 50 with an average length of 10 cm. giving about 5 metres per bulb. This huge network earns the plant **Hoehne**'s title of "nest-epiphyte" since the colony looks like a bird's nest and indeed, larger colonies may often contain them.

Inflorescence and Flowers: There are from 4 to 12 flowers on a tough erect, basal, spicate inflorescence, 30-60 cm. long with generally two spikes to each pseudobulb. This is a promiscuous species, flowering more profusely in large colonies than on individual plants due to the colonies capacity to trap large quantities of leaf and other detritus which decays and adds to the nutrient store available to the massive root system. The lip, 3 cm. long by 2.7 cm. wide is a strong white with two distinct parallel ridges at the throat and with two dark purple spots. The sepals and petals are similar, chocolate brown with wavy cream stripes and the apex is always cream coloured; they measure about 3.6. cm. long x 1.3 cm wide. The lateral sepals curve and touch at their tips behind the lip.

Flowering Periods: In September and October, when the flowers last for up to three weeks.

Pollination: Frequent, possibly pollinated by bees.

Cultivation: In large pots or on large tree fern slabs, allowing space for much increased growth, in good dappled light, good air movement and a high constant humidity.

Miltonia russelliana
Lindl. | 2058 (Plate 40)

Etymology: Named for **John Russell**, 6th Duke of Bedford (1766-1839) whose family largely financed the travels of **George Gardner**.

Habitat: Found in transitional forest in section V, at 550-650 M.asl, at mid-tree level enjoying dappled light, a wet short summer and the remainder of the year dry.

Occurrence: Rare.

Plant: A somewhat more delicate variation of the other plants in the **M. clowesii** alliance; a clustered epiphyte with pseudobulbs up to 6 cm. long by 2 cm. wide, typically laterally compressed, narrowly oblong-ovate, and light green. The leaves are narrowly linear and lingulate, up to 30 cm. long by 1.5 cm. broad, dark green, floppy, keeled, constricted at their bases into conduplicate, longish false petioles. There are two long-lived apical and opposite leaves, while two or three subtending basal leaves last but a year or so. The root system is the usual hessian type mass of fine white long roots, both adherent and capable of catching all nutrients that pass.

Inflorescence and Flowers: One or two basal and opposite inflorescences emerge in April rising to ± leaf height. These are up to 15 cm. long, arching with 1 to a few alternate and opposite flowers with short pedicels and very short scapes. The sepals and petals are ± equal, an elliptic lanceolate, 3 cm. long by 0.6-0.8 cm. The lip is lingulate, very vaguely bilobed, protrudes and measures 3 cm. by 1.0 cm. The basal half is a light magenta while the apical lobe is a clear milk-white and slightly acuminate at the apex. Faint parallel calli descend from the throat to mid-lip.

Flowering Periods: May and June; the flowers last for 3 weeks.

Pollination: Infrequent, **Dressler** suggests oil-gathering bees. However these are not very active in May/June at these altitudes.

Cultivation: As for *Miltonia candida*.

The genus
Aspasia Lindl.

Raised by **Lindley** in 1832 and the name refers to *Aspasia* the beautiful mistress of **Pericles**, an important Athenian general and statesman (±495 - 429 BC), who abandoned his aristocratic wife to marry this beauty. The genus is composed of six species from Central and South America of which three are attributed to the Organ Mountain Range.

Aspasia lunata
Lindl | 2080 (Plate 40)

Etymology: *Lunata* is Latin for crescent-shaped probably referring to the shape of the tepals.

Habitat: Found in large colonies in transitional forest either as a low tree epiphyte or as a lithophyte at 500 M.asl, in sections V, VI and VIII. The plants are usually in deep litter and leaf mould but with good drainage, high light and high humidity to the roots.

Occurrence: Occasional to rare, like its habitat.

Plant: A small to medium-sized plant with the Miltonia habit of forming crowded colonies collecting detritus. The pseudobulbs are borne on a 0.6 cm. thick stiff, sheathed and woody rhizome at intervals of 3 cm. and are bilaterally compressed, narrowly oval and measure 7 cm. high by 2 cm. New growths have five leaves, two apical and long-lived and three smaller which subtend the pseudobulb and last only one or two years. Apical leaves are lanceolate, slightly keeled, pale green and floppy, measuring 20 cm. long by 3 cm. The root system is vigorous and consists of stiff, white roots, 0.2 cm. diameter which are up to 20 cm. long, seldom branching and arising from all parts of the rhizome.

Inflorescence and Flowers: One or two flowers on short pedicels emerge from either side of the previous years mature pseudobulb, sometimes opening within the leaf bracts. The sepals and petals are more or less equal in size and shape at 2.5 cm. long by 0.4 cm., narrowly lanceolate, acuminate and with brown blotches on a light green background. The lip is violin-shaped, trilobed and white with pale purple staining from the mid-lobe to the throat. It has a fine 'V'shaped callus at the throat and measures 3.0 cm. long by 1.5 cm. at the broadest part.

Flowering Periods: November, when the flowers last for ten days or so but a good mature plant may bear flowers for a month.

Pollination: Seems to be rare and possibly *Euglossine* bees are responsible.

Cultivation: The root system is always in deep litter and keeps the plant active. This suggests a rich loose and well-drained substrate in fairly deep baskets and grown in good light and air movement with year-long high humidity.

The genus
Warmingia Rchb. f.
Heinrich Reichenbach dedicated this two species genus in 1881 to **Eugene Warming**, a Danish botanist, the founder of plant ecology, who collected in Minas Gerais.

Warmingia eugenii
Rchb. f. | *2090* (Plate 40)

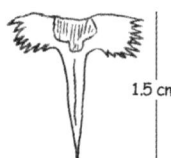

Etymology: Named in dedication to **Eugene Warming** a Danish botanist who specialised in Greenland flora.

Habitat: A low tree epiphyte found in transitional forest between 400 and 500 M.asl, in section V, living in low light, low air movement and humidity and with a long dry winter.

Occurrence: Locally frequent like many pioneer plants.

Plant: A small twig epiphyte with minute, cylindrical pseudobulbs 1.5 cm. high and 0.5 cm. wide. A mature clump will have about 15 pseudobulbs, each with a single apical, long-lived, dark green leaf which is elliptical, keeled and measures up to 16 cm. long by 0.3 cm. and is slightly coriaceous, with a 1.5 cm. false petiole. The root system is not extensive and is composed of thin pearly-white roots which embrace the small twigs on which this plant is commonly found.

Inflorescence and Flowers: A branched, pendent raceme up to 30 cm. long arises from the pseudobulb base bearing up to 70 clear translucent white flowers which are extraordinarily attractive. The sepals are equal in size and shape, lanceolate, acuminate and 1.3 cm. long by 0.3 cm. The petals are also lanceolate but broader, 1.3 cm. long by 0.4 cm. and for the distal three quarters of their length are grossly serrated. The lip is trilobed with two basal round, marginally serrated side-lobes which together measure 0.75 cm. across and 0.3 cm. wide and has serrated margins for its apical half with a tiny V-shaped callus at the throat.

Flowering Periods: November and December are the main months although flowers may be found until April.

Pollination: We have observed a pair of large bees with orange-abdomens (probably *Eulaema* sp.) possibly collecting oil from the lip centre and throat and in so doing, removing the pollinia from all 40 of the flowers visited. However, only two flowers were fertilized, five pollinia were stuck on petals and sepals and the remaining 33 were mostly lost during the transfer of the oil from the forelegs to the hind leg pouches.

Cultivation: Mount on pieces of rough bark and hang in shade with daily spraying during growth in summer.

The genus
Trichocentrum Poepp. & Endl.

This genus was established by **Poeppig** and **Endlicher** in 1838. The Greek compound word *tricho* meaning hairy or hair-like and *kentron* a spur refers to the long, thin spur in some species. This is a Latin American species of which nine are attributed to Brazil and one to the Organ Mountain Range.

Trichocentrum fuscum
Lindl. | *2097* (Plate 40)

Etymology: *Fuscum* is Latin for sombre brown, referring, somewhat unfairly, to the flower colour.

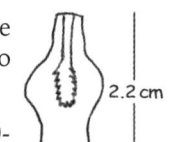

Habitat: Found in section VIII at 300-400 M.asl on relict trees enjoying a hot wet summer, a cooler and dryer winter with high light and good air movement.

Occurrence: Locally common.

Plant: A small to medium-sized, low-to-mid-tree epiphyte which can be found in large clumps. The minute light green sheathed pseudobulbs are closely packed and arise at tiny intervals from a short wandering rhizome. Each holds a single, apical, long-lived, light green, stiff, leathery, leaf. This is a broad lanceolate, up to 7 cm. long by 2.5 cm. broad, lightly keeled on emergence but becoming more so with age. The root system is strong and penetrating, composed of many 0.2 cm. diameter, long, white, apparently non-branching roots.

Inflorescence and Flowers: One to four flowers emerge on short, green, 0.2 cm. thick and 1.5 cm. long, basal inflorescences. The flower's predominant feature is a long, slightly bilobed protuberant lip backed by an equally long, 0.25 cm. thick nectary. The petals and sepals are more or less equal in size, shape and colouring: a broad lanceolate, 1.5 cm. long by 0.7 cm. wide, centred with orange-brown, fringed with yellow, and accentuated at their apices. The lip is almost rectangular, 2.2 cm. long by 1.5 cm. wide, crisped at its margins, ridged vertically at its centre with the tight basal ridges tipped in wine. The nectary, 2.2 cm. long, is butter-yellow. The column head has significant wings.

Flowering Periods: March through April and the flowers last for up to two weeks, but as the flowers are sequential the plant may show open blooms for up to two months.

Pollination: The sweet scent, long white lip and a significant nectary, indicate pollination by night-flying Lepidoptera.

Cultivation: We have seen this plant cultivated with serious success at sea level on tree fern slabs in both high light and shade, by **Rita Gripp** in Macaé.

General: Although named by **Lindley** in the early 19th century, we have found very little relevant material in our orchid literature. It seems to be a very local species, though **Pabst & Dungs** have examined specimens from as far off as Mexico.

The genus
Centroglossa **Barb. Rodr.**

This genus is made up of a small group of 6 species confined to eastern Brazil of which **Pabst & Dungs** attribute all to the Atlantic rain forest and 4 to the Organ Mountain Range, of which we have found all 4. All are small scrub epiphytes and these we have found are pretty, and all in section VI between 800-1500 M.asl.

Centro is from he Latin for at the centre, and *glossa* is Greek for tongue, thus the implication that the lip is very reminiscent of a central tongue which it is in most species, and as usual, well named by **Barbosa Rodrigues**.

Centroglossa castellensis
Brade | 2104

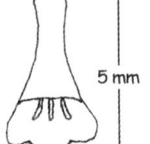

Etymology: Refers to the city of Castelo, E.S. which is the region where the plant was collected.

Habitat: Grows at 1100 M.asl in original forest in section VI, on scrub trees or low trunks, enjoying constant high humidity, low light and wind movement.

Occurrence: Occasional.

Plant: A small tufted epiphyte with no apparent stem swellings. The lanceolate, sessile leaves are keeled, light green, arising in a loose whorl from the stem and are up to 5 cm. long by 1 cm. The stem emerges from a node in the centre of the root system which is composed of a number of thick, greenish furry roots.

Inflorescence and Flowers: This is apical, round, green

and up to 4 cm. long with single pale yellow flowers. The sepals are ± equal in size and shape, a broad lanceolate, 4 cm. long by 0.3 cm., while the petals are slightly more spathulate and measure 3.8 cm. by 2 cm. The lip is trumpet-shaped, with the margins of the basal 2/3 fused to form a tube and the wider apex is vaguely trilobed with an inverted arrow-head of three green lines at its centre. The background is yellow.

Flowering Periods: March and August and the flower lasts for a week or so.

Pollination: Unknown.

Cultivation: As for *Chytroglossa marileone*.

Centroglossa macroceras
Barb. Rodr. | 2107 (Plate 41)

Etymology: *Macro-ceras* is Latin for long horn-like projection, referring to this projection behind the anther.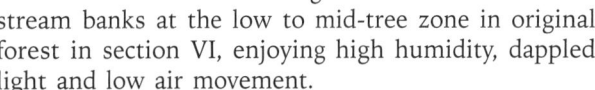

Habitat: **Hoehne** describes this plant as a child of the pluvial forest and so it is at 800-1000 M.asl along river or stream banks at the low to mid-tree zone in original forest in section VI, enjoying high humidity, dappled light and low air movement.

Occurrence: Occasional.

Plant: A small pendent epiphyte easily mistaken for *Zygostates multiflora* or *Chytroglossa marileoniae*, when not in flower. It has a single, linear, markedly keeled grey green leaf, 3 cm. long by 0.3 cm. wide, emerging from the apex of a minute emerald-green almost round pseudobulb. The new pseudobulb, 0.4 cm. long by 0.3 cm. in diameter, is subtended by two opposite leaves of similar size and shape as the apical leaf. A typical plant will have six or seven such pseudobulbs. The root system is weak, composed of up to a dozen or so fine, short, whitish green roots up to 4 cm. long, which are densely covered by short hairs.

Inflorescence and Flowers: These are held on a thin basal inflorescence some 1 cm. long. One or two flowers are borne on 1 cm. long pedicels which stem from a trumpet-shaped bract at the stem's apex. The flowers are milk-white with emerald green veins at the base of the tepals. The flowers are surprisingly large for such a small plant, 1 cm. long by 0.75 cm. wide and are quite distinctive. The dorsal sepal is elliptic, 0.7 cm. by 0.4 cm broad. The lateral sepals are the same shape but smaller. The petals are almost round, 0.7 cm. by 0.6 cm. broad, the apical half having a finely serrated margin, but the flower's really distinctive feature is the lip. This is a funnel or tube 0.8 cm. long by 0.6 cm. wide with an 0.8 cm. long, 0.2 cm. wide nectary at its base. The leading edges of the tube's mouth are faintly serrated, while the inside base is intensely veined with emerald green. Just behind the anther cap are two opposite sickle-like horns and just behind the pollinia is a third horn-like protuberance facing downwards. There are four pollinia on a long caudicle.

Flowering Periods: In September and October and the flowers last for 10 days or so.

Pollination: **Dressler** suggests small Anthophorid bees gathering oil to feed their larvae, but the colour and the depth of the nectary might indicate night-flying moths.

Cultivation: Best grown on barks or in small pots on a well drained fibrous compost, plenty of water in summer, misting in the rest of the year but never becoming water-logged or drying out.

Centroglossa nunes-limae
Pôrto & Brade | 2108 (Plate 41)

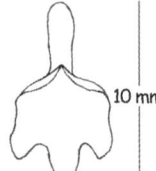

Etymology: Named in honour of **Nunes-Lima**.

Habitat: Found in Section III at 700-800 M. asl in old regrowth forest by streams in low light and very high humidity in the company of *Cirrhaea*, *Stanhopea* and other humidity-loving species, on low trunks, in similar situations to *Zygostates multiflora*.

Occurrence: Locally common.

Plant: A small epiphyte with surprisingly luxurious foliage made up of a tight tuft of linear, dark green, keeled leaves up to 6 cm. long by 0.4 cm. arising from the centre of the root mass. There was no sign of pseudobulbs. The root mass is strongly developed and composed of numerous, white, thinnish unbranched roots, a little longer than the leaves.

Inflorescence and Flowers: A slightly fleshy, erect inflorescence arises from the centre of the leaves and a cm. or so higher. It bears up to 8 small flowers which resemble *Zygostates* but are yellowish and the horns on either side of the midrib are quite distinctive. The sepals and petals are of similar size, yellowish-white, more or less linear, 0.6 cm. long by 0.2 cm. wide and the sepals are greener than the petals. The lip is trilobed with horns, green striped towards the throat, white at the tip and 0.5 cm. long by 0.4 cm. wide at the central shoulders. The dorsal sepal is tinged red at the base.

Flowering Periods: September and the flowers last for a week or so.

Pollination: Unknown.

Cultivation: Grow as *Zygostates* but with higher humidity and lower light.

Centroglossa tripollinica
Barb. Rodr. | 2109

Etymology: *Tri-pollinica* is Greek for three pollinia.

Habitat: Found in section VI at 1500 M.asl in scrub forest, requiring reasonable indirect light, high constant humidity and good air movement.

Occurrence: Occasional.

Plant: A small tufted epiphyte found around the bases and low trunks of stunted forest trees. The leaves arise on a short false petioles which in turn emerge from a root node. Leaves are sheathed, lanceolate, keeled, covered densely with minute warts giving them a shiny appearance and measure up to 5 cm. long by 1 cm. The roots are relatively thick, long and densely pilose.

Inflorescence and Flowers: A 4 cm. inflorescence arises from a new pseudobulb base and bears up to 4 small, resupinate, light green flowers. The petals and sepals are ± equal, a concave oval, 0.3 cm. by 0.2 cm. The lip is a trumpet-like tube, 0.4 cm. by 0.2 cm. wide at the apex. The column is flanked by two horn-like protuberances.

Flowering Periods: June and August; the flowers last a week or so.

Pollination: Unknown.

Cultivation: As for *Chyrtroglossa marileoniae*.

The genus
Comparettia Poepp. & Endl.

Poeppig a German, travelled in Chile, Peru and Brazil, mainly in the Amazon region, and together with **Endlicher** published the new species found. *Comparettia* was named by them and dedicated to **Comparetti**, professor of botany in Padua, Italy. Only one species of this small Latin American genus is to be found in the Organ Mountain Range.

Comparettia coccinea
Lindl. 2110 (Plate 41)

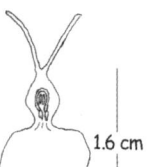

Etymology: *Coccineus* is Latin for deep red, from scarlet to carmine, referring to the flower's colour.

Habitat: We have never found this plant in original forest but frequently come across it at mid-tree level in 30 year old, open, regenerating forest, within citrus orchards in large colonies, and even as a terrestrial on cleared road-side banks. It requires good light, some wind movement and reasonable humidity at 800-1000 M.asl in sections V and VI.

Occurrence: Common.

Plant: A small variable tufted plant, essentially a twig epiphyte. The pseudobulbs are compressed narrow ovoids on a narrow convoluted rhizome, each up to 3 cm. long by 1 cm. wide, and generally wine-coloured. There is a single deeply-keeled, brittle, dark green, pointed oval, apical leaf, up to 11 cm. long by 1 cm, and long-lived. The root system is quite extensive and adherent. The fine grey-brown roots burrow into bark cracks for some distance around the plant. The plant does not transplant well and is relatively short-lived on its original roost.

Inflorescence and Flowers: The flowers are held on a thin, semi-erect, red, racemose, basal inflorescence up to 30 cm. long and may be many flowered but normally has between 2 and 10. The predominant feature of the flower is its colour, a deep cochineal orange, and its wide square trilobed lip, 1.5 cm. wide x 1.4 cm. long, with a distinct yellow vertical dividing line, and two backward and downward facing narrow pointed lateral lobes. The dorsal sepal and the petals are more or less equal, pointed, keeled ovals, 0.8 cm. long by 0.5 cm. wide, light yellow to orange in colour. The lateral sepals are fused and form at their base a long spur, looking like a nectary.

Flowering Periods: March and April and the flowers last for two weeks.

Pollination: Pods are frequent with up to 50% of flowers fertilized. The pollinator is not known, but, the colour would attract humming birds.

Cultivation: Difficult; grow on bark or in a small pot on a coarse fibrous compost, only misting, in good light and air movement; but a caveat; twig epiphytes are short-lived.

The genus
Rodriguezia Ruiz & Pavón

A Central and South American genus of up to 35 species mainly pertaining to Brazil and of which, seven should be found in the Organ Mountain Range. Raised by **Ruiz & Pavón**, two Spanish botanists, who were looking for medicinal plants in Central America and Peru during the 1790's, and who dedicated this genus to **Manuel Rodriguez**, a Spanish botanist and pharmacist.

Rodriguezia obtusifolia
(Lindl.) Rchb. f. | *2123*

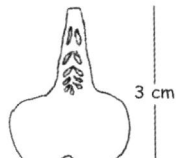

Etymology: Obtusi-folia is Latin for blunt, obtuse leaves, referring to the blunt leaf apices.

Habitat: In secondary forest scrub and old orchards, requiring high light and humidity with low wind movement in section VI around 900 M.asl.

Occurrence: Occasional.

Plant: This plant's principal characteristic is the length of the rhizome between pseudobulbs which can be up to 50 cm, thin, round and brittle with internodes. A new shoot arises from the base of the most recent pseudobulb and develops the new growth at its apex, up to 50 cm. on. Here several leaves develop and the stem swells at this point to form the new pseudobulb. One apical and one subtending leaf remain and are oval, keeled, stiff, light green and up to 8 cm. long by 3 cm. and obtuse at their apices. The pseudobulb is light green, a bilaterally-compressed oval, 4 cm. tall by 2.5 cm. wide. The root system emerges from the pseudobulb base and is composed of a number of long, non-branching, wiry white roots which remain aerial unless contacting a suitable substrate.

Inflorescence and Flowers: Single 15-35 cm. long inflorescences arise from the bases of the most recent pseudobulbs, and may show a raceme of up to 10 flowers at 1 cm. intervals at its apex. They are overall milky-white when viewed from the front while from the sides they show a pink tinge. The dorsal sepal is a narrow lanceolate 1.5 cm. long by 0.4 cm. while the lateral sepals are the same shape and size but fused for 90% of their length. The petals are oval, 2 cm. long by 0.8 cm. The bilobed lip is by far the most dominant feature with two almost round apical lobes, 2 cm. across by 1.5 cm. deep and held on a long tapering neck 1.5 cm. long, itself with a long central callus in a series of inverted "V"s.

Flowering Periods: June and July and the flowers last for a week or so.

Pollination: Probably Euglossine bees according to **Dressler**.

Cultivation: One could consider fixing the basal rooted pseudobulb in a large pot with a rich fibrous compost and allow the plant to grow up the side or end of a greenhouse, allowing some pseudobulbs to root on strategically placed tree fern or bark plaques.

Rodriguezia bracteata
(Vell.) Hoehne | *2130* (Plate 41)

Etymology: Bracteata is Latin and refers to the bracts.

Habitat: An epiphyte found in all sections between 200-800 M.asl, and often in large colonies in abandoned coffee plantations and orchards. A low tree epiphyte or lithophyte needing good air movement, strong dappled light and high humidity.

Occurrence: Common and clearly an opportunist shown by the enormous numbers of plants we have found in abandoned orchards.

Plant: A compact and long-lived epiphyte, typically with up to eight bilaterally compressed pseudobulbs arising tightly from a 1 cm. diameter rhizome, each pseudobulb measuring 2.4 cm. high by 1.1 cm. wide and sheathed by the old leaf bases for three years or so. New growths have one apical and two basal leaves, all coriaceous, deeply keeled, lanceolate, light green and retained for two to three years. The apical leaf is up to 12 cm. long by 1.8 cm. wide. The confused root system is part adherent and part aerial with numerous thin, wiry, branched, white roots often extending for a metre or more towards both ends of a smallish branch.

Inflorescence and Flowers: One or two inflorescences arise from the most recent pseudobulb base with round stems, 2.0 cm. diameter and up to 15 cm. long and the apical 9.0 cm. bears a raceme of up to eight alternate, outward facing, showy flowers. The concave dorsal sepal is a broad lanceolate measuring 1.8 cm. by 1.0 cm. while the laterals are completely fused and together measure 2.0 cm. by 0.75 cm. and form a deeply concave rowboat shape. The petals are ovate and measure 1.5 cm. long by 0.9 cm. Both sepals and petals are milk-white. The lip is flared into two lobes at the tip, narrow waisted towards the throat where there are two elevated parallel, sail-like protuberances. The lip measures 2.5 cm. long by 1.7 cm. at the lower lobes and is milk-white, but the protuberances and the tip of the lip are light butter-yellow. These are flowers of exceptional beauty.

Flowering Periods: November and December and the flowers last for two weeks.

Pollination: They are probably pollinated by *Euglossine* bees.

Cultivation: On bark plaques or in hanging baskets with a well-aerated compost. As the roots are untidy and partly aerial, a good humid environment is needed.

Rodriguezia venusta
Rchb. f. | *2133* (Plate 41)

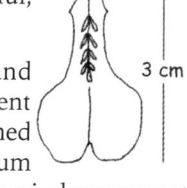

Etymology: Venusta is Latin for beautiful, graceful referring to the flowers.

Habitat: Found on both trunks and branches of scrub trees over permanent swamps in section V. Also in abandoned orchards at 600 M.asl, enjoying medium light, high constant humidity and low wind movement.

Occurrence: Occasional.

Plant: A medium-sized, untidy epiphyte found usually in large colonies in its specific habitat. The pseudobulbs are held on a sheathed woody rhizome arising at 0.5-1.5 cm. intervals and are narrowly elliptic, strongly bilaterally compressed, up to 3 cm. long by 1.4 cm. There are one apical and two basal and opposite leaves, light green, leathery, deeply keeled, lanceolate and measuring up to 15 cm. by 2 cm. broad. They are long lasting. The root system tends to show a tangled mass of wiry, long, white roots of which part are aerial and part adhere strongly to the substrate.

Inflorescence and Flowers: Inflorescences, up to 8 cm. long, round and pendulous arise from the bases of the most recent pseudobulbs and are partly sheathed by the subtending leaves. They bear up to seven very pretty white flowers which are sometimes lightly blushed in pink. The dorsal sepal is a broad lanceolate, 1.9 cm. long by 0.7 cm. and concave. The lateral sepals are 100% fused as a conduplicate unit and together measure 2 cm. long by 0.7 cm. The petals are elliptic, measuring 2.0 cm. by 1.0 cm. broad. The lip is bilobed, 3 cm. long by 2 cm. at the apical lobes. The apical lobes almost form a circle which narrows towards the throat to show a yellow, several streaked callus, 2.2 cm. long and reaching to the indentation separating the apical lobes.

Flowering Periods: December and January and the flowers last for two weeks or so.

Pollination: **Dressler** suggests probably *Euglossine* bees.

Cultivation: In pots on a coarse fibrous substrate taking care not to let the plant become water-logged or dried out.

General: It differs from **R. bracteata** in that the flowers are larger and the callus is much more defined.

The genus
Ornithophora Barb. Rodr.

Barbosa Rodrigues raised this one-species genus and used the Greek compound word meaning looking like a bird and referring to the flower.

Ornithophora radicans
(Rchb. f.) Garay & Pabst | 2142

Etymology: *Radicans* is Latin for putting forth aerial roots which we have not noticed.

Habitat: Found along riverside banks in regrowth, almost scrub forest, at mid- tree level, enjoying high constant humidity, reasonable light and air movement in section VI at 900 M.asl

Occurrence: Rare.

Plant: A small, somewhat delicate epiphyte. The pseudobulbs are almost cylindrical arising alternately at 1.5 cm. intervals from a sheathed woody rhizome and are 2 cm. tall by 0.8 cm. broad, slightly laterally compressed, and sheathed by the bases of dead subtending leaves. The leaves on new growths are both subtending and apical, two of the former and up to four of the latter. Only a single apical leaf survives for several years and this is linear-lanceolate, some 10 cm. long by 0.3 cm. broad, light green, erect and somewhat stiff. The root system is not strong, composed of few short white roots which emerge from both the pseudobulb base and the rhizome.

Inflorescence and Flowers: Basal, stiff, thin inflorescences arise from the most recently matured pseudobulbs, each up to 8 cm. long with up to six alternate flowers on the apical 4 cm. These are borne on 1 cm. pedicels and are resupinate. The dorsal sepal is lanceolate, 0.35 cm. by 0.2 cm. broad and is flanked by the two similarly sized and shaped petals while the lateral sepals are a boomerang-shape and much smaller. All are a very pale lime-green. The lip is relatively large, milk-white, ± a broad anchor-shape, vaguely bilobed with a central four-lobed, short-nosed callus, partly butter-yellow. The prominent column is dark purple while the anther cap is butter-yellow. This mélange of colour induced **Barbosa Rodrigues** to name the plant *O. quadricolor* – now a synonym. The lip is 0.5 cm. long by 0.4 cm.

Flowering Periods: March and the flowers last for 10 days or so.

Pollination: **Dressler** affirms that it is pollinated by small oil-gathering bees.

Cultivation: Unknown.

The genus
Rodrigueziopsis Schltr.

Rudolf Schlechtler named this genus so because it was similar to *Rodriguezia*. It is a small Brazilian genus of two species of twig epiphytes. Both are to be found in the Organ Mountain Range.

Rodrigueziopsis eleutherosepala
(Barb. Rodr.) Schltr. | 2145

Etymology: *Eleuther-sepala* is a Greek compound for free, not united - sepals.

Habitat: A high treetop plant from original forest in section VI at 1000-1500 M.asl, enjoying high light and wind movement and an underlying constant humidity.

Occurrence: Occasional.

Plant: A sprawling epiphyte which appears to migrate from branch to branch and even tree to tree using its spider-like aerial root system and very long rhizome between pseudobulbs. The pseudobulbs are ovate, light green, bilaterally compressed, 2.5 cm. tall by 1 cm. broad, and are held on a long rhizome, up to 50 cm. in length and 0.15 cm. thick, which is reproduced for the following years growth. The young pseudobulb has two apical and two subtending leaves on either side. The apical leaves are long-lasting, lanceolate, deeply keeled, dark green, coriaceous and measure 6.5 cm. by 2.5 cm. broad. The subtending leaves are short-lived, the same shape, colour and texture but smaller. The root system emerges from the base of the pseudobulbs and is composed of some half dozen initially aerial roots. These are white, pliable and measure up to 10 cm. long. If they come up against a substrate they will adhere and thus accomplish a first stage of migration.

Inflorescence and Flowers: A short, hair-like inflorescence up to 5 cm. long, emerges from the base of the pseudobulb, appearing at the junction of the upper subtending leaf, with one or two flowers on 1 cm. pedicels. The dorsal sepal is spathulate, 0.5 cm. long by 0.3 cm. broad and somewhat concave. The lateral sepals are an asymmetric lanceolate, more or less the same size while the petals are an asymmetric oblong, 0.65 cm. long by 0.4 cm. broad. The lip is 1.4 cm. long and four-lobed. The two apical lobes are together almost round, 0.6 cm. in diameter and take up half the lip's length. The basal half, which has two horn-like, opposite side-lobes, each 0.25 cm. long by 0.1 cm. wide, also shows two parallel nose-like calli which run centrally for a third of the lips length. The overall flower colour is white.

Flowering Periods: In November to January and the

flowers last for two weeks.

Pollination: Unknown.

Cultivation: Unknown.

Rodrigueziopsis microphyta (Barb. Rodr.) Schltr. | *2146* (Plate 41)

Etymology: *Micro* is Greek for little or small and *phyta* is Greek for plants 1.2 cm which is a fair description.

Habitat: A true aerophyte of the original forest at 900-1500 M.asl in section VI, which dangles in strands from the tree tops and so manages to pass from tree to tree across the canopy. The twigs, where it is anchored, are densely covered with moss and lichen. The pseudobulbs' dense bracts may protect against water loss.

Occurrence: Common.

Plant: This is truly a treetop aerophytic species which may be found hanging from the smallest branches of the tallest trees. Strings of 5 to 7 pseudobulbs may be seen hanging in lengths of up to half a metre joined by a shiny, wiry rhizome. The small oval pseudobulbs measure 2.5 cm. x 1.1 cm. and are separated by the rhizome in lengths of up to 25 cm. Young, developing pseudobulbs have five leaves, two apical and three sheathing the bulb. All but one of the sheathing leaves dry out during the second year and remain as bulb sheaths. Some apical leaves are retained until the fourth year. On the free, hanging pseudobulbs, some 2-4 unbranched text book aerial roots of about 6 cm. are produced. On anchored pseudobulbs, a more extensive root network penetrates the surface moss and adheres to branch surfaces.

Inflorescence and Flowers: A 6-8 cm. inflorescence arises from within the sheath of the previous years pseudobulb. The individual flowers are pale yellow and measure 1.5 cm. x 0.6 cm. and do not open widely. The dorsal sepal is a broad lanceolate, 0.7 cm. long by 0.4 cm. The laterals are narrower, fused for 10% of their length and 0.7 cm. by 0.3 cm wide. The trilobed lip is relatively long, 1.3 cm. by 0.6 cm., and dominated by the roundish apical lobe. The basal, opposite triangular lobes surround a 4-nosed purple-haired callus.

Flowering Periods: December and January; the flowers last for two weeks.

Pollination: Infrequent and the pollinator is unknown.

Cultivation: Anchor the base bulb to almost any substrate in high light, strong air movement and reasonable humidity, allowing support for its sprawling habit

The genus Capanemia Barb. Rodr.

This genus was dedicated to the Baron of Capenema, **Gustavo Schuch**, who was one of the first to warn that the railroads would be one of the key factors in the ultimate destruction of the Atlantic Rain Forest. **Barbosa Rodrigues** dedicated this genus to him.

The genus is represented by some 15 small epiphytes from Brazil, Argentina and Paraguay of which around six could be attributed to the Organ Mountain Range. These are characterised by tiny pseudobulbs and their capacity to colonize defunct coffee plantations and run down orchards.

Capanemia fluminensis Pabst | *2151*

Etymology: *Fluminensis* Latin for river or in this case, from the state of Rio.

Habitat: A twig epiphyte, found in abandoned orchards, coffee plantations and suburban woods in section V and VI, at 800-1000 M.asl, enjoying hot humid summers and dry winters.

Occurrence: Common.

Plant: This plant is surprisingly large for a twig epiphyte. A typical plant has up to 30 linear-lanceolate, fleshy, deeply keeled, dark green leaves, one of which is basal, measuring up to 2.4 cm. long by 0.5 cm. and one apical, 3.5 cm. long by 0.7 cm. These leaves are long-lived and it is rare to find an old pseudobulb without leaves. The pseudobulbs are dark green, ovate and measure 1 cm. long by 0.35 cm. in diameter and arise from the base of a previous pseudobulb leaving no space to show a distinctive rhizome. The root system is composed of numerous long, thin white, exposed roots which completely envelop the thin twigs that form the plant's perch.

Inflorescence and Flowers: The plant produces both basal and apical inflorescences, sometimes both on the same pseudobulb. The flowers, between two and four, appear as a raceme on the apical 1 cm. section of a 2 cm. stem and rarely rise above the leaf tip. A typical plant will have up to 10 inflorescences, eight basal and two apical, holding some 30 flowers, yellow green in colour, with tepals ± equal in shape and size, measuring 0.4 cm. long by 0.1 cm. wide. The lip is distinctive, single-lobed, measuring 0.5 cm. long by 0.2 cm. broad. The basal central section has two 0.3 cm. long tear-drop-like, fused calli, 0.3 cm long.

Flowering Periods: In June and the flowers last for three

weeks

Pollination: Frequent; probably by nectar-seeking, wasps.

Cultivation: Only seems to thrive on a live twig substrate.

General: What separates this plant from *C. therezae* is its size and that its calli are far grosser than the latter's.

Capanemia thereziae
Barb. Rodr. | 2152 (Plate 41)

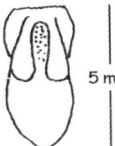

Etymology: Named by **Barbosa Rodrigues** in homage to Her **Majesty the Empress S. Tereza Christina Maria**. Since this is a small, insignificant plant with tiny flowers, it hardly seems a compliment.

Habitat: Found on the tiny twigs of **Tibouchina** species along ridges and exposed high mountain fields where it is experiences a cool, dry winter and heavy spring rains; also found in regrowth forest of small size and on fruit trees from 800-1500 M.asl in sections V and VI.

Occurrence: Common but not easy to find.

Plant: A diminutive twig epiphyte with clusters of ovoid pseudobulbs measuring 1 cm. long x 0.3 cm. wide and borne on a very short stem. The fine roots anchor the plant to tiny twigs, along which they run for some distance.

Inflorescence and Flowers: Flowers arise from a 2-3 cm. long basal inflorescence which bears 3-5 flowers. These are spiky-looking with fine tapered lanceolate greeny-yellow tepals about 0.6 cm. long and 0.2 cm. wide. The lip, 5 mm. long by 2 mm. broad is lanceolate, with two parallel nose-like calli taking up half its length and is curled back and under.

Flowering Periods: July and August and the flowers last for two weeks.

Pollination: Pods are frequent, up to 50%; possibly fertilized by oil-gathering bees.

Cultivation: We have only managed to transplant this orchid onto live branches. On any other medium, transplants have quickly perished, and remember, twig epiphytes have short lives, which is probably why they are so precocious and fructiferous.

General: Growth in vitro has been fast, the seedlings coming to flower in flask at nine months after sowing.

Capanemia carinata
Barb. Rodr. | 2156

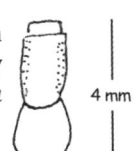

Etymology: *Capanemia* Refers to the Baron of Capanema, an early 19th century Brazilian conservationist, while *carinata* means keeled, referring to the keeled lip.

Habitat: A twig epiphyte from 400 M.asl, in section VIII, enjoying reasonable shade and humidity with some air movement.

Occurrence: Occasional.

Plant: Distinguished from most other **Capanemia** species by its erect terete leaves and small, almost fusiform light green pseudobulbs, which are tightly clumped, 1 cm. tall by 0.3 cm. each bearing a 4 cm. tall, 0.2 cm. broad, grooved, darker green, long-lived, terete leaf. The root system, as with most twig epiphytes, is surprisingly extensive and composed of innumerable longish white roots, 0.1 cm. thick and up to 6 cm. long which run up and down the substrate.

Inflorescence and Flowers: Short, thin, wiry inflorescences, up to a third of the leaf height arise from the bases of most pseudobulbs and bear up to five small alternate flowers on short, bracted pedicels. The sepals are a narrow lanceolate, 0.5 cm. long by 0.15 cm, the petals are shorter but thicker and the lip is shaped like the petals with a ¾ length, tear-drop like callus which starts at the throat. The overall flower colour is greenish-white.

Flowering Periods: September, when the flowers last for 10 days or so and surprisingly, remain on the inflorescence when withered.

Pollination: Infrequent and probably by small flies.

Cultivation: **Hoehne** suggests that these plants should be tied to Azalea twigs.

The genus
Dipteranthus **Barb. Rodr.**

A small Brazilian genus of some seven species which are confined to eastern Brazil. Four are attributed to the Organ Mountain Range, three of which we have found. In **João Barbosa Rodrigues**'s description, to elevate the genus he uses the Greek word to indicate an insect with two clear wings and looking like a mosquito. There seems recently to be some polemic amongst taxonomists between this genus and *Zygostates*.

Dipteranthus aff. octavio-reisii
(C. Pôrto & Brade)
Pabst. | 2173 (Plate 41)

Etymology: Not known.

Habitat: A low to mid-tree trunk epiphyte in stream-side original forest at 1200 M.asl in section VII enjoying high humidity, medium light and fair wind movement.

Occurrence: Occasional and in small colonies.

Plant: A very small epiphyte. When not in flower is quite indistinguishable from certain small *Pleurothallis* species. There is a hint of a minute pseudobulb. There are 4 or 5 leaves, both basal and apical, long-lived, dark green, linear-lanceolate, deeply keeled, curved downwards and 4.5 cm. long by 0.3 cm. The root system fits the plant; small and delicate, a few short, small white roots which do not exceed the circumference of the plant.

Inflorescence and Flowers: A short basal inflorescence, up to 2 cm. long, bears up to 5 alternate small flowers clustered around the apex and opening sequentially. The overall colour is a light greenish yellow. The sepals are ± equal, the dorsal 0.4 cm. by 0.2 cm. is ± oval while the laterals are slightly shorter and more wedge-like. The lip is shaped like a long-stemmed wine glass, 0.3 cm. deep, 0.2 cm. wide, green tinged inside, and whitish yellow outside.

Flowering Periods: In mid May and the inflorescence shows flowers for up to two weeks.

Pollination: Unknown, but the spur and the colour would indicate night flying Lepidoptera.

Cultivation: As for any small *Pleurothallis* species, always kept humid and in low light but with air movement.

Dipteranthus pellucidus
(Rchb. f.) Cogn. | 2176 (Plate 41)

Etymology: *Pellucidus* is Latin for translucent but not colourless referring to the tepals.

Habitat: A low to mid-tree epiphyte found around 100 M.asl in original or regrowth forest and requiring medium light, humidity, and air movement in section VI.

Occurrence: Occasional.

Plant: A small epiphyte with pseudobulbs rather larger than others in the genus, each a narrow ovoid measuring 2 cm. by 0.6 cm. wide, greyish-green, faintly ridged and tightly held on the thin, woody, branching

rhizome. There is a single, grey-green, long-lived, apical leaf on a 0.3 cm. false petiole, elliptical, mildly keeled and up to 10 cm. long by 2.2 cm. wide with a tendency to reflex at the margins. The root system is strongly developed and composed of long greenish, very hairy roots which sometimes branch.

Inflorescence and Flowers: A 10-15 cm. basal or apical inflorescence arises with a loose cone of up to 40 small white and green flowers on the 8-12 cm. apical section. The sepals are an obtuse-apexed rectangle, 0.4 cm. long by 0.2 cm. wide. The petals are a blunt club-shape and measure 0.45 cm. long by 0.25 cm. wide. The dorsal sepal and petals are milk-white. The lip is horned at its base and cup-shaped at its apex, single lobed, triangular and measures 0.4 cm. long by 0.25 cm. at the horns. The column and anther are complex.

Flowering Periods: November and the flowers last for two weeks.

Pollination: Fruits are rarely seen and pollination is probably by oil-gathering bees.

Cultivation: On bark plaques or in small pots with a coarse bark substrate keeping humid all year round and watering well in spring.

Dipteranthus grandiflorus
Pabst | 2177 (Plate 42)

Etymology: *Grandiflorus* is Latin for large-flowered which in Dipteranthus terms, fits.

Habitat: Grows in tall original forest at 600-800 M.asl in section VI. This is a low-to-mid-trunk epiphyte enjoying deep shade, some wind movement and high humidity.

Occurrence: Common.

Plant: A compact fleshy-leafed epiphyte, know locally as 'The Fan' orchid, with small pseudobulbs, 1.5 cm. x 1.0 cm. and surrounded by previous years' leaf bases which remain as brown sheaths. The rhizome is very short. There are 4 leaves, one long-lived apical leaf, 2 cm. x 1 cm, a broad oval, and 3 leaves subtending the pseudobulb. These are larger at 5.0 cm. x 2.0 cm, pale green and retained for two years. The leaves are set in a fan-shaped arrangement. The roots are thick and white with a shiny velamen, adherent and reaching about 10 cm, each pseudobulb producing about six.

Inflorescence and Flowers: The inflorescences, produced from the base of the mature pseudobulbs, are about 10 cm. long, bearing 6-10 white flowers with green markings. The lateral sepals are circular, 0.6 cm. in diameter. The petals are 1 cm. long, fan-shaped and extended towards the base, and white with a deep green tinge on the basal quarter. The lip is 1.4 cm. long with one upper and two basal lobes and notched on the lateral edge of the basal lobes.

Flowering Periods: January and the flowers last for three weeks.

Pollination: This is frequent and **Dressler** suggests probably by small oil-gathering bees.

Cultivation: In small pots on a rich fibrous compost, in shade with some air movement and high humidity.

The genus *Zygostates* Lindl.

John Lindley raised this genus of eight species, mostly from the Brazilian Atlantic rain forest using the Greek compound word meaning joined scales or balance and referring to the long balance-like appearance of the staminodes joined at the base of the column.

Zygostates cornuta Lindl. | *2179* (Plate 42)

Etymology: *Cornuta* is Latin for horned, referring to the horn-like column tip.

Habitat: Found between 1200-1500 M.asl, in section VI, growing on low tree trunks in dark, original montane forest with dappled light, low wind movement and constant high humidity

3 mm

Occurrence: Rare.

Plant: A most curious miniature epiphyte easily mistaken for a *Dryadella* or small *Pleurothallis* when not in flower. The light green leaves are linear-lanceolate and are up to 6 cm. long by 0.6 cm. wide. They are apical, single, long-lived, faintly keeled and borne on a minute cylindrical pseudobulb. The weak root system is composed of numerous short fine greenish-white roots which do not branch nor exceed the plant's circumference.

Inflorescence and Flowers: A pendulous inflorescence, up to 10 cm. long arises from the base of the most recent leaf with a loose cone of curious flowers on the apical half. The sepals are green, concave and lanceolate, 0.5 cm. long by 0.2 cm. wide. The petals are spade-shaped, smoothly rounded at the apices and deeply toothed at their orange margins, greenish-orange in their centres and they measure 0.5 cm. by 0.5 cm. The lip is concave, single-lobed, lanceolate and 0.3 cm. long by 0.15 cm, a clear white with indented margins. There is a club-like incurved horn, 0.2 cm. long at each side of the column.

Flowering Periods: December and January and the flowers last for a week or so but the inflorescence may bear flowers for up to three weeks.

Pollination: Probably pollinated by **Anthophorid**, oil-gathering bees.

Cultivation: In small pots or on bark plaques in low light, low air movement and high humidity.

Zygostates multiflora (Rolfe) Schltr. | *2183* (Plate 42)

Etymology: *Multi-flora* is Latin for many flowered.

Habitat: Almost exactly the same as *Oncidium cogniauxianum*; top-tree twigs in original forest at 1000-1500 M.asl in section VI. However, it also seems perfectly happy in the deep gloom beloved by *Pabstia jugosa* provided there is a certain air movement.

4.5 mm

Occurrence: Common.

Plant: A small, attractive, miniature epiphyte forming quite large colonies in a variety of situations. The pseudobulbs are a deep green oval, slightly laterally compressed, 1 cm. high x 0.4 cm. wide, tightly clustered on a thick green rhizome. A single dark green apical leaf is retained and this is 5 cm. long by 0.6 cm. wide, lanceolate and deeply keeled. The young pseudobulb has two to six basal sheathing leaves which are the same shape and colour as the apical leaf, but vary in length and width. The root system is vigorous and adherent, and 1 mm. diameter green roots insinuate into bark cracks for significant distances up and down the branch to total 60-80 cm. for each plant.

Inflorescence and Flowers: One or 2 pendulous, 8 cm. long, spicate, basal inflorescences are produced, each bearing up to 12 alternate resupinate flowers. The petals are large, axe-head shaped, 3-4 mm. long by 3 mm. wide. The sepals are oval, 3 mm. long by 2 mm. wide; all parts are milk-white. The lip is curious with a complex callus and two quite distinct backward facing horns which are about 4 mm. long by 2.5 mm. wide. The column is a light green.

Flowering Periods: In October through to December and the flowers last for three weeks.

Pollination: Frequent seed set, **Dressler** suggests possibly effected by small oil-gathering bees.

Cultivation: Grow on twigs or in small pots on a fibrous compost, in dappled light, good air movement, and high humidity.

| 323 |

The genus
Ornithocephalus Hooker

Only one of this New World genus of about 50 species is attributed to the Organ Mountain Range. Sir. **W. J. Hooke**r, was the director of the *Royal Botanic Gardens* at Kew, England, and used the compound Greek word for like a bird's head alluding to the column.

Ornithocephalus myrticola
Lindl. | *2192* (Plate 42)

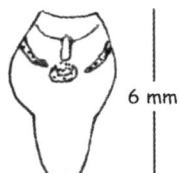

6 mm

Etymology: *Myrti-cola* is Latin for myrtle-dwelling as the first plants, found in Bananal, were growing on Myrtle trees.

Habitat: Found on mossy tree trunks in both original and maturing regrowth forest in sections III and VI at 700-1000 M.asl, where it enjoys high humidity, dappled light and low wind movement.

Occurrence: Occasional, but generally found in colonies.

Plant: A small fan-shaped epiphyte with up to 12 lanceolate, curved acuminate leaves which measure up to 8 cm. long by 0.8 cm. broad. These are infolded, Iris-like, fused for half their length and completely fused for the remaining apical quarter. They arise alternately, tightly and sequentially from a hidden central stem and are light green, tinged grey as they mature, and slightly succulent. The root system is weak, arising from the base of the fan, and is composed of a number of short white roots usually under moss.

Inflorescence and Flowers: A short erect inflorescence measuring up to 4 cm. and with 10 bracted, alternate flowers at 0.4 cm. intervals, emerges from the base of one of the most recent, central leaves. All the stems, the 0.5 cm. pedicels and bracts are hairy and the flowers open sequentially from the base of the inflorescence. The dorsal sepal is spathulate, 0.4 cm. long by 0.25 cm., concave, white and with green basal veins. The lateral sepals are similar in size, shape and colour but are pilose at the margins. The petals are an irregular square shape and white with three green central veins and they measure 0.5 cm. long by 0.4 cm. wide. The lip is a complex structure, vaguely trilobed and longitudinally green-veined at its elevated base and shaped like a narrow spade. It measures 0.6 cm. long by 0.3 cm. wide at the side-lobes. The column and the rostellum are equal in size and, proportionally to the flower, very long.

Flowering Periods: April and May and individual flowers may last for a week but the inflorescence lasts for a month.

Pollination: Probably pollinated by oil-gathering *Anthophorid* bees.

Cultivation: Best grown on a small piece of hard wood, like vine or on bark plaques.

The genus
Chytroglossa Rchb. f.

Named by **Reichenbach Jnr.**, a German botanist who worked in tandem with **Lindley** in England, and who named many orchids, some of them with almost incomprehensible names such as this one, which **Pabst&Dungs** inform us means that the lip is joined to the stigmatic cavity forming a pouch. Four species are attributed to the Atlantic rain forest and three to the Organ Mountain Range including one new one which our friend and colleague, **Helmut Seehawer** discovered in Macaé de Cima, Nova Friburgo.

Chytroglossa marileoniae
Rchb. f. | *2195* (Plate 42)

Etymology: Not known.

Habitat: We have found this extraordinarily beautiful small orchid in 40-year-old regrowth, on relict original forest trees and once on the 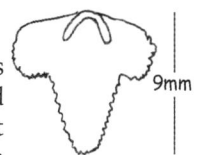 trunk of *Cupressus*, all between 1000-1100 M.asl in section VI, from mid to upper tree, enjoying good dappled light, fair air movement and humidity.

9mm

Occurrence: Occasional.

Plant: A small epiphyte, easily confused with a *Pleurothallis* when not in flower. The leaves are dark green, spreading, erect to pendulous, 4-6 cm. in length, 1 cm. wide and deeply keeled, growing from a central point or much reduced rhizome. The roots are numerous, thin, white and radiating from the leaf bases for up to 6 cm. around.

Inflorescence and Flowers: Flowers arise on thin, basal, pendulous, and spicate inflorescences, often several to a plant. We have found up to 56 alternate flowers on one plant. The sepals are lime-yellow. The dorsal is 0.7 cm. long by 0.3 cm. wide and bluntly pointed and the laterals curve behind the lip and are 0.3 cm. long x 1.5 cm. wide. The petals are flat, also lime-yellow, and are 0.7 cm. long by 0.2 cm. wide and obtuse at the apices. The trilobed lip is a broad anchor-shape, 1 cm. wide at the basal side-lobes by 0.9 cm. long towards the triangular apical lobe, white, with claret blotches on each lobe's centre and the margins are faintly serrated.

Flowering Periods: June - September and the flowers last for 15 days or so.

Pollination: Infrequent; **Dressler** suggests probably oil-gathering bees.

Cultivation: Best mounted and kept in good light and air movement, always slightly damp. It detests transplanting.

Chytroglossa seehaweri
Bock | 2195b

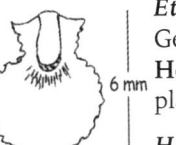

Etymology: Dedicated to the noted German specialist in Pleurothallidinae **Helmut Seehawer,** who discoverd this plant in Macaé de Cima, RJ.

Habitat: Found in section VI in the upperbranches in original forest trees needing good light and air movement with a constant underlying humidity.

Occurrence: Occasional.

Plant: Very similar to *C. Marileoniae*. A small epuphyte showing minute clustered emerald green pseudobulbs 0.2 cm. tall. These arise from a short, wandering, wiry rhizome. The leaves, one apical and two subtending are held on short whitish green false petioles, elliptic keeled up to 2 cm. long by 0.3 cm, light green and fleshy. Only the apical leaf is long lived. The root system is strong, composed of 6-7 short thick white roots from the pseudobulbs base.

Inflorescence and Flowers: The inflorescence, up to 3 cm. long, holds one to seven alternate flowers on short pedicels, and arises from a mature pseudobulb base. The sepals and petals are erect, laterally reflexed, ± the same size and shape, lanceolate, 0.5 cm. long by 0.12 cm. and a greenish white with a wine blob at the bases of the dorsal sepal and petals. The lip is the principal feature, differentiating this flower from *C. marileoniae*. It is single-lobed with a high, proportionally thick neck. The lobe is almost round to spade-shape, white, while the neck is orange heavily blotched in wine. All margins are intensely serrated, while a cuplike, green callus is shown at mid-throat. The overall size is 0.6 cm. long by 0.5 cm.

Flowering Periods: January and the flowers last for 10 days or so.

Pollination: Unknown, but probably as for *C. marileoniae*.

Cultivation: As for *C. marileoniae*.

The genus
Phymatidum Lindl.

John Lindley used this Greek word, a diminutive of *phyma* for growth, or small plant - small growth, which aptly describes most species of this 15 strong small Brazilian genus. Six are attributed to the Organ Mountain Range, mostly from high altitudes. **John Lindley** was perhaps the most prolific namer of Brazilian orchids.

Phymatidium aquinoi
Schltr. | 2196

Etymology: Not known.

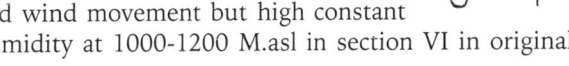

Habitat: A small epiphyte at low tree trunk level in under forest, found in very damp conditions similar to *P. tillandsoides* and requiring low light and wind movement but high constant humidity at 1000-1200 M.asl in section VI in original forest.

Occurrence: Occasional.

Plant: A very small epiphyte, quite insignificant when not in flower, and as with others in the genus, more like a lichen. It has three to five short stems up to 6 cm. long, which arise from nodules in the centre of a root mass. These bear five to 10 closely alternate 1 cm. long, very narrow, light green, 0.1 cm. wide, fleshy leaves, giving a fish-bone skeleton-like appearance. The root system, composed of up to a dozen, white, unbranching, brittle roots run up and down the liana or small branch and in fact give away its orchid status. A gypsy type plant, short-lived normally, but flowering and producing seed quickly.

Inflorescence and Flowers: A short raceme of three to five small white flowers arise at 0.4 cm. intervals at the apex of each stem of 6 cm. Sepals and petals are more or less equal, lanceolate and measure 0.4 cm. by 0.2 cm. broad, a clear white. The lip is distinctive as the base is a round bowl, light green with two small horn-like projections at its front. The single lobe is an arrow-head, green at the centre and white at the margins and measures 0.45 cm. long by 0.25 cm. broad.

Flowering Periods: April and May and the flowers are long-lasting for up to three weeks.

Pollination: **Dressler** suggests possibly small night-flying moths or probably small *Anthophorid* oil-gathering bees.

Cultivation: As for *P. mello-barretoi*.

Phymatidium delicatulum Lindl. | *2197* (Plate 42)

3 mm

Etymology: *Delicatulum* is from Latin *delicatu* - delicate, weak, feeble, referring to this sturdy little plant's apparent fragility.

Habitat: Deep shady, original or regrowth, riverside forest on twigs, small branches and lianas, in section VI, with low wind movement, high humidity and good light, at 1000 M.asl.

Occurrence: Occasional, but may be more common as being so small and camouflaged. It is easy to overlook.

Plant: A minute and delicate twig epiphyte whose stems rarely exceed 3 cm. with up to eight alternate, terete, pale green leaves, up to 0.4 cm. long. The root system is extensive and often the only way to differentiate the plant from one of the many lichens that are its neighbours.

Inflorescence and Flowers: The flowers arise alternately on minute stems from the leaf bases, often as many inflorescences as leaves. Petals and sepals are milk white, ± equal in size and shape, which is a broad lanceolate. The lip is larger, 0.4 cm. by 0.3 cm. wide, single-lobed, almost square with one of the corners being the apex, and white with a three-knobbed, green callus at its throat.

Flowering Periods: March/April and October/November when the flowers last for 10 days or so.

Pollination: Unknown.

Cultivation: As for *P. mello-barretoi*.

Phymatidium hysteranthum Barb. Rodr. | *2199* (Plate 42)

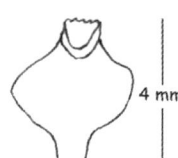

4 mm

Etymology: *Hysteranthum* is a Greek compound for flowers that appear before the leaves.

Habitat: Epiphytic on twigs and scrub tree trunks, in regrowth forest on mountain ridges at 1400-1600 M.asl, in section VI, enjoying high light and wind movement with constant mountain misting, characteristic of these regions. **Barbosa Rodrigues** reported that it almost always grows on *Myrtaceae* spp.

Occurrence: Common.

Plant: A very small tufted epiphyte similar to most other *Phymatidium* species except perhaps it is slightly larger and marginally longer-lived. When not in flower, only its roots distinguish it as an orchid species. The numerous leaves are a light translucent green, up to 2 cm. long by 0.2 cm. broad, a fleshy, awl-shape and a typical plant may have up to 60. They arise from nodes in the root system, itself composed of few surprisingly thick, short, whitish green roots which stretch for about 5 cm. beyond the plant's circumference. It lacks pseudobulbs.

Inflorescence and Flowers: Long spicate inflorescences emerge from nodes at the leaf bases, reaching up to 9 cm. with up to 12 flowers. A typical plant may have up to 15 such inflorescences and over 100 white flowers, which make a fine sight in early spring. The flowers are milk-white with an emerald green callus. The sepals and petals are roughly the same size and shape, 0.4 cm. long by 0.2 cm. broad, linear-lanceolate and milk-white. The lip is very distinctive, trilobed, appearing like a white cross below a cuplike emerald-green callus. When pressed or flattened and the back folded serrated margins exposed, the lip assumes a more conventional shape, 4 mm. long by 0.4 cm. wide at the side-lobes.

Flowering Periods: In September and October and the flowers last for two weeks.

Pollination: **Dressler** suggests that probably fertilized by small *Anthophorid* bees that gather oil as food for their larvae.

Cultivation: Unknown. However, we believe that this plant has a very short life. This is borne out by long term observation and the fact that it is a prodigious and swift seed producer.

Phymatidum mello-barretoi Hoehne & Williams | *2201* (Plate 42)

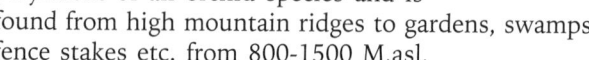

6 mm

Etymology: Dedicated to **Mello Barreto** who collected in Brazil in 1935.

Habitat: Said to be epiphytic on *Eugenia* species, however, in sections V and VI it is perhaps the most polyvalent of all orchid species and is found from high mountain ridges to gardens, swamps, fence stakes etc. from 800-1500 M.asl.

Occurrence: Common.

Plant: A very small epiphyte and easy to pass by when not in flower, not only because of the small size of the plant but also for its lack of similarity to any other orchid species. It resembles tufts of lichen among moss on very small twigs. Apart from the flowers, the only orchid-like features are the typical long roots. The leaves are thin, terete and pale grey and are gathered into a loose fan-like structure, 1.5 cm.-1.8 cm. long and less than 1 mm. wide and about 8 to a plant. The roots distinguish this plant as an orchid when seen on a trunk or branch, as they are white with green tips and run in either direction from the plant along twigs. A sample plant had a total length of 40 cm. of roots.

Inflorescence and Flowers: For such a small plant, the inflorescences are disproportionately large. They are spicate and up to 4 cm. in length with 5-9 flowers, each subtended by a small leaf-like bract. The flowers are white, 0.5 cm. x 0.6 cm. high. The tepals are equal in size and shape, a narrow lanceolate 0.6 cm. long by 0.3 cm, while the broad pointed, white recurved lip is 0.7 cm. long by 0.4 cm. wide. The callus is large, fleshy and green and has 2 upper lobes of darker green

Flowering Periods: February through to March and the flowers last for three weeks or so.

Pollination: Seed is set frequently and up to 50% of flowers are with fruit; **Dressler** suggests small oil-gathering bees.

Cultivation: Best grown mounted on twigs in shady, moist conditions all year.

General: By any standards this plant has no right to exist and we can only conclude that further study will reveal the mechanisms by which such an apparently vulnerable, slight plant with no visible storage capacity, copes with the environmental stresses.

Phymatidium myrtophilum
Barb. Rodr. | 2202

Etymology: *Myrto-philum* is Latin for *Myrtaceae*-loving describing where the plant was first found.

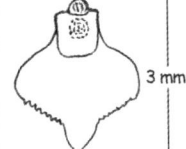

Habitat: Found on twigs of old orchard trees, in low scrub in young regrowth forest, in humid conditions but with high light and reasonable air movement at 900-1000 M.asl in section VI.

Occurrence: Occasional.

Plant: Very similar to *P. hysteranthum*: a small epiphyte, quite insignificant when not in flower. The plant has up to four short stems, each pale yellow green, compressed and with up to seven thin, horn-like leaves up to 0.5 cm. long and emerging alternately at 0.5 cm. intervals. The root system is not extensive, composed of brittle, short white roots. The plant, as with many twig epiphytes is short-lived and produces much fruit.

Inflorescence and Flowers: The flowers are held alternately on single short pedicels on the last 3 cm. from the leaf axils. They are a milk-white except for an emerald green horseshoe-shape callus which covers the labellum throat. The single-lobed labellum is a broad heart-shape, with serrate margins at the midlobe, the sepals and petals are ± equal, a reflexed lanceolate, 0.3 cm. long by 0.15 cm. broad.

Flowering Periods: November and the flowers last for up to three weeks.

Pollination: Unknown but frequent.

Cultivation: As for *P. mello-barretoi*.

Phymatidium tillandsoides
Barb. Rodr. | 2204 (Plate 43)

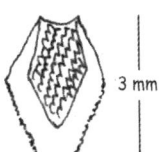

Etymology: *Tillandsoides* means *Tillandsia*-like, a genus of bromeliads.

Habitat: It is found in original under forest in humid, gloomy conditions near to running water and exposed to some air movement and diffuse light. Its perch is usually a small branch rarely more than 10 metres above the forest floor. It is common in its specific environment at altitudes between 1000 and 1300 in section VI. The major requirement is for high humidity.

Occurrence: Common.

Plant: This is an extraordinary orchid which looks precisely like a bromeliad, **Tillandsia sp.**, which it mimics not only in its growth habit but also with its "air-plant" features. It is a tufted epiphyte with a myriad of light green, linear-lanceolate leaves which, in perfect conditions can grow into a ball-shaped colony up to 30 cm. diameter. The root system is not extensive but is fiercely adherent and composed of thin, white wiry roots which seem to function more as an anchor than nutritionally.

Inflorescence and Flowers: Many thin, green racemose inflorescences are produced, each bearing up to 8 flowers arranged alternately. The flowers are an overall watery milk-white with a distinct green centre to the lip. Both sepals and petals are lanceolate and more or less equal in size, measuring 0.5 cm. x 0.2 cm. The lip is a blunt, serrated heart-shape, green but fringed with white and 0.3 cm. long by 0.2 cm.

Flowering Periods: November through March and the flowers last for two weeks but as all inflorescences do not flower together the plant may bear flowers for over a month.

Pollination: Frequent, up to 30% of flowers are pollinated and Dressler suggests probably by oil-gathering bees.

Cultivation: Though essentially a twig epiphyte, this plant is long-lived and easy to grow successfully. Anchor on a piece of tree fern in shade with low air movement and high humidity. Local success has been achieved by suspending plants above a constantly moist gravel tray.

General: Interesting for the bromeliad-like habit and the inflorescences are a bonus.

Phymatidium vogelii
Pabst | 2205

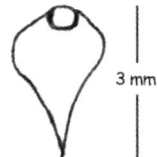

Etymology: Not known.

Habitat: A small epiphyte at low tree trunk level in high mountain scrub forest, found in dryish conditions similar to *P. delicatulum* in section VI, requiring medium light and air movement and year round humidity furnished by bromeliad undercover and constant misting at 1500 M.asl.

Occurrence: Occasional.

Plant: A very small twig epiphyte, quite insignificant when not in flower and, like all in the genus, unorchid-like. The plant has up to five 7 cm long stems. These are light green and somewhat flattened and hold up to 10 alternate leaves which are a compressed terete shape up to 1 cm. long by 0.1 cm. broad and arise at 0.4 cm. intervals. The root system is surprisingly extensive, composed of unbranched brittle, white roots which run for long distances up and down the thin branches and are often the factor that calls one to the plant. Like most twig epiphytes the plant is short-lived and produces much fruit.

Inflorescence and Flowers: The flowers are on short inflorescences on the apical half of the 7 cm. stem. They arise alternately on 0.5 cm. pedicels from the leaf axils and are milk-white except for the green callus. Tepals are ± equal in size and shape, lanceolate, 0.4 cm. long by 0.15 cm. broad. The lip is broad at the base and shoulders gradually tapering to an acuminate apex and measures 3 mm. long by 0.2 cm. broad at the shoulders. The green callus is an elevated horseshoe-shape, small and appears at the throat.

Flowering Periods: October and November and the flowers last for three weeks.

Pollination: Unknown.

Cultivation: As for *P. mello-barretoi*.

The genus
Thysanoglossa Pôrto & Brade

This Brazilian genus of one or two enchanting miniature species was named from the Greek compound word meaning lip with tassels and refers to the fringed lip.

Thysanoglossa jordanensis
Pôrto & Brade | 2207
Thysanoglossa organensis
Brade | 2208 (Plate 43)

Etymology: *Jordanensis*, from Campos do Jordão mountains. *Organensis*, from the Organ Mountain Range.

Habitat: Epiphytic from sections IV and VI at 900-1000 M.asl on relict trees in original forest. A tree-top, twig epiphyte, which will not grow in regrowth forest. Enjoying its specific perch with high light, vigorous air movement and a constant underlying humidity.

Occurrence: Occasional.

Plant: A twig epiphyte with almost microscopic pseudobulbs. These have both basal and apical leaves, normally up to five in number, and are lanceolate, deeply keeled, light green and measure from 1-2 cm. long by 0.2 cm.-0.5 cm. in width. The few roots which arise from the base of the pseudobulb are relatively thick, almost always exposed and deep green except at the white growing tip. The root system is not extensive; three or four of these roots measure up to 10 cm. on a mature plant.

Inflorescence and Flowers: Up to eight surprisingly large *Oncidium*-like flowers are held alternately on an apical inflorescence of up to 5 cm. The sepals and petals are similar in size and shape and are a bluntish lanceolate measuring 0.5 cm. long by 0.2 cm, a deep butter-yellow. The bilobed lip is violin shaped, bright yellow and 0.9 cm. long by 0.5 cm. at the apical lobes, narrow-waisted with a warty, green callus at the base, the basal half being intensely cilate.

Flowering Periods: Flowers appear in late November and December and last for two weeks or so.

Pollination: **Dressler** suggests probably by oil-gathering bees of the genus *Paratetrapedia*.

Cultivation: In our experience most twig epiphytes tend to be short-lived and prolific in seed production. Their cultivation perforce requires singular devotion and an acceptable substrate to assure the continuum. Knowing their natural propensities will guide you to determine the best medium to use as substrate. We have put the plants on the twigs of an orange or lemon tree with success.

General: The two species are so similar that one suspects that one is a geographical variety of the other. *T. organensis* found in the Serra dos Órgãos while *T. jordanensis* comes from the Campos do Jordão some 500 Km. to the south west. We believe that we have found both on the same tree.

The genus
Lockhartia Hooker

Hooker dedicated this Central and South America genus of about 30 species to the first superintendent of the Trinidad Botanic Gardens, Sir **David Lockhart**. Of the five species **Pabst & Dungs** attribute to Brazil, only one should be found in the Organ Mountain Range.

Lockhartia lunifera
(Lindl.) Rchb. f. | *2217*

Etymology: *Lunifera* is Latin for moon-shape, probably referring to the two crescent-shaped horns at either side of the lip base.

Habitat: This plant was brought to us, having been collected in the west of section III. The habitat described was gallery forest at 400-500 M.asl on a large fallen tree trunk and growing in dappled light, reasonable air movement and with a cool dry winter. We also found It in Pará State.

Occurrence: Occasional.

Plant: A small non-pseudobulbous epiphyte, vaguely reminiscent of a *Dichaea* or *Isochilus* species when not in flower. A mature plant will show over 30 leafy stems which arise from the base of mature stems to form a dense clump of an average 20 cm. height. The long-lived leaves arise closely and alternately from the base to the apex of the stem, each leaf enfolding the stem and the following leaf for over half its length. They are equitant, conduplicate and fused for the apical half, measuring up to 2.5 cm. long by 0.5 cm. broad, in their doubled-up iris-like way, and are a bluntish lanceolate, light green and stiff. The root system is prolific, composed of many fine white roots which emerge from the base of the new growth, mounting up on each other but not spreading for more than a few centimetres beyond the plant's circumference.

Inflorescence and Flowers: Short, delicate, hair-like inflorescences, 1.5 cm. long arise from the leaf axils of the six to eight apical leaves producing one to three separate flowers. Each flower is preceded by a delicate light green cup-like bract at the base of the pedicel. The flowers are reminiscent of some *Oncidium* species' flowers, both showing similar complex calli and butter-yellow petals, sepals and lips. However the column is very short, which is the main reason for setting this genus apart. The sepals are almost round, slightly pointed at their apices and measure 0.5 cm. long by 0.45 cm. broad and are a clear yellow. The petals are the same shape and colour but slightly larger at 0.6 cm. long by 0.55 cm. The lip is complex and depending how one looks at it, either three or six-lobed, with a long central callus. The upper two lobes are acuminate and appear as horns, measuring 0.6 cm. long by 0.1 cm. and are orange-brown with yellow margins. Two central lobes which flank the brown central calli appear as two small yellow triangles, while the flared, skirt-like apical lobe is yellow, divided, and measures 0.4 cm. long by 0.7 cm. wide when flattened.

Flowering Periods: May and the flowers last for a week, but as they are to a degree sequential, the plant may show flowers for over a month.

Pollination: Male *Euglossine* bees, according to **Dressler**.

Cultivation: On bark or in pots, on a dense fibrous compost, well drained and aerated, requiring watering during the spring and summer, misting thereafter.

The genus
Saundersia Rchb. f.

Heinrich Reichenbach raised this genus, containing two species only, and dedicated it to his friend **Saunders**, a keen orchidist. Both species should be found in the Organ Mountain Range. **Reichenbach**, a German, spent many years working in England, studying under **Lindley**. He named many orchid genera and species including 16 Brazilian genera and 308 species; one of the greatest of orchid specialists.

Saundersia paniculata
Brade | *2219* (Plate 43)

Etymology: *Paniculata* is Latin for a loose, irregularly branched indeterminate flower cluster; a compounded raceme, referring to the inflorescence.

Habitat: We have found plants in original under forest at between 450-550 M.asl, where they enjoy an intensely humid atmosphere, low light and minimal wind movement. An epiphyte from section VI.

Occurrence: Occasional to rare.

Plant: A small pendent epiphyte with small cylindrical pseudobulbs resembling a swollen petiole. These arise, tightly clustered and sheathed from a thick, wandering rhizome and are 1.5 cm. long by 0.4 cm. diameter. The leaves are apical, single, long-lived, dark green, linear-lanceolate, coriaceous and deeply keeled, measuring 25 cm. long by 2.3 cm. A well established plant will have up to eight such pendent leaves. The root system is somewhat weak and consists of a number of 1 mm. thick, grey hairy roots which are not very adherent and spread for about 10 cm. under moss.

Inflorescence and Flowers: As many as 15 small flowers are borne on alternate branches on the apical half of a short, hairy, pendent, paniculate inflorescence up to 10 cm. long. The dorsal sepal is heart-shaped and 0.5 cm. long and broad. The lateral sepals are the same shape but slightly larger. The petals are rectangular but with bluntly curved apices and measure 0.6 cm. long by 0.4 cm. wide. All parts are muddy-white and the petals have strong brown blotches. The bilobed lip is 0.9 cm. long x 0.7 cm. at the widest point, and a broad anchor-shape with a gross, longitudinal, central, bilobed callus. The basal halves of all the flower parts are densely pilose.

Flowering Periods: November and the flowers last for a week.

Pollination: Probably pollinated by small bees.

Cultivation: On bark or tree fern plaques in a damp, dark part of the greenhouse; not easy to flower and very slow to re-root.

The genus *Notylia* Lindl.

John Lindley raised this genus of about 40 species. In Greek this means a dorsal hump, referring to the recurved apical part of the column. **Pabst & Dungs** attribute some 27 species of this Central and South American genus to Brazil, and five to the Organ Mountain Range. These are plants which colonize regrowth and old orchards on the relatively dry anticline of the Organ Mountain Range.

Notylia hemitricha
Barb. Rodr. | *2231*

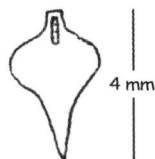

Etymology: *Hemi-tricha* from the Greek *hemi*, half and *tricha*, hairy referring to the pubescence on the lip.

Habitat: In humid regrowth forest at low tree level on branches and shrubs in shade with some wind movement and a reasonable constant humidity in section V at 600 M.asl.

Occurrence: Occasional.

Plant: A small twig epiphyte only slightly different from *N. lyrata*. The small, ovate, pseudobulbs are tightly clustered on an imperceptible rhizome and 0.5 cm. by 0.3 cm. wide. They hold one apical, keeled, light green, long-lived elliptical leaf, 6 cm. long by 1.5 cm. wide. The root system is extensive, composed of long,

sometimes branching, thin white roots some of which adhere to the substrate but the majority are aerial.

Inflorescence and Flowers: The inflorescence is basal, sheathed, round, green, thin, up to 8 cm. long and pendent with a cone of up to 30 flowers, very typical of the *Notylia* genus. The sepals are a light yellow while the petals, lip and column are milk-white. The dorsal sepal is lanceolate, 0.55 cm. long by 0.15 cm. wide. The lateral sepals are fused for 80% of their length and reflexed at their tips, 0.5 cm. long by 0.2 cm. at the fused section. The petals are narrow linear-lanceolate, 0.5 cm. long by 0.1 cm. The lip is an acuminate, arrow-head shape, longer than the slightly upturned column and is 0.5 cm. long and pubescent.

Flowering Periods: January and the flowers last for two weeks or so.

Pollination: **Dressler** suggests small wasps, or male *Euglossine* bees.

Cultivation: As for *N. lyrata*.

Notylia longispicata
Hoehne and Schltr. | *2236*

Etymology: *Longi-spicata* is a Latin compound for long flower spike.

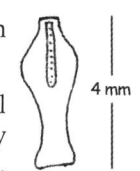

Habitat: Found in secondary transitional regrowth forest and aptly named by countrymen "Parasitas da capoeira", "orchids of regrowth forest", as being the first orchids to colonize regrowth, well before the *Gomesa spp*, in dry forests. We have found this plant in quantities in two fragments of 30 year old regrowth forest. In section V, accepting a long dry winter in deep shade with reasonable air movement, at 600-800 M.asl.

Occurrence: Locally common.

Plant: A twig or small branch epiphyte from low forest. More than half of its extensive root system is aerial and its anchorage always seems precarious. The pseudobulbs are small and sheathed, 1.2 cm. by 0.4 cm. wide and bilaterally compressed. They have a single long-lived, dark green apical leaf. This is keeled and elliptic, 11 cm. long by 3 cm. broad and with a false petiole. The root system as noted, is extensive, and composed of long green to grey branching roots, many of which are aerial.

Inflorescence and Flowers: Produced in February and March on long, basal branching, racemose inflorescences up to 25 cm. long. There may be up to 100 small yellow to white flowers, borne closely and alternately, and forming a veritable show, faintly reminiscent of *Epidendrum armeniacum*. The dorsal sepal is a blunt lanceolate 0.4 cm. long by 0.15 cm.

broad. The partially fused lateral sepals, which curve upwards at their apices are ± the same size and shape. The petals, vaguely lingulate, are 0.4 cm. long by 0.1 cm. broad, while the narrow violin-shaped lip is 0.42 cm. long by 0.15 cm. broad at its shoulders with a lean callus at its throat.

Flowering Periods: February and March and the flowers last for two weeks.

Pollination: Infrequent; unknown.

Cultivation: In baskets, on a fibrous compost in low light and wind movement. Cease watering when the inflorescences start to appear.

Notylia lyrata
Sp. Moore. | *2237* (Plate 43)

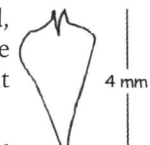

Etymology: *Lyrata* is Latin for lyre-shaped, referring to the shape of the lip. The lyre was a Greek stringed musical instrument used to accompany singers.

4 mm

Habitat: An epiphyte from section V and VI at 600-900 M.asl, growing at low tree level in original mountain forest receiving good filtered light, reasonable air movement and a constant high humidity. More frequently found in section V in steep river valleys.

Occurrence: Occasional in section VI. However when not in flower is easily confused with species of *Gomesa* or *Rodrigueziella*.

Plant: A small clustered epiphyte with minute laterally-compressed light green pseudobulbs which measure 0.5 cm. long by 0.5 cm. wide. There is a single broad lanceolate apical leaf measuring up to 7 cm. long by 3 cm. wide, deeply keeled, leathery and a light green. The root system is weak, composed of a number of thick white roots which are not particularly adherent and are sometimes aerial. A typical plant will have up to five long-lived leaves.

Inflorescence and Flowers: Thick round, light green, fleshy inflorescences emerge from the bases of the most recent pseudobulbs, each 10 cm. long and 0.1 cm. in diameter and thickly sheathed only at the bases. They bear a raceme, for 7 cm. of the apical section, in the form of a cone of up to 30 alternate small flowers held on short, 0.1 cm. pedicels. The lanceolate dorsal sepal is both convex and bent downwards towards the apex and measures 0.45 cm. by 0.23 cm. broad. The lateral sepals are fused for up to half their length. Their loose apices curve upwards and outwards and they measure together 0.5 cm. by 2.2 cm. broad. The petals are also lanceolate though slightly asymmetrical, measuring 0.45 cm. by 0.1 cm. broad. The broad-shouldered lip is a pointed wedge-shape and measures 4 cm. long by 0.2 cm. broad at the shoulder showing a faint nose-like

callus at the central base.

Flowering Periods: In late January and early February when the flowers which open simultaneously, last for two weeks. They are yellow to start with and orange towards maturity.

Pollination: **Dressler** suggests probably small wasps or male Euglossine bees.

Cultivation: As for *Gomesa* spp with perhaps a little less light. It is comfortable on bark plaques or in baskets using a well aerated substrate. During winter it requires higher light to induce the inflorescence.

The genus
Campylocentrum Bentham.

This American genus of 30 or so species in the principally African sub tribe Angraecinae, and is found from Florida through to Brazil. It is a monopodial genus and in Brazil's case always with tiny flowers and with a large nectary. The flowers are almost always pollinated, probably by night flying moths. It is not the sort of orchid one would include in your bride's bouquet or as a present to your girlfriend. However the tiny flowers are interesting in as much as the labellum of an orchid, which is normally a highly modified petal, is hardly modified at all in a number of *Campylocentrum* species. This could be explained by the exaggerated nectary, an adaptation for night flying moths, the scent and a white colour together which would obviate the radical changes in the labellum needed in most orchids to attract pollinators.

We have found and described 11 species out of a possible 14, attributed by **Pabst&Dungs** to the Organ Mountain Range. We feel sure that there are at least 6 more to be found.

Campylocentrum organense
(Rchb. f.) Rolfe | *2257* (Plate 43)

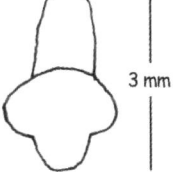

Etymology: *Organense* means from the Organ Mountain Range.

3 mm *Habitat*: The few colonies we have found were at mid-tree at over 1400 M.asl in original forest in section VI, within epiphytic gardens of bromeliads, presupposing high humidity, reasonable light and air movement.

Occurrence: Occasional.

Plant: This is a mid-tree to tree-top monopodial epiphyte, forming a long, 0.2 cm diameter stiff and continuously growing rhizome, off which, at intervals of around 2 cm., emerge thick long alternate aerial roots. The leaves are short, 1.2 cm. x 0.5 cm. wide, lanceolate, light green, deeply keeled and fleshy, arising alternately at 0.5 cm. intervals from the top 6 cm. of the stem. The stem is anchored to the substrate by roots not dissimilar to their aerial counterparts.

Inflorescence and Flowers: One to three inflorescences emerge opposite a leaf where the root node is found. Up to 10, 0.1 cm. thick inflorescences may be found, always as long as, or longer than the leaf, bearing up to 25 closely aligned tiny yellow alternate flowers. There could be as many as 250 on a typical plant. Under a dissecting microscope the sepals appear broadish, pointed, 0.1 cm. long by 0.1 cm. wide. The petals are of similar size but obtuse. The lip, infurled at its base, is trilobed and when opened is as wide as it is long at 3 mm. The tapering centre lobe is flat at the apex.

Flowering Periods: January and the flowers last for 10 days.

Pollination: Unknown, but almost 100% of flowers set seed and the significant nectary would suggest *Lepidoptera*.

Cultivation: Anchor the plant base to a piece of tree fern and let it run, always leaving space to fix additional growths to the substrate. Grow in dappled light with reasonable air movement and night time humidity.

Campylocentrum lansbergii (Rchb. f.) Schltr. | 2260

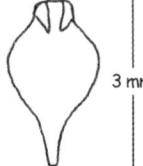

Etymology: Lansbergii. Named in honour of **Lansberg**.

3 mm *Habitat*: Found at 1000 M.asl on small branches in regrowth forest receiving good light, reasonable humidity and wind movement in section VI.

Occurrence: Occasional, but could be common as it is so insignificant.

Plant: A small monopodial epiphyte with most of the characteristics of the truly leafy group except for its small size and capability of flowering when still minute. The leaves are leathery, dark green, arising alternately at very short intervals from the stem and 2 cm. long by 0.5 cm. broad, keeled and lanceolate. The root system is strong, composed of relatively thick white roots running up and down the substrate and these are usually the first clue one sees before following them to find the plant.

Inflorescence and Flowers: A short compact inflorescence emerges from a node opposite the more mature leaves, approximately 1 cm. long with a raceme of up to six tightly packed, minute white flowers. Under a dissecting microscope the sepals and petals are a ±

equal shape, a narrow lanceolate 0.22-0.29 cm. long by 0.06 cm. broad, while the arrow-head looking, slightly trilobed lip is just over 0.2 cm. long by 0.1 cm. wide, backed by a 0.2 cm. long nectary.

Flowering Periods: March and the flowers last for a week or so.

Pollination: Probably moths and pollination is frequent.

Cultivation: As for *C. organensis*.

Campylocentrum linearifolium Schltr. ex Mansf. | 2261 (Plate 43)

Etymology: *Linear-folius* is Latin for linear-leaved, referring to the leaves of the plant.

Habitat: Found in deep dark original forest between 900-1300 M.asl in section VI, frequently over running water or humid ponds where the atmosphere is charged with moisture. The roots soon shrink when taken into the dry.

Occurrence: Occasional.

Plant: A monopodial, hanging epiphyte found in the most dark, moisture-laden forest. It is of interest for its photosynthetic roots, which, in some leafless species of **Campylocentrum** take over the role of leaves and stem. The plant reaches 25 cm. long on a 3 mm. diameter stem and appears dark green. The linear leaves which are alternate and opposite, 3.8 cm. long x 1 cm. wide, are a glossy dark green and have a midrib and a sheath enclosing the stem. The roots arise from behind the fifth youngest leaf and behind most of the older leaves. The young roots are short, with a snow-white velamen and have green tips. Older roots of 25 cm. maximum length have very fine ridging along their length and appear dark green with pale tips. A stem of 25 cm. had 1.08 metres of root.

Inflorescence and Flowers: Flowers arise from inflorescences immediately below each root and borne on 1.2 cm. spikes, each with up to 20 small white, opposite flowers on thick, short pointed pedicels. The flowers are sweetly scented and have a 3 mm. long spur. The tepals are more or less equal, lanceolate, 0.4 cm. long by 0.1 cm. broad while the arrow-head shaped lip is 0.4 cm. by 0.2 cm. wide.

Flowering Periods: In February and March and the flowers last for 10 days or so.

Pollination: As for *C. organense*.

Cultivation: As for *C. organense*.

Campylocentrum pauloense
Hoehne & Schltr. | 2264

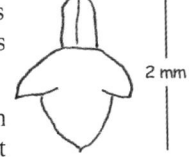

Etymology: *Pauloensis* probably refers to São Paulo where **Hoehne** was Director of the Botanical Institute.

Habitat: Found at mid-tree level on shrubs in regrowth and original forest at 1000 M.asl in section VI, enjoying a reasonable constant humidity, shade and wind movement.

Occurrence: Occasional to common.

Plant: A very small monopodial epiphyte forming a continuously growing, stiff rhizome which may branch. Aerial roots emerge irregularly from the rhizome and may be up to 6 cm. in length, relatively thick and greenish-white. A few, long-lasting, dark olive-green, alternate leaves appear towards the apex of the principal or branch rhizomes, at 0.1 cm. intervals. These are strongly conduplicate, lanceolate, stiff and measure up to 1.5 cm. by 0.4 cm. wide.

Inflorescence and Flowers: A very short inflorescence emerges from the apical three or four leaf axils. These are up to 0.5 cm. long and hold up to 6 minute white flowers. The dorsal sepal is broad, obtuse and lanceolate, 2.1 mm. long by 0.8 mm. The slightly asymmetrical lateral sepals are 1.9 mm. by 0.8 mm. wide. The petals are smaller and tending to acuminate and 1.6 mm. long by 0.8 mm. The lip is typically gross and trilobed, as wide as it is long 1.9 mm. by 1.6 mm. and has a shortish but thick nectary.

Flowering Periods: January and the flowers last for a week or so.

Pollination: Is infrequent and probably by *Lepidoptera*.

Cultivation: As for *C. organense*.

Campylocentrum robustum
Cogn. | 2265 (Plate 43)

Etymology: *Robustum* is Latin for strong - robust, referring to the habit of this plant.

Habitat: Found in the transitional forest in section V at 450-600 M.asl, at mid-tree level requiring good light and air movement with high humidity during the summer and a long dry winter.

Occurrence: Occasional in its habitat.

Plant: Very similar to all the monopodial species in the **C. micranthus** alliance, but three times the size of any other. Once again the roots are the dominant factor of the genus; long, unbranched, thick, to a real extent aerial, but quite capable of adhering to the bark

substrate and enveloping medium-sized branches at the mid-tree level. The leaves which are long-lived, are linear-lanceolate, emarginate at the apex and up to 8 cm. long by 1.5 cm., light green and keeled, arising alternately at 2 cm. intervals from the continuing stem.

Inflorescence and Flowers: Sepals and petals are more or less equal, lanceolate, white and minute, up to 0.8 cm. by 0.2 cm. broad, and arising in two tight parallel racemes of up to 10 flowers each emerging from a node opposite the leaf bases on the previous years growth. The lip is an acuminate-lanceolate, 0.6 cm. long by 0.2 cm. and is subtended by a small nectary. Significantly the lip, or rather the third petal, has hardly evolved in any extravagant manner and merely is slightly more acuminate and longer than the other two.

Flowering Periods: May/June and the flowers last for up to 20 days.

Pollination: Probably moths and seed is frequent.

Cultivation: Adpressed to a large bark or tree fern plaque taking care to keep the plant slightly moist.

Campylocentrum calostachyum
Cogn. | 2266

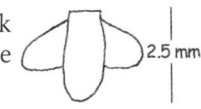

Etymology: *Calo* and *stachyo* are Greek for beautiful spike, referring to the inflorescence.

Habitat: Found growing on twigs in lower growth original forest, over streams and rivers in year-round, highly humid conditions at 250-300 M.asl, in section VIII.

Occurrence: Rare but locally frequent.

Plant: A small monopodial epiphyte difficult to differentiate from the *Notylia* species that also inhabit these regions, until you see the flowers or seed pods. The leaves, which emerge in a fan-like spray are closely alternate, papery, lanceolate, keeled, apically indented, dark green and up to 6 cm. long by 1.2 cm. wide, arising from a very short continuous stem. The roots, arising from the stem base are profuse and run up down and around small branches, both penetrating rotting twigs and holding them together.

Inflorescence and Flowers: The inflorescence emerges from the base of one of the older leaves and is up to 8 cm. long and can bear over 70 closely alternate flowers which are white and have a significant bulbous nectary. The minute petals are a narrow oval, the sepals are slightly larger and rounder and the lip is trilobed with a nectary of 0.3 cm. attached to the throat.

Flowering Periods: May and June.

Pollination: Most probably by night flying moths.

Cultivation: On twigs or cork bark but always grown in high humidity.

Campylocentrum aff. ulei Cogn. | *2268* (Plate 44)

Etymology: Not known.

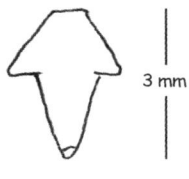

Habitat: On twigs in original forest at mid to tree-top level in section VI at 500-600 M.asl., enjoying year round high humidity, good light and wind movement.

Occurrence: Occasional.

Plant: A small monopodial twig epiphyte. The long-lasting, light green, lanceolate, leathery, keeled leaves measure 2.5 cm. long by 0.7 cm. and arise at 0.6 cm. intervals alternately from a flattish stem which they sheath with a broad petiole. The root system is extensive, running up and down the substrate for some fair distance with numerous long greyish green, non-branching roots indicating a capacity to photosynthesise. Roots that are mainly aerial, also emerge from internodes along the stem.

Inflorescence and Flowers: Up to five spicate inflorescences emerge at irregular intervals from nodes on the stem. These are up to 5 cm. long and hold up to 25 minute forward-facing alternate white, star-like flowers. Sepals and petals are similar in size and shape, lanceolate, 0.2 cm. long by 0.1 cm. The lip is trilobed with a dagger-shaped apical lobe, shouldered by two smaller triangular side-lobes and a small, fat, light green nectary at the base.

Flowering Periods: August and September and the flowers last for two weeks or so.

Pollination: As for *C. organensis*.

Cultivation: As for *C. organensis*.

Campylocentrum parahybunense (Barb.Rodr.) Rolfe | *2273* (Plate 44)

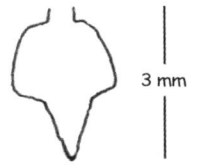

Etymology: Found on the banks of the Parahybuna river.

Habitat: Grows in sections V and VI at 800-1000 M.asl in original and regrowth forest and in montane scrub, enjoying dappled light, good wind movement and a high constant humidity.

Occurrence: Occasional.

Plant: A monopodial epiphyte forming 30 cm. long strings. The leaves are acicular, up to 6 cm. long by

0.15 cm. thick, dark green, with a distinct groove on the upper side and long-lived. They arise at 1-1.5 cm. intervals, sometimes alternately but more often in a line. The roots, initially very pilous and a greenish white, show a dark green smooth tip, which penetrates a mossy substrate with rapidity. These roots emerge from an internode, a fraction above the leaf base and do not appear to branch.

Inflorescence and Flowers: A very short raceme of up to six small white alternate flowers arises from an internode opposite where the root emerges. The lanceolate dorsal sepal is 0.2 cm. by 0.1 cm. broad. The laterals are the same shape but longer while the petals are the same shape and size as the dorsal sepal. The trilobed lip shows two wedge-like opposite basal side-lobes for half its length and an acuminate apical lobe taking up the other half. The overall size is 0.25 cm. by 0.2 cm. broad at the side-lobes. A significant nectary sits behind the lip, 0.2 cm. long by 0.1 cm. wide.

Flowering Periods: August and the flowers last for two weeks but the plant may bloom for up to a month.

Pollination: Frequent; probably *Lepidoptera*.

Cultivation: As for *C. sellowii*.

General: This plant differs from *C. sellowii*, having thinner, longer and a more regular leaf structure and in addition, its roots are far more pilous.

Campylocentrum sellowii (Rchb. f.)Rolfe | *2274* (Plate 44)

Etymology: Dedicated to **F. Sellow** who collected plants from Uruguay to Bahia in the 1820s.

Habitat: An epiphyte found at mid tree level in original and regrowth forest at 900-1200 M.asl enjoying high humidity, medium light and air movement in section VI.

Occurrence: Occasional.

Plant: A small monopodial epiphyte which forms quite large groups. A central stem, which branches, grows for up to 40 cm. This is thin, 0.2 cm. diameter, and throws out alternate leaves at 1-2 cm. intervals which are terete with a central groove and measure 3.5 cm. long by 0.2 cm. in diameter. Single, thick, 0.3 cm. diameter roots emerge from nodes on the main stem, normally opposite the leaves. These are greenish-white, somewhat rough, and only up to 3 cm. long until they find a substrate when they adhere and branch.

Inflorescence and Flowers: Various racemes of flowers emerge from opposite the new leaves' emergence. These are short, up to 2 cm. long, with up to 12 small, creamy-white, minute flowers. The dorsal sepal is a broad lanceolate, 2 mm. long by 1 mm. broad. The lateral sepals are slightly longer and narrower, while the petals are shorter and narrow, 2 mm. long by 0.9 mm. broad. The lip is vaguely trilobed, 3 mm. long by 2 mm. at the shoulders of the side-lobes and has a broad nectary, 1.5 mm. long, at its base. Needless to say these delicacies can only be admired under a good hand lens or even a microscope.

Flowering Periods: June and July and the flowers last for up to 10 days.

Pollination: Probably moths; the deep nectary would indicate this. Nearly all flowers bear fruit.

Cultivation: On tree fern or large bark slabs. If transplanted, make sure that all parts of the plant are in close contact with the substrate. This plant requires a daily misting both in winter and summer.

Campylocentrum burchellii Cogn. | *2278*

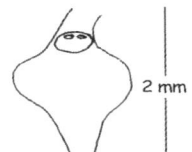

Etymology: Named after **William Burchell** (1781-1863) who collected in Brazil from 1825-30.

Habitat: Found at mid-tree in original forest enjoying variable light, good humidity and wind movement in section VI at 400-1100 M.asl.

Occurrence: Occasional, but it may be more common as its almost leafless habit and small size make it difficult to see.

Plant: A small epiphyte notable for its large and extensive root system. The roots are thickish, long and occasionally branching, intimately involved with the substrate which is usually a small, thin branch. They tend to be grey-green which may indicate a tendency to photosynthesise. The tiny, infrequent leaves are elliptic, and up to 1 cm. long by 0.4 cm. broad. The rhizome is so short as to make it difficult to conceive it being monopodial. Both roots and leaves arise from around a central, indistinct point.

Inflorescence and Flowers: Up to six short tidy inflorescences arise from a central point. These are up to 1.5 cm. long with to 12 tiny white alternate flowers. At first glance these remind one of a rachitic *C. ulei*, but the lack of leaves and distinct monopodial growth puts paid to this potential confusion. The sepals and petals are equal in size, shape and colour, lanceolate, 0.2 cm. long by 0.1 cm. and milk-white. The lip is very vaguely trilobed with an accentuated acuminate apex and the usual fat posterior.

Flowering Periods: November and the flowers last for two weeks.

Pollination: Frequent and probably by moths.

Cultivation: Place on a tree fern or bark plaque with good light, high humidity and air movement.

Campylocentrum aff. hirtellum Cogn. | *2281* (Plate 44)

Etymology: *Hirti* is Latin for hairy probably referring to the apparent hairs on the roots.

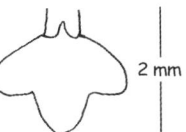

Habitat: Low to high tree and on undercover twigs in original forest at 800-1200 M.asl in sections V and VI, in the intensely humid, low lit and still ambiance favoured by **Pabstia jugos**a.

Occurrence: Common.

Plant: Practically leafless, but the plant has an enormous root network composed of medium thick roots, some of which adhere to the support while others hang in the air. The few leaves tend to be green, spathulate and centrally keeled, 1.5 cm. by 0.4 cm. and emerge fan-like from a 1 cm. stem which in turn arises from a nodule in the roots.

Inflorescence and Flowers: A 4.5 cm spicate, green, erect, inflorescence arises from the leaf-stem joint below the lowest leaf. Up to 22 white flowers form two very symmetrical alternate lines. Petals and sepals are white and slightly fleshy, pointed and equal at 0.2 cm. long by 0.1 cm. wide and appear like tiny stars. The white lip has a heart-shaped central lobe and two rounded side-lobes that furl and funnel back to the base, and it measures 0.25 cm. long by 0.15 cm. wide.

Flowering Periods: In November at December and the flowers last for 10 days or so.

Pollination: As for **C. organense**.

Cultivation: As for **C. organense**.

Campylocentrum aff. pubirhachys Schltr. | *2283*

Etymology: *Pubi-rhachys* is a Latin-Greek compound referring to the weak, slender hairs on the stem.

Habitat: Found in section VI at 900-1000 M.asl in regrowth forest and old orchards requiring good light and air movement with reasonable humidity.

Occurrence: Occasional, but it may be more common as it is almost invisible when not flowering.

Plant: A most curious, monopodial, leafless, epiphyte from the section, **Dendrophyllopsis**. It has no apparent rhizome and the flowering stems emerge from a nodule in the centre of the root system. The roots are thick and overall green, indicating their photosynthetic role.

The root system is extensive, running for a good distance, either side of the plant on its thin branch substrate.

Inflorescence and Flowers: Several inflorescence stems emerge from nodules around the central root system base, each up to 4 cm. long with a loose raceme of alternate small white flowers. The sepals are ± equal 0.2-0.24 cm. long by 0.09 cm. broad and lanceolate. The petals are the same shape but smaller. The lip is distinctly trilobed, 2 mm. long by 0.20 cm. wide at the side-lobes and with a 0.25 cm. nectary at its throat.

Flowering Periods: November and December and the flowers last for a week or so.

Pollination: Frequent, probably by moths.

Cultivation: As for other *Campylocentrum* species.

ERRATA from first printing.

Below are the missing captions for Plates 01-14

Plate 01 (page 339)
Phragmipedium vittatum – 4

Plate 02 (page 340)
Habenaria fastor – 9 Habenaria trifida – 27 Habenaria angulosa – 28

Habenaria parviflora – 29 Habenaria josephensis – 38 Habenaria leptoceras – 39

Habenaria montevidensis – 41 Habenaria achnantha - 64 Habenaria repens – 73

Plate 03 (page 341)
Habenaria riedelii – 74 Habenaria rupicola – 75 Habenaria petalodes 87

Habenaria rodeiensis – 106 Habenaria itatiayae – 161 Vanilla edwallii – 206

Cleistes brasiliensis – 232 Cleistes gracilis – 233 Cleistes ionoglossa – 235

Plate 04 (page 342)
Cleistes itatiaiae – 236 Cleistes lepida – 237 Cleistes pleuriflora – 241

Cleistes metallina (lip) - 252 Cleistes calantha – 257 Cleistes speciosa – 270

Cleistes vinosa – 271 Psilochilus modestus – 281 Elleanthus brasiliensis – 283

Plate 5 (page 343)
Elleanthus crinipes – 286 Elleanthus linifolius – 287 Cranichis candida – 310

Prescottia stachyodes – 317 Prescottia montana – 320 Prescottia nivalis – 321

Prescottia rodeiensis – 324 Prescottia epiphyta – 329 Prescottia octopollinica - 331

Plate 6 (page 344)
Prescottia polyphylla – 332 Sauroglossum nitidum – 353 Sauroglossum sp. – 353 a

Cyclopogon argyrifolius – 358 Cyclopogon elegans – 365 Cyclopogon variegatus – 367

Cyclopogon congestus – 371 Cyclopogon iguapensis – 381 Cyclopogon candidus – 388

Plate 7 (page 345)
Eurystyles actinosophila – 421 Eurystyles cotyledon – 422 Eurystyles cogniauxii – 423

Pelexia laminata – 427 Pelexia itatiayae – 434 Pelexia novafriburgensis – 453 a

Lankesterella ceracifolia – 468 Lankesterella gnomus – 470 Lankesterella aff. longicollis – 471

Plate 8 (page 346)
Mesadenella esmeraldae – 473 Stenorrhynchos lanceolatus – 476 Erythrodes commelinoides – 528

Erythrodes nobilis – 530 Erythrodes bidentifera – 532 Liparis nervosa – 540

Malaxis excavata – 544 Malaxis parthonii – 547 Polystachya caespitosa – 533

Plate 9 (page 347)
Eltroplectris roseo-alba – 501

Plate 10 (page 348)
Polystachya estrellensis – 554 Galeandra beyrichii – 577 Bulbophyllum atropurpureum – 585

Bulbophyllum campos-portoi – 586 Bu. aff. campos-portoi – 586a Bulbophyllum cantagallense – 587

Bulbophyllum luederwaldtii – 592 Bulbophyllum paranaense – 594 Bulbophyllum mirandaianum – 621

Plate 11 (page 349)
Bulbophyllum napellii – 634 Encyclia bracteata – 648 Encyclia flabellifera – 655

Encyclia odoratissima – 663 Encyclia oncidioides – 664 Encyclia pygmaea 674

Encyclia calamaria – 679 Encyclia kautskyi – 681

Plate 12 (page 350)
Encyclia inversa – 687 Encyclia suzanensis – 694

Encyclia vespa – 695 Encyclia fragrans – 696 Epidendrum armeniacum – 699

Epidendrum addae – 702 Epidendrum chlorinum – 711 Epidendrum aff. geniculatum – 714

Plate 13 (page 351)
Epidendrum aff. hololeucum – 716 Epidendrum mantiqueiranum – 718 Epidendrum robustum – 725

Epidendrum ecostatum – 726 Epidendrum janeirense – 727   Epidendrum ochrochlorum – 729

Epidendrum parahybunense – 730 Epidendrum proligerum – 731 Epidendrum aff. tenue – 734

Plate 14 (page 352)
Epidendrum setiferum - 737

13. Photos by Izabel Moura Miller

- 4

P
L
A
T
E
02

161

PLATE 04

Foto: Akos Litsek

P
L
A
T
E
05

PLATE 07

PLATE
08

Photo: Helmut Seehawer

PLATE
09

PLATE 11

PLATE 12

695

PLATE 14

Foto: Ákos

Epidendrum denticulatum - 753

Epidendrum xanthinum - 761

Epidendrum latilabre - 767

Epidendrum aquaticum - 774

Epidendrum paniculatum - 779

Epidendrum saxatile - 783

P
L
A
T
E
15

Epidendrum paranaense - 787

Epidendrum pium - 789

Epidendrum ramosum - 790

Epidendrum rodriguesii - 791

Epidendrum saximontanum - 792

Epidendrum vesicatum - 797

Lanium avicula - 802

Cattleya bicolor - 809

Cattleya velutina - 810

Cattleya dormaniana - 815

Cattleya guttata - 816

Cattleya harrisoniana - 822

Laelia crispa - 863

Laelia perrinii - 867

Laelia virens - 870

P
L
A
T
E
17

Laelia pumila - 876

Laelia cinnabarina - 895

Schomburgkia crispa - 931

Pseudolaelia corcovadensis - 934

Brassavola flagellaris - 942

Brassavola tuberculata - 945

Isabelia virginalis - 949

Sophronitis brevipedunculata - 950

Sophronitis cernua - 951

Sophronitis coccínea - 952

Sophronitis coccínea var. laranja - 952d

Sophronitis coccínea, raceme with 3 flowers - 952

Sophronitis coccinea var. pigmea - 952a

Sophronitis coccínea var. rosea - 952b

P
L
A
T
E
19

Sophronitis coccínea var. flava - 952c

Sophronitis coccínea var. acuensis - 952e

PLATE 20

Sophronitis wittigiana var. brevipedunculata - 956a

Sophronitella violacea - 957

Leptotes bicolor - 963

Leptotes tenuis - 964

Loefgranianthus blanche-amesii - 966

Tetragamestus modestus - 980

Ponera striata - 981

Isochilus linearis - 987

Bletia catenulata - 1540

Catasetum aff. macrocarpum - 1557

P
L
A
T
E
21

Catasetum cernuum - 1567

Catasetum hookeri - 1577

Foto: Marta Moraes

Cycnoches pentadactylon - 1614

Eulophia alta - 1616

Cyrtopodium glutiniferum - 1618

Grobya amherstiae - 1647

Promenaea stapelioides - 1664

Houlletia brocklehurstiana - 1671

Foto: Inghra Scart

Grobya galeata - 1649

Promenaea xanthina - 1662

Stanhopea guttulata - 1678

Gongora bufonia - 1685

Cirrhaea dependens - 1699

Cirrhaea longiracemosa - 1702

Cirrhaea saccata - 1703

Xylobium variegatum - 1708

Bifrenaria atropurpurea - 1709

Bifrenaria aurea - 1710

Bifrenaria calcarata - 1711

Foto: Sula Ramos

Bifrenaria harrisoniae - 1712

Bifrenaria inodora - 1713

Bifrenaria tetragona - 1717

Bifrenaria melanopoda - 1722

Bifrenaria aureo-fulva - 1724

PLATE 25

Bifrenaria wendlandiana - 1726

Bifrenaria vitellina - 1729

| 363 |

Bifrenaria mellicolor - 1715

Bifrenaria stefanae - 1729a

Pabstia jugosa - 1736

Zygopetalum brachypetalum - 1741

Zygopetalum crinitum - 1742

Zygopetalum crinitum var. major - 1742a

Zygopetalum intermedium - 1744

Zygopetalum mackayi - 1745

Zygopetalum maxillare - 1746

Zygopetalum pedicellatum - 1748

P
L
A
T
E
27

| 365 |

Neogardneria murrayana - 1751

Warrea warreana - 1776

Huntleya meleagris - 1782

P L A T E 28

Cochleanthes candida - 1784

Maxillaria leucaimata - 1791

Maxillaria modesta - 1792

Maxillaria caparaoensis - 1798

Maxillaria cleistogama - 1799

Maxillaria rufescens al. - 1803a

Maxillaria rufescens - 1803

Maxillaria consanguinea - 1808

Maxillaria phoenicanthera - 1810

Maxillaria picta - 1811

Maxillaria porphyrostele - 1813

P
L
A
T
E
29

Maxillaria rupestris - 1814

Maxillaria ubatubana - 1815

Maxillaria gracilis - 1817

Maxillaria chrysantha - 1819

Maxillaria marginata - 1820

Maxillaria piresiana - 1821

Maxillaria bradei - 1825

Maxillaria ochroleuca - 1827

Maxillaria brasiliensis - 1833

Maxillaria aff. discolor -1834

Maxillaria cogniauxiana - 1844

Maxillaria ferdinandiana - 1849

Maxillaria minuta - 1851

Maxillaria plebeja - 1855

Maxillaria pumila - 1856

P L A T E 31

Maxillaria aff. paulistana - 1858

Maxillaria acicularis - 1859

Maxillaria (Alliance subulata) - 1863a

Maxillaria valenzuelana - 1867

Maxillaria cerifera - 1871

Maxillaria notylioglossa - 1871a

Maxillaria jenischiana - 1874

Maxillaria loefgrenii - 1876

Maxillaria rigida - 1879

Ornithidium parviflorum - 1880

Scuticaria hadwenii - 1881

Scuticaria hadwenii var. strictifolia - 1881a

Dichaea muricata - 1892

Foto: Inghra Scart

Dichaea pendula - 1893

Dichaea cogniauxiana - 1899

Dichaea mosenii - 1909

Gomesa crispa - 1914

PLATE 33

Gomesa barkeri - 1917

Gomesa fischeri - 1919

P
L
A
T
E
34

Gomesa laxiflora - 1921

Gomesa recurva - 1923

Gomesa sessilis - 1924

Gomesa glaziovii - 1935

Binotia brasiliensis - 1936

Rodrigueziella gomezoides - 1938

Rodrigueziella handroi - 1939

Oncidium pumilum - 1949

Oncidium divaricatum - 1951

Oncidium sphegiferum - 1954

Oncidium harrisonianum - 1955

Oncidium harrisonianum var. concolor - 1955

Oncidium cornigerum - 1960

Oncidium aff. cruciatum - 1961

Oncidium lietzei - 1965

P
L
A
T
E
35

| 373 |

Oncidium truncatum - 1969

Oncidium waluewa - 1976

Oncidium trulliferum - 1977

P L A T E 36

Oncidium hians - 1983

Foto: Marta Moraes

Oncidium kraenzlinianum - 1984

Oncidium hookeri - 1985

Oncidium concolor - 1989

Oncidium cogniauxianum - 1992

Oncidium longipes - 1994

Foto: Inghra Scart

Oncidium uniflorum - 1995a

Oncidium barbatum - 1996

Oncidium aff. ciliatum - 1998

P L A T E 37

Oncidium crispum - 2006

Oncidium marshallianum - 2008

| 375 |

Natural hybrid - *O. marshallianum* × *O. crispum* - 2008a

Oncidium enderanum - 2011

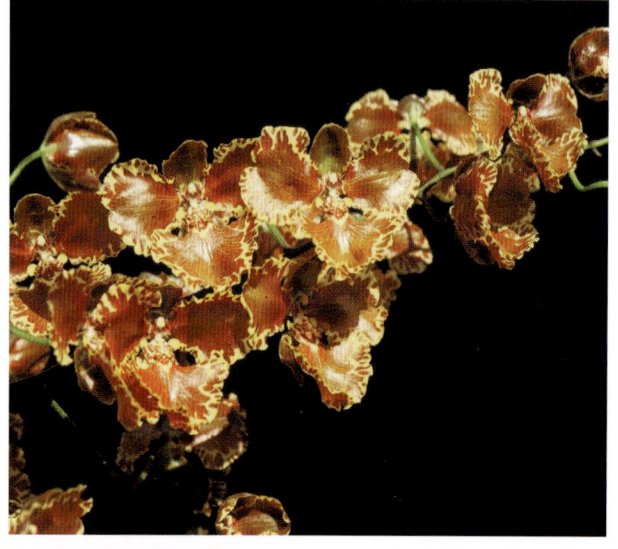

Oncidium forbesii - 2012

Oncidium gardneri - 2013

Oncidium flexuosum - 2024

Oncidium batemannianum - 2031

Oncidium blanchetii - 2032

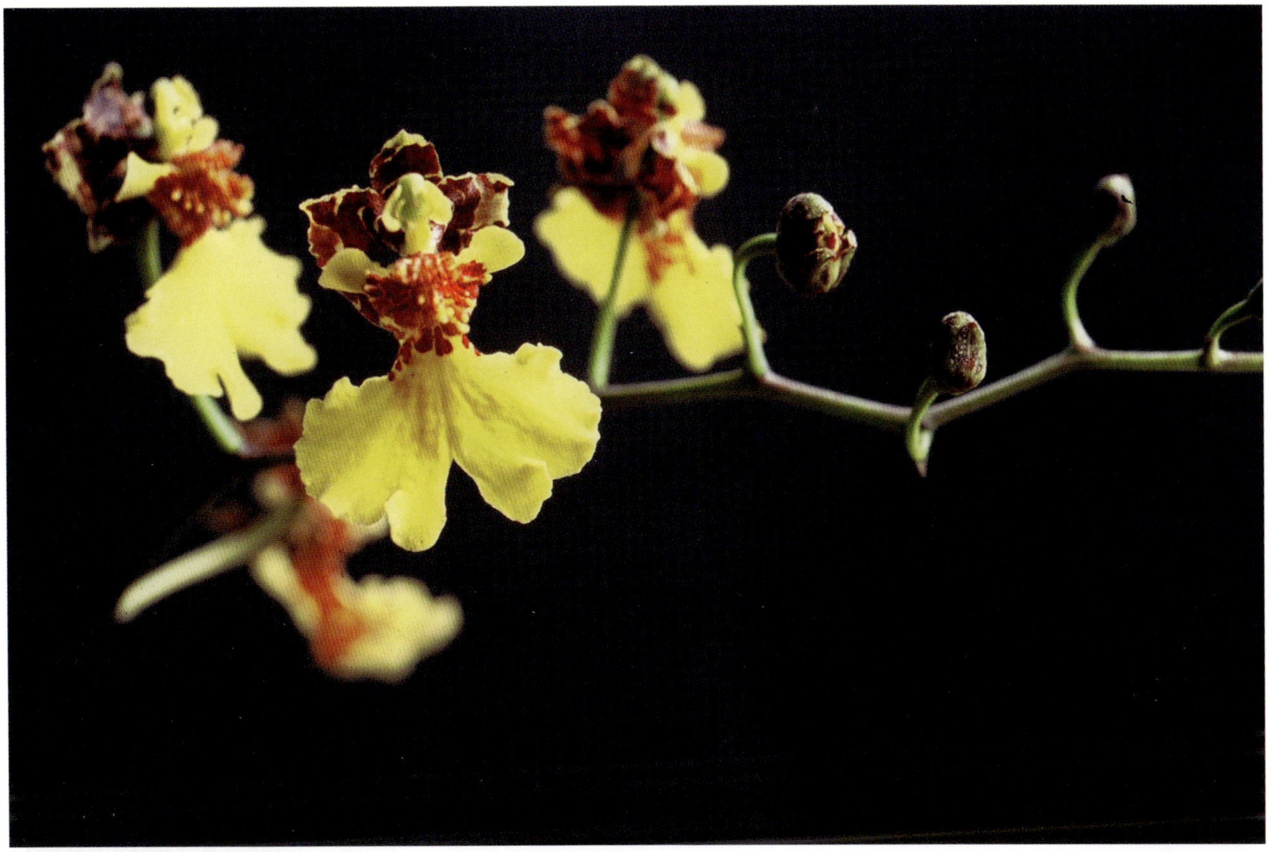

Oncidium hydrophilum - 2033a

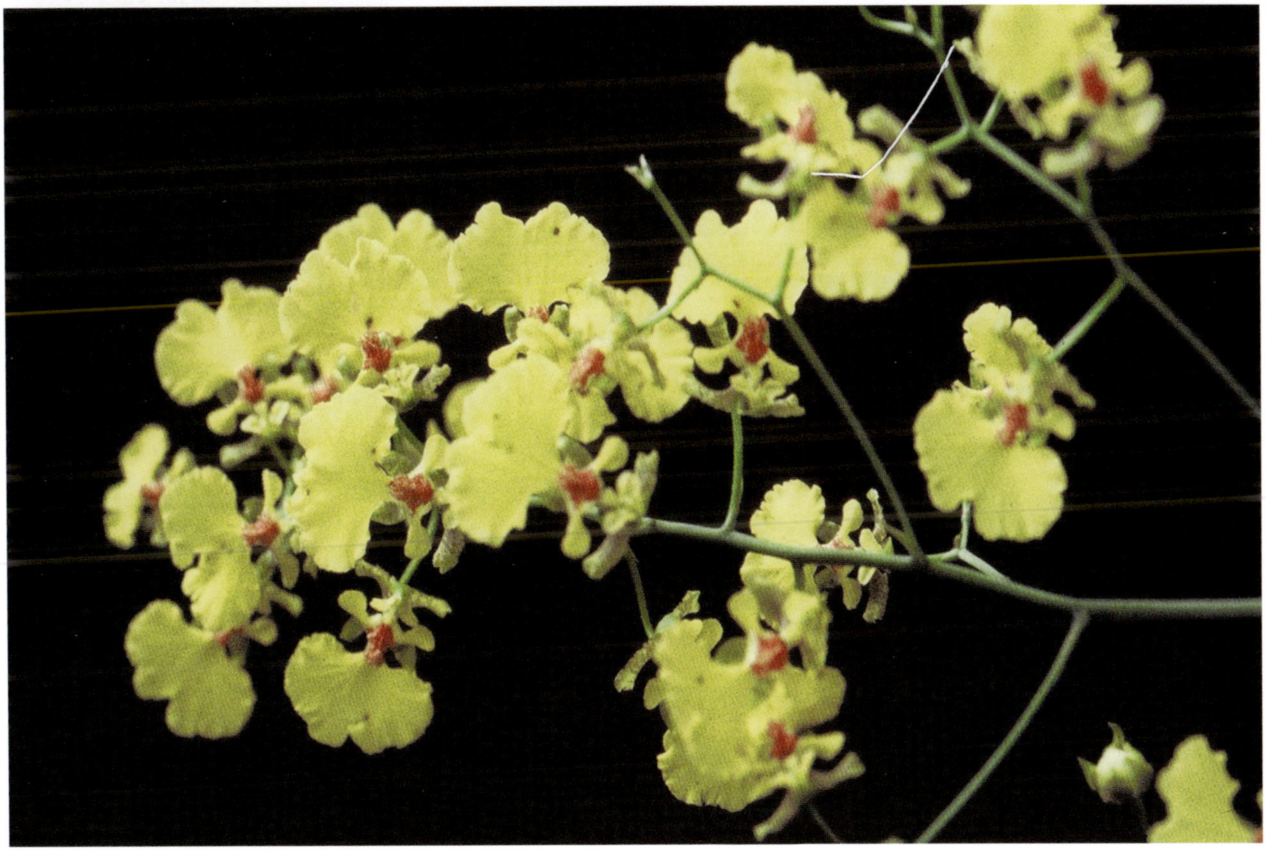

Oncidium ramosum - 2034

Miltonia flavescens - 2051

Miltonia spectabilis - 2052

Miltonia candida - 2053

Miltonia clowesii - 2054

Miltonia cuneata - 2055

Miltonia russelliana - 2058

Aspasia lunata - 2080

Warmingia eugenii - 2090

Trichocentrum fuscum - 2097

P L A T E 40

Centroglossa macroceras - 2107

Centroglossa Nunes-limae - 2108

Comparettia coccinea - 2110

Foto: Sula Ramos

Rodriguezia bracteata - 2130

Rodriguezia venusta - 2133

Rodrigueziopsis microphyta - 2146

Capanemia thereziae - 2152

Dipteranthus aff. octavio-reisii - 2174

Dipteranthus pellucidus - 2176

P L A T E 41

Dipteranthus grandiflorus - 2177

Zygostates cornuta - 2179

Zygostates multiflora - 2183

Ornithocephalus myrticola - 2192

Chytroglossa marileoniae - 2195

Phymatidium delicatulum - 2197

Phymatidium hysteranthum - 2199

Phymatidium mello-barretoi - 2201

Phymatidium tillandsoides - 2204

Thysanoglossa organensis - 2208

Saundersia paniculata - 2219

Notylia lyrata - 2237

Campylocentrum organense - 2257

Campylocentrum linearifolium - 2261

Campylocentrum robustum - 2265

PLATE 43

| 381 |

Campylocentrum aff. ulei - 2268

Campylocentrum parahybunense - 2273

Campylocentrum sellowii - 2274

Campylocentrum aff. hirtellum - 2281

14. Systematic Table

GENUS	SPECIES	N° PABST	HABITAT E	HABITAT T	HABITAT L	FREQ C	FREQ O	FREQ R	SEG 1	SEG 2	SEG 3	SEG 4	SEG 5	SEG 6	SEG 7	SEG 8	ALT 0–400	ALT 400–800	ALT 800–1200	ALT 1200–1600	ALT 1600–2000
Phragmipedium	vittatum	4		X	X	X		X			X								X		
Habenaria	fastor	9		X	X	X								X					X		
	trifida	27		X	X		X								X				X		
	angulosa	28		X	X		X							X					X		
	parviflora	29		X	X		X							X						X	
	hexaptera	36		X	X		X							X					X		
	josephensis	38		X	X		X						X	X						X	X
	leptoceras	39		X	X		X							X							
	montevidensis	41		X	X		X							X					X		
	crassipes	50		X	X		X							X					X		
	minarum	56		X	X		X						X						X		
	achalensis	63		X	X		X							X					X		
	achmantha	64		X	X		X							X						X	
	aff. marupaana	70		X	X			X	X						X						
	repens	73		X	X		X							X					X		
	riedelii	74		X	X		X							X					X		
	rupicola	75		X	X	X						X		X					X	X	
	petaloides	87		X	X		X				X	X	X	X					X		
	rodeiensis	106		X	X		X							X					X	X	
	itatiayae	161		X	X	X								X					X	X	
Vanilla	edwallii	206	X	X			X							X				X	X		
Epistephium	portellanum	219	X	X				X						X					X		
Cleistes	brasiliensis	232		X	X			X						X						X	
	gracilis	233		X	X			X						X						X	
	ionoglossa	235		X	X			X						X						X	
	itatiaiae	236		X	X		X							X					X	X	
	lepida	237		X	X		X							X						X	
	lepida (alliance)	237a		X	X			X						X						X	
	pluriflora	241		X	X		X							X						X	
	metallina	252		X	X	X			X			X		X					X		

GENUS	SPECIES	Nº PABST	HABITAT			FREQUENCY			SEGMENT OF THE MOUNTAINS								ALTITUDE m.ASL				
			E	T	L	C	O	R	1	2	3	4	5	6	7	8	0/400	400/800	800/1200	1200/1600	1600/2000
Cleistes	calantha	257		x		x			x	x		x		x					x	x	
	speciosa	270		x			x		x	x		x		x						x	
	vinosa	271		x			x		x					x					x	x	
Psilochilus	modestus	281		x			x		x					x				x	x	x	
Elleanthus	brasiliensis	283	x	x		x						x		x				x	x		
	crinipes	286	x	x		x						x		x				x	x	x	
	linifolius	287	x	x	x		x							x				x		x	
Cranichis	candida	310	x	x		x	x							x		x		x	x	x	
Prescottia	glazioviana	315		x			x							x				x		x	x	
	stachyodes	317		x		x							x	x				x		x	
	montana	320		x			x						x	x				x		x	x
	nivalis	321		x			x					x	x					x			
	rodeiensis	324		x			x					x	x					x			
	epiphyta	329	x	x			x					x	x					x		x	
	lancifolia	330		x			x							x				x			
	octopollinica	331		x		x									x			x	x	x	
	polyphylla	332		x	x		x						x	x			x				
Wullschlaegelia	aphylla	336		x		x			x		x			x		x		x	x		
Sauroglossum	nitidum	353		x		x	x		x	x	x	x	x	x	x	x			x		
	sp.	353a		x			x						x		x						x
Cyclopogon	argyrifolius	358		x			x					x	x	x				x	x		
	longibracteatus	360		x			x	x				x	x	x			x	x			
	calophyllus	363		x			x							x	x				x		
	elegans	365		x			x		x						x				x		
	variegatus	367		x		x								x	x					x	
	congestus	371		x			x							x	x			x			
	iguapensis	381		x		x							x	x	x			x		x	
	candidus	388		x		x								x	x					x	
Hapalorchis	lineatus	389	x	x				x						x			x		x		
	micranthus	390		x			x							x	x					x	

Genus	Species	No.
Sarcoglottis	granatflora	403
Eurystyles	acthnosophila	421
	cotyledon	422
	cogniauxii	423
	lorenzii	424
Pelexia	larvinata	427
	itatiayae	434
	hypnophila	453
	novafriburguense	453a
Cogniauxiocharis	glazioviana	466
Lankesterella	ceracifolia	468
	gnomus	470
	aff. longicollis	471
Mesadenella	esmeraldae	473
Stenorrhynchos	lanceolatus	476
Eltroplectris	roseo-alba	501
	arietina	511
Erythrodes	commelinoides	528
	nobilis	530
	bidentifera	532
Liparis	nervosa	540
Malaxis	excavata	544
	parthonii	547
Polystachya	caespitosa	553
	estrellensis	554
	micrantha	559
Galeandra	beyrichii	577
Bulbophyllum	atropurpureum	585
	campos-portoi	586
	aff. campos-portoi	586a
	cantagallense	587
	liuederwaldtii	592
	paranaense	594
	ricaldonei	596

GENUS	SPECIES	N° PABST	HABITAT E	HABITAT T	HABITAT L	FREQ C	FREQ O	FREQ R	SEG 1	SEG 2	SEG 3	SEG 4	SEG 5	SEG 6	SEG 7	SEG 8	ALT 0/400	ALT 400/800	ALT 800/1200	ALT 1200/1600	ALT 1600/2000
Bulbophyllum	cribbianum	620	X				X							X					X		
	mirandaianum	621	X				X							X			X				X
	regnellii	622	X			X								X				X		X	
	napellii	634	X			X								X					X	X	
Amblostoma	tridactylum	635	X					X						X	X				X		
Encyclia	bracteata	648	X		X			X								X		X			
	flabellifera	655	X			X								X					X		
	odoratissima	663	X				X							X					X		
	oncidioides	664	X			X							X	X					X		
	pygmaea	674	X			X								X					X		
	calamaria	679	X		X	X							X	X					X	X	
	kautskyi	681	X			X							X	X					X	X	
	inversa	687	X			X			X				X	X					X	X	
	suzanensis	694	X					X					X	X					X		
	vespa	695	X			X							X	X						X	
	fragrans	696	X					X					X		X			X	X		
Epidendrum	armenacum	699	X			X							X	X				X	X	X	
	addae	702	X			X								X					X	X	
	chlorinum	711	X			X								X					X		
	aff. geniculatum	714	X					X						X					X		
	henschenii	715	X			X						X		X			X	X			
	aff. hololeucum	716	X			X						X		X			X	X			
	mantiqueiranum	718	X			X								X			X	X			
	robustum	725		X	X	X								X	X			X	X		
	ecostatum	726	X					X						X					X		
	janeirense	727	X			X								X					X		
	obergii	728	X				X						X					X			
	ochrochlorum	729	X				X							X						X	X
	parahybunense	730	X				X							X						X	
	proligerum	731	X			X								X						X	

Genus	Species	No.
Epidendrum	aff. cadense	732
	filicaule	733
	aff. tenue	734
	setiferum	737
	purpureum	748
	ansiferum	751
	denticulatum	753
	elongatum	755
	xanthinum	761
	difforme	766
	latilabre	767
	aquaticum	774
	dendrobioides	775
	nutans	778
	paniculatum	779
	saxatile	783
	paranaense	787
	pium	789
	ramosum	790
	rodriguesii	791
	saximontanum	792
	vesicatum	797
	infaustum	798
Lanium	avicula	802
Cattleya	bicolor	809
	velutina	810
	dormaniana	815
	guttata	816
	harrisoniana	822
Laelia	crispa	863
	perrinii	867
	virens	870
	pumila	876
	cinnabarina	895

GENUS	SPECIES	Nº PABST	HABITAT			FREQUENCY			SEGMENT OF THE MOUNTAINS								ALTITUDE m.ASL				
			E	T	L	C	O	R	1	2	3	4	5	6	7	8	0–400	400–800	800–1200	1200–1600	1600–2000
Schomburgkia	crispa	931	X	X	X		X		X		X		X	X			X	X			
Pseudolaelia	corcovadensis	934	X			X			X		X		X		X			X	X		
Brassavola	flagellaris	942	X	X			X				X						X				
Brassavola	tuberculata	945	X	X	X	X			X		X		X		X		X	X	X		
Isabelia	virginalis	949	X				X				X	X	X	X				X	X		
Isabelia	brevipedunculata	950	X			X				X				X					X	X	
Sophronitis	cernua	951	X	X			X		X		X		X					X		X	
Sophronitis	coccinea	952	X	X		X				X		X		X						X	
Sophronitis	coccinea v. pygmea	952a	X				X							X						X	
Sophronitis	coccinea v. rosea	952b	X				X							X						X	
Sophronitis	coccinea v. flava	952c	X					X						X						X	
Sophronitis	coccinea v. laranja	952d	X					X						X						X	
Sophronitis	coccinea v. acuensis	952e	X					X		X										X	
Sophronitis	cernua v. pterocarpa	955	X					X					X					X			
Sophronitis	wit. v. brevipedunculata	956a	X				X							X						X	
Sophronitella	violacea	957	X				X	X						X				X			
Leptotes	bicolor	963	X				X						X	X				X			
Leptotes	tenuis	964	X					X					X	X					X		
Loefgrenianthus	blanche-amesii	966	X					X						X					X		
Tetragamestus	modestus	980	X	X		X						X		X					X	X	
Ponera	striata	981	X	X	X	X					X		X						X		
Isochilus	linearis	987	X			X					X		X	X				X			
Hexadesmia	sessilis	989	X			X				X		X		X				X			
Zootrophion	schenkii	992	X			X								X			X			X	
Cryptophoranthus	fenestratus	993	X					X						X					X		
Cryptophoranthus	jordanensis	994	X				X						X						X		
Cryptophoranthus	langeana	996	X					X				X		X					X		
Cryptophoranthus	punctatus	999	X					X						X				X		X	
Cryptophoranthus	spictatus	1000	X					X						X						X	
Stelis	drosophila	1004/1004b	X			X								X		X			X	X	
Stelis	hoehnei	1006	X				X							X					X	X	
Stelis	megantha	1008	X			X								X		X			X	X	

Genus	Species	No.
Stelis	parvifolia	1009
	megantha v. robusta	1010
	rubrechtiana	1011
	thermophila	1012
	triangularis	1013
	microcaulis	1017
	campos-portoi	1020
	chlorantha	1021
	paquerensis v. fraterna	1024
	itatiayae	1025
	modesta	1026
	omalosantha	1027
	puberula	1028
	grandiflora	1030
	dusenii	1037
	loefgrenii	1040
	aprica	1045
	binotii	1046
	paraensis	1051
	argentata v. pterostele	1052
	paquerensis v. porschiana	1053
	reflexisepala	1054
	paquerensis v. tweediana	1057
	paquerensis	1058
	parahybunensis	1059
	paquerensis v. guttifera	1061
	paquerensis v.	1063
	vinosa	1064
Phloeophila	pubescens	1069
	bradei	1070
	echinantha	1071
Masdevallia	curtipes	1083
	infracta	1084
Dryadella	edwallii	1088
	zebrina	1096

| 391 |

GENUS	SPECIES	N° PABST	HABITAT E	HABITAT T	HABITAT L	FREQ C	FREQ O	FREQ R	SEG 1	SEG 2	SEG 3	SEG 4	SEG 5	SEG 6	SEG 7	SEG 8	ALT 0-400	ALT 400-800	ALT 800-1200	ALT 1200-1600	ALT 1600-2000
Lepanthopsis	aff. densiflora	1101b						X						X		X				X	
	floripecten	1102					X							X	X	X			X		
Platystele	oxyglossa	1103				X		X						X		X			X		
Myoxanthus	punctatus	1119				X						X		X	X	X			X	X	
	aff. lonchophyllum	1120					X						X				X		X		
Pleurothallis	aff. barbulata	1106b						X	X												
	parvifolia	1108	X				X							X						X	
	scabripes	1114	X			X								X		X			X		
	collina	1125	X					X						X						X	
	quartzicola	1126	X			X								X		X			X	X	
	curti-bradei	1130	X			X								X					X	X	
	mattinhensis	1131	X				X							X					X	X	
	miragliae	1132	X				X							X		X			X		
	punctatifolia	1134	X					X						X		X			X		
	quadridentata	1135	X			X						X		X					X		
	bradei	1137	X					X						X		X		X	X		
	carinifera	1138	X					X						X		X			X		
	podoglossa	1139	X					X						X				X			
	sordida	1140	X			X						X	X	X		X			X	X	
	calcarata	1141	X			X						X		X		X				X	X
	conspersa	1142				X								X					X		
	avenacea	1146				X							X	X		X			X	X	
	tripterantha	1152				X							X	X		X			X	X	
	imbeana	1153											X	X				X			
	fluminensis	1154						X						X		X			X		
	colorata	1155				X								X					X		
	cordilabia	1157	X					X						X		X				X	
	gehrtii	1158	X			X							X	X		X			X	X	
	lineolata	1162	X					X						X		X			X	X	
	ramphastorhyncha	1163	X				X								X			X			

Pleurothallis	No.
rubro-lineata	1164
trifida	1165
uniflora	1166
hypnicola	1167
hypnicola v. cuneifolia	1167b
hypnicola v. flava	1167c
hypnicola v. major	1167d
piraquarensis	1158
seriata	1159
sp. not identified	1172a
arcuata	1174
granulosa	1177
henrique-aragonii	1178
hians	1179
pellifeloides	1180
wacketii	1185
teres	1195
leptotifolia	1197
sonderana v. longicaulis	1200b
bidentata	1205
capanemiae	1208
cearensis	1209
saundersiana	1215
serpentula	1216
spilantha	1218
translucida	1220
farinosa	1223
ramosa	1230
aff. ramosa	1230b
aphtosa	1231
panduripetala	1236
ramosa	1238
limae	1239
modestissima	1240

GENUS	SPECIES	N° PABST	HABITAT			FREQUENCY			SEGMENT OF THE MOUNTAINS								ALTITUDE m.ASL				
			E	T	L	C	O	R	1	2	3	4	5	6	7	8	0 400	400 800	800 1200	1200 1600	1600 2000
Pleurothallis	prolifera	1241		X		X								X						X	
	exarticulata	1242	X					X						X				X			
	heringeri	1243	X				X		X				X	X					X		
	luteola	1244	X	X			X						X	X		X			X		
	acuminatipetala	1245	X			X			X											X	
	binotii	1247	X			X						X		X				X			
	pantherina	1251b	X					X						X					X		
	cryptophoranthoides	1252	X					X						X						X	
	dracula	1252b	X					X						X				X			
	klotzschiana	1255	X																		
	porphyrantha	1257	X					X						X					X		
	albo-rosea	1260	X					X						X						X	
	macropoda	1262	X				X							X				X			
	octophrys	1273	X				X									X			X		
	strictophylla	1275	X				X		X			X		X					X	X	
	lingua	1279	X			X			X				X	X					X		
	crinita	1280	X				X							X					X		
	karlii	1281	X				X							X	X	X				X	
	recurva	1286	X				X							X	X	X		X	X		
	leucopyramis	1289	X			X								X		X			X		
	platystachya	1293	X			X								X		X			X		
	auriculata	1295	X			X						X		X		X		X	X		
	capillaris	1298	X				X							X	X				X		
	hygrophila	1301	X					X						X					X		
	longicaulis	1303	X				X							X					X	X	
	malachantha	1304	X				X							X					X		
	saurocephala	1314	X				X				X			X		X		X	X	X	
	marginalis	1319	X			X								X					X	X	
	viridiflora	1321b	X					X						X					X		
	grobyi	1323	X			X							X	X		X			X	X	

Pleurothallis	Code																							
aff. grobyi	1323b	X	X						X	X	X			X	X		X
subpicta	1325	X	X						X	X	X			X			X	X
edvallii v. major	1329a	X	X			X			X								X	X
edvallii v. pallida	1329b	X									X		X			X		
bocainensis	1335	X			X						X			X			X
depauperata	1338	X		X									X			X	X
eugenii	1339	X		X					X					X	X		X	X
linearifolia	1341	X	X		X				X	X				X	X		X	X
matmeana	1342	X		X						X				X				X
adenochila	1348	X	X		X				X	X	X			X	X		X	X
tigridens	1357	X			X				X	X							X	X
microgemma	1362	X			X				X		X			X	X		X
modesta	1363	X			X				X		X			X	X		
recurvipetala	1365	X		X					X	X	X		X	X	X		X
aff. laciniata	1371	X			X				X		X			X	X		X
microphyta	1374	X			X				X		X			X	X		X	X
aff. microphyta	1374b	X	X		X					X	X			X	X		X	X
rubro-limbata	1375	X	X		X					X				X	X		X	X
aff. rubro-limbata	1375b	X	X		X									X	X		X	X
corticicola	1376	X			X				X		X			X	X		X	X
aff. corticicola	1376b	X			X				X		X			X	X		X	X
foliata	1379	X			X				X					X	X		X	X
sparnageliana	1384	X			X									X	X		X	X
pulvinata	1385	X			X				X					X	X		X	X
fasciculata	1386	X			X				X			X					X	X
fasciculata v. densiflora	1385b	X							X		X			X	X	X		X
obovata	1387	X			X				X		X			X	X	X		X
dryadum	1388	X	X		X									X	X		X	X
ferdinandiana	1389	X			X				X					X	X		X	X
montipelladensis	1394	X			X				X					X	X		X	X
pabstii	1355	X			X				X		X			X	X		X	X
simpliciglossa	1357	X			X				X		X			X	X		X	X
sclerophylla	1399	X			X				X		X			X	X		X	X
radialis	1404	X			X						X			X	X		X	X

GENUS	SPECIES	Nº PABST	HABITAT E	T	L	FREQ C	O	R	SEG 1	2	3	4	5	6	7	8	ALT 0-400	400-800	800-1200	1200-1600	1600-2000
Pleurothallis	rubens	1405	X			X			X	X	X	X	X	X	X	X				X	
	heterophylla	1407	X			X			X	X	X			X	X	X		X	X	X	
Barbosella	australis	1427	X			X						X		X						X	
	gardneri	1429	X			X							X	X					X	X	
	gardneri v. microphylla	1432	X			X								X		X			X	X	
Barbrodia	miersii	1433	X			X								X					X	X	
Brachionidium	dungsii	1437	X					X						X		X			X		
Octomeria	alpina	1440	X			X								X						X	
	crassifolia	1443	X					X				X		X					X	X	
	densiflora	1445	X			X							X	X		X		X	X	X	
	ementosa	1446	X				X							X		X			X		
	fasciculata	1448	X				X							X		X					
	robusta	1454	X			X								X						X	
	sarcophylla	1455	X					X							X					X	
	serrana	1457	X				X							X				X			
	anceps	1458	X				X								X				X		
	grandiflora	1460	X			X								X		X			X	X	
	gracilicaulis	1462	X			X								X					X	X	
	oxichela	1465	X				X							X		X			X		
	concolor	1469	X			X							X	X	X	X		X	X	X	
	micrantha	1471	X			X							X	X	X	X			X		
	aff. micrantha	1471b	X					X							X	X	X			X	X	
	bradei	1476	X			X								X	X	X		X		X	
	linearifolia	1478	X				X					X				X			X	X	
	aff. linearifolia	1478b	X			X								X					X	X	
	aff. linearifolia	1478c	X				X							X					X		
	aff. linearifolia	1478d	X					X						X				X	X		
	albopurpurea	1480	X				X							X		X			X	X	
	crassilabia	1483	X				X							X					X		
	diaphana	1484	X				X								X			X			

Genus	Species	Nº
Octomeria	glazioviana	1486
	aff. glazioviana	1486b
	rechiana	1487
	reitzii	1488
	rodriguesii	1489
	rotundiglossa	1490
	tricolor	1491
	estrellensis	1493
	minuta	1495
	aff. ochroleuca	1497b
	rhodoglossa	1498
	aloefolia	1506
	fimbriata	1508
	chamaeleptotes	1515
	riograndensis	1517
	alexandri	1518
	sancti-angeli	1519
	decumbens	1521
	geraensis	1522
	aff. geraensis	1522b
	juncifolia	1523
	praestans	1527
	truncicola	1528
	wilsoniana	1529
	gracilis	1530
	campos-portoi	1536
Bletia	catenulata	1540
Catasetum	aff. macrocarpum	1557
	cernuum	1567
	hookeri	1577
Cycnoches	pentadactylion	1614
Oeceoclades	maculata	1615
Eulophia	alta	1616
Cyrtopodium	glutniferum	1618

GENUS	SPECIES	N° PABST	HABITAT			FREQUENCY			SEGMENT OF THE MOUNTAINS								ALTITUDE m.ASL				
			E	T	L	C	O	R	1	2	3	4	5	6	7	8	0–400	400–800	800–1200	1200–1600	1600–2000
Cyrtopodium	gigas	1624	X	X			X					X		X			X				
Govenia	gardneri	1646	X	X			X						X	X							X
Grobya	amherstiae	1647	X	X		X			X			X	X	X	X	X		X	X	X	
Grobya	galeata	1649	X					X	X										X		
Promenaea	xanthina	1662	X	X		X								X						X	
Promenaea	stapelioides	1664	X			X						X		X		X		X			
Houlletia	brocklehurstiana	1671		X				X						X							
Stanhopea	guttulata	1678	X	X	X		X				X					X		X		X	
Gongora	bufonia	1685	X	X			X					X		X		X	X		X		
Cirrhaea	dependens	1699	X			X						X	X	X	X		X		X		
Cirrhaea	loddigesii	1700	X				X						X	X				X		X	
Cirrhaea	longiracemosa	1702	X					X						X				X		X	
Cirrhaea	saccata	1703	X					X								X	X				
Xylobium	variegatum	1708	X	X	X	X						X		X			X	X			
Bifrenaria	atropurpurea	1709	X			X				X		X		X			X		X	X	
Bifrenaria	aurea	1710	X				X						X						X		
Bifrenaria	calcarata	1711	X					X						X					X	X	
Bifrenaria	harrisoniae	1712	X		X		X			X		X	X	X					X		
Bifrenaria	inodora	1713	X				X							X		X			X		
Bifrenaria	mellicolor	1715	X				X							X	X				X		
Bifrenaria	tetragona	1717	X					X						X					X		
Bifrenaria	thyrianthina	1718	X		X			X						X	X				X		
Bifrenaria	melanopoda	1722	X				X							X					X	X	
Bifrenaria	racemosa	1723	X			X								X					X	X	
Bifrenaria	aureo-fulva	1724	X					X						X					X		
Bifrenaria	wendlandiana	1726	X			X								X						X	
Bifrenaria	vitellina	1729	X				X							X					X	X	
Bifrenaria	stefanae	1729a	X				X							X					X	X	
Pabstia	jugosa	1736	X	X		X			X	X	X	X	X	X	X	X		X	X	X	
Zygopetalum	brachypetalum	1741			X	X			X	X	X	X	X	X		X		X	X		

Genus	Species	No.
Zygopetalum	crinitum	1742
	intermedium	1744
	mackayi	1745
	maxillare	1746
	pedicellatum	1748
Neogardneria	murrayana	1751
Warrea	warreana	1776
Huntleya	meleagris	1782
Cochleanthes	candida	1784
Maxillaria	desvauxiana	1787
	leucaimata	1791
	modesta	1792
	caparaoensis	1798
	cleistogama	1799
	monantha	1801
	aff. osmantha	1802
	rufescens	1803
	rufescens (alliance)	1803a
	consanguinea	1808
	phoenicanthera	1810
	picta	1811
	porphyrostele	1813
	rupestris	1814
	ubatubana	1815
	gracilis	1817
	kautskyi	1818
	chrysantha	1819
	marginata	1820
	piresiana	1821
	bradei	1825
	ochroleuca	1827
	rodriguesii	1829
	brasiliensis	1833
	aff. discolor	1834

GENUS	SPECIES	N° PABST	HABITAT E	HABITAT T	HABITAT L	FREQ C	FREQ O	FREQ R	SEG 1	SEG 2	SEG 3	SEG 4	SEG 5	SEG 6	SEG 7	SEG 8	ALT 0–400	ALT 400–800	ALT 800–1200	ALT 1200–1600	ALT 1600–2000
Maxillaria	cogniauxiana	1844	X			X										X				X	
	madida	1846	X				X							X	X				X	X	
	neuwiedii	1847	X				X						X					X	X		
	ferdinandiana	1849	X		X		X						X	X				X			
	minuta	1851	X				X						X					X			
	plebeja	1855	X			X								X	X			X	X		
	pumila	1856	X				X							X	X			X			
	spannagelii	1857	X				X							X	X				X		
	aff. paulistana	1858a			X			X			X								X			
	acicularis	1859	X			X								X	X				X	X	
	subulata	1863	X			X								X	X				X		
	subulata (alliance)	1863a	X					X						X	X				X		
	verrucosa	1864	X				X							X	X				X		
	valenzuelana	1867	X				X						X	X	X			X	X		
	johannis	1870			X		X						X	X	X	X		X	X		
	cerifera	1871	X			X								X	X				X	X	
	notylioglossa	1871a	X				X							X	X				X	X	
	jenischiana	1874	X			X								X	X				X		
	loefgrenii	1876	X		X	X								X	X				X		
	rigida	1879	X		X		X		X					X	X				X		
Ornithidium	parviflorum	1880	X					X	X					X	X			X			
Scuticaria	hadwenii	1881	X				X		X					X	X	X		X			X
Dichaea	muricata	1892	X			X			X					X	X				X	X	
	pendula	1893	X			X						X		X	X	X	X	X			
	aff. bryophyla	1898	X			X						X		X	X	X	X	X			
	cogniauxiana	1899	X			X			X	X	X	X		X	X	X		X	X	X	
	mosenii	1909	X			X								X			X	X			
	anchorifera	1911	X					X								X		X			
Gomesa	crispa	1914	X	X		X			X	X	X	X		X	X	X		X	X	X	
	barkeri	1917	X	X			X		X	X	X	X		X	X			X	X	X	

Gomesa	*fischeri*	1919	X				X			X		X		X					X		
	laxiflora	1921	X				X							X					X		
	recurva	1923	X	X		X			X	X	X	X	X	X	X	X			X	X	
	sessilis	1924	X				X						X					X			
	glaziovii	1935	X	X		X								X						X	
Binotia	*brasiliensis*	1936	X					X						X			X	X			
Rodrigueziella	*gomezoides*	1938	X				X							X					X	X	
	handroi	1939	X				X							X						X	
	jacunda	1940	X				X							X					X		
	petropolitana	1941	X					X						X							X
Oncidium	*pumilum*	1949	X				X						X				X	X			
	divaricatum	1951	X				X							X					X		
	pulvinatum	1952	X				X							X					X		
	robustissimum	1953	X					X				X					X				
	sphegiferum	1954	X				X							X					X		
	harrisonianum	1955	X			X						X		X		X		X	X		
	cornigerum	1960	X				X					X	X	X					X		
	aff. cruciatum	1961	X				X							X				X	X		
	kautskyi	1963	X					X						X						X	
	lietzei	1965	X				X						X	X					X		
	truncatum	1969	X					X						X		X			X	X	
	fimbriatum	1973	X				X							X					X		
	waluewa	1976	X					X				X								X	
	trulliferum	1977	X					X			X		X	X				X			
	hians	1983	X				X							X		X			X		
	kraenzlinianum	1984	X					X								X			X		
	hookeri	1985	X			X							X	X					X	X	
	loefgrenii	1986	X			X							X	X					X	X	
	raniferum	1988	X			X							X	X					X	X	
	concolor	1989	X				X					X		X		X				X	
	cogniauxianum	1992	X			X								X						X	
	longipes	1994	X				X						X	X					X	X	
	uniflorum	1995a	X			X								X					X	X	
	barbatum	1996	X				X							X		X		X			

GENUS	SPECIES	N° PABST	HABITAT			FREQUENCY			SEGMENT OF THE MOUNTAINS								ALTITUDE m.ASL				
			E	T	L	C	O	R	1	2	3	4	5	6	7	8	0-400	400-800	800-1200	1200-1600	1600-2000
Oncidium	aff. ciliatum	1998	X				X		X	X	X	X	X	X	X	X		X			
	crispum	2006	X			X			X	X	X	X	X	X	X	X			X	X	
	marshallianum	2008	X			X					X	X	X		X				X	X	
	praetextum	2009	X				X					X								X	
	curtum	2010	X			X									X					X	
	enderanum	2011	X				X			X		X		X	X		X	X		X	
	forbesii	2012	X			X				X		X		X	X		X		X	X	
	gardneri	2013	X				X							X	X		X			X	
	flexuosum	2024	X			X				X		X		X			X			X	X	
	batemannianum	2031		X	X	X			X		X		X	X			X		X		
	blanchetii	2032		X	X	X			X		X		X	X			X		X		
	hydrophilum	2033a	X	X				X						X			X			X	
	ramosum	2034	X	X				X						X		X	X			X	
	baueri	2045	X	X				X								X	X				
Miltonia	flavescens	2051	X			X				X		X		X		X	X				
	spectabilis	2052	X				X			X				X	X		X				
	candida	2053	X			X						X	X	X	X	X	X			X	
	clowesii	2054	X			X						X	X	X		X	X			X	
	cuneata	2055	X			X								X		X	X				
	russelliana	2058	X				X						X	X		X		X		X	
Aspasia	lunata	2080	X				X		X	X	X	X	X	X	X		X	X			
Warmingia	eugenii	2090	X						X	X			X	X		X	X				
Trichocentrum	fuscum	2097	X			X	X		X						X	X					
Centroglossa	castellensis	2104	X				X							X					X		
	macroceras	2107	X				X							X					X		
	nunes-limae	2108	X			X					X				X						
	tripollinica	2109	X				X							X			X				
Comparettia	coccinea	2110	X			X			X	X	X	X	X	X	X	X		X			
Rodriguezia	obtusifolia	2123	X			X			X	X	X	X		X	X		X			X	
	bracteata	2130	X		X	X			X	X	X	X	X	X	X	X	X				

Genus	Species	No.																		
Rodriguezia	venusta	2133	X				X			X			X					X		
Ornithophora	radicans	2142	X			X		X		X	X							X		X
Rodrigueziopsis	eleutherosepala	2145	X			X				X					X			X		X
	microphyta	2146	X		X					X	X	X			X			X		X
Capanemia	fluminensis	2151	X		X		X	X	X						X			X		
	thereziae	2152	X		X		X	X	X	X					X		X	X		X
	carinata	2156	X			X	X				X									
Dipteranthus	octavio-reisii	2173	X			X	X			X			X		X			X		
	pellucidus	2176	X			X	X				X						X			
	grandiflorus	2177	X		X					X	X				X			X		
Zygostates	cornuta	2179	X			X	X			X	X							X		
	multiflora	2183	X		X		X			X	X							X		X
Ornithocephalus	myrticola	2192	X			X		X		X	X							X	X	X
Chytroglossa	marileoniae	2195	X			X	X			X	X							X	X	X
	seehaweri	2195b	X		X		X			X	X				X			X	X	X
Phymatidium	aquinoi	2196	X			X				X	X							X	X	X
	delicatulum	2197	X			X	X			X	X							X	X	
	hysteranthum	2199	X		X					X	X							X		X
	mello-barretoi	2201	X		X					X	X		X					X	X	X
	myrtophilum	2202	X			X	X			X	X							X	X	
	tillandsoides	2204	X		X		X			X	X							X	X	X
	vogelii	2205	X			X	X			X	X						X	X	X	X
Thysanoglossa	jordanensis	2207	X			X		X		X	X							X	X	
	organensis	2208	X			X		X		X	X							X	X	
Lockhartia	lunifera	2217	X			X		X		X	X			X				X	X	
Saundersia	paniculata	2219	X			X				X	X			X				X		
Notylia	hemitricha	2231	X		X	X		X		X	X	X		X				X		
	logispicata	2236	X		X	X		X		X	X	X		X						
	lyrata	2237	X		X	X		X		X	X	X		X						
Campylocentrum	organense	2257	X			X	X			X	X							X		
	lansbergii	2260	X			X	X			X	X							X	X	X
	linearifolium	2261	X			X	X			X	X							X	X	
	pauloense	2264	X		X			X		X	X							X	X	X
	robustum	2265	X			X	X			X	X	X	X	X			X	X	X	

| 403 |

GENUS	SPECIES	N° PABST	HABITAT			FREQUENCY			SEGMENT OF THE MOUNTAINS								ALTITUDE m.ASL				
			E	T	L	C	O	R	1	2	3	4	5	6	7	8	0 / 400	400 / 800	800 / 1200	1200 / 1600	1600 / 2000
	callostachyum	2266	X					X								X	X				
	aff. ulei	2268	X				X							X				X			
	parahybunense	2273	X				X						X	X					X		
Campylocentrum	sellowii	2274	X				X							X					X		
	burchellii	2278	X		X								X	X					X		
	aff. hirtellum	2281	X		X								X	X					X		
	aff. pubirrhachys	2283	X				X							X					X		

15. Pleurothallidinae
Descriptions

The genus
Zootrophion Luer

Zootrophion was established in 1982 by **Carlyle Luer** to cover about 11 Central and South American species of *Cryptophoranthus* unrelated to *C. fenestratus* **Barb. Rodr.** from Brazil. The name is from the Greek *zootrophion* meaning a menagerie and refers to the similarity of the flowers to animal heads.

Key: The species have two waxy pollinia with a visci-dium. The sepals are fused at their tips to form two lateral windows. The lip has two basal side lobes.

Zootrophion schenkii
(Rchb) Luer | *992* (plate I)

Etymology: Named in honour of the German Botanist **J. A. Schenk.**

Habitat: Epiphytic or terrestrial from section VI at 800-1200 M.asl in humid rain forest growing up to 3 meters above ground, also on dead wood. L 3-5; H 5-8; W 3-5.

Occurrence: Common.

Plant: A clustered epiphyte, 120-180 mm. high with 10-40 leaves and a moderate root system composed of white or light yellow-green, 1 mm. diameter roots. The round leaf stems arise alternately at 10-15 mm. intervals from a 2.5 mm. diameter branching rhizome and are enveloped by up to 5 loose, long-lived, trumpet-shaped sheaths with recurved margins. The leaf stems are 50-100 mm. long by 1-1.5 mm. wide at their bases, widening to 2-2.5 mm. diameter at the leaf junctions. The leaves are lanceolate, leathery, distinctively conduplicate and slightly infolded and measure 50-90 mm. long, 20-30 mm. wide and 1-1.5 mm. thick.

Inflorescence and Flowers: One or two 5 mm. inflorescences emerge simultaneously from the apices of both young and old leaf stems and bear up to 3 flowers. A slit between the dorsal sepal and the fused lateral sepals opens in the middle third of the acutely pointed flower. The pedicel is 7-9 mm. long and the ovary, 6 mm. The concave sepals are distinctively keeled. The lanceolate dorsal sepal is 16-20 mm. long by 4-8 mm and is 50% fused to the completely fused laterals. The dorsal sepal is yellow at the base and turns to blood red towards the apex with 5 darker longitudinal veins. The blood red lateral sepals have 2 yellow points at the apex and together measure 12-17 mm. long by 9-14 mm. wide. The petals are spathulate and cuspidate, 4.5-5.0 mm. long by 1.4-2.8 mm. wide and light leather-yellow with 4 purple, longitudinal veins. The lip is arrow-shaped, light purple with 2 faint yellowish ear-like appendages at its base and 2 narrow, erect purple side lobes and is 5.1 mm. long by 3.3 mm wide. The narrow main lobe and the basal third of the lip

are warty. The pale purple and light yellow column is 2.6 mm. long by 1.3 mm. wide with narrow yellow wings.

Flowering Periods: January to June and the flowers last for 14 days.

Cultivation: It loves humus in nature. We recommend cultivating it like *Paphiopedilum.*

The genus
Cryptophoranthus Barb. Rodr.

Crypto and *anthus* are Greek words for hidden flower and refer to the fact that the inner parts of the flower are hidden by the partially fused sepals. **Barbosa Rodrigues** was the most famous of Brazilian orchid botanists. The genus was effectively eliminated by **Carlyle Luer** and its species transferred to *Pleurothallis* and *Zootrophion.* However, while recognising the logic behind these changes, we have kept the genus because of the distinct characteristics of the flower and our referring all Organ Mountain Range species to **Pabst&Dungs'** *Orchidaceae Brasiliensis.* All the species from this genus are found in under forest or on tree trunks in original cloud forest at high altitudes. Growers should bear this in mind and never expose the plants to direct sunlight.

Key: Two waxy pollinia with a viscidium. The sepals are fused at their tips forming two lateral windows between the dorsal and lateral sepals. The lip is bilobed above the middle.

Cultivation: On bark with moss and the substrate should be kept moist. Do not expose to direct sunlight.

Cryptophoranthus fenestratus
Barb. Rodr. | *993* (plate I)

Etymology: Fenestratus is Latin for window, referring to the eye between the fused sepals.

Habitat: Epiphytic from section VI, at 850 M. asl, on a southern facing slope, in very humid original cloud forest, on trunks up to 5 M. above the ground. L 3; H 8; W 3.

Occurrence: Rare. We only know of one colony.

Plant: The root system is moderate in intensity and composed of white or ochre, 0.5 mm. diameter roots. The leaf stems are 10-20 mm. long and 2 mm. diameter and emerge alternately at 6 mm. intervals from a 2-3 mm. diameter creeping and sheathed rhizome. The undersides of the dark green, fleshy leaves turn purple in high light. The keeled, lanceolate leaves are slightly infolded and 100-120 mm. long, 20-26 mm. wide and 2.5 mm. thick, finely attenuated towards their bases forming 5-10 mm. long false petioles and are enveloped by the loose, short-lived sheaths of the leaf stems.

Inflorescence and Flowers: Two pendent 15-25 mm. inflorescences emerge from the apex of young and older leaf stems with up to 7 flowers on 3-5 mm. pedicels. A split between the dorsal sepal and the fused lateral sepals opens in the middle two thirds of the acutely pointed flower. The short hairy ovary is speckled purple. The fleshy, concave sepals are irregularly undulate on the inside, with hairy margins and keels, olive-green with purple speckles on the back and light beige with dark purple blotches, which turn to solid dark purple towards the apices on the shiny insides. The dorsal sepal is 13 mm. long and 4.7 mm. wide and is 10% fused to the lateral sepals, which are themselves 90% fused. The apical two thirds of the lateral sepals are dark purple and together measure 12 mm. long by 9.5 mm. wide. The spoon-shaped, concave outside, acute, 4.7 mm. long petals are 1.4 mm. wide, light ochre, speckled with dark purple with serrated apical margins. The lip is arrow-shaped, rose with wine-red or purple speckles and 6.5 mm. long by 3.4 mm. wide and has 2 ear-like appendages and a lateral callus at the light green base. There are 2 side-lobes at the mid-section of the lip, where 2 long wine-red, elevated lines originate and the lip is recurved by 60°. The elliptical main lobe has a rounded apex with serrate margins. The column is light purple with darker coloured margins and is 4.5 mm. long and 2.5 mm. wide.

Flowering Periods: April to May and the flowers last for 14 days.

Similar Species: See *C.langeana - 996*.

Cryptophoranthus jordanensis Brade | *994* (plate I)

Etymology: Not known.

Habitat: Epiphytic from section IV, at 1100-1300 M. asl, in humid original forest, on lower trunks, branches and lianas. L 3-5; H 4-7; W 2-4.

Occurrence: Occasional.

Plant: A typical plant has 7 to 10 leaves and a moderately intensive root system composed of white or light green, 1 mm. diameter roots. The leaf stems are enveloped by loose, short-lived sheaths, 1.5 mm. diameter, 4-7 mm. long, including the 3-5 mm. false petiole, and emerge alternately at 5 mm. intervals from a creeping, 2 mm. diameter rhizome. The leaves are 25-40 mm. long, 13-18 mm. wide and 2.5 mm. thick, lanceolate, fleshy, conduplicate and dark green.

Inflorescence and Flowers: Up to 2 pendent, 4 mm. inflorescences emerge from the apex of young and older leaf stems with up to 2 flowers on 1.5 mm. pedicels. A split opens at the mid-section of the flower between the partly fused dorsal sepal and the completely fused lateral

sepals. The spathe is 2.5 mm. long; the bracts are 2 mm. and the ovary is 2.5 mm long, short-haired and speckled dark wine-red. The fleshy, concave sepals are granular on the dark wine-red inside and olive-green with purple speckles on the short-haired outside. The lanceolate dorsal sepal is 9-10 mm. long, 4-4.5 mm. wide and 20% fused with the lateral sepals, which are 8.8 mm. long and 4 mm. wide together. The acute petals are 3.5 mm. long by 1.4 mm. wide, spoon-shaped, concave outside and dark wine-red. The broadly wedge-shaped lip is 5.2 mm. long and 3.5 mm. wide, ochre with dark purple longitudinal callosities, a transverse callosity and 2 hook-shaped appendages at the base. The main lobe is faintly warty. The dark wine-red column is 3.5 mm. long and 1.4 mm. wide with an orange, hairy foot.

Flowering Periods: January. The flowers last for 14 days.

Cryptophoranthus langeana (Krzl.) Garay | *996* (plate I)

Etymology: Not known.

Habitat: Epiphytic from section VI at 1450 M. asl, found 50 M below the crest of a southern facing slope, in humid original forest, on a 50 mm. diameter branch. L 3; H 9; W 7.

Occurrence: Rare. We have found only one plant to date.

Plant: The 10 mm. long and 3 mm. diameter leaf stems emerge alternately at 2 mm. intervals from a 1 mm. diameter, creeping rhizome. The root system is moderate, composed of white, 0.5 mm. diameter roots. The faintly keeled, lanceolate leaves are up to 45 mm. long, 10-15 mm. wide, 1.5-2 mm. thick, slightly infolded, dark green, fleshy, turning purple in high light and finely attenuated towards their bases to form 5-9 mm. false petioles. The young leaf stems are enveloped by loose, short-lived sheaths.

Inflorescence and Flowers: The 5-10 mm. inflorescences emerge from the apices of adult leaf stems and produce up to 33 flowers on 3-5 mm. pedicels. A slit opens at the middle third between the dorsal and the lateral sepals. The short hairy ovary is speckled purple and the flower has an acute tip. The surfaces of the fleshy, concave sepals are irregularly undulate or granular on the inside with a rougher outside. The margins and the keels are hairy. The sepals are light olive-green with purple speckled longitudinal veins on the back, and on the shiny inside they are wine-red. The lanceolate, dark wine dorsal sepal is 10.5 mm. long, 4 mm. wide and is 15% fused to the lateral sepals which are 85% connate and together measure 9 mm. long and wide. The spoon-shaped, concave, pointed, 4.6 mm. long petals are 1.8 mm. wide and wine-red. The light yellow lip, 6.5 mm. long and 3 mm. wide, has 2 ear-like lateral appendages at its base and a lateral callus recurved towards its apex.

There are 2 erect, rose side-lobes short of the centre, where 2 wine-red, long elevated lines originate and where the lip is recurved by 90°. The apex of the elliptical and convex, orange-yellow main lobe is rounded. The orange and wine-red column is 5 mm. long by 1 mm. wide with very faintly gnawed narrow wings and apex. The base of the column is hairy on the front and back.

Flowering Periods: January to February and April and the flowers last for 14 days.

Cryptophoranthus punctatus
Barb. Rodr. | *999* (plate I)

Etymology: Punctatus is from Latin meaning dotted or spotted referring to the flower.

Habitat: Epiphytic from section IV and VI at 800-1100 M. asl, on a southern facing slope, on trunks and strong branches in very humid original forest L 3; H 7-9; W 2.

Occurrence: Rare. We know of only three populations.

Plant: A typical plant has 7 to 10 leaves, held on 10-15 mm. long by 1.5 mm. diameter leaf stems which emerge alternately at 2 mm. intervals from a 1-1.5 mm. diameter creeping rhizome. The root system is moderately intensive, composed of white or ochre, 1 mm. diameter roots. The dark green, fleshy leaves are 60-70 mm. long, 20-25 mm. wide and 2 mm. thick and are purple in high light, keeled, lanceolate, fleshy and attenuated to their bases where they are enveloped by the loose, short-lived sheaths of the leaf stems.

Inflorescence and Flowers: Up to 2 pendulous, 10-15 mm. inflorescences emerge from the apex of young and older leaf stems and each bears 3 flowers on 2-3 mm. pedicels and may bloom together. A wide slit opens at the flower's mid-section between the partly fused dorsal sepals and the completely fused lateral sepals. The fleshy, concave sepals are irregularly undulate and light to dark purple on the shiny inside. The rougher outside is light olive-green with purple speckles and the margins and keels are hairy. The dorsal sepal is lanceolate, 11 mm. long by 4.1 mm. wide and 25% fused to the lateral sepals which are 85% fused, 10.5 mm. long and 7.2 mm. wide together and dark purple. The spoon-shaped, concave outside, acute petals are 2.5 mm. long by 0.7 mm. wide with faintly toothed margins. The broadly wedge-shaped lip is yellow and orange with darker coloured speckles and is 4.7 mm. long by 3.5 mm. wide with 2 ear-like appendages at the light purple base. The lip broadens beyond the lateral callus into the wide side-lobes and long undulate elevated lines extend far into the large, wedge-shaped convex main lobe. The apex is rounded.

Flowering Periods: January to February. The flowers last for 14 days.

Cryptophoranthus spicatus
Dutra | *1000* (plate I)

Etymology: Spicatus is Latin for bearing a spike.

Habitat: Epiphytic from section VI, at 1400-1600 M. asl, on a southern facing slope, on trunks and strong branches in very humid original forest. L 3; H 8; W 4.

Occurrence: Rare. We know of only one habitat but with quite a number of plants.

Plant: A typical plant has 7 to 10 leaves and a moderately intensive root system composed of white or ochre, 1 mm. diameter roots. The leaf stems are 7-13 mm. long by 1.5 mm. diameter, thickening to 3 mm. at their bases and emerge alternately at 5 mm. intervals from a creeping, 1.5 mm. diameter rhizome. The leaves are 50-60 mm. long, 25-30 mm. wide and 2 mm. thick and lanceolate, fleshy, keeled, dark green with purple undersides in high light and finely attenuated to their bases. The 5-10 mm. long false petiole is enveloped by the loose, short-lived sheaths of the leaf stems.

Inflorescence and Flowers: The pendent, 7 mm. inflorescences emerge from the apex of young and older leaf stems with up to 3 flowers on 2-3 mm. pedicels, which may bloom together. A split opens at the mid-section of the obtuse flower between the partly fused dorsal sepal and the completely fused lateral sepals. The spathe is 4 mm. long; the bracts are 2-2.5 mm. and the ovary 1.5 mm. The fleshy, concave sepals are irregularly undulate on the shiny and light to dark purple inside and are light olive-green with purple speckles on the rough outside with hairy margins and keels. The lanceolate dorsal sepal is 11 mm. long, 4.7 mm. wide and 25% fused with the lateral sepals, which are dark purple, 85% fused and together, 10.5 mm. long and 9.4 mm. wide. The petals are spoon-shaped, concave outside, acute, with faintly toothed margins and are 3.5 mm. long by 1.3 mm. wide. The broadly wedge-shaped lip is yellow and orange with darker speckles and 2 ear-like appendages at the light purple base. It measures 4.7 mm. long and 3.5 mm. wide and broadens out at the lateral callus into the wide side-lobes. Long undulate elevated lines extend far into the large, wedge-shaped, convex main lobe, which has a rounded apex. The dark wine column is 4.5 mm. long and 2 mm. wide with acuminate wing tips.

Flowering Periods: January to February and the flowers last for 14 days.

Similar Species: See *C. langeana - 996.*

The Genus
Stelis Sw.

In Greek, the word *Stelis* means small pillar and was used for a kind of mistletoe which, like Stelis, grows on trees. **Olof Swartz**, who first described Stelis in 1799, was the most famous Swedish botanist after **Linnaeus** and was in fact, a student of **Linnaeus's** son. **F. C. Hoehne** states that all *Stelis* species are the legitimate children of hygrophilic forests and that it is not difficult to find more than five species on the same tree. It is rare to find *Stelis* species outside original forest and as such, the genus is as threatened as its environment.

The descriptions of the *Stelis* species that follow were compared with the original Latin descriptions where possible. However, our descriptions are more detailed in terms of habitat and micro-environment. Some species are so similar to others that the reference material available did not allow for sure identification and we further suspect that some given species are merely variations of others.

The material used to produce *"Flora Brasiliensis"*, which was a principal reference source for the genus *Stelis*, was put together by a number of famous botanists and taxonomists and completed in 1906. These specimens were collected in the middle of the 19th. Century. Environmental conditions have changed radically since then. In addition 150 years have passed which means that up to 150 generations of these plants have occurred. Furthermore the original species were almost inevitably collected in areas different from our collections. All of which leads us to speculate that we should not worry too much should our plants' measurements differ somewhat from those cited in *"Flora Brasiliensis"*, in particular the plant sizes and flower parts. As such, we beg pardon in advance for any identification errors that perchance our readers discover.

Key: Two waxy pollinia with a viscidium. The sepals are fused at their bases in a dish-shaped display. The petals and lip are fleshy and minute and the petals are wider than long. The callus is at the apex and the lip is articulate with the column foot.

Stelis drosophila
Barb. Rodr. | *1004 e 1004b* (plate I)

Etymology: Droso-phila is Greek for dew-loving and refers to the moist habitat.

Habitat: Epiphytic from section VI and VIII, at 1000-1600 M. asl in original cloud forest on branches of the upper half of trees. L5-8; H 6-8; W 5-8.

Occurrence: Common.

Plant: This 50-80 mm. high epiphyte forms clusters with 10-50 leaves. The leaf stems are enveloped by 2 light ochre sheaths and are 10-20 mm. long and 0.6-0.8 mm. diameter. The leaves are narrowly oval, finely attenuated to their bases, faintly conduplicate, fleshy and 30-50 mm. long, 5-7 mm. wide and 1-1.5 mm. thick.

Inflorescence and Flowers: The 50-70 mm, slightly bent inflorescence emerges from the apex of the leaf stem with 15-25 flowers on 1-2 mm. pedicels. The spathe is 4 mm. long and the bracts are 1.5 mm. The light yellow-green flowers open wide in humid conditions and are 3.6 mm. by 3.2 mm. The slightly concave sepals are almost equal and covered in small swellings. The slightly concave petals are fleshy, yellow-green and 0.7 mm. long and 0.8 mm wide. The succulent, yellow-green lip is 0.5 mm. long, 0.5 mm. wide and 0.35 mm. thick. The light green column is 0.8 mm. long, 0.6 mm. wide and 0.6 mm. thick.

Flowering Periods: November to December.

Stelis hoehnei
Schltr. | *1006* (plate I)

Etymology: Named in honour of the Brazilian Taxonomist **F. C. Hoehne**.

Habitat: Epiphytic from section VI, 1000-1400 M. asl, in original cloud forest on mossy trunks and low branches. L 2-3; H 6-9; W 2-3.

Occurrence: Occasional.

Plant: A 20-25 mm. high epiphyte forming clusters with 10-40 leaves and light green 0.5 mm. diameter roots. The leaf stems are covered by 2 loose, short-lived, light ochre sheaths and are 0.5 mm. diameter by 5 mm. long, including the 2 mm. pseudopetiole. The fleshy leaves are narrowly lanceolate, grass green, finely attenuated to their bases, faintly conduplicate, fleshy and measure 20 mm. long, 2.7-3.5 mm. wide and 1 mm. thick.

Inflorescence and Flowers: The 50-70 mm, slightly curved inflorescence arises from the apex of the most recent leaf stem and produces 10-20 flowers on 1.5 mm. pedicels. The spathe is 3 mm. long, the bracts are wide open and 1.5 mm. and the ovary is 0.5 mm. long. The light yellow-green flower only opens in very humid conditions, does not open wide and is 2.3 mm. by 2.1 mm. when spread. The concave sepals are almost equal, covered in small swellings when dry and short-haired when wet. The slightly concave petals are fleshy, yellow-green 0.5 mm. long and 0.7 mm. wide. The succulent, yellow-green lip is 0.5 mm. long, 0.6 mm. wide and 0.5 mm. thick. The light green column is 0.5 mm. long, 0.5 mm. wide and 0.4 mm. thick. The apex, rostellum and the wings are longer than average.

Flowering Periods: November to January.

General: This is the smallest Stelis species we have found and so is easily overlooked.

Stelis megantha
Barb. Rodr. | *1008* (Plate II)

Etymology: Mega - antha form a Greek compound word for large - flower and this species has the largest flowers in the genus found in the Organ Mountain Range.

Habitat: Epiphytic from section VI and VIII, at 1000-1500 M. asl in humid original forest, on mid-tree branches. L 4-7; H 6-9; W 2-6.

Occurrence: Common.

Plant: This 150-250 mm. high epiphyte forms clusters with 7 to 25 leaves. The leaf stems are 80-120 mm. long and 3-4 mm. diameter and enveloped by 1 short, brown and 2 light ochre sheaths. The leathery leaves are elliptic, abruptly attenuated to their bases, faintly conduplicate, and 80-130 mm. long, 25-30 mm. wide and 1-1.5 mm. thick.

Inflorescence and Flowers: A 150-200 mm, slightly bent inflorescence emerges from the apex of the leaf stem with 20-25 flowers on 2.5-3.5 mm. pedicels. The spathe is 10-15 mm. and the bracts are 6 mm. long. The light yellow-green and red-brown flowers open wide and measure 9 mm. long by 8 mm. wide. The sepals are almost equal, glabrous, yellow-green and red-brown at the base. The petals are slightly concave, fleshy, olive-green and dark wine-red and 1.2 mm. long, 1.8 mm. wide and 0.35 mm. thick. The succulent, dark wine-red lip is 1.1 mm. long, 1.2 mm. wide and 0.6 mm. thick. The column is light green and wine-red and 1.1 mm. long, 1.2 mm. wide and 0.8 mm. thick.

Flowering Periods: November to January and May. The inflorescence lasts for a month.

Cultivation: The young leaves have to be kept dry in the greenhouse.

Stelis parvifolia
Garay | *1009* (Plate II)

Etymology: Parvi-folia is Latin for small leaf.

Habitat: Epiphytic in section VI and VIII, at 1200-1600 M. asl on the upper and mid-tree branches in original forest. L 5-8; H 6-8; W 4-7.

Occurrence: Common.

Plant: A small epiphyte, 20-25 mm. high, forming clusters of 15-40 leaves. The leaf stems are 2-5 mm. long and 0.5-0.7 mm. diameter and are enveloped by 2 loose, 2-4 mm. sheaths. The lanceolate leaves are faintly conduplicate, fleshy, yellow-green, narrowing to their

bases and measure 10-20 mm. long, 2.5-5.5 mm. wide and 1.3-1.6 mm. thick.

Inflorescence and Flowers: The 40-60 mm. inflorescence emerges from the apex of the leaf stem with 5 to 10 flowers. The spathe is 2.5-3 mm. and the bracts 2 mm. long. The light green and wine-red flower opens wide in very humid conditions and is 2.7 mm. long and 2.5 mm. wide. The sepals are almost equal and warty on the inside. The petals are kidney-shaped, 0.5 mm. long by 0.7 mm. wide, yellowish-green and have 0.25 mm. fleshy tips. The succulent, yellowish-green lip is 0.45 mm. long, 0.6 mm. wide and 0.45 mm. thick. The trilobed, yellow-green column is 0.4 mm. long, 0.55 mm. wide and 0.45 mm. thick with an elongated apex.

Flowering Periods: November to January.

Stelis megantha Pabst var. *robusta*
Schltr. | *1010* (Plate II)

Etymology: Robusta means strong and refers to the sturdy nature of this plant.

Habitat: Epiphytic from section VI, at 1100-1500 M. asl in humid original forest, on mid-tree trunks and branches. L 4-7; H 5-7; W 4-6.

Occurrence: Occasional. It is interesting that **Pabst and Dungs** classify this plant as "Brasil sine loco" which implies that only one herbarium consulted had a specimen and even the curator did not know from whence it came.

Plant: A 200-300 mm. tall epiphyte forming clusters with 7 to 15 leaves. The leaf stems are enveloped by 2 long, light ochre sheaths and are 80-120 mm. long by 3-4 mm. diameter. The leaves are elliptical, faintly conduplicate, leathery and measure 80-130 mm. long, 25-30 mm. wide and 1.5-2.5 mm. thick and abruptly attenuated to their bases.

Inflorescence and Flowers: Up to 2, 150-200 mm. inflorescences emerge from the apices of the leaf stems with 20-25 flowers on 2-3 mm. pedicels. The spathe is up to 28 mm. and the bracts are 4 mm. long. The light yellow-green and red-brown flowers open wide and measure 9.0 mm. by 7.5 mm. The glabrous sepals are yellow-green and red-brown at their bases and almost equal. The slightly concave, fleshy, olive-green and dark wine-red petals are 1.5 mm. long, 2.8 mm. wide and 0.4 mm. thick. The lip is succulent, dark wine-red, 1.1 mm. long, 1.3 mm. wide and 0.8 mm. thick. The light green, wine-red column is 1.0 mm. long and 1.3 mm wide, 0.8 mm. thick with small teeth between the wings and the apex.

Flowering Periods: December to January.

Similar Species: See *S. megantha - 1008*.

Stelis rubrechtiana
Rchb. f. | *1011* (Plate II)

Etymology: Unknown.

Habitat: Epiphytic from section IV, VI and VIII, at 1100-1600 M. asl in humid original forest, on upper-tree branches. L 4-7; H 6-9; W 2-6.

Occurrence: Common.

Plant: This 30-45 mm high epiphyte forms clusters with 7 to 30 leaves. The leaf stems are 10-15 mm. long and 0.6-0.8 mm. in diameter and enveloped by 2 tight, light brown sheaths up to 10 mm. long. The leaves are 20-35 mm. long, 7-9 mm wide. and 1-1.5 mm. thick, obovate, stiff, faintly conduplicate and attenuated to their bases forming 3 mm. false petioles.

Inflorescence and Flowers: The 50-75 mm, slightly bent inflorescence emerges from the apex of the leaf stem bearing 10-15 flowers on 1.5 mm. long pedicels. The spathe is 3-4 mm. and the relatively large bracts 4-6 mm. long. The light green and pale wine-coloured flower seldom opens wide and is 3.5-4.5 mm. by 3.7-4.6 mm. The slightly concave sepals are smooth or slightly grainy and almost equal. The yellow-green sepals are wine-red at their bases. The concave, fleshy, green and dark wine-red petals are 0.7-1.0 mm. long, 1.1-1.3 mm. wide and 0.25 mm thick. The lip is succulent, greenish, wine-red and measures 0.8-1.1 mm. long, 1-1.4 mm. wide and 0.6 mm. thick. The light green to dark wine-red column is 0.8 mm. long, 1.4 mm. wide and 0.8 mm. thick.

Flowering Periods: June to July and November.

General: There is a var. *latifolia* in which the leaves are 25-30 mm. long and 12 mm. wide; also var. major were the leaf stems are 20-30 mm. long and the leaves are 30-45 mm. long and 8-16 mm. wide.

Stelis thermophila
Schltr. | *1012* (Plate II)

Etymology: Latin *thermae* means warm and *philus* is Greek for loving.

Habitat: Epiphytic from section VI and VIII, at 1100-1500 M. asl in humid original forest, on low and mid-tree branches. L 5-7; H 5-8; W 2-6.

Occurrence: Common.

Plant: A 60-110 mm high epiphyte forming clusters with 7 to 20 leaves. The leaf stems are covered by 2 short-lived, light brown sheaths and are 20-40 mm. long and 1-1.7 mm. diameter. The leaves are obovate, stiff, dark green, faintly conduplicate, finely attenuated and slightly infolded to their bases forming 3-7 mm. false petioles and they are 30-60 mm. long, 5-8 mm. wide and 2 mm. thick.

Inflorescence and Flowers: The 60-150 mm. inflorescences arise from the apices of the most recent leaf stems and produce 15-25 flowers on 2.5-3.5 mm. pedicels. The spathe is 7-10 mm, the acute, funnel-shaped bracts are 3-4 mm. and the ovary is 2 mm. long. The wine and light yellow-green flowers only open wide in high humidity and measure 5.5 mm. by 4.8 mm. The light yellow-green, smooth sepals are light-wine coloured at their bases and are almost equal. The concave, fleshy, wine-red petals are 1 mm. long, 1.3 mm. wide and 0.3 mm. thick. The lip is succulent, round, greenish and wine-red and 1 mm. long, 1.2 mm. wide and 0.75 mm. thick. The transverse callus is indistinctly divided. The light green and wine-red column is 0.9 mm. long, 1.3 mm. wide and 0.6 mm. thick.

Flowering Periods: November and May.

Stelis triangularis
Barb. Rodr. | *1013* (plate II)

Etymology: *Triangularis* is Latin and refers to the triangular flower shape.

Habitat: Epiphytic from section VI, at 800-1500 M. asl in humid original forest, on upper-tree branches. L 6-8; H 5; W 7.

Occurrence: Occasional.

Plant: A 150-220 mm. high epiphyte forming clusters of 7 to 30 leaves with white or greenish 0.7 mm. diameter roots. The leaf stems are enveloped by 2 tight, light brown, long-lived sheaths and are 30-60 mm. long by 2-3 mm. diameter. The leaves are obovate, stiff, faintly conduplicate, slightly infolded, attenuated to their bases forming 6-9 mm. false petioles and measure 90-150 mm. long, 20-35 mm. wide and 1.5-2 mm. thick.

Inflorescence and Flowers: The 70-150 mm. inflorescences emerge from the apex of the leaf stem of both young and older leaves. They produce 15-25 flowers on 3-4 mm. long pedicels. The spathe is 10-15 mm. long, the bracts are 3-7 mm. and the ovary is 1.5 mm. long. The light wine and light yellow-green flowers open wide and are 6.5 mm. by 5 mm. The sepals are acute, smooth with the dorsal slightly longer. The concave, fleshy, green and wine-red petals are 1.5 mm. long, 1.7 mm. wide, 0.4 mm. thick with granular apices. The lip is succulent, greenish, wine-red and 1 mm. long, 1.2 mm. wide and 0.8 mm. thick with an indistinctly divided transverse callus. The light green and wine-red column is 1.3 mm. long, 1.2 mm. wide and 0.7 mm. thick.

Flowering Periods: November to January and August.

Stelis microcaulis
Barb. Rodr. | *1017* (Plate II)

Etymology: *Micro-caulis* is Greek for with a very short stem.

Habitat: Epiphytic from section VI and VIII, at 700-1000 M. asl in humid original forest, on upper and mid-tree branches. L 4-7; H 5-7; W 4-6.

Occurrence: Occasional. We have only identified a few plants.

Plant: A 20-40 mm. high epiphyte forming clusters with 7 to 20 leaves and 0.5 mm. diameter whitish roots. The leaf stems are covered by 2 tight, 4-7 mm. long, light brown sheaths and measure 8-12 mm. long, including a 2 mm. pseudo-petiole, and 0.8-1.2 mm. diameter. The dark olive-green elliptical leaves are 20-30 mm. long, 7-11 mm. wide, 1 mm. thick, fleshy, stiff, faintly conduplicate and attenuated abruptly to their bases.

Inflorescence and Flowers: The 60-90 mm, slightly curved inflorescence arises from 2 mm. below the leaf base with 12-18 flowers on 1-1.5 mm. pedicels. The spathe is 4-6 mm, the bracts are 1.5-2.0 mm. and the ovary is 1 mm. long. The flowers open wide and are 4.6 mm. by 4 mm. The sepals are smooth or very short haired at the margins and almost equal. The concave, fleshy petals are 0.5-0.75 mm. long, 0.75-1 mm. wide and 0.15 mm. thick. The succulent, round lip is light green and wine-red and 0.6 mm. long, 0.75 mm. wide and 0.4 mm. thick with an indistinctly divided transverse callus. The column is 1 mm. long, 0.7 mm. wide and 0.5 mm. thick.

Flowering Periods: November.

Stelis campos-portoi
Garay | *1020* (Plate II)

Etymology: **Paulo de Campos Pôrto**, the grandson of **Barbosa Rodrigues**, was director of the Rio de Janeiro Botanical Garden and wrote various articles on Orchids together with **A. C. Brade**.

Habitat: Epiphytic from section VI and VIII, at 1000-1400 M. asl, in humid original forest, on low-tree branches. L 4-5; H 6-8; W 4-5.

Occurrence: Common.

Plant: This 40-80 mm. high epiphyte forms clusters with 10-50 leaves. The leaf stems are covered by 2 tight, light brown sheaths and are 20-35 mm. long and 0.8-1.0 mm. diameter. The leaves are narrowly obovate, finely attenuated to their bases forming 0.7-10 mm. false petioles, faintly conduplicate, leathery and 30-40 mm. long, 6-10 mm. wide and 1 mm. thick.

Inflorescence and Flowers: The 100-150 mm, slightly bent inflorescence emerges from the apex of the leaf stem and bears 30-40 flowers on 1.5 mm. pedicels. The spathe is 8 mm. and the bracts are 2 mm. long. The light or dark wine and light yellow-green flowers open wide and are 3.4 mm. by 3.3 mm. The sepals are almost equal and have long white hairs along their margins and are olive-green to light wine at their bases. The concave, fleshy, green and wine-red petals have rough apices and are 0.6 mm. long, by 0.8 mm. wide. The succulent, square, yellow-green and wine-red lip is 0.5 mm. long, 0.7 mm. wide and 0.4 mm.thick, concave at the apex only and the margins have short hairs. The transverse callus is indistinctly divided. The light green and wine-red column is 1 mm. long, 0.9 mm. wide and 0.8 mm. thick.

Flowering Periods: November to December.

Stelis chlorantha
Barb. Rodr. | *1021* (Plate II)

Etymology: *Chlor - antha* is a Greek compound word for green-flowered.

Habitat: Epiphytic from section VI, at 1000-1200 M. asl, in humid original forest, on upper-tree branches. L 4-7; H 6-8; W 5-7.

Occurrence: Occasional.

Plant: A 80-100 mm. high epiphyte forming clusters with many leaves. The leaf stems are covered by 2 tight, light brown sheaths and are 20-25 mm. long and 0.7-1.0 mm. diameter. The leaves are narrowly obovate, attenuated to their slightly infolded bases forming 0.7-10 mm. false petioles, faintly conduplicate, leathery and 60-75 mm. long, 12-14 mm. wide and 1 mm. thick.

Inflorescence and Flowers: The 100 mm, slightly bent inflorescence emerges from the apex of the leaf stem with an annulus and develops 15-25 flowers. The spathe is 10-15 mm. and the bracts are 3 mm. The light yellow-green flowers open wide and are 3.2 x 3 mm. The flat, glandular sepals are wider than long and unequal. The dorsal sepal is 1.75 mm. long, while the lateral sepals are 1.5 mm. long. The concave, fleshy petals are 0.7 mm. long by 0.9 mm. wide. The lip is succulent, slightly concave and is 0.6 mm. long, 0.7 mm. wide and 0.5 mm. thick, with an indistinctly divided transverse callus. The column is 0.6 mm. long, 0.7 mm. wide and 0.4 mm. thick.

Flowering Periods: March.

General: The flower in our drawing is more wine than green and the flower part measurements differ slightly.

Stelis paquerensis var. fraterna
Lindl. | *1024* (Plate III)

Etymology: Fraterna is Latin for brotherly, closely allied.

Habitat: Epiphytic from section VI, at 800-1400 M. asl, in original cloud forest, on lower and mid-tree branches. L 4-7; H 6-8; W 2-6.

Occurrence: Common.

Plant: This 90-120 mm. high epiphyte forms clusters with many leaves. The leaf stems are 30-45 mm. long by 1.5-2 mm. diameter and are covered by 2 or 3 tight, light brown sheaths up to 40 mm. long. The leaves are narrowly obovate and attenuated to their bases forming 12-15 mm. false petioles, faintly keeled, furrowed, slightly infolded at their bases, leathery or stiff and 80-100 mm. long, 15-18 mm. wide and 1-2 mm. thick.

Inflorescence and Flowers: The 200 mm, slightly bent inflorescence emerges from the apex of the leaf stem and bears 30-40 flowers on 1.5 mm. pedicels. The spathe is 13 mm, the bracts are 2-4 mm, and the ovary is 0.75-1.0 mm. long. The flowers open wide and are 4-6 mm. by 2.5-4 mm. The slightly convex, slender sepals are unequal and hairy, these being longer and denser at the margins. The light yellow-green sepals are light wine at their bases and the dorsal is 2.8-3.3 mm. long and 1.8-2.5 mm. wide, while the laterals are 1.6-2.7 mm. long and 1.3-2.5 mm. wide. The concave, fleshy, green and wine-red petals are 0.5-0.8 mm. long and 0.9-1.2 mm. wide. The lip is succulent, concave, yellow-green and dark wine-red and is 0.7 mm. long, 0.8 mm. wide and 0.5 mm. thick, with an indistinctly divided transverse callus. The light green and wine-red column is 0.9 mm. long, 0.9 mm. wide and 0.7 mm. thick.

Flowering Periods: November.

Stelis itatiayae
Schltr. | *1025* (Plate III)

Etymology: Itatiaia is an area between São Paulo and Rio de Janeiro where the plant was probably first collected.

Habitat: Epiphytic from section VI, at 1000-1400 M. asl, in original cloud forest, on lower and mid-tree branches. L 4-7; H 6-8; W 2-6.

Occurrence: Common.

Plant: This 100-230 mm. high epiphyte forms clusters with 10-40 leaves and 0.7 mm. diameter white and light green roots. The leaf stems are 50-100 mm. long and 1.5-2.0 mm. diameter and enveloped by 2 or 3 tight, light brown sheaths, up to 60 mm. long. The leaves are narrowly obovate and attenuated to their bases forming 7-15 mm. false petioles, faintly conduplicate, slightly infolded at their bases, leathery and measure 90-110 mm. long, 14-18 mm. wide and 1 mm. thick.

Inflorescence and Flowers: A 70-130 mm, slightly curved, sparse inflorescence emerges from the apex of the leaf stem and produces 15-20 flowers on 3-4 mm. pedicels. The spathe is 8-12 mm. and the bracts 2-3 mm. long. The flowers open wide and are 8-9 mm. long by 7 mm. wide. The faintly concave sepals are slightly unequal and have long white hairs at the margins but are short haired along the margins. The light yellow-green sepals are light wine at their bases. The concave, fleshy, light green and dark wine-red petals are 0.9 mm. long, 1.3 mm. wide and 0.15 mm. thick. The lip is succulent, concave, yellow-green, dark wine-red and measures 0.8 mm. long, 1 mm. wide and 0.5 mm. thick, with an indistinctly divided transverse callus. The light green and dark wine-red column is 1.2 mm. long, 1.0 mm. wide and 0.8 mm. thick.

Flowering Periods: October to January.

Stelis modesta
Barb. Rodr. | *1026* (Plate III)

Etymology: Modestus refers to the modest, almost shy aspect of the flower.

Habitat: Epiphytic from all sections, at 600-1300 M. asl in original cloud and regrowth forest, on mid and upper-tree branches. L 4-8; H 6-8; W 5-7.

Occurrence: Common.

Plant: This 50-70 mm. high epiphyte forms clusters with 10-40 leaves. The leaf stems are 10-30 mm. long and 0.7-1 0 mm. diameter and enveloped by 2 or 3 tight, light brown sheaths, up to 15 mm. long. The leaves are narrowly obovate, attenuated to their base forming 7 mm. false petioles, fleshy, faintly conduplicate, slightly infolded at their bases and 35-50 mm. long, 7-9 mm. wide and 1-2 mm. thick.

Inflorescence and Flowers: A 70-100 mm, slightly bent inflorescence emerges from the apex of the leaf stem with 15-25 flowers. The spathe is 4-6 mm, the bracts are 2 mm. and the bent pedicel of the flower is 1.5 mm. long. The light yellow-green and light wine-red flowers open wide and are 3.9 mm. long by 3.4 mm. The concave sepals are almost equal, and densely covered by short hairs on the inside. The concave, fleshy, yellow-green and wine petals are 0.6 mm. long, 0.8 mm. wide and 0.25 mm. thick at their apices. The lip is succulent, slightly concave, yellow-green and wine and 0.6 mm. long, 0.7 mm. wide and 0.5 mm. thick with a divided transverse callus. The light yellow-green column is 0.75 mm. long, 0.75 mm. wide and 0.6 mm. thick.

Flowering Periods: November to January.

Stelis omalosantha
Barb. Rodr. | *1027* (late III)

Etymology: Omalos-antha is a Greek compound word for a symmetrical flower.

Habitat: Epiphytic from section VI, at 1000-1300 M. asl in original cloud forest, on mid-tree branches. L 4-7; H 6-8; W 4-7.

Occurrence: Common.

Plant: This 50-80 mm. high epiphyte forms clusters with 7 to 20 leaves. The leaf stems are 25-35 mm. long and 1 mm. diameter and enveloped by 2 tight, light brown sheaths, up to 20 mm. long. The leaves are narrowly obovate, attenuated to their bases forming 12-15 mm. false petioles, faintly conduplicate, slightly infolded at their bases and measure 25-50 mm. long, 6-10 mm. wide and 1-2 mm. thick

Inflorescence and Flowers: A 50-80 mm, slightly bent, sparse inflorescence emerges from the apex of the leaf stem and develops 10-20 flowers on 2 mm. pedicels. The spathe is 4-6 mm. and the bracts, 1.5 mm. long. The light yellow-green and wine-red flowers open wide and are 3-5 mm. by 3.5-4 0 mm. The slightly concave sepals are almost equal, covered densely by short hairs and are light green and wine towards the centre. The concave petals are square, fleshy, dark wine-red and 0.4 mm. long, 0.6 mm. wide. The lip is succulent, concave, dark wine-red and 0.4 mm. long, 0.6 mm. wide and 0.4 mm. thick with an indistinctly divided transverse callus. The olive-green and dark wine-red column is 0.5 mm. long, 0.7 mm. wide and 0.5 mm. thick.

Flowering Periods: January to April.

Stelis puberula
Barb. Rodr. | *1028* (Plate III)

Etymology: Puberulus is Latin for downy or with short, soft hairs referring to the hairy sepals.

Habitat: Epiphytic from section VI and VIII, at 1000-1300 M. asl in original forest, on mid-tree branches. L 4-7; H 5-8; W 4-7.

Occurrence: Common.

Plant: An epiphyte, 100-150 mm. high forming clusters with 7 to 20 leaves. The leaf stems are 50-70 mm. long and 1.5-2 mm. diameter and enveloped by 2 tight, light brown sheaths, up to 40 mm. long. The leaves are narrowly elliptical, attenuated to their bases forming 10 mm. false petioles and faintly conduplicate, slightly infolded at their bases and measure 50-80 mm. long, 17-20 mm. wide and 1-2 mm thick.

Inflorescence and Flowers: A 150-170 mm, slightly bent, sparse inflorescence emerges from the apex of

the leaf stem and bears 15-25 flowers which open wide and are 3.2 mm. by 3 mm. The spathe is 15 mm. long. The slightly concave sepals are almost equal, covered densely by short hairs and are light wine-red except the light olive margins. The concave, fleshy petals are 0.5 mm. long, 0.9 mm. wide. The succulent, concave lip is 0.5 mm. long, 0.7 mm. wide, 0.5 mm. thick with an indistinctly divided transverse callus. The column is 0.5 mm. long, 0.7 mm. wide and 0.5 mm. thick.

Flowering Periods: March.

Stelis grandiflora
Lindl. | *1030* (Plate III)

Etymology: Grandiflora in Latin means a large flower which, in *Stelis* terms, it is.

Habitat: Epiphytic from section VIII, at 900-1000 M. asl in humid original cloud forest, on mid-tree branches. L 5-8; H 6-8; W 4-7.

Occurrence: Rare. We have only found a few plants to date.

Plant: This 200-270 mm. high epiphyte forms clusters with 7 to 25 leaves and 0.75 mm. diameter roots which form a profuse system. The leaf stems are enveloped by 3 light ochre sheaths, up to 60 mm. long and measure 70-110 mm. long and 2.5-4 mm diameter. The leaves are narrowly obovate and attenuated to their bases, leathery, faintly conduplicate, slightly infolded at their bases and measure 120-150 mm. long, 25-35 mm. wide and 1-2 mm. thick.

Inflorescence and Flowers: A 150-220 mm, slightly bent inflorescence emerges from the apex of the leaf stem and produces 25-30 flowers on 2-3 mm. pedicels. The spathe is 15-20 mm, the bracts 3 mm. and the ovary 2-2.5 mm long. The light yellow-green flowers open wide and are 9 mm. by 7.8 mm. The smooth sepals are almost equal and have 5 distinctive longitudinal veins. The concave, fleshy petals are 1.4 mm. long, 1.8 mm. wide and 0.5 mm. thick and the lip is succulent, concave, 0.9 mm. long, 1.2 mm. wide and 1 mm. thick. The pale green column is 0.9 mm. long, 1 mm. wide and 0.7 mm. thick.

Flowering Periods: October to December and March to April.

Stelis dusenii
Garay | *1037* (Plate III)

Etymology: Named in honour of the Swedish Botanist **P. K. Dusen.**

Habitat: Epiphytic from section III, V, VI, VII and VIII, at 1000-1500 M. asl, in original cloud forest on upper-tree branches. L 6-8; H 4-5; W 6.

Occurrence: Common.

Plant: An 80-120 mm. high epiphyte forming clusters of 10-50 leaves. The leaf stems are 30-45 mm. long by 1-1.5 mm. diameter and enveloped by 2 light ochre sheaths, up to 20 mm. long. The fleshy leaves are 50-75 mm. long, 4-5 mm. wide and 1-2 mm. and are thick, almost linear, faintly conduplicate, slightly infolded and attenuated to their bases forming 5-10 mm. long false petioles.

Inflorescence and Flowers: A 60-100 mm, slightly bent inflorescence emerges from the apex of the leaf stem, bearing 25-40 flowers on 1.5 mm. pedicels. The spathe is 4-6 mm. and the bracts are 1-2 mm. long. The light yellow-green flowers open wide and are 2.8 mm. by 2.5 mm. The papillate, slightly concave sepals are almost equal while the petals are an equilateral triangle with one point cut off, concave, fleshy, yellow-green and 0.6 mm. long, 0.65 mm. wide and are short haired on the apical outer surface. The lip is succulent, yellow-green, slightly concave and 0.5 mm. long, 0.6 mm. wide and 0.5 mm. thick. The transverse callus is divided and the light green column is 0.7 mm. long, 0.6 mm. wide and 0.5 mm. thick.

Flowering Periods: December to February.

Similar Species: See ***S. drosophila - 1004.***

Stelis loefgrenii
Cogn. | *1040* (Plate III)

Etymology: Named in honour of the Swedish botanist **A. Loefgren.**

Habitat: Epiphytic from section VI, at 1200-1300 M. asl, in original cloud forest on upper-tree branches. L 5-8; H 5-8; W 4-7.

Occurrence: Occasional.

Plant: A 100-130 mm. high epiphyte forming clusters of 10-30 leaves. The leaf stems are 30-40 mm. long by 1.5 mm. diameter and enveloped by 2 light brown sheaths, up to 20 mm. long. The fleshy leaves are 60-80 mm. long, 16-24 mm. wide and 1 mm. thick, obovate, leathery, conduplicate, slightly infolded and attenuated to their bases forming 3 mm. false petioles. The roots are 0.5 mm. diameter.

Inflorescence and Flowers: A 70-90 mm. inflorescence emerges from the apices of young and older leaf stems with 10-15 flowers on 1 mm. pedicels. The spathe is 8 mm. and the bracts are 2 mm. long. The light yellow-green flowers open wide and are 9-10 mm. by 7 mm. The slightly convex sepals are a little unequal in size with short-haired margins and the petals are kidney-shaped, concave, slightly fleshy, light green and 1.1 mm. long, 1.7 mm. wide and 0.2 mm. thick. The lip is succulent, yellow-green, slightly concave and 0.9 mm. long, 1.1 mm. wide and 0.8 mm. thick with an indistinctly divided transverse callus. The light green column is 1 mm. long, 1.1 mm. wide and 0.7 mm. thick.

Flowering Periods: December to February.

Stelis aprica
Lindl. | *1045* (Plate III)

Etymology: *Apricus* is Latin for uncovered, growing in the sunshine referring to the plant's habit.

Habitat: Epiphytic from section V, VI and VIII, at 1000-1600 M. asl, in original forest on the upper-tree branches. L 5-8; H 3-5; W 5-7.

Occurrence: Common.

Plant: This 80-100 mm. high epiphyte forms clusters with 10-50 leaves. The leaf stems are 25-45 mm. long and 0.8-1.2 mm. diameter and enveloped by 3 light ochre sheaths, up to 30 mm. long. The leaves are linear-lanceolate, leathery or succulent, attenuated to their bases to form 5-7 mm. false petioles and slightly infolded at the base, faintly conduplicate and measure 50-60 mm. long, 5-8 mm. wide and 1.5-2 mm. thick.

Inflorescence and Flowers: A 60-80 mm, inflorescence emerges from the apex of the leaf stem and bears 20-30 flowers on 1 mm. pedicels. The spathe is 4-7 mm; the bracts and ovary are 1 mm. long. The light yellow-green flowers open wide and are 3.0 by 2.8 mm. The sepals are not quite equal, smooth, slightly convex and very slightly fused at their bases. The petals are concave, kidney-shaped, fleshy, light green to yellow and 0.45 mm. long, 0.7 mm. wide. The lip is 0.6 mm. long, 0.45 mm. wide and 0.35 mm. thick and succulent, hook-shaped and green with a divided transverse callus. The light green column is 0.6 mm. long, 0.7 mm. wide and 0.4 mm. thick.

Flowering Periods: February to April. The inflorescence flowers for one month.

General: We also found plants with spathulate leaves, 65 mm. long and 14 mm. wide in section I at 1200 M. asl. The inflorescence was 60 mm. tall. This is the only Stelis species in the Organ Mountain Range in which the lateral sepals are longer than the dorsal sepal.

Stelis binotii
De Willd. | *1046* (Plate IV)

Etymology: Named in honour of the Brazilian orchid grower **Binot**.

Habitat: Epiphytic from section V, VI and VIII, at 1000-1500 M. asl, in original cloud forest on the upper-tree branches. L 5-8; H 6-8; W 4-7.

Occurrence: Common.

Plant: A 90-140 mm. high epiphyte forming clusters with 10-50 leaves. The leaf stems are 50-70 mm. long and 1-1.5 mm. diameter and are enveloped by 3 light ochre sheaths, up to 40 mm. long. The leathery leaves are linear-lanceolate and attenuated to their bases to form 10-12 mm. false petioles. They are slightly infolded at the base, faintly conduplicate and 50-70 mm. long, 6-12 mm. wide and 2-3 mm. thick.

Inflorescence and Flowers: Up to three 110 mm. slightly bent inflorescences arise from the apices of each leaf stem and each produces up to 60 flowers on 1-1.5 mm. pedicels. The spathe is 9-12 mm, the bracts are 2-3 mm. and the ovary is 1-1.5 mm. long. The light yellow-green flower opens wide and is 3 mm. long by 2.7 mm. wide. The slightly convex, smooth sepals are almost equal and slightly fused at their bases. The concave, kidney-shaped, fleshy, green petals are 0.45 mm. long, 0.75 mm. wide and have distinct apices. The succulent, light green lip is 0.7 mm. long, 0.5 mm. wide, 0.35 mm thick and has a hook-shaped apex. The transverse callus is divided. The light green column is 0.6 mm. long, 0.7 mm. wide and 0.35 mm. thick.

Flowering Periods: All months of the year but mainly in October. The inflorescence lasts for one month.

Similar Species: See **S. aprica - 1045**. Our measurements are slightly larger than those of **Cogniaux**.

Stelis paraensis
Barb. Rodr. | *1051* (Plate IV)

Etymology: The first plant was found in the state of Pará.

Habitat: Epiphytic from section VIII, from 900-1500 M. asl, in humid original forest, on mid-tree branches. L 4-7; H 6-8; W 4-6.

Occurrence: Common.

Plant: An epiphyte, 150-250 mm. high, found in groups of 7-20 leaves. The leaf stems are 70-120 mm. long and 1.5-3.5 mm. diameter and enveloped by 3 greyish sheaths for up to 50% of their length. The faintly conduplicate leaves are obovate, narrow and leathery and narrow towards the base, forming false petioles of about 5 mm, and measure 80-130 mm. long by 20-35 mm, wide and 1.5 mm. thick.

Inflorescence and Flowers: One or 2, 150-300 mm. high inflorescences emerge from the apex of the leaf stem bearing 50-100 small flowers borne on 1.0-1.5 mm. pedicels. The spathe is 12 mm. long, the bracts 2.5 mm. and the ovary 1.5 mm. The flowers are yellowish-green, open widely and measure 4.0-4.5 mm. high by 3.0 mm. The dorsal sepal is longer than the laterals and all have smooth surfaces with 3 longitudinal veins and are partly fused. The petals are concave, broadly oval, fleshy and 0.6 mm. by 1.0 mm. wide. The succulent lip is 0.6 mm. long by 0.8 mm. and 0.5 mm. thick and the transverse callus is divided. The column is 0.6 mm. long by 1.0 mm, including the two wings, and 0.5 mm. thick.

Flowering Periods: January to mid-February.

Stelis argentata var. *pterostele*
Hoehne & Schltr. | *1052* (Plate IV)

Etymology: *Ptero-stele* is Greek. for winged-column which this plant clearly shows.

Habitat: Epiphytic from section VI and VIII, at 900-1500 M. asl, in original forest, on mid-tree branches. L 4-7; H 6-8; W 4-6.

Occurrence: Common.

Plant: A 150-250 mm. high epiphyte forming clusters with 7 to 20 leaves. The leaf stems are 70-120 mm. long and 1.5-3.5 mm. diameter and are enveloped by 3 tight, light brown sheaths, up to 50 mm. long. The leathery leaves are narrowly obovate and attenuated to their bases to form 5 mm. false petioles. They are faintly conduplicate, and are 80-130 mm. long, 20-35 mm. wide and 1-1.5 mm. thick.

Inflorescence and Flowers: One or two 150-300 mm, slightly bent inflorescences arise from the apices of the leaf stems of young and older leaves and bear 50-100 flowers on 1 1.5 mm. pcdiccls. The spathe is 12 mm, the bracts 2.5 mm. and the ovary is 1.5 mm. long. The pure light yellow-green flowers open wide and measure 4-4.5 mm. by 3 mm. The dorsal sepal is longer than the laterals and all are smooth with 3 distinct longitudinal veins and are partly fused. The concave, broadly oval, fleshy petals are 0.6 mm. long and 1 mm. wide. The succulent lip is 0.6 mm. long, 0.8 mm. wide and 0.5 mm. thick with a divided transverse callus. The column is 0.6 mm. long, 1 mm. wide, including the two wings, and 0.5 mm. thick.

Flowering Periods: January to mid-February.

Stelis paquerensis var. *porschiana* Schltr. | *1053* (Plate IV)

Etymology: Named in honour of the German Botanist **Otto Porsch**.

Habitat: Epiphytic from section VI and VIII, at 1200-1500 M. asl, in original forest, on mid-tree branches. L 4-7; H 6-8; W 4-7.

Occurrence: Common.

Plant: This 160-220 mm. high epiphyte forms clusters with 7 to 30 leaves. The leaf stems are 80-110 mm. long and 2.5-3.5 mm. diameter and enveloped by 3 tight, light brown sheaths, up to 40 mm. long. The leathery leaves are obovate, attenuated to their bases to form 5 mm. false petioles, faintly conduplicate, slightly infolded at their bases and 80-120 mm. long, 20-32 mm. wide and 2 mm. thick.

Inflorescence and Flowers: One or two, 80-200 mm, slightly bent inflorescences arise from the apices of each leaf stem with 20-40 flowers on 2 mm. pedicels. The spathe is 12-15 mm. and the bracts are 2 mm. long. The light yellow-green or wine-red flowers open wide and measure 6.7 mm. by 4.9 mm. The dorsal sepal is longer than the laterals and all are smooth with 3 distinct longitudinal veins, totally green or wine-red at their bases but yellow-green towards their apices. The concave, kidney-shaped, fleshy, wine-red or green petals are 1.1 mm. long, 1.5 mm. wide and 0.3 mm. thick. The succulent, concave, green or dark wine-red lip is 0.7 mm. long, 0.8 mm. wide and 0.8 mm. thick. The transverse callus is divided and the light green or wine-red column is 0.9 mm. long, 1.2 mm. wide and 0.7 mm. thick.

Flowering Periods: February to March and October.

General: There is a smaller variety with smaller flowers.

Stelis reflexisepala Garay | *1054* (Plate IV)

Etymology: Reflexus is Latin for bent abruptly backwards referring to the recurved sepals.

Habitat: Epiphytic from section V, VI and VIII, at 1200-1500 M. asl, in original cloud forest, on mid-tree branches. L 4-8; H 4-8; W 4-8.

Occurrence: Common.

Plant: This 120-180 mm. high epiphyte forms clusters with 7 to 30 leaves. The leaf stems are 60-85 mm. long and 1.8-2.4 mm. diameter and enveloped by 3 tight, light brown sheaths, up to 40 mm. long. The leathery leaves are obovate and attenuated to their bases to form 6 mm. false petioles. They are faintly conduplicate and

measure 70-90 mm. long, 20-25 mm. wide and 2 mm. thick.

Inflorescence and Flowers: A 120-160 mm. slightly bent inflorescence emerges from the apex of the leaf stem and produces 20-40 flowers on 2 mm. pedicels. The spathe is 10-12 mm; the bracts and the ovary are 2 mm. long. The very dark wine-red flowers open wide and are 4.7 mm. by 4 mm. The dorsal sepal is longer than the laterals and all are smooth, dark wine-red and have recurved margins. The petals are concave, kidney-shaped, fleshy, purple, dark wine-red and measure 0.8 mm. long, 1.6 mm. wide. The succulent, concave, dark wine-red lip is 0.8 mm. long and 1 mm. wide. The transverse callus is indistinctly divided. The light yellow-green and dark wine-red column is 0.8 mm. long and 1.2 mm wide.

Flowering Periods: January to February and October.

General: This species shows the darkest coloured flowers in the Organ Mountain Range.

Stelis paquerensis var. *tweediana* Lindl. | *1057* (Plate IV)

Etymology: Named in honour of the British Botanist **J. Tweedie**.

Habitat: Epiphytic from section VI, at 1400 M. asl, in humid original cloud forest on mid tree branches. L 5, H 8, W 5.

Occurrence: Occasional.

Plant: A 100-200 mm. high epiphyte forming clusters with 8-20 leaves and roots up to 40 cm. long. The light green rhizome is short and 0.4 mm. diameter. The leaf stems are 50-100 mm. long and 1.7-2.3 mm. diameter and enveloped by 2 tight, light brown sheaths, up to 45 mm. long. The leathery leaves are narrowly elliptical and attenuated to their bases to form 10-12 mm. false petioles, faintly conduplicate, slightly infolded at the base and 55-100 mm. long, 18-24 mm. wide and 1.5 mm. thick.

Inflorescence and Flowers: Up to two, 100-140 mm. inflorescences arise from the apex of the leaf stem and bend away from it and bear 20-35 flowers on 1.5 mm. pedicels. The spathe is 12-14 mm; the bracts are 1.5-2.0 mm. and the ovary 1 mm. long. The light yellow-green flowers open wide and measure 4.0 mm. by 3-4 mm. The dorsal sepal is longer than the laterals and all are slightly concave and smooth. The petals are kidney-shaped, fleshy, dark wine and purple and 0.7 mm. long, 1 mm. wide. The lip is succulent, concave, light green and dark purple, 0.7 mm. long, 1.0 mm. wide and 0.5 mm. thick. The transverse callus is indistinctly divided. The light yellow-green and dark wine-red column is 0.7 mm. long, 1.0 mm. wide and 0.5 mm. thick.

Flowering Periods: January to February.

Stelis paquerensis var.*inaequalisepala* Hoehne & Schltr. | *1058* (Plate IV)

Etymology: Inaequali-sepala is Latin for unequal sepals.

Habitat: Epiphytic from section VI and VIII, at 800-1100 M. asl, in original cloud forest, on mid-tree branches. L 4-6; H 6-8; W 4-7.

Occurrence: Occasional.

Plant: This 130-200 mm. high epiphyte forms clusters with many leaves and 0.5 mm. diameter whitish roots. The leaf stems are 35-70 mm. long and 1.5-2.3 mm. diameter and enveloped by 2 tight, light brown sheaths, up to 40 mm. long. The leathery leaves are narrowly obovate and attenuated to their bases to form 10 mm. false petioles and are faintly conduplicate, slightly infolded at the base and 90-130 mm. long, 13-18 mm. wide and 1.5 mm. thick.

Inflorescence and Flowers: One or two, 110-180 mm, slightly bent inflorescences emerge from the apices of each leaf stem with 35-50 flowers on 1.5 mm. pedicels. The spathe is 10 mm, the bracts 2-3 mm. and the ovary 1.5 mm. long. The yellow, green and purple flowers open wide and are 5 mm. by 3.8 mm. The dorsal sepal has 5 longitudinal veins and is distinctly longer than the laterals which have 3 longitudinal veins. The smooth sepals have light wine-red markings and are fused for almost half their length. The petals are concave, kidney-shaped, fleshy, light green and light purple and 0.75 mm. long, 1.2 mm. wide with short hairs on their obtuse. 0.2 mm. thick apices. The succulent, concave, light green and dark purple lip is 0.8 mm. long, 0.8 mm. wide and 0.6 mm. thick. The light yellow-green and dark wine-red column is 0.8 mm. long, 1 mm. wide and 0.6 mm. thick.

Flowering Periods: April and November.

General: This species has the most unequal sepals of those found in the Organ Mountain Range.

Stelis parahybunensis Barb. Rodr. | *1059* (Plate IV)

Etymology: Found by the Parahybuna River.

Habitat: Epiphytic from section VI and VIII, at 1300-1600 M. asl, in original cloud forest, on mid- and lower tree branches. L 3-5; H 6-8; W 2-5.

Occurrence: Common.

Plant: This 150-220 mm. high epiphyte forms clusters with 7-20 leaves. The 2.0-2.5 mm. diameter leaf stems are 80-110 mm. long and enveloped by 3 tight, brown sheaths, up to 65 mm. long. The leathery leaves are 80-120 mm. long, 20-30 mm. wide and 1.5 mm. thick, elliptic, conduplicate, slightly infolded and attenuated to

their bases to form 5 mm. false petioles. The numerous, 0.8 mm. diameter roots are whitish.

Inflorescence and Flowers: The 150-200 mm. inflorescences emerge from young and older leaf stems with 20-35 flowers on 1.5-2.5 mm. pedicels. The spathe is 15 mm, the bracts 3 mm. and the ovary 4 mm. long. The light green to dark wine-red flowers open wide and are 6-8.5 mm. by 4-6.2 mm. The dorsal sepal is longer than the laterals and all are slightly convex with short haired margins. The concave petals are kidney-shaped, fleshy, light green to wine-red and 1.0 mm. long, 1.5-2 mm. wide and 0.2 mm. thick. The succulent, concave, light green to wine-red lip is 0.8 mm. long, 1.1 mm. wide and 1 mm. thick. The light yellow-green and pale wine-red column is 0.8 mm. long, 1.2 mm. wide and 0.7 mm. thick.

Flowering Periods: November.

Stelis paquerensis var. *guttifera* Porsch | *1061* (Plate IV)

Etymology: Gutta and *fer* make a Latin compound word meaning drop-bearing suggesting the humid conditions these plants enjoy.

Habitat: Epiphytic from section VI and VIII, at 1000-1200 M. asl, in original forest, on lower tree branches. L5; H 8; W 4.

Occurrence: Common.

Plant: This 180-250 mm. high epiphyte forms clusters with 10-30 leaves. The leaf stems are 90-120 mm. long and 2-3 mm. diameter and enveloped by 3 tight, light brown, short-lived sheaths, up to 40 mm. long, The leathery leaves are narrowly obovate and attenuated to their bases to form 6-8 mm. false petioles and are faintly conduplicate, slightly infolded at the base and 100-150 mm. long, 16-24 mm. wide and 2-2.5 mm. thick.

Inflorescence and Flowers: A 120-150 mm, slightly bent inflorescence emerges from the apex of the leaf stem and produces 20-25 flowers on 3 mm. pedicels. The spathe is 10-15 mm; the bracts are 2 mm. and the ovary 2.5 mm. long. The very light green and wine-red flowers open wide and are 8 mm. by 6 mm. The slightly convex, smooth sepals are almost equal with short-haired margins and are light wine at their bases turning to light green towards the apices. The petals are concave, kidney-shaped, fleshy, wine-red and 1 mm. long, 1.8 mm. wide. The succulent, concave, light green and dark wine-red lip is 0.75 mm. long, 0.75 mm. wide and 0.75 mm. thick with a divided transverse callus. The light yellow-green and wine-red column is 1 mm. long, 1.2 mm. wide and 0.7 mm. thick.

Flowering Periods: March to April.

Stelis paquerensis var. *plurispicata* Barb. Rodr. | *1063* (Plate V)

Etymology: Pluris and *spicatus* are a Latin for many spikes.

Habitat: Epiphytic from section VI and VIII, at 1000 M. asl, in original cloud forest, on lower tree branches. L 4-6; H 6-8; W 4-5.

Occurrence: Occasional.

Plant: This 80-100 mm. high epiphyte forms clusters with 10-30 leaves. The leaf stems are 30-45 mm. long and 1.5 mm. diameter and are enveloped by 2 tight, light brown sheaths, up to 25 mm. long. The leathery leaves are elliptic and attenuated to their bases to form 10 mm. long false petioles. They are conduplicate, slightly infolded at the base and 50-65 mm. long, 10-20 mm. wide and 1-2 mm. thick.

Inflorescence and Flowers: One or two, 80-130 mm. inflorescences emerge from the apical leaf stem with 20-25 flowers on 1.5 mm. long pedicels. The spathe is 8-10 mm; the bracts are 1 mm. and the ovary 1.5 mm. long. The light green and wine-coloured flowers open wide and measure 3.4 mm. by 2.5 mm. The dorsal sepal is slightly longer than the laterals; all are slightly convex, warty, with hairy margins and are green or wine-coloured at their bases but light green towards their apices. The concave petals are kidney-shaped, fleshy, wine-red and are 0.65 mm. long, 1 mm. wide. The succulent, concave, light green and wine-red lip is 0.55 mm. long, 0.75 mm. wide and 0.4 mm. thick. The transverse callus is indistinctly divided. The light yellow-green and light wine-red column is 0.9 mm. long, 0.9 mm. wide and 0.6 mm. thick.

Flowering Periods: December to January

Stelis vinosa Barb. Rodr. | *1064* (Plate V)

Etymology: Vinosa is Latin for wine-red and refers to the lilac-wine-red flowers.

Habitat: Epiphytic from section VI and VIII, at 1000-1300 M. asl, in original forest, on lower tree branches. L 3-5; H 6-8; W 2-5.

Occurrence: Common.

Plant: This 90-120 mm high epiphyte forms clusters with 10-30 leaves. The 1.5-2.0 mm. diameter leaf stems are 25-35 mm. long and enveloped by 2 tight, light brown sheaths, up to 25 mm. long. The leathery leaves are elliptical and attenuated to their bases to form 5 mm. false petioles. They are faintly conduplicate, slightly infolded at the base and 50-65 mm. long, 15-25 mm. wide and 2 mm. thick.

Inflorescence and Flowers: One or two, 80-130 mm, slightly bent inflorescences emerge from the apices of each leaf stem with 25-40 flowers on 1.5-2.5 mm. pedicels. The spathe is 7-10 mm; the bracts are 1.5-2.5 mm. and the ovary 1.5 mm. long. The light green and dark wine-red flowers open wide and are 6 mm. by 5 mm. The dorsal sepal is slightly longer than the laterals and all are slightly convex, papillate with short-haired margins and green or wine-red at their bases turning to light green towards their apices. The concave petals are kidney-shaped, fleshy, wine-red and 1.0 mm. long, 1.8 mm. wide and 0.2 mm. thick. The succulent, concave, light green and wine-red lip is 0.9 mm long, 1.1 mm wide and 0.7 mm thick. The transverse callus is indistinctly divided. The light yellow-green and light wine-red column is 0.8 mm long, 1.5 mm wide and 0.8 mm thick.

Flowering Periods: October to January and April.

The genus *Phloeophila* Hoehne & Schltr.

This small Brazilian genus was raised by **Hoehne & Schlechter** because of the curious and distinctive flower shape and the plant's growth habit. **Carlyle Luer** transferred it to Pleurothallis based on the similarity of the flower parts to that genus. We have left the genus separate because we consider the flower parts distinctive and we wish readers to refer to **Pabst&Dungs'** *Orchidaceae Brasiliensis* Vol. 1, where it is found under *Phloeophila*. *Phloeo* and *phila* are Greek words which mean bark-loving which described these plants' preference.

Key: Two waxy pollinia with viscidium. Plants are small with a creeping rhizome and the sepals are fused more or less tube-like at the base.

Phloeophila pubescens (Barb. Rodr.) Garay | *1069* (Plate V)

Synonym: Pleurothallis sarracenia (Garay; Luer).

Etymology: Pubescens is Latin and refers to the hairy inside of the sepals.

Habitat: Epiphytic from section V, VI and VIII, at 700-1000 M. asl, in humid original mountain cloud forest on shady trunks, branches and rotten wood, usually not above 5 metres above ground. L 4-6; H 7-8; W 5.

Occurrence: Occasional, but locally common.

Plant: This creeping epiphyte covers large areas of a trunk or branch. Membranous, short-lived sheaths cover the green 1.5-2.0 mm. diameter rhizome. The leaves arise alternately after 2 or 3 internodes of the rhizome at 5 mm. intervals on 2 mm. long and 0.5 mm. diameter leaf stems covered by a membranous sheath. The yellow-green leaves are faintly conduplicate, fleshy and straight, narrowly elliptical, almost linear, 2.5 mm. thick and are 15-24 mm. long and 5 mm. wide. The plant is loosely attached to the bark by white or light greenish, 0.5-0.7 mm. diameter roots.

Inflorescence and Flowers: The very short, almost sessile inflorescence bears 1 or 2 opposite flowers, which emerge from the apex of approximately 1-year-old leaf stems. Both the peduncle and the pedicel are less than 1 mm. and are enveloped by the membranous spathe and bracts. The ovary is 1 mm. long. The flower opens only at the apical section, which is curved forward by 90-180°. The dark wine and yellow-green sepals are fleshy and 80% fused and covered inside with up to 1.5 mm, whitish hairs. The acute sepals are 10-18 mm. long and 4-6 mm. wide. The lanceolate, slightly asymmetrical petals are 4-4.5 mm. long and 1.5-1.9 mm. wide and have long acuminate apices. The yellow and dark wine lip has 2 wide, erect side lobes and the main lobe has a rounded apex with serrate margins. The longitudinal callosities are relatively large and the lip is 3.2 mm. by 2.3 mm. The yellowish-wine column is 2.8 mm. long and 1.3 mm. wide.

Flowering Periods: December to April and the flowers last for 10 to 20 days.

Cultivation: It will do well on bark, which should not dry out completely.

Phloeophila bradei (Schltr.) Garay | *1070* (Plate V)

Synonym: Pleurothallis neobradrei (Garay) **Luer**.

Etymology: Named in honour of the German Botanist **A. Brade**.

Habitat: Epiphytic in section VI, 800-950 M asl, in humid original mountain forest, on rocks, bare trunks, roots and branches up to 3 meters above ground. L 2-4; H 7-8; W 3.

Occurrence: Occasional but locally common.

Plant: This creeping epiphyte covers large areas of a trunk or branch resembling a Barbosella species. The green 0.5-0.7 mm. diameter rhizome is completely covered by membranous sheaths. The leaves are alternate arising after about 3 internodes of the rhizome at 5 mm. intervals on a 3 mm. long and 0.5 mm. diameter leaf stems. A membranous sheath also envelops it. The faintly conduplicate, leathery leaf is dark, shiny green and rotated by 45-90° and has a 3 mm. false petiole. The narrowly elliptical, 0.8 mm. thick leaves are 15-20 mm. long and 5 mm. wide. The plant is attached to the bark by a few white or light greenish, 0.5-0.7 mm diameter roots.

Inflorescence and Flowers: The single flowered inflorescence emerges 2 mm. below the leaf base. The peduncle and the pedicel are 1.5 mm. each and are enveloped by a membranous, 1.3 mm. spathe and bract. The ovary is 1.3 mm. long. The flower only opens at the apical half. The sepals are dark wine colour and have some short hairs on the back. The 4.5 mm. long, acute ovate dorsal sepal is 2 mm. wide and 40% fused to the lateral sepals which are 95% fused and 5.5 mm. long and 4.4 mm. wide together. The lanceolate petals are 4 mm. long and 1.5 mm. wide. The tongue-shaped lip has two hook-shaped side lobes, a lanceolate apex with serrate margins and is 3.5 mm. by 1.6 mm. The light to dark wine-red column is 2.5 mm. long and 1.3 mm. wide. The wings are faint and the column foot is dark wine.

Flowering Periods: October to November and January. The flower lasts for 7-10 days.

Cultivation: We also found it on dead branches; it should do well on bark with some moss, which should never dry out completely.

Phloeophila echinantha (Barb. Rodr.) Hoehne & Schltr. | *1071* (Plate V)

Synonym: Pleurothallis echinantha **Barbosa Rodrigues**.

Etymology: From the Latin *echinatus* for spiny and *antha* a flower referring to the hairs on the sepals.

Habitat: Epiphytic from section VI and VIII, 900-1250 M. asl, in humid original mountain forest on bare trunks and branches at low tree level. L 4-7; H 7; W 3-7.

Occurrence: Common.

Plant: The plant is closely addpressed to bark and as the rhizome is many branched it can cover large areas of a trunk or branch. The rhizome is 0.7-1.0 mm. thick and completely covered by funnel-shaped hairy sheaths. The leaves arise alternately after about 3 sections of the rhizome at 2-4 mm. intervals on very short hairy, sheathed leaf stems. The faintly conduplicate leaf is dark shiny green and its surface is completely covered by more or less equal sized dimples. The fleshy, 0.5-1.0 mm. thick leaves are ovate or nearly round and 5-7 mm. x 4-5.5 mm. and the underside may be purple. The few, very adherent, white or light greenish roots are 0.5-0.7 mm. diameter.

Inflorescence and Flowers: Very short, single flowered inflorescences emerge from the leaf bases and the flowers open only a little and have a pronounced chin and faint keels. The spathe, the bract and the 1.5 mm. long ovary are hairy. The outside of the sepals has strong red hairs, disorderly arranged. The sepals are reddish at the base becoming yellowish towards the apex. The broad obovate dorsal is 40% fused to the laterals, has an obtuse apex, 3 wine-red longitudinal veins and is 4.0

| 421 |

mm. long by 3.6 mm. wide. The lateral sepals are 80% fused, and have 3 wine-red veins, a row of wine-red dots at each margin and are 5 mm. long by 5.2 mm. wide together. The yellow petals are spathulate, 3 mm. long by 0.7 mm. wide with 3 yellow longitudinal veins and an obtuse apex. The lip is tongue-shaped, white with purple blotches and 2 indistinct side lobes with serrated margins. It is 3.5 mm. long by 1.3 mm with an obtuse apex.

Flowering Periods: October to December and the flower lasts for 14 days.

The genus *Masdevallia* Ruiz & Pavon

Ruiz and **Pavon** were Spanish scientists who visited various Spanish-American countries between 1778-88, searching principally for medicinal plants. They named this large genus of over 350 species, which are concentrated in the Andean countries of Peru, Ecuador and Colombia in honour of Dr. **Jose Masdeval**, another Spanish botanist and physician. Very few species have strayed as far as the Organ Mountain Range.

Key: Two waxy pollinia with a viscidium. All sepals are fused variously below the apex and are very finely acuminate. The petals have a callus on the lower part.

Masdevallia curtipes Barb. Rodr. | *1083* (Plate V)

Etymology: Curti and *pes* are Latin for short foot referring to the very short peduncle.

Habitat: Epiphytic from section VIII, at 700-1000 M. asl, in the shade of very humid original cloud forest, on trunks and thick lower branches. L 3-5; H 8; W 2.

Occurrence: Common in original forest but absent from many apparently suitable areas.

Plant: This 100-150 mm. high epiphyte forms clusters with 8 to 30 leaves and a moderately intensive root system composed of white, 1.5 mm. diameter roots. The 7-15 mm. long, round leaf stems are 1.5 mm. diameter and enveloped by 2 loose, light ochre sheaths. The leaves are 100-140 mm. long, 14-17 mm. wide and 1.5-2 mm. thick, linear-lanceolate, conduplicate, fleshy, verdigris-green, slightly infolded and finely attenuated towards their bases forming 10-15 mm. false petioles.

Inflorescence and Flowers: The inflorescences emerge 1 to 2 mm. above the bases of the leaf stems of the most recent leaves. The peduncles are 20-40 mm. with 1.5 mm. equilateral triangle cross-sections with up to 3 consecutive flowers. The pedicels are 15-20 mm, the

bracts 25 mm. and the ovaries are 8 mm. long. The flower is yellow-green and wine and all sepals are fused at their bases to form an asymmetric cup with 10-25 mm. finely acuminate tips. The tongue-shaped, fleshy 5.5 mm. long petals are 1.9 mm. wide, light yellow-green with obtuse 3-pointed apices. The tongue-shaped lip is 6 mm. long, 1.9 mm. wide, fleshy, light green and ochre at the more fleshy base and apex. The distinct lamellae end in the undulate surface of the apical half and the side-lobes are close to the lip base. Both lip and petals are usually dotted purple. The light green, 4.3 mm. column is 1.5 mm. wide with a purple dotted column foot.

Flowering Periods: November to March. Individual flowers last for 15 days.

Cultivation: Temperate climate with a maximum temperature below 30°C. and minimum 10-15°C., no direct sunlight, relative humidity above 75%, substratum should be kept moist; a recommended mixture is fine bark, fibre and some shredded foliage.

General: Our drawing seems to show a hybrid between **M. infracta** and **M. curtipes**. The peduncle should be shorter than shown.

Masdevallia infracta Lindl. | *1084* (Plate V)

Etymology: Infracta is Latin for bent or sharply incurved probably referring to the flower habit.

Habitat: Epiphytic or lithophytic from sections I, V, VI and VIII, at 900-1100 M. asl, in the shade of very humid original forest, on trunks and thick lower branches. L 3-5; H 8; W 2.

Occurrence: Common in original forest.

Plant: This 100-160 mm, high epiphyte forms clusters of 8 to 40 leaves and a moderately intensive root system composed of white, 1.5 mm. diameter roots. The 10-20 mm. long, round leaf stems are 2 mm. diameter and enveloped by 2 loose, light green sheaths. The leaves are 100-160 mm. long, 15-20 mm. wide and 1.5-2 mm. thick and linear-lanceolate, conduplicate, fleshy, verdigris-green, slightly infolded and finely attenuated towards the bases to form 13-15 mm. false petioles.

Inflorescence and Flowers: The inflorescences emerge about 1 to 3 mm. above the bases of the most recent leaf stems on 80-120 mm. peduncles, with 2 mm. equilateral triangle cross-sections and up to 3 consecutive flowers. The pedicels are 15-20 mm, the bracts 15 mm. and the ovaries 5 mm. long. The normal colour of the flower is yellow-green and wine, but there are colour forms from pure yellow to deep purple. All sepals are fused at the base to form an asymmetric cup with 35-45 mm. finely acuminate tips. The asymmetric, spathulate, fleshy 6.2

mm. long petals are 1.5 mm. wide and light yellow with strong yellow pointed apices. The tongue-shaped lip is 6.5 mm. long, 2.2 mm. wide, light yellow and fleshy with distinct lamellae ending in the undulate surface of the apical half and side-lobes close to the lip base. The lip and the petals are usually dotted purple. The light green, 6 mm. column is 2 mm. wide and has a yellow base.

Flowering Periods: November to March. An individual flower lasts for 15 days.

Cultivation: Temperate climate with a maximum below 30°C., and minimum of 10-15°C., no direct sunlight and relative humidity above 75%. The substratum should be kept moist; and a recommended mixture is fine bark, fibre and some shredded foliage.

The genus *Dryadella* Luer

Dryadella is a small Central and South American genus recently separated from *Masdevallia* by **Carlyle Luer**. Its appropriate name is the diminutive of dryad, the mythological nymphs of trees and forests.
Key: Two waxy pollinia with a viscidium. Plants are small; the sepals are fused at the base and the lateral sepals have a transverse callus above their bases; the lip is faintly although intensely serrate at the margin.
Cultivation: Temperate to cool with a maximum below 30°C., and minimum of 10°C; no direct sunlight, relative humidity above 75% and the substrate should be kept moist. The recommended mixture is fine bark, fibre and some shredded foliage.

Dryadella edwallii (Cogn.) Luer | *1088* (Plate V)

Etymology: Named in honour of the Brazilian Botanist **G. Edwall**.

Habitat: Epiphytic and lithophytic from section VI, at 900-1500 M. asl, in the shade of very humid original forest, on trunks, thick branches shrubs and rocks. L 4-6; H 7; W 5.

Occurrence: Common.

Plant: A 50-70 mm high epiphyte forming clumps with 10-50 leaves and a very intensive and dense root system composed of white, 0.8 mm. diameter roots. The leaf stems are 5-8 mm. long and 0.8 mm. diameter, round and covered by 2 loose, light green sheaths. The leaves are linear-lanceolate, conduplicate, verdigris-green, fleshy, infolded and narrowed to their bases to form 3 mm. false petioles and are 40-50 mm. long, 5-7 mm. wide and 1.5-2 mm. thick.

Inflorescence and Flowers: Up to 3 individual flowers emerge consecutively from the base of the most recent leaves on 8 mm. long pedicels, which are partly enveloped by loose sheaths and the spathe. The distinctively keeled ovary is 3 mm. long. All parts of the flower are yellow with purple speckles or spots. The finely acuminate, ovate sepals are distinctively keeled. The concave dorsal sepal is 13-15 mm. long and 4.5 mm. wide and the laterals are 5% fused, convex and 14 mm. long by 4.5 mm. wide. The asymmetric petals are rhomboid, fleshy and 4.2 mm. long by 4 mm wide. The lip is spoon-shaped, 4.8 mm. long and 2.8 mm. wide with 2 small hook-shaped ear-like appendages at its light green base and widens after a narrow 2 mm. long basal section into the hook-shaped side-lobes. There are 2 erect, triangular lamellae at the base of the almost round apical lobe. The light green and yellow, 4 mm. long column is 2 mm. wide. The wings and the apex of the column are acuminate.

Flowering Periods: October to February and the individual flowers last for 15 days.

Dryadella zebrina (Porsch) Luer | *1096* (Plate V)

Etymology: The flowers have zebra-like markings.

Habitat: Epiphytic in sections VI, 1100-1500 M. asl, in the shade of very humid montane cloud forest, on trunks and branches, frequently exposed to drifting fog. L 2-4; H 9; W 4.

Occurrence: Rare.

Plant: This 20-30 mm. high epiphyte forms clusters with 10-50 leaves and a densely packed root system composed of white, 0.5 mm. diameter roots which build a kind of stem, like a tree fern, reaching 30 mm. after about 20 years. The leaf stems are 4-5 mm. long and 0.7 mm. diameter, round and enveloped by up to 4, light green sheaths. The leaves are 20-30 mm. long, 2.5-3 mm. wide and 1.5-2 mm. thick, linear, conduplicate, verdigris-green, fleshy and infolded and narrowed towards their bases forming 3 mm. false petioles and are enveloped by the 4 loose sheaths of the leaf stems.

Inflorescence and Flowers: One or 2 single flowers emerge consecutively from the apices of the most recent leaf stems on 5 mm. peduncles and 2 mm. pedicels. The distinctively keeled ovary is 2 mm. long. All sepals are speckled purple on a whitish background and are concave and distinctively keeled. The margins of the apical half are faintly toothed. The finely acuminate apices are yellow-green. The 13-16 mm. long dorsal sepal is 4 mm. wide and the lateral sepals, which are 5% fused are 13-15 mm. long and 3.8 mm. wide. The asymmetric, square, fleshy petals are 3 mm. long by 3 mm. wide and coloured like the sepals with darker bases. The spoon-shaped, 4.3 mm. long and 2.3 mm. wide lip has 2 small ear-like appendages at the whitish base. The lip widens after a narrow, 1.5 mm. long basal section into the sickle-shaped side-lobes. There are 2 erect, toothed lamina on the plate of the almost round apical lobe, which is recurved by 90°.

Flowering Periods: January to April and the flowers last for 15 days.

The genus *Lepanthopsis* (Cogn.) Ames

Lepanthopsis means like a Lepanthes, both genera that were earlier considered sections of Pleurothallis. **Oakes Ames** raised it to generic rank in 1933 because of the structure of the minute, complex flowers. There are about 25 species from Florida to Mexico and south to Brazil, distinguished by the trumpet-like sheaths on the stems.

Key: Two waxy pollinia with a viscidium. The column is short in comparison to its width. The lip is round and the petals are much smaller than the sepals; there is a rudimentary column foot and an apical, bilobed stigma.

Lepanthopsis aff. densiflora (Barb. Rodr.) possible new species Ames | *1101b* (Plate VI)

Etymology: Densi-flora is Latin for densely-flowered.

Habitat: An epiphyte from section VI and VIII, at 1250-1350 M. asl, in very humid original mountain forest, on mid-tree branches. L 3-5; H 9; W 4.

Occurrence: Rare.

Plant: The plant is 30-45 mm. high and forms clumps with 5-20 leaves and greenish 0.8 mm. diameter roots and a very short rhizome. Three to 5 trumpet-shaped hispid sheaths with recurved, hairy margins envelop the 10-22 mm. long and 0.5 mm. diameter leaf stems. The sheaths are olive with dark purple veins, turning brown with age. The stiff leaves are 12-20 mm. long, 5-6 mm. wide and 1.5 mm. thick, olive green, elliptical, flat, conduplicate and narrowed abruptly into 1.5-2.0 mm. false petioles.

Inflorescence and Flowers: The 40-75 mm. inflorescences arise from the apex of young and older leaf stems with 20-40 flowers in 2 dense, opposite ranks on 0.5 mm. pedicels and ovaries. The flowers open in quick succession and bloom simultaneously. The floral bracts are 1.5 mm. and the spathe is 2.5 mm long. The wide open flowers are 3-4 mm. long and 1.3-1.5 mm. wide. The ovate sepals are slightly concave, keeled, light green to yellow green and stained purple at the apices. The

dorsal sepal is 1.5-1.7 mm. long and 0.6-0.9 mm. wide and the laterals are 50% fused and 1.3-1.7 mm. long by 0.6 mm. wide. The round and slightly concave petals are transparent light green with green dots and 0.8 mm. long and wide. The tongue-shaped lip is concave, fleshy and light purple with a light green base and is 0.9 mm. long, 0.5 mm. wide. The yellow green and faintly purple column is 0.5 mm. long, 0.6 mm. wide and 0.4 mm. thick. The anther is apical and the stigma bilobed.

Flowering Periods: December to January and all leaf bases may repeat flowering. We have seen it blooming twice in December. The flowers last for two weeks.

General: Our plant differs from that drawn by **Barbosa Rodrigues** and **Carlyle Luer** as the leaves are wider, the sepals more obtuse and the lip is more slender and tongue-shaped, not ovate as in their drawings.

Lepanthopsis floripecten (Rchb. f.) Ames | *1102* (Plate VII)

Etymology: Flori and *pectinatis* are Latin for comb-like, referring to the densely arranged flowers.

Habitat: Epiphytic in sections VI and VIII, at 1000-1200 M. asl, in humid original montane forest, on mid-tree branches. L 4-6; H 8; W 5.

Occurrence: Occasional.

Plant: The plant is 80-140 mm. high, forming clumps with 10-20 leaves. Six to 8 tight, trumpet-shaped bristly sheaths with recurved, hairy margins envelop the 60-90 mm. long and 1 mm. diameter leaf stems. The sheaths are dark purple, which turns to dark brown and black with age. The stiff leaves are light green, elliptical to obovate, flat, conduplicate and narrowed abruptly into 2-5 mm. false petioles and are 35-50 mm. long, 18-22 mm. wide and 1.5-2 mm. thick.

Inflorescence and Flowers: The 50-80 mm. inflorescences arise from the apex of young and older leaf stems with 15-20 flowers in 2 opposite ranks on 0.75 mm. pedicels and ovaries. The flowers are 7 mm. long and 2 mm. wide, open wide in quick succession and bloom simultaneously. The floral bracts are 1 mm. and the spathe is 4 mm. long. The ovate, slightly concave, light green and light purple sepals are keeled. The dorsal is 3.2-4 0 mm. long and 1.7 mm. wide. The lateral sepals are 50% fused and are 3.6-4.5 mm. long by 2 mm. wide together. The round and slightly concave petals are 0.8 mm. long and wide and purple with increasing intensity from their green bases to their apices. The fleshy, round and slightly concave lip has 2 lobes at the base and is 0.9 mm. long, 0.8 mm. wide, 0.35 mm. thick and yellow and orange from the green base to the red apex, while the back is wine-red. The basal transverse callus is faintly divided and the light green and purple column is 0.5 mm. long, 0.6 mm. wide and 0.4 mm. thick. The anther is apical and the stigma bilobed.

Flowering Periods: All months of the year. All leaf bases may repeat flowering during the year. The flowers last for two weeks.

The genus
Platystele Schltr.

Platystele was named by **Rudolph Schlechter** in 1910 because of its broad column, platys and stele in Greek. Although closely related to *Pleurothallis*, it differs in having an entire lip without a callus and the petals being equal to the sepals in length. There are about 50 species through Central and South America, frequent and widespread from Mexico to Bolivia but infrequent in Brazil with only 2 species in the Organ Mountain Range.

Key: Two waxy pollinia with a viscidium, a lanciform lip lacking callosities on the plate. The petals are as long as the sepals and both are free. The column is short in comparison with its width; hooded at the apex, the stigma is bilobed and the foot is rudimentary.

Platystele oxyglossa (Schltr.)
Garay | *1103* (Plate VII)

Etymology: Oxy and *glosso* is a Greek compound word for a pointed lip.

Habitat: Epiphytic from section VI and VIII, at 1000-1200 M. asl, in humid original cloud forest, on moss-covered mid-tree branches. L 4-7; H 5-7; W 4-6.

Occurrence: Rare.

Plant: A 10-20 mm. high epiphyte forming clusters of 5 to 50 leaves and few, 0.5 mm. diameter, white roots. The leaf stems are 1-4 mm. long and 0.5 mm. diameter and enveloped by 3, tiny, green imbricated sheaths. The stiff leaves are 8-20 mm. long, 2.5-4.0 mm. wide and 1 mm. thick, light green, obovate, fleshy, conduplicate and finely narrowed to the base.

Inflorescence and Flowers: Up to two, 15-30 mm. inflorescences emerge from the apex of the most recent leaf stem with up to 3 alternate, consecutive flowers. The peduncle is very thin, 9 mm. long, the thin floral bracts are 0.5-1 0 mm., the pedicels 1-3.0 mm. and the ovary 0.5 mm. long. The flower opens wide and is 4 mm. long by 3.5 mm. The acuminate sepals are ovate, keeled, light green and light purple. All sepals are free and 2-4 mm. long and 0.8 mm. wide. The narrowly ovate, acuminate petals with finely serrate or hairy margins are 1.8 mm. long and 0.3 mm. wide, and the fleshy, lanceolate, red-purple lip is 1.2-2.0 mm. long, 0.5-0.7 mm. wide and 0.25 mm. thick with an acuminate apex

and a triangular or heart-shaped callosity at the base. The light yellow-green and reddish column has 2 wide wings, an indistinct apex and is approximately 0.7 mm. long and 0.9 mm. wide. The margins of the anther bed are undulate, almost entire and the column foot is rudimentary.

Flowering periods: January to April.

General: The plants and flowers are remarkably variable in size and the lip colour varies from yellow-green to orange, pure red or purple.

The genus
Myoxanthus Poepp. & Endl.

Poeppig and **Endlicher** originally established the genus in 1835 since when **Barbosa Rodrigues** named it *Chaetocephala* from the Latin for bristle-headed referring to the hairy stem sheaths, and **Cogniaux** placed the species in *Pleurothallis*. **Luer** resurrected *Myoxanthus* as these species differ from *Pleurothallis* because of their bristly sheaths and the single flowers with thickened petal tips. There are about 40 species found in Central and South America. Myoxos and *anthos* are Greek for dormouse and flower, referring obscurely to the hairy stems and anther cap.

Key: Two waxy pollinia with a viscidium. The sheaths of leaf stems have bristles and the leaf stems are without an annulus. The inflorescence is single-flowered and the peduncles are much shorter than the leaf. The petals are fleshy at the apex.

Myoxanthus punctatus (Barb. Rodr.)
Luer | *1119* (Plate VII)

Etymology: Punctatus means spotted which these stems and flowers are.

Habitat: Epiphytic and terrestrial from section V, VI, VII and VIII, at 900-1500 M. asl, in mountain cloud forest on trunks or thick branches, in good light. It also survives on the ground. L 5-7; H 5; W 3.

Occurrence: Common, occasionally forming large colonies.

Plant: The 100-250 mm. long by 1.5-2.5 mm. diameter leaf stems arise alternately at 10 mm. intervals from a 7 mm. diameter, creeping rhizome with 1.5-2.0 mm. diameter, orange and brown roots. The stems are totally enveloped by 5 sheaths of which the 3 lower ones have black, stiff hairs or trichomes. The narrowly lanceolate, green to light green, conduplicate leaves are 150 mm. long by 18 mm. wide, flat and leathery, with slightly back-folded margins.

Inflorescence and Flowers: One or 2 pendent flowers

arise from the apices of all leaf stems, hidden by loose papery sheaths. The flower hangs on a 20 mm. peduncle with 3 hairy bracts, a 2 mm. pedicel and a 10 mm. bristly spathe. After some months rest a new inflorescence may appear. All parts of the flower are shiny and smooth. The 7-veined sepals are ovate, convex and the margins slightly recurved. The bases are greenish, the mid-sections beige or maroon and the apices yellow-green. There are dark brown marks on the lower 2/3. The dorsal sepal is 14 mm. long and 7 mm. wide, while the laterals together measure 13 mm. by 7.5 mm. The petals are 13 mm. long and the margins of the mid-section are recurved to form a tube with elevated, dark purple dots and purple dotted longitudinal veins on the lower widened and flat part of the petals, while the spindle-shaped apex is brown or greenish. The wine-red lip has two whitish and round ear-like appendages at its green base. The mid-section has undulate margins, which are rolled backwards forming a round club-like lip tip. The mid-section is dominated by 2 orange callosities. The lip is 8 mm. long by 3.8 mm. wide and the short, sturdy column is pale yellow to green with small red spots on the back and 4 mm. long, 3 mm. wide with rounded wings and indented apex. The column foot is 2 mm. long with 2 thick callosities at the end.

Flowering Periods: The main periods are April to June and September to October and the flowers last for 14 days.

Myoxanthus aff. lonchophyllum (Pabst) Luer | *1120* (Plate VII)

Etymology: Loncho-phyllum is a Greek compound word meaning lance-like leaf.

Habitat: Epiphytic or terrestrial from section V, at 700-800 M. asl, in original forest on trunks, thick branches and shady rocks. L 3-5; H 5; W 3:

Occurrence: Occasional.

Plant: A 150-250 mm. high plant forming clusters of 20-50 leaves. The 80-150 mm. long, 1.5-2.5 mm. diameter leaf stems arise alternately at 10 mm. intervals from a 5-7 mm. diameter, creeping rhizome. The stems are enveloped by 5 sheaths of which the 3 lower, have black or dark brown to purple stiff hairs or trichomes. The narrowly lanceolate, olive-green, conduplicate leaves are 120-150 mm. long by 15-18 mm. wide and are flat, leathery with slightly back-folded margins. The 1.5-2.0 mm. diameter roots are orange or brown.

Inflorescence and Flowers: One or 2 flowers emerge from the apices of all leaf stems, hidden by their loose papery sheaths. The flower hangs on a 20 mm. peduncle with 3 hairy bracts, a 2 mm. pedicel and a 10 mm. bristly spathe. After some months rest a new inflorescence may appear. All sepals and petals are shiny and smooth. The 7-veined sepals are ovate and convex with slightly recurved margins, light green bases, the midsections beige or maroon and the apices yellow-green. The dorsal sepal is 10-12 mm. long and 5-6 mm. wide, while the laterals are 10 mm. by 4-5 mm. The petals are 9-10 mm. long and the mid-section margins are recurved to form a tube. The spindle-shaped apex is brown or greenish. The dark wine-red lip is 6-7 mm. long by 2.5-3 mm. wide and has a short-haired mid-section and apex. The margins of the apical half are rolled backwards forming a round club-like lip-tip. Parallel to the indistinct side-lobes are 2 longitudinal calli. The short, sturdy column is light yellow to light green and is 4 mm. long by 3 mm. wide with rounded wings and a gnawed apex.

Flowering Periods: April to June and September to October. The flowers last for 14 days.

Similar Species: See *M. punctatus 1119.*

The Genus *Pleurothallis* R. Br.

Both **Frederic Hoehne** and **Pabst&Dungs** estimate that Brazil is host to over 350 species of this genus with 136 in the State of Rio de Janeiro. That is to say they have examined 136 herbarium specimens collected in the state of Rio. We have found 124 species, excluding *Cryptophoranthus* which **Luer** has included in *Pleurothallis*. In addition, for 31 of the 124 species in the Organ Mountain Range, **Pabst&Dungs** only examined specimens in herbaria from other states. Finally, we have not found 43 species that **Pabst&Dungs** attribute to the State of Rio. What does all this mean?

It is difficult to say and serious speculation is involved. The missing 43 species represent 31% of **Pabst&Dungs** total and we have not found them after seven years of searching. The first explanation that springs to mind is that most of these species came from the anticline forests that are now virtually extinct after the holocaust of the coffee cycle. Secondly one must assume, that certainly up until 1975, the year when *Orchidaceae Brasiliensis* was published, research into the Pleurothallidinae was limited which would explain why we have found 31 species not attributed to the State of Rio by **Pabst&Dungs** and a number of new species and varieties. This latter factor more or less eliminates the third possibility, that we are poor field researchers and have overlooked 43 species.

Almost all the *Pleurothallis* species we have found, with the dramatic exception of *P. teres 1195* and to a lesser extent, *P. leptotifolia 1197*, are adapted to living in permanently humid forest with high light to deep shade. *P. teres* is the arch rebel and has rejected all laws governing the Pleurothallidinae by being a succulent and living on exposed rock faces. There is almost a complete absence of the terete leaf form which is relatively common in *Octomeria* but lacking in *Stelis* and most *Pleurothallis* species. This leads us to speculate that both *Pleurothallis* and *Stelis* experienced their main speciation after the climax of the most recent glaciation; a mere 25 thousand years ago.

Orchids from the genus *Pleurothallis* are amongst the smallest of all orchidaceous plants while being one of the largest genera. Surprisingly little is known about them, largely because up to the present they have had very little commercial value and commercial value is the key factor to the interest in almost all plants. **Watson&Chapman** in Orchids: *Their Culture and Management* (1903 edition) describes them as 'botanically interesting inconspicuous orchids'. Botanically interesting means, of course, of interest to botanists only. And botanists in general tend to concentrate on plants of commercial value, quite understandably, as the study of such plants will raise funds for research grants. This, in turn, is why the majority of the rural population do not even consider the Pleurothallidinae as orchids but, at best, small parasites. This is a serious benefit to the conservationist and in today's terms, the botanist. In the first place, no one collects Pleurothallidinae from the forest so if you are studying orchids in original forest you may be reasonably sure that the Pleurothallidinae found are the same and in the same places as they were 500 years ago when the European colonists arrived. Secondly, from the point of view of the restoration ecologist examining regrowing forest: if the trunks and branches do not hold a population of Pleurothallidinae then the constant high humidity layer and the depth of humus and organic soils are not yet adequate for the more noble forest trees to enter naturally. This applies to areas between 900-1600 M. asl on the scarp slopes of the Organ Mountain Range and assumes a Pleurothallidinae seed bank in neighbouring original forest. Most species in most genera of the Pleurothallidinae have small flowers and identification is very difficult. A hand lens is normally insufficient and a good microscope is an absolute necessity. To a very great extent one must first use the morphological approach to categorize the plant into a general group and only then concentrate on the flower. This is why we have given such detailed descriptions of the plants, the inflorescences, the flowers and the habitats as well as providing detailed water colours instead of photographs. Every one of these water colours is based on plants found in the Organ Mountain Range as are the plants' descriptions. As previously mentioned, research material is mostly sparse, somewhat historic and ancient. Many of the plants have synonyms in double digits which again confuses the issue and many of the

species are variable. As such, we apologise in advance for any errors in identification that our readers may perchance discover.

Etymology: From the Greek *pleuron*, a rib and thallus a branch, referring to the tufted, rib-like leaf stems.

Pleurothallis is a genus with more than 350 brazilian species and it differs from the genus *Octomeria*, in the number of pollinia.

Key: Two waxy pollinia with a viscidium. The dorsal sepal is not fused to the lateral sepals – **Carlyle Luer** changed this to reintroduce some genera into *Pleurothallis*; the petals are distinctly smaller than the sepals, longer than wide and not callused at the apex; the column slender in comparison with its width; the anther and stigma ventral and the column has a short or well-developed foot; the lip is articulate with the column foot.

Pleurothallis aff. barbulata Lindl. | *1106b* (Plate VII)

Etymology: Latin for with a small beard referring to the bearded lip.

Habitat: Epiphytic from section II, at 900-1000 M. asl, in humid original forest, on lower trunks and branches of trees. L 5; H 5-7; W 3.

Occurrence: Rare.

Plant: This epiphyte is 20-25 mm. high and forms clusters with 10 to 40 leaves and whitish 0.5 mm. diameter dense roots. The leaf stems are 1.5-2.5 mm. long, 0.6 mm. diameter and enveloped by 3 sheaths of which the lowest is very short and the apical is up to 4 mm. long. The dark olive-green, obovate, conduplicate leaves are attenuated towards their bases forming 1.5 mm. false petioles and are 12-18 mm. long, 4-5.5 mm. wide and 1.5 mm. thick.

Inflorescence and Flowers: Up to 2, 10-20 mm. inflorescences emerge from the apices of the younger leaf stems and develop their first wide open flower at leaf height followed by up to 6 consecutive flowers on 2.5 mm. racemes. The spathe is 1.5 mm., the pedicels are up to 3 mm., the bracts are 2 mm. and the ovary is 1 mm. long. The sepals are ovate with a dark wine base and an olive-green apex. The dorsal sepal is 4.3 mm. long and 1.6 mm. wide and the laterals are 80% fused, recurved by 90° and 3.8 mm. long by 2.9 mm. wide together. The lanceolate petals have hairy margins and are asymmetrically coloured; one side yellow and the other dark wine and are 3.1 mm. long and 1.3 mm. wide. The dark wine lip has hairy margins, 2 small appendages at the base and 2 longitudinal callosities which join at the fleshy apex and enclose a lighter coloured depression. The lip is 2.7 mm. long by 0.9 mm. wide. The whitish to dark wine, faintly curved column is 1.8 mm. long, 1.5 mm. wide and has lacerate wings and apex.

Flowering Periods: November to February. The flowers last for 2 weeks and the inflorescence for 2 months.

Similar Species: See *P. gehrtii - 1158.*

Geral: Collected by a friend in the habitat as described above.

Pleurothallis parvifolia
Lindl. | *1108* (Plate VII)

Etymology: *Parvi-folia* is a Latin compound word for small leaf.

Habitat: Epiphytic from section VI, at 1100-1350 M. asl, in humid original cloud forest on moss-covered branches in good light. L 6; H 8; W 3.

Occurrence: Occasional.

Plant: The plant is under 20 mm. high and forms small clusters of 5 to 15 leaves with 5-10 mm. long and 0.3 mm diameter, round leaf stems, enveloped by 3 funnel-shaped sheaths. The shiny, obovate leaves are very dark bottle-green, indistinctly conduplicate, attenuated to their bases to form 1 mm. false petioles. The leaves are 6-12 mm. long, 4-7 mm. broad and 1-1.5 mm. thick. The undersides are usually purple. The few white or green roots are 1 mm. diameter

Inflorescence and Flowers: The 20-35 mm. inflorescence arises from the apex of the leaf stem. Up to 6 flowers open consecutively, one at a time, out of a basket-shaped sheath on 5 mm. pedicels. The flower has a pronounced chin. The basic colour is pale green or cream. The ovate dorsal sepal is 7 mm. by 3.2 mm. with 3 purple longitudinal veins. The indistinctly keeled laterals are 95% fused and are 5.5 mm. long by 4.6 mm. wide together with one purple longitudinal vein. The spathulate petals have 3 purple longitudinal veins, an acute tip and are 4.5 mm. long by 2.3 mm. wide. The lip is arrow-head-shaped with round corners and has undulate margins on the apical half. The light yellowish column is 2.3 mm. long and with wings spread, 1.4 mm. wide.

Flowering Periods: January to May. The flowers last for 14 days.

Pleurothallis scabripes
Lindl. | *1114* (Plate VII)

Etymology: *Scaber* is Latin for rough and probably alludes to the rough column.

Habitat: Epiphytic from section VI at 900-1200 M. asl, in original humid forest, on high tree branches. L 4-7; H 5; W 5.

Occurrence: Common.

Plant: A 50-130 mm. high epiphyte forming loose groups of 5-50 leaves. The leaf stems emerge from the 1.0-2.0 mm. diameter rhizome and measure 25-70 mm. long, are round and enveloped for 80% of their length by 3 sheaths covered with purple spots, pointed warts and hairs. The leaves are linear, slightly recurved, olive-green and measure 60-90 mm. by 5-6 mm. wide and 1.5 mm. thick. The undersides turn purple in strong light.

Inflorescence and Flowers: One to 3 flowers emerge consecutively on an inflorescence from the leaf stem apex. The ovary and the stem sheaths have purple spots and hairs. The under surfaces of the sepals are granular or hairy. The dorsal sepal is lanceolate and 8.8 mm. long by 2.4 mm. wide. All tepals are striped purple over a base of pale yellowish-green. There are 3 stripes on the dorsal sepal and the petals while there are 5 on the laterals which are 95% fused and together measure 7.3 mm. long by 4.3 mm. The oblanceolate petals are 3.8 mm. long by 1.3 mm. The rectangular lip is purple and 4.0 mm. long by 1.6 mm. with two lateral lobes and indistinct callosities. The internal sides of the callosities are granular. The column is purple-yellow and 3 mm. long by 1.5 mm. with open wings.

Flowering Periods: October to May with flowers open for ten days.

Pleurothallis collina
Cogn. | *1125* (Plate VII)

Etymology: Not known.

Habitat: Epiphytic from section VI, at 1100-1300 M. asl, in very humid original cloud forest on trunks and lower tree branches. L 4; H 8; W 2.

Occurrence: Rare.

Plant: This 100-130 mm. high epiphyte forms clusters with 7 to 10 leaves and white, 0.7 mm. diameter roots. The round leaf stems are 50-75 mm. long, 0.8 mm. diameter, arising alternately from a short 1 mm. diameter rhizome, enveloped by 5 to 7 tight trumpet-shaped sheaths with short stiff hairs on recurved margins. The longitudinal veins of the sheaths are purple when young, brown and later black with age. The leaves are 40-55 mm. by 10 mm., and 1-1.5 mm. thick, almost flat, lanceolate, leathery, slightly twisted at their bases, light green, and conduplicate.

Inflorescence and Flowers: The 100 mm. inflorescences arise from the apex of the younger leaf stems and bear the first flower at half leaf height followed by 5 to 8 consecutive, alternate flowers on 4-6 mm. pedicels and up to 5 flowers bloom at once. The yellow sepals are finely acuminate, with green keels which turn cinnamon with age. The dorsal sepal is 15-20 mm. long by 3.5-5.0

mm. wide, while the laterals are 40% fused and together measure 15-18 mm. long by 3.5-5.0 mm. wide. The spathulate yellow petals have pointed, blunt apices. The elliptical yellow lip is 5 mm. long by 3 mm. wide with 2 erect, narrowly sickle-shaped side-lobes at its base. Two light red and green longitudinal lamellae enclose a light yellow-green depression, extending from the base to near the rounded apex. The lip is recurved by 90° at its mid-section and the margins of the main lobe are also recurved. The narrow column is light green and 2 mm. long by 0.6 mm. wide with its faint wings spread.

Flowering Periods: September to December. The flowers last for 14 days; the inflorescence for a month.

Similar Species: See *P. quartzicola - 1126*.

Pleurothallis quartzicola
Cogn. | *1126* (Plate VII)

Etymology: Not known.

Habitat: Epiphytic from sections VI and VIII, at 950-1400 M. asl, in humid original cloud forest on mid-tree branches and trunks. L 4; H 8; W 3.

Occurrence: Common.

Plant: This 80-110 mm. high epiphyte forms clusters with 5 to 10 leaves and white, 0.6 mm. diameter roots. The round leaf stems are 40-60 mm. long and 0.8 mm. in diameter, arising alternately from a short, 1 mm. diameter rhizome and are enveloped by 4 to 8 tight trumpet-shaped sheaths with recurved margins and short stiff hairs. The longitudinal veins of the sheaths are purple when young, brown and later black with age. The almost flat leaves are lanceolate, leathery, dark green, slightly resupinate at their bases, conduplicate and 40-55 mm. long by 10 mm. wide and 1-1.5 mm. thick.

Inflorescence and Flowers: The 150-250 mm. inflorescences arise from the apices of younger leaf stems with the first flower at half leaf height followed over 6 months by up to 30 consecutive, alternate flowers on 4-5 mm. pedicels and up to 2 bloom at once. The yellow sepals are finely acuminate with cinnamon keels which turn maroon with age. The dorsal sepal is 10-16 mm. long by 3-4 mm. wide, while the laterals are 40% fused and together measure 15-18 mm. long by 3.5-5.0 mm. wide. The obovate petals are yellow, with pointed, blunt apices and are 3.5 mm. long and 2 mm. wide. The elliptical, yellow lip is 4.5 mm. long and 2.5 mm. wide with 2 erect, narrowly sickle-shaped side-lobes at its base. Two light red and yellow longitudinal lamellae enclose a light yellow to ochre depression, extending from the base to close to the rounded apex. The lip is 90° recurved at its mid-section and the reddish margins of the main lobe are recurved. The narrow column is light green and 1.7 mm. long and 0.5 mm. wide with the

faint wings spread.

Flowering Periods: September to March. The flowers last up to 2 weeks; the inflorescence for half a year.

Pleurothallis curti-bradei
Pabst | *1130* (Plate VIII)

Etymology: Named after the German systematic botanist **A. Curt Brade** (1881-1971), who worked in the National Museum and the Rio de Janiro Botanic Garden.

Habitat: Epiphytic from section VI, at 900-1600 M. asl, in humid original forest on lianas and low and mid-tree branches. L 3-5 ; H 7-9; W 5.

Occurrence: Common.

Plant: This 40 mm. high epiphyte forms clusters of 10-50 dark green leaves with white, 0.4 mm. diameter roots. The leaf stems are 15-22 mm. long and 0.3 mm. diameter and enveloped by 3 trumpet-shaped sheaths with short, stiff hairs and recurved margins. The sheaths have longitudinal veins, purple when young and brown with age. The stiff leaves are spathulate, faintly conduplicate, attenuated to their bases to form false petioles, and are 10-15 mm. long by 4-6 mm. wide and 1.5 mm. thick.

Inflorescence and Flowers: Inflorescences of 3 to 6 flowers, which may bloom together, emerge from the apices of young and old leaf stems and grow in a zig-zag pattern to 30-40 mm., with the first flower above leaf height. All flower parts are yellow-green and keeled. The dorsal sepal is 4.3 mm. by 1.5 mm. and the laterals are 90% fused and together are 4 mm. long by 2.2 mm. wide. The obovate petals are slightly bent downwards and outwards and are 3 mm. long by 1.1 mm. wide. The lip is 2.7 mm. long by 1.2 mm. wide with 2 indistinct side lobes and 2 faint longitudinal callosities parallel to them. The round apex is faintly warty and the base is undulate. The column is 1.3 mm. long and wide.

Flowering Periods: November to January. The flowers last for one to two weeks.

Pleurothallis mattinhensis
Hoehne | *1131* (Plate VIII)

Etymology: Refers to a municipality on the Parana coast.

Habitat: Epiphytic from section VI, at 900-1300 M. asl, on the lower half of humid original cloud forest trees and on lianas. L 3-5; H 7-9; W 3-5.

Occurrence: Occasional.

Plant: This 35-45 mm high epiphyte forms clusters of 10-30 slightly conduplicate light green leaves and white, 0.4 mm. diameter roots. Two trumpet-shaped sheaths with short, stiff hairs and recurved margins envelop the leaf stems, which are 15-30 mm. long and 0.2-0.4 mm. diameter. The third and apical sheaths have a normal tube-shape. The obovate leaves are attenuated to their bases to form false petioles and are 12-15 mm. long, 6-7 mm. wide and 0.5-1.0 mm. thick.

Inflorescence and Flowers: The erect inflorescences arise from the apex of young and older leaf stems and grow in a zig-zag pattern to 25-40 mm. producing 3 to 4 flowers of which 3 may bloom at once. The lowest flower opens at leaf height. The pedicel and the ovary are 5-6 mm. long together. All parts of the flower are keeled and yellow- or light green, occasionally with a touch of purple. The dorsal sepal is 6.5 mm. by 2.0 mm. The laterals are 90% fused and are 6.0 mm. long and 2.6 mm. wide. The narrowly elliptic petals are slightly bent downwards and outwards and are 4.7 mm. long by 1.5 mm. wide and the apical section is faintly granular with crenate margins. The lip is 3 mm. long by 1.6 mm. wide with indistinct side lobes and the surface of the base is undulate while the rounded or cut apex is warty. The light green column is 2.0 mm. long and 1.2 mm. wide and the wings are extended above the height of the apex and the base is purple.

Flowering Periods: September to December. The flowers last for one to two weeks.

Pleurothallis miragliae
J. E. Leite | *1132* (Plate VIII)

Etymology: Not known.

Habitat: Epiphytic from section VI, at 1000-1300 M. asl, on trunks and thicker branches in humid original cloud forest on mountain crests with frequent wind and fog. L 3-5; H 8-9; W 7.

Occurrence: Occasional.

Plant: This epiphyte is 25-35 mm. high and forms clusters of 5 to 30 leaves with whitish, 0.3 mm. diameter roots. Two trumpet-shaped sheaths with short stiff hairs and recurved margins envelop the leaf stems, which are 10-18 mm. long and 0.3 mm. diameter. The apical sheaths are tube-shaped with slightly unrolled margins. The stiff leaves are 15 mm. long, 5-6 mm. wide and 1.5 mm. thick, lanceolate, dark green and conduplicate with distinctively purple-dotted backs.

Inflorescence and Flowers: The 60 mm. bent inflorescences arise from the apex of young and older leaf stems with up to 15 alternate and consecutive flowers. The first flower develops above leaf height. All sepals have distinct keels. The ovate dorsal sepal is light rose to light green and yellow-green at the apex and 5.2 mm. long by 1.9 mm. wide, while the slightly granular laterals are 80% fused and 5.6 mm. long by 3.8 mm. wide, pink at the bases deepening to dark wine

towards their apices. The lanceolate petals are light rose and 2.8 mm. long by 0.8 mm. wide with slightly serrate apical margins. The dark wine-red lip is 4 mm. long by 1.3 mm. wide and has recurved margins. Two rectangular lamellae enclose a depression at the lip's mid-section and the rounded apex is crater-shaped. The green column has a purple base and is 3 mm. long and 1.3 mm. wide with spread wings. The wings and apex are faintly gnawed.

Flowering Periods: All months of the year. The flowers last for a week and the inflorescence up to 5 months.

Pleurothallis punctatifolia
(Barb. Rodr.) Pabst | *1134*
(Plate VIII)

Etymology: *Punctati-folia* is Latin for spotted leaves referring to the purple spots on the leaf backs.

Habitat: Epiphytic from section VI and VIII, at 900-950 M. asl, in humid original mountain forest on trunks and low-tree branches. L 5; H 8; W 5:

Occurrence: Rare.

Plant: This epiphyte is 25-50 mm. high and forms clusters with 10-25 leaves and 0.4 mm. diameter whitish roots. Three tight, trumpet-shaped sheaths usually with short, stiff hairs and recurved margins envelop the leaf stem, which is 15-30 mm. long and 0.3 mm diameter. The almost round leaves are dark green, fleshy, faintly conduplicate, attenuated abruptly to their bases to form 3 mm. false petioles with distinctly purple-dotted undersides and measure 15-18 mm. long, 9-12 mm. wide and 2-3.5 mm. thick. Older leaves have yellowish margins.

Inflorescence and Flowers: The inflorescences arise from the apex of young and older leaf stems and grow in very short zigzags to half leaf height with 4 to 6 alternate, consecutive flowers, 2 of which bloom at once. All the sepals are ovate and faintly keeled. The dorsal sepal is light yellow-green with a rose mid-section and is 5 mm. long by 1.8 mm. wide. The partly granular lateral sepals are 80% fused and together are 5 mm. long by 4 mm. wide and pink at the bases deepening to dark wine towards the tips. The lanceolate petals are acute, translucent yellow with faint traces of purple and are 3 mm. long by 0.8 mm. wide and the margins of the apical halves are serrate. The lip is wine-red, 2.5 mm. long by 1.1 mm. wide and has recurved margins, while 2 lamina enclose a depression at the mid-section. The rounded apex may also be olive-green. The yellow-green column is 2.5 mm. long and 1.2 mm. wide, with finely gnawed wings and apex and a purple base.

Flowering Periods: Flowers all months of the year and the flowers last for 10-15 days.

Pleurothallis quadridentata (Barb. Rodr.) Cogn. | *1135*

(Plate VIII)

Etymology: Latin for with four teeth referring to the column.

Habitat: Epiphytic from sections V, VI and VIII, at 900-1300 M. asl, in humid original cloud forest on lower branches and lianas. L 3; H 8; W 2.

Occurrence: Common.

Plant: This small, 20-30 mm. high epiphyte forms clusters of 5 to 20 leaves and whitish, 0.4 mm. diameter roots. Three trumpet-shaped sheaths with short, stiff hairs and recurved margins envelop the leaf stems, which are 10-17 mm. long and 0.2 mm. diameter. The elliptic to lanceolate leaves are light green, faintly conduplicate, 10-17 mm. long, 4-6 mm. wide and 1-1.5 mm. thick and attenuated to their bases to form 1.5 mm. false petioles.

Inflorescence and Flowers: The 50-60 mm. inflorescences arise from the apex of young and older leaf stems with up to 15 alternate and consecutive flowers and 2 flowers may be open at once. The keeled sepals are light green and ovate. The dorsal is 5.2 mm. long and 2.3 mm. wide, while the laterals are 90% fused and together are 4.4 mm. long by 2.4 mm. wide. The petals are spathulate with rounded apices, light yellow-green and 2.5 mm. by 1.1 mm. wide. The yellow-green lip has 2 wide but short side-lobes and a wine-red, tuberculate main lobe with serrate margins and is 2.5 mm. long by 1.3 mm. wide. The slender column is light green, 1.8 mm. long, 0.8 mm. wide with 2 narrow, short wings and a narrow, gnawed tip.

Flowering Periods: All months of the year. The flowers last for 15-25 days and the inflorescence up to 6 months.

General: We have also found plants with normal *Pleurothallis*-type sheaths.

Pleurothallis bradei Schltr. | *1137* (Plate VIII)

Etymology: Named in honour of the German Botanist **A. Brade**.

Habitat: Epiphytic from section VI, at 800-900 M. asl, in humid but open original mountain forest, on trunks and mid-tree branches. L5; H 8-9; W 5.

Occurrence: Rare.

Plant: This epiphyte is 25-45 mm high and forms clusters of 5 to 25 leaves with light green, 0.4 mm. diameter roots. The leaf stems are 10-25 mm. long and 0.2-0.4 mm. diameter, covered by 3 tight, trumpet-

shaped green sheaths with recurved hairy margins and purple veins and margins. The broadly lanceolate leaves are dark olive-green, stiff, erect, indistinctly conduplicate, attenuated to their bases to form 2 mm. false petioles and are 15-18 mm. long, 8-10 mm. wide and 1.2 mm. thick.

Inflorescence and Flowers: The 35 mm., erect inflorescences arise from the apex of young and older leaf stems and produce the first flower at leaf height, followed by 3 to 6 alternate and consecutive flowers at 4 mm. intervals, 2 of which are open at once. All sepals are ovate, keeled, light green to golden yellow and have faintly serrated apical margins. The dorsal sepal is 5.5 mm. long by 2.6 mm. wide and the laterals are 80% fused and 4.9 mm. long by 3 mm. wide together. The spathulate petals are pointed, with yellow or wine-red central veins on a light yellow background and are 2.6 mm. long and 1.2 mm. wide. The trilobed lip is 2.3 mm. long by 1.6 mm. wide with faintly crenate margins and is yellow-green with a dark purple base and apex. The surface of the fleshy base is rough, the side-lobes are faintly hook-shaped and the keel and the 2 longitudinal callosities are crenate. The light green column is slightly curved with a hairy light purple base and is 1.5 mm. long by 0.7 mm. wide.

Flowering Periods: January to March and the flowers last for 8 to 12 days, the inflorescence 2 months.

Pleurothallis carinifera (Barb. Rodr.) Cogn. | *1138* (Plate VIII)

Etymology: Latin for bearing keels referring to the petals, sepals and lip.

Habitat: Epiphytic from section VI and VIII, at 1100-1300 M. asl, in humid original cloud forest along river banks or swampy forest, on moss-covered trunks and mid- to low-tree branches. L 3-5; H 7-9; W 3.

Occurrence: Rare.

Plant: This 30-50 mm high epiphyte forms clusters with 5 to 20 leaves and an intense root system of 0.5 mm. diameter whitish roots. Three trumpet-shaped sheaths with short, stiff hairs and recurved margins envelop the leaf stems, which are 10-20 mm. long and 0.5 mm. diameter. The leaves are lanceolate, light green to green, faintly conduplicate, attenuated to their bases to form short false petioles and are 20-25 mm. long, 8-10 mm. wide and 0.7 mm. thick.

Inflorescence and Flowers: The 50-60 mm. inflorescences arise from the apex of young and older leaf stems with up to 7 alternate and consecutive flowers on a zig-zag raceme. Up to 5 flowers may bloom at once. The pedicels are 6-8 mm. long and the ovary 1.3 mm. All parts of the flower except the column are distinctly keeled and white with small purple spots. The drop-shaped sepals have long tapered points and the apices are recurved by 90°. The apices of the sepals and petals are yellowish and not dotted. The dorsal sepal is 11 mm. long by 3.3 mm. wide, while the laterals are 80% fused and together are 11 mm. long by 3.8 mm. wide. The petals are 4.8 mm. long by 0.7 mm. wide with thread-like elongated apices, 3 mm. long by 0.2 mm. The anchor-shaped, trilobed lip is 2.8 mm. long and 1.7 mm. wide, with almost triangular side lobes and 2 longitudinal, crenate callosities, and ends in a broader, fan-shaped, rounded, warty apex. The column is rose, slightly curved, with purple flanks, a yellow-green mid-section and is 2 mm. long and 1.2 mm. wide with the 2 spread acuminate wings.

Flowering Periods: November to December. The flowers last for three weeks.

Pleurothallis podoglossa
Hoehne | *1139* (Plate VIII)

Etymology: Podo and *glossa* are Greek for foot-like tongue, referring to the lip shape.

Habitat: Epiphytic from section VI, at 800-900 M. asl, in humid, open original mountain cloud forest, on trunks and mid-tree branches. L5; H 8-9; W 5.

Occurrence: Rare.

Plant: This epiphyte is 30-50 mm. high and forms clusters of 5 to 30 leaves and whitish 0.5 mm. diameter roots. The leaf stems are enveloped by 3 tight trumpet-shaped sheaths with hairy recurved margins and are 15-25 mm. long and 0.4 mm. diameter. The dark to bluish-green leaves are 18-25 mm. long, 6-8 mm. wide and 1.5 mm. thick, stiff, indistinctly conduplicate, attenuated to their bases to form short false petioles and occasionally have the backs dotted with fine purple marks.

Inflorescence and Flowers: The 60-100 mm. inflorescences arise from the apex of young and older leaf stems with the first flower at leaf height followed by 10-15 alternate, consecutive flowers at 4 mm. intervals. Up to 3 flowers are open at once and the longer inflorescences bend. All sepals are ovate and keeled, with white hairs on the inner, apical halves and are light green to yellow but wine-red along the central veins broadening towards their apices. The dorsal sepal is 5.1 mm. long by 2.2 mm. wide and the 85% fused laterals are 4.6 mm. long by 2.6 mm. wide together and very pale green at their common centre. The spathulate petals are 2.7 mm. long and 0.9 mm. wide with brown, rounded apices with an acute point, while the central vein is wine-red on a light yellow background. The trilobed lip is 1.8 mm. long by 1.4 mm. wide, dark wine-red with yellow margins with a V-shaped callus surrounded by white hairs at its base, a serrated keel and 2 longitudinal callosities. There is a white hairy U-shaped line on the main lobe parallel to the gnawed margins of the rounded apex. The wine-red

column is slightly curved and 1.2 mm. long by 0.5 mm. wide.

Flowering Periods: February to April. The flowers last for 8 to 12 days and the inflorescence for two to three months.

Pleurothallis sordida
Krzl. | *1140* (Plate VIII)

Etymology: Sordida is Latin for dirty-looking, and refers to the colour of the sepals and petals.

Habitat: Epiphytic from sections V, VI and VIII, at 900-1600 M. asl, in humid original cloud forest, on lianas and mid- to lower branches. L 2-5; H 7-8; W 4-7.

Occurrence: Common.

Plant: This 20-35 mm high epiphyte forms clusters with 5 to 40 leaves and whitish 0.4 mm. diameter roots. The leaf stems are 10-25 mm. long and 0.5 mm. diameter, enveloped by 3 tight trumpet-shaped stiffly hairy sheaths with recurved margins. The faintly conduplicate leaves are broadly lanceolate, stiff, dark bluish-green, speckled purple on the sea-green undersides and 10-12 mm. long, 5 mm. wide and 1.5 mm. thick.

Inflorescence and Flowers: The 60-100 mm. inflorescences emerge from the apex of young and older leaf stems with the first flower at leaf height followed by 20-30 consecutive flowers at 3-4 mm. intervals, bending with length and one flower blooming at a time. The ovate sepals are keeled with yellow hairs on the inner, apical halves. The concave dorsal sepal is 3.3 mm. long and 1.5 mm. wide, yellow with a wine-red base and centre, while the veins and the margins are dark wine-red. The laterals are 80% fused and together measure 3 mm. long by 2.4 mm. wide with light yellow centres, dirty yellow apices and the rest is dark wine-red. The lanceolate, acute petals are 2.1 mm. long by 0.7 mm. wide and dark wine-red at their bases, central veins and apices, while the background, including the hairs at the margins, is yellow. There is a shiny, yellow-green blister on the outsides of the petals. The trilobed, yellow lip is dark wine-red along the central vein and 1.5 mm. long, 0.9 mm. wide and the narrow main lobe widens to a rounded, warty apex. The slightly curved column is light green, rose towards the base, and is 1.5 mm. long and 0.9 mm. wide.

Flowering Periods: All months of the year; the flowers last for 12 days and the inflorescence for up to 9 months.

Pleurothallis calcarata
Cogn. | *1141* (Plate IX)

Etymology: Calcarata refers to the spur.

Habitat: Epiphytic from sections V, VI and VIII, at 900-

1500 M. asl, in humid original cloud forest, on lianas, trunks and branches of the lower two thirds of forest trees. L 3-5; H 5-7; W 5.

Occurrence: Common.

Plant: The plant is 45 mm. high and forms clumps with 4 to 20 leaves. The 5-10 mm. long and 0.3 mm. diameter leaf stems are enveloped by 3 sheaths of which the lowest is very short and the apical is loose. The light green, linear, conduplicate leaves are attenuated towards their bases forming false petioles and are 30-40 mm. long, 2.5-3 mm. wide and 1 mm. thick.

Inflorescence and Flowers: The 40-60 mm. inflorescences arise from the apex of young and older leaf stems. Three to 4, long-lasting flowers with very pronounced chins, bloom consecutively and alternately, 1 flower at a time on the 20 mm. raceme. The sepals are narrowly lanceolate, concave, light yellow to light green and keeled. The dorsal is 10 mm. long and 2.4 mm. wide and the laterals are 95% fused and together measure 9.5 mm. long by 3.3 mm. wide. The lanceolate, light yellow petals are 6 mm. long and 1.5 mm. wide with acuminate apices, serrate margins and 3 orange-yellow longitudinal veins. The light green, strap-shaped lip is 5 mm. long and 1.4 mm. wide, with 2 erect but faint, faintly sawed side-lobes. The surface along the central vein is undulate near the orange base and tuberculate at the apex. The light green, slightly curved column is 3.3 mm. long and 1.6 mm. wide. The margins of the wings and the apex are faintly gnawed.

Flowering Periods: September to April. The flowers last for 2 to 3 weeks and the inflorescence for 2 months.

Pleurothallis conspersa
Hoehne | *1142* (Plate IX)

Etymology: *Conspersus* is Latin for sprinkled, perhaps referring to the numerous tiny warts or hairs sprinkled on the flower parts.

Habitat: Epiphytic from section V, at 900-1000 M. asl, in humid original gallery forest, on lower trunks and branches of the forest trees. L 3; H 5-7; W 3.

Occurrence: Rare.

Plant: The plant is 25 mm. high and forms clusters with 20 to 40 leaves and 0.5 mm. diameter light green, densely packed roots. The leaf stems are 4.5-7.5 mm. long, 0.4 mm. diameter and enveloped by 3 sheaths of which the lowest is very short and the apical is up to 9 mm. long. The light green, obovate, conduplicate leaves are attenuated towards their bases forming 1.5 mm. false petioles, and are 15-20 mm. long, 4-5.5 mm. wide and 0.8 mm. thick.

Inflorescence and Flowers: The 30-50 mm. inflorescences emerge from the apices of the younger leaf stems with

their first flower at leaf height followed by 1 to 3 consecutive flowers. The spathe is 1.3 mm., the bracts are 1.5 mm. and the ovary 1 mm. long. The flower has a pleasant scent and a very pronounced chin. The sepals, petals and lip are keeled, tongue-shaped and wine-red with a lighter coloured base and darker veins and margins. The sepals are concave with hairy apical margins. The dorsal sepal is 5.2 mm. long and 2.2 mm. wide and the laterals are 95% fused and together are 5.0 mm. long by 3.4 mm. wide and hairy on their inner apical section. The petals are 3.5 mm. long, 1.2 mm. wide with finely toothed margins at the rounded warty apices. The lip is 4 mm. long and 1.7 mm. wide, with 2 erect hook-shaped, faintly sawed side-lobes. The margins of the longitudinal callosities and the warty apex are toothed. The whitish and light wine, faintly curved column is 2.5 mm. long, 1.2 mm. wide with dark wine-red flanks and the 2 mm. long foot. The margins of the faint wings and the apex are gnawed.

Flowering Periods: November to April. The flowers last for 2 weeks and the inflorescence for 1 or 2 months.

Similar Species: This plant looks similar to *P. colorata*, *P. gehrtii* and *P. rubro-lineolata* but when in flower it is easily identified.

Pleurothallis avenacea
Ames | *1146* (Plate IX)

Synonym: P. mathildae Brade.

Etymology: *Avenacea* means oat-like and refers to the inflorescence form.

Habitat: Epiphytic from section V, VI and VIII, at 900-1600 M. asl, in humid original cloud forest, on lianas, trunks and mid to lower branches. L 3-6; H 7; W 3-5.

Occurrence: Common and in large colonies.

Plant: A 50-70 mm. high epiphyte forming clusters with 10-30 leaves and an intense root system composed of 0.6 mm. diameter whitish roots. Three faintly trumpet-shaped, hairless sheaths with recurved margins envelop the leaf stems, which are 25-35 mm. long and 0.4 mm. diameter. The lowest segments of the leaf stems are 2 mm. long. The leaves are light green, oblanceolate, faintly keeled, stiff and attenuated to their bases to form 3 mm., mostly resupinate, false petioles and have rounded apices and are flat but concave in dry conditions and are 25-35 mm. long, 9-13 mm. wide and 1-2 mm. thick.

Inflorescence and Flowers: The 70 mm. inflorescences emerge from the apex of the most recent leaf stem. Four to 6 long-lasting, slightly pendulous flowers on 8 mm. pedicels open alternately in rapid succession and bloom together. The flower has a pronounced chin and an unpleasant scent and are pollinated by fruit- or fungus flies. The ovate sepals are concave, keeled and whitish to light green. The dorsal has 3 distinctive purple longitudinal veins on the apical half and is 7.5 mm. long by 3.1 mm. wide while the laterals are 90% fused and together measure 6.7 mm. long by 5.5 mm. wide. Each has a purple outer vein. The acute obovate petals have 3 purple longitudinal veins and are 5.3 mm. long and 2.5 mm. wide. The light green, trilobed lip is 6 mm. long and 2.5 mm. wide. The margins of the apical half are serrate and the warty apex is rounded, while the surface along the central vein is undulate. The slightly curved column is light green with purple margins and is 4.3 mm. long by 1.5 mm. wide. The narrow wings and the apex are faintly gnawed.

Flowering Periods: September to December, with occasionally a second flowering period in April. The flowers last for four to five weeks and the inflorescence for two months.

Pleurothallis tripterantha
Rchb. f. | *1152* (Plate IX)

Synonym: P. procumbens Lindl.

Etymology: Tripterantha means three-winged flower.

Habitat: Epiphytic from section V, at 650 M. asl, on a remnant original forest tree on an almost horizontal trunk near the riverbank. L 3; H 7; W 3.

Occurrence: Rare.

Plant: This epiphyte tends to lie almost flat on the substratum and forms clusters with many leaves covering the trunk. The intense root system is composed of light yellow-green, adherent, 0.6 mm. diameter roots. The 12-18 mm. long and 1.2-1.5 mm. diameter leaf stems emerge alternately from the 2-3 mm. diameter rhizome at 5 mm. intervals, enveloped by 2 short-lived sheaths. The leaves are olive-green, narrowly lanceolate, folded, slightly recurved, leathery, conduplicate and are finely attenuated to their bases into a 9 mm. false petiole and measure 140-180 mm. long, 17-22 mm. wide and 1 mm. thick.

Inflorescence and Flowers: The 80-100 mm. slightly pendent inflorescence emerges from the apex of the third oldest leaf stem. Up to 6 flowers on 5 mm. pedicels bloom consecutively and alternately with up to 2 flowers at a time on the 40-50 mm. apical half; the bracts are 3 mm. and the ovary 1.5 mm. long. The sepals are narrowly lanceolate with acuminate apices, concave at the base, distinctively keeled, light yellow-green, dotted purple mainly along the veins and the margins. The dorsal is 12 mm. long by 3 mm. wide and has 3 longitudinal veins and the laterals, which are 50% fused, together measure 11 mm. long by 4 mm. wide. The very light yellow, spathulate, petals have acute apices are speckled purple along the veins and are 3.5

mm. long and 1.7 mm. wide. The light yellow-green, tongue-shaped lip is fleshy and is 5 mm. long and 2 mm. wide. There are 2 small ear-like appendages at the base and 2 erect, indistinct side lobes. The acute apex is warty and has crenate margins. The light yellow-green column is 3.2 mm. long and 1.6 mm. wide. The anther is apical and the stigma ventral.

Flowering Periods: April. The flowers last for two weeks.

Pleurothallis imbeana
Brade | *1153* (Plate IX)

Etymology: Not known.

Habitat: Epiphytic from section VI and VIII, at 600-1100 M. asl, in original cloud forest found on trunks and lower tree branches. L 3; H 7; W 3.

Occurrence: Rare.

Plant: An epiphyte, 40-80 mm. high forming clusters with 10 to 20 leaves and an intense root system composed of light yellow-green, adherent, 0.7 mm. diameter roots. The 5-12 mm. long and 0.5-0.7 mm. diameter leaf stems are enveloped by 2 short-lived sheaths, up to 9 mm. long. The leaves are olive-green, linear, slightly recurved, leathery, conduplicate, finely attenuated to their bases into a 2.5 mm. false petiole and measure 60-75 mm. long, 4.5-6 0 mm. wide and 1-1.5 mm. thick.

Inflorescence and Flowers: Up to 3, 5-10 mm. inflorescences emerge from the apices of the younger leaf stems. Up to 4, long-lasting flowers bloom consecutively and alternately, one at a time. The spathe is 1-2 mm., the peduncle 3-5 mm, the pedicels 2 mm. and the ovary 1.5 mm. long. The keeled, lanceolate, concave, light yellow-green and very light rose sepals have a yellow-green apex. Their margins and longitudinal veins are partly wine-red. The dorsal is 7-10 mm. long by 3.3 mm. wide and has 3 longitudinal veins. The laterals, which together measure 7-9.5 mm. long by 4.8 mm. wide, are 95% fused and have 6 to 7 wine-red longitudinal veins. The apices of the spathulate, very light yellow petals are dark purple with more or less gnawed margins and are 3.5-3.8 mm. long and 1.8-2 mm. wide. The light yellow-green, strap-shaped lip is 4.5-5 mm. long and 1.3-1.8 mm. wide. There are 2 small ear-like appendages at the base and 2 erect, slightly undulate gills extend from the base to the convex apex, which is wine-red, rounded and covered with tubercles up to 0.5 mm. long. The plate is rough to warty. The light green column is 4 mm. long with a long foot and a purple granular tip.

Flowering Periods: December to February. The flowers last for two to three weeks.

Similar Species: See *P. ramphastorhyncha - 1163*.

Pleurothallis fluminensis
Pabst | *1154* (Plate IX)

Etymology: Fluminensis Latin for river or in this case, from the state of Rio.

Habitat: Epiphytic from section VI, at 900-1100 M. asl, in humid original cloud forest, on lianas, trunks and branches of the lower quarter of forest trees. L 2, H 7; W 2.

Occurrence: Common, occasionally in large colonies.

Plant: The plant is 30-50 mm. high and forms clusters with 10-30 leaves. The 5-10 mm. long and 0.3-0.4 mm. diameter leaf stems are enveloped by 3 thin, short-lived sheaths. The leaves are light green, linear, conduplicate and 25-35 mm. long, 2.5 mm. wide and 1 mm. thick.

Inflorescence and Flowers: The 5-10 mm. inflorescences emerge from the apex of young and older leaf stems and produce up to 3 consecutive flowers. One flower blooms at a time followed by the next after a month's rest. The ovary and the bracts are 1 mm. long. The flower has a pronounced chin. The lanceolate, concave, light ochre sepals have green keels and are covered with purple speckles. The dorsal is on the lower side and is 4 mm. long by 2.4 mm. wide, coloured dark wine from the base and narrowed up to two thirds of its length. The laterals, which are 95% fused, are 4.3 mm. long by 3.3 mm. wide together. The light yellow to ochre, spathulate petals are dark wine-red at the base and at the pointed apices with dark wine-red speckles at their margins and measure 3 mm. long by 1.4 mm. wide. The light green lip is 3 mm. long and 1.5 mm. wide and has 2 white ear-like appendages at its base, and the margins and apex are dark wine-red. The lip surface turns from undulate at the base to tuberculate at the rounded fleshy apex. The triangular side lobes are erect and wine speckled and the lip is recurved by 70° at the mid-section and the green main keel is partly serrate. The light green, slightly curved column is 2.7 mm. long and 1.4 mm. wide with wine-red margins, gnawed wings and apex and wine-red flanks.

Flowering Periods: All months of the year. The flowers last for two weeks.

General: There is a variety with inflorescences up to 30 mm. long.

Pleurothallis colorata
Pabst | *1155* (Plate IX)

Etymology: Coloratus is Latin for coloured.

Habitat: Epiphytic from sections VI and VIII, at 100-1300 M. asl, in the shade of humid original forest, on trunks and lianas. L 2-3; H 6-8; W 2.

Occurrence: Common, occasionally in large colonies.

Plant: A variable, 40-70 mm. high plant forms clusters with 7 to 30 leaves and whitish, 0.3 mm. diameter roots and mostly protected by short moss. Two loose, short-lived sheaths envelop the 10-15 mm. long and 0.4-0.6 mm. diameter leaf stems. The leaves are light green, linear, conduplicate, finely attenuated to their bases and 30-50 mm. long, 4-5 mm. wide and 0.5 mm. thick.

Inflorescence and Flowers: The 40 mm. inflorescences arise from the apex of young and older leaf stems. The first flower blooms at less than half the leaf height followed by 4 to 6 consecutive and alternate flowers. The lanceolate sepals are keeled, concave, yellow, and golden-yellow towards the margins and the apices, while the margins themselves are speckled light wine. The 4.5-6.0 mm. long dorsal is 1.6-2.0 mm. wide and has 3 wine-red longitudinal veins. The laterals are 95% fused and are 4.3-5.9 mm. long by 2.5-3.3 mm. wide together and have 6 wine-red longitudinal veins. The spathulate petals are coloured like the sepals and have 3 wine-red longitudinal veins and are 2.7-3.2 mm. long and 0.8 mm. wide. The light yellow and wine-red lip is 3.0-4.0 mm. long by 1.3 mm. with 2 faint whitish ear-like appendages at its base and crenate margins from the side lobes to the rounded, fleshy, warty apex. The side lobes are erect and the central vein is speckled wine-red. The light green, slightly curved column is 2.3 mm. long, 1.1 mm. wide and has wine-red margins. The wings and apex are faintly gnawed.

Flowering Periods: All months of the year and the flowers last for two weeks.

Pleurothallis cordilabia
Pabst | *1157* (Plate IX)

Etymology: Cordilabia is a Latin compound word meaning heart-shaped lip.

Habitat: Epiphytic from section IV 1500 M. asl, in the shade of humid original cloud forest, on trunks and mid-tree branches. L 4; H 7; W 4.

Occurrence: Rare.

Plant: This 30-40 mm. high plant forms clusters with 5-15 leaves and 0.7 mm. diameter light green roots. The round leaf stems are 4-6 mm. long by 0.5-0.7 mm. diameter and are enveloped by 2 light brown, short-lived sheaths. The narrowly lanceolate, light green leaves are leathery, conduplicate, 20-30 mm. long 4-5 mm. wide and 1.0 mm. thick and are attenuated to their bases to form 1.5 mm. long false petioles.

Inflorescence and Flowers: The 30 mm. inflorescences emerge from the apex of the leaf stems to about three quarters of the leaf height with up to 5 consecutive flowers. The spathe is 1mm. long, the bracts 2 mm. and the ovary 1.3 mm. The distinctly keeled, concave sepals are ovate, light wine-red with darker speckles and yellow margins. The dorsal sepal is 5 mm. long and 2.5 mm. wide while the laterals are 80% fused and 4.4 mm. long by 3.1 mm. wide together. The petals are spathulate and 2.5 mm. long by 1.3 mm. with arrowhead-like apices and dark purple to almost black bases and apices. The dark wine-red-blotched lip is 3.0 mm. long by 2 mm. wide with a paler base and 2 distinct erect side lobes with strong crenate, longitudinal callosities and a fleshy apex. The light yellow-green, slightly curved column is 2.3 mm. long and 1.3 mm. wide with purple margins and a dark purple, granular column foot.

Flowering Periods: March to April. The flowers last for 3-4 weeks.

Pleurothallis gehrtii Hoehne & Schltr. | *1158* (Plate IX)

Etymology: Not known.

Habitat: Epiphytic from section V and VI at 1100 M. asl, in humid original cloud forest, along riverbanks and on trunks with sufficient light. L 5; H 8; W 3.

Occurrence: Rare.

Plant: The 25-35 mm. high plant forms clusters with 10-30 leaves. The 1-2 mm. long and 0.4 mm. diameter leaf stems are enveloped by 2 loose sheaths, up to 4 mm. long. The leaves are light green, narrowly lanceolate, conduplicate and finely attenuated to their bases, 20-30 mm. long, 4-5 mm. wide and 0.5-1.0 mm. thick. The yellow-green roots are 0.5 mm. in diameter.

Inflorescence and Flowers: The 45 mm. inflorescences emerge from the apex of young and older leaf stems. The first flower blooms at half the leaf height followed in short zigzags by 5 to 7 consecutive, alternate flowers. One or 2 flowers are open at a time. The lanceolate, keeled, flat, lightly hairy sepals are dark saffron, paler towards the hairy margins and green towards the apices. The slightly recurved dorsal sepal is 5 mm. long and 1.4 mm. wide while the laterals are 90% fused and 4.1 mm. long by 2.2 mm. wide together and are recurved by up to 90°. Dark wine-red hairs between their keels cover them densely. The petals are finely acuminate, ovate, red-brown to orange-brown, bent down and outwards and measure 3.5 mm. by 0.85 mm.. The margins of the apical half are serrate. The lip is dark wine-red, with hairy margins and 2 small, erect, narrow ear-like appendages at its base. Two dark wine gills extend from the base alongside the deepened and light yellow central vein to the orange coloured apex. The basal half of the lip is concave, the apical half is convex, the side-lobes are erect and it measures 2.2 mm. by 1.0 mm.. The light yellow, curved column is 2.0 mm. long and 1.1 mm. wide with the wings slightly turned in towards the gnawed apex.

Flowering Periods: December to April. The flowers last

for two weeks.

General: In section V we found plants with shorter, wider leaves and almost pure yellow flowers.

Pleurothallis lineolata (Barb. Rodr.) Cogn. | *1162* (Plate X)

Etymology: *Lineolata* is Latin and means with fine parallel lines referring to the markings on the tepals.

Habitat: Epiphytic from section VI, at 300-700 M. asl, in humid original forest, growing on moss-covered lower tree trunks, branches and lianas. L 2-3; H 6-8; W 2.

Occurrence: Common.

Plant: The 80-140 mm. high plant forms clusters of 7 to 20 leaves and 0.5 mm diameter whitish roots. The round leaf stems are 25-35 mm. long and 0.6-0.8 mm. diameter and enveloped by 3 tight, short-lived sheaths. The leaves are dark green, linear lanceolate, keeled, furrowed and finely attenuated to their bases. They are slightly recurved and 60-110 mm. long, 5-7 mm. wide and 1 mm. thick.

Inflorescence and Flowers: The 25-50 mm. inflorescences arise from the apices of young and older leaf stems and produce 3 flowers consecutively, one at a time, each on 3.5 mm pedicels. The narrowly ovate sepals are lemon-coloured, deepening towards the margins and the recurved apices. The dorsal is 9 mm. long and 2.3 mm. wide with 5 wine-red longitudinal veins. The laterals are 90% fused and together measure 9.6 mm. by 4.1 mm. with a greenish central vein, 2 dark purple longitudinal veins and beaked apices. The arrow-shaped petals are 4.1 mm. long by 1.3 mm. wide and are coloured like the sepals but have 5 dark purple longitudinal veins. The pale yellow, arrow-headed lip is covered with dark speckles, 4.1 by 2.2 mm. with 2 faint, white ear-like appendages at its base. The side lobes are erect, almost triangular, while the apex is obtuse. The light green, slightly curved column is 3.2 mm. by 1.8 mm.

Flowering Periods: January to April. The flowers last for up to three weeks.

Pleurothallis ramphastorhyncha (Barb. Rodr.)Cogn. | *1163* (Plate X)

Etymology: *Rhyncha* is from the Greek meaning provided with a snout or beak referring to the flower's chin.

Habitat: Epiphytic from section IV, V, VI and VIII, at 300-1100 M. asl, in the shade of humid original forest, on trunks and lianas. L 2-3; H 7-9; W 2.

Occurrence: Common and we have found it in large colonies.

Plant: The plant is 60-75 mm. high and forms clusters

with 10-40 leaves and an intense root system consisting of whitish roots, less than 0.5 mm. diameter. The 10-20 mm. long and 0.5 mm. diameter leaf stems are enveloped by 3 loose sheaths. The leaves are dark green, narrowly lanceolate, almost linear, conduplicate, finely attenuated to the base and measure 50-60 mm long., 4-6 mm. wide and 0.5-1 0 mm. thick.

Inflorescence and Flowers: The 30-40 mm. inflorescences arise from the apex of young and older leaf stems. The first flower blooms at half the leaf height followed by 4 to 6 consecutive, alternate flowers. There may be a second inflorescence from the same leaf base in the same year. The flower has a pronounced chin. The lanceolate sepals are keeled, concave, light yellow-green and turn to green at their apices. The dorsal is 7.5 mm. long by 3.3 mm. wide with 5 wine-red longitudinal veins. The lateral sepals are 95% fused and are, together, 7 mm. long by 4.5 mm. wide with 10 wine-red longitudinal veins. The spathulate petals have arrow-head-like apices and are 3.6 mm. long, 2.2 mm wide and coloured like the sepals. The light green and wine-red lip has 2 faint, whitish ear-like appendages at its base, crenate margins from the side lobes to the rounded, fleshy and warty apex and erect side lobes. The light yellow-green slightly curved column is 3 mm. long and 1.6 mm. wide with wine-red margins and a wine-red and warty column foot.

Flowering Periods: All months of the year. The flowers last for three to four weeks.

General: The sepals may be light green with dark green veins.

Pleurothallis rubro-lineata
Hoehne | *1164* (Plate X)

Etymology: Rubro and *lineatus* are Latin for with red parallel lines and refer to the tepals.

Habitat: Epiphytic from section VI and VIII, at 500-800 M. asl, in humid original cloud forest on upper and mid-tree moss-covered branches in good light. L 3-4; H 5-8; W 2.

Occurrence: Common.

Plant: A 20-25 mm. high plant forming small clusters of 4 to 20 leaves with adherent, 0.5 mm. diameter beige roots. The round, 5-10 mm. long, 0.5 mm. diameter leaf stems are enveloped by 3 sheaths. The narrowly lanceolate, light green leaves are faintly conduplicate and 10-12 mm. long, 2.5-3.5 mm. wide by 1-1.5 mm. thick and attenuated to their bases to form 1-2 mm. false petioles.

Inflorescence and Flowers: The inflorescence arises from the leaf base to about half the leaf height. Four to 5 flowers follow consecutively at very short intervals in a zig-zag and 1 or 2 flowers bloom at once. The distinctly keeled, ovate sepals are orange to light red-brown with darker longitudinal veins and saffron margins. The dorsal is 4.5 by 2.2 mm. and the laterals are 95% fused and together measure 4.3 mm. long by 3.2 mm wide. The asymmetric lanceolate petals are orange with 3 red-brown longitudinal veins and measure 2.3 mm. by 0.9 mm. The orange, elliptical lip is 2.7 mm. long and 2 mm. wide and has 2 robust, half round, erect side lobes. The margins of the rounded apex are toothed and the central vein is elevated increasingly towards the base. The light green column is 2 mm. long and with the narrow wings spread, 1 mm. wide.

Flowering Periods: All months of the year. The flowers last for 10-15 days.

Similar Species: See *P. colorata - 1155*.

Pleurothallis trifida
Lindl. | *1165* (Plate X)

Etymology: Trifidus is Latin for divided into three.

Habitat: Epiphytic from sections V, VI and VIII, at 900-1600 M. asl, in humid original cloud forest, growing at all levels of the forest on lianas, trunks and branches. The main occurrence is on lianas and branches less than 80 mm. diameter, usually with moss and Barbosella species. L 3-6; H 5-8; W 3.

Occurrence: Common, occasionally in large colonies.

Plant: The plant is 20-40 mm. high and forms clusters with 5 to 30 leaves. The 7-15 mm. long and 0.3 mm. diameter leaf stems are enveloped by 3 thin sheaths. The light green, narrowly lanceolate and conduplicate leaves are attenuated to their bases to form 2 mm. false petioles, and are 25 mm. long, 3.5 mm. wide and 0.5-0.8 mm. thick. The white roots form a normal root system mostly covered by moss.

Inflorescence and Flowers: The 10-25 mm. inflorescence emerges from the top of the most recent leaf stem. The first flower blooms at half leaf height followed by 5 to 10 consecutive and alternate flowers in very short zigzags. The inflorescence reaches leaf height and 1 or 2 flowers are open at a time. The sepals are lanceolate, concave, yellow to orange and keeled with orange margins and longitudinal veins. The dorsal is 5.2 mm. long by 1.8 mm. wide and the laterals are 95% fused and 4.5 mm. long by 2.5 mm. wide together. The translucent yellowish petals are spathulate, 2.6 mm. long by 1 mm. wide and have 3 orange longitudinal veins. The light yellow-green lip, turning orange with age, is 3 0 mm. long by 1.2 mm. wide and has 2 faint ear-like appendages at its base. The margins are crenate from the indistinct side lobes to the rounded and papillate apex and the light green, slightly curved column is 2.4 mm. long and 1.5 mm. wide. The acuminate wings and the apex are gnawed.

Flowering Periods: All months of the year. The flowers

last for two to three weeks.

Similar Species: See *P. rubro-lineata - 1164.*

Pleurothallis uniflora
Lindl. | *1166* (Plate X)

Etymology: *Uni* and *flora* are Latin for single flowered.

Habitat: Epiphytic from section IV, VI, 500-900 M. asl, in humid original forest, on thin mid- and upper-tree branches. L 6, H 7, W 5.

Occurrence: Occasional but in some locations found in large colonies.

Plant: This 50-75 mm. high epiphyte forms clusters with 5 to 15 leaves and an intense root system of 0.5 mm. diameter roots. The 8-18 mm. long and 0.5 mm. diameter leaf stems arise at 1.5 mm. intervals from a 0.7 mm. diameter rhizome and are enveloped by 3 short-lived sheaths of 0.5 mm., 4 mm. and 8 mm. length. The linear to narrowly lanceolate, conduplicate, papery leaves are finely attenuated to their bases forming 2 mm. false petioles between the annulus and the abscission layer and are 50-60 mm. long, 5-6 mm. wide and 0.5 mm. thick.

Inflorescence and Flowers: The 25-40 mm. inflorescences arise from the apex of young and older leaf stems. The first flower blooms at half the leaf height followed by 1 to 3 consecutive, alternate flowers. There may be a second inflorescence in the same year from the same leaf base. The narrowly ovate, distinctively keeled, concave sepals are light yellow-green and deeper coloured at their acute apices. The dorsal is 7.2 mm. long, 2.8 mm. wide with 5 wine-red longitudinal veins. The laterals are 95% fused and together are 7 mm. long by 4.8 mm. wide with 10 wine-red longitudinal veins. The lanceolate petals are 4 mm. long by 2 mm. wide and coloured like the sepals with 5 wine-red longitudinal veins. The light green and dark wine-red lip is 4.3 mm. long by 2 mm. wide with 2 faint light yellow ear-like appendages at its base. The margins are slightly crenate from the erect side lobes to the rounded, fleshy, papillate, dark purple apex. The light yellow-green slightly curved column is 3.3 mm. long and 1.5 mm. wide and has a wine-red, papillate base.

Flowering Periods: November through April. The flowers last for three weeks.

Similar Species: See *P. ramphastorhyncha - 1163.*

Pleurothallis hypnicola
Lindl. | *1167* (Plate X)

Etymology: A Latin compound word from *hypnum* and *cola* meaning moss-dwelling.

Habitat: Epiphytic from section IV, V, VI and VIII, at 300-1600 M. asl, in humid original forest, on low-tree trunks, branches and lianas. L 3-5; H 5-8; W 3-5.

Occurrence: Common.

Plant: This epiphyte is 50-100 mm. high and forms clusters with 7 to 25 leaves and whitish 0.5 mm. diameter roots. Three sheaths envelop the 15-30 mm. long and 1 mm. diameter leaf stems. The lanceolate leaves are light green, conduplicate, finely attenuated towards their bases and measure 60-80 mm. long, 15-20 mm. wide and 0.8 mm. thick.

Inflorescence and Flowers: The 70-90 mm. inflorescence arises from the apex of young and older leaf stems. The first flower blooms at less than half leaf height followed by 5 to 7 consecutive and alternate flowers. Up to 2 flowers are open at once. The lanceolate sepals are keeled and their basal halves are concave and pale yellow. The apical halves are convex and bright yellow. The dorsal is 7 mm. long by 3.1 mm. wide and speckled in the lower half transversely in deep purple. The laterals are 60% fused, measure 7 mm. long and 3.8 mm. wide together and have purple speckles in the basal halves from the margins to the keels. Both dorsal and lateral sepals are shielded by purple-green, lanceolate, addpressed bracts for nearly half their length. The spathulate petals are 3.2 mm. long by 1.7 mm. wide and dark wine at their bases and the pointed, fleshy apices. The arrow-shaped, fleshy lip is light greenish and has 2 faint whitish ear-like a-ppendages at its base. The margins are purple and the apex is rounded. The lip is speckled wine-red between the margins and the faintly warty, green centre. The light green, slightly curved column is 2.5 mm. long and 1.5 mm. wide with purple-speckled wings and apex.

Flowering Periods: All months of the year. The flowers last for two weeks.

Pleurothallis hypnicola
var. *cuneifolia*
Lindl. | *1167b* (Plate X)

Etymology: Cunei-folia is Latin for wedge-shaped leaves, probably referring to the leaf bases.

Habitat: Epiphytic from section V, VI and VIII, at 400-1600 M. asl, in humid original forest on mid-tree branches. L 2-6; H 5-9; W 3-6.

Occurrence: Common.

Plant: The plant is 70-100 mm. high and forms clusters with 7 to 25 leaves and plentiful thin whitish roots, usually running under moss. The round, 15-30 mm. long and 1 mm. diameter leaf stems arise alternately from the branching, 1.5 mm. diameter rhizome and are enveloped by 3 tube-shaped sheaths. The lanceolate leaves are papery, conduplicate, light green, attenuated towards their bases and 50-70 mm. long, 15-18 mm. wide and 0.8 mm. thick.

Inflorescence and Flowers: The inflorescences arise from the apex of young and older leaf stems. The first flower blooms at half the leaf height followed by 4 to 7 consecutive and alternate flowers. Up to 2 flowers are open at a time. The slightly bent inflorescence reaches a length of 60-70 mm. The lanceolate, wine-red and yellowish sepals are keeled and the lower halves concave and speckled purple horizontally from the central vein to the margins. The convex and granular upper halves are speckled red on an orange base. The dorsal sepal is 7 mm. long and 3.5 mm. wide. The lateral sepals are 60% fused and are 6.5 mm. long by 3.8 mm. wide together and light yellow between the keels. Both dorsal and lateral sepals are shielded by purple-green, lanceolate, addressed bracts for nearly half their length. The spathulate petals are 3.3 mm. long by 1.8 mm. wide with pointed apices and are dark wine-red at their bases and light wine towards the fleshy and almost black apices. The arrow-shaped lip is fleshy, light green with 2 faint, white ear-like appendages at its base, a granular centre and it measures 3 mm. long and 1.5 mm. wide and is purple from its base to the erect side-lobes. The margins and the rounded apex are purple, while the rest of the lip is speckled wine on a light green background. The light green, slightly curved column is 2.5 mm. long and 1.6 mm. wide. The rose-coloured wings and the apex are speckled with wine-red dots.

Flowering Periods: All months of the year. The flowers last for two to three weeks.

Pleurothallis hypnicola var. *flava* 1167c (Plate X)

Habitat: Epiphytic from section V and VI, at 700-1400 M. asl, in humid original forest on mid-tree branches. L 2-6; H 5-9; W 3-6.

Occurrence: Rare.

Plant: The measurements of the plant and flowers are identical to 1167 and 1167b.

Flower: The flowers are pure light green and yellow.

Pleurothallis hypnicola var. *major* possible new species 1167d (Plate X)

Habitat: Epiphytic from section VIII, at 700 M. asl, in a gorge within remnant original forest on low-tree trunks, branches and lianas. L 4; H 5-8; W 4.

Occurrence: Rare.

Plant: This epiphyte is 160-180 mm. high and forms clusters with 5 to 15 leaves and whitish 0.5 mm. diameter roots. The leaf stems are 80-100 mm. long and are enveloped by 3 sheaths. The lanceolate leaves are 120-140 mm. long, 25 mm. wide and 0.8 mm. thick, conduplicate, light green and finely attenuated towards their bases to form a 15 mm. false petiole.

Inflorescence and Flowers: The inflorescence is 140-170 mm. long and arises from the apex of the leaf stems. The first flower blooms at three quarters of the leaf height followed by 4 to 5 consecutive and alternate flowers. The bracts are 2.5-3 mm. long, the spathe is 8 mm. and the ovary, 5 mm. The lanceolate sepals are keeled with concave, pale yellow basal parts and convex, golden-yellow, short-haired apices. The dorsal sepal is 8 mm. long and 4 mm. wide with dark wine blotches at the concave section. The yellow laterals are 70% fused and are 8 mm. long by 4.4 mm. wide together. The dark wine-red petals are arrow-shaped, 4.1 mm. long by 1.6 mm. wide. The arrow-shaped, fleshy lip is 4.5 mm. long and 2 mm. wide, light green, with two faint whitish ear-like appendages at the base. The main lobe and the round wine-speckled apex are granular. The light green and red, slightly curved column is 3.5 mm. long and 1.8 mm. wide with a dark purple column foot.

Flowering Periods: All months of the year. The flowers last for two weeks.

Pleurothallis piraquarensis Hoehne | 1168 (Plate XI)

Etymology: A municipality in Parana; Piraquara.

Habitat: Epiphytic from section VI, at 1000-1100 M. asl, in humid original cloud forest, on lower trunks and horizontal branches. L 3; H 6; W 3.

Occurrence: Rare. We have only found a few solitary plants.

Plant: This 60-120 mm. high epiphyte forms clusters with 7 to 25 leaves and whitish, 0.5 mm. diameter roots forming an intensive system. The 15-30 mm. long by 0.7 mm. diameter leaf stems are enveloped by 3 sheaths. The leaves are oblanceolate, olive-green, conduplicate, attenuated towards their bases and are 40-80 mm. long, 8-14 mm. wide and 0.8 mm. thick.

Inflorescence and Flowers: The 65 mm. inflorescences

arise from the apex of young and older leaf stems. The first flower blooms at almost leaf height followed by 2 to 3 consecutive and alternate flowers and up to 2 may be open at once. The lanceolate sepals are keeled, olive-green or orange-brown and are speckled purple along the convex margins; the longitudinal veins have smaller purple speckles. The sepals are hairy on the inside along the margins and close to the apices. The 5.6 mm long. dorsal is 2.4 mm. wide and the laterals are 60% fused and 6 mm. long and 3.4 mm. wide together. The fleshy petals are 3.8 mm. long by 2 mm. wide, spathulate with rhomboid apices and are light rose with dark wine speckles. The inside surface is undulate. The arrow-shaped, light greenish and rose, fleshy lip has two faint whitish ear-like appendages at the base and is speckled purple with increasing intensity towards the warty apex and is 2.4 mm. long by 1.4 mm. wide. The light green, slightly curved, narrow column is 3 mm. long and 1 mm. wide The wings and the apex are gnawed.

Flowering Periods: October to January. The flowers last for three weeks.

Pleurothallis seriata
Lindl. | *1169* (Plate XI)

Etymology: Not known.

Habitat: Epiphytic from section VI, at 500-1600 M. asl, in humid original cloud forest, on trunks, lianas and mid-tree branches. L 3-6; H 3-6; W 3-5.

Occurrence: Common.

Plant: This 50-70 mm. high epiphyte forms clusters with 6 to 40 leaves. Two tight sheaths envelop the 10-25 mm. long and 0.6-1 0 mm. diameter, round leaf stems. The leaves are light green, lanceolate, leathery or stiff, faintly conduplicate and finely attenuated towards their bases forming 2 mm. false petioles. They are flat or concave and 30-50 mm. long, 9-11 mm. wide and 2-3 mm. thick and usually speckled black or dark purple.

Inflorescence and Flowers: The 150 mm. inflorescences arise from the apex of young and older leaf stems. The first flower blooms well above leaf height, followed at 1.5 mm. intervals by 4 to 7 consecutive and alternate flowers on 10 mm. pedicels. Up to 4 flowers are open at a time. The slender ovary is 3 mm. long. The lanceolate, keeled sepals have concave, greenish lower halves and convex light ochre, apical ends and are speckled purple. The 9 mm. long dorsal sepal is 4 mm. wide and the margins have short, purple hairs. The laterals, with entire margins, are 95% fused and together are 8 mm. long by 5 mm. wide. The whitish petals are broadly spathulate, 2.8 mm. long and 2.2 mm. wide and have a purple base, speckled dark purple with increasing intensity towards the round or cut, tuberculate apex. The light greenish, tongue-shaped, fleshy lip has 2 faint, pointed ear-like

appendages at its base. Two lamina originate at the greyish green, erect hook-shaped side-lobes and end at the spherical almost black apex. The fleshy, ball-shaped, dark purple apex encloses a depression, which opens towards the base. The keels of the undersides of the lip are warty for the apical half. The lip is 4.5 mm. long and 2 mm. wide. The light green, slightly curved column is 2.3 mm. long and 2 mm. wide. The wings are finely acuminate and the base is purple.

Flowering Periods: March to May and August to December. The flowers last for 7 to 10 days.

Pleurothallis sp.
Possible new species
1172a (Plate XI)

Etymology: Open.

Habitat: Epiphytic from section VI, at 1300-1550 M. asl, in humid original cloud forest, on trunks, low branches and low branches. L 3-6; H 6-8; W 3-5.

Occurrence: Rare.

Plant: The plant is 100-150 mm. high and forms clusters with 5-12 leaves and whitish 0.7 mm. diameter, light green roots. The round leaf stems are 50-65 mm. long by 0.7-1 mm. diameter and arise alternately at 1 mm. intervals from a 1.5 mm. diameter branched rhizome, enveloped by 3 tube-shaped sheaths. The light green, lanceolate, leathery, keeled leaves are infolded at the base and are 60-90 mm. long, 15-18 mm. wide, 1 mm. thick and attenuated to the base to form a 6 mm. false petiole.

Inflorescence and Flowers: The slightly bent 80 mm. inflorescences arise from the apices of young and older leaf stems with the first flower blooming at half leaf height and followed by 2 to 4 consecutive, alternate flowers on 7 mm. pedicels and up to 2 may bloom at once. The ovate sepals are faintly keeled, concave and greenish-yellow to orange and the margins are speckled red with 3 red longitudinal veins and red speckled margins. The upper halves have yellow hairs on the inside, increasing in density to the apices. The dorsal sepal is 8.6 mm. long, 3.3 mm. wide and the laterals are orange yellow, light yellow between the keels, are 95% fused and measure 8.4 mm. long by 4.3 mm. wide together. The spathulate petals are faintly pointed, bright orange inside with an olive-green double keel on the granulose outside and measure 3.9 mm. long by 1.8 mm. wide. The arrow-shaped, fleshy lip has two hook-shaped, cream, ear-like appendages at its base and a pointed apex. It is red from the pale yellow base to the side-lobes, yellow along the central vein and 4 mm. long by 2.4 mm. wide with red to red-brown speckles on the side-lobes and on the tuberculate, elliptic main lobe. The light yellow, slightly curved column is 3.5 mm. long and 1.5 mm. wide with red or red-brown margins and flat, faintly gnawed wing tips and apex.

Flowering Periods: November to January; the flowers last for 2 weeks and the inflorescence 1 month

General: This new species has since been named *Pleurothallis gracilicaulis* **Seeh**. The holotype is in the Rio de Janeiro Botanic Garden, # RB-410773.

Pleurothallis arcuata
Lindl. | *1174* (Plate XI)

Etymology: From the Latin arcuatus, curved like a bow referring to the curved sepals.

Habitat: Epiphytic from section V, VI and VIII, at 900-1600 M. asl, in humid original cloud forest, on lower trunks, branches and lianas. L 3-5; H 4-7; W 2.

Occurrence: Common.

Plant: This 100-150 mm. high epiphyte forms clusters with 7 to 30 leaves and a prolific, wiry root system penetrating bark. The round leaf stems are 30-50 mm. long and 1-2 mm. diameter, and enveloped by 3 sheaths. The leaves are light green, broadly lanceolate, leathery, conduplicate, attenuated towards their bases, slightly infolded and are 60-100 mm. long, 25-35 mm. wide and 1.8 mm. thick.

Inflorescence and Flowers: The slightly bent 250-300 mm. inflorescences arise from the apex of young and older leaf stems. The first flower blooms at one and a half times the leaf height followed by 10-20 consecutive and alternate flowers on 10-20 mm. bent pedicels. Up to 4 flowers are open at once. The lanceolate sepals are keeled, concave with granular flanks at their bases and convex on the recurved apical halves. The dorsal sepal is 13 mm. long by 4.2 mm. wide and is speckled purple with diminishing intensity towards the light yellow-green apex and towards the margins on the rose-coloured lower half. The light yellow-green lateral sepals are faintly speckled purple between the margins and the keels and are 100% fused, together measuring 12 mm. long by 4 mm. wide. The arrow-shaped petals have dark purple bases and apices, purple speckled whitish midsections and are 4.2 mm. long, 2.6 mm. wide. The light greenish, tongue-shaped, fleshy lip is 3.3 mm long and 1.2 mm wide and has 2 faint ear-like appendages and 2 wine-red speckles at its base and the margins of the erect side-lobes are speckled purple. The surface of the central vein and the rounded apex are undulate or papillate. The whitish, slightly curved column is 2.7 mm. long by 1.5 mm. wide and speckled purple. The wings and the apex are finely acuminate and the column base is hairy.

Flowering Periods: October to June. The flowers last for 10-15 days and the inflorescence for 3 months.

Pleurothallis granulosa
Barb. Rodr. | *1177* (Plate XI)

Etymology: The Latin granulatus refers to little knobs on the sepals and lip.

Habitat: Epiphytic from section IV, at 1100-1300 M. asl, in humid original mountain cloud forest, on lower trunks, branches and lianas. L 3-5; H 4-7; W 2-4.

Occurrence: Common.

Plant: A 110-170 mm. high epiphyte forming clusters with 7 to 20 leaves and yellow-green or whitish, wiry roots less than 1 mm. diameter. The round leaf stems have an elliptical upper cross-section and are 40-60 mm. long by 1-2.5 mm. diameter and enveloped by 2 or 3, orange-brown sheaths. The leaves are light green, broadly lanceolate, leathery, conduplicate, slightly infolded, attenuated towards their bases, and 60-110 mm. long, 25-35 mm. wide and 1.8 mm. thick.

Inflorescence and Flowers: The slightly curved 250-350 mm. inflorescences arise from the apex of both young and older leaf stems. The first flower blooms at one and a half times the leaf height followed by 10-15 consecutive and alternate flowers on 10-20 mm. pedicels. Up to 4 flowers are open at once. The spathe is 8 mm. and the bracts are 7-12 mm. long. The lanceolate sepals are keeled, concave at their bases and convex with granular surfaces on the apical two thirds. The dorsal is light yellow and 11.4 mm. long by 4 mm. wide with 3 dark red-brown veins. The light yellow laterals are 50% fused and together measure 10.5 mm. long by 3.5 mm. wide. The arrow-shaped petals have dark purple central veins and are 3.2 mm. long and 2.2 mm. wide. The light yellow-green, tongue-shaped, fleshy lip is 3.2 mm long and 1.5 mm wide and has 2 faint ear-like appendages at its base. The surface of the plate and the rounded apex are undulate to warty. The wine-red, slightly curved column is 3 mm. long by 1.5 mm. and the column foot is granulate. The wings and the apex are finely acuminate.

Flowering Periods: October to March. The flowers last for 10-15 days and the inflorescence for three months.

General: P. arcuata - 1174 and *P. granulosa* are so similar that they may be varieties of one species.

Pleurothallis henrique-aragonii
Pabst | *1178* (Plate XI)

Etymology: Named in honour of Dr. **Henrique Aragao**.

Habitat: Epiphytic from section IV and V, at 900-1200 M. asl, in humid original forest, on lower, and mid-tree branches and lianas. L 3-5; H 4-7; W 2-4.

Occurrence: Common.

Plant: This 110-140 mm high epiphyte forms clusters

with 7 to 20 leaves and yellow-green or whitish, wiry roots less than 1 mm. diameter. Two or 3, orange-brown sheaths envelop the 20-25 mm. long and 1-2 mm. diameter, round leaf stems which have elliptical upper cross-sections. The leaves are obovate, papery, conduplicate, slightly infolded at their bases and are 80-110 mm. long, 20-25 mm. wide and 1 mm. thick.

Inflorescence and Flowers: Slightly curved 200-250 mm. inflorescences arise from the apex of both young and older leaf stems. The slightly pendent first flower blooms at one and a half times the leaf height, followed by 5-7 consecutive, alternate flowers on 10-20 mm. pedicels and up to 4 flowers are open at once. The spathe is 6 mm. long and the bracts are 5-6 mm. All parts of the flower are granulate, fleshy, keeled and olive green with purple speckles The lanceolate sepals are concave at their bases and slightly convex on the apical quarters. The dorsal is 15 mm. long by 4 mm. wide and the 95% fused laterals together measure 14.5 mm. long by 7 mm. wide. The arrow-head-shaped petals are paler and 9 mm. long by 3.8 mm. wide. The broadly tongue-shaped lip is 5 mm. long and 3.2 mm. wide with 2 faint ear-like appendages and a lateral callosity at its base. The surface of the plate is dark purple and the rounded, fleshy apex is warty. The slightly curved, light green column is 4.5 mm. long by 2 mm. wide with a purple column foot and finely lacerate wings and apex.

Flowering Periods: April to May. The flowers last for 15 days; the inflorescence for 2 months.

General: This species has the largest *Pleurothallis* flowers in the Organ Mountain Range. There is also a yellow variety with a wine-red lip.

Pleurothallis hians
Lindl. | *1179* (Plate XI)

Etymology: Hians is Latin for gaping, with open mouth referring to the open flowers.

Habitat: Epiphytic from section VI at 900-1400 M. asl, in humid original montane forest on moss-covered, lower tree branches. L 5; H 7; W 3-7.

Occurrence: Occasional.

Plant: A 150-180 mm high epiphyte forming clusters with 7 to 20 leaves and a prolific wiry root system of whitish roots which penetrate bark and the mossy strata around. The 1.5-2.5 mm. diameter leaf stems are alternate on a branching, 2 mm. diameter rhizome. The leaf stems are 40-60 mm. long and enveloped for 60% of their length by 3 tube-shaped sheaths. The lower leaf stem cross-section is round and the apical is oval. The leaves are obovate, conduplicate, shiny olive-green and are 80-120 mm. long, 25-35 mm. wide and 1 mm. thick. Younger leaves are speckled with wine-red spots.

Inflorescence and Flowers: The 200-250 mm. inflorescences emerge from the apices of young and older leaf stems. The first pendulous flower blooms on a 15 mm. pedicel at almost double the leaf height followed by 10-15 consecutive and alternate flowers at intervals of 7-20 mm. Up to 6 flowers are open at once. The sepals are obovate, yellow-green, faintly keeled, with purple longitudinal veins. The concave lower halves are rose-coloured and the apical halves are convex, light olive-green and densely covered by short hairs. The dorsal sepal is 10-13 mm. long by 5.0-6.0 mm. wide and the laterals are 90% fused and 9.5-12 mm. long and 5.0-8.0 mm. wide together and very light green between the keels. The arrowhead-like petals have fleshy, pointed apices, very light green with wine-red speckles and are 3.5-4.0 mm. long by 2-2.6 mm. wide and hairy on the inside. The wedge-shaped lip is fleshy, light yellow-green, hairy on the inside with 2 faint ear-like appendages at its base, purple margins from the base to the erect side-lobes and light purple from the side-lobes to the wedge-shaped apex. The lip is 4 mm. long by 2.0 mm. wide and covered with wine-red speckles. The light green, slightly curved column is 3.0 mm. long, 2.0 mm. wide and is speckled dark wine-red and has a hairy column-foot.

Flowering Periods: December to March and the flowers last for 2 to 3 weeks; the inflorescence for 4 to 5 months.

Pleurothallis pellifeloides
(Barb. Rodr.)Cogn. | *1180* (Plate XI)

Etymology: Pellifel-oides is Latin for skin-like. In Barbosa Rodrigues' original description, Lepanthes pellifeloidis, he stated that the sepals are spotted like Jaguar skin.

Habitat: Epiphytic from section VI and VIII, 500-1200 M. asl, in humid original cloud forest, mainly on moss-covered mid- and lower tree trunks and branches. L 5; H 7; W 5.

Occurrence: Rare.

Plant: A 140-160 mm. high epiphyte forming clusters with 7 to 20 leaves and 0.5 mm. diameter roots The round leaf stems are 40-60 mm. long and 1.5-2.5 mm. diameter and emerge alternately from a branching, 2 mm. diameter rhizome. They are enveloped for 70% of their length by 3 tube-shaped sheaths. The leaves are obovate, fleshy, faintly conduplicate, shiny green and are 80-120 mm. long, 30-35 mm. wide and 1.0 mm. thick.

Inflorescence and Flowers: The 200 mm. inflorescences arise from the apex of young and older leaf stems. The first pendulous flower blooms on a 20 mm. pedicel at leaf height followed by 6-10 consecutive, alternate flowers at intervals of 7.0-20 mm. Up to 3 flowers may be open at once. The obovate, yellow-green, purple spotted sepals are keeled with concave and granular bases and convex, hairy apices. The dorsal sepal is 10.0 mm. long and 5.0 mm. wide. The laterals are 80% fused, very light green between the keels and measure 9.8 mm. long by 6.0 mm. wide together. The arrow-shaped petals have fleshy, pointed apices, very light green with purple speckles and 4.2 mm. long, 3.0 mm. wide. The wedge-shaped lip is 4.0 mm. long by 3.0 mm. wide, fleshy, light yellow-green and purple dotted, with 2 faint ear-like appendages at its base. The light green, purple dotted column is narrow, slightly curved, 3.3 mm. long, 2.0 mm. wide and the column foot is hairy.

Flowering Periods: November to February and the flowers last for 2 to 3 weeks: the inflorescence for 4 months.

Similar Species: See *P. hians - 1179*.

Pleurothallis wacketii
Handro & Pabst | *1185* (Plate XI)

Etymology: Not known.

Habitat: Epiphytic from section VI and on rocks from section I, at 1200-1300 M. asl, in humid original forest on trunks and on low branches along riverbanks. L 2-4; H 7-9; W 3.

Occurrence: Occasional.

Plant: A 40-150 mm. high, variable plant forming clusters with 7 to 20 leaves and 0.7 mm. diameter whitish roots. The round leaf stems are 30-50 mm. long and 0.7-1.0 mm. diameter and arise alternately at intervals of 3-5 mm. from a branching, 1.5 mm. diameter rhizome, enveloped by 3 tube-shaped sheaths. The light green to dark olive leaves are 20-100 mm. long, 5-18 mm. wide and 0.7-1.5 mm. thick, narrowly lanceolate, papery to fleshy, faintly conduplicate, flat and resupinate near their bases and their backs may be speckled dark purple.

Inflorescence and Flowers: The 10-150 mm. inflorescences arise from the apex of young and older leaf stems. The first flower blooms on a 7 mm. pedicel at half the leaf height, followed by 3 to 5 consecutive, alternate flowers and up to 4 flowers are open at once. The ovate sepals are keeled, concave, greenish yellow to light golden yellow with the margins and veins tending to red. The apical halves are convex and have light yellow hairs on the inside. The dorsal is 3.5-6.5 mm. long and 1.8-2.5 mm. wide and the laterals are 90% fused and measure 3.3-6.0 mm. long by 2-3.5 mm. wide together. The petals are light green to bright orange, spathulate and the central veins end with a point at the apical underside. They are 2-3.7 mm. long, 1-2 mm. wide. The lip is 2-3.5 mm. long by 1.5-2 mm. wide, wine-red to dark purple with a lighter coloured central vein and plate. The side lobes are round or a wide hook-shape, the margins of

the main lobe are serrate and the surface is undulate to rough. The light yellow green column is narrow, slightly curved, 2.5-3.5 mm. long and 1.8 mm. wide with red or dark purple margins. and faintly gnawed wings and column apex and the dark wine column-foot warty.

Flowering Periods: November to February. The flowers last for 2 weeks and the inflorescence for a month.

General: The size of the plant and flowers vary over a wide range, as does the colour.

Pleurothallis teres
Lindl. | *1195* (Plate XII)

Synonym: Pleurothallis rupestris.

Etymology: Refers to the terete leaves.

Habitat: A terrestrial lithophyte from sections V and VIII, at 800-1400 M. asl, in humus filled cracks of sun exposed rock faces with adequate water. Very often on the lower and north-facing side of other plant islands thus ensuring a fairly constant moisture run-through. L 5-9; H 1-3; W 7-9.

Occurrence: Occasional.

Plant: The plants are 35-80 mm. high and form clumps with 5 to over 100 leaves with whitish, 0.75 mm. diameter roots. The round leaf stems are 5-15 mm. long and 3-9.0 mm. wide and emerge from the sheath-covered, 2-3 mm. diameter rhizome at 3-4 mm. intervals, enveloped by two scale-shaped sheaths, up to 10 mm. long. The terete, very succulent, leaves are linear or narrowly ovate with a point and slightly curved sideways, light green to dark wine-red and 20-60 mm. long, 6-10 mm. wide and 5-7 mm. thick.

Inflorescence and Flowers: The 50-100 mm. inflorescences arise from the apices of younger leaf stems. The first, almost sessile pendulous flower develops slightly below leaf height followed by 3 to 10 alternate flowers blooming together. The ripening seed pod turns the long-lasting flower upright. The peduncle is 20-30 mm. long and 1-2 mm wide, the spathe is 5-8 mm., and the bracts are 4 mm. long. The lanceolate sepals have acute apices, wine coloured with a yellowish centre and an orange apex, keeled, fleshy, with 3 darker longitudinal veins. The dorsal sepal is 5-7 mm. long by 2-3 mm. wide. The laterals are 90% fused and 5-7 mm. long and 3.5-4.5 mm. wide together. The spathulate petals are dark purple with pale ochre centres and are 3-4 mm. long and 1.4 mm. wide. The tongue-shaped lip is wine-red, 3.5-4.5 mm. long, 1.5-1.8 mm. wide with 2 yellowish, ear-shaped appendages at its base, indistinct round side-lobes and 2 longitudinal callosities. The round apex has a point. The slightly curved column is light green and wine-red and 2.5 mm. long by 2 mm. wide.

Flowering Periods: December to February.

General: We can find no discernable differences between herbarium specimens of *P. teres* and *P. rupestris*.

Pleurothallis leptotifolia
Barb.Rodr. | *1197* (Plate XII)

Etymology: Named because of the similarity of the leaf to those in the genus *Leptotes*.

Habitat: Epiphytic from section VI, at 900-1000 M. asl, in original forest remnants on the upper branches of relict trees. L 6-8; H 5; W 6-7.

Occurrence: Rare, but in large colonies on a few relict trees.

Plant: A 20-30 mm high, branching epiphyte forming clusters with many leaves, clinging tightly to the bark with 0.4 mm. diameter, white to green or beige roots. The round leaf stems are 3-5 mm. long and 0.4-2.0 mm. diameter and arise alternately at 2-4 mm. intervals from a branching, 1-2 mm. diameter, creeping, sheath-covered rhizome, enveloped by scale-shaped, short-lived sheaths, up to 5 mm. long,. Three to 10 sequential leaves are produced annually from the tip of each rhizome branch and are indistinctly conduplicate, succulent, olive-green, but the lower halves of the younger leaves are lighter-coloured and blotched purple. They stand upright, are slightly recurved and are 20-25 mm. long by 2.5-3.5 mm. diameter.

Inflorescence and Flower: The 30-35 mm. inflorescences arise from the apex of the younger leaf stems. The first flower develops at leaf height followed by 2 to 3 alternate flowers blooming simultaneously. The spathe is 5 mm. long, the bracts 2 mm. and the ovary 1.5 mm. The lanceolate sepals are light yellow, faintly keeled with 3 light green longitudinal veins. The dorsal is 5.5 mm. long by 2 mm. wide and the 30-70% fused laterals are 5.5 mm. long and 3 mm. wide together. The lanceolate, pale yellow petals are 3.8 mm. long and 1.4 mm. wide. The lip is narrowly rectangular, light yellow and 4.1 mm. long by 1.6 mm. wide with discrete side-lobes and no visible callosities. The margins of the rectangular apex are toothed. The light green, slightly curved column is 2.5 mm. long, 0.7 mm. wide with rounded and faintly gnawed wings.

Flowering Periods: May. The flowers last for two weeks.

Pleurothallis sonderana
var. *longicaulis*
Rchb. f. | *1200b* (Plate XII)

Etymology: *Longi-caulis* is Latin for long stem referring to the proportionately long stems.

Habitat: Epiphytic from sections VI and VIII, at 600-800 M. asl, in open original cloud forest on upper branches of trees exposed to sunshine and wind. L 8; H 5; W 8.

Occurrence: Rare.

Plant: The 50-60 mm. high, branched epiphyte forms tight clusters with many leaves and clings to the bark with 0.3 mm. diameter, white to green or beige roots. The round leaf stems are 25-35 mm. long by 0.8-2 0 mm. diameter and enveloped by 2 tube-shaped, short-lived sheaths up to 7 mm. long. The upright, linear leaves are fleshy, infolded, conduplicate, light olive-green and are 25-35 mm. long by 3-4 mm. diameter and 2 mm. thick.

Inflorescence and Flowers: The 30-35 mm. tall inflorescences arise from the apex of young and older leaf stems. The first flower develops at ¾ of the leaf height followed by 2 to 4 alternate flowers blooming simultaneously. All parts of the flower are lemon-yellow. The obovate sepals are keeled and have 3 light green longitudinal veins. The dorsal sepal is 6.3 mm. long by 2.3 mm. wide. The laterals are 30% fused and each measures 6-8 mm. long and 1.6 mm. wide. The lanceolate petals are 2.9 mm. long and 0.8 mm. wide. The lip is tongue-shaped, trilobed, and 2.5 mm. long by 1.4 mm. wide. The hook-shaped apices of the side lobes point towards the apex of the lip whose margins are serrate. The longitudinal callosities are undulate. The lightly curved column is 2.2 mm. long and 1.2 mm. wide with faintly gnawed wings and apex.

Flowering Periods: April. The flowers last for two weeks.

Pleurothallis bidentata
Lindl. | *1205* (Plate XII)

Etymology: *Bidentata* is Latin for with two teeth referring to the side lobes of the lip.

Habitat: Epiphytic from section VI, at 250 M. asl, on trunks in remnant gallery forest. L 6; H 7-8; W 4-6.

Occurrence: Rare.

Plant: This 220-420 mm. high epiphyte forms loose clusters of 5 to 50 leaves with relatively few 1.5 mm. diameter reddish roots, whitish in moss. The 100-230 mm. leaf stems arise alternately at 15 mm. intervals from a 4-6 mm. diameter, creeping rhizome covered by 5 short-lived sheaths. The lower sections of the leaf stems are 2 mm. diameter widening at the leaf junctions to 3.5-4.5 mm. by 3 mm. The narrowly lanceolate leaves are conduplicate, leathery to stiff, light olive-green, erect and 150-200 mm. long, 15-18 mm. wide and 1.5-2 mm. thick.

Inflorescence and Flowers: One to 3 single flowered inflorescences emerge consecutively on 2-4 mm. peduncles from the apices of the most recent leaf stems. The ovary is 2 mm. long and the spathe slightly longer. The fleshy, tongue-shaped sepals are concave at their bases, slightly convex at the fleshy apices with inside surfaces undulate to warty and orange with 3 wine-red longitudinal veins and red margins. The dorsal sepal is 8 mm. long by 2.8 mm wide and the 30% fused laterals are 7.2 mm. long by 2.9 mm. wide. The lanceolate petals are light orange with wine-red central veins and margins, serrate on their apical halves and ending in one or two pointed apices. The petals are 4.5 mm. long and 1.1 mm. wide, bent slightly downwards and outwards. The fleshy, narrowly elliptical lip is 4.9 mm. long, 2.1 mm. wide and has 2 erect, hook-shaped side-lobes. The oval, fleshy, hairy main lobe has faintly toothed or warty and revolved margins and ends in a double pointed or round apex. The light yellow column is 3.6 mm. long and, with spread wings, 1.1 mm. wide with moderately gnawed apex and wings. The column foot may be granular and wine-red.

Flowering Periods: February and March. The flowers last for ten days.

General: This species is variable in size, lip shape and flower colour as shown in the drawings.

Pleurothallis capanemiae
Barb. Rodr. | *1208* (Plate XII)

Etymology: Named in honour of the **Baron of Capanema**, the first to forecast that the railways would be the key factor of the final destruction of the Atlantic Rainforest.

Habitat: Epiphytic or terrestrial from section VIII, at 700-800 M. asl, in low-growing, swampy forest on shady trunks and branches, usually not higher than 2 M. above the ground. L 2-4; H 6; W 2.

Occurrence: Occasional, but locally common.

Plant: The 20-50 mm. long, pendent leaf stems are 1 mm. diameter, partly covered by short-lived sheaths and emerge alternately at 10 mm. intervals from a 1 mm. diameter, creeping, branching rhizome. The lanceolate leaves are 80-110 mm. long, 18-22 mm. wide by 1 mm. thick, flat, leathery, olive-green turning wine-red in high light. One or 2 whitish or green, 0.8 mm. diameter roots emerge from each leafless internode of the rhizome.

Inflorescence and Flowers: The 20 mm. inflorescences emerge from the apices of the leaf stems with 5 simultaneously blooming flowers. The pedicel is 1.5 mm. long, the bract 1.5 mm. and the ovary 3 mm. The ovate, keeled sepals are fleshy, concave at the basal half and convex at the apical end. They are light yellow with dark wine-red dots and blotches. The dorsal sepal is 9 mm. long by 2.5 mm. wide and the laterals are 70% fused and 7.5 mm. long by 5 mm. wide together. The arrow-head shaped petals are translucent with wine-red markings along the yellow central veins and slashed apical margins and 2.6 mm. long by 1.4 mm. wide. The

tongue-shaped lip is 3.4 mm. long by 2.2 mm. wide with 2 small, ear-like appendages at its base. The margins of the purple side lobes and the copper-coloured longitudinal callosities are tuberculate and the margins of the warty, rounded apex are faintly toothed. The light green and dark purple column is 3 mm. long and, with spread wings, 1.4 mm. wide with a dark purple and hairy foot.

Flowering Periods: February to April and September and the flowers last for 14 days.

Pleurothallis cearensis
Schltr. | *1209* (Plate XII)

Etymology: *Cearensis* possibly means that the plant examined came from Ceara State.

Habitat: Epiphytic from section VI, at 600-900 M. asl, in shady, humid forest on trunks and mid-tree branches and in abandoned orchards. L 3-5; H 8; W 2-3.

Occurrence: Common.

Plant: The round leaf stems arise alternately at 20-30 mm. intervals from a 2-2.5 mm. diameter, creeping and branching rhizome and enveloped for 60% of their length by 2 brown sheaths, while the apical segments are furrowed deeply on the front. The leaf stems are 50-70 mm. long by 1.5 mm. diameter at their bases widening to 1.8 by 2.5 mm. diameter at the leaf junctions. The ovate leaves are flat, leathery, dark olive-green, conduplicate, 50-65 mm. long, 20-25 mm. wide and 1 mm. thick. The yellow-green or whitish roots are 0.7 mm. diameter.

Inflorescence and Flowers: Two to 3, 15-20 mm. inflorescences emerge from the apices of younger and older leaf stems, with up to 3 consecutive flowers on 1.5 mm. pedicels and up to two blooms at once. The spathe is 6 mm. and the ovary 1.5 mm. long. The concave, fleshy, keeled sepals are faintly verrucous on the outside and undulate on the apical inside, while the margins are faintly toothed. The dorsal sepal is 9 mm. long by 4.1 mm. wide, and appears longer than the laterals with wine-red margins and 5 longitudinal veins. The row-boat-shaped laterals are 100% fused and 7.8 mm. long by 7.2 mm. wide. The margins are yellow-green, the centre is purple and the apices are reddish with olive-green or dark brown. The petals are lanceolate, faintly yellow, and 2.6 mm. long and 1.2 mm. wide with 3 purple longitudinal veins and lacerate margins. The elliptical lip is 3.3 mm. long by 1.7 mm. wide with 2 thread-like appendages at its green base and is dark purple between the base and the green side-lobes. The purple margins of the side lobes and the green longitudinal callosities are undulate and turn to toothed and fringed towards the apex. The light green column is 2.2 mm. long and with spread wings 1.1 mm. wide with both margins and granular base wine-red. The wings and the apex of the column are faintly lacerate.

Flowering Periods: October to February. The flowers last for three to six weeks.

Pleurothallis saundersiana
Rchb. f. | *1215* (Plate XII)

Etymology: unknown.

Habitat: Epiphytic from section VI, at 800 M. asl, covering large branches and trunks in original forest. L 4-6; H 6; W 5.

Occurrence: Occasional but locally common.

Plant: The 20-25 mm long leaf stems emerge alternately at 20 mm. intervals from a 1.5-2 mm. diameter, creeping, branching sheath-covered rhizome and are 1-1.5 mm. diameter, enveloped for half their length by 2 sheaths with deeply furrowed apical segments on the front side. The elliptical leaves are leathery or stiff, dark olive-green, conduplicate, and 30-45 mm. long, 20-25 mm. wide and 2 mm. thick.

Inflorescence and Flowers: The single-flowered 6 mm. inflorescences emerge from the apex of young and older leaf stems with flowers on 2-3 mm. pedicels. The ovary is 3 mm; the spathe and the bract are 3-4 mm long. The keeled sepals are fleshy and concave at their light yellow-green bases and convex at their orange apices. The ovate dorsal sepal is 12 mm. long and 4.3 mm. wide with 5 longitudinal veins. The laterals are 70% fused and are 11 mm. long by 7 mm. wide together with a light purple speckled centre. The petals are lanceolate, faintly convex, light yellow, with 3 yellow-green longitudinal veins and 3 mm. long by 1.3 mm. wide. The margins of the apical halves are lacerate. The lip is 4 mm. long, 1.9 mm. wide, green, fleshy, elliptical with 2 hair like pointed appendages at the base and speckled wine on the main lobe and the margins of the small, lacerate side lobes and longitudinal callosities. The surface is rough at the base and the apex and the margins of the main lobe are toothed. The light green and purple column is 2.7 mm. long and with spread wings 1.2 mm. wide with a purple base and column foot.

Flowering Periods: January to February and April. The flowers last for 14 days.

Pleurothallis serpentula
Barb. Rodr. | *1216* (Plate XII)

Etymology: Serpentula is Latin for snake-like, referring to the snake-like rhizome habit.

Habitat: Epiphytic from section IV, at 1400 M. asl, in original forest covering large branches and trunks. L 4-6; H 6; W 5.

Occurrence: Occasional, but locally common.

Plant: The 25-35 mm. leaf stems emerge alternately at 20-35 mm. intervals from a 1.5-2 mm. diameter, creeping, branching, sheath-covered rhizome and are 1.5-2.0 mm. diameter, enveloped for 50% of their length by 2 sheaths and their apical segments are deeply furrowed on the front. The elliptical leaves are leathery or stiff, dark olive-green, conduplicate, and 35-50 mm. long, 20-25 mm. wide and 2-2.5 mm. thick.

Inflorescence and Flowers: The single-flowered 10 mm. inflorescences emerge from the apex of young and older leaf stems with flowers on 3 mm. pedicels. The ovary 3-4 mm. and the spathe and the bract are 4-5 mm. long. The keeled sepals are fleshy, concave at their light yellow-green bases and convex at their apices. The ovate dorsal sepal is 15 mm. long and 4.8 mm. wide with a purple apex and 5 green or wine longitudinal veins. The laterals are 70-90% fused and 14 mm. long by 8 mm. wide together and speckled purple. The petals are lanceolate, faintly convex, light yellow, with 3 yellow-green longitudinal veins and are 5.5 mm. long by 1.6 mm. wide. The margins of the apical halves are lacerate. The lip is fleshy, elliptical and purple with darker speckles and is 5.2 mm. long, 2.6 mm. wide with 2 hair-pointed appendages at its light base. The surface is rough at the base and the apex. The margins of the rudimentary side-lobes are crenate and dark purple; the margins of the main lobe are toothed. The light green and purple column is 4 mm. long and, with spread wings, 1.4 mm. wide. The base and the column foot are purple.

Flowering Periods: January to February. The flowers last for 14 days.

Pleurothallis spilantha
Barb. Rodr. | *1218* (Plate XII)

Etymology: Spil and *antha* are from Greek meaning hairy flower, referring to the hairy ovary and keels to the sepals.

Habitat: Epiphytic from section VIII, at 1300 M. asl, in shady, humid original forest on trunks and mid-tree branches. L 3-5; H 8; W 2-3.

Occurrence: Occasional.

Plant: The leaf stems emerge alternately at 20-30 mm. intervals from a 2-2.5 mm. diameter, creeping, branching rhizome, and are 40-50 mm. long by 1 mm. diameter at their bases widening to 1.5-2 mm. at the leaf junctions. The apical sections are deeply furrowed and they are enveloped for 60% of their length by 2 brown sheaths. The ovate leaves are almost flat, leathery, dark olive-green and 40-50 mm. long, 20-25 mm. wide and 1 mm. thick. The roots are yellow-green to whitish and 0.7 mm. Diameter.

Inflorescence and Flowers: Up to two, 15 mm. inflorescences emerge from the apices of young and older leaf stems with up to 2 flowers on 10 mm. pedicels. The spathe is 4 mm. and the long-haired ovary is 1.5 mm. long. The very fleshy, concave, keeled sepals are hairy on the outer bases and rose-coloured with dark wine speckles. The dorsal sepal is 6.4 mm. long by 3.2 mm. wide. The laterals are 80% fused and 5.9 mm. long by 6 mm. wide. The petals are 3.5 mm. long and 2 mm. wide, lanceolate, whitish, with lacerate apical margins and 3 purple longitudinal veins. The tongue-shaped lip has 2 faint side-lobes, is rose-coloured with dark wine sepals and 4.2 mm. long by 2.4 mm. wide. The pale whitish column is 3 mm. long and, with spread wings, 1.5 mm. wide. The fleshy column-foot is orange to yellow with some dark wine spots.

Flowering Periods: April to May and the flowers last for 14 days.

Pleurothallis translucida
Barb. Rodr. | *1220* (Plate XIII)

Etymology: Translucidus is Latin for allowing light to shine through referring to the clear dorsal sepal.

Habitat: Epiphytic from section VI, at 900-1300 M. asl, in humid original cloud forest on trunks and lower branches. L 2-4; H 8-9; W 2-3.

Occurrence: Occasional. We have found large colonies at a few sites.

Plant: The 40-60 mm. long leaf stems emerge alternately at 15-30 mm. intervals from a 1.5 mm. diameter, creeping, branching rhizome and are 0.6 mm. in diameter at their bases and 2 mm. by 1.5 mm. at their apices. The apical segments are furrowed deeply on the front, while 2 sheaths envelop the lower halves of the younger leaf stems. The oblong leaves are fleshy, dark green, conduplicate, 50 mm. long, 12 mm. wide and 1 mm. thick.

Inflorescence and Flowers: The inflorescences emerge from the apices of young and older leaf stems. The first flower develops at 10-15 mm. up the inflorescence on a 1-2 mm. pedicel followed by up to 2 consecutive flowers and they may bloom at the same time. The concave sepals have indistinct keels and the narrowly ovate dorsal is 16 mm. long by 4.7 mm. wide, light green at the base turning ochre and yellow at the apex with brown margins and 5 brown longitudinal veins. The heart-shaped laterals are 70% fused and are 14 mm. long by 11.5 mm. wide together. They are grass-green between the keels and yellow-green from the keels to the margins. The 6 longitudinal veins are a deeper colour. The lanceolate petals are faintly convex, ochre, with 3 olive-green longitudinal veins and 5.5 mm. long by 1.8 mm. wide with purple serrate margins on the

apical halves. The fleshy, ovate, grass-green lip is 6.2 mm. long, 4 mm. wide with 2 linear, yellow thread-like appendages at its base. The margins of the basal quarter are divided into a crenate and a dark purple cauliflower-like third, while the centre of the lip is undulate. The light green column has a purple-dotted flat apex and is 3.2 mm. long and with spread wings, 1.5 mm. wide. The stigma is olive-green and the flanks are wine-red.

Flowering Periods: April to June. The flowers last for 14 days.

Pleurothallis farinosa
Pabst | *1223* (Plate XIII)

Etymology: Farinosa is Latin for mealy or powdery and probably refers to the lip margins.

Habitat: Epiphytic or terrestrial from sections VI and VIII, at 800-900 M. asl, in very humid original cloud forest on shady trunks, roots or rocks, less than 2 M. above the ground. L 2-3; H 8; W 2.

Occurrence: Occasional, but locally common.

Plant: The 40-60 mm. long, pendent leaf stems emerge alternately at 15-20 mm. intervals from a 1 mm. diameter, creeping and branching rhizome. The leaf stems are 0.7-1.0 mm. diameter at their bases and 0.8-1.5 mm. at the leaf junctions and enveloped by 2 sheaths for 60% of their length. The conduplicate leaves are lanceolate, leathery, dark olive-green and measure 80-100 mm. long, 25-40 mm. wide by 1 mm. thick. One or 2 whitish green, 1 mm. diameter roots emerge from each internode of the rhizome.

Inflorescence and Flowers: The 15-25 mm. inflorescences emerge from the apices of the 1 year old leaf stems and 2 to 3 flowers bloom simultaneously. The lanceolate, fleshy sepals have indistinct keels and are warty inside the tips and rose to light yellow with dark wine-red longitudinal veins and margins. The dorsal sepal is 8 mm. long by 2.8 mm. wide and the laterals are 99% fused and 7.7 mm. long by 7 mm. wide together. The spathulate, whitish petals have purple central veins and gnawed, purple apical margins and are 2.8 mm. long by 1.5 mm. wide. The elliptic, wine-red lip is 4.8 mm. long by 3 mm. wide with 2 small erect ear-like appendages at its base and 2 longitudinal lamina originate from the erect side-lobes. The whitish margins of the verrucous, rounded apex are faintly toothed. The rose to wine-red column is 3 mm. long and with spread wings 1.5 mm. wide. The pear-shaped seedpods are 10 mm. long and 7 mm. diameter.

Flowering Periods: February to April. The flowers last for 14 days.

Similar Species: See *P. cryptophoranthoides - 1252.*

General: The drawing shows the oldest leaf with the inflorescence of *P. cryptophoranthoides - 1252*, to demonstrate the difference.

Pleurothallis ramosa
Barb. Rodr. | *1230* (Plate XIII)

Etymology: Ramosa is Latin for branching and refers to the branched rhizome habit.

Habitat: Epiphytic from section V and VII, at 900-1400 M. asl, on relict trees, covering large, mossy branches and lianas. L 5-7; H 5; W 5.

Occurrence: Rare, but probably abundant in former times. We have found large colonies on fallen relict trees.

Plant: A small, creeping epiphyte forming large colonies on trunks and lower branches with 0.5-0.8 mm. diameter light yellow-green roots, usually moss-covered. The leaf stems are 70% enveloped by 1 or 2 sheaths, their apical segments are furrowed on the front and they are 10-25 mm. long by 0.5-1.0 mm. diameter and emerge alternately at 5-15 mm. intervals from a 0.7-1.2 mm. diameter, creeping and branching sheath-covered rhizome. The ovate or elliptic leaves with reflexed apices are conduplicate, leathery, dark olive-green and tarnished wine. They measure 25-30 mm. long, 10-14 mm. wide and 1-1.5 mm. thick.

Inflorescence and Flowers: The inflorescences are up to 8 mm. and emerge from the apex of young and older leaf stems with 1 or 2 flowers on 3 mm. pedicels. The ovary is 1.0 mm; the spathe and the bract are 2.0 mm. long. The keeled sepals are concave at their yellow bases and convex at their yellow-green apices. The narrowly ovate dorsal is 9.0 mm. long and 3.3 mm. wide with 3 dark wine longitudinal veins. The laterals are 90% fused, 8.3 mm. long by 4.9 mm. wide together and are lighter coloured at the common centre. The petals are lanceolate, faintly curved outwards, lemon-yellow, with 3 green longitudinal veins and are 3.3 mm. long by 1.0 mm. wide with lacerate margins on the apical halves. The tongue-shaped lip is fleshy, dark wine and 3.4 mm. long, 1.9 mm. wide with 2 hair-pointed appendages at its base and a rough surface at the base and the apex. The margins of the side-lobes are crenate and dotted dark purple, but the margins of the light purple dotted main lobe are hairy. The light green column is 2.5 mm. long and with spread wings 1.0 mm. wide with a purple base and flanks.

Flowering Periods: May to June The flowers last for 14 days.

Pleurothallis aff. ramosa
Barb. Rodr. | *1230b* (Plate XIII)

Etymology: See *1230*.

Habitat: Epiphytic from section V, at 700-850 M. asl, in original forest on trunks and lower thick branches. L 3-4; H 6; W 2-4.

Occurrence: Rare. We have found some colonies on fallen relict trees.

Plant: A small, creeping and branching epiphyte forming large colonies on trunks and lower branches with 0.5 mm. diameter light yellow-green roots, usually moss-covered. The leaf stems are 15-23 mm. long by 0.5-1.0 mm. diameter and emerge alternately at 5-10 mm. intervals from a green 1 mm. diameter rhizome and are covered by 3 short-lived sheaths. The leaves are 20-30 mm. long, 8-10 mm. wide and 1.5-2 mm. thick, elliptic with an acute apex, conduplicate, stiff, dark olive-green and tarnished wine-red in high light.

Inflorescence and Flowers: Up to 2, 10 mm. inflorescences with up to 3 flowers emerge from the apices of both young and older leaf stems. The spathe is 2 mm., the pedicel 2 mm., the bracts 2 mm. and the ovary 1 mm. long and all are scaly or short haired. All parts of the flower are fleshy with hairy keels and, excepting the petals, have hairy margins. The sepals are narrowly lanceolate, light yellow-green with dark wine blotches and 7.8 mm. long by 2 mm. wide. The laterals are 80% fused, 7.2 mm. long by 5 mm. wide and coloured as the dorsal. The petals are lanceolate, yellow-green with some wine dots at the base, 2.8 mm. long by 1 mm. wide with faintly serrate apical margins. The elliptic lip is 3.1 mm. long by 1.6 mm. wide with long hairs at the margins of the main lobe and the side-lobes are reduced to a thorn-shape. The light yellow column is 2.6 mm. long and has a dark wine, short-haired column-foot.

Flowering Periods: March to May and the flowers last for 14 days.

Pleurothallis aphtosa
Lindl. | *1231* (Plate XIII)

Etymology: Aphtosa is Greek for inaccessible perhaps referring to the difficulty of collecting.

Habitat: Epiphytic from section V, at 700-850 M. asl, in original cloud forest on lower, thick branches. L 3-4; H 6; W 2-4.

Occurrence: Rare.

Plant: A 250-400 mm. high epiphyte forming loose clusters of 10-50 leaves with 0.6 mm. diameter whitish roots. The 150-200 mm. long leaf stems are alternate at 10-14 mm. intervals from a 4-6 mm. diameter, creeping rhizome and are covered by 2 short-lived sheaths, up to 70 mm. long,. The round lower sections of the leaf stems are 3 mm. diameter widening at the leaf junctions to 7-8 mm. by 3 mm., where they are deeply furrowed. The leaves are narrowly lanceolate, leathery, dark olive-green, erect, conduplicate and 150-180 mm. long, 35-45 mm. wide and 2-3 mm. thick.

Inflorescence and Flowers: One or 2, 20 mm. inflorescences emerge from the apex of young and older leaf stems with 3 to 7 flowers and 4 may bloom at once. The hairy ovary is 2 mm., the bracts are 4 mm., the pedicels 5 mm. and the spathe is 5 mm. long. The fleshy, lanceolate sepals are concave at their bases, convex at their apices with undulate to warty inside surfaces and greenish, hairy backs. The yellow-green dorsal has a light brown apex with dark brown speckles and is 10.5 mm. long by 3.9 mm. wide. The dark wine-red laterals are 30% fused and are 10 mm. long by 4.5 mm. wide. The wedge-shaped petals are ivory with dark wine markings and 3.5 mm. long by 1.4 mm. wide. The fleshy, tongue-shaped lip is dark wine-red, 4.8 mm. long, 2.4 mm. wide with 2 erect side-lobes and 2 predominant, undulate longitudinal callosities. There are 2 thin appendages at the yellow, light purple base and the apex of the fleshy warty main lobe is rounded. The light green, wine column is 3 mm. long and, with spread wings, 2 mm. wide. Both apex and wings are faintly gnawed.

Flowering Periods: April. The flowers last for fourteen days.

General: Sometimes the flowers have an unpleasant smell.

Pleurothallis panduripetala
Barb. Rodr. | *1236* (Plate XIII)

Etymology: Panduratus-petala is Latin for violin-shaped and petals.

Habitat: Epiphytic from section VI, at 1300 M. asl, in humid original mountain cloud forest, on lower trunks and branches. L 3-4; H 7; W 4.

Occurrence: Rare.

Plant: The plant is fixed to the bark by a few, adherent, whitish or light green roots. The round leaf stems bend slightly under the weight of the succulent leaves and are 60-80 mm. long by 1-1.5 mm. diameter and arise at 2 mm. intervals from a creeping 1-1.5 mm. diameter rhizome, enveloped by 3 short-lived sheaths. The acute ovate, shell-like leaves are dark olive-green to purple, conduplicate and 35-45 mm. long, 17-20 mm. wide and 2-3 mm. thick.

Inflorescence and Flowers: Up to 3, 7 mm., single-flowered inflorescences, emerge from the apex of young and older leaf stems and bloom together. The whitish, keeled sepals have a yellow base, a bright purple apex

and the dorsal is tongue-shaped, 11 mm. long by 2 mm. wide, concave at its base and convex at the apex. with 3 striking purple central veins. The lateral sepals are 100% fused and together are 9.5 mm. long by 6.5 mm. wide, concave and speckled purple with increasing intensity from the yellowish base to the apex. The violin-shaped, whitish petals are 5.5 mm. long by 2 mm. wide and widen from the narrow, yellowish base. The central veins are purple and the wider centres of the petals are speckled purple, as are the deeply fringed margins of the wider section. The yellow, violin-shaped lip is 3.8 mm. long by 1.8 mm. wide and has a heart-shaped callus and 2 narrow ear-like appendages at the base. The side lobes are thread-shaped followed by 2 triangular greenish sail-like protuberances. The main lobe spreads like a fan towards the obtuse apex with toothed margins. The light green, slightly curved column is 2.7 mm. long by 1.0 mm. wide with light purple margins, wings and apex.

Flowering Periods: February to June. The flowers last for two weeks.

Pleurothallis hamosa
Barb. Rodr. | *1238* (Plate XIII)

Etymology: Hamosa is Latin for hooked and refers to the hook-like leaf apices.

Habitat: Terrestrial or lithophytic from sections VI and VIII, at 750-850 M. asl, in very humid original mountain cloud forest in humus on almost vertical and shady rocks. L 2-3; H 9; W 1.

Occurrence: Rare.

Plant: The plant has plenty of light yellow-green, long, 1.5 mm. diameter roots. The round leaf stems are enveloped for half their length by up to 3, tube-shaped sheaths and emerge at 10 mm. intervals from a creeping 2.5 mm. diameter rhizome. The leaf stems are 100-140 mm. long by 1.5-2 mm. in diameter at their bases and elliptic in cross-section; 4.5 mm. by 3 mm. at the leaf junctions. The broadly ovate leaves are fleshy, conduplicate, dark green, shell-like, tilted down by at least 45° with reflexed, hook-shaped, acute apices and are 50-100 mm. long, 40-63 mm wide. and 2.5-3 mm. thick.

Inflorescence and Flowers: One or two 20-25 mm. inflorescences emerge from the apex of both young and older leaf stems with up to 7 alternate flowers blooming at once. All parts of the flower are wine-red. The sepals are lanceolate, concave, fleshy and faintly keeled with ochre-coloured centres and orange-brown to dark brown apices with faintly undulate or granular inner surfaces and granular outer surfaces. The dorsal sepal is 9.5 mm. long by 2.8 mm. wide and the laterals, fused for 70% of their length, together are 9 mm. long by 7.5 mm. wide. The petals are lanceolate or spear-head shaped and 4.6 mm. long by 1.4 mm. wide with lacerate apical margins. The wine-red, tongue-shaped lip is 5.3 mm. long by 2.8 mm. wide with ochre-coloured margins; the side lobes are deeply gnawed and the margins of the rounded apex are serrate. The wine-red surface of the lip is rough. The light wine-red, slightly curved column is 3.3 mm. long and 1.8 mm. wide with light purple margins to the column, wings and apex.

Flowering Periods: March to April and November are the main flowering periods. Some plants flower repeatedly during the year and the flowers last for three weeks.

Similar Species: See P. *Prolifera - 1241. P. limae - 1239.* Both are terrestrials growing in half shade and 500 M. higher altitude. The leaves are narrower and thinner.

Plurothallis limae
Pôrto & Brade | *1239* (Plate XIII)

Etymology: Limae is Latin for file or queue and refers to the file-like flower arrangement.

Habitat: Terrestrial from section VI, at 1200-1500 M. asl, in low regrowing montane or elfin ridge forest, in leaf litter or rotting wood, often surrounded by moss. L 5-7; H 3-5; W 4-6.

Occurrence: Common.

Plant: The round, 60-180 mm. long leaf stems arise at 10-20 mm. intervals from a creeping 1.5-2.5 mm. diameter rhizome and are 1.5 mm. diameter at their bases widening to 3-5 mm. by 2-3 mm. in a 2-edged cross-section at their leaf junctions. Two tube-shaped sheaths envelop the leaf stems for a third of their length. The leaves are ovate to narrowly ovate, conduplicate, dark green or purple, horizontal or vertically inclined with hook-shaped, pointed apices and 40-75 mm. long, 20-42 mm. wide and 2 mm. thick.

Inflorescence and Flowers: One or two 20 mm. inflorescences emerge from the apex of young and older leaf stems. Up to 9 alternate flowers bloom together on a short raceme. All parts of the flower are wine-red. The sepals are lanceolate, fleshy, very faintly keeled, concave and have a granular to warty outer surface. The inner surface is smooth but the apex is faintly undulate. The dorsal sepal is 5-7 mm. long by 2-2.5 mm. wide with dark wine-red longitudinal veins and margins, and the laterals, fused for 70% of their length, together are 4.5-6 mm. long and 4-5.5 mm. wide. The petals are lanceolate or arrow-headed, rose, and 2.7-4.0 mm. long by 1.1-2 mm. wide with purple longitudinal veins. The margins of the wider sections are slashed. The tongue-shaped lip is 3-4 mm. long and 1.4-2.0 mm. wide, purple with slashed margins at the second fifth of the lip and a mid-section with 2 round, toothed lamellae. The margins of the round, tuberculate apex are faintly toothed. The column is light purple, slightly

curved, 2.2 mm. long by 1.2 mm. wide and the apex is light purple and yellow.

Flowering Periods: January to March is the main flowering period. Some plants produce several inflorescences during the year. The flowers last for three weeks.

Pleurothallis modestissima
Rchb. f. & Warm. | *1240* (Plate XIII)

Etymology: Modestissima is a Latin superlative of modest referring to the small flowers.

Habitat: Terrestrial from section VI, at 1300-1400 M. asl, in low growing montane forest, in moss or rotting wood. The whitish, or if uncovered, wine-red roots do not enter the soil. L 6-8; H 3-5; W 5-7:

Occurrence: Occasional.

Plant: The round, 60-110 mm. long leaf stems arise at 10-15 mm. intervals from a creeping 1.5 mm. diameter rhizome and are 1 mm. diameter at their bases widening to 2 mm. by 1.5 mm. at the leaf junctions with an oval cross-section. They are enveloped for a third of their length by 2 tube-shaped sheaths. The leaves are ovate to narrowly ovate, conduplicate, olive-green or purple, horizontal or slightly inclined, with hook-shaped, pointed apices and are 30-45 mm. long, 18-22 mm. wide and 2 mm. thick.

Inflorescence and Flowers: One or 2, 10-15 mm. inflorescences emerge from the apex of young and older leaf stems with up to 5 closely packed, alternate flowers blooming together. All parts of the flower are purple to wine-red. The lanceolate, concave sepals are fleshy, faintly keeled with granular outer surfaces, smooth inner surfaces and faintly undulate close to the margins and the apices. The dorsal sepal is 4 mm. long by 1.7 mm. wide with dark purple longitudinal veins and margins. The laterals are 90% fused and together are 3.8 mm. long by 4 mm. wide. The petals are lanceolate, light yellow and 2.3 mm. long by 1.2 mm. wide and the bases and central veins are purple with slashed margins at the wider section. The margins of the second fifth of the purple, tongue-shaped lip are slashed and the middle section has 2 semi-circular, tuberculate, orange-yellow longitudinal callosities. There are toothed margins on the round, tuberculate apex and the lip is 2.6 mm. long and 1.6 mm. wide. The column is light purple and greenish, slightly curved and 2.2 mm. long by 1.3 mm. wide.

Flowering Periods: January to March is the main flowering period. Some plants produce several inflorescences during the year. The flowers last for three weeks.

Pleurothallis prolifera
Herb. ex Lindl. | *1241* (Plate XIV)

Etymology: Prolifera is Latin for producing off-shoots, and refers to the keikis often produced.

Habitat: Terrestrial from section VI, at 1300-1600 M. asl, in low growing montane ridge forest, in moss or rotting wood. L 4-8; H 3-5; W 5.

Occurrence: Common.

Plant: The round, 60-150 mm. long leaf stems arise at 10-15 mm. intervals from a creeping 2.5 mm. diameter rhizome and are 1-1.5 mm. diameter at their bases widening to an oval, 3 mm. by 2 mm. cross-section at the leaf junctions and enveloped for a third of their length by 2 tube-shaped sheaths. The leaves are ovate to narrowly ovate, keeled, dark green or purple, horizontal or slightly inclined, have hook-shaped pointed apices and are 40-70 mm. long, 20-35 mm. wide and 1.5-2 mm. thick. The roots are whitish or wine-red if exposed to intense light.

Inflorescence and Flowers: One or two 10-15 mm. inflorescences emerge from the apex of young and older leaf stems and up to 7 closely packed, alternate, wine-red flowers may bloom together. The sepals are lanceolate, fleshy, faintly keeled, and concave with a granular to verrucous outside. The inner surface is smooth, but faintly undulate near the apex. The dorsal sepal is 5.2 mm long. by 1.9 mm. wide with dark wine-red longitudinal veins and margins. The laterals are 75% fused and 4.4 mm. long by 4.1 mm. wide together. The lanceolate petals are rose-coloured, 2.7 mm. long by 1.2 mm. wide with a purple central vein and purple-dotted side veins. The margins of the wider section are slashed. The tongue-shaped lip is purple, 3 mm. long by 1.7 mm. wide. The margins of the second fifth are slashed and on the middle section are 2 half round, toothed lamellae and the margins of the rounded, warty apex are toothed. The slightly curved column is light purple, 2.4 mm. long and 1.3 mm. wide and the apices and wings are light yellow.

Flowering Periods: January to March is the main flowering period. Some plants show several inflorescences during the year. The flowers last for three weeks.

Pleurothallis exarticulata
Barb. Rodr. | *1242* (Plate XV)

Etymology: Ex-articulatus is Latin for without joints referring to the smooth leaf stems.

Habitat: Epiphytic from section VI, at 700-800 M. asl. We found an immense colony on an original forest relict tree on horizontal mid- to upper-tree branches. L 7-8; H 5; W 5:

Occurrence: We classify the plant as rare now but in

earlier times it must have been common.

Plant: The branching, 100 mm. high plant forms loose clusters of 10 to many leaves and an extensive root system composed of white, wiry, adherent roots, less than 1 mm. diameter. The round leaf stems emerge alternately at 12-18 mm. intervals from a 2.5 mm. diameter, green, creeping and branching rhizome and are up to 100 mm. long, and 1.5 mm. diameter at their bases, thickening to 2 mm. at the furrowed, unsheathed top half. The leaves are ovate with the apex slightly recurved, slightly infolded, fleshy, stiff, conduplicate, grass-green or yellow in bright sun and are 40-50 mm. long, 15-18 mm. wide and 1.5-2.0 mm. thick.

Inflorescence and Flowers: One or two, 25 mm. inflorescences emerge from the apices of young and old leaf stems and up to 5 flowers bloom together. The spathe is 5 mm. and the sheathed stem, the pedicel and the granular ovary are all 0.5 mm. long. The flowering colony has an overall orange-yellow hue. The dorsal sepal is fleshy, elliptical, erect, slightly convex, orange-yellow and 6 mm. long, 2 mm. wide and turns yellow towards the base where the longitudinal veins are wine-red. The slightly concave, totally fused, broadly elliptical lateral sepals are 4.8 mm. long and 3.6 mm. wide together, and orange-yellow with a light green centre at their common base. The longitudinal veins and the margins are wine-red at the basal half. The light yellow, spathulate petals have a green central vein and are purple at the base. The margins of the apical half are lacerate. The petals are 2 mm. long by 0.8 mm. wide and slightly curved inside and downwards. The yellow, fleshy, tongue-shaped lip is 2.5 mm. long by 1.4 mm. wide and has 2 small, white, ear-like appendages at the base and 2 very dark wine-red side-lobes. The surfaces of the plate and apex are warty and all margins are toothed. The yellow-green column is 2 mm. long and with spread wings, 1 mm. wide, with dark wine flanks and base.

Flowering Periods: December to January. The flowers last for two weeks.

Pleurothallis heringeri
Hoehne | *1243* (Plate XV)

Etymology: Not known.

Habitat: Epiphytic from sections I, V, VI, at 900-1200 M. asl, in the shade of humid original cloud forest on low trunks and branches. L 3; H 8; W 2.

Occurrence: Occasional.

Plant: A 150-350 mm. high epiphyte forming clumps of 7 to 20 leaves. The 100-200 mm. long leaf stems arise alternately at 3-5 mm. intervals from the 2 mm. diameter rhizome and are round, 1.5 mm. diameter at their bases widening to 3-5 mm. by 2-3 mm. with a Y-shaped cross-section at the leaf junctions and 2 tube-shaped sheaths envelop their lower halves. The leathery leaves are ovate to narrowly ovate, conduplicate, dark green and 60-80 mm. long, 20-30 mm. wide and 1.5 mm. thick.

Inflorescence and Flowers: One or two 25-40 mm. inflorescences emerge from the apex of young and older leaf stems with up to 7 alternate consecutive flowers on short pedicels on the apical 15 mm., and up to 4 flowers bloom together. The keeled, concave sepals are yellow with dark wine-speckled margins and longitudinal veins and greenish apices. The tongue-shaped dorsal sepal is 9 mm. long by 2.6 mm. wide and the laterals are 90% fused and together are 8 mm. long and 4.5 mm. wide. The spathulate petals are light yellow and 4 mm. long by 1.5 mm. wide with wine-speckled longitudinal veins and serrate apical margins. The yellow tongue-shaped lip is 5.3 mm. long by 2.2 mm. wide with 2 yellow ear-like appendages at the base and 2 sickle-shaped, erect side-lobes with serrate margins. The longitudinal veins are speckled wine-red. The apex of the violin-shaped main lobe is rounded, fleshy, warty, green, wine and yellow. The light yellow-green, slightly curved column is 4.2 mm. long and 2 mm. wide with a light purple and yellow apex and the golden-yellow base is hairy.

Flowering Periods: April to June. The flowers last for two weeks.

Pleurothallis luteola
Lindl. | 1244 (Plate XV)

Etymology: Luteolus is Latin meaning pale yellow; this flower's colour.

Habitat: An epiphyte or occasional terrestrial from sections V, VI and VIII, at 1000-1200 M. asl, in very humid original cloud forest, on low-tree branches, mainly in deep shade along river banks. L 2; H 9; W 1.

Occurrence: Occasional.

Plant: This 120-160 mm high epiphyte forms clusters of 5 to 15 leaves with 1 mm. diameter whitish roots. The 90-120 mm. long leaf stems emerge alternately at 5-10 mm. intervals from a creeping, 2 mm. diameter rhizome, enveloped on the lower halves by 2 tube-shaped sheaths. The basal 15 mm. of the leaf stems are round and 1-2 mm. diameter but at the leaf base they are 3 mm. by 2.5 mm. and deeply conduplicate on the front, forming a Y in cross-section. The leaves are ovate to narrowly heart-shaped, keeled, yellow-green to dark green, leathery, 65-80 mm. long by 20-30 mm. wide and 1 mm. thick with acute apices and the false petioles hide the source of the inflorescences.

Inflorescence and Flowers: One or 2, 15 mm. inflorescences emerge from the apex of young and older leaf stems. Up to 4 flowers bloom alternately and simultaneously on the apical 8 mm. All parts of the flower are yellow with orange-yellow longitudinal veins. The narrowly ovate sepals are keeled and the dorsal is 8.5 mm. long by 2.5 mm. wide with a recurved, convex apical half. The laterals are 90% fused and 7.5 mm. long by 4 mm. wide together. The petals are spathulate, 3.3 mm. long by 1.1 mm. wide with faintly toothed margins at their rounded apices. The narrowly violin-shaped lip has 2 rows of dark wine-red speckles along the deepened central vein and is 5 mm. long, 1.8 mm. wide crenate margins to its rounded apex. The 3.5 mm, long, slightly curved column is 1.6 mm. wide and only faintly gnawed.

Flowering Periods: April to May and July through August. The flowers last for two weeks.

General: We found a variety in sector VIII with rust-coloured veins on the sepals and petals. The lip was light green with two small ear like appendages at its base and faint side lobes while the apical half of the lip is not widened and the apex is pointed or acute; type B on plate XV.

Pleurothallis acuminatipetala
A. Samp. | 1245 (Plate XV)

Etymology: Latin for with a pointed petal referring to the flowers.

Habitat: Epiphytic from section I, at 1100-1400 M. asl, in gallery forest on thick, moss-covered mid – to lower tree branches. L 3-5; H 5-7; W 4.

Occurrence: Locally common.

Plant: A 200-400 mm. epiphyte forming clusters with 7 to 15 leaves. The round leaf stems are 120-210 mm. long by 3-4 mm. diameter, arising from a short, 5 mm. diameter rhizome and are 60% enveloped by 3 sheaths. The leaves are elliptic, stiff, conduplicate, dark olive-green and are 140-190 mm. long, 35-45 mm. wide and 2-3 mm. thick.

Inflorescence and Flowers: The 150-200 mm. spicate inflorescences arise from the apex of young and older leaf stems with up to 35 erect alternate flowers, which open in quick succession on 3.5 mm. pedicels. The first flower blooms at half leaf height. The short-haired spathe is 35-45 mm., the ovary is 1.5 mm. and the bracts are 3-4 mm long. The flowers are orange and fleshy and only open a little. The lanceolate sepals are hairy on their outsides. The dorsal is 9.5 mm. long by 2.8 mm. wide and the laterals are 70% fused and 9.8 mm. long by 4.6 mm. wide together. The narrowly spathulate petals have distinctive points and are 2.4 mm. long and 1.5 mm. wide. The tongue-shaped lip is 2.7 mm. long by 1.3 mm. wide with 2 small ear-like appendages at its base, 2 predominant callosities on the plate, short side-lobes and a round apex. The light green, slightly curved column is 2.2 mm. long by 1.5 mm. wide with gnawed margins on the apex and wings.

Flowering Periods: December to January. The flowers last for two to three weeks.

Pleurothallis binotii
Regel | *1247* (Plate XV)

Etymology: Named in honour of the Brazilian orchid grower **Binot**.

Habitat: Epiphytic from sections IV and VI at 300-700 M. asl, in humid original forest on thick branches of the lower canopy, in relative poor light conditions. L 4-6; H 6; W 4.

Occurrence: Common.

Plant: The plant is 250-500 mm. tall and forms loose clusters with 10-25 leaves and a vigorous root system composed of green, 1.5 mm. diameter roots which brown with age. The 150-300 mm. leaf stems emerge alternately at 12-15 mm. intervals from a 5-7 mm. diameter, creeping, sheath-covered rhizome and are enveloped by 3 sheaths for half of their length. They are 4 mm. in diameter and round at the basal section, but furrowed and elliptical in cross-section at the apical section where they are 7.5 mm. by 4 mm. The leathery, erect, conduplicate and shiny olive-green leaves are ovate with slightly recurved apices and 110-170 mm. long, 40-55 mm wide and 1.5-2 mm. thick.

Inflorescence and Flowers: One or 2, 100-140 mm., slightly bent inflorescences emerge from the apices of 1 to 3 year-old leaf stems with 9 to 12 flowers, which bloom simultaneously. The spathe is 15-18 mm. long, the bracts 4-5 mm. and the ovary only 1.5 mm. The strongly keeled, lanceolate sepals are concave at their bases, slightly convex at their apices and yellow turning green towards the apices. The 10 mm. long dorsal sepal is 2.3 mm. wide with small wine-red spots. The laterals are 90% fused and together, 9.8 mm. long by 5.5 mm. wide with dense wine-red spots. The yellow, lanceolate petals are 4.4 mm. long and 1.2 mm. wide and have 3 green longitudinal veins and some purple points at their bases and faintly serrate margins to the apical halves. The fleshy, tongue-shaped lip is 5 mm. long and 2.1 mm. wide with 2 purple ear-like appendages at its purple base. The 2 small triangular side-lobes and the margins at the rounded apex are yellow. The fleshy plate is slightly undulate, greenish with purple points. The yellow column has a purple dotted foot and is 4 mm. long and with spread wings 2 mm. wide.

Flowering Periods: March. The flowers last for two to three weeks.

Pleurothallis pantherina
Seeh. New species | *1251b* (Plate XV)

Etymology: *Pantherina* refers to the blotched colouring of the flower.

Habitat: Epiphytic from section VI, at 1050 M. asl, in relatively dry regenerating forest on the trunk of a mid-size tree, 5-15 M. below the canopy. L 6; H 4; W 5.

Occurrence: Rare. We have found few plants to date.

Plant: The 100-150 mm. high plant forms clusters with 10-30 leaves and an intense root system composed of whitish, 0.5 mm. diameter roots. The round leaf stems are alternate at 1 mm. intervals from the 1 mm. diameter, branching rhizome and are 40-80 mm. long by 1-1.8 mm. diameter and 2 tight, rugose, brown sheaths envelop 70% of their length. The flat leaves are narrowly lanceolate, olive-green, conduplicate and 50-90 mm. long, 10-15 mm. wide, 1.5-2 mm. thick with the upper halves up to 90° recurved.

Inflorescence and Flowers: Two 50 mm. inflorescences arise simultaneously from the apices of 1 or 2 year-old leaf stems with the first flower at 25 mm. from the base, followed by 3 to 5 consecutive, alternate flowers but only 2 flowers bloom together. The lanceolate, strongly keeled, concave sepals are light yellow, turning darker towards the margins and the obtuse apices. They have wine-red markings along the longitudinal veins, which are smaller and paler towards the margins. The dorsal sepal is 7.2 mm. long by 2.7 mm. wide and the laterals are 80% fused and together measure 6.3 mm. long and 4.1 mm. wide with a lighter coloured area between the keels. The spathulate petals are pale rose and cream towards the apices, speckled with wine-red dots and 4 mm. long by 1.6 mm wide with serrate margins at the apex. The elliptical lip is also speckled with wine-red dots and is 4.4 mm. long by 2.1 mm. wide with a yellow-green base, turning pink towards the side-lobes and ochre towards the apex. The wine-red margins are serrate from the indistinct side-lobes to the apex. The faintly serrate longitudinal callosities are parallel to the margins. The plate of the lip is faintly warty and the apex tuberculate. The keels of the lip's underside have green and purple warts and the yellow-green column has wine-red markings and is 3.5 mm. long by 2 mm. wide.

Flowering Periods: October to April. The flowers last for two to three weeks and the inflorescence for two to three months.

Pleurothallis cryptophoranthoides
Loefgr. | *1252* (Plate XV)

Etymology: From the Greek *Cryptophoranthus*-like as the flower opens like those in that genus (now named *Zootrophion*).

Habitat: Epiphytic from section VI, at 1300-1500 M. asl, in humid original forest on mid- and low-tree branches. L 6; H 5; W 6.

Occurrence: Rare. We have found only a few solitary plants and a small colony.

Plant: The 35-80 mm. long, pendent, 1 mm. diameter leaf stems emerge alternately at 3 mm. intervals from a 1-1.5 mm. diameter, creeping, branching rhizome and are slightly compressed at their apical end and 60% covered by 2 wide sheaths. The flat leaves are lanceolate, conduplicate, slightly asymmetric, fleshy, dark olive-green and 80-110 mm. long, 7-15 mm. wide and 2 mm. thick. Their undersides turn purple in high light.

Inflorescence and Flowers: One or 2, 30-60 mm. inflorescences emerge from the apex of young and older leaf stems with up to 8 flowers which bloom simultaneously. In typical Cryptophoranthus pattern, the flower opens only a slit in the middle third between the dorsal and the lateral sepals which are lanceolate, keeled, fleshy, light yellow at their bases, purple in the middle and greenish at their apices. The 5.5-7 mm. long dorsal sepal is 2.2 mm. wide with 3 purple longitudinal veins fading at the apices. The laterals are 90% fused and together are 5-8 mm. long by 4.0-5.5 mm. wide. The light greyish-yellow petals are spathulate with faintly pointed, blunt apices and 2.0-2.5 mm. long by 1.7-2 mm. wide. The tongue-shaped lip is yellow, 2.8 mm. long by 1.2 mm. wide with 2 ear-like appendages at its base and, on the apical half, 2 semicircular, erect lamellae with undulate margins. The plate is also undulate and the margins of the acute apex are toothed. The light green and purple marked column is 2 mm. long and 1.2 mm. wide and has a yellow base.

Flowering Periods: February through April. The flowers last for three weeks.

Pleurothallis dracula
Seeh. New Species | *1252b* (Plate XV)

Etymology: Refers to the dark, hairy and strangely shaped flower.

Habitat: Epiphytic from section VI, at 700-900 M. asl, in humid original cloud forest on the upper branches of mature trees. L 6; H 5; W 6.

Occurrence: Rare. We have found numerous plants, but restricted to two fallen trees.

Plant: This 70-100 mm. high epiphyte forms clusters with 10-30 leaves and an intense root system composed of white or greenish roots, 0.7 mm. diameter. The round leaf stems are 30-40 mm. long and 0.7-1.0 mm diameter and are 70% enveloped by 3 tight, short-lived, brown sheaths. The leaves are obovate, leathery, dark olive-green, indistinctly keeled, furrowed and measure 50-70 mm. long, 12-16 mm. wide and 1 mm. thick.

Inflorescence and Flowers: One or 2, 35 to 45 mm. inflorescences arise from the most recent leaf stems,

with the first flower at half leaf height, and up to 7 consecutive flowers after several weeks rest. The acute spathe is 2-4 mm., the ovary 3-4 mm. and the pedicel 4-8 mm long. The fleshy, distinctively keeled, light olive-green sepals are hairy on the inside, concave at the base and convex at the apex. The obovate dorsal sepal is 9.5 mm. long by 4.8 mm. wide with longitudinal veins and the margins light olive-green with dark wine-red speckles. The laterals are similarly coloured and are 75% fused, together measuring 8 mm. long by 6 mm. wide. The spathulate petals are pale green, 4 mm. long by 2.7 mm. wide and have very fleshy, papillate, dark purple apices with serrate apical margins. The longitudinal veins and the margins are coloured or speckled dark purple. The tongue-shaped lip is whitish to light green, 2.7 mm. long by 1.1 mm. wide, has 2 ear-like appendages at its base and at the mid-section 2 erect side lobes with serrate margins to the rounded, fleshy, tuberculate and dark purple apex. The lip surface is undulate along the central vein. The light green, slightly curved column is 2.5 mm. long by 1.4 mm. wide with dark purple apical margins and a granular base.

Flowering Periods: February through April. The flowers last for 15-20 days.

Pleurothallis klotzschiana
Rchb. f. | *1255* (Plate XV)

Etymology: **Johann Friedrich Klotzsch** was a German botanist.

Habitat: Epiphytic from section VIII, at 1300 M. asl, in humid original cloud forest. L 5; H 5; W 5.

Occurrence: Rare. We have found only one plant.

Plant: This 80-110 mm. high epiphyte forms a loose cluster with 8 leaves and a wiry root system of 1.2 mm. diameter whitish roots. Pairs of 20-30 mm. long and 2.5-3 mm. diameter leaf stems arise at 10 mm. intervals from the 4 mm. diameter rhizome and are enveloped by 3 loose, brown dotted sheaths The apical segments of the leaf stems are furrowed. The leaves are elliptic, stiff, conduplicate, erect, dark olive-green and measure and 45-70 mm. long, 20-27 mm. wide and 3 mm. thick.

Inflorescence and Flowers: Up to two 50-70 mm. long inflorescences arise from the apex of young and older leaf stems with 10-15 closely packed, alternate flowers on 0.5 mm. long pedicels. They open in quick succession and bloom simultaneously. The spathe is 12 mm. long, and the bracts are 3 mm. The lanceolate sepals are fleshy, keeled, with olive-green and dark wine-red hairy backs, while the warty inside is yellow-green with dark wine-red veins and apices. The 7.5 mm. long dorsal sepal is 3.6 mm. wide and the laterals, which are 90% fused, are 8 mm. long and 5.6 mm. wide together. The arrow-head shaped petals are 2.5 mm. long by 1.3 mm.

wide with acute, pointed apices and undulate margins, light yellow-green and speckled light wine-red apices. The fleshy, tongue-shaped lip is 2.5 mm. long by 1.5 mm. wide with a smooth, transverse callus, 2 hook-shaped appendages at its light green base and 2 large, darker longitudinal callosities arise beside the 2 small and round side-lobes. The rounded apex with recurved faintly toothed margins is faintly warty and dark purple. The sturdy column is 2 mm. long by 1.5 mm. wide, olive-green and purple and the column foot has short-hairs.

Flowering periods: January. The flowers last for two weeks.

Pleurothallis porphyrantha
Krzl. | *1257* (Plate XVI)

Etymology: *Porphyro* and *antha* are Latin for purple flower.

Habitat: Epiphytic from section VI, at 1050 M. asl, in regenerating forest, 3 M. above ground on the trunk of a medium size tree. L 4; H 5; W 3.

Occurrence: Rare. We have found only one plant to date.

Plant: This epiphyte formed a cluster with 30 pendent leaves with whitish or light green 1 mm. diameter wiry roots. The leaf stem and the leaf are 200-300 mm. long together. The round leaf stems are 150-180 mm. long and 1.3-3.5 mm. in diameter and emerge alternately at 5-10 mm. intervals from the 2-4 mm. diameter, green rhizome bent 40°downwards, and 3 sheaths envelop 40% of the leaf stem whose upper section is furrowed and elliptical in cross-section. The ovate, flat or slightly convex leaves are leathery, dark olive-green, conduplicate and are 80-100 mm. long, 25-35 mm. wide and 1.5-3 mm. thick.

Inflorescence and Flowers: One or 2, 35-45 mm. inflorescences arise from the apex of young and older leaf stems. Up to 6 flowers on 2-3 mm. pedicels open in rapid succession and bloom together. The acute spathe is 25 mm long and the hairy ovary 1.5 mm. The sepals are fleshy, keeled, light yellow-green and hairy on the outside. The tongue-shaped dorsal sepal is 10.5 mm. long by 2.7 mm. wide with purple longitudinal veins and margins. The ovate laterals are 95% fused, 9.8 mm. long and 6 mm. wide together and light yellow-green, speckled purple with less intensity towards the acute, green apices. The spathulate petals are 4.8 mm. long by 2.4 mm. wide, pale yellow-green, with serrate margins at the acute apices. The base, the longitudinal veins and margins are speckled dark purple. The light yellow-green, 5 mm. long and 3.5 mm. wide lip has 2 small, hook-shaped appendages at its base and 2 erect, sickle-shaped side-lobes with serrate margins. The margins of

the obtuse, fleshy, tuberculate and dark purple-coloured apex are toothed. The light green, slightly curved, 4 mm. long and 1.7 mm. wide column has a purple base.

Flowering Periods: October to November. Single flowers last for 10 days.

Pleurothallis albo-rosea
(Krzl.) Brade | *1260* (Plate XVI)

Etymology: Latin for white and pink - the flower's colours.

Habitat: Epiphytic from section VI, at 1200-1500 M. asl, in original cloud forest on south facing slopes or along river banks on low and mid-tree branches. L 2-4; H 7-8; W 2-3.

Occurrence: Rare.

Plant: The round, pendent 100-200 mm. long and 1.2-2 mm. diameter leaf stems emerge at 3-8 mm. intervals from the creeping and branching 1-2 mm. diameter rhizome, enveloped for 50% of their length by 3 tube-shaped, short-lived sheaths, up to 40 mm. long. The fleshy leaves are narrowly lanceolate, almost linear, conduplicate, dark olive-green, turning purple in high light and are 100-200 mm. long, 7-12 mm. wide and 2-3 mm. thick. The whitish roots are 0.7 mm. diameter.

Inflorescence and Flowers: One or 2, inflorescences emerge from the apex of young and older leaf stems. The 50-75 mm. inflorescences bear 8 to 18 flowers, which open in rapid succession and bloom simultaneously. The fleshy, lanceolate, keeled sepals are concave at their light yellow-green bases and slightly convex at their apices, hairy on the backs and with 3 wine-red longitudinal veins. The dorsal sepal is 9.5 mm. long by 2.7 mm. wide and the laterals are 90% fused and 8.5 mm long by 5 mm wide together. The spathulate petals are ivory-white, 4 mm. long by 1.3 mm. wide, with a light wine-red central vein and speckled purple on the acute apex. The tongue-shaped lip has an acute apex and is 5.5 mm. long by 2 mm. wide, light yellow with dark wine-red speckles and a yellow transverse callus at the base; the plate of the lip is granular to warty and the side-lobes are small and erect. The yellow-green column is wine-red on the back and 3.7 mm. long by 2 mm. wide with spread wings.

Flowering Periods: February through to April. The flowers last for three weeks.

Pleurothallis macropoda
Barb. Rodr. | *1262* (Plate XVI)

Etymology: *Macro* and *poda* are Greek. for large foot and may refer to the swollen stem base.

Habitat: Epiphytic from section V, at 700-850 M. asl,

in original forest on lower, thick branches. L 3-4; H 6; W 2-4.

Occurrence: Occasional.

Plant: This 250-500 mm. high epiphyte forms clusters of 10-20 leaves and whitish or light green, 0.8-1.0 mm. diameter, wiry roots. The round, erect leaf stems are 120-230 mm. long, alternate at 5-15 mm. gaps from the 4-6 mm. diameter, sheathed rhizome which is 3 mm. diameter at the bases, widening to 4 by 5 mm. at their apices and the upper section is furrowed for half its length. Their acute sheaths are light green with dark purple dots when young. The leaf stems are partly covered by 2 short-lived sheaths up to 60 mm. long, light brown, dark brown dotted and granulose. The leaves are elliptic, acute, slightly infolded at their bases, leathery, dark olive-green, conduplicate and 90-120 mm. long, 28-40 mm. wide and 2-3.5 mm. thick.

Inflorescence and Flowers: One or two 80-130 mm. inflorescences arise from the apex of young and older leaf stems with 12 to 20 flowers, on very short pedicels, which open in rapid succession, blooming simultaneously. The acute spathe is 20 mm.; the bracts are 1-3 mm. and the hairy ovary 2-3 mm. All parts of the flower are fleshy and strongly keeled. The sepals are lanceolate, light yellow-green, undulate to granular inside with hairy backs. The obovate dorsal sepal is 8.5 mm. long by 3.3 mm. wide with light green longitudinal veins. The laterals are 85% fused, 8.3 mm. long and 5.8 mm. wide together, yellow green and sometimes purple-speckled. The spathulate petals are pale yellow-green, 2.5 mm. long by 1 mm. wide with serrate margins at their acute apices. The broadly elliptic lip is 3.8 mm. long and 2.7 mm. wide, yellow-green, with 2 ear-like appendages at its base. Two longitudinal callosities extend from the base to the centre and the plate is undulate to warty, the golden yellow margins are serrate and the apex is acute. The light green, slightly curved column is 3 mm. long and 1.8 mm. wide and has a granular or hairy foot and lacerate wings.

Flowering Periods: April to May. A single flower lasts for 10-14 days.

Pleurothallis octophrys
Rchb. f. | *1273* (Plate XVI)

Etymology: Octophrys is Greek for eight eyebrows possibly referring to the rows of hairs on the lip.

Habitat: Epiphytic from sections VI and VIII, at 900-1200 M. asl, in original cloud forest, on mid-tree branches of large trees. L 4-7; H 3-5; W 4.

Occurrence: Occasional.

Plant: This 120-250 mm. high epiphyte forms loose clusters with 7 to 12 leaves, slightly pendent when in flower, and wiry, whitish 0.7 mm. diameter roots.

The round leaf stems arise at 2 mm. intervals from the creeping, branching 1.5-2.0 mm. diameter rhizome and are 50-70 mm. long and 0.8 mm. diameter and 80% enveloped by 3 tube-shaped, rugose, hairy, purple-speckled sheaths. The leaves are linear to narrowly lanceolate, fleshy, faintly conduplicate, dark olive-green and 70-100 mm. long, 5-6 mm. wide and 1.5-2.5 mm. thick. In high light, the backs of the leaves are more or less purple-speckled.

Inflorescence and Flowers: Up to 4, 70-100 mm., curved to pendent inflorescences emerge from the apex of young and older leaf stems with 5 to 12 consecutive flowers and up to 5 bloom at once. The 6 mm. spathe, the 3 mm. pedicels, the bracts and the 2 mm. long ovary are hairy. The sepals and petals are hairy on the outsides and along the margins and the lanceolate, faintly keeled sepals are light yellow-green, with red-brown margins and middle section, but green longitudinal veins. The dorsal is 10.5 mm. long by 2.7 mm. wide with a golden yellow apex. The laterals, always on the upper side, are 85% fused and together are 9 mm. long by 6 mm. with yellow-green apices. The petals are hammer-shaped, pale yellow-green and 3 mm. long by 2 mm. wide. The central veins are speckled red and the apices are flat and pointed with serrate and red speckled margins. The square, fleshy, hairy lip is 3 mm. long by 2 mm. wide, yellow-green with dark purple speckles and an obtuse apex with 2 dark purple, hairy dots and hairy margins to the longitudinal callosities. The yellow-green, wine speckled column is 3.5 mm. long and with spread wings 1.5 mm. wide and with almost entire margins.

Flowering Periods: April to June and the flowers last for two weeks.

Pleurothallis strictophylla
Schltr. | *1275* (Plate XVI)

Etymology: Stictus and *phylla* are Latin and mean very straight leaf.

Habitat: Epiphytic from sections I and IV, at 900-1300 M. asl, in humid original cloud forest, on trunks and lower and mid-tree branches. L 4-6; H 6; W 4.

Occurrence: Occasional.

Plant: This 70-120 mm. high epiphyte forms clusters with 10-50 leaves and a root system composed of 0.5 diameter green roots. The leaf stems arise alternately at 1 mm. intervals from a 2 mm. diameter rhizome and are 50-70 mm., long, 0.6 mm. diameter, round at their bases and enveloped by a membranous short-lived sheath and their apical segments are furrowed and 1 mm. by 1.5 mm. in cross-section. The erect leaves are narrowly lanceolate to linear, stiff, conduplicate, 40-60 mm. long, 5-6 mm. wide, 1.5-2 mm. thick and olive- to yellow-green with dark purple points depending on light exposure.

Inflorescence and Flowers: The 30-35 mm. erect inflorescence arises from the apex of the most recent leaf stem and produces the first flower at 15 mm. from the base followed by 5-7 flowers in quick succession at 4 mm. intervals. Up to 5 flowers may bloom simultaneously. The spathe is 5 mm., the bracts are 2.5 mm., both have purple points. The pedicels and the ovary are 1 mm. The concave sepals are lanceolate, distinctly keeled, light green and golden-green, speckled with purple dots. The 7 mm. long dorsal sepal is 2 mm. wide. The laterals are 95% fused and 6.7 mm. long by 3.7 mm. wide together. The lanceolate, 5.5 mm. long petals are 2 mm. wide with finely serrate apical margins. The lip is 3-lobed, 4.1 mm. long and 2 mm. wide with a yellow-green transverse callus at the base. The light green, slightly curved column is 2.7 mm. long by 1.1 mm. wide with lacerate wings and apex and the foot tip is hairy.

Flowering Periods: October and February to March; in good conditions it will flower several times a year. The flowers last for up to 15 days and the inflorescence for a month.

Pleurothallis lingua
Lindl. | *1279* (Plate XVI)

Etymology: Lingua is Latin for tongue and refers to the tongue-shaped lip.

Habitat: Epiphytic from sections V, VI, VII and VIII, at 700-1200 M. asl, in the shade of humid original cloud forest, on trunks and lianas but rare on branches. L 2-4; H 8; W 2.

Occurrence: Common.

Plant: A 25-40 mm. high epiphyte forming clusters of 10-60 leaves. The leaf stems are 0.4 mm. diameter, 5-10 mm. long and enveloped by 3 tube-shaped, membranous sheaths. The linear-lanceolate leaves are faintly conduplicate, light green, finely attenuated to their bases and are 25-30 mm. long, 3-5 mm. wide and 1 mm. thick.

Inflorescence and Flowers: A single 3-5 mm. inflorescence arises from the apex of young and older leaf stems and bears 1 or 2 flowers on 2-3 mm. pedicels. The ovary is 1 mm. long. The ovate, keeled sepals are light ochre at their bases and light yellow-green at their acute apices. The margins and 3 longitudinal veins are wine-red. The 4.8 mm. long dorsal sepal is 1.8 mm. wide and the laterals are 95% fused and 4.5 mm. long by 2.2 mm. wide together. The spathulate, acute petals are pale ochre, 2.9 mm. long by 0.8 mm. wide with purple margins and central veins. The tongue-shaped lip is light ochre, with 2 indistinct side lobes, 2 crenate, dark wine longitudinal callosities and is 4 mm. long by 1.3 mm. wide. The crenate margins, the apex and the

longitudinal veins are wine-red. The light green column is 3 mm. long and 1.3 mm. wide with purple margins and faintly gnawed wings and apex which are almost the same length.

Flowering Periods: October to June. The flowers last for two weeks and a leaf stem apex may produce a second inflorescence during the year.

General: We have also found plants with greenish-yellow flowers.

Pleurothallis crinita
Barb. Rodr. | *1280* (Plate XVI)

Etymology: Crinita is Latin for with long weak hairs referring to the hairs on all flower parts.

Habitat: Epiphytic from section VI, at 800-900 M. asl, in very humid original mountain cloud forest, on trunks, lianas and on mid- to low-tree branches. L 3; H 8; W 2.

Occurrence: Occasional.

Plant: This creeping, branching epiphyte forms plants with 10-30 leaves addpressed to bark. The whitish or light green roots are 0.5 mm. in diameter. The leaf stems are 8-12 mm. long by 1-1.5 mm. diameter, totally enveloped by sheaths and emerging at 4-5 mm. intervals from the 1.5-2.5 mm. diameter, totally sheathed rhizome. The leaves are obovate, conduplicate, yellowish olive-green, speckled purple on both sides and are 40-50 mm. long, 20 mm. wide and 1.5 -2 mm. thick. They are attenuated abruptly to their bases to form 4 mm. twisted false petioles causing the inflorescences to appear under the leaves.

Inflorescence and Flowers: The 15-17 mm. inflorescences emerge from the apex of the most recent leaf stem. Three to 5 consecutive flowers develop on the apical 12 mm. and up to 2 may bloom together. The flower has a pronounced chin. The fleshy, keeled, triangular sepals are concave with slightly incurved margins and whitish hairs up to 2 mm. long cover the outsides. The light yellow-green sepals are speckled purple with diminishing intensity towards the apices. The dorsal is 6.2 mm. long by 2.1 mm. wide and the laterals are 90% fused and 5.1 mm. long by 5.1 mm. wide together. The spathulate, acute petals are pale ochre with yellow margins and central veins and are 2.2 mm. long by 1.3 mm. wide. The light yellow, obovate lip is 4 mm. long by 2 mm. wide with 2 indistinct sickle-shaped side lobes. The main lobe is elliptic, fleshy, yellow with purple speckles and serrate margins. The yellow column is 2.7 mm. long, 1.6 mm. wide and the purple apex is gnawed.

Flowering Periods: February. The flowers last for two weeks.

Pleurothallis karlii
Pabst | *1281* (Plate XVI)

Etymology: Not known.

Habitat: Epiphytic from section VIII at 1300-1400 M. asl, in original cloud forest on branches of older trees, exposed to wind and sunlight. L 7-9; H 4-5; W 7.

Occurrence: Occasional. We have only found one location but with many plants.

Plant: A creeping, branching epiphyte with five to many leaves and relatively few greenish to beige roots, 0.5-0.8 mm. diameter. The leaf stems are 5-7 mm. long and 1 mm. diameter and emerge alternately at 4-8 mm. intervals from a 2-2.5 mm. diameter sheathed rhizome. The elliptic leaves are 25-40 mm. long, 15-28 mm. wide and 1 mm. thick with wide wedge-shaped bases and are leathery, conduplicate, dark olive-green with purple-speckled undersides.

Inflorescence and Flowers: One or two 5-10 mm. inflorescences emerge from the apices of young and older leaf stems with 2 to 4 flowers on 1.2 mm. pedicels and may bloom simultaneously. The spathe is 2.5 mm., the pedicels are 1.2 mm. and the ovary is less than 1 mm. long. The sepals are acutely ovate, keeled and smooth with purple bases and light green apical parts with purple freckles. The dorsal sepal is 15% fused to the basal margins of the laterals and measures 7-9 mm. long by 3 mm. wide. The laterals are 95% fused and 6.5-7.5 mm. long and together, 4.5-5.5 mm. wide. The lanceolate petals are 3 mm. long and 1.2-1.4 mm. wide, dark purple with a lighter apex. The tongue-shaped lip is 3 mm. long by 2.2 mm. wide, slightly fleshy and warty with 4 erect, finger-shaped appendages forming a transverse callosity at the base. The lip is light yellow with purple freckles, purple longitudinal lamina and purple central margins and the round main lobe is convex. The purple column is 3 mm. long and, with spread wings, 1.5 mm. wide with a lighter apex and column foot.

Flowering periods: November to February and the flowers last for ten days.

Pleurothallis recurva
Lindl. | *1286* (Plate XVI)

Etymology: *Recurvatus* is Latin for curved backwards.

Habitat: Epiphytic from sections IV, VI and VIII, at 600-1400 M. asl, in original cloud forest on branches of older trees, exposed to wind and sunlight. L 7-9; H 8; W 7.

Occurrence: Occasional.

Plant: This creeping and branching epiphyte has 5 to 15 leaves and relatively few greenish to beige roots, 1.5 mm. in diameter. The 4-7 mm. long and 1.5-2 mm. diameter leaf stems arise alternately at 4-5 mm. intervals from a 2-2.5 mm. diameter, rhizome, entirely covered by sheaths for their first year. The ovate leaves are fleshy, conduplicate, dark olive-green with purple speckled undersides and measure 25-40 mm. long, 15-22 mm. wide by 3-5 mm. thick and are only slightly elevated above the substrate.

Inflorescence and Flowers: One or two 25-50 mm. inflorescences arise from the apices of the most recent leaf stems with 3 to 9 flowers on 2-4 mm. pedicels and up to 5 may bloom at once. The ovary is 2.5 mm. long and the bracts 3 mm. The fleshy, obovate sepals are concave at their bases, convex at their apices, faintly keeled and hairy on the back. The dorsal sepal is 9.5 mm. long and 3.5 mm. wide with a light green basal half turning wine-red towards the granular, hairy apex. The 8.5 mm. long laterals are 95% fused and together are 6.3 mm wide and the surface of the purple base is granular turning hairy and wine-red towards the apex. The spathulate petals are pale green, 3.6 mm. long and 1.6 mm. wide with serrate margins on their apical halves. The fleshy, tongue-shaped lip is purple or wine-red, 3.5 mm. long by 2 mm. wide with 2 erect side-lobes and darker coloured plate and fleshy lamellae. The round main lobe is convex. The cream coloured column is 3.3 mm. long and with spread wings 2.8 mm. wide. The wings are relatively large and only faintly gnawed.

Flowering Periods: December to April. The flowers last for ten days.

General: The flower colour varies from dark ruby red to almost white and the side-lobes may be more or less distinct. We also found plants with 70 mm. long leaves.

Pleurothallis leucopyramis
Rchb. f. | *1289* (Plate XVII)

Etymology: *Leuco-pyramis* are Greek for white-pyramid, referring to the triangular dorsal sepals.

Habitat: Epiphytic from sections VI and VIII, at 1100-1300 M. asl, in humid original forest on mid-tree branches and in swampy forests on low stems. L 5; H 8; W 4-6.

Occurrence: Common.

Plant: This 80-100 mm. high epiphyte forms clusters of 7 to 25 leaves with an intense root system of 0.5 mm. diameter whitish roots. Two trumpet-shaped sheaths with short, stiff hairs and recurved margins envelop the leaf stems, which are 40-50 mm. long and 0.5-1.0 mm. diameter. The faintly conduplicate leaves are obovate, light green with faint purple blotches and 40-50 mm. long, 13-15 mm. wide and 0.7 mm. thick. The backs of the leaves are blotched purple.

Inflorescence and Flowers: One or two 80-110 mm.

inflorescences arise from the apex of young and older leaf stems with their first flower at leaf height followed in rapid succession by up to 13 pure white, fragrant flowers which bloom at the same time. All sepals have very distinct keels. The lanceolate dorsal sepal is concave at the base and convex at the apex and 10 mm. long by 1.2 mm. wide. The 95% fused laterals are 9 mm. long and 4.9 mm. wide together, while the lanceolate petals are white and 3.4 mm. long by 1 mm. wide. The tongue-shaped lip is 2.5 mm. long by 1.2 mm. wide, light yellow-green, fleshy, with 2 erect side lobes, 2 indistinct ear-like appendages at its base and a blunt apex with a point. The sturdy column is light green and 2.0 mm. long by 1.4 mm. wide including spread wings.

Flowering Periods: December to February and April. The flowers last for 3 to 4 weeks and the inflorescence for more than a month.

Pleurothallis platystachya
Regel | *1293* (Plate XVII)

Etymology: Platy- and *stachy-* are Greek for a broad spike describing the flattened inflorescence.

Habitat: Epiphytic, occasionally terrestrial from sections V, VI and VIII, at 900-1300 M. asl, in original and regenerating cloud forest, in humus and on low to mid-tree branches and trunks. L 3-6; H 5-7; W 3-5.

Occurrence: Common.

Plant: The plant is up to 300 mm. high and forms clusters with 7 to many leaves and a root system composed of 1.5 mm. diameter wiry, adherent roots. The leaf stems are laterally compressed and arise alternately at 3 mm. intervals from a branching, 4-5 mm. diameter rhizome. They are 60% enveloped by 3 sheaths and are 150-200 mm. long with a cross-section of 1.5 by 2 mm. at their bases widening to 9 mm. by 2 mm. with 2-edged cross-section at the leaf junctions. Their apical segments are twisted by 45-90°. The leaves are narrowly ovate, leathery, conduplicate, erect, grass to yellow-green with purple margins, depending on light, and are 100-140 mm. long, 20-30 mm. wide and 1.5-2 mm. thick.

Inflorescence and Flowers: The 200 mm. inflorescence is spicate, laterally compressed and arises from the apex of the most recent leaf stem. The first flower blooms at three quarters of the leaf height, on a very short pedicel, almost sitting in the acuminate bract. It is followed in rapid succession by 10-15 alternate flowers, which are directed upwards, and bloom simultaneously. The spathe is 15 mm. long and the peduncle 75-90 mm. The dorsal sepal is 15 mm. long and 3 mm. wide. The laterals are completely fused and 14 mm. long by 3.7 mm. wide. All are concave, lanceolate and speckled with wine-red blotches on a yellow-green base and

distinctly laterally compressed, thus showing sharp keels. The 7.5 mm. long, sword-shaped petals are 1.4 mm. wide with faintly pointed apices and faintly sawed margins. The narrowly lanceolate lip is 10 mm. long by 3 mm. wide, fleshy, light yellow-green with 2 erect, indistinct side-lobes close to its base. The lowered main vein and the apex have wine-red speckles. The 5 mm. long by 2 mm. wide column is speckled wine-red and has indistinct wings.

Flowering Periods: August to October. The flowers last for two to three weeks.

Pleurothallis auriculata
Lindl. | *1295* (Plate XVII)

Etymology: Auriculata is Latin for ear-like appendages, referring to the lip side-lobes.

Habitat: Epiphytic from sections V, VI, VII and VIII, at 600-1300 M. asl, in humid original cloud forest, on trunks and lower and mid-tree branches. L 4-6; H 6; W 4.

Occurrence: Common.

Plant: This 120-250 mm high epiphyte forms clusters with 6 to 20 leaves and a profuse root system, composed of 0.7 mm. diameter whitish roots. The leaf stems arise alternately at 1 mm. intervals from a 0.8 mm. diameter rhizome and are 70-140 mm. long and 0.6 mm. in diameter round their bases and 60% enveloped by 3 membranous sheaths. Their apical segments are deeply furrowed and 2 mm. by 1 mm. at their Y-shaped cross-sections. The narrowly ovate, leathery, conduplicate, erect leaves are 60-100 mm. long, 7-12 mm. wide, 1-1.5 mm. thick and are olive-green to dark wine-red depending on the light conditions.

Inflorescence and Flowers: Up to three 35-50 mm., erect or pendent inflorescences emerge from the apices of the most recent leaf stems with 5 to 10 laterally compressed flowers on 2 mm. pedicels which open in quick succession and may bloom simultaneously. The concave sepals are lanceolate, keeled and yellow-green, speckled with wine-red blotches. The 5 mm. long dorsal sepal is 2.3 mm. wide and the laterals are 95% fused and 5 mm. long by 3.7 mm. wide together. The lanceolate, 3.8 mm. long petals are 1.6 mm. wide, light green and purple from their bases to their midsections, while their apical margins are serrate. The lip is tongue-shaped, 3.8 mm. long and 1.8 mm. wide with a purple centre. The margins, the erect sickle-shaped side-lobes and the sharply pointed apex are light green with a smooth, transverse callus at the base. The light green, slightly curved column is 2.2 mm. long by 1.4 mm. wide with light purple speckles on the wings and apex and a hairy foot. The stigma has a purple margin.

Flowering Periods: November to March. The flowers last for up to 20 days and the inflorescence for two months.

Pleurothallis capillaris
Lindl. | *1298* (Plate XVII)

Etymology: Not known.

Habitat: An epiphyte from section VI, at 1000-1100 M. asl, in humid original mountain forest on trunks and low, thick, moss-covered branches. L 5; H 8; W 4.

Occurrence: Occasional.

Plant: This 80-130 mm. high epiphyte forms clusters with 7 to 20 leaves and 0.7 mm. diameter roots. The round leaf stems are 70% enveloped by 3 overlapping sheaths and are 50-70 mm. long and 0.7 mm. diameter at the base. Their mostly twisted apical segments are furrowed and 2 mm. by 1.5 mm. with Y-shaped cross-sections. The leaves are narrowly oval, stiff, conduplicate, erect, light green and are 50-65 mm. long, 10-13 mm. wide and 1 mm. thick.

Inflorescence and Flowers: Up to 3 erect or slightly pendent 100-120 mm. inflorescences arise from the apex of young and older leaf stems and bear 15-20 flowers. The first flower blooms at three quarters of the leaf height followed sequentially by others at 1 or 2 day intervals. Half of the flowers may bloom at once. The keeled, lanceolate sepals and petals are light yellow-green. The dorsal sepal is 6.6 mm. long by 2 mm. wide, while the laterals are 90% fused and 6.4 mm. long by 3.0 mm. wide together. The lanceolate, light yellow-green petals are 3.9 mm. long by 1.6 mm. wide and have faintly serrate margins at their acute apices. The tongue-shaped lip is light green, 3.3 mm. long and 2 mm. wide with a smooth, transverse callus at its base and 2 erect hook-shaped side-lobes. The surface of the lip's plate is slightly undulate, while the apex is acute. The light green, slightly curved column is 2.5 mm. long and 1.5 mm. wide and has wide and finely acuminate wings.

Flowering Periods: October to December. The flowers last for a week and the inflorescence 2 months.

Pleurothallis hygrophila
Barb. Rodr. | *1301* (Plate XVII)

Etymology: *Hygro-philus* is Greek for moist-loving describing its moist mossy habitat.

Habitat: Epiphytic from section VI, at 1000-1300 M. asl, in humid original forest on moss-covered mid-tree branches. L 3-6; H 7-9; W 4.

Occurrence: Rare.

Plant: This 30-70 mm high epiphyte forms small clumps with 6 to 10 leaves and 0.7 mm. diameter

whitish roots, mostly covered by short moss. The leaf stems are alternate at intervals of less than 1 mm. on a 0.7 mm. diameter rhizome and are 20-35 mm. long, 60% enveloped by 2 or 3 sheaths and round, 0.5 mm. diameter at their bases, while the apical furrowed segments widen to 0.8-2.0 mm. by 0.6-1.0 mm. in a Y-shaped cross-sections. The narrowly ovate leaves are fleshy, conduplicate, erect and 25-40 mm. long, 5-6 mm. wide and 1-2.5 mm. thick and dark olive-green with purple-speckled backs in high light.

Inflorescence and Flowers: The erect or slightly bent, 40 mm. inflorescences arise from the apex of young and older leaf stems bearing 4 to 7 flowers on 0.5 mm. pedicels which open in quick succession and may bloom simultaneously. The lanceolate sepals are keeled, have a fleshy apex and are light green with faintly purple splotches. The 5 mm. long dorsal sepal is 1.5 mm. wide, while the laterals are 90% fused and 4.2 mm. long by 2.3 mm. wide together. The lanceolate petals are 3.6 mm. long by 0.8 mm. wide and have acuminate apices with serrate margins and are spotted purple on a very light green background. The light green, tongue-shaped lip is 2.7 mm. long and 1.3 mm. wide with a smooth, transverse callus at its base and 2 erect hook-shaped side-lobes and a light red or purple pointed apex. The light green, slightly curved column is 2 mm. long and 1.2 mm. wide with a light purple centre.

Flowering Periods: November to March. The flowers last for two to three weeks and the inflorescence for two months.

Pleurothallis longicaulis
Lindl. | *1303* (Plate XVII)

Etymology: *Longi-caulis* are Latin for long stem, referring to the leaf stem.

Habitat: Epiphytic from section VI, at 1000-1400 M. asl, in humid original forest on thick branches. L 3-5; H 7; W 2.

Occurrence: Occasional.

Plant: This 100-220 mm. high epiphyte forms clusters with 5 to 10 leaves. The leaf stems are 70-150 mm. long and 0.7 mm. diameter, arising alternately at 1 mm. intervals from a 0.8 mm. diameter rhizome. They are round at their bases and deeply furrowed at the apical segments forming a Y-shaped cross-section, 2 mm. by 1 mm. and are 60% enveloped by 3 membranous sheaths. The leaves are narrowly ovate, leathery or stiff, conduplicate, erect, yellow-green to olive-green depending on the light, and are 40-70 mm. long, 8-13 mm. wide and 1-2 mm. thick.

Inflorescence and Flowers: Up to three 60 mm., erect or pendent inflorescences emerge from the apex of the most recent leaf stem bearing 8 to 12 laterally compressed flowers on 2 mm. pedicels. The flowers open in quick succession and may bloom together. The keeled, lanceolate sepals are yellow-green or lemon. The 5.5 mm. long dorsal sepal is 1.8 mm. wide and the laterals are 90% fused and 4.5 mm. long by 3.5 mm. wide together. The lanceolate petals are 3.2 mm. long by 1.4 mm. wide and light yellow with green central veins. The margins of the acuminate apex are slightly undulate or serrate. The ovate lip is 3 mm. long by 2 mm. wide, light yellow-green or lemon with a smooth transverse callus at its base, 2 hook-shaped side-lobes and a pointed, granular apex occasionally with purple speckles. The light green, slightly curved column is 2.2 mm. long and 1.6 mm. wide and the wings and apex are pale rose.

Flowering Periods: November to May. The flowers last for two to three weeks and the inflorescence for six weeks.

Pleurothallis malachantha
Rchb. f. | *1304* (Plate XVII)

Etymology: Malacantha in Greek means yellow-green flower.

Habitat: An epiphyte from section VI at 1300 M. asl, in original forest found on low to mid-tree branches. L 4; H 7; W 4.

Occurrence: Occasional.

Plant: A 100-170 mm. high epiphyte found in groups of 10-15 leaves and with a root system composed of 0.7 mm. diameter, yellow roots. The leaf stems emerge alternately at intervals of 1.0 mm. from a 1.5 mm. diameter rhizome and are 70-120 mm. high and 1.0 mm. diameter at the base. The apical segments are grooved, 3 mm. long by 2.0 mm. in transverse section and are enveloped by 3 short-lived, brown-spotted sheaths. The erect leaves are 35 mm. long, 15-17 mm. wide, 2 mm. thick and are lanceolate, narrow and stiff.

Inflorescence and Flowers: Various, 80 mm. high, erect inflorescences emerge from the apices of the most recent leaf stems. The first flower opens at 40 mm. from the base of the inflorescence followed in rapid succession by 4-6 flowers at 8 mm. intervals. All seven flowers can be open at once. The spathe is 3 mm. long and splashed wine-red. The pedicel and ovary are 1.0 mm. long. The sepals are narrowly oval, keeled, pale green and splashed with rose. The dorsal is 7.0 mm. long by 2.9 mm. while the laterals are 95% fused and together measure 7.6 mm. long by 4.5 mm. The petals are spathulate, 3.6 mm. long by 1.9 mm. with lightly serrated apical margins. The tongue-shaped lip is 4.0 mm. long by 1.9 mm., pale yellowish-green with a transverse callus, which is green and purple at the base. The slightly curved column is 3.5 mm. long by 1.5 mm.

with lacerate wings. The apex and the base of the foot are purple and hairy.

Flowering Periods: January. The flowers are open for 15 days.

Pleurothallis saurocephala
Lodd. | *1314* (Plate XVII)

Etymology: Sauro-cephala is Greek for lizard head, referred to the shape of the flower.

Habitat: Epiphytic from sections III, VI and VIII, at 500-1300 M. asl, in humid original cloud forest along river banks on moss-covered mid-tree branches. L 5; H 7; W 4.

Occurrence: Occasional.

Plant: This 150-400 mm. high epiphyte forms loose clusters with 7 to 15 leaves and a wiry root system, composed of 1 mm. diameter whitish roots. The 80-180 mm. long leaf stems arise from the 2-4 mm. diameter rhizome, 60% enveloped by 3 loose sheaths and are 2-3.5 mm. in diameter just above the pseudobulb-like widened bases, which are 5 mm. diameter and 8 mm. long. The apical segments of the leaf stems are furrowed and 4-6 mm. by 3-5 mm. cross-section. The leaves are elliptic, stiff, conduplicate, erect, dark olive-green and 70-130 mm. long, 30-45 mm. wide and 3 mm. thick.

Inflorescence and Flowers: The 80-150 mm., shortly racemose inflorescences arise from the apex of young and older leaf stems with 12-18 closely packed, alternate flowers, which open in quick succession and bloom simultaneously. The first flower blooms at one third of the leaf height. The spathe is 20 mm. long. The lanceolate sepals are fleshy, keeled, with olive-green, hairy backs, while the inside is yellow-green with dark brown to almost black margins to the longitudinal veins and apices. The surfaces of their apical halves are undulate. The 10 mm. long dorsal sepal is 3 mm. wide and the laterals, which are 95% fused, are 10 mm. long and 5.4 mm. wide together with greenish to brown basal halves. The light yellow-green petals are speckled wine colour with increasing intensity towards their apices and are 2.4 mm. long by 1.5 mm. wide with pointed apices and undulate margins. The fleshy, tongue-shaped lip is 2.7 mm. long by 1.7 mm. wide with a smooth transverse callus, 2 hook-shaped ear-like appendages at its yellow-green base and 2 large darker longitudinal callosities arise beside the 2 small round side-lobes, enclosing an ochre depression. The rounded apex with recurved margins is granular and densely speckled purple. The sturdy column is light green and purple and 2 mm. long by 1.3 mm. wide. The narrow wings and the apex together are almost square.

Flowering Periods: November to January. The flowers last for two weeks.

Pleurothallis marginalis
Rchb. f. | *1319* (Plate XVII)

Etymology: Marginalis is Latin for with a margin referring to we know not what.

Habitat: Epiphytic from section VI, at 900-1600 M. asl, in humid original forest on mid and lower moss-covered branches and lianas. L 5; H 7; W 5.

Occurrence: Common.

Plant: This 30-35 mm. high epiphyte forms clusters with 10-80 leaves. The leaf stems are 2-3 mm. long by 0.3-0.4 mm. diameter and hidden in the profuse root system, of 0.6 mm diameter roots. The leaf stems are enveloped by 3 membranous sheaths, of which the 2 lower are less than 2 mm. long together. The obovate leaves are dark-green, faintly keeled, 20-30 mm. long, 5-6 mm. wide and 0.7 mm. thick, finely attenuated to their bases and mostly covered by purple markings on the back.

Inflorescence and Flowers: The 35-140 mm., erect inflorescences arise from the apex of the most recent leaf stems and produce up to ten flowers successively. The first flower opens at leaf height and up to 3 flowers bloom at once. The lanceolate sepals are concave, keeled, light yellow-green and the dorsal is 5.6 mm. long, 2.7 mm. wide with 3 red-brown longitudinal veins. The laterals are 60% fused and together are 5 mm. long by 3 mm. wide. The spathulate petals are 2.5 mm. long by 1 mm. wide, light green with red-brown central veins and acute apices. The tongue-shaped lip is 3.2 mm long by 1.3 mm wide, light yellow-green, fleshy with 2 very indistinct side lobes. These and the side veins are wine-red at the mid-section. The column is light yellow-green, 3 mm. long and 2 mm. wide with 2 drop-shaped protuberances at the column foot.

Flowering Periods: September to November and March to April. The flowers last for 7 to 10 days, and the inflorescence for more than a month.

Pleurothallis viridiflora
Seeh. | *1321b* (Plate XVIII)
New species

Etymology: Viridi and *flora* are Latin for green flower which describes this plant.

Habitat: Epiphytic from section VI, 900-1000 M. asl, on a lower horizontal branch in very humid original mountain cloud forest. L 7; H 7; W 6.

Occurrence: Rare. We have found a dozen plants on one hill side.

Plant: This epiphyte is 20 mm. high and forms clusters with 10-30 leaves and relatively few light brown roots, 0.7 mm in diameter. The 0.4 mm. diameter leaf stems are 4-6 mm. long and 2 membranous, short-lived, loose sheaths, up to 5 mm. long, cover them. The obovate, faintly conduplicate leaves are 15-18 mm. long, 3-4 mm. wide, 0.5 mm. thick and finely attenuated to their bases forming 3 mm. false petioles.

Inflorescence and Flowers: The 25-35 mm. inflorescence arises from the apex of the most recent leaf stem. The first flower develops at half leaf height followed by 2 to 5 flowers in rapid succession on very short pedicels and all bloom together. Both spathe and bracts are 1 mm. The sepals are keeled, wedge-shaped, concave, light yellow-green with fleshy apices. The dorsal is 4.6 mm. long and 1.8 mm. wide with 3 yellow-green veins. The laterals are 90% fused and are 4.3 mm. long by 2 mm. wide together. The light green, lanceolate petals are 2 mm. long by 0.6 mm. wide and have an acute apex. The light yellow-green, tongue-shaped lip is 2 mm. long by 0.65 mm. wide with 2 indistinct side lobes and a fleshy, rough base. The light yellow-green column is 1.8 mm. long by 1.3 mm. wide and the apex is lower than the tips of the wings. There are 2 small, drop-shaped protuberances at the column foot.

Flowering Periods: February to April and the flowers last seven to ten days.

Pleurothallis grobyi
Lindl. | *1323* (Plate XVIII)

Etymology: Groby is in Leicestershire, England where Lord Grey, a patron of orchids, lived.

Habitat: Epiphytic from sections V, VI and VIII, at 900-1600 M. asl, in humid original cloud forest on lower moss-covered branches and lianas. L 4-7; H 6; W 5.

Occurrence: Common.

Plant: This 50-70 mm high epiphyte forms clusters with 10-100 leaves and the bases of the 3-5 mm. long and 0.8-1.2 mm. diameter leaf stems are hidden in the profuse root system, and are enveloped by 3 membranous sheaths, up to 7 mm. long, of which the basal sheath is less than 1 mm. long. The obovate, faintly keeled leaves are finely attenuated towards their bases, occasionally speckled purple on their backs and are 30-65 mm. long, 10-15 mm. wide and 1 mm. thick.

Inflorescence and Flowers: The 150-200 mm., erect inflorescences emerge from the apex of the most recent leaf stem with up to 25 flowers on 6-15 mm. pedicels. The first flower opens at twice the leaf height and up to 15 flowers bloom together. The ovary is 1 mm. long. The lanceolate sepals are distinctively keeled, concave and yellow-green to beige. The dorsal sepal is 9-11 mm. long and 3.5-4.0 mm. wide with 3 wine-red longitudinal veins. The apex is slightly convex. The laterals are 95% fused and together are 10-15 mm. long by 4-5 mm. wide with olive to red-brown central and outer veins and apices. The very light green or beige, lanceolate petals are 2.2 mm. long by 0.8 mm. wide and have dark wine-red central veins. The articulate, rectangular lip is light yellow-green, fleshy and 3.3 mm. long by 1.3 mm. wide with 2 brown longitudinal side veins, framed in red-brown. The green central vein is lowered for the basal half, also framed red-brown on the obtuse, wider apical section. The lip is recurved by 90°at the fleshy base. The light yellow-green column is 2.8 mm. long by 2.2 mm. wide with hook-shaped wings pointing to the column apex and the base has 2 drop-shaped protuberances.

Flowering Periods: October to January. The flowers last for 7 to 10 days and the inflorescence for more than a month.

General: The plant and flower seem to be very variable in size and colour. See 1323b.

Pleurothallis aff. grobyi Lindl. | *1323b* (Plate XVIII)

Habitat: Epiphytic from section V and VIII, at 700-800 M. asl, in humid original forest on lower moss-covered trunks and shady rocks. L 2-4; H 6; W 3.

Occurrence: Occasional but locally common.

Plant: This 50 mm. high epiphyte forms clusters with 5-15 leaves and 4-8 mm. long and 0.8-1.2 mm. diameter leaf stems, enveloped by 3 membranous sheaths, up to 7 mm. long, of which the basal sheath is less than 1 mm. The obovate leaves are conduplicate, furrowed, finely attenuated towards their bases and 30-40 mm. long, 5-7 mm. wide and less than 1 mm. thick.

Inflorescence and Flowers: The 120-150 mm., erect inflorescences emerge from the apex of the most recent leaf stems with up to 25 erect flowers held on 3-5 mm. pedicels. The first flower opens at leaf height and up to 15 flowers bloom together. The ovary is 1.5 mm. long. The lanceolate sepals are distinctly keeled, concave and light yellow-green. The dorsal is 5-7 mm. long and 2.5-3 mm. wide with 3 green longitudinal veins and a slightly convex apex. The laterals are 75% fused and together are 6-7.5 mm. long by 3.5-4 mm. wide. The light yellow-green, lanceolate petals are 2.2 mm. long by 1 mm. wide with darker coloured central veins. The articulate, rectangular lip is light yellow-green, fleshy and has a wider basal half with 2 brown longitudinal side veins and is 2.3 mm. long by 1.1 mm. wide. The apex is obtuse and the fleshy base recurved by 90°. The light yellow-green column is 1.9 mm. long by 1.3 mm. wide. The wings are hook-shaped and pointing to the column apex. The column base has 2 protuberances, which are less pronounced than those of P. grobyi.

Flowering Periods: November through March. The flowers last for seven to 10 days and the inflorescence for more than a month.

Pleurothallis subpicta Schltr. | *1325* (Plate XVIII)

Etymology: Sub and *pictus* are Latin for somewhat coloured, referred to the lip colours.

Habitat: Epiphytic from section VI, at 900-1600 M. asl, in humid original cloud forest on mid and lower, moss-covered branches and lianas. L 4; H 7; W 4.

Occurrence: Common.

Plant: A 20-30 mm. high epiphyte forming clusters with 10-100 leaves and a profuse root system of 0.5 mm. diameter, whitish roots, usually covered by moss. The leaf stems are enveloped by 2 membranous sheaths hidden in the mass of the root system and are 3-4 mm. long by 0.4 mm. in diameter. The basal sheath is under 2 mm. and the upper reaches far above the leaf base. Leaves are obovate, faintly keeled, attenuated towards their bases, dark green, occasionally purple-backed and are 20-25 mm. long, 5-7 mm. wide and 1-1.5 mm. thick.

Inflorescence and Flowers: The 55-110 mm., erect inflorescences emerge from the apex of the most recent leaf stems with up to 18 flowers. The first flower blooms at 30 mm. from the base followed by the others in quick succession and all may bloom at once. The lanceolate sepals are keeled, concave, yellow-green to yellow with green longitudinal veins. The dorsal is 6 mm. long by 2.3 mm. wide with 3 wine-red longitudinal veins and a slightly convex apex. The laterals are 80% fused and together measure 5.8 mm. long by 2.9 mm. wide. The spathulate petals have acuminate apices, are light yellow with green or slightly wine-red central veins and are 2.2 mm. long by 1.1 mm wide. The articulate, rectangular lip is fleshy, light yellow-green with 2 indistinct, dark purple side-lobes and is 2.1 mm. long by 1.1 mm. wide. The light yellow-green column is 1.7 mm. long by 1 mm. wide with hook-shaped wings pointing towards the short column apex.

Flowering Periods: October to November. A flower lasts up to 10 days; the inflorescence for a month.

Pleurothallis edwallii var. *major* | *1329a* (Plate XVIII)

Etymology: Named in honour of the Brazilian Botanist **G. Edwall**.

Habitat: Found near Petropolis, section II, 1000-1200 M. asl, in humid original forest on moss-covered mid-tree branches. L 4; H 8; W 4.

Occurrence: Common.

Plant: This 120-170 mm. high epiphyte forms clusters with 10-25 leaves and a profuse root system composed of 1 mm. diameter, whitish roots. The tough leaf stems are

80% enveloped by 3 tight, dark green or brown sheaths and are 50-85 mm. long and 0.8-1.0 mm. diameter. The flat leaves are 50-80 mm. long, 15-18 mm. wide and 1 mm. thick, lanceolate, faintly conduplicate, dark olive-green and narrowed to their bases to form short false petioles.

Inflorescence and Flowers: The 200-250 mm., erect inflorescences arise from the apex of young and older leaf stems with up to 10 flowers on 10-15 mm. pedicels. The first flower blooms 100 mm. up the inflorescence followed by the rest successively and 4 may bloom together. The lanceolate sepals are keeled, concave, hairy, yellow-green to yellow with wine-red longitudinal veins and apices. The dorsal sepal is 10.5 mm. long and 3.6 mm. wide with a slightly convex apex. The laterals are 95% fused and together are 10 mm. long and 5 mm. wide with wine-red bases. The petals are light green, spathulate, blunt, pointed, concave and 4 mm. long by 2.5 mm. wide with green or light wine-red central veins. The articulate lip leans back for 90° close to its base and is light yellow and purple, trilobed, 5 mm. long by 3 mm. wide with a rounded apex and gnawed margins. The basal half of the lip and the side-lobes are purple and the plate is undulate with increasing intensity towards the granular to warty yellowish apex. The 3.5 mm. long column is 2.5 mm. wide and wine-red deepening towards the margins and the apex and wings are slightly gnawed.

Flowering Periods: January to April. The flowers last for 15-20 days and the inflorescence for more than a month.

Pleurothallis edwallii var. *pallida* Hoehne&Schltr. | *1329b* (Plate XVIII)

Description and drawing is to be found under *1185 Pleurothallis wacketii*.

Etymology: Pallidus is Latin for pale.

General: The differences in size of the plant and flower of the variety pallida, as well as the lack of hairy sepals and the differences in shape of the lip should justify species status.

Pleurothallis bocainensis Pôrto & Brade | *1335* (Plate XVIII)

Etymology: Bocaina is a location west south west of Rio de Janeiro.

Habitat: Epiphytic and lithophytic from section VI, 1300-1400 M. asl, in humid original cloud forest on lower and mid-tree branches and on a steep rock face in half shade. L 6; H 7; W 6.

Occurrence: Rare.

Plant: This 30-70 mm. high epiphyte forms clusters with 10-30 leaves and an adherent root system composed of 0.6 mm. diameter, whitish roots. The leaf stems are 5-20 mm. long by 1 mm. diameter and alternate at 2 mm. intervals from a 1 mm. diameter, branching rhizome enveloped by 3 tight, dark brown sheaths. The flat leaves are lanceolate, faintly conduplicate, yellow-green, narrowed to the base and are 25-50 mm. long, 8-10 mm. wide and 1.5-2 mm. thick.

Inflorescence and Flowers: The bent, 120 mm. inflorescences emerge from the apex of the most recent leaf stems. The first flower blooms at leaf height, followed by up to 9 flowers on 5 mm. pedicels, all blooming simultaneously. The sepals are almost linear, concave at the base, keeled, light yellow-green to yellow and wine-red along the central veins. The dorsal is 20 mm. long and 2.6 mm. wide with recurved margins at the apex. The free laterals are 18 mm. long by 2.6 mm. wide and hairy at the inner, basal margins. The petals are linear, slightly bent down, and 11 mm. long by 1.3 mm. wide with wine-red central veins and very fine hairs at the margins. The light yellow-green lip is narrowly elliptical, 2.9 mm. long and 1 mm. wide, with indistinct side-lobes and crenate margins. The central vein is sunken and granular at the base and tuberculate at the apex. The sturdy, wine-red column is 3 mm. long by 2.5 mm. wide with spread wings. The apex and the wings are lacerate.

Flowering Periods: March to April. The flowers last for 14 days.

Pleurothallis depauperata Cogn. | *1338* (Plate XVIII)

Etymology: Depauperatus is Latin for undeveloped, and refers to the flowers which are usually only partly open.

Habitat: Epiphytic from section VIII, at 950-1000 M. asl, in humid original cloud forest on the lower side of mid-tree branches. L 7: H 7: W 7.

Occurrence: Rare. We have only found a few plants to date.

Plant: This creeping epiphyte is 50-120 mm. high and forms clusters with many leaves and a profuse root system, composed of 0.4 mm. diameter light yellow-green or golden-yellow roots. The leaf stems emerge at 5 mm. intervals from the 1.2 mm. diameter rhizome and are 5-18 mm. long and 0.8 mm. diameter, each enveloped by a single orange-brown sheath. The yellow-green leaves are narrowly obovate, almost linear, faintly conduplicate, finely narrowed to their bases and are 30-90 mm. long, 6-8 mm. wide and up to 1 mm. thick.

Inflorescence and Flowers: One or 2, 100-120 mm. inflorescences emerge from the apex of the most recent leaf stems, with up to 7 simultaneously blooming, partly-opened flowers, on 3-4 mm. pedicels. The first flower blooms above leaf height followed by the rest in quick succession. The ovary is 1 mm. long. The linear sepals and petals are acuminate, keeled and very light yellow-green. The dorsal sepal is 9-12 mm. long by 1.4-1.8 mm. wide. The free laterals are slightly asymmetric and 8-11.5 mm. long by 1.2-1.4 mm. wide with a purple base. The petals are 6.5-8 mm. long and 1.0 mm. wide with faintly serrate margins and a purple base. The tongue-shaped, convex lip is whitish to very light yellow-green, with 2 indistinct side-lobes and is 2.7 mm. long by 0.6 mm. wide. The surface of the basal half is hairy and the low-lying central vein has a row of pearl-like callosities. The sturdy, curved column is light yellow-green and 1.8-2.2 mm. long by 1.5 mm. wide with finely acuminate apex and wings.

Flowering Periods: November to April. The flowers last for 7 days.

Pleurothallis eugenii
Pabst | *1339* (Plate XVIII)

Etymology: Named in honour of **Eugene Warming**, the Danish botanist.

Habitat: Epiphytic from section VI, at 1100-1400 M. asl, in humid original forest on mid-tree branches. L 5; H 7; W 5.

Occurrence: Occasional.

Plant: This 100-140 mm. high epiphyte has clusters with 10-30 leaves and an intensive root system of 0.8 mm. diameter, whitish roots. The 40-60 mm. long and 1.5-2 mm. diameter leaf stems are covered by 3 loose brown or cinnamon sheaths. The lanceolate leaves are erect, faintly conduplicate, slightly infolded and twisted at their bases and 60-80 mm. long, 13-17 mm. wide and 1.5 mm. thick.

Inflorescence and Flowers: The 150-200 mm. bent inflorescences emerge from the apex of young and older leaf stems bearing 9 to 12 pendent flowers which bloom simultaneously on 3-4 mm. pedicels. The first flower develops at 100 mm. up the inflorescence. The white sepals are narrowly ovate, almost linear, keeled, with the central veins wine-red at their concave bases. The dorsal sepal is 14 mm. long and 3.1 mm. wide. The free lateral sepals are 13.5 mm. long by 2.7 mm. wide and slightly asymmetric. The acuminate petals are similarly coloured, slightly bent downwards and to the outside with serrate margins and 6.6 mm. long by 2.3 mm. wide. The indistinctly trilobed lip is 3.0 mm. long by 1.2 mm. wide and white, narrowly oval, almost rectangular, recurved at the mid-section by 180° and has 2 indistinct side-lobes. The margins are crenate near the base. The low lying central vein has a row of pearl-like callosities

and the centre of the lip is undulate, while the overall surface is rough. The column is white, 3 mm. long and 3 mm. wide including the spread wings, sturdy, slightly curved with lacerate wings and apex.

Flowering Periods: August to September and April to June. The flowers last for 20 days.

Pleurothallis linearifolia
Cogn. | *1341* (PlateXVIII)

Etymology: Lineari-folia is Latin for linear leaves which this plant almost shows.

Habitat: Epiphytic from section VI, at 900-1400 M. asl, in humid original forest on trunks, mid- and lower tree branches and lianas. L 7; H 6; W 6.

Occurrence: Common.

Plant: This 50-70 mm. high epiphyte has clusters of 10-100 leaves and a profuse root system of 0.5 mm. diameter, whitish roots. The round leaf stems are 5-15 mm. long by 0.5 mm. diameter and enveloped by 3 brown or cinnamon, loose sheaths. The linear-lanceolate leaves are 40-55 mm. long, 3-4 mm. wide and 1.5 mm. thick, erect, faintly conduplicate, infolded and finely attenuated to their bases to form false petioles.

Inflorescence and Flowers: The 70-90 mm. inflorescences arise from the apex of young and older leaf stems with 9 to 12 flowers on 2 mm. pedicels. The first flower is 45 mm. from the inflorescence base, followed by the rest in rapid succession and all may bloom at once. All parts of the flower are light yellow-green and the narrowly lanceolate sepals are concave and keeled. The dorsal is 7 mm. long by 1.5 mm. wide, while the asymmetric, free laterals are 6.5 mm. long by 1.5 mm. wide. The petals are narrowly lanceolate with faintly serrate margins, an acuminate apex and are 5 mm. long by 1.1 mm. wide. The light green lip is 2.9 mm. long and 0.75 mm. wide and recurved by 180° at the mid-section, narrowly elliptical or tongue-shaped, with a rough surface and faintly crenate margins. The low-lying central vein has a row of pearl-like callosities. The light yellow-green column is sturdy, slightly curved and 2.4 mm. long by 2.2 mm. wide with spread wings. The apex of the column and the relatively wide wings are distinctly lacerate.

Flowering Periods: October to November and March to April. The flowers last for 10 days.

Pleurothallis malmeana
Dutra ex. Pabst | *1342* (Plate XIX)

Etymology: Named in honour of **Dr. Malme**.

Habitat: Epiphytic from section VI, at 600-1000 M. asl, in humid original mountain forest on trunks and mid-

and upper tree branches. L 7; H 8; W 7.

Occurrence: Occasional.

Plant: This 60-125 mm. high epiphyte forms clusters with 5 to 15 leaves and a root system composed of 0.7 mm. diameter, whitish roots. The round leaf stems are 25-45 mm. long, 1 mm. diameter and are enveloped by 3 to 5 brown or cinnamon, tube-shaped sheaths. The leaves are flat, lanceolate, erect, conduplicate, leathery or stiff and are 35-80 mm. long, 13-15 mm. wide and 1.5-2 mm. thick.

Inflorescence and Flowers: The 70-120 mm. inflorescences, with 5 to 11 flowers on 3 mm. pedicels, arise from the apex of one-year-old leaf stems. The first flower develops at leaf height followed by the rest in quick succession, all blooming at once. The ovary is 1.5 mm. long. All parts of the flower are light yellow-green and the margins of the sepals and petals are slightly recurved. The narrowly wedge-shaped sepals are keeled with acuminate apices. The 6.4 mm. long dorsal sepal is 1.4 mm. wide, while the asymmetric lateral sepals are 6.5 mm. long by 1.5 mm. wide. The narrowly lanceolate petals with acuminate apices are 5.0 mm. long by 1.1 mm. wide and have faintly serrate apical margins. The light green lip is narrowly oval, hairy on the front and back, and 2.5-3.2 mm. long by 1.0 mm. wide with 2 very indistinct side-lobes, slightly recurved and the low-lying central vein has a row of pearl-like callosities. The sturdy column is slightly curved, light yellow-green, and is 2.5 mm. long and 2.0 mm. wide including the spread wings. The wings and the apex are distinctively lacerate.

Flowering Periods: November and occasionally January to February; the flowers last for 10 days.

Pleurothallis adenochila
Loefgr. | *1348* (Plate XIX)

Etymology: Adenos and *cheilos* are Greek for gland-lipped, referring to the tips of the petals and the lip.

Habitat: Epiphytic from section V, VI and VIII, at 1100-1500 M. asl, in humid original forest on trunks, lianas and low-tree branches. L 2-4; H 8; W 2-3.

Occurrence: Common.

Plant: This 40-120 mm. high epiphyte forms clusters with 10-30 leaves and a root system composed of 0.5 mm. diameter, whitish roots. The round leaf stems are 5-50 mm. long by 0.6-0.8 mm. diameter and enveloped by 3 to 5 membranous, green or brown sheaths. The leaves are narrowly lanceolate, dark green, faintly conduplicate, infolded and finely attenuated to their bases forming 2 mm. false petioles and 25-40 mm. long, 5-8 mm. wide and 0.5 mm. thick.

Inflorescence and Flowers: Up to 4, 10 mm. inflorescences arise simultaneously from the apex of young and older

leaf stems with up to 8 alternate, consecutive flowers on 1 mm. pedicels. The first flower is 3 mm. from the base of the inflorescence. The sepals are keeled, concave and lanceolate with entire margins, yellow at their bases turning to cinnamon towards the yellow-orange apices. The concave dorsal sepal is 4.2 mm. long by 2 mm. wide, while the asymmetric laterals are 10% fused at the base and 3.8 mm. long by 1.4 mm. wide. The petals are lanceolate with bright yellow-green, pearl-shaped blobs at their apices and 3 mm. long by 1 mm. wide with hairy and slightly recurved margins, pale-yellow at their bases but red-brown and dark liver-brown towards their apices. The dark blood-coloured, fleshy lip is 2.3 mm. long and 1.0 mm. wide, narrowly elliptical with indistinct hairy side-lobes, a bright yellow-green, pearl-shaped blob at the apex and 2 white ear-like appendages at its green base. The central vein is elevated at the lip's base but lowered towards the apex. The sturdy column is light yellow or white and is 2 mm. long and 1.5 mm. wide including the spread wings and has large flanks and a reddish back at the apex.

Flowering Periods: September through to May. The flowers last for 14 days.

Pleurothallis tigridens
Loefgr. | *1357* (Plate XIX)

Etymology: Tigrinus and *dens* are Latin for tiger-toothed, referring to the lip shape.

Habitat: Epiphytic from section V, at 800 M. asl, in original mountain forest on trunks and mid-tree branches. L 7; H 8; W 7.

Occurrence: Rare.

Plant: This 60-100 mm. high epiphyte forms clusters with 5 to 15 leaves and has a root system of 0.9 mm. diameter, whitish or light brown roots. The round leaf stems are 25-35 mm. long., 1.2-1.5 mm. diameter and are enveloped by 3 to 5 overlapping, wide, orange-brown, tube-shaped long-lived sheaths. The leaves are 60-70 mm. long, 18-20 mm. wide and 1 mm. thick, flattish, elliptical, leathery, faintly conduplicate and narrowed to a 10 mm. petiole-like base.

Inflorescence and Flowers: The 100-150 mm. inflorescences, which produce 5 to 7 flowers on 6-8 mm. pedicels, arise from the younger leaf bases. The first flower is above leaf height followed consecutively by the rest but only one flower blooms at a time. The bracts are 5 mm. and the ovary 2 mm. long. From afar the flower seems to be black. The narrowly lanceolate sepals and petals are fleshy, keeled, dark olive-green and wine with lighter bases, short-haired on the backs with acuminate apices. The dorsal sepal is 10.5 mm. long and 3 mm. wide, while the almost free laterals are 10.5 mm. long by 2 mm. wide. The petals are 10 mm. long by 1.6

mm. wide and are also hairy on the inside. The green to bluish-green lip is trilobed, narrowly lanceolate, fleshy, short-haired on the base and back, and 3.6 mm. long by 1.5 mm. wide with 2 narrow side-lobes with wine-red, short-haired margins. On the base are 2 light yellow-green, narrow ear-like appendages and the acute apex is slightly bent upwards. The sturdy column is slightly curved, yellow-green except the dark wine-red column foot, and 2.2 mm. long and 1.8 mm. wide including the spread wings. The wings and the apex are distinctly lacerate.

Flowering Periods: January to March. The flowers last for 10 days.

Pleurothallis microgemma
Schltr. | *1362* (Plate XIX)

Etymology: Micro is Greek for small and *gemma* Latin for bud-like.

Habitat: Epiphytic from section VI, at 1100-1400 M. asl, in humid original cloud forest on trunks, lianas and low-tree branches. L 3; H 7; W 3.

Occurrence: Occasional.

Plant: This 30-60 mm. high epiphyte forms clumps with 10-30 leaves with a root system composed of 0.6 mm. diameter, whitish roots. The round leaf stems are 10-30 mm. long by 0.7 mm. diameter and are enveloped by 3 overlapping, green or brown, membranous sheaths. The leaves are 15-40 mm. long by 7-10 mm. wide and 1-1.5 mm. thick, obovate, dark olive-green, conduplicate, slightly infolded and attenuated to the base to form 2 mm. false petioles.

Inflorescence and Flowers: The 5-10 mm. inflorescences arise from the apex of young and older leaf stems with 2 to 6 alternate and consecutive flowers on 1.5 mm. pedicels. The first flower develops at 2 mm. from the inflorescence base. The ovary is less than 1 mm. long. The sepals are ovate, keeled and concave with entire margins and are yellow at their bases and turn to cinnamon towards the orange-yellow apices. The concave dorsal sepal is 4.5 mm. long by 1.9 mm. wide and the asymmetric laterals are 20% fused at the base and 4.2 mm. long by 1.8 mm. wide. The petals are narrowly lanceolate with hairy apical margins and apices and are 3.2 mm. long by 0.8 mm. wide, whitish but dark wine at their bases and along the central veins. The fleshy, elliptical lip has 2 small white ear-like appendages at its darker base and a slightly retuse apex. The dark wine lip is 1.5 mm. long by 0.8 mm, the side-lobes have hairy margins leaning upwards and extending for less than half the lip's length. The curved column is light wine except the yellow or white apex and 1.8 mm. long and 1.4 mm. wide including the spread wings. The wings and

the apex are deeply gnawed.

Flowering Periods: April to June. The flowers last for 14 days.

Pleurothallis modesta
Cogn. | *1363* (Plate XIX)

Etymology: Modestus refers to the modest, almost shy aspect of the flower.

Habitat: Epiphytic from section VI, at 1200-1400 M. asl, in humid original cloud forest on trunks, lianas and low-tree branches. L 4; H 7; W 5.

Occurrence: Occasional.

Plant: This 40-50 mm. high epiphyte has clusters with 10-30 leaves and a root system of 0.6 mm. diameter, whitish roots. The round leaf stems are enveloped by 3 to 5 membranous, green or brown sheaths and are 5-15 mm. long by 0.6-0.7 mm. diameter. The leaves are 25-35 mm. long, 7-11 mm. wide and 1.5 mm. thick and narrowly obovate, obtuse, dark olive-green, conduplicate, slightly infolded and finely attenuated to their bases to form 2 mm. false petioles.

Inflorescence and Flowers: The 10 mm. inflorescences arise from the apex of young and older leaf stems with 4 to 6 alternate, consecutive, wide open flowers on 1.5 mm. pedicels. The first flower is 5 mm. from the inflorescence base. The ovary is less than 1 mm. long. The sepals are ovate, concave, faintly keeled with hairy margins, light yellow at their bases, turning cinnamon towards the yellow-orange apices. The dorsal sepal is 7 mm. long by 2.5 mm. wide, while the asymmetric laterals are 6.8 mm. long, by 2.3 mm. wide. The narrowly lanceolate petals are 4.8 mm. long and 1.3 mm. wide and have recurved, hairy margins at the apical sections and are rose turning to dark wine towards the central veins. The lip is 3.5 mm. long by 1.2 mm. wide, narrowly elliptical, slightly waisted, wine to dark purple with 2 white, ear-like appendages at its whitish base and its margins are hairy. The rose-coloured side-lobes on the basal half of the lip lean upwards and the fleshy tip is rounded. The basal third of the central vein is elevated. The column is light yellow or white, sturdy, slightly curved and 2 mm. long by 1.5 mm. wide including the spread wings which are deeply gnawed.

Flowering Periods: April and October to November. The flowers last for 14 days.

Pleurothallis recurvipetala
Cogn. | *1365* (Plate XIX)

Etymology: Recurvi-petala is Latin for recurved petals.

Habitat: Epiphytic from section VI and VIII, at 900-1000 M. asl, in humid original mountain forest on low tree trunks. L 4-6; H 8; W 5.

Occurrence: Rare.

Plant: An epiphyte, 25-55 mm. high forming clumps with 5 to 15 leaves and a root system of 0.7 mm. diameter, whitish roots. The round leaf stems are 7-20 mm. long by 0.5-0.75 mm. diameter and are enveloped by 3 to 5 light green or yellow-brown, loose, membranous sheaths. The leaves are 20-35 mm. long, 8-13 mm. wide and 0.5-1.0 mm. thick, leathery, obovate, faintly conduplicate, dark olive-green, slightly infolded and finely attenuated to their bases to form 2 mm, resupinate false petioles. The backs are nearly dark purple.

Inflorescence and Flowers: The 5-7 mm. inflorescences arise from the apex of young and older leaf stems with 3 to 4 alternate and consecutive flowers on 1.5 mm. pedicels and the first flower is 3 mm. from the inflorescence base. The ovary is 1.5 mm. long. All parts of the flower are fleshy and keeled. The sepals are ovate, concave, keeled, white to very light yellow-green and may turn purple at their apices, which have hairy margins. The dorsal sepal is 3.8 mm. long and 2 mm. wide, while the asymmetric laterals are 3.4 mm. long and 1.4 mm. wide. The lanceolate petals are 2.4 mm. long and 0.9 mm. wide with finely acuminate, slightly recurved apices, dark wine-red and hairy along their faintly serrate margins. The trilobed lip is elliptic and 1.8 mm. long by 1 mm. wide, golden-yellow with hairy, dark purple margins and has 2 recurved ear-like appendages at its base. The yellow side-lobes lean upwards and extend from the base to the middle, while the central vein is distinctively elevated on the basal half and extends to the fleshy, occasionally purple apex. The sturdy column is slightly curved, light green and is 1.8 mm. long by 1.6 mm. wide including the spread wings. The wings and the apex are deeply lacerate and the foot of the column is purple.

Flowering Periods: November through to April. The flowers last for 7 days.

Pleurothallis aff. laciniata
Barb. Rodr. | *1371b* (Plate XIX)

Etymology: Laciniatus is Latin and means slashed with pointed incisions and refers to the petals.

Habitat: Epiphytic from section VIII, in original mountain forest at 800-900 M. asl. L 5; H 6; W 8.

Occurrence: We are not certain if we have found the real *P. laciniata.* The plant described here occurs occasionally.

Plant: This epiphyte forms tight clusters with 10-30 overlapping leaves, more or less addpressed to the substratum. The root system is moderately intensive, sticking tightly to bark and the 0.5 mm. diameter roots are whitish or yellow-green. The round leaf stems are very short, 0.3-0.5 mm. diameter and enveloped by

2 light green or brown, membranous sheaths. The leaves are 12-20 mm. long, 4-6 mm. wide and 1 mm. thick, obovate with obtuse apices, shiny green to dark olive-green and occasionally covered with purple dots, faintly conduplicate and attenuated towards their bases forming short false petioles.

Inflorescence and Flowers: The 5-10 mm. inflorescences arise from the apices of young and older leaf stems with 3 consecutive flowers on 1.5 mm. pedicels. The ovary is less than 1 mm. long. The sepal margins are entire. The sepals are ovate, faintly keeled, concave at the whitish or light yellow bases, turning yellow with some wine red speckles towards their apices. The dorsal sepal is 4.3 mm. long by 1.8 mm. wide. The asymmetric laterals are 20% fused, 4.3 mm. long by 1.3 mm. wide. The petals are narrowly lanceolate, wine-red with light yellow apices, finely serrate margins and 3.5 mm. long by 0.7 mm. wide. The trilobed lip is 1.9 mm. long by 0.9 mm. wide, tongue-shaped, dark wine-red, fleshy along the central vein with 2 indistinct, obtuse, white recurved ear-like appendages at its yellowish base and a fleshy yellow apex. The flesh-coloured side-lobes lean upwards and extend over less than half of the lip's length. The sturdy column is whitish, slightly curved and 2 mm. long by 1.8 mm. wide including the spread wings. The wings and the apex are lacerate.

Flowering Periods: January through to March. The flowers last for 10 days.

Pleurothallis microphyta
(Barb.Rodr.) Cogn. | *1374* (Plate XIX)

Etymology: Micro and *phyta* are Greek for small plant, relating to the growth habit.

Habitat: Epiphytic from section VI and VIII , at 800-1000 M. asl in humid original mountain forest, on trunks, lianas and low tree branches. L 5-7; H 7-8; W 5-7.

Occurrence: Rare.

Plant: This 10-20 mm high epiphyte forms clusters with 5 to 20 leaves and a profuse root system, composed of whitish or yellow-green, 0.4 mm. diameter roots. The very short rhizome is 0.8 mm. diameter and the round leaf stems are 1-2 mm. long by 0.3-0.5 mm. diameter, enveloped by 3 light green or brown, membranous, short-lived sheaths. The leaves are fleshy, elliptical, dark olive-green and 10-13 mm. long, 4-6 mm. wide and 1-1.5 mm. thick.

Inflorescence and Flowers: One or two 4-6 mm. inflorescences arise from the apices of young and older leaf stems with 1 or 2 consecutive flowers on 2 mm. pedicels. The ovary is less than 1 mm .long, the bracts and the spathe are 1.2 mm. long. The cuneate sepals have entire margins, are concave, keeled, very light wine at their bases and turning wine and golden-yellow towards their apices. The dorsal sepal is 3.6 mm. long and 1.6 mm. wide, while the asymmetric laterals are 20% fused at their bases, 3.4 mm. long and 1.3 mm. wide each. The petals are narrowly lanceolate, wine-red, slightly asymmetric, and 2.7 mm. long by 0.8 mm. wide with faintly serrate margins at the apical halves. The tongue-shaped lip has a fleshy, rounded apex and is 2 mm. long by 0.8 mm. wide and dark wine-red, almost black, with 2 white ear-like appendages at its black base. The narrow side-lobes with hairy margins, lean upwards and extend over less than half the lip's length. The light ochre central vein is sunk from the base to the apex. The white, sturdy, slightly bent, 1.8 mm. long column is 1.4 mm. wide with spread wings and both wings and apex are lacerate.

Flowering Periods: October to April. The flowers last for 14 days.

Pleurothallis aff. microphyta (Barb.Rodr.)Cogn. | *1374b* (Plate XIX)

Etymology: as for *1374*.

Habitat: Epiphytic from section VI and VIII, at 800-1000 M. asl in humid original mountain forest, on trunks, lianas and low tree branches. L 5-7; H 7-8; W 5-7.

Occurrence: Occasional.

Plant: This epiphyte varies from 10-20 mm. and forms dense clusters with 5 to 20 leaves and a root system composed of whitish or yellow-green, 0.4 mm. diameter roots and a very short 0.7 mm. diameter rhizome. The round leaf stems are enveloped by 3 light green or brown, membranous, short-lived sheaths and are 2-6 mm. long by 0.3-0.7 mm. diameter. The leaves are fleshy, stiff, obovate or elliptic, faintly conduplicate, dark olive-green and 10-15 mm. long, 4-6 mm. wide and 1-1.5 mm. thick. Younger leaves tend to be purple.

Inflorescence and Flowers: The 2-3 mm. inflorescences arise from the apices of young and older leaf stems with up to 3 consecutive flowers on 2 mm. pedicels. The ovary is under 1 mm. long. The ovate sepals are concave with granulate to short-haired insides and hairy margins, very light yellow at their bases and rose or wine towards their orange-yellow apices. The dorsal sepal is 3.6 mm. long and 1.5 mm. wide, and the asymmetric laterals, 20% fused at their bases, are 3.3 mm. long by 1.1 mm. wide each. The lanceolate petals are 2.2 mm. long by 0.7 mm. wide, slightly asymmetric, wine-red with hairy margins. The tongue-shaped lip is 1.4 mm. long by 0.8 mm. wide, very dark wine and has 2 white ear-like appendages at its even darker base. The narrow side-lobes with long-haired margins lean upwards and extend over half the lip's length. The central vein is elevated at the basal half and the 2 longitudinal callosities end in the fleshy, rounded apex. The light wine-red, slightly bent column is 2.5 mm. long and 1.3 mm. wide, with spread wings,

and both apex and wings are lacerate.

Flowering Periods: November to March. The flowers last for 14 days.

Pleurothallis rubro-limbata Hoehne | *1375* (Plate XX)

Etymology: Rubro and *limbata* are Latin for bordered in red probably referring to the lip edge.

Habitat: Epiphytic from section VI, at 1000-1400 M. asl in humid original cloud forest on trunks, lianas and low-tree branches. L 3-5; H 6-8; W 2-5.

Occurrence: Common.

Plant: This 20-30 mm. high epiphyte forms clusters with 10-40 leaves and an intensive root system of whitish or yellow-green, 0.5 mm. diameter roots. The round leaf stems are enveloped by 3 overlapping, light green or brown, membranous sheaths and are 2.5 mm. long and 0.7-0.8 mm. diameter. The leaves are 20-25 mm. long by 7-10 mm. wide and 2 mm. thick, fleshy, obovate with obtuse apices, olive-green with purple spots, faintly keeled, furrowed and finely attenuated to their bases to form 2 mm. false petioles.

Inflorescence and Flowers: The 7 mm. inflorescences arise from the apices of young and older leaf stems with 2 to 5 alternate, consecutive flowers on 2-3 mm. pedicels and the first flower is 1-2 mm. from the base. Up to 2 flowers bloom at once and a leaf base may produce several consecutive inflorescences per year. The ovary is less than 1 mm. long. All margins of the sepals are entire. The sepals and petals are cuneate, keeled, concave and light yellow at their bases and red-brown towards their apices. The dorsal sepal is 4 mm. long and 1.8 mm. wide, while the asymmetric laterals are 3.7 mm. long and 1.8 mm. wide. The lanceolate petals are 2.8 mm. long and 1.0 mm. wide, with hairy margins to the dark brown apical halves. The lip is narrowly elliptical with a round apex, hairy margins and is 2.5 mm. long by 1.1 mm. wide. It is dark red-brown to dark purple with 2 white ear-like appendages at the darker base. The narrow side-lobes lean upwards and extend over more than half of the lip's length. The basal half of the dark purple central vein is elevated and lowered on the apical half of the lip. The column is light yellow or white, slightly curved, 2 mm. long and, including the spread wings, 1.2 mm. wide. The wings and the apex are lacerate.

Flowering Periods: October to April. The flowers last for 7 days.

Pleurothallis aff. rubro-limbata
1375b (Plate XXI)

Etymology: See *1375.*

Habitat: Epiphytic from section VI, at 1000-1400 M. asl. in humid original forest on trunks, lianas and low-tree branches. L 3-5; H 6-8; W 2-5.

Occurrence: Common.

Plant: This 20-25 mm. high epiphyte forms clusters with 10-40 leaves and an intensive root system of whitish or yellow-green, 0.5 mm. diameter roots. The round leaf stems are enveloped by 3 overlapping, light green or brown membranous sheaths and are 2.5 mm. long and 0.7-0.8 mm. diameter. The leaves are 20-25 mm. long by 7-10 mm. wide and 2 mm. thick, fleshy, obovate with obtuse apices, olive-green with purple spots, faintly keeled, furrowed and finely attenuated to the base forming 2 mm. false petioles.

Inflorescence and Flowers: The 7 mm. inflorescences arise from the apices of young and older leaf stems with 2 to 5 alternate, consecutive flowers on 2-3 mm. pedicels. The first flower is 1-2 mm. from the inflorescence base and up to 2 flowers bloom at once. Leaf bases may produce several consecutive inflorescences per year. The ovary is less than 1 mm. long. All sepal margins have very short hairs. The sepals and petals are wedge-shaped, keeled, concave and light yellow at their bases and red-brown towards their apices. The dorsal sepal is 5.7 mm. long and 1.9 mm. wide, while the asymmetric laterals are 5.1 mm. long and 1.5 mm. wide. The narrowly lanceolate petals are 4 mm. long and 1 mm. wide, wine-red with hairy margins to the dark brown apical halves. The dark wine-red lip is 1.7 mm. long by 0.9 mm. wide, tongue-shaped with a round apex, hairy margins and 2 white ear-like appendages at the base. The narrow side-lobes lean upwards and extend over three quarters of the lip's length. The dark purple central vein is elevated at the basal half and lowered on the apical half. The column is light yellow or white, slightly curved and 2 mm. long and, including the spread wings, 1.2 mm. wide. The wings and the apex are lacerate.

Flowering Periods: October to April. The flowers last for 7 days.

Pleurothallis corticicola
Schltr. | 1376 (Plate XXI)

Etymology: Cortex and *-cola* are Latin for bark-dwelling describing this plant's habit.

Habitat: Epiphytic from section VIII, at 800-900 M. asl in humid original mountain forest, on trunks and low tree branches. L 4-7; H 7-8; W 5-7.

Occurrence: Occasional.

Plant: A creeping epiphyte which usually lies flat on the substrate with leaves arranged like a hand of cards. We also found plants in clusters with 10 to 20 leaves. The root system is profuse and composed of whitish or yellow-green, 0.5 mm. diameter roots. The round leaf stems, under 1 mm. long by 0.5-0.75 mm. diameter, emerge alternately at 1 mm. intervals from a green, 0.7-0.9 mm. diameter rhizome and are enveloped by 1 or 2 light green or brown, membranous, short-lived sheaths. The dark olive-green leaves are 15-25 mm. long, 5-8 mm. wide and 1-1.5 mm. thick, acute, fleshy, narrowly obovate, faintly conduplicate and finely attenuated to the bases to form up to 2 mm. false petioles.

Inflorescence and Flowers: One or two, 6-10 mm. inflorescences arise from the apices of young and older leaf stems, and produce 2 to 3 alternate, consecutive flowers on 3 mm. pedicels and the first flower is 3 mm. from the base. The spathe is 3 mm. the bracts, 2 mm. and the ovary 1.0 mm. long. The sepal margins are entire and the narrowly wedge-shaped sepals and petals are concave, keeled, light yellow-green at their bases and yellow-orange towards their apices with some fine purple speckles. The dorsal sepal is 7 mm. long by 1.9 mm. wide, while the asymmetric laterals are 20% fused at their bases and are 7 mm. long by 1.7 mm. wide each. The slightly asymmetric petals are 5.8 mm. long by 0.9 mm. wide with faintly serrate margins. The narrowly elliptic lip is yellow-green and purple with a yellow to ochre centre and a fleshy, hairy base and has a fleshy, pointed apex and is 2.5 mm. long by 1.2 mm. wide. The narrow, erect and purple side-lobes with serrate margins extend over half the lip length. The very light green, slightly bent, 2 mm. column is 1.5 mm. wide with spread wings and both wings and apex are lacerate.

Flowering Periods: January to March. The flowers last for 14 days.

Pleurothallis aff. corticicola
Schltr. | 1376b (Plate XXI)
Possible new species

Etymology: see *1376.*

Habitat: Epiphytic from section VIII, at 800-900 M. asl in humid original mountain forest, on trunks and low tree branches. L 4-7; H 7-8; W 5-7.

Occurrence: Rare.

Plant: This epiphyte forms tight clusters with 10 to 15 leaves and a root system of yellow-green, 0.4 mm. diameter roots. The round leaf stems, which are less than 1-2 mm. long by 0.5-0.75 mm. diameter, including a 1 mm. pseudopetiole, emerge alternately at under 1 mm. gaps from the green, 0.5 mm. diameter rhizome and are enveloped by 1 or 2 membranous, short-lived sheaths. The leaves are 8-12 mm. long, 3-5 mm. wide,

0.8 mm. thick, dark olive with more or less dark purple dots and obovate, fleshy, faintly conduplicate and attenuated to their bases.

Inflorescence and Flowers: The inflorescences arise from young and older leaf stems with 2 to 3 consecutive flowers on 2 mm. pedicels and the first flower is 3 mm. from the base. The spathe is 1.5 mm., the bracts 1 mm. and the ovary under 1 mm. long. The sepals and petals are narrowly wedge-shaped with entire margins, concave, keeled, light yellow-green at their bases and dark purple towards their apices. The dorsal sepal is 7 mm. long and 1.9 mm. wide, while the asymmetric laterals are 10% fused at their bases and 7 mm. long and 1.6 mm. wide each. The slightly asymmetric petals are 5 mm. long by 0.8 mm. wide and are slightly rough at their apices. The narrowly elliptical lip is 2.8 mm. long by 0.9 mm. wide, has a pointed apex, is fleshy, dark purple with a lighter coloured base and central vein. It has 2 ear-like appendages at its rough base and the margins have short, very dark hairs from the base to the middle of the lip. The side lobes and the longitudinal callosities are very faint. The very light green, slightly bent column has purple flanks and a dark purple column foot and is 2.4 mm. long and 1.5 mm. wide.

Flowering Periods: January to March. The flowers last for 10 to 14 days.

Pleurothallis foliata
Griseb. | *1379* (Plate XXI)

Etymology: Foliata is Latin for leafy which this plant is.

Habitat: Epiphytic from section VI, at 1100-1200 M. asl, in very humid original cloud forest on lower and mid-tree branches, usually woven into moss or other epiphytes and hard to find. L 3; H 8; W 3.

Occurrence: Occasional.

Plant: A creeping, branching epiphyte with 5 to 30 leaves reaching 200 mm. diameter and whitish roots emerging only from the leaf-bearing or branching internodes, close to the substrate. Only a few of the 0.3 mm. diameter roots are fixed to the bark of the host. The internodes are 5-10 mm apart. The sheaths of the very short leaf stems are trumpet-shaped with recurved margins and short, stiff hairs and longitudinal veins, purple when young, brown and later black with age. The conduplicate leaves are dark green with wine-red backs and emerge from the second to fifth internode of the 0.25 mm. diameter, sheathed rhizome. The flat leaves, attenuated to their bases to form false petioles, are 10-14 mm. long and 6-8 mm. wide.

Inflorescence and Flowers: One or two, 30-45 mm. inflorescences arise from the apex of the most recent leaf stems. The first flower blooms at 15-20 mm. from the base followed by 4 to 7 flowers in rapid succession,

all blooming at once. A second inflorescence may follow 1 or 2 months later. All parts of the flower are light yellow-green and distinctly keeled. The sepals are narrowly lanceolate and the dorsal is 6.5 mm. long by 1.6 mm. wide, while the free laterals are 5.6 mm. long and 1.1 mm. wide. The spathulate petals have faintly serrated margins to the apical sections and are 2 mm. long by 0.8 mm. wide. The keeled lip is tongue-shaped. light yellow-green with indistinct side-lobes, 2 crenate longitudinal callosities and is 2.1 mm. long and 0.9 mm. wide with rounded, hairy, apical margins and an undulate green central vein. The light green column is 1.2 mm. long by 0.5 mm. wide, with 2 ear-like wings, a faint apex and an orange base.

Flowering Periods: September to December. A flowers lasts for 5 to 10 days.

Pleurothallis spannageliana
Hoehne | *1384* (Plate XXI)

Etymology: Not known.

Habitat: Epiphytic from section VI, at 1100 M. asl, in humid original mountain forest, on moss-covered mid-tree branches with more than 3 M. precipitation per year. L;6; H 6-8; W 6.

Occurrence: Rare.

Plant: This 40-65 mm. high epiphyte forms clusters with 5 to 15 leaves and white or brownish roots, 0.4-0.7 mm. diameter. The round leaf stems are enveloped by 3 membranous sheaths 4, 7 and 9 mm. long and are themselves 20-28 mm. long and 0.7-1.0 mm. diameter. The leaves are obovate, leathery or stiff, flat, dark green, conduplicate, attenuated to their bases to form 4 mm. false petioles and 25-35 mm. long, 7-9 mm. wide and 1-2 mm. thick.

Inflorescence and Flowers: The flowering plant looks like an Octomeria. Up to four, 3 mm. inflorescences emerge from the apex of young and older leaf stems, each developing 2 flowers simultaneously. The spathe is 3.5 mm. long and the bracts 2 mm. All parts of the flower are keeled, light yellow and speckled more or less with wine-red blotches. The inner surfaces of the sepals are granular. The dorsal sepal is acute, ovate and is 6.5 mm. long by 2.5 mm. wide. The almost free laterals are acute, narrowly ovate and each is 6.4 mm. long by 1.7 mm. wide. The almost linear petals are 5.3 mm. long and 0.8 mm. wide with acuminate apices and serrate margins. The yellow lip is tongue-shaped and 2.7 mm. long by 1.2 mm. wide with a rounded apex and 2 indistinct ear-like appendages at its base. The hairy plate is white and framed a dark wine colour. The column is whitish to light green, slightly curved and 2 mm. long, 1.6 mm. wide including the spread wings, while the apex and the wings are gnawed.

Flowering Periods: October to November. Flowers last for five to ten days.

Pleurothallis pulvinata
(Barb.Rodr.)Cogn. | *1385* (Plate XXI)

Etymology: Pulvinata is Latin for cushion-shaped, probably referring to the lip's structure.

Habitat: Epiphytic from section VIII at 800-900 M. asl, in original cloud forest on moss-covered, mid-tree branches. L 5: H 5: W 5.

Occurrence: Rare.

Plant: A 140-220 mm. tall epiphyte forming loose clusters with 10-30 leaves and wiry white to brownish 1.3 mm. diameter roots. The round leaf stems are 70-110 mm. long and 1.5-3.5 mm. diameter and arise from the creeping 3.5 mm. diameter rhizome, enveloped by 5 sheaths with rust-brown short hairs when young. The leaves are elliptic, stiff and 80-120 mm. long, 25-30 mm. wide and 2 mm. thick.

Inflorescence and Flowers: The flowering plant looks like an *Octomeria.* Up to 20 flowers emerge from the apices of young and older leaf stems on 3 mm. long pedicels and bloom simultaneously. The spathe is 3 mm. and the ovary is 4.5 mm. long. The sepals and petals are obtuse, ovate and whitish to very light green. The dorsal sepal is 6.7 mm. long by 2.7 mm. wide. The laterals are 40% fused and 6.2 mm. long by 4.8 mm. wide together. The petals are 6 mm. long and 1.8 mm. wide. The tongue-shaped, fleshy lip is 3.1 mm. long by 1.5 mm. wide, light yellow-green with a rose basal half and faint side lobes. The surface of the main lobe is papillate or short-haired. The whitish column is 2.5 mm. long and 1.5 mm. wide with a wine-red column foot. The anther cap is orange with wine-red short hairs.

Flowering Periods: April and the flowers last for ten days.

Pleurothallis fasciculata
Cogn. | *1386* (Plate XXI)

Etymology: Fasciculata is Latin for bundles, referring to the clustered inflorescences.

Habitat: Epiphytic from sections VI and VIII at 900-1000 M. asl, in very humid original cloud forest on moss-covered, mid-tree branches. L 5; H 8; W 5;

Occurrence: Occasional.

Plant: This 170-250 mm. tall epiphyte forms loose clusters with 10-15 leaves and wiry, white or brownish branched roots of 1 mm. diameter. The round leaf stems are 50-130 mm. long and 2-3 mm. diameter and arise from each second or third internode at 10-15 mm. intervals from the creeping, sheath-covered,

2.5 mm. diameter rhizome and are 70% enveloped by 3 membranous short-lived brown sheaths. The leaves are oval, stiff, yellow-green, indistinctly conduplicate and 50-120 mm. long, 26-30 mm. wide and 2.5 mm. thick.

Inflorescence and Flowers: The flowering plant resembles an Octomeria. Up to 10, 7-15 mm. inflorescences arise from the apex of both young and older leaf stems, each producing up to 3 flowers on 2 mm. pedicels and blooming simultaneously. All parts of the flower are keeled and light yellow. The inner sepal surfaces are granular. The dorsal is acute, narrowly ovate and 6 mm. long by 1.4 mm. wide. The slightly asymmetric laterals are almost linear and each is 5.5 mm. long by 1.2 mm. wide. The narrowly lanceolate, almost linear petals are 4.8 mm. long and 0.8 mm. wide. The tongue-shaped lip is 1.9 mm long by 0.6 mm. wide with a rounded apex. The column is light yellow, slightly curved and 1.8 mm. long by 1.5 mm. wide including the spread wings, and both apex and round wings are faintly gnawed.

Flowering Periods: April and the flowers last for 10 days.

Pleurothallis fasciculata
var. densiflora
Cogn. | *1386b* (Plate XXI)

Etymology: Densi and *flora* are Latin for densely packed flowers.

Habitat: Epiphytic from section VIII at 600-800 M. asl, in humid original mountain forest on moss-covered mid-tree branches. L 5; H 8; W 5;

Occurrence: Occasional.

Plant: This 150-220 mm. tall epiphyte forming loose clusters with 10-20 leaves and a few wiry, white or brownish branched roots of 0.7 mm. diameter. The round leaf stems are 80-110 mm. long and 2-3.5 mm. diameter and arise from each second or third internode at 8-12 mm. intervals from the creeping, sheath-covered, 3.5 mm. diameter rhizome and are 70% enveloped by 3 membranous short-lived brown sheaths. The leaves are acute, stiff, flat, and 80-110 mm. long, 25-40 mm. wide and 2 mm. thick.

Inflorescence and Flowers: The flowering plant resembles an Octomeria. Up to 20, 15-20 mm. inflorescences arise from the apex of both young and older leaf stems, each producing up to 3 flowers on 5 mm. pedicels and blooming simultaneously. The ovary is 1 mm. long and the flowers have a pleasant scent. All parts of the flower are keeled and light yellow. The sepals and petals are narrowly ovate with long tapering apices.. The dorsal sepal is 7.4 mm. long mm. long by 1.6 mm. wide. The slightly asymmetric laterals are almost linear and each is 7 mm. long by 1.1 mm. wide. The narrowly lanceolate, almost linear petals are 5.3 mm. long and 0.9 mm. wide.

The tongue-shaped lip is 2 mm long by 0.4 mm. wide with a rounded apex. The column is 1.8 mm. long by 1.5 mm. wide.

Flowering Periods: April and the flowers last for 10 days.

Pleurothallis obovata
Lindl. | *1387* (Plate XXI)

Etymology: Latin for inverted egg-shaped referring to the leaves.

Habitat: Epiphytic from section VIII, at 600-800 M. asl, in humid original mountain forest on moss-covered lower and mid-tree branches. L 6-7; H 6-8; W 6-8.

Occurrence: Occasional.

Plant: This 150-200 mm tall epiphyte forms loose clusters with 10-20 leaves and wiry, white or brownish branching roots, 0.8 mm. in diameter. The round leaf stems are 50-110 mm. long and 2-2.5 mm in diameter, arising from each second or third internode at 10-15 mm. intervals from the creeping, sheath-covered, 2.5 mm. diameter rhizome and are 70% enveloped by 3 membranous, short-lived, brown sheaths. The leaves are indistinctly keeled, obovate, stiff, olive or yellow-green, and 50-100 mm long, 20-25 mm wide and 2 mm thick.

Inflorescence and Flowers: In flower the plant resembles an Octomeria. Up to four 7-10 mm. inflorescences arise from the apex of young and older leaf stems, each producing up to 3 flowers, blooming simultaneously on 2 mm. pedicels. The ovary is 2 mm. and the bracts are 2 mm. long. All parts of the flower are keeled and light yellow and the inner surfaces of the sepals are smooth. The dorsal sepal is acute, ovate and 4.5 mm. long by 1.9 mm. wide. The slightly asymmetric laterals are each 4.8 mm. long by 1.2 mm. wide. The lanceolate petals are 3.5 mm. long and 1.2 mm. wide. The tongue-shaped lip is 2.2 mm. long, 0.9 mm. wide with a blunt, pointed apex. The column is light yellow, slightly curved and 2.2 mm. long, 1.8 mm. wide, including the spread wings, while the apex and the roundish wings are faintly gnawed.

Flowering Periods: April. A flower lasts for 10 days.

Pleurothallis dryadum
Schltr. | *1388* (Plate XXII)

Etymology: Dryads were Greek forest nymphs and the name refers to the plants small and delicate beauty in the forest.

Habitat: Epiphytic from section VI and VIII, at 900-1600 M. asl, in humid original cloud forest on moss-covered mid-tree branches and trunks. L 4-6; H 7; W 5.

Occurrence: Common.

Plant: This 40-60 mm. high epiphyte forms loose clusters with 10-100 leaves and white or brownish 0.4 mm. diameter roots. The round thin leaf stems are 90% enveloped by 4 membranous sheaths and are 20-35 mm. long and 0.3 mm. in diameter. The leaves are 15-25 mm. long, 5-6 mm wide, 1.5-2.0 mm thick, obovate, stiff, dark olive-green, indistinctly conduplicate and finely attenuated to their bases to form 3 mm. false petioles. Young leaves have a more or less purple-coloured back.

Inflorescence and Flowers: One or two 50-60 mm. hair-like, wiry inflorescences arise from the apices of young and older leaf stems, each with 4 to 7 flowers. The first sessile flower blooms 25 mm. from the base; the others follow in rapid succession in a straight line at 2 mm. intervals and remain open simultaneously. The spathe is 1 mm., the bracts are 1.5 mm. and the ovary 1.5 mm long. The narrowly wedge-shaped sepals are light green to yellow and keeled. The dorsal is 7 mm. long by 1.5 mm. wide, while the laterals are 6.6 mm. long and 1.1 mm. wide. The slightly asymmetric petals are 5.5 mm. long, 0.8 mm. wide and with or without serrate margins at the finely acuminate apices. The tongue-shaped lip is vaguely violin-shaped, keeled and 2.2 mm. long by 0.7 mm. wide, has a rough surface at the yellow base, 2 very indistinct side-lobes and a faintly acute apex with occasionally two wine-red longitudinal stripes along the side-lobes. The column is light green, 2.2 mm. long by 1 mm. wide with some wine-red markings and the wings and apex are faintly gnawed.

Flowering Periods: April to May and October to November and the flowers last for 15-20 days.

Var. major: We found some plants in section VI at 1600 M. asl, which have 70 mm. leaf stems with 6 sheaths and the 80 mm. inflorescence flowered in May. It is shown in the drawing as type B.

Var. minor: We found some plants in section VI at 1350 M. asl, which are 25 mm. tall and have 10 mm. leaf stems with 3 sheaths. The 60 mm. inflorescences produce up to 10 light green flowers in December. The sepals are 2.6 mm. long and 0.3-0.5 mm. wide, the petals are 1.6 mm. by 0.3 mm. and the lip is 0.8 mm. long by 0.35 mm. wide.

Pleurothallis ferdinandiana
(Barb.Rodr.) Cogn. | *1389* (Plate XXII)

Etymology: Not known.

Habitat: Epiphytic from section VI, at 1100-1400 M. asl, in humid original forest on moss-covered mid-tree branches and trunks. L 5; H 7; W 5.

Occurrence: Occasional.

Plant: This 80-100 mm. high epiphyte forms clusters

with 6 to 20 leaves and white, 0.5 mm. diameter roots. The round leaf stems are enveloped by 4 brown sheaths and are 35-45 mm. long and 1 mm. diameter. The leaves are elliptical, leathery, olive-green, conduplicate, finely attenuated to their bases and 45-65 mm. long, 10 mm. wide and 0.5-1.0 mm. thick.

Inflorescence and Flowers: One or two 80 mm. inflorescences arise from the apices of young and older leaf stems with up to 10 flowers on 2-3 mm. pedicels. The first flower blooms at half leaf height and the rest open in rapid succession and remain open together. The spathe is 5 mm. and the ovary is 1.5 mm. long. All sepals and petals are light yellow, narrowly wedge-shaped with finely acuminate apices. The dorsal sepal is 9.5 mm. long by 2 mm. wide, while the laterals are 9.5 mm. long and 1.7 mm. wide. The slightly asymmetric petals have finely serrated margins and are 9 mm. long by 0.9 mm. wide. The lip is narrowly elliptical with faint side lobes and 2.5 mm. long by 1 mm. wide. The central vein is lowered at the fleshy base and the apex is obtuse. The column is light yellow, slightly curved, and 2 mm. long by 1.5 mm. wide

Flowering Periods: November to January and the flowers last for 7 to 14 days.

Pleurothallis montipelladensis Hoehne | *1394* (Plate XXII)

Etymology: A bare mountain. Where?

Habitat: Epiphytic from section VI, at 1100-1400 M. asl, in humid original cloud forest on moss-covered lower and mid-tree branches and trunks. L 5; H 8; W 4.

Occurrence: Rare.

Plant: This 15-25 mm. high epiphyte forms loose clusters with 10-100 leaves and white, 0.5 mm. diameter roots. The round leaf stems are enveloped by 2 membranous sheaths and are 8-15 mm. long and 0.3-0.5 mm. diameter. The stiff leaves are obovate, olive-green, faintly conduplicate, finely attenuated to their bases and are 10-12 mm. long, 4-5 mm. wide and 0.7 mm. thick. Young leaves occasionally have purple speckles and blotches on their backs.

Inflorescence and Flowers: The 25-30 mm. inflorescences arise from the apex of young and older leaf stems with up to 4 flowers on 2-3 mm. pedicels. The first flower blooms 10-15 mm. from the base of the inflorescence and 3 flowers may be open at once. The ovary is 0.7 mm. long. All sepals and petals are light lemon and narrowly ovate with finely acuminate apices. The dorsal sepal is 4.5 mm. long by 1.2 mm. wide while the laterals are 4.6 mm. long by 1.2 mm. wide. The slightly asymmetric petals are 4 mm. long by 1 mm. wide, with entire margins. The keeled lip is narrowly heart-shaped, with a rounded apex and is 2.2 mm. long by 1.4 mm.

wide. Its broad base and the short, fleshy side-lobes are short haired. The light green column is slightly curved and 1.8 mm. long, 1.3 mm. wide with distinctly gnawed wings and apex.

Flowering Periods: June to July and the flowers last for 7 to 14 days.

Pleurothallis pabstii Garay | *1395* (Plate XXII)

Etymology: Named in honour of the taxonomist **F. J. Pabst.**

Habitat: Epiphytic from section VI, at 1100-1400 M. asl, in humid original cloud forest on moss-covered mid-tree branches and trunks. L 6; H 8; W 5.

Occurrence: Occasional.

Plant: This 80-150 mm. high epiphyte forms clusters with 6 to 30 leaves and white, 0.5 mm. diameter roots. The round leaf stems are covered by 4 brown sheaths and are 35-65 mm. long and 1 mm. diameter. The leaves are obovate, leathery, dark olive-green, conduplicate, finely attenuated to their bases and are 45-70 mm. long, 12-18 mm. wide and 0.5-1.0 mm. thick. Young leaves occasionally have purple speckles on the back.

Inflorescence and Flowers: One or two 100 mm. inflorescences arise from the apices of young and older leaf stems with up to 6 fragrant flowers on 2-3 mm. pedicels. The first flower blooms 25 mm. up the inflorescence and the rest open in rapid succession and stay open together. The ovary is 1 mm. long. All sepals and petals are light yellow, narrowly cuneate with finely acuminate apices. The dorsal sepal is 11.5 mm. long by 1.8 mm. wide, while the laterals are 11 mm. long and 1.5 mm. wide. The slightly asymmetric petals are lighter and 8-9 mm. long by 1.2 mm. wide. The lip is narrowly elliptical and 3 mm. long by 1 mm. wide. The central vein is lowered at the fleshy base and the apex is obtuse. The column is light yellow, slightly curved, and 2.4 mm. long by 2 mm. wide and has lacerate wings and apex.

Flowering Periods: November to December and June and the flowers last for 7 to 14 days.

Pleurothallis simpliciglossa Loefgr. | *1397* (Plate XXII)

Etymology: Latin for simple tongue, referring to the regular shape of the lip.

Habitat: Epiphytic from section VI, at 1100-1300 M. asl, in humid original forest on moss-covered mid-tree branches and trunks. L 5; H 7; W 5.

Occurrence: Occasional.

Plant: This 40-50 mm. high epiphyte forms clusters

with 6 to 30 leaves and white, 0.5 mm. diameter roots. The round leaf stems are enveloped by 5 membranous, red-brown sheaths, which widen at the tips and are 10-35 mm. long and 0.3-0.5 mm. diameter. The stiff leaves are 15-20 mm. long, 7-8 mm. wide and 1-2 mm. thick, olive-green, obovate, faintly conduplicate and finely attenuated to their bases. Young leaves occasionally have purple speckles on their backs.

Inflorescence and Flowers: One or two 35-40 mm. inflorescences emerge from the apices of the most recent leaf stems with up to 6 wide open flowers on 2 mm. pedicels. The first flower is 15 mm. from the base and the remainder open in rapid succession and bloom simultaneously. The ovary is 1 mm. long. All sepals and petals are light yellow, keeled, narrowly wedge-shaped with finely acuminate and 90° recurved apices. The dorsal sepal is 6-8.5 mm. long by 1.8 mm. wide, while the laterals are 7.6 mm. long and 1.4 mm. wide. The slightly asymmetric petals are paler and 6.5 mm. long by 1.2 mm. wide with faintly serrate margins. The lip is narrowly tongue-shaped with a round or obtuse apex, and 2.4 mm. long, 0.8 mm. wide with faintly undulate margins and faint red longitudinal callosities. The light yellow column is slightly curved, 2.4 mm. long, 2 mm. wide, with a red back, gnawed wings and apex.

Flowering Periods: July and December and the flowers last for 20 days.

Pleurothallis sclerophylla
Lindl. | *1399* (Plate XXII)

Etymology: Sclero is Latin for hard, and *phyllus* Greek for leaf, referring to leaf texture.

Habitat: Epiphytic from section VI and VIII, at 1100-1500 M. asl, in humid original cloud forest, near mountain ridges, on upper-tree branches and trunks. L 6-9; H 5-8; W 5-9.

Occurrence: Common.

Plant: This 150-200 mm. tall epiphyte forms clusters with 10-30 leaves and white, 0.7 mm. diameter roots. The round leaf stems are 80-120 mm. long and 1.5-2.5 mm. diameter and arise alternately at 3 mm. intervals from the 2-3 mm. diameter rhizome and are 50-70% enveloped by 2 or 3, green or brown sheaths. The leaves are 60-90 mm. long, 15-20 mm. wide and 1.5-2.0 mm. thick, elliptic, stiff, shiny yellow-green, infolded at their bases and slightly convex at the rounded apices, conduplicate and attenuated to their bases to form 3 mm. false petioles.

Inflorescence and Flowers: One or two 150-250 mm. inflorescences arise from the apices of young and older leaf stems with up to 50 wide open, alternate and fragrant flowers on 2.5 mm. pedicels. The lengthening inflorescence bends. The first flower blooms 15

mm. from the base followed by the remainder in rapid succession at 1.5 mm. intervals and all bloom simultaneously. The ovary is 2 mm. long. The sepals are keeled, narrowly wedge-shaped, almost linear with finely acuminate apices, light yellow-green and hairy inside except for their concave bases. The dorsal sepal is 12-18 mm. long by 1.8 mm. wide, while the laterals are also 12-18 mm. long but 1.7 mm. wide. The slightly asymmetric petals are paler, spathulate with pointed apices and are 2.3 mm. long and 1 mm. wide. The central vein is fleshy and purple or dark brown at the apical half. The oval, slightly trilobed lip is light green and 2.5 mm. long and 1.3 mm. wide. The central vein and the keels are undulate, while the margins and the longitudinal callosities are toothed. The light green column is purple at the foot and apex, slightly curved, 2.2 mm. long and 0.6 mm. wide with the faint wings spread.

Flowering Periods: September to November and January to March. The flowers last for 20 days.

Pleurothallis radialis
Pôrto & Brade | *1404* (Plate XXII)

Etymology: Radialis is Latin for spreading straight out and refers to the tepals.

Habitat: Epiphytic from sections VI and VIII, at 1200-1300 M. asl, in humid original forest, on trunks 1-5 metres from the ground in humid habitats with good light. L 5; H 7; W 5.

Occurrence: Occasional.

Plant: This 150-250 mm. tall epiphyte forms loose clusters with 7-50 leaves and white, 1.3 mm. diameter roots. The round leaf stems arise alternately at 5-10 mm. intervals from the 2 mm. diameter rhizome and are 70-120 mm. long by 2 mm. diameter, 60% enveloped by 3 short-lived sheaths. The upper segment is compressed, deeply furrowed and 2.5 mm. by 5 mm. in cross-section. The stiff leaves are elliptic, erect, dark olive-green and 80-130 mm. long, 30-45 mm wide. and 1.5-2 mm. thick.

Inflorescence and Flowers: Almost all leaves produce up to 3 curved, 100-140 mm. inflorescences with 20-30 light lemon-yellow flowers which smell unpleasant, and are on very short pedicels. The peduncle is 30-35 mm., the spathe 20 mm. and the ovary is 2.5 mm. long. The whitish sepals and petals are keeled and almost linear with an obtuse apex. The dorsal sepal is 6.3 mm. long by 2 mm. wide and the free laterals are 6.3 mm. long by 1.7 mm. and the petals 5 mm. long by 1 mm. wide. The whitish, three-lobed lip has 2 hook-shaped erect side-lobes and is 3 mm. long by 1.9 mm. wide. The surface is warty at the base and apex. The light green, slightly curved column is 2.6 mm. long by 1.8 mm. wide and the apex and wings are lacerate.

Flowering Periods: February though to April and the flowers last 3 weeks.

Pleurothallis rubens
Lindl. | *1405* (Plate XXII)

Etymology: Rubens is Latin for reddish possibly referring to the reddish sepal veins.

Habitat: Epiphytic in all sections, at 900-1600 M. asl, in humid original mountain forest, near mountain ridges, in regrowing forest and even on suburban trees. L 7-9; H 5-7, W 5-7.

Occurrence: Common.

Plant: This 150-240 mm high epiphyte forms clusters with 10 to 30 leaves and white, 1 mm. diameter roots. The round leaf stems arise alternately at 3 mm. intervals from the 2-3 mm. diameter rhizome and are 70-110 mm. long by 2.5-4 mm. diameter, 50-70 % enveloped by 3 green or brown sheaths which widen at their apices. The upper sheaths are short-lived, and the upper segment of the leaf stem is furrowed on both sides. The leaves are elliptic, erect, leathery or stiff, conduplicate and 100-140 mm. long, 30-40 mm,. wide and 2 mm. thick.

Inflorescence and Flowers: The 250-350 mm. inflorescences emerge from the apices of young and older leaf stems with up to 25 wide open fragrant flowers on 6 mm. pedicels. The first flower blooms at 60 mm. from the base and the rest open in rapid succession and bloom simultaneously. The spathe is 8 mm. and the ovary is 2 mm long. The acute ovate sepals are keeled, light yellow and have small pearl-like protuberances at the margins. The dorsal sepal is 11 mm. long by 4 mm. wide, the laterals are 10.5 mm. long and 3.6 mm. wide with pearl-like protuberances on the apical section. The lighter coloured petals are spathulate with an indented apex and 5.5 mm. long by 2.5 mm. wide. The keeled lip is vaguely violin-shaped with a widened main lob, light yellow-green with 2 faint side-lobes and 5.3 mm. long by 2.5 mm. wide with 3 rough-surfaced greenish veins. The central vein may be purple in the lip mid-section. The column is light green, slightly curved, and 4 mm. long by 0.75 mm. wide and is dotted purple between the flanks and at the margins of the faint wings. The column is acuminate with entire margins.

Flowering Periods: End of November to January and the flowers last for 7 to 14 days.

Pleurothallis heterophylla
(Barb. Rodr.)Cogn. | *1407* (Plate XXII)

Etymology: Hetero - phylla is Greek meaning different leaves as they vary in shape.

Habitat: Epiphytic from section VI and VIII, at 700-900 M. asl, in humid original mountain forest on mid-tree horizontal branches. L 6; H 6, W 5.

Occurrence: Common.

Plant: This 60-90 mm high epiphyte forms clusters with 6 to 30 leaves and whitish, 0.7 mm. diameter roots. Three brown sheaths envelop the round leaf stems, which are 25-35 mm. long and 1 mm. diameter. Leaf shape varies from lanceolate to elliptical and they are 35-60 mm. long, 10-15 mm. wide and 1-1.5 mm. thick, leathery, dark olive-green to wine-red, conduplicate and attenuated to their bases into 2 mm. pseudo-petioles.

Inflorescence and Flowers: One or two 60-80 mm. inflorescences emerge from the apices of young and older leaf stems with up to 8 fragrant flowers on 2-3 mm. pedicels. The first flower blooms at over half the leaf height and the others open in rapid succession and remain open simultaneously. The ovary is 1 mm long. All sepals and petals are light yellow with wine-red patches, narrowly wedge-shaped with finely acuminate apices and keeled. The dorsal sepal is 10-13 mm. long by 2 mm. wide, while the laterals are 10-13 mm. long and 1.8 mm. wide. The slightly asymmetric petals are light green and 8-11 mm. long by 1.2 mm. wide. The lip is narrowly elliptical and 3 mm. long by 1 mm. wide with 2 indistinct side-lobes. The central vein is lowered at the fleshy, rough base and the apex is obtuse. The column is light green, slightly curved, and 2.8 mm. long by 1.8 mm. wide with slightly lacerate wings and apex.

Flowering Periods: November to December and the flowers last for 7 to 14 days.

The genus
Barbosella Schltr.

Barbosella has been included in *Pleurothallis* but **Rudolf Schlechter** created the genus in 1918, naming it in honour of **Dr. João Barbosa Rodrigues**. The suffix *-ella* hints at the diminutive nature of these plants which, although small, may occur in vast quantities completely covering branches and twigs in very humid forest and by riversides. There are about 20 species of these creeping epiphytes found from Costa Rica to Argentina with 10 listed for Brazil. They differ from *Pleurothallis* in having four, rather than 2 pollinia and usually, single flowers.

Key: Four pollinia with sticky viscidia at the base. The flowers are individual with ventral anthers and stigmas.

Barbosella australis (Cogn.) Schltr. | *1427* (Plate XXIII)

Etymology: Australis is Latin for southern which could mean anything.

Habitat: Epiphytic from sections IV, V and VI, at 1100-1400 M. asl, in the shade of very humid original forest on trunks and branches. L 1-3; H 8-9; W 2.

Occurrence: Common.

Plant: A creeping, branching, 25-35 mm. high epiphyte forming plants with 10-40 leaves. A pair of leaf stems with a whitish root up to 1 mm. diameter, or a branching rhizome, emerge from the 1 mm. diameter, light green rhizome at 2-6 mm. intervals. The 4 mm. long, round leaf stems are 0.75 mm. diameter and enveloped by up to 3 loose light green sheaths, up to 3 mm. long. The lanceolate leaves are succulent, conduplicate, yellow-green or flesh-coloured and 25-40 mm. long, 5-11 mm. wide, 1 mm. thick and narrowed towards their bases to form 3 mm. false petioles.

Inflorescence and Flowers: The single flowered inflorescences are 25-30 mm. and emerge from the apex of the most recent leaf stems. The spathe is 3 mm. and the ovary is 2 mm long. The narrowly ovate dorsal sepal is light yellow-green and 11 mm. long by 2 mm. wide with an acuminate golden-yellow apex. The ovate laterals are also yellow-green, 60% fused and together are 13 mm. long by 8.3 mm. wide. The narrowly lanceolate petals are 7 mm. long and 1.8 mm. wide with faintly serrated apical margins. The broad tongue-shaped lip has a blunt apex with 3 indistinct points and is 4.3 mm. long by 2.9 mm. wide with 2 yellow and wine-red protuberances enclosing the light yellow-green, bowl-shaped base. The elliptic main lobe is concave, yellow or yellow-green and has wine-red stripes parallel to the margins. The column is very straight, light green and 4 mm. long by 1.3 mm. wide and the base has a spherical foot which corresponds with the bowl-shaped lip base.

Flowering Periods: February to March. The flowers last for 15 days.

Barbosella gardneri Schltr. | *1429* (Plate XXIII)

Etymology: Named in honour of the Scottish Botanist **George Gardner**.

Habitat: Epiphytic from section VI, at 900-1500 M. asl, on trunks and branches in the shade of very humid original cloud forest. L 1-3; H 8-9; W 2.

Occurrence: Common.

Plant: This creeping and branching epiphyte forms large colonies and may cover branches or trunks in a short time. A pair of leaf stems, only 0.5 mm. apart and

a 0.4-0.5 mm. diameter root or a branching rhizome emerge from the 0.5 mm. diameter, light green rhizome. The next pair of leaf stems follows after 2 internodes at 3-7 mm. intervals. The 0.5 mm. long, round leaf stems are 0.3 mm. diameter and enveloped by a loose, 1 mm. long, light green sheath. The elliptic leaves are fleshy, faintly conduplicate, grass-green and 5-6.5 mm. long by 3.5-4 mm. wide and 1 mm. thick with mucronate tips.

Inflorescence and Flowers: The single flowered inflorescences are 25-30 mm. and emerge from the apex of the most recent leaf stems. The ovary is 1 mm. long. The sepals are narrowly ovate, light yellow-green with finely acuminate, old-gold-coloured apices and short-haired or toothed margins. The dorsal sepal is 8 mm. long and 1.6 mm. wide, and the 25% fused laterals are 8 mm. long by 3.8 mm wide together. The narrowly ovate petals have serrate margins at their finely acuminate apices and are 6.5 mm. long by 1.8 mm. wide. The lip is trilobed, elliptic, has a finely acuminate apex and is 2 mm. long by 1.3 mm. wide with 2 yellow-green longitudinal callosities at its base connected to the dark purple main callosities at the lip's centre where the 2 light purple side-lobes end. The longitudinal veins of the main lobe are purple. The column is light green and 1.6 mm. long by 0.4 mm. wide.

Flowering Periods: February through to May. The flowers last for 15 days.

General: We also found plants with linear leaves 20 mm. long and 2-3 mm. wide.

Barbosella gardneri var. *microphylla* Schltr. | *1432* (Plate XXIII)

Etymology: Micro and phylla together mean small leaves.

Habitat: Epiphytic from section VI, at 900-1500 M. asl, on trunks and branches in the shade of very humid original forest. L 1-3; H 8-9; W 2.

Occurrence: Common.

Plant: This creeping and branching epiphyte forms large colonies and may cover branches or trunks in a short time. A pair of leaf stems only 0.5 mm. apart and a 0.4-0.5 mm. diameter root or a branching rhizome emerge from the 0.5 mm. diameter light green rhizome. The next pair of leaf stems follows after 2 internodes at 3-7 mm. intervals. The 0.5 mm. long, round leaf stems are 0.3 mm. diameter and enveloped by a loose, 1 mm. long, light green sheath. The elliptical leaves are fleshy, faintly conduplicate, grass-green and are 5-6.5 mm. long by 3.5-4.0 mm. wide and 1 mm. thick. The apices are sharply pointed.

Inflorescence and Flowers: The single flowered inflorescence is 25-30 mm. and emerges from the apex of the most recent leaf stems. The ovary is 1 mm. long. The sepals are narrowly ovate, light yellow-green with finely acuminate, old gold apices with short-haired or toothed margins. The dorsal sepal is 7 mm. long and 1.5 mm. wide, while the 40% fused laterals are 6.3 mm. long by 3.8 mm. wide together. The narrowly ovate petals have serrate margins at their finely acuminate apices and are 5.5 mm. long by 1.6 mm. wide. The lip is trilobed, elliptic, has an acuminate apex and is 2.1 mm. long and 1.4 mm. wide with 2 yellow-green longitudinal callosities at the base, connected to the dark purple main callosities at the lip's centre where the 2 light purple side-lobes end. The longitudinal veins of the main lobe are purple. The column is light green and 1.6 mm. long by 0.4 mm. wide.

Flowering Periods: October to November. The flowers last for 15 days.

The genus *Barbrodia* Luer

A single species genus named by **Carl Luer** in honour of the Brazilian Botanist **Barbosa Rodrigues**. The plants were previously included in *Barbosella*. It was transferred as it was much smaller than the other *Barbosella* species. The column is short and robust; the anther is apical and the stigma entire.

Key: Plant minute; leaves less than 3 mm. long; column short; anther apical and stigma entire; four pollinia.

Barbrodia miersii (Lindl.) Luer | *1433* (Plate XXIII)

Etymology: Species named in honour of the British Botanist **John Miers**.

Habitat: Epiphytic from section VI, at 900-1500 M. asl, on trunks and branches in the shade of very humid original cloud forest. L 1-3; H 8-9; W 2.

Occurrence: Common.

Plant: This creeping and branching epiphyte forms large colonies and may cover branches or trunks in a short time. A pair of leaf stems only 0.5 mm. apart and a 0.4 mm. diameter root or a branching rhizome emerge from the 0.3 mm. diameter light green rhizome. The next pair of leaf stems appears after 2 internodes at 4-6 mm. intervals. The 0.4 mm. long, round leaf stems are 0.25 mm. diameter and enveloped by a loose light green sheath, 1 mm long. The elliptical leaves are fleshy, faintly conduplicate, grass-green and 3-5 mm. long by 2.5-3 mm. wide and 0.5 mm. thick with sharply pointed apices.

Inflorescence and Flowers: The single flowered inflorescence is 10-20 mm. and emerges from the apex of the most recent leaf stems. The ovary is 1 mm. long. The sepals are ovate, light olive-green and short haired on the apical halves. The dorsal sepal is 2.2 mm. long by 1.1 mm wide and the 30% fused laterals are 2 mm. long by 2.1 mm. wide together. The lanceolate petals have serrate margins at the acuminate apices and are 1 mm. long by 0.35 mm. wide. The fleshy lip is tongue-shaped, green with wine-red stains at its apex and back folded margins. The column is light green and 0.8 mm. long by 0.4 mm. wide.

Flowering Periods: October and the flowers last for 15 days.

The genus *Brachonidium* Lindl.

The name comes from the compound Greek word *brachionidium* for short and refers to the arms of the stigma. *Brachonidium* is genus of about thirty five species, all from tropical America. Two species are listed from Brazil.

Key: The species have six pollinia, leaf stems which are shorter than the rhizome, lacking an annulus, single flowers and a column with 2 lateral cavities.

Brachonidium dungsii Pabst | *1437* (Plate XXIII)

Etymology: Named in honour of the German botanist **F. Dungs** who lived in Nova Friburgo and died in 1977.

Habitat: Epiphytic from section VI and VIII, at 1000-1250 M. asl, in original forest on thin branches of large trees. L 5; H 7; W 5.

Occurrence: Rare. To date we have only found a few plants.

Plant: The very short, 0.5 mm. diameter leaf stems emerge at 5-10 mm. intervals from the 0.5-1.0 mm. diameter, creeping, branching, sheathed rhizome, enveloped by up to 3 greenish or yellow-green sheaths. The stiff leaves are fleshy, lanceolate, faintly keeled, furrowed, olive green and 12-20 mm. long, 6-10 mm. wide and 1.5-2.0 mm. thick and are only slightly elevated above the substrate.

Inflorescence and Flowers: We have only seen inflorescences with buds which fell off before flowering. The 20 mm. single-flowered inflorescences arise from the apex of young and older leaf stems. The ovary is 2.5 mm. and the bracts 3 mm long. The following description is according to **Barbosa Rodrigues** and **Hoehne** and measurements are taken from drawings. The acuminate, broadly ovate sepals are concave and light ochre. The dorsal sepal is 4.5 mm. long and 3 mm. wide. The laterals are 100% fused and together are 5 mm. long and wide. The acuminate, lanceolate, wine-red petals are 5 mm. long and 2.5 mm. wide with cuspidate apices. The fleshy, transverse, reniform lip is wine-red, 1.3 mm. long by 2 mm. wide and has a triangular callosity on its plate. The light green and rose column is 1.3 mm. long and 1 mm. wide with acuminate wings and short apex.

Flowering Periods: December to January. The flowers last for ten days.

The genus
Octomeria R. Br.

Robert Brown named this genus from the Greek *okto* - eight and *meros* - a part referring to the eight pollinia that all its species have. The genus is distributed from Central America and the Caribbean down through South America to Argentina. About 100 species are attributed to Brazil of which perhaps 50 to the Organ Mountain Range. Like the genus *Stelis*, almost all *Octomeria* species in these mountains are inhabitants of high altitude, original cloud forest, happiest in shade and high constant humidity. However, with some of the species, it appears this was not always the case. Here we refer to those plants that have terete or narrow and succulent leaves. Botanical theoreticians have speculated that these species perhaps evolved to restrict water loss during the glacial periods when the mountain chain passed through many thousands of years with climates much drier than that of today. The most dramatic example of this phenomenon is *Octomeria aloefolia - 1506*, which also has the most interesting and beautiful flowers. Flowers of the terete-leaved species interestingly, are consistently more attractive or striking than those of species more appropriate to the current wet forest conditions.

Identification of a number of the following species was difficult due to the variability of some, the similarity of others and the lack or inadequacy of good source material. Once again we apologise in advance for errors in identification which perchance our readers discover.

Key: Eight waxy pollinia with stipes and viscidium. The flowers are in bundles or single on short peduncles. The sepals and petals are almost equal and the column is without lateral cavities. The inflorescence stems – peduncles, are usually slightly shorter than the bracts and the flower stems – pedicels, are distinctly longer.

Octomeria alpina
Barb. Rodr. | *1440* (Plate XXIII)

Etymology: *Alpina* is Latin for Alps, referring to the higher altitudes this plant inhabits.

Habitat: Epiphytic from section VI, at 1300-1600 M. asl, in very humid original cloud forest on moss-covered trunks and low- to mid-tree branches. L 4; H 7-8; W 4.

Occurrence: Frequent where the habitats still exist.

Plant: This 200-250 mm. high epiphyte forms clusters of 10-30 leaves with a profuse root system of 0.8 mm. diameter, tough, adherent, whitish roots. The round leaf stems are alternate at 2-3 mm. intervals on the 3.5 mm. diameter rhizome and are 120-150 mm. long by 2-4 mm. diameter, enveloped by 3 to 5 greyish sheaths. The slightly convex leaves are lanceolate, fleshy, dark olive-green, conduplicate and 100-120 mm. long, 20-25 mm. wide and 2-2.5 mm. thick.

Inflorescence and Flowers: Two to 5, wide open, yellow flowers emerge simultaneously from the apex of young and older leaf stems. The ovary is 3 mm. long. The ovate sepals and petals are acuminate. The dorsal sepal is 9-12 mm. long by 3.8 mm. wide, while the laterals are 8.5-12 mm. long by 3.8 mm. wide and the petals are 8-11 mm. long by 2.5-3.5 mm. wide. The lip is yellow with a red blotch between the lamellae and is 5.5-6.5 mm. long by 3.5-3.7 mm. wide. The margins are entire and the 3 points of the apex are indistinct. The light yellow column is 3 mm. long and 0.8 mm. wide.

Flowering Periods: April to May. The flowers last for 10 days.

Similar Species: See *O. densiflora - 1445. O. fasciculata - 1448*.

Octomeria crassifolia
Lindl. | *1443* (Plate XXIII)

Etymology: *Crassi* and *folia* are Latin for thick leaf.

Habitat: Epiphytic from sections IV and VI, at 1100 M. asl, in original mountain cloud forest on mid-tree branches. L 6; H 8; W 7.

Occurrence: Rare.

Plant: This 120-160 mm. high epiphyte forms clusters of 7 to 15 leaves with a root system of 0.8 mm. diameter, whitish roots. The round leaf stems are alternate at 3 mm. intervals on the 1.5-2.0 mm. diameter rhizome and are 70-90 mm. long and 1.2 mm. diameter at their bases widening to 2.5 mm. at the leaf junctions and enveloped by 5 to 6 tight, short-lived, tube-shaped sheaths. The slightly convex leaves are narrowly lanceolate, erect, dark olive-green, fleshy, faintly conduplicate and 65-80 mm. long, 12-14 mm. wide and 2.5-3 mm. thick.

Inflorescence and Flowers: Two to 6 unscented, bell-shaped, yellow flowers emerge simultaneously from the apex of both young and older leaf stems. The spathe, the bracts, the pedicels and the ovary are each 2.5- 3.0 mm. long. The ovate sepals and petals are acuminate. The dorsal sepal is 8.5 mm. long by 2.7 mm. wide, the laterals 8.2 mm. by 2.3 mm., while the petals are 8 mm. long and 2.4 mm. wide. The trilobed lip is 3.5-4.8 mm. by 2.5-2.7 mm. and light yellow with 1 or 2 purple blotches between the longitudinal callosities. The distinct side-lobes are narrow and close to the base. The main lobe is narrow and has a 0.5 mm. elongated mid point, while the margins are slightly crenate. The light yellow column is 2.3 mm. long and 0.8 mm. wide.

Flowering Periods: November to December. The flowers last for 15 days.

Similar Species: See *O. densiflora - 1445*.

Octomeria densiflora
Barb.Rodr. | *1445* (Plate XXIII)

Etymology: Densi and *flora* are Latin referring to the closely packed flowers.

Habitat: Epiphytic from sections V, VI and VIII, at 500-1500 M. asl, in original cloud forest on branches of large trees. L 4-6; H 5-7; W 5-7.

Occurrence: Common.

Plant: This 150-250 mm high epiphyte forms clusters of 10-30 leaves with an intense root system of 0.8.mm. diameter, whitish roots. The round leaf stems are 100-130 mm. long by 1.5 mm. diameter at their bases widening to 3-3.5 mm at their leaf junctions and enveloped by 5 to 7 tight, short-lived, tube-shaped sheaths arising alternately at 2 mm. intervals from the 1.5-2 mm. diameter rhizome. The slightly convex leaves are narrowly lanceolate, erect, dark olive-green, fleshy, faintly conduplicate and 80-120 mm. long, 16-22 mm. wide and 2-2.5 mm. thick.

Inflorescence and Flowers: Ten-35 strongly scented bell-shaped, light yellow to yellow flowers emerge simultaneously from the apex of young and older leaf stems. The ovate sepals and petals are acuminate. The dorsal sepal is 7-8 mm. long by 2.5-3.0 mm. wide, the laterals 7-8.5 mm. by 2.3-2.7 mm. wide, while the petals are 7-7.5 mm. long and 2.4-2.7 mm. wide. The lip is yellow with 1 or 2 red blotches between the lamellae and is 4-4.5 mm. by 2.4-3 mm. The margins are entire and the main lobe is narrowed to a faintly 3-pointed apex. The light yellow column is 2.5 mm. long and 1 mm. wide.

Flowering Periods: December. The flowers last for 15 days.

General: In section VI, at 1350 M. asl, we found some 50

mm. tall plants of the species flowering in November.

Octomeria ementosa
Barb. Rodr. | *1446* (Plate XXIII)

Etymology: Ementosa refers to the flower's column which lacks a mentum.

Habitat: Epiphytic from section VI and VIII, at 900-1200 M. asl, in humid original forest on low and mid-tree branches and trunks. L 5; H 8; W 5-7.

Occurrence: Occasional.

Plant: This 170-240 mm. high epiphyte forms clusters of 10-25 leaves with a profuse root system of 0.6 mm. diameter, whitish roots. The round leaf stems are alternate at 2 mm. intervals from the 2.5 mm. diameter rhizome and enveloped by 4 to 5 tight, tube-shaped sheaths and 60-120 mm. long and 2-2.5 mm. diameter at the base widening to 4 mm. at the leaf junctions. The narrowly lanceolate leaves are slightly bent back, mildly convex, fleshy, olive-green, conduplicate and 100-120 mm. long, 20-24 mm. wide and 2 mm. thick.

Inflorescence and Flowers: Up to ten flowers emerge simultaneously from the apex of young and older leaf stems. The pedicels and the ovaries are each 2 mm. long. The wide-open, bell-shaped flowers are light yellow to yellow-green and both sepals and petals are lanceolate. The dorsal is 10.5 mm. long by 3.7 mm. wide and the laterals are 10.5 mm. by 3.1 mm., while the petals are 10 mm. long by 3 mm. wide. The pale lemon lip is 5-6 mm. long by 3-4.1 mm. wide and has a dark purple blotch between the longitudinal callosities; 2 knife-shaped calli extend far into the main lobe from the 2 larger callosities in the lip centre. The lip margins are undulate and the apex is indistinctly 3-pointed. The light yellow-green column is 4 mm. long and 1 mm. wide.

Flowering Periods: March to May and the flowers last for 10 days.

Octomeria fasciculata
Barb. Rodr. | *1448* (Plate XXIV)

Etymology: From the Latin *fasciculatus* meaning clustered, growing in bundles referring to the clustered flowers.

Habitat: Epiphytic from section VI, at 1100-1400 M. asl, in very humid original cloud forest on low and mid-tree branches and trunks. L 5; H 8-9; W 5-7.

Occurrence: Occasional.

Plant: This 250-350 mm. high epiphyte forms clusters of 10-25 leaves and a profuse root system of 1 mm. diameter, tough, adherent, whitish roots. The round leaf stems arise alternately at 2 mm. intervals from the 2.5-3.0 mm. diameter rhizome, enveloped by 5 to 7 tight, tube-shaped sheaths 100-200 mm. long by 2-2.5 mm. diameter at their bases widening to 5 mm. at the leaf junctions. The narrowly lanceolate, dark olive-green leaves are 100-160 mm. long, 22-33 mm. wide, 2.5-3.5 mm. thick, slightly bent back, slightly convex, fleshy and conduplicate.

Inflorescence and Flowers: Five to 25 wide-open, bell-shaped, light yellow flowers emerge simultaneously from the apex of young and older leaf stems. The pedicels and the ovary are 3-4 mm. long. The tips of the ovate sepals and lanceolate petals are obtuse. The dorsal sepal is 10-12 mm. long by 3-4 mm. wide and the laterals are 9.5-12 mm. by 2.5-3.8 mm., while the petals are 9-11 mm. long by 2-3.5 mm. wide. The yellow lip, 5.3-6.0 mm. long by 3.5-4.0 mm. wide, has a light red blotch between the lamellae, entire margins and the apical points are rounded. The light yellow column is 2.2 mm. long and 0.8 mm. wide.

Flowering Periods: April to May. The flowers last for 10 days.

Octomeria robusta
Rchb. f. & Warm. | *1454* (Plate XXIV)

Etymology: Robusta is Latin for strong, referring to the plant's robust habit.

Habitat: Epiphytic from section VI, at 1400-1600 M. asl, in very humid original cloud forest on branches of large trees. L 6-7; H 8-9; W 7.

Occurrence: Common.

Plant: This 150-250 mm high epiphyte forms clusters of 10-25 leaves with a profuse root system of whitish, wiry, adherent, 1 mm. diameter roots. The round leaf stems are enveloped by 4 to 6 tight, tube-shaped sheaths which arise alternately at 5 mm. intervals from the 3-5 mm. diameter rhizome and are 50-110 mm. long and 2-2.5 mm. diameter close to the rhizome but 3-5 mm. diameter and oval at the leaf junctions. The leaves are 70-140 mm. long, 12-18 mm. wide and 2-2.5 mm. thick, narrowly lanceolate, erect, fleshy, yellow-green, faintly conduplicate.

Inflorescence and Flowers: Three to 20 wide open, bell-shaped, light yellow or ochre, single flowers emerge together from the apex of young and older leaf stems. The pedicel and the ovary are both 3 mm. long. The apices of the ovate sepals and lanceolate petals are obtuse and recurved. The dorsal sepal is 7 mm. long by 3 mm. wide, the laterals are 6.8 mm. by 2.9 mm. and the petals are 7.2 mm. long and 2.6 mm. wide. The lemon-yellow lip is 4 mm. by 2.9 mm. with small side-lobes, short lamellae, a robust main lobe with faintly gnawed margins and the centre may be maroon. The

light yellow-green column is 2.7 mm. long and 0.8 mm. wide.

Flowering Periods: March to April. The flowers last for 10-15 days.

Octomeria sarcophylla
Barb. Rodr. | *1455* (Plate XXIV)

Etymology: Sarco and *phylla* are Greek for fleshy leaves.

Habitat: Epiphytic from section VII, at 750-800 M. asl, in original mountain cloud forest, on upper and mid-tree branches. L 8; H 7; W 7.

Occurrence: Rare.

Plant: This 130-150 mm high epiphyte forms clusters of 5-15 leaves and a whitish, wiry root system of adherent, 1 mm. diameter roots. The round leaf stems are enveloped by 4 to 5 tight, tube-shaped sheaths and alternate at very short intervals on the 1 mm. diameter rhizome and are 70-85 mm. long and 1 mm. diameter close to the rhizome and 2.5 mm. diameter at the leaf junctions. The leaves are 50-65 mm. long, 5-7 mm. wide and 2.5 mm. thick, linear-lanceolate, fleshy, dark olive-green, conduplicate and indistinctly keeled. The backs are usually blotched dark wine-red.

Inflorescence and Flowers: One to 3 flowers emerge consecutively from the apex of young and older leaf stems. The pedicel and the ovary are both 3 mm. long. All parts of the flower are whitish, occasionally with pale purple speckles on the central veins. The flower has a distinctive chin. The sepals and petals are narrowly lanceolate with acute apices and the dorsal is 10 mm. long by 3.5 mm. wide; the laterals are 10 mm. long by 3.2 mm. wide, and the petals are 9 mm. long and 2.7 mm. wide. The wide lip is 4 mm. long and 3.6 mm. wide and light lemon with light purple speckles along the centre vein. The side-lobes are wide and the main lobe faintly trilobed. The yellow-green column is 2.3 mm. long, 1.1 mm. wide with a proportionally massive foot.

Flowering Periods: October, November and February to April. The flowers last for 14 days.

Octomeria serrana
Hoehne | *1457* (Plate XXIV)

Etymology: Serra is Latin for saw and in Portuguese means a saw-like mountain range referring to the habitat.

Habitat: Epiphytic from section VI, at 900-1300 M. asl, in original forest, on moss-covered mid-tree branches. L 5-7; H 7-8; W 5-7.

Occurrence: Occasional.

Plant: This 300-350 mm high epiphyte forms clusters

of 10-30 leaves. The erect round, leaf stems arise from the slightly creeping, 3-6 mm. diameter rhizome and are 200-250 mm. long by 2.5 mm diameter above their swollen bases widening to 3-4 mm. at the leaf junctions and enveloped by 5 to 7 greyish sheaths. The apical section is elliptical in cross section. The leaves are narrowly lanceolate, erect, conduplicate, leathery, olive-green and 150-180 mm. long, 20-35 mm. wide and 2.5 mm. thick.

Inflorescence and Flowers: Up to 18, light yellow-green, bell-shaped flowers emerge simultaneously from the apex of young and older leaf stems. The pedicels are 10 mm. and the ovary is 5 mm. long. The light yellow green sepals and petals are narrowly lanceolate with acuminate apices; the sepals are 9-15 mm. long by 3-3.5 mm. wide. The petals are 8.5-14 mm. long by 2.5-3 mm. wide. The tri-lobed lip is intensely purple between the longitudinal callosities and is 6-7 mm. long by 3.5-4 mm. wide with the apex of the main lobe rounded and pointed with faintly crenate margins. The yellow-green column has some purple at the apex and is 3.5 mm. by 1 mm.

Flowering Periods: February. The flowers last for 3-8 days.

Similar Species: See *O. fasciculata - 1448.*

Octomeria anceps
Pôrto & Brade | *1458* (Plate XXIV)

Etymology: From the Latin anceps meaning two-edged referring to the flattened stems.

Habitat: Epiphytic from Section VII at 1200-1400 M. asl, in original cloud forest on moss-covered mid-tree branches. L 5-7; H 7; W 5-7.

Occurrence: Occasional.

Plant: This 150-200 mm high epiphyte forms clusters of 10-20 leaves. The round leaf stems are enveloped by 4 to 5, tube-shaped sheaths, which arise alternately at short intervals from the 2-3 mm. diameter rhizome and are 70-90 mm. long by 0.8 diameter at the base and 2 mm by 0.8 mm at the leaf junctions. The leathery to stiff leaves are narrowly lanceolate, slightly recurved, conduplicate, dark green and 60-90 mm. long, 5-7mm. wide and 1 mm. thick.

Inflorescence and Flowers: Two to 3 whitish, wide-open flowers emerge simultaneously from the apices of the most recent leaf stems. The spathe is 5-6 mm., the bracts are 1.5-2 mm. and the ovary is 2.5 mm. long. The long acuminate sepals and petals are narrowly ovate. The sepals are 8.5-9 mm. long by 2.3 mm. wide, while the petals are 8.3 mm. long and 1.8 mm. wide. The trilobed lip is 3.8 mm. long and 2.6 mm. wide, yellow with a dark purple centre. The main lobe margins are faintly

crenate and the apex is pointed. The whitish column has a purple foot and is 2.5 mm. long by 0.7 mm. wide.

Flowering Periods: January to February and November.

Similar Species: This is one of only three *Octomeria* species with laterally compressed leaf stems in the region and these are easily differentiated.

Octomeria grandiflora
Lindl. | *1460* (Plate XXIV)

Etymology: *Grandi* and *flora* are Latin for large flower and this species has the largest in the genus.

Habitat: Epiphytic from section VI and VIII at 900-1600 M. asl, in humid original cloud forest on branches and trunks, tolerant but prefers shade. L 2-5; H 5-8; W 2-5.

Occurrence: Common.

Plant: This 150-230 mm. high epiphyte forms clusters of 10-30 leaves with round leaf stems enveloped by 5 tube-shaped sheaths, arising alternately at 2 mm. intervals from the 2-3 mm. diameter rhizome. They are 70-100 mm. long by 1-1.5 mm. diameter at their oval bases and 2 by 3 mm. at the leaf junctions. The leaves are narrowly lanceolate, slightly recurved, leathery, distinctly conduplicate and 90-140 mm. long, 13-17 mm. wide and 1-1.5 mm thick.

Inflorescence and Flowers: One to 3 wide open, white, flowers emerge simultaneously from the apex of young and older leaf stems. The apices of the ovate sepals and lanceolate petals are obtuse and slightly recurved. The dorsal sepal is 15 mm. long by 5.5 mm. wide and the laterals are 14.5 mm. long by 4.5 mm. wide, while the petals are 13.5 mm. long and 3.6 mm. wide. The lip is 7.5 mm. long by 6 mm. wide, light yellow, while the narrow side lobes and the lamellae are wine-red. The yellow main lobe has 2 distinct points, faintly gnawed margins and the centre is sometimes maroon. The column is 5 mm. long by 1 mm. wide and wine-red, while the apex is yellow-green.

Flowering Periods: November to April. The flowers last for 2 weeks but flowering plants may be found all months of the year.

Variety: We found a var. *robusta* in section VI, at 500 M. asl, which is a 550 mm. tall plant with 50 leaves. The leaf stems are up to 330 mm. long, 3 mm. diameter at their bases and 6 mm. by 4 mm. near the leaf junctions with a drop-shaped cross-section and up to 7 segments. The narrowly lanceolate leaves are leathery to stiff and are up to 250 mm. long, 35 mm. wide and 2 mm. thick. The plant blooms in January with 1 or 2 light yellow flowers at the most recent leaf bases. The sepals are 10 mm. long and 3.8 mm. wide, the petals are 9.8 mm. by 3 mm and the lip 6 mm, long and 5.4 mm, wide.

Octomeria gracilicaulis
Schltr. | *1462* (Plate XXIV)

Etymology: From the Latin *gracilis* and *caulis* for thin stem referring to the thin leaf stem.

Habitat: Epiphytic from section VI, at 700-1600 M.asl, in original montane cloud forest on moss-covered upper-tree branches. L 7; H 7; W 6.

Occurrence: Common.

Plant: This epiphyte is 150-220 mm. high and forms clusters with 10-20 leaves and 60-140 mm. long by 1.2-1.8 mm. diameter leaf stems enveloped by 4 to 6 short-lived sheaths and arising alternately at 1-2 mm. intervals from the 1.2 mm. diameter rhizome. The leaves are narrowly lanceolate, leathery to stiff, erect, conduplicate, very faintly keeled and 60-90 mm. long by 7-9 mm. wide and 3 mm. thick.

Inflorescence and Flowers: Two to 4 wide open flowers emerge from the apex of young and older leaf stems. The short bracts are acuminate and the pedicel is 2 mm., while the ovary is 1 mm. long. The light yellow sepals and petals are ovate with obtuse apices. The dorsal is 5.5-6.5 mm. long by 2-2.7 mm. wide, the laterals 5.8-6.0 mm. long by 1.8-2.5 mm. wide and the petals are 5.7-6 mm. long by 1.8-2.5 mm. wide. The broad yellow lip is 3-3.5 mm. long by 2.3-2.7 mm. wide, and the main lobe has only one distinct point at its apex. The yellowish column is 2 mm. long and 1 mm. wide.

Flowering Periods: November to April. The flowers last for about 14 days.

Octomeria oxichela
Barb. Rodr. | *1465* (Plate XXIV)

Etymology: *Oxy* and *chela* are Greek for pointed lip referring to the 3-pointed lip.

Habitat: Epiphytic from section VI and VIII, at 900-1200 M. asl, in humid original mountain cloud forest on branches and trunks, preferring shady habitats. L 2-5; H 5-8; W 2-5:

Occurrence: Occasional.

Plant: This 150-240 mm high epiphyte forms clusters of 10-20 leaves with 0.9 mm diameter roots. The round leaf stems are enveloped by 6 to 8 short-lived sheaths, are alternate at 2 mm. intervals from the 2-4 mm. diameter rhizome and are 100-140 mm. long by 1-1.5 mm. diameter at their bases but 2 by 3.2 mm. at the leaf junctions with an oval cross section. The leaves are narrowly lanceolate, slightly recurved, leathery, conduplicate and 65-120 mm. long, 11-23 mm. wide and 1-1.5 mm. thick.

Inflorescence and Flowers: Up to 10 pure white to very light yellow, bell-shaped flowers emerge from the apex of young and older leaf stems, opening in rapid succession and staying open simultaneously. The pedicel is 1 mm., the whitish ovary is 2 mm and the greyish brown spathe and the bracts are 2-2.5 mm. long. The apices of the ovate, light yellow or amber sepals and lanceolate petals are acute and slightly fleshy. The dorsal sepal is 6 mm. long by 2.5 mm. wide and the laterals are 6 mm. long by 2.4 mm. wide, while the petals are 5.1 mm. long and 2.2 mm. wide. The lip is 3.4 mm. long by 2.5 mm. wide and light yellow with a short base, the side lobes not widely extended and the relatively long main lobe distinctly 3-pointed. The whitish column is 2.3 mm. long by 0.8 mm. wide.

Flowering Periods: November to January. The flowers last for two weeks.

General: The 8 segments of the leaf stems combined with the numerous flowers and the lip shape makes it easy to recognise.

Octomeria concolor
Barb. Rodr. | *1469* (Plate XXIV)

Etymology: From the Latin: uniformly-coloured referring to the flower.

Habitat: Epiphytic from sections V, VI, VII and VIII, at 600-1600 M. asl, in original mountain cloud forest on upper-tree branches. L 5-7; H 6-7; W 6.

Occurrence: Common.

Plant: This epiphyte is 80-150 mm. high and forms clusters of 10-20 leaves. The leaf stems are enveloped by 5 to 7 sheaths and are alternate at 1-2 mm. intervals from a 1.5-2.0 mm. diameter rhizome and are 50-90 mm. long and 1-1.8 mm. diameter. The flat or faintly folded leaves are narrowly lanceolate, leathery, erect, keeled, furrowed and are 60-70 mm. long, 7-9 mm. wide and 1 mm. thick.

Inflorescence and Flowers: Two to 4 wide open flowers emerge from the apex of young and older leaf stems with a short, acute spathe. The pedicels are 2 mm. and the ovary is 1 mm. long. The light yellow sepals and petals are ovate with obtuse apices and the dorsal is 6.4 mm. long and 2.4 mm. wide, while the laterals are 6.2 mm. long by 2.2 mm. wide and the petals, 5.5 mm. by 2.0 mm. The yellow lip is a typical *Octomeria* shape and 3.9 mm. long and 2.6 mm. wide. The yellowish column is 2.5 mm. long by 0.6 mm. wide.

Flowering Periods: November to April. The flowers last for 5-10 days but flowering plants may be found during all months of the year.

Octomeria micrantha
Barb. Rodr. | *1471* (Plate XXV)

Etymology: *Micra* and *antha* are Greek for small flower.

Habitat: Epiphytic from sections V, VI, VII, VIII, at 800-1300 M. asl, in original forest on trunks and mid- and upper-tree thick branches. L 5-8; H 6; W 5-7.

Occurrence: Common.

Plant: This epiphyte is 60-120 mm high and forms clusters with 10-40 leaves and 0.5 mm diameter whitish roots. The slightly curved leaf stems are alternate at 1-2 mm. intervals from the 1.2-2 mm. diameter rhizome, covered by 4 to 6 sheaths and 30-60 mm. long by 0.8-1.8 mm. diameter. The olive-green leaves are lanceolate, slightly conduplicate and 50-60 mm. long, 6-8 mm. wide and 1 mm. thick.

Inflorescence and Flowers: One to 5 flowers emerge from the apex of young and older leaf stems on 3 mm. pedicels and open wide. The spathes are short and acute and the ovary is 1 mm. long. The light yellow-green sepals and petals are ovate with obtuse apices. The dorsal sepal is 3.5 mm. long by 1.7 mm. wide, the laterals are 3.5 mm. by 1.5 mm. and the petals are 3.5 mm. long by 1.3 mm. wide. The yellow lip is a typical *Octomeria* shape and 3.0 mm. long and 1.7 mm. wide. The yellow-green column is 2.0 mm. by 1.0 mm. long.

Flowering Periods: January-February and the flowers last for 7 days.

Octomeria aff. micrantha
Barb. Rodr. | *1471b* (Plate XXV)

Etymology: See *1471*.

Habitat: Epiphytic from sections VI, VIII, at 1100-1300 M. asl, in original forest on thicker mid- and lower-tree branches. L 5; H 8; W 5.

Occurrence: Rare.

Plant: This epiphyte is 60-90 mm high and forms clusters with 5-15 leaves and 0.4 mm diameter whitish roots. The slightly curved leaf stems are alternate at 1 mm. intervals from the 1-1.5 mm. diameter rhizome, covered by up to 5 sheaths and 30-50 mm. long by 1.5 mm. diameter. The olive-green leaves are narrowly lanceolate, almost linear, slightly conduplicate and 50-60 mm. long, 7 mm. wide and under 1 mm. thick.

Inflorescence and Flowers: Up to 2 flowers emerge from the apices of young and older leaf stems on 2.5 mm. pedicels and do not open wide. The spathes are 2.5 mm. and the ovary is 2 mm. long. All parts of the flower are light yellow-green and the sepals and petals are ovate with obtuse apices. The dorsal sepal is 4.5 mm. long by 2.5 mm. wide, the laterals are 4.3 mm. by 2.3 mm.

and the lanceolate petals are 4.1 mm. long by 1.9 mm. wide. The concave lip is 3-lobed and 2.9 mm. long by 2.2 mm. wide and the rounded apex has a single point. The column is 2 mm. long by 1 mm.

Flowering Periods: January and the flowers last for 7 days.

General: This is possibly a variety of *O. concolor*.

Octomeria bradei
Schltr. | *1476* (Plate XXV)

Etymology: *Bradei* is named after the German **A. Brade** (1881-1971), a systematic botanist who worked in the National Museum and the Rio de Janeiro Botanic Garden.

Habitat: Epiphytic from section VI, at 900-1100 M. asl, in humid original cloud forest, on mid- and upper-tree trunks and thick branches. L 5-7; H 6-7; W 6.

Occurrence: Rare.

Plant: This epiphyte forms clusters with 7 to 30 leaves. The round, pendent leaf stems are 80-150 mm. long by 1-2 mm. diameter and alternate at 1 mm. intervals from a 1.5 mm. diameter rhizome, enveloped by 3 loose sheaths. The leaves are terete, finely attenuated towards their apices, with one longitudinal furrow, olive-green or purple and 150-200 mm. long, 3-3.5 mm. wide and 2.5-3 mm. thick.

Inflorescence and Flowers: One to 5 light yellow to yellow, wide open flowers on 3-4 mm. pedicels emerge from the apex of young and older leaf stems. The concave sepals and petals are ovate with obtuse apices and the dorsal is 6.5 mm. long by 2.2-2.4 mm. wide, the laterals are 6.4 mm. long by 2.2-2.3 mm. wide and the petals are 6 mm. long by 1.9-2.0 mm. wide. The trilobed lip is 3.6 mm. long by 2.4 mm. wide, lemon-yellow and the main lobe has undulate margins and a faintly 3-pointed apex. The column is 2 mm. long by 0.6 mm. wide and yellow.

Flowering Periods: September to December. The flowers last for 10-14 days.

Octomeria linearifolia
Barb. Rodr. | *1478* (Plate XXV)

Etymology: *Lineari* and *folia* are Latin meaning linear or straight and narrow leaves.

Habitat: Epiphytic from section IV, VI and VIII, 900-1300 M. asl, in original mountain cloud forest, on upper and mid-branches of large trees. L 6; H 6; W 6.

Occurrence: Occasional.

Plant: This 70-90 mm high epiphyte forms clusters of 10-100 leaves and a profuse root system of whitish, wiry, adherent, 0.7 mm. diameter roots. The round leaf stems are enveloped by 4 to 5 tight, tube-shaped sheaths and alternate at very short intervals on the 0.8 mm. diameter rhizome and are 10-25 mm. long and 0.7 mm. diameter close to the rhizome but 1.5 mm. and elliptical at the leaf junctions. The leaves are linear, terete, dark olive-green, indistinctly conduplicate and 40-65 mm. long, 3.5-4.5 mm. wide and 2-2.5 mm. thick.

Inflorescence and Flowers: One to 7 flowers emerge simultaneously from the apex of young and older leaf stems. The pedicel and the ovary are both 2 mm. long. All parts of the flower are light yellow-green, occasionally with very light purple speckles along the central veins. The sepals and petals are narrowly ovate with acute apices and the dorsal sepal is 7 mm. long by 2.3 mm. wide, the laterals are 6.5 mm. long by 2.2 mm. wide, while the petals are 6.7 mm. long and 2 mm. wide. The narrow lip is 4.2 mm. long and 2.6 mm. wide and light yellow with pale purple speckles along the centre vein and wide side-lobes not much extended and a relatively long main lobe. The light green column is 2.2 mm. long and 0.6 mm. wide.

Flowering Periods: April to June and September to October and the flowers last 10-15 days.

Octomeria aff. linearifolia
1478b (Plate XXV)

Etymology: See *1478*.

Habitat: Epiphytic from section VI, 900-1500 M. asl, in original mountain forest, on upper and mid-tree branches of large trees. L 5; H 6; W 5.

Occurrence: Common.

Plant: This 60-110 mm high epiphyte forms clusters of 10-30 leaves and an intense root system of 0.8 mm diameter white roots. The round leaf stems are 10-30 mm. long and 0.7-1.2 mm. diameter at their bases widening to 2.2 mm. diameter at the leaf junctions, enveloped by 3 to 4 loose sheaths and arise at very short intervals from the rhizome. The linear leaves are dark green, faintly conduplicate with V-shaped cross-sections and are 40-80 mm. long, 3.5-4 mm. wide and 2-2.5 mm. thick.

Inflorescence and Flowers: Up to 6 flowers emerge simultaneously from the apices of young and older leaf stems. The pedicel is 2.5 mm. and the ovary is 1.5 mm. long. The lanceolate sepals and petals are whitish to light yellow-green and speckled purple along their central veins. The dorsal sepal is 7.6 mm. long and 3.1 mm. wide and the lateral sepals are 8 mm. long by 2.7 mm. wide, while the petals measure 7 mm. by 2.5 mm. The trilobed lip is 4.3 mm. long and 2.9 mm. wide and is light yellow to light yellow-green with purple speckles along its central vein and a three-pointed apex.

Flowering Periods: January. The flowers last for 10 days.

General: This plant seems to be a variety of *O. linearifolia - 1478* but the latter plant is larger, the flowers open wider and the lip is narrower.

Octomeria aff. linearifolia
1478c (Plate XXV)

Etymology: See *1478*.

Habitat: Epiphytic from section VI, at 900-1300 M. asl, in original mountain cloud forest, on moss-covered mid-tree branches. L 5-7; H 7-8; W 5-7.

Occurrence: Occasional.

Plant: This 70-90 mm. high epiphyte forms clusters of 10-30 leaves. The round leaf stems are enveloped by 3 to 4 long-lived sheaths and are 30-40 mm. long and 0.7 mm. diameter at their bases widening to 2 mm. at their leaf junctions. The leaves are linear, conduplicate, erect, olive-green and 40-50 mm. long, 3-4 mm. wide and 1.5 mm. thick.

Inflorescence and Flowers: One to 6 bell-shaped flowers emerge from the apex of young and older leaf stems. The pedicels are 5 mm. long and the ovary is 1 mm. The white sepals and petals are narrowly lanceolate with finely acuminate apices and the flower is more or less marked purple along its central veins, depending on the light. The dorsal sepal is 10 mm. long by 2.3 mm wide, the laterals are 9 mm. by 2.1 mm. and the petals are 8.6 mm. long by 1.8 mm. wide. The light yellow lip has purple marks along its central vein and is 5 mm. long by 2.4 mm. wide. The side-lobes are broad and the apex has only one point. The yellow-green column is 2.1 mm by 0.6 mm long.

Flowering Periods: February to June and September. The flowers last 10-14 days.

General: This plant and its flowers are larger and more graceful than the normal *O. linearifolia* and apex of the lip has only one point. It is possibly a new species.

Octomeria aff. linearifolia
1478d (Plate XXV)
Possible new species

Etymology: See *1478*.

Habitat: Epiphytic from section VI, at 700-900 M. asl, in original mountain cloud forest, on moss-covered mid-tree branches. L 5-7; H 7-8; W 5-7.

Occurrence: Rare.

Plant: This 70-85 mm high epiphyte forms clusters of 10-30 leaves. The round leaf stems are enveloped by 3 sheaths and are 20-30 mm. long and 0.7 mm. diameter at their bases widening to 1.5 mm at the leaf junctions. The leaves are linear, erect, dark olive-green, conduplicate and 40-50 mm. long, 3-5 mm. wide and 2.5 mm. thick.

Inflorescence and Flowers: Up to 4 flowers emerge simultaneously from the apex of young and older leaf stems. The pedicels are 3 mm and the ovary is 1.5 mm long. The flower opens wide and all parts are very light green. The sepals and petals are narrowly lanceolate. The dorsal sepal is 6-7 mm. long by 2 mm. wide, the laterals are slightly fused at the base and are 6 mm. long by 2 mm. wide and the petals are 6 mm. long by 1.6 mm. wide. The lip is 4 mm. long and 2 mm. wide with wide but short side-lobes and the main lobe has a faintly 3-pointed apex.

Flowering Periods: April to June and September. The flowers last for two weeks.

General: The shape of the plant and the flower details are so different to *O. linearifolia* that it is possibly a new species.

Octomeria albopurpurea
Barb. Rodr. | 1480 (Plate XXV)

Etymology: Albo and *purpurea* mean white and purple referring to the lip colours.

Habitat: Epiphytic from section VI and VIII, at 900-1500 M. asl, in original mountain forest near river banks or on south facing slopes, on moss-covered lower branches and trunks. L 4; H 8; W 4.

Occurrence: Occasional. It may be found in colonies.

Plant: This 100-130 mm. high epiphyte forms clusters of 4 to 15 leaves with a root system of whitish, adherent, 0.7 mm. diameter roots. The round leaf stems are enveloped by 4 to 5 loose, long-lived sheaths alternating at 2 mm. intervals from the 1.2 mm. diameter rhizome and are 45-60 mm. long by 0.8 mm, round in cross-section at their bases, but 1.5 mm. diameter and elliptical at the leaf junctions. The papery leaves are 45-70 mm. long by 10-17 mm. wide, narrowly lanceolate, conduplicate, olive-green, occasionally tarnished wine in high light, and attenuated to the bases forming 5-8 mm. false petioles, mostly twisted up to 180°, causing the flowers to appear on the backs of the leaves.

Inflorescence and Flowers: One to 3 whitish, wide-open flowers emerge from the apex of young and older leaf stems. The sepals and petals are lanceolate and acute. The dorsal sepal is 7-9.5 mm. long by 3-3.2 mm. wide, the lateral sepals 8-9.8 mm. long and 2.5-2.8 mm. wide and the petals are 7-8.5 mm. long by 2.5-2.8 mm. wide. The lip is whitish and light yellow with a purple centre and 4.7-5.2 mm. long by 3.2-4.0 mm. wide. The margins

are gnawed from the side lobes to the 3-pointed apex. The slightly curved column is light yellow-green and 2.5 mm. long by 1.1 mm. wide.

Flowering Periods: October to November. We also found a colony flowering in January through to April in a separate habitat. The flowers last for about 14 days.

Octomeria crassilabia
Pabst | 1483 (Plate XXV)

Etymology: From the Latin *crassus* and *labia* meaning thick lipped.

Habitat: Epiphytic from section VI, at 900-1200 M. asl, in original cloud forest on moss-covered trunks and mid-tree branches. L 5; H 8; W 5.

Occurrence: Occasional.

Plant: This 100-160 mm high epiphyte forms clusters with 5 to 15 leaves. The leaf stems are enveloped by 7 to 9 loose, long-lived sheaths and are alternate at 2-3 mm intervals on the 1.5-2 mm. diameter rhizome and 50-80 mm. long, 0.8 mm. diameter and round in cross-section at their bases but elliptical and 1.8 mm. diameter at the apical segments. The leaves are narrowly lanceolate, olive-green, occasionally tarnished wine-red in high light, conduplicate, attenuated to the base forming 5-10 mm false petioles, which are usually resupinate up to 180°, causing the flowers to appear on the backs of the leaves. The flat leaves are 50-80 mm. long by 11-16 mm. wide and 1 mm. thick.

Inflorescence and Flowers: One to 3 bell-shaped flowers emerge simultaneously from the apex of young and older leaf stems. The sepals and petals are acuminate and ovate, concave and white to light yellow. The dorsal sepal is 9 mm. long by 3.5 mm. wide, the laterals 8.5 mm. long by 3.1 mm. wide, while the petals are 7.7 mm. long and 2.8 mm. wide. The round lip is 3.6 mm. long by 3 mm. wide, light yellow and wine-red with a darker, fleshy centre and an elevated central vein on the main lobe which has serrate margins. The light green column is 2 mm. long and 1.2 mm. wide.

Flowering Periods: November to February and April and the flowers last for about 12-18 days.

Octomeria diaphana
Lindl. | 1484 (Plate XXVI)

Etymology: Diaphanus is from the Latin meaning colourless or quite translucent which describes many *Octomeria* flowers.

Habitat: Epiphytic from section VI and VIII, at 900-1200 M. asl, in original forest on moss-covered mid-tree branches and trunks. L 5; H 8; W 5.

Occurrence: Occasional but often in large numbers locally.

Plant: This 100-150 mm. high epiphyte forms clusters with 5 to 15 leaves. The leaf stems are alternate at 2-3 mm. intervals from the 1.5-2.0 mm. diameter rhizome, are enveloped by 7 to 9 loose, long-lived sheaths and measure 50-85 mm. long by 0.8 mm. and round in cross-section at their bases, but 1.8 mm. diameter and elliptical at the leaf junctions. The leaves are narrowly lanceolate, conduplicate, attenuated to their bases to form 5-10 mm. false petioles, which are usually up to 180° resupinate, causing the flowers to appear on the backs of the leaves. The slightly recurved leaf is 40-65 mm. by 13-18 mm. and 1-2 mm. thick and is green, occasionally tarnished red in sunny situations.

Inflorescence and Flowers: Three to 7 wide open, bell-shaped flowers arise simultaneously from the apex of young and older leaf stems. The narrowly ovate sepals and petals are finely acuminate, concave and white. The dorsal sepal is 15 mm. long by 4.5 mm. wide, the laterals 15 mm. by 4 mm. and the petals are 13.5 mm. long by 3.5 mm. wide. The white lip has a 3-pointed elliptical main lobe with indented margins, while the base is light beige to yellow and has slightly undulate margins. The longitudinal callosities are dark wine-red. The lip is 5.5 mm. long by 3.2 mm. wide and the light yellow column 3 mm. by 1 mm.

Flowering Periods: November to January. The flowers last for about 12-18 days.

Octomeria glazioviana
Regel | *1486* (Plate XXVI)

Etymology: Named in honour of the French Botanist **A. F. M. Glaziou**.

Habitat: Epiphytic from section VI, at 500-700 M. asl, in original cloud forest on moss-covered mid-tree branches and trunks. L 5; H 8; W 5.

Occurrence: Occasional.

Plant: This 140-180 mm. high epiphyte forms clusters of 5 to 15 leaves. The leaf stems are alternate at 3 mm. intervals from the 2.0 mm. diameter rhizome and enveloped by 5 to 7 loose, long-lived sheaths. They are 50-75 mm. long by 1 mm. diameter and round in cross-section at their bases, but 1.5 by 2 mm. diameter and elliptical at the leaf junctions. The leaves are 60-75 mm. long, 15-16 mm. wide and 1 mm. thick, olive-green to wine-red, lanceolate, conduplicate and attenuated to their bases to form 1 mm. false petioles.

Inflorescence and Flowers: Three, 3 mm wide-open bell-shaped flowers emerge in rapid succession from the apices of the leaf stems. The spathe is 2 mm., the bract and the ovary are 2.5 mm. long. The acute ovate sepals and petals are concave and white. The dorsal sepal is 10.5 mm. long by 4.2 mm. wide, the laterals 9.5 mm. by 3 mm. and the petals are 9.5 mm. long and 3.3 mm. wide. The trilobed lip is 5 mm. long by 3.2 mm. wide, light yellow-green and dark purple between the longitudinal callosities with a yellow centre vein. The light yellow column is 3 mm. long and 0.8 mm. wide.

Flowering Periods: January. The flowers last for 12 to 18 days.

Octomeria aff. glazioviana
1486b (Plate XXVI)

Etymology: See *1486*.

Habitat: Epiphytic from section VI, at 1000-1500 M. asl, in original cloud forest on moss-covered mid-tree branches and trunks. L 5; H 8; W 5.

Occurrence: Common.

Plant: This 90-120 mm. high epiphyte forms clusters of 5 to 15 leaves. The leaf stems are alternate at 2-3 mm. intervals from the 1.5-2.0 mm. diameter rhizome and enveloped by 5 to 7 loose, long-lived sheaths. They measure 50-70 mm. long by 0.8 mm. diameter and are round in cross-section at their bases, but 1.8 mm. diameter and elliptical at the leaf junctions. The leaves are 50-60 mm. long, 13-16 mm. wide and 1 mm. thick, olive-green, conduplicate, lanceolate, and attenuated to their bases to form 3-6 mm. false petioles.

Inflorescence and Flowers: Three to 7 bell-shaped flowers emerge in rapid succession or simultaneously from the apices of leaf stems. The ovary is 2.5 mm. long. The lanceolate sepals and petals are concave, acute and white. The dorsal sepal is 8.5 mm. long by 4 mm. wide, the laterals 9 mm. by 3.5 mm. and the petals are 8 mm. long and 3 mm. wide. The trilobed lip is 4.5 mm. long by 3 mm. wide, yellow, with the mid-section dark wine-red and the main lobe 3-pointed with faintly indented margins. The light yellow column is 2.5 mm. long and 1 mm. wide.

Flowering Periods: December to January. However, flowering plants may be found in all months of the year. The flowers last for 12 to 18 days.

Octomeria rechiana
Hoehne | *1487* (Plate XXVI)

Etymology: Not known.

Habitat: Epiphytic from section VI and VIII, 1000-1600 M. asl, in original cloud forest on moss-covered trunks and mid-tree branches. L 4; H 8; W 4.

Occurrence: Common.

Plant: This 90-120 mm. high epiphyte forms clusters of

5 to 15 leaves and a root system of whitish, adherent, 0.6 mm. diameter roots. The round leaf stems are enveloped by 5 to 7, loose, long-lived sheaths and alternate at 2-3 mm. intervals, from the 1.6 mm diameter rhizome. They measure 50-70 mm. long by 0.8 mm. wide and are round in cross-section at their bases widening to 1.8 mm. and an elliptical cross-section at the leaf junctions. The papery leaves are 40-50 mm. long by 13-18 mm. wide and 1.5 mm. thick, narrowly lanceolate, conduplicate, olive-green, occasionally tarnished wine in high light and attenuated to the base forming 5 mm. false petioles.

Inflorescence and Flowers: Three to 8, white, bell-shaped flowers on 3-5 mm. pedicels emerge in rapid succession or simultaneously from the apex of young and older leaf stems. The ovary is 2.5 mm long. The narrowly ovate sepals and petals are finely acuminate and concave. The dorsal sepal is 11 mm. long by 3.3 mm. wide, the laterals 11 mm. by 2.6 mm. and the petals are 10 mm. long by 2.3 mm. wide. The lip is 5.2 mm. long by 3 mm. wide, light ochre with an oval, faintly verrucous, yellowish, 3-pointed main lobe with toothed margins. The base and the central vein are undulate, while the longitudinal callosities are dark wine-red. The light yellow column is 3 mm. by 1.3 mm.

Flowering Periods: December to January. However flowering plants may be found during all months of the year. The flowers last for 12-18 days.

Octomeria reitzii
Pabst | *1488* (Plate XXVI)

Etymology: Named in honour of **R. Reitz**, botanist, of Santa Catarina.

Habitat: Epiphytic from section VIII, 900-1100 M. asl, in original cloud forest on moss-covered trunks and mid-tree branches. L 4; H 8; W 4.

Occurrence: Occasional.

Plant: This 90-150 mm. high epiphyte forms clusters of 5 to 15 leaves and a root system of whitish, adherent, 0.8 mm. diameter roots. The round leaf stems are enveloped by 4 loose, overlapping, long-lived sheaths and arise alternately at 2 mm. intervals from the 1.5 mm. diameter rhizome. They measure 40-80 mm. long by 0.5 mm. wide with a round cross-section at their bases and widen to 1.5 mm. with an elliptical cross-section at the leaf junctions. The leaves are 50-70 mm. long by 10-15 mm. wide and 1.5 mm. thick, narrowly lanceolate, papery to leathery, conduplicate, olive-green occasionally tarnished wine in high light and attenuated to the base forming 5 mm. false petioles.

Inflorescence and Flowers: Up to 5 white, wide-open flowers emerge in rapid succession from the apex of the leaf stems. The ovary is 2-3 mm. long. The narrowly

ovate sepals and petals are acuminate and concave. The dorsal sepal is 9.5 mm. long by 2.5-3.2 mm. wide, the laterals 9.5 mm. by 2.2-2.4 mm. and the petals 8.5 mm. long by 2 mm. wide. The lip is 4.5 mm. long by 3 mm. wide, light yellow with an orange base and purple longitudinal callosities. The wide main lobe has a 3-pointed apex and undulate margins and the very light green column is 2.5 mm. by 0.8 mm.

Flowering Periods: October to November. The flowers last for 12-15 days.

Octomeria rodriguesii
Cogn. | *1489* (Plate XXVI)

Etymology: One more species named in honour of the botanist **João Barbosa Rodrigues**.

Habitat: Epiphytic from section VI and VIII, 1000-1300 M. asl, in humid original cloud forest on moss-covered trunks by river banks and low-tree branches. L 2-4; H 8; W 2-4.

Occurrence: Occasional.

Plant: This 80-150 mm. high epiphyte forms clusters of 5 to 15 leaves and a root system of whitish, adherent, 1 mm. diameter roots. The round leaf stems are enveloped by 5 loose, long-lived sheaths and arise alternately at 2 mm. intervals, from the 1.5 mm. diameter rhizome. They measure 30-80 mm. long by 0.5-1.0 mm. wide, are round in cross-section at their bases and widen to 1.5 mm, diameter with an elliptical cross-section at the leaf junctions. The leaves are 45-80 mm. long, 8-10 mm. wide and 1.5 mm. thick, narrowly lanceolate, leathery, conduplicate, dark olive-green and attenuated to the base forming 3 mm. false petioles.

Inflorescence and Flowers: Single white, bell-shaped flowers emerge from the apices of young and older leaf stems and are directed more or less upward but do not open wide. The ovary is 2 mm long. The ovate sepals and petals are obtuse and concave. The dorsal sepal is 4.5 mm. long by 2-3 mm. wide, the lateral sepals 4.5 mm. by 1.8-2.6 mm. and the petals are 4.5 mm. long by 1.6-2.2 mm. wide. The lip is 2.4-2.6 mm. long by 1.8-2.3 mm. wide, light yellow and 3-lobed and its faintly 3-pointed main lobe is oval with a wine-red centre and lacerate margins. The light yellow-green column is 1.6 mm. by 0.5 mm.

Flowering Periods: November to February. The flowers last for 12-15 days.

Octomeria rotundiglossa
Hoehne | *1490* (Plate XXVI)

Etymology: Rotundi and *glossa* mean round lip, which is almost as long as wide.

Habitat: Epiphytic from section VIII, at 700-900 M. asl, in original mountain forest with over 3 meters of rainfall per year, on thicker branches. L 4-6; H 6-8; W 4-7.

Occurrence: Rare.

Plant: This 30-70 mm. high epiphyte forms clusters of 5 to 15 leaves and a root system of whitish or green, adherent, 0.6 mm. diameter roots. The round leaf stems are enveloped by 4 to 8 loose, long-lived sheaths and arise alternately at very short intervals from the rhizome and are 15-50 mm. long by 1 mm. diameter. The narrowly elliptical leaves are flat, stiff, erect, light green, conduplicate and 20-35 mm. long, 6-8 mm. wide and 1.2 mm thick.

Inflorescence and Flowers: Up to 4 single-flowered inflorescences emerge together from the apex of young and older leaf stems on very short, 1 mm., pedicels. The ovary is 1.5 mm. and the spathe is 3 mm. long. The ovate sepals and petals are concave and canary-yellow. The sepals are 4-7.2 mm. long and 1.8-2.7 mm. wide and the petals are 4-6.8 mm. long and 1.5-2.5 mm. wide. The lemon-yellow lip is 2.4-3.6 mm. long, 2-3.3 mm. wide with 2 small side-lobes and 2 significant, undulate longitudinal callosities. The apex of the main lobe is rounded. The light yellow column is 1.4-1.8 mm. long and 0.6 mm. wide The apex and the wings have entire margins.

Flowering Periods: February and November The flowers last for 10 days.

General: The plants and flowers vary in size.

Octomeria tricolor
Rchb. f. | *1491* (Plate XXVI)

Etymology: Tri and *color* relate to the three colours of the lip: yellow, wine-red and white.

Habitat: Epiphytic from section V, VI and VIII, at 1000-1600 M. asl, in original cloud forest on moss-covered, mid-tree branches. L 3-5; H 6-8; W 4-6.

Occurrence: Common.

Plant: A normal cluster is 120-200 mm. high with 5 to 15 leaves. The leaf stems are alternate at 5 mm. intervals on the 2 mm. diameter rhizome and are enveloped by 5 to 7 loose, long-lived sheaths and are 80-110 mm. long by 1 mm., round in cross-section at their bases, but 1.8 mm. and elliptical at the leaf junctions. The leaves are 50-70 mm. long, 7-8 mm. wide and 1 mm. thick, flat, lanceolate, olive-green, occasionally wine-tarnished in high light, conduplicate and attenuated to their bases to form short false petioles.

Inflorescence and Flowers: One to 3 white, bell-shaped flowers emerge in rapid succession or simultaneously from the apex of young and older leaf stems. The ovate

sepals and petals are acute and concave. The dorsal sepal is 8 mm. long by 3 mm. wide, the lateral sepals 8.2 mm. by 2.8 mm. and the petals are 7.2 mm. long by 2.3 mm. The golden-yellow, trilobed lip is 3.4 mm. long by 2.1 mm. wide with an elliptical main lobe, light ochre and speckled wine-red with slashed margins and a 3-pointed apex. The centre of the lip is dark wine-red. The light yellow column is 2 mm. by 1.1 mm..

Flowering Periods: October to January. However, flowering plants may be found in all months of the year. The flowers last for 7-10 days.

Octomeria estrellensis
Hoehne | *1493* (Plate XXVI)

Etymology: Estrellensis means star-like and refers to the flower shape and colour.

Habitat: Epiphytic from sections IV and VI and VIII, at 800-1100 M. asl, on trunks and thick branches in relatively dry and sunny positions. L 7; H 5; W 5.

Occurrence: Rare.

Plant: This epiphyte is 20-30 mm. high and forms clusters of 10-30 leaves and a root system of 0.7 mm. diameter, white, very adherent roots. The leaf stems are alternate at 1 mm. intervals from the 1.5-2 mm. diameter rhizome and measure 5-8 mm. long and 0.7-1.2 mm. diameter and are enveloped by 3 loose, long-lived sheaths. The erect leaves are 15-25 mm. long, 10-13 mm. wide and 2-3 mm. thick, concave, obovate or elliptical, fleshy, faintly conduplicate and olive-green with dark wine-red backs.

Inflorescence and Flowers: One to 3 single-flowered inflorescences on 3-6 mm. pedicels emerge consecutively or simultaneously from the apices of young and older leaf stems. The ovary is 8 mm long. The flower opens wide. The narrowly lanceolate, finely acuminate, concave sepals and petals are white or light olive. The dorsal sepal is 12-14 mm. long and 3.1 mm. wide, the laterals 11.5-14 mm. by 3.2 mm. and the petals are 11-13 mm. long and 2.5-3 mm. wide. The lip is light yellow-green and 7-7.5 mm. long by 3.4 mm. wide with a faintly 3-pointed apex, while the side lobes, the central vein and the lamellae are dark purple to wine-red. The margins of the main lobe are crenate. The column is light yellow-green, and 3 mm. long by 1.5 mm. wide and has acuminate, purple wings and an indistinct apex. At the end of the purple blotched column foot are two 0.2 mm. long horn-shaped appendages.

Flowering Periods: November to January. The flowers last for 10 days.

Octomeria minuta
Cogn. | *1495* (Plate XXVII)

Etymology: Minuta means small which this attractive plant is.

Habitat: Epiphytic from section VI, at 1100 M. asl, in cloud forest, on thick branches in relatively dry and sunny positions. L 7; H 5; W 6-8.

Occurrence: Rare.

Plant: A 30-40 mm high epiphyte forming clusters of 5-15 leaves and a root system of 0.5 mm. diameter, white, very adherent roots. The leaf stems are alternate at 1 mm. intervals on the 1.5 mm. diameter rhizome and are 3.0-7.0 mm. long and 0.5-1.2 mm. diameter and enveloped by 2 loose, long-lived sheaths. The erect leaves are 15-25 mm. long, 7-10 mm. wide and 1.5-2 mm. thick, obovate, slightly infolded, fleshy, conduplicate and dark olive-green.

Inflorescence and Flowers: The single-flowered inflorescences emerge from the apex of the most recent leaf stems. The ovary is 3.5 mm. and the bracts 2 mm. long. The white, bell-shaped flower points upwards. The narrowly lanceolate, acuminate, concave sepals and petals are white or very light yellow green. The dorsal sepal is 8 mm. long and 2.4 mm. wide. The 33% fused lateral sepals measure 8 mm. by 2.3 mm. The petals have faintly serrate margins and are 8.1 mm. long by 2 mm. wide. The 3-lobed lip is 5.8 mm. long by 2.3 mm. wide, white and has 2 dark purple lamellae. The apex has one point and the margins of the main lobe are faintly gnawed. The column is light yellow-green, and 2.2 mm. long by 1 mm. wide.

Flowering Periods: December to January. The flowers last for 10 days.

Octomeria aff. ochroleuca
1497b (Plate XXVII)
Possible new species

Etymology: Ochro-leuca is Greek for pale ochre colour referring to the flower.

Habitat: Epiphytic from section VI, at 1100-1350 M. asl, in original montane cloud forest, near ridges in good light, on mid-tree branches and trunks. L 5-8; H 7-9; W 5-8.

Occurrence: Occasional.

Plant: This 40-120 mm. high epiphyte forms clusters with 7 to 20 leaves. The plant lies flat on the substrate if it grows on the upper side of horizontal branches, but is otherwise erect. The leaf stems are enveloped by 3 to 5 long-lived sheaths and are 15-80 mm. long by 0.8-1.5 mm. diameter at their bases but 2.5 mm. diameter at the leaf junctions. The leaves are 30-45 mm. long,

8-19 mm. wide and 1-2.5 mm. thick, lanceolate, fleshy, faintly conduplicate, dark olive-green or wine-red and attenuated to their bases to form short false petioles.

Inflorescence and Flowers: One or 2, usually pendent flowers on 20-30 mm. pedicels emerge from the apices of the most recent leaf stems and there may be up to 12 inflorescences on each stem. The ovaries are 10-15 mm. long. The flowers open wide and are very light yellow-green or white. The concave sepals and petals are narrowly ovate, finely acuminate with faintly purple dotted bases. The dorsal sepal is 15-20 mm. long by 3.5-4 mm. wide. The slightly asymmetric laterals are 15-20 mm. long by 3.3-3.6 mm. wide and the petals measure 13-18 mm. long by 3.5 mm. wide. The elliptical lip is ochre with a purple centre, faint side lobes and a broad main lobe with a 3-pointed apex. and is 5-6 mm. long by 3.5-4 mm. wide. The green column is 2.7 mm. long and 1.2 mm. wide.

Flowering Periods: October to November and June. The flowers last for 10 days.

General: This new species has since been named *Octomeria longopedicellata* Seeh. The holotype is in the Rio de Janeiro Botanic Garden.

Octomeria rhodoglossa
Schltr. | *1498* (Plate XXVII)

Etymology: From the Greek *rhodo-* for rosy-red and *glossa* for lip referring to the red lip, although it was yellow in our specimens.

Habitat: Epiphytic from section VI, at 1100 M. asl, in original mountain forest on trunks and branches. L 6; H 8; W 6.

Occurrence: Rare.

Plant: This 40-60 mm high epiphyte forms clusters with 3 to 10 leaves. The leaf stems have furrowed apical segments and are enveloped by 3 loose, 4-6 mm. long sheaths and arise alternately at 1 mm. intervals from the 1 mm. diameter rhizome and are 10-20 mm. long by 1-1.5 mm. diameter. The erect leaves are 30-40 mm. long, 17-20 mm. wide and 2 mm. thick, obovate, concave, fleshy, wine-red or purple, distinctly conduplicate and attenuated to the bases forming 3 mm. false petioles which are enveloped by the apical sheaths of the leaf stems.

Inflorescence and Flowers: One or 2 wide open flowers arise consecutively from the apices of the most recent leaf stems. The ovary is 2 mm. long by 0.7 mm. wide. The sepals and petals are narrowly ovate, finely acuminate, faintly keeled and white with light yellow apices. The concave dorsal sepal is 10-11 mm. long by 3.5-4.0 mm. wide and the laterals are 11-11.5 mm. long by 3-3.2 mm. wide with slightly recurved apices. The petals are 9-9.5 mm. long and 3-3.3 mm. wide. The lip is light yellow-green, 5 mm. long by 3 mm. wide and has 2 small side lobes. The wide main lobe ends in a narrow 3-pointed apex with recurved and faintly gnawed margins. The white column is 3 mm. long and 1.3 mm. wide.

Flowering Periods: May to June, occasionally in October and November. The flowers last for 10 days.

General: The plant described above is probably a colour variety.

Octomeria aloefolia
Barb. Rodr. | *1506* (Plate XXVII)

Etymology: Aloe and *folia* refer to the aloe-like leaves.

Habitat: Epiphytic from section VI, at 900-1050 M. asl, in relatively dry and sunny habitats, on the upper branches of large trees in original forest. L 6-8; H 4-6; W 5.

Occurrence: In very limited locations but in colonies.

Plant: A normal cluster with 10-100 leaves is 15 mm. high. The purple-speckled leaf stems are alternate at 1 mm. intervals from the 0.8 mm. diameter rhizome and 1-2 mm. long by 1.2 mm. diameter, covered by fibres of the short-lived sheaths. The terete, faintly conduplicate, dark olive-green, succulent leaves are 10-15 mm. long, 3-4 mm. wide and 3-3.5 mm. thick.

Inflorescence and Flowers: Up to 3 wide open flowers arise consecutively from the apices of the younger leaf stems. The ovary is 1.5-2 mm. long. The narrowly ovate, narrowly acuminate, concave sepals and petals are light yellow and have 3 wine-red longitudinal veins. The sepals are 12-15 mm. long by 3 mm. wide and the petals are 11-13 mm. long by 2.4 mm. wide. The 3.2 mm. long, lemon-yellow lip is 2.5 mm. wide with narrow, erect side lobes and wine-red plate, while the wide heart-shaped main lobe has indented margins. The flanks of the lamellae are wine-red to dark purple. The light yellow-green column is 2.1 mm. long and 0.7 mm. wide and has acuminate, purple wings and a faint apex.

Flowering Periods: September to December and June. The flowers last for 10 days. A plant may produce flowers over a period of three months.

Octomeria fimbriata
Pôrto & Brade | *1508* (Plate XXVII)

Etymology: Referring to the fimbriate (fringed) petals and lip.

Habitat: Epiphytic from section VI, at 1300-1400 M. asl, in original mountain cloud forest, on branches at 5 to 15 M. below the canopy. L 5-7; H 6-8; W 5.

Occurrence: Rare.

Plant: This 30-40 mm. high epiphyte forms clusters with 7 to 20 leaves and whitish 0.8 mm. diameter roots. The leaf stems are enveloped by 2 short-lived sheaths, 10 mm. long and 0.7 mm. diameter at their bases but 1.2-3 mm. at the leaf junctions. The leaves are terete, faintly furrowed longitudinally, dark olive-green and 20-30 mm. long, 2-2.5 mm. wide and 2 mm. thick.

Inflorescence and Flowers: Up to 3 wide open, lemon-yellow flowers on 4 mm. pedicels arise consecutively at month-long intervals from the apex of young and older leaf stems. The ovary is 1.5 mm long. The concave sepals are narrowly ovate, finely acuminate with faintly serrate margins and measure 12 mm. long by 3.1 mm. wide and have brown apices. The petals have fringed margins and are 11 mm. long and 3 mm. wide. The triangular lip is lemon-yellow, 3.2 mm. long by 2.5 mm. wide, with fringed margins and a narrow 3-pointed apex. The light yellow-green column is 2.5 mm. long and 1 mm. wide.

Flowering Periods: November to January. The flowers open in dry weather and last 5 to 10 days.

Octomeria chamaeleptotes
Rchb. f. | *1515* (Plate XXVII)

Etymology: From the Greek *chamae-* meaning creeping and refers to the creeping, terete, *Leptotes*-like plant.

Habitat: Epiphytic from section VI, at 900-1300 M. asl, in original cloud forest, on mid-tree trunks and thick branches. L 5-7; H 5-6; W 5-7

Occurrence: Rare.

Plant: This 130-160 mm. high epiphyte forms clumps with 7 to 25 leaves. The round leaf stems are enveloped by 3 to 5 short-lived sheaths and are 70-90 mm. long and 1.5-2 mm. diameter at their bases and widen to 3-4 mm. at their apices. The leaves are 50-80 mm. long, 3.5-5.0 mm. wide and 2.5-3.0 mm. thick, terete with relatively obtuse apices, faintly conduplicate, olive-green.

Inflorescence and Flowers: One to 6 wide open, bell-shaped flowers emerge simultaneously from the apex of young and older leaf stems. The ovary is 2 mm. long. The narrowly ovate and acute sepals and petals are white with 4 to 5 light purple speckled longitudinal veins. The dorsal sepal is 10 mm. long by 3.0 mm. wide, the lateral sepals are 9.5 mm. long by 3.0 mm. wide and the petals are 9.5 mm. long by 2.7 mm. wide. The trilobed lip is 5 mm. long by 3 mm. wide, purple along both sides of the light green central vein and has round side-lobes and a faintly 3-pointed, almost acute apex. The yellow-green column is 3 mm. long and 1.2 mm. wide.

Flowering Periods: January to November. The flowers last for 10 days.

General: We found this species with flower parts larger than those described.

Octomeria riograndensis
Schltr. | 1517 (Plate XXVII)

Etymology: First found in Rio Grande do Sul.

Habitat: Epiphytic from section VI, at 1200 M. asl in original cloud forest, on thick branches. L 5-8: H 5-8; W 5-7.

Occurrence: Rare.

Plant: This 55-75 mm. high epiphyte forms clusters with 10 to 20 leaves and whitish to light brown roots. The erect leaf stems are 23-45 mm. long by 0.4 to 0.8 mm. diameter and emerge at short intervals from a branching 1 mm. diameter rhizome, enveloped by 3 to 5 light brown sheaths. The leaves are terete and finely acuminate, faintly conduplicate, dark olive-green to purple and are 30-55 mm. long, 1.2 mm. wide and 1 mm. thick.

Inflorescence and Flowers: Few flowers on very short pedicels emerge simultaneously from the apex of young and older leaf stems. The ovary is 1-2 mm. long. The whitish or light yellowish sepals and petals are lanceolate with acute apices. The dorsal sepal is 6.5-8 mm. long by 2.5 mm. wide, the laterals 6.2-7.5 mm. long by 2.1 mm. wide and the petals are 6-8 mm. long by 2.1 mm. wide. The yellow, 3-lobed lip is 3.8 mm. long by 2.4 mm. wide and has a long main lobe with an almost cut, flat apex and small side lobes, relatively close to the base. The light yellow-green column is 2 mm. long by 0.8 mm. wide.

Flowering Periods: March and occasionally November. The flowers last for five to 10 days.

Octomeria alexandri
Schltr. | 1518 (Plate XXVII)

Etymology: Alexander is the fist name of Dr. **Curt Brade**.

Habitat: Epiphytic from section VI, at 1100-1300 M. asl, in humid original mountain cloud forest, on mid-tree branches and trunks. L 4-6: H 6-8; W 5.

Occurrence: Common.

Plant: This 90-150 mm. high epiphyte forms clusters with 5 to 20 leaves. The leaf stems are enveloped by 2 to 4 sheaths and are 25-65 mm. long and 0.6-1.0 mm. diameter. The erect leaves are linear, terete, succulent, faintly conduplicate, dark olive-green and 35-90 mm. long, 3-4 mm. wide and 2-3 mm. thick.

Inflorescence and Flowers: Up to 3 wide open, bell-shaped flowers emerge simultaneously from the apex of young and older leaf stems. The ovary is 2 mm. long. The ovate dorsal sepal, the obovate lateral sepals and the lanceolate petals are yellow with obtuse apices. The dorsal sepal is 6.5-7.0 mm. long by 3.5-3.7 mm. wide,

the slightly asymmetric laterals are 6.7 mm. long by 2.8-3.4 mm. wide and the petals, 6.5 mm. long by 2.7-3.5 mm. wide. The lip is 3.6 mm. long by 2.8 mm. wide and yellow with red longitudinal callosities. The side lobes are round and the main lobe has entire margins and 2 rounded points at the apex. The slim column is 2.2 mm. long by 0.5 mm. wide, green and has a wine-red front.

Flowering Periods: November to January. The flowers last for 7 to 10 days. It may have a second flowering period after two months rest.

Octomeria sancti-angeli
Krzl. | 1519 (Plate XXVII)

Etymology: Santo-angelo is a municipality in Rio Grande do Sul.

Habitat: Epiphytic from section VI, at 900-1100 M. asl, in humid original forest, on mid to upper-tree branches and trunks. L 5; H 8; W 5.

Occurrence: Rare. We have found it in only one habitat, but with numerous plants.

Plant: This 90 mm. high epiphyte forms clusters with 10-30 leaves. The leaf stems are 20-30 mm. long and 0.6-1.3 mm. diameter and arise at 1 mm. intervals from a branching rhizome, enveloped by 5 sheaths. The leaves are linear, terete, succulent, faintly conduplicate, olive-green and 40-60 mm. long, 2.2-2.7 mm. wide and 2 mm. thick.

Inflorescence and Flowers: One to 3, bell-shaped, wide open, yellow flowers emerge simultaneously from the apex of young and older leaf stems. The ovary is 1 mm long. The ovate sepals and the lanceolate petals have obtuse apices. The dorsal sepal is 5.3 mm. long by 2.2 mm. wide, the laterals 5.4 mm. long by 2.1 mm. and the petals are 5.1 mm. long by 1.9 mm. The lip is 3 mm. long by 2.2 mm. wide and has 2 rounded points at its apex. The slim column is green and 1.7 mm. long and 0.5 mm. wide.

Flowering Periods: November to January. The flowers last for 4 to 7 days.

Octomeria decumbens
Cogn. | 1521 (Plate XXVIII)

Etymology: Decumbens is Latin for prostrate, lying flat and refers to the plant's habit.

Habitat: Epiphytic from section VI and VIII, 900-1300 M. asl, in original forest, on upper-tree trunks and branches near ridges. L 5-7: H 6-8; W 6-8.

Occurrence: Occasional.

Plant: The creeping, branching plant can cover whole trunks and branches of trees exposed to wind and light. The roots are whitish and 0.6 mm diameter. The pendent leaf stems are 40-80 mm. long and 1-2 mm. diameter and emerge at 5-7 mm. intervals from the green or purple, 2-3 mm. diameter, sheath-covered rhizome, enveloped by 5 to 7 short-lived sheaths. The leaves are linear, terete, finely acuminate, fleshy, faintly conduplicate, olive-green with occasional wine blotches and are 70-150 mm. long, 4-6 mm. wide and 3-4.5 mm. thick.

Inflorescence and Flowers: Three to 6 yellow, bell-shaped flowers emerge together from the apices of young and older leaf stems. The ovary is 2 mm. long. The apices of the ovate sepals and petals are obtuse. The dorsal sepal is 7-8 mm. long by 2.5-3.5 mm. wide, the laterals 7-8 mm. long by 2.2-2.7 mm. wide and the petals are 6-7 mm. long by 2.2-2.5 mm. wide. The trilobed lip has well developed side-lobes, a wide main lobe with an indistinctly 3-pointed apex and is 4-4.5 mm. long by 2.7-3.0 mm. wide. The yellow-green column is 2.5 mm. long and 0.6 mm. wide.

Flowering Periods: June.

Octomeria geraensis
Barb. Rodr. | *1522* (Plate XXVIII)

Etymology: First found in Minas Gerais in the Serra do Lenheiro.

Habitat: Epiphytic from section VI, 1100-1400 M. asl, in humid original cloud forest, on mid and upper trunks and thick branches. L 4; H 7; W 5.

Occurrence: Rare.

Plant: This 150-220 mm. high epiphyte forms clusters with 7 to 30 leaves. The round leaf stems are enveloped by 3 to 5 short-lived sheaths and are 50-80 mm. long by 1-1.5 mm. diameter close to their bases but 2-3 mm. at the leaf junctions. The leaves are 100-140 mm. long, 5-6 mm. wide and 4-6 mm. thick, terete, erect, olive-green, vaguely conduplicate, finely attenuated to the base and have finely acuminate apices. The leaves are slightly concave at the basal halves and the longitudinal side veins are elevated on the lower halves and lowered in the apical halves.

Inflorescence and Flowers: Five to 15 wide open flowers emerge simultaneously from the apex of young and older leaf stems. The pedicels are 5 mm. and enveloped by the 2 mm. spathes and bracts. The smooth ovary is 3-4 mm long. The light yellow sepals and petals are narrowly ovate and obtuse. The dorsal and lateral sepals are 8-11 mm. long by 2-3 mm. wide, while the petals are 7.8-10 mm. long by 2.4-2.5 mm. wide. The narrow lip is lemon-yellow and 4-5 mm. long by 2.5-3 mm. wide with a faintly three-pointed apex. The yellow-green column is 2.0 mm. long.

Flowering Periods: June. The flowers last for 10 days.

Octomeria aff. geraensis
1522b (Plate XXVIII)

Etymology: See *1522*.

Habitat: Epiphytic from section VI and VIII, at 900-1200 M. asl in original cloud forest, on upper-tree trunks and thick branches. L 5-8; H 5-8; W 5-7.

Occurrence: Occasional.

Plant: This epiphyte forms clusters with 10 to many leaves. The pendent leaf stems emerge at 6-9 mm. intervals from a branching, creeping, purple, 2 mm. diameter rhizome and are 60% enveloped by 3 to 5 sheaths and measure 80-120 mm. long by 0.8-2.0 mm. diameter. The leaves are terete, finely acuminate, faintly conduplicate, olive-green to purple and 80-120 mm. long, 3-4 mm. wide and 2.5-3.0 mm. thick.

Inflorescence and Flowers: Three to 12 pale yellow, bell-shaped flowers emerge simultaneously from the apices of young and older leaf stems. The ovary is 2 mm. long and 0.6 mm. wide. The sepals and petals are obovate with obtuse apices and translucent margins. The dorsal sepal is 6.5 mm. long by 2.7 mm. wide, the laterals 6.2 mm. long by 2.2 mm. wide and the petals are 6 mm. long by 2 mm. wide. The 3-lobed yellow lip is 3.8 mm. long by 2.4 mm. wide with a long main lobe with an almost cut, flat apex. The light yellow-green column is 1.8 mm. long by 0.8 mm. wide.

Flowering Periods: June. The flowers last for five to 10 days.

Octomeria juncifolia
Barb. Rodr. | *1523* (Plate XXVIII)

Etymology: Junci a rush and *folia,* leaf are Latin meaning with rush-like leaves.

Habitat: Epiphytic from section VI and VIII, at 600-1100 M. asl in original cloud forest or remnant trees, on upper-tree trunks and thick branches. L 5-8; H 5-8; W 5-7.

Occurrence: Common.

Plant: This epiphyte forms clusters with 10 to many leaves. The pendent leaf stems emerge at 3-6 mm. intervals from a branching, creeping, purple and 2.5-3.0 mm. diameter rhizome, 30% enveloped by 3 to 5 fleshy, obtuse sheaths and measure 80-250 mm. long by 1-2.5 mm. diameter. The leaves are terete, finely acuminate, vaguely conduplicate, olive-green to purple and 200-400 mm. long, 3-4 mm. wide and 2.5-3.0 mm. thick.

Inflorescence and Flowers: Three to 12 golden-yellow, wide open, bell-shaped flowers emerge simultaneously from the apices of young and older leaf stems. The ovary is 3.5 mm long and 1 mm wide. The sepals and petals are obovate with obtuse apices and translucent margins. The dorsal sepal is 8-13 mm. long by 3-4 mm. wide, the laterals are 7.5-12.5 mm. long by 3-4 mm. wide and the petals 6-11 mm. long and 3-3.8 mm. wide. The lip has wine dots between the base and the longitudinal callosities and is 5-5.8 mm. long by 4-5 mm. wide. The side-lobes are close to the base and the 2 points of the apex are triangular, while the margins of the wide main lobe are laterally folded. The yellow-green column has wine-red flanks and base and is 3.8 mm. long by 1.1 mm. wide.

Flowering Periods: March to April and August to September. The flowers last for five to 10 days. A flowering plant is very showy.

General: We found a plant showing leaf stems 200 mm long and leaves 600 mm long at 700 M. asl, in Section VI with smaller, yellow-green flowers.

Octomeria praestans
Barb. Rodr. | *1527* (Plate XXVIII)

Etymology: Praestans is Latin, meaning superior or excellent probably referring to the showy flowers.

Habitat: Epiphytic from section VI, at 1200-1300 M. asl, in original cloud forest on mid-tree branches and trunks. L 5-7; H 7-8; W 5-6.

Occurrence: Occasional and usually it is solitary.

Plant: This 150-220 mm. high epiphyte forms clusters with 7 to 20 leaves and whitish, wiry, adherent and 1.5 mm. diameter roots. The leaf stems are enveloped by 5 short-lived sheaths and are 70-100 mm. long by 2 mm. diameter at their bases but 4-5 mm. at the leaf junctions. The erect leaves are terete, conduplicate, dark olive-green, occasionally stained wine-red and measure 80-120 mm. long, 4-6 mm. wide and 3-4 mm. thick.

Inflorescence and Flowers: Three to 9 flowers emerge simultaneously from the apices of young and older leaf stems. The ovary is 1.5 mm. long. The flowers only open a little and are ivory coloured or very light yellow-green with slightly rosy veins on the back. The concave sepals and petals are acute, narrowly ovate with faintly purple dotted apices. The sepals are 10-14 mm. long by 3-3.2 mm. wide and the petals 9.5-13 mm. long by 2.7-2.9 mm. wide. The elliptical lip has a 3-pointed apex, is light yellow-green and 5.8-6.8 mm. long by 3.3-3.5 mm. wide with yellow faint side-lobes and centre, while the longitudinal callosities are purple. The light green column is 3 mm. long and 1.2 mm. wide and has a purple foot.

Flowering Periods: November. The flowers last for six to ten days.

Octomeria truncicola
Barb. Rodr. | *1528* (Plate XXVIII)

Etymology: Truncus and *cola* are Latin for trunk-dweller.

Habitat: Epiphytic from section VIII, at 900-1100 M. asl, in humid original cloud forest, on mid- and upper-tree trunks and thick branches. L 5-7; H 6-7; W 6.

Occurrence: Occasional.

Plant: This 140-200 mm. high epiphyte forms clusters with 10-20 leaves and a profuse system of 0.5-0.7 mm. diameter roots. The round leaf stems are alternate at 1-2 mm. intervals from a 2 mm. diameter rhizome and are enveloped by up to 6 brown sheaths. The leaf stems are 50-65 mm. long by 1.8-3.5 mm. diameter. The leaves are terete, furrowed longitudinally, dark olive-green or purple and 80 mm. long, 3-4 mm. wide and 4-5 mm. thick.

Inflorescence and Flowers: Up to 10 light yellow, bell-shaped flowers on 4 mm. stems, covered by the spathes, emerge from the apex of young and older leaf stems and do not open wide. The light yellow sepals and petals are narrowly ovate with obtuse apices and have wine spotted veins. The dorsal sepal is 10.5 mm. long by 2.8 mm. wide, the laterals are 10.5 mm. long by 2.7 mm. wide and the petals are 8.5 mm. long by 2.2 mm. wide. The lip is 4.5 mm. long, 2.8 mm. wide and is light lemon-yellow with 2 side-lobes. The main lobe is relatively narrow with slightly undulate margins and a faintly 3-pointed apex. The yellow-green column is 2 mm. long and 0.7 mm. wide with some wine-red longitudinal markings.

Flowering Periods: June and October to November. The flowers last for 10 days.

General: This species varies greatly in plant and flower size and lip shape.

Octomeria wilsoniana
Hoehne | *1529* (Plate XXVIII)

Etymology: Not known.

Habitat: Epiphytic from sections VI, VIII, at 1100-1300 M. asl, in original forest on thicker mid- and lower-tree branches. L 5; H 8; W 5.

Occurrence: Rare.

Plant: This epiphyte is 60-90 mm high and forms clusters with 5-15 leaves and 0.4 mm diameter whitish roots. The slightly curved leaf stems are alternate at 1 mm. intervals from the 1-1.5 mm. diameter rhizome, covered by up to 5 sheaths and 30-50 mm. long by 1.5 mm. diameter. The olive-green leaves are narrowly lanceolate, almost linear, slightly conduplicate and 50-60 mm. long, 7 mm. wide and under 1 mm. thick.

Inflorescence and Flowers: Single bell-shaped flowers on 4-5 mm. stems covered by the spathes, emerge from the apices of young and older leaf stems and do not open wide. The white sepals and petals are ovate with obtuse apices. The dorsal sepal is 6.8 mm. long by 3.2 mm. wide, the laterals are 6.5 mm. long by 3 mm. wide and the petals are 6.2 mm. long by 2.5 mm. wide. The lip is 3.8 mm. long, 2.6 mm. wide and very light lemon-yellow, with 2 small side-lobes, while the main lobe is wide with entire margins and a 2-pointed apex. The column is yellow-green.

Flowering Periods: July to August and December to January. We found only one plant with flowers and others have not flowered within five years. The flowers last for 10 days.

Octomeria gracilis
Lodd. | *1530* (Plate XXVIII)

Etymology: Gracilis is Latin, meaning slender or graceful and refers to the plant.

Habitat: Epiphytic from section VI and VIII, at 900-1600 M. asl, in original cloud forest on trunks and thick branches of the mid- to upper-tree, sometimes creeping. L 7-9; H 6; W 5-7.

Occurrence: Common, often in huge colonies.

Plant: This epiphyte is 150-250 mm. high and forms clusters of 20-100 leaves. The slightly curved leaf stems arise alternately at 1-2 mm. intervals from the 1.2-2 mm. diameter rhizome, enveloped by 4 to 6 sheaths and measure 50-100 mm. long and 0.3-0.7 mm. diameter. The olive-green leaves are linear, terete, slightly recurved, finely attenuated towards the apices, 60-150 mm. long, 3-4 mm. wide and 1-2 mm. thick.

Inflorescence and Flowers: One to 6 wide open flowers emerge from the apex of young and older leaf stems on 3 mm. pedicels. The spathes are short and acute and the ovary is 1 mm. long. The light yellow sepals and petals are ovate with obtuse apices. The dorsal sepal is 6 mm. long by 2.2 mm. wide, the lateral sepals are 6.2 mm. by 2.2 mm. and the petals are 5.5 mm. long by 1.8 mm wide. The yellow lip has the typical Octomeria shape and measures 3.3 mm long and 2.4 mm. wide. The yellow-green column is 1.8 mm. long by 0.8 mm.

Flowering Periods: January. The flowers last for 10 days.

Octomeria campos-portoi
Schltr. | *1536* (Plate XXVIII)

Etymology: **Paulo de Campos Pôrto**, the grandson of **Barbosa Rodrigues**, was a director of the Rio de Janeiro Botanic Garden. He wrote various articles about orchids together with **A. C. Brade**.

Habitat: Epiphytic from section VI, at 1000-1300 M. asl, in humid original cloud forest, close to the mountain peaks or ridges, on trunks and mid-tree branches. L 7; H 8; W 8.

Occurrence: Common, but in low numbers.

Plant: This 120-140 mm. high epiphyte forms clusters with 7 to 15 leaves. The round leaf stems are alternate at 1 mm. intervals on a 0.7 mm. diameter branching rhizome, enveloped by 3 to 5 loose, short-lived sheaths and are 45-65 mm. long and 0.7-1.5 mm. diameter. The leaves are terete, erect, finely attenuated to their apices, faintly conduplicate, olive-green and 60-80 mm. long, 3.5-4 mm. wide and 2.5-3 mm. thick.

Inflorescence and Flowers: Single flowers emerge on 5.5 mm. stems from the apex of young and older leaf stems. The ovary is 4 mm. long. The bell-shaped flowers are white, tinged wine-red or very light lemon and open wide but close in high humidity. The concave sepals and petals are ovate with obtuse apices. The dorsal sepal is 9 mm. long by 3.5-4.0 mm. wide with purple marks along the longitudinal veins. The asymmetric laterals are 9 mm. long by 3.5-4.0 mm. wide and are more distinctly purple between the 2 inner longitudinal veins. The petals are coloured like the dorsal sepal and 7.8-8.5 mm. long by 2.7-3.0 mm. wide. The lip is 3.7-4.7 mm. long by 2.8 mm. wide, axe-shaped, lemon-yellow and the 3 veins are dark wine-red. The surface of the convex basal half is undulate with 2 golden yellow, longitudinal callosities. The yellow-green column is 3.3 mm. long and 0.9 mm. wide.

Flowering Periods: December, January and May. The flowers last for seven days.

16. Botanic Drawings
by Helmut Seehawer

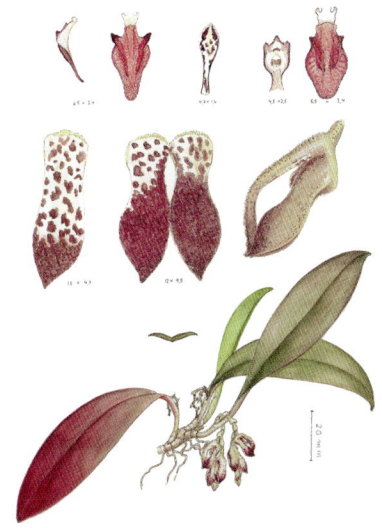

Zootrophion schenkii   992

Cryptophoranthus fenestratus   993

Cryptophoranthus jordanensis   994

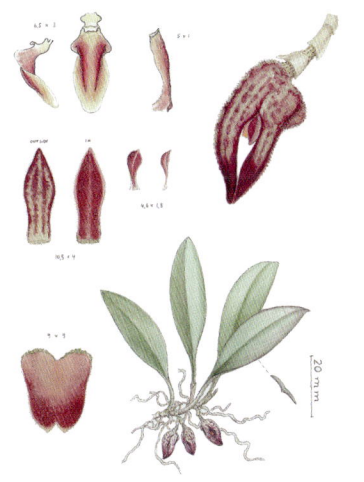

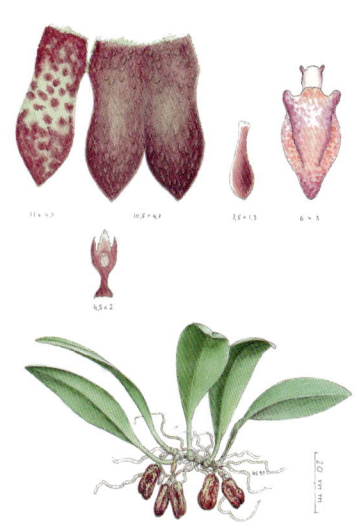

Cryptophoranthus langeana   996

Cryptophoranthus punctatus   999

Cryptophoranthus spicatus   1000

P L A T E   I

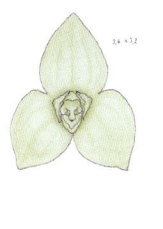

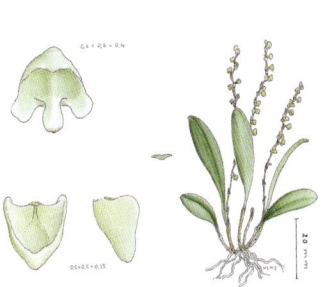

Stelis drosophila   1004

Stelis drosophila   1004b

Stelis hoehnei   1006

| 499 |

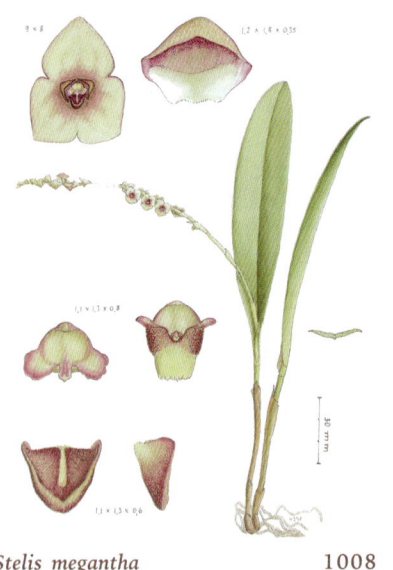

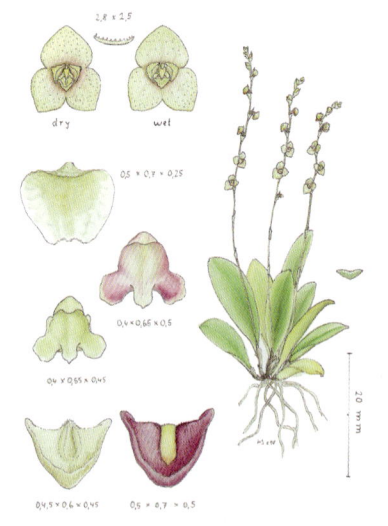

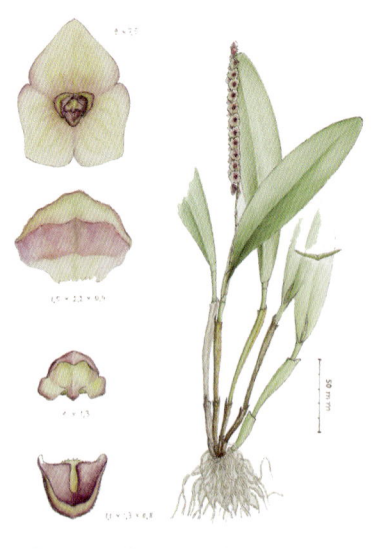

Stelis megantha 1008

Stelis parvifolia 1009

Stelis megantha var. *robusta* 1010

P L A T E   II

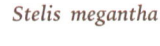

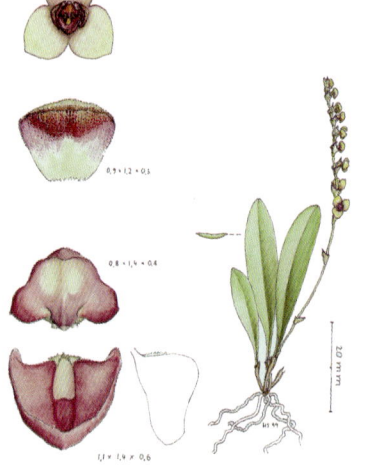

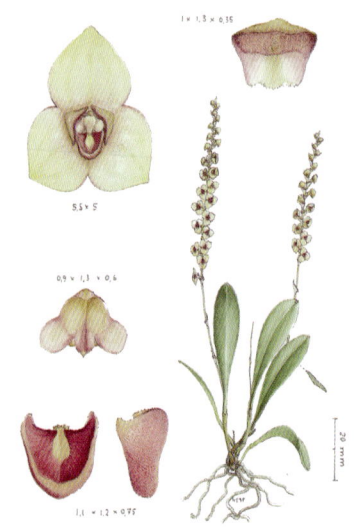

Stelis ruprechtiana 1011

Stelis thermophila 1012

Stelis triangularis 1013

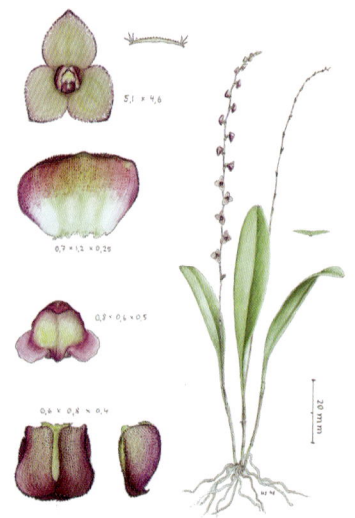

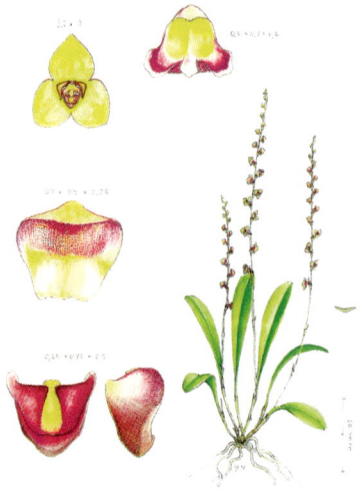

Stelis microcaulis 1017

Stelis campos-portoi 1020

Stelis chlorantha 1021

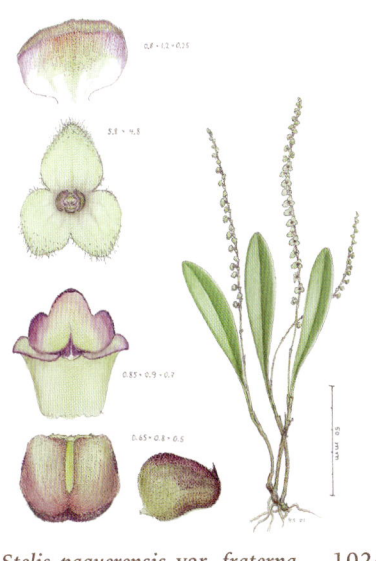

Stelis paquerensis var. *fraterna*   1024

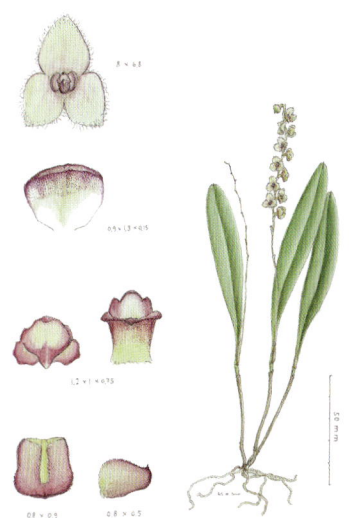

Stelis itatiayae   1025

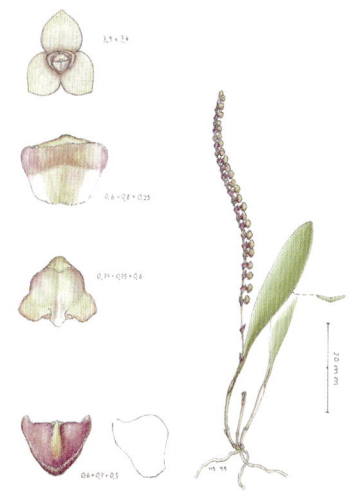

Stelis modesta   1026

Stelis omalosantha   1027

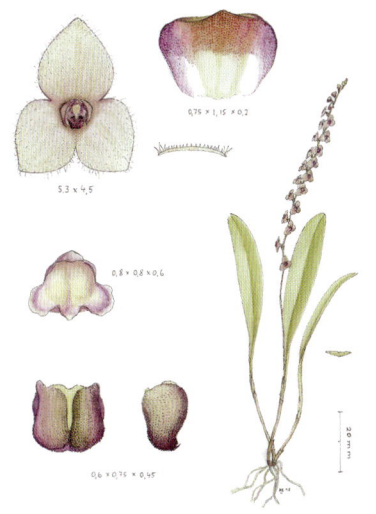

Stelis puberula   1028

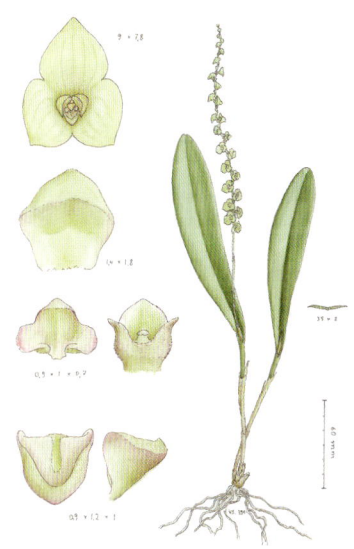

Stelis grandiflora   1030

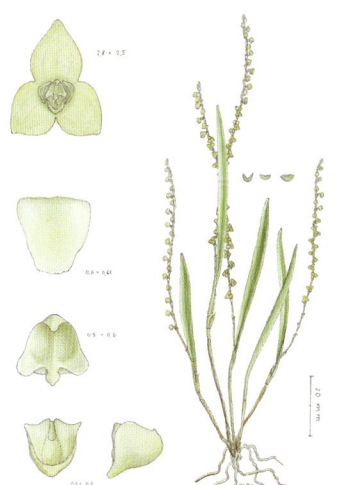

Stelis dusenii   1037

Stelis loefgrenii   1040

Stelis aprica   1045

P
L
A
T
E

III

| 501 |

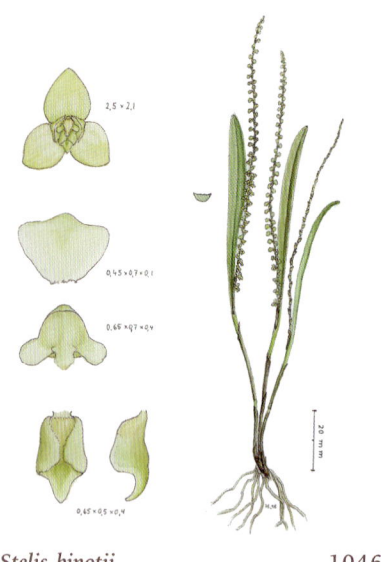

Stelis binotii   1046

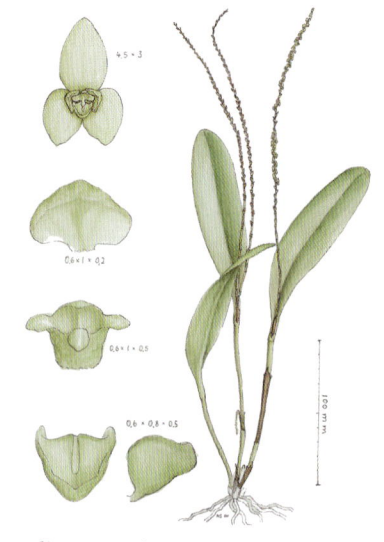

Stelis paraensis   1051

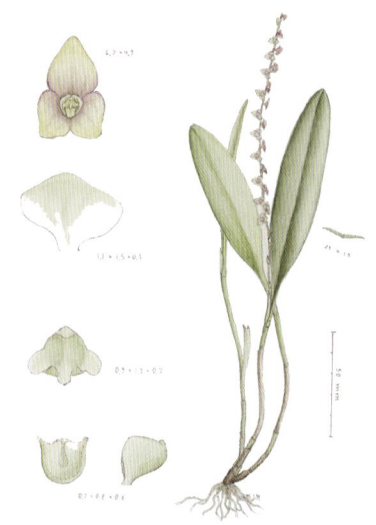

Stelis argentata var. *pterostele*   1052

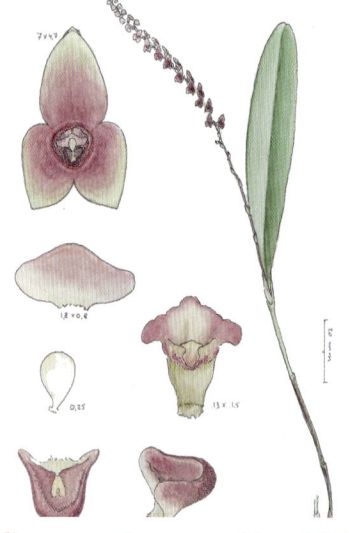

Stelis paquerensis var. *porschiana* 1053

Stelis reflexisepala   1054

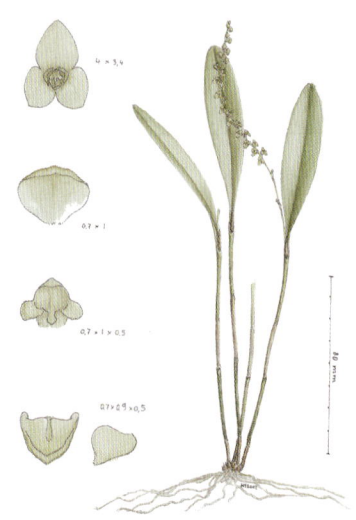

Stelis paquerensis var. *tweediana*   1057

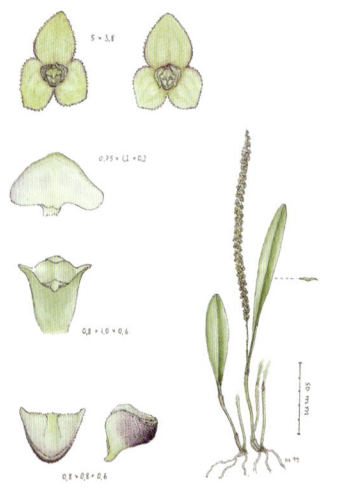

Stelis paquerensis var.   1058
inaequalisepala

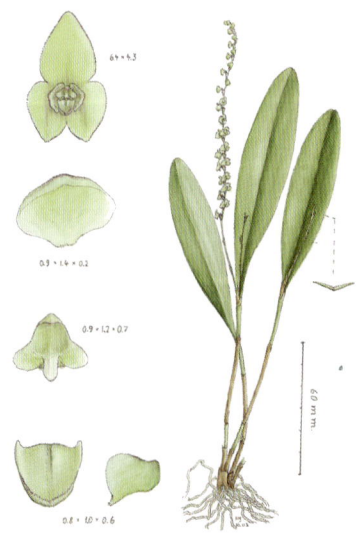

Stelis parahybunensis   1059

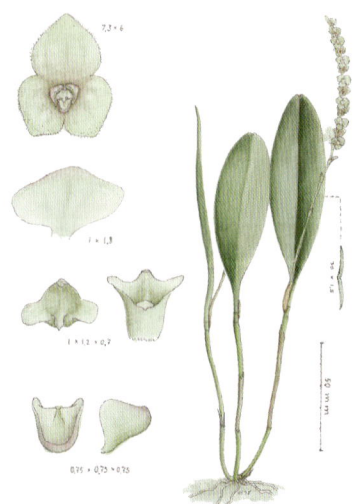

Stelis paquerensis var. *guttifera*   1061

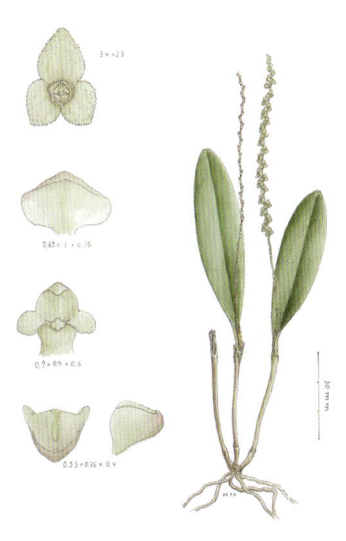

Stelis paquerensis var.   1063
plurispicata

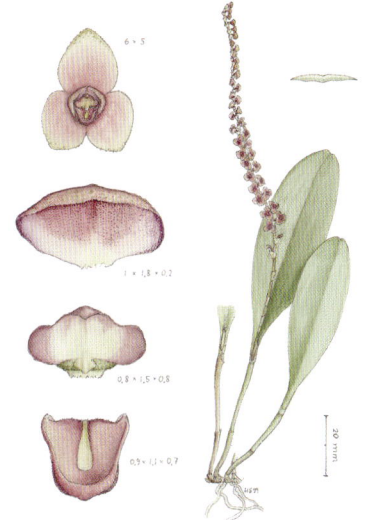

Stelis vinosa   1064

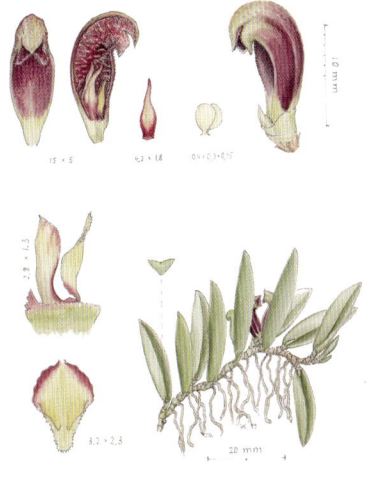

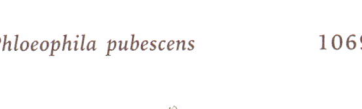

Phloeophila pubescens   1069

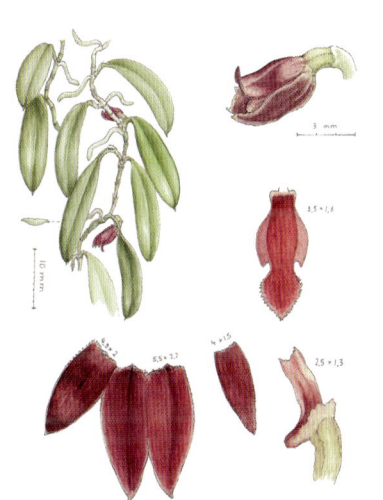

Phloeophila bradei   1070

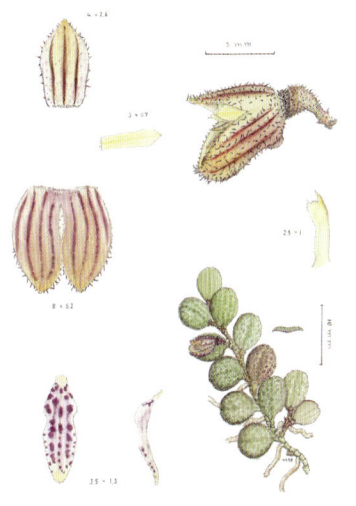

Phloeophila echinantha   1071

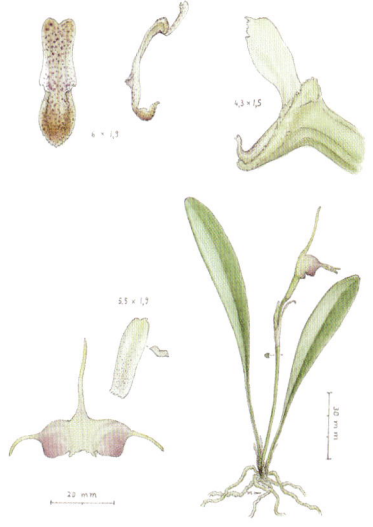

Masdevallia curtipes   1083

**P
L
A
T
E
V**

Masdevallia infracta   1084

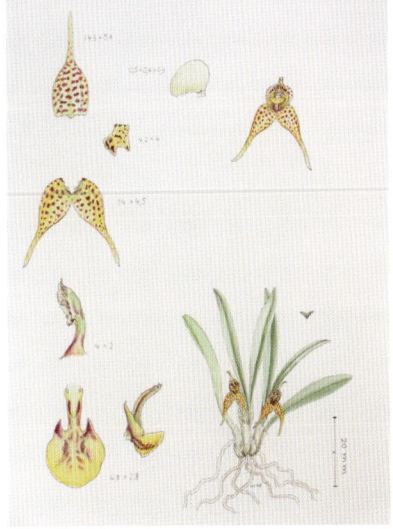

Dryadella edwallii   1088

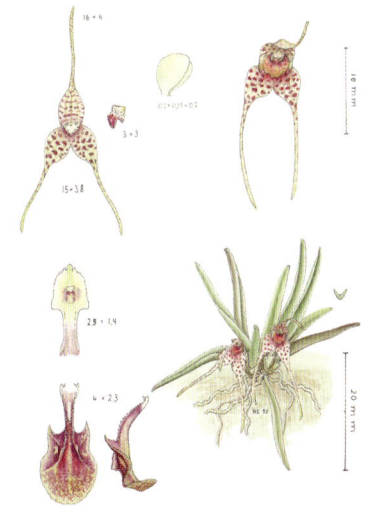

Dryadella zebrina   1096

| 503 |

1.7 × 0.9

0.75 × 0.8

1.3 × 0.6

0.6 × 0.4

0.9 × 0.5

2.0 mm

HS 1.03

P
L
A
T
E

VI

Lepanthopsis aff. densiflora

1101b

| 504 |

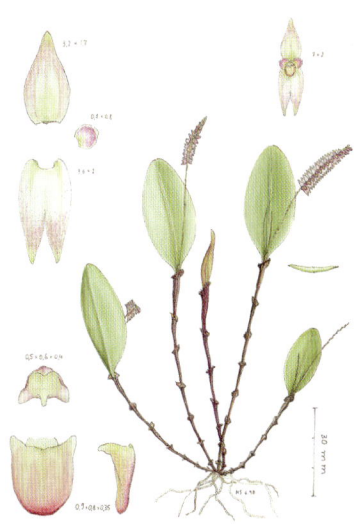

Lepanthopsis floripecten   1102

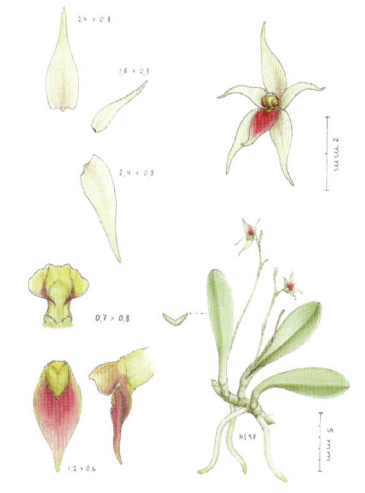

Platystele oxyglossa   1103

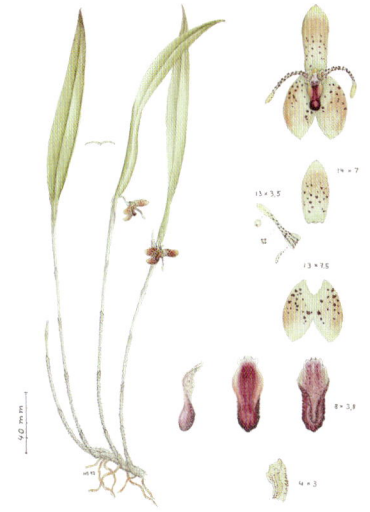

Myoxanthus punctatus   1119

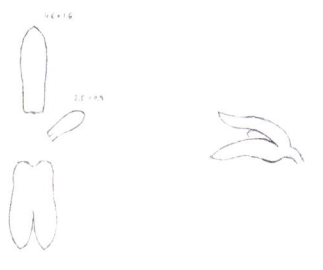

Myoxanthus aff. lonchophyllum   1120

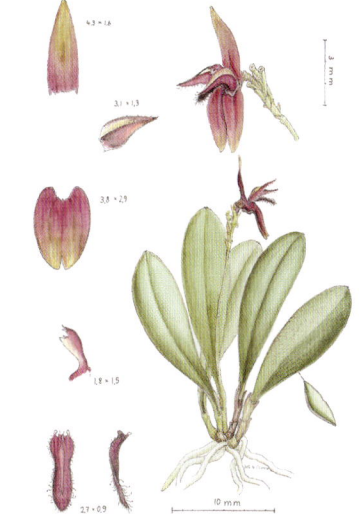

Pleurothallis aff. barbulata   1106b

Pleurothallis parvifolia   1108

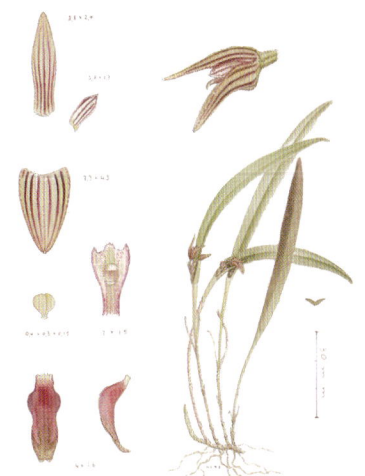

Pleurothallis scabripes   1114

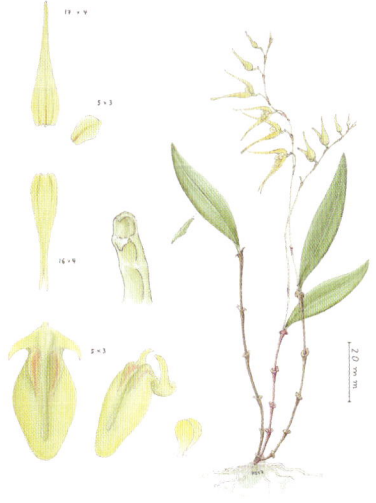

Pleurothallis collina   1125

Pleurothallis quartzicola   1126

P
L
A
T
E

VII

Pleurothallis curti-bradei   1130

Pleurothallis mattinhensis   1131

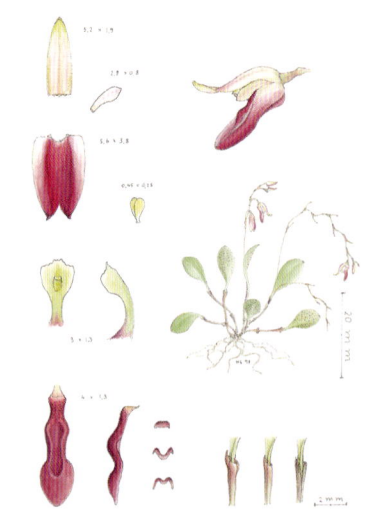

Pleurothallis miragliae   1132

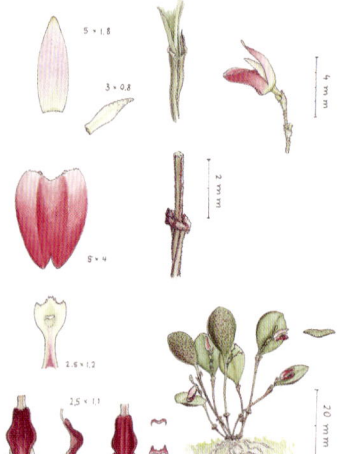

Pleurothallis punctatifolia   1134

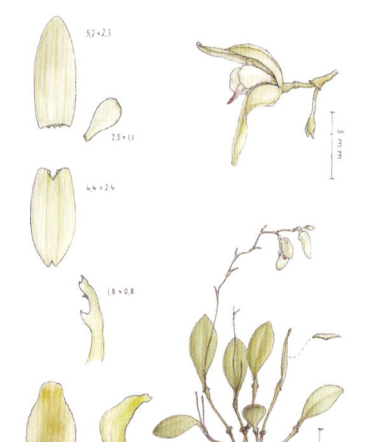

Pleurothallis quadridentata   1135

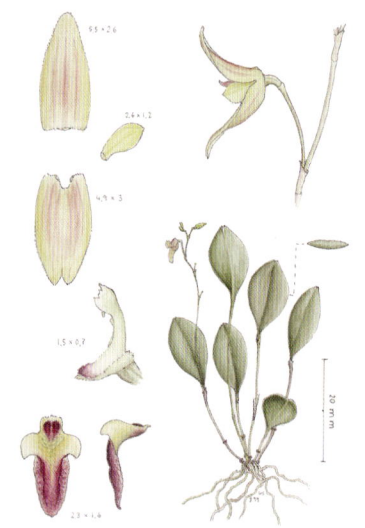

Pleurothallis bradei   1137

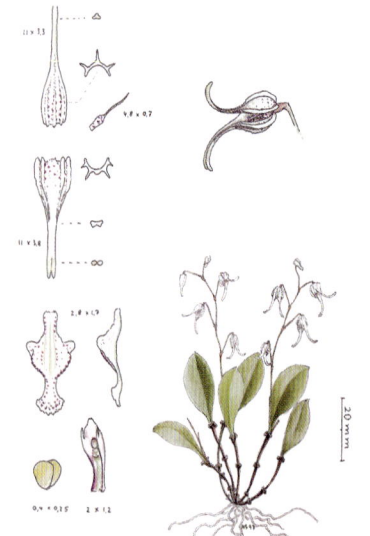

Pleurothallis carinifera   1138

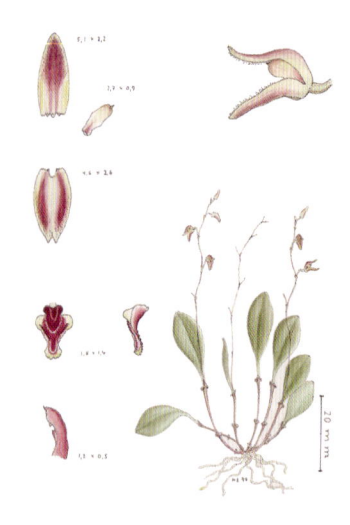

Pleurothallis podoglossa   1139

Pleurothallis sordida   1140

P
L
A
T
E

VIII

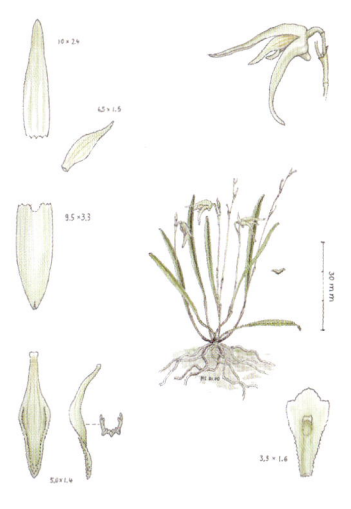

Pleurothallis calcarata   1141

Pleurothallis conspersa   1142

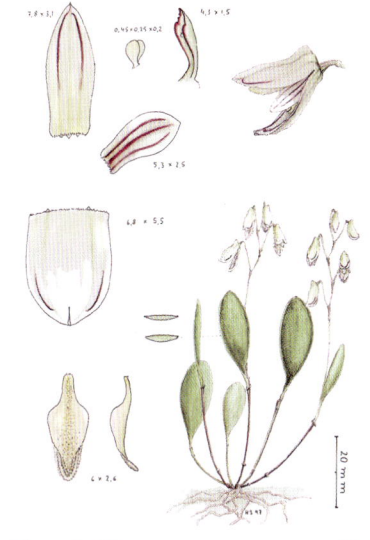

Pleurothallis avenacea   1146

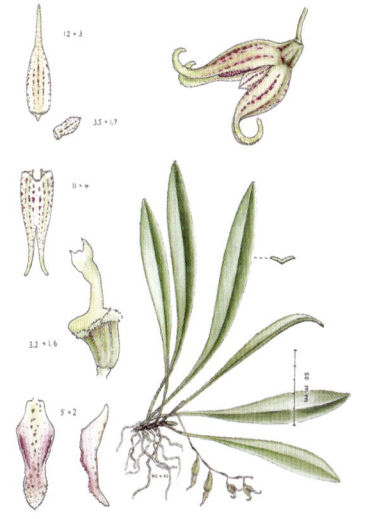

Pleurothallis tripterantha   1152

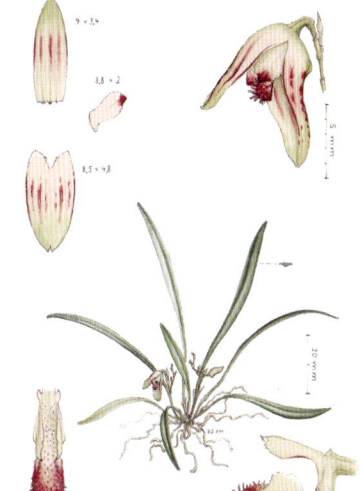

Pleurothallis imbeana   1153

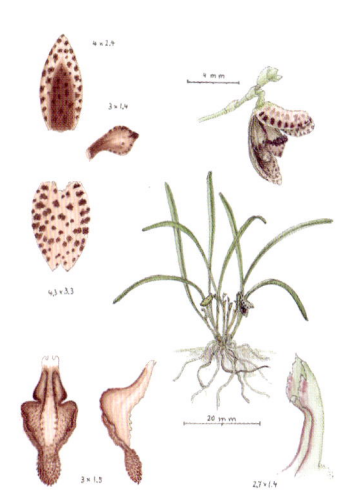

Pleurothallis fluminensis   1154

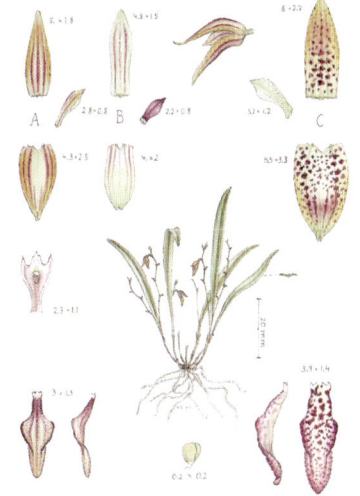

Pleurothallis colorata   1155

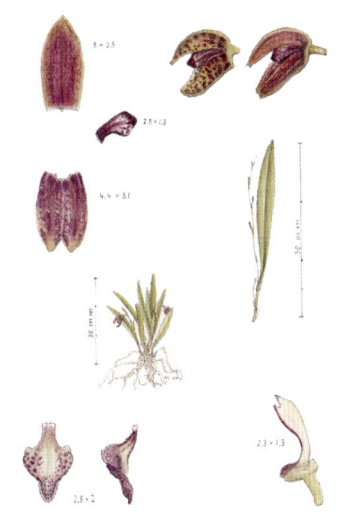

Pleurothallis cordilabia   1157

Pleurothallis gehrtii   1158

P L A T E IX

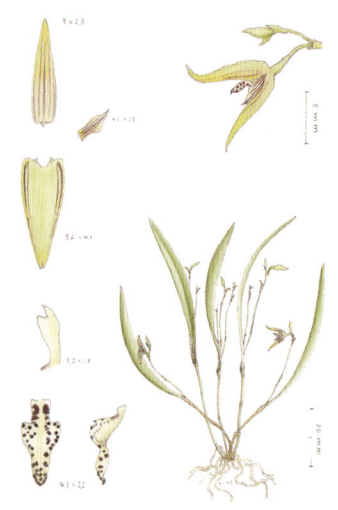

Pleurothallis lineolata 1162

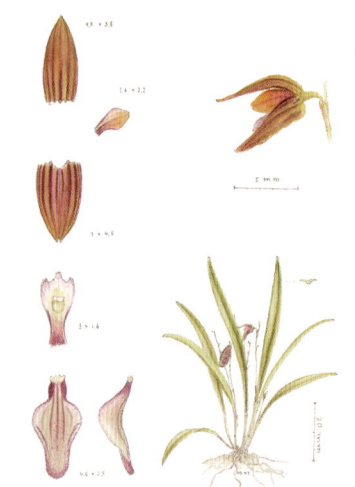

Pleurothallis ramphastorhyncha 1163

Pleurothallis rubro-lineata 1164

Pleurothallis trifida 1165

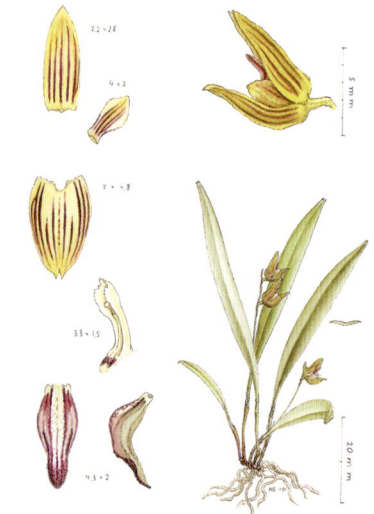

Pleurothallis uniflora 1166

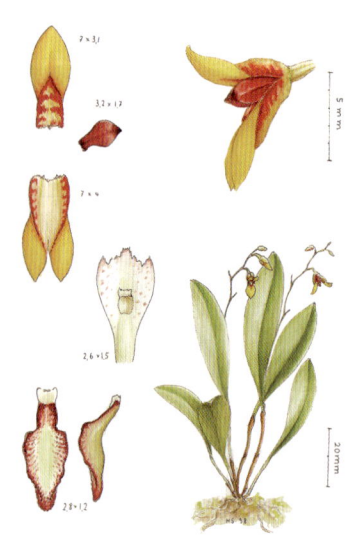

Pleurothallis hypnicola 1167

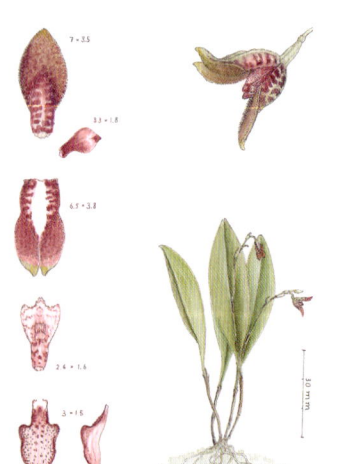

Pleurothallis hypnicola
var. *cuneifolia* 1167b

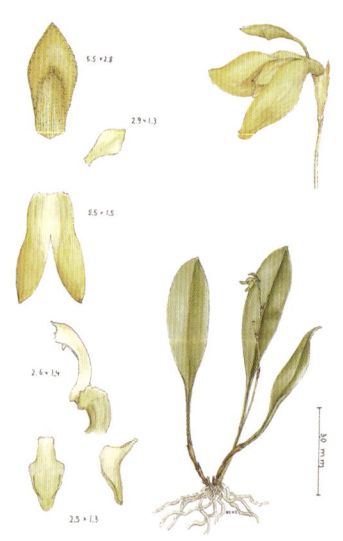

Pleurothallis hypnicola var. *flava* 1167c

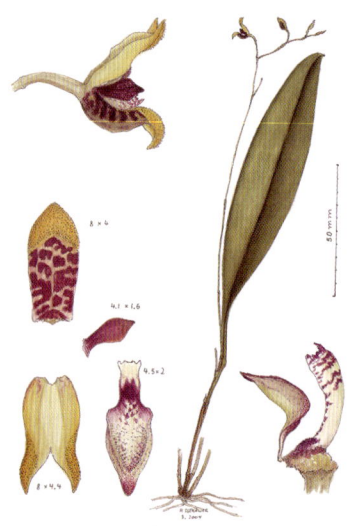

Pleurothallis hypnicola
var. *major* 1167d

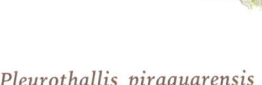

Pleurothallis piraquarensis   1168

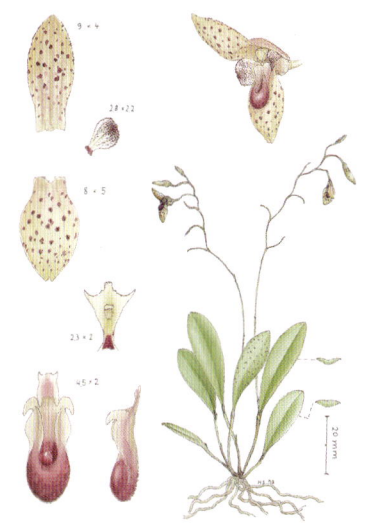

Pleurothallis seriata   1169

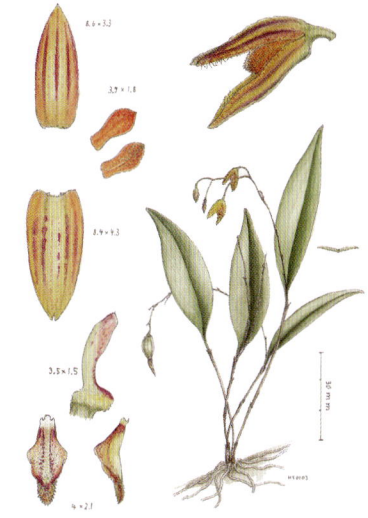

Pleurothallis sp.
não identificada   1172a

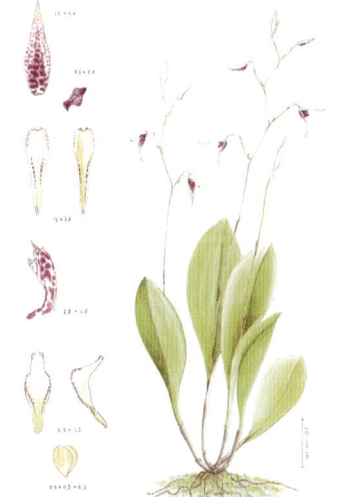

Pleurothallis arcuata   1174

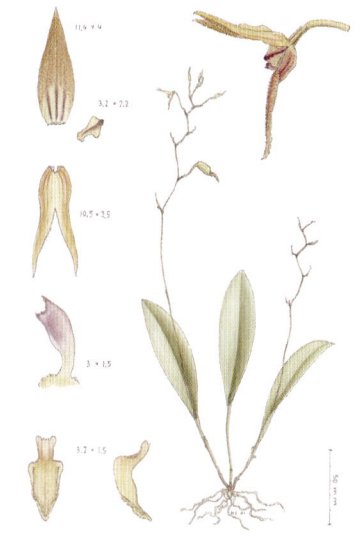

Pleurothallis granulosa   1177

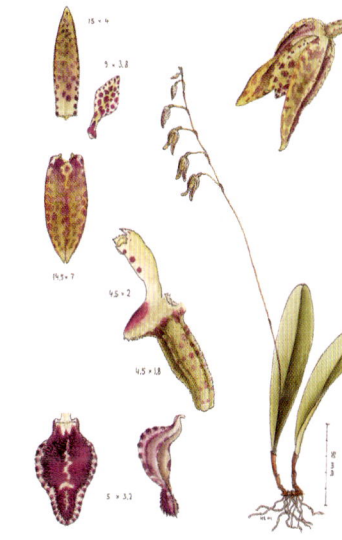

Pleurothallis henrique-aragonii   1178

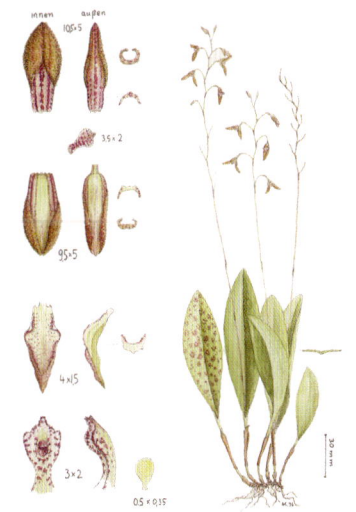

Pleurothallis hians   1179

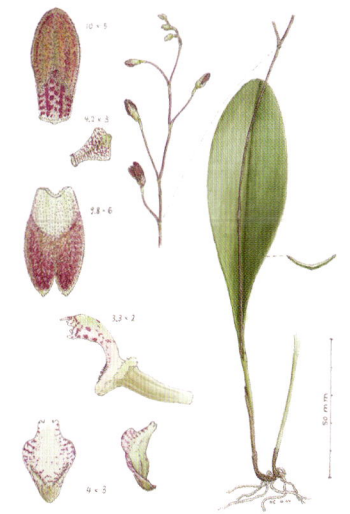

Pleurothallis pellifeloides   1180

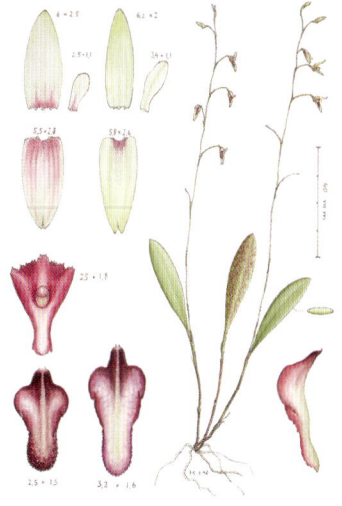

Pleurothallis wacketii   1185

P
L
A
T
E

XI

P L A T E

XII

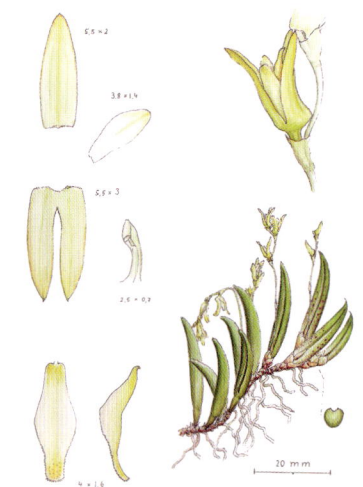

Pleurothallis teres   1195

Pleurothallis leptotifolia   1197

Pleurothallis sonderana
var. longicaulis   1200b

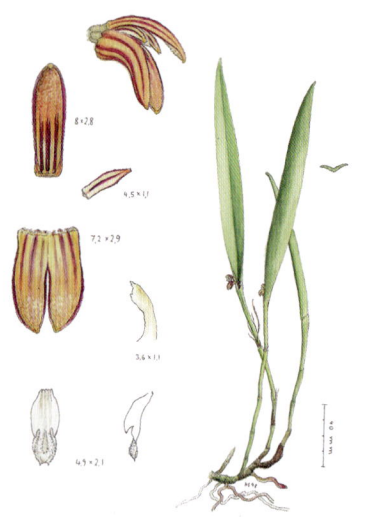

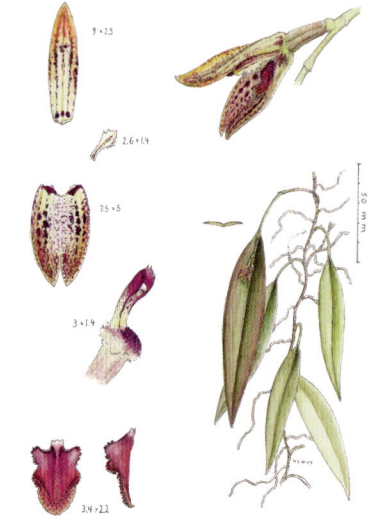

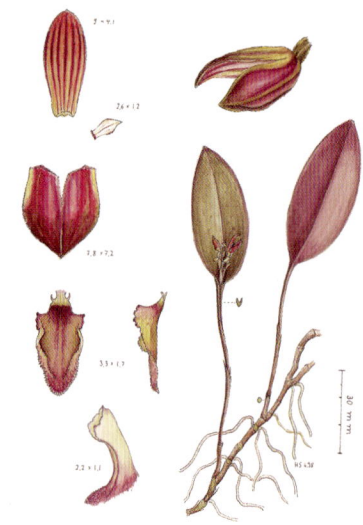

Pleurothallis bidentata   1205

Pleurothallis capanemiae   1208

Pleurothallis cearensis   1209

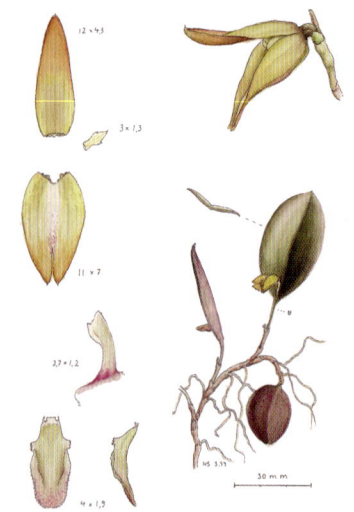

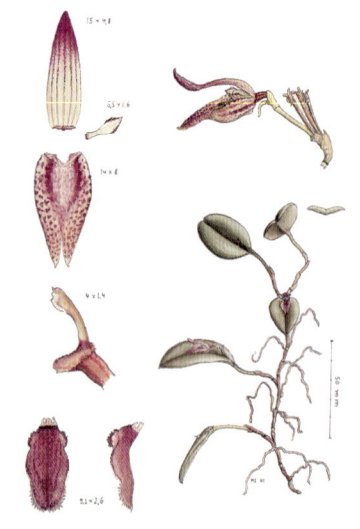

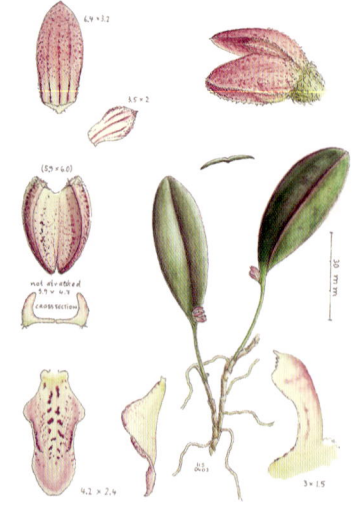

Pleurothallis saundersiana   1215

Pleurothallis serpentula   1216

Pleurothallis spilantha   1218

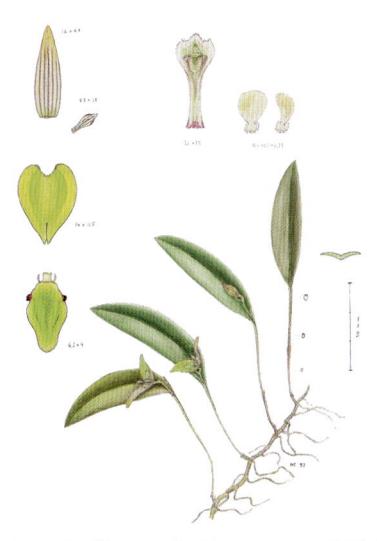

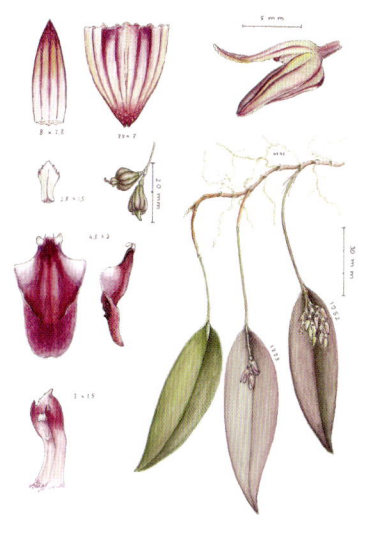
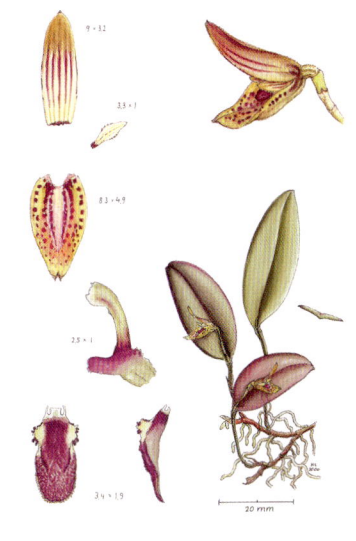

Pleurothallis translucida   1220

Pleurothallis farinosa   1223

Pleurothallis ramosa   1230

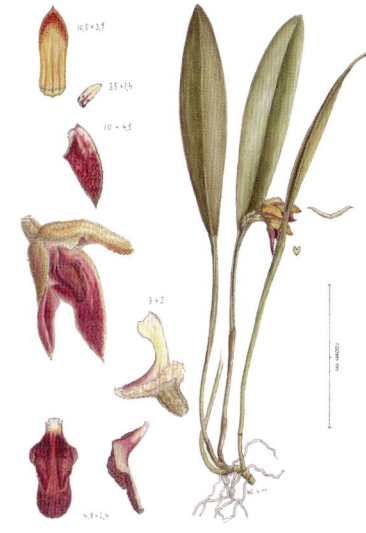

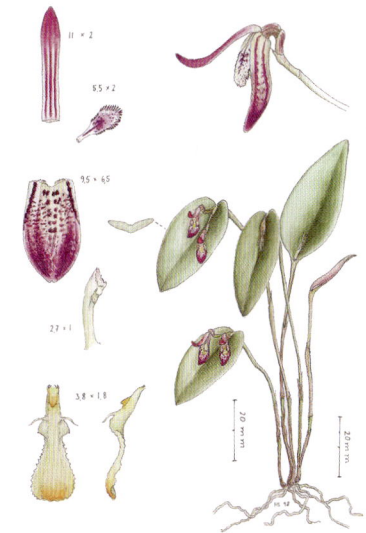

Pleurothallis aff. ramosa   1230b

Pleurothallis aphtosa   1231

Pleurothallis panduripetala   1236

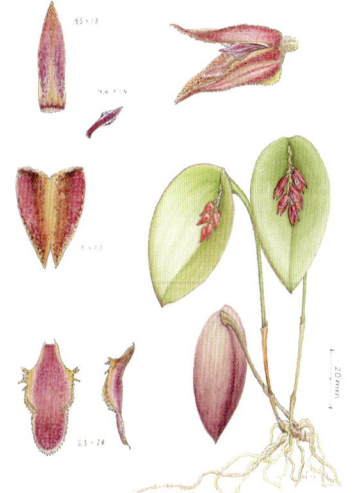

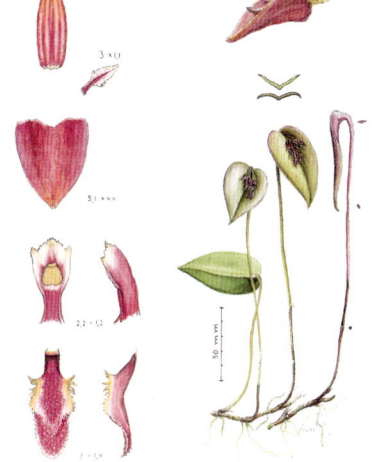

Pleurothallis hamosa   1238

Pleurothallis limae   1239

Pleurothallis modestissima   1240

P
L
A
T
E

XIII

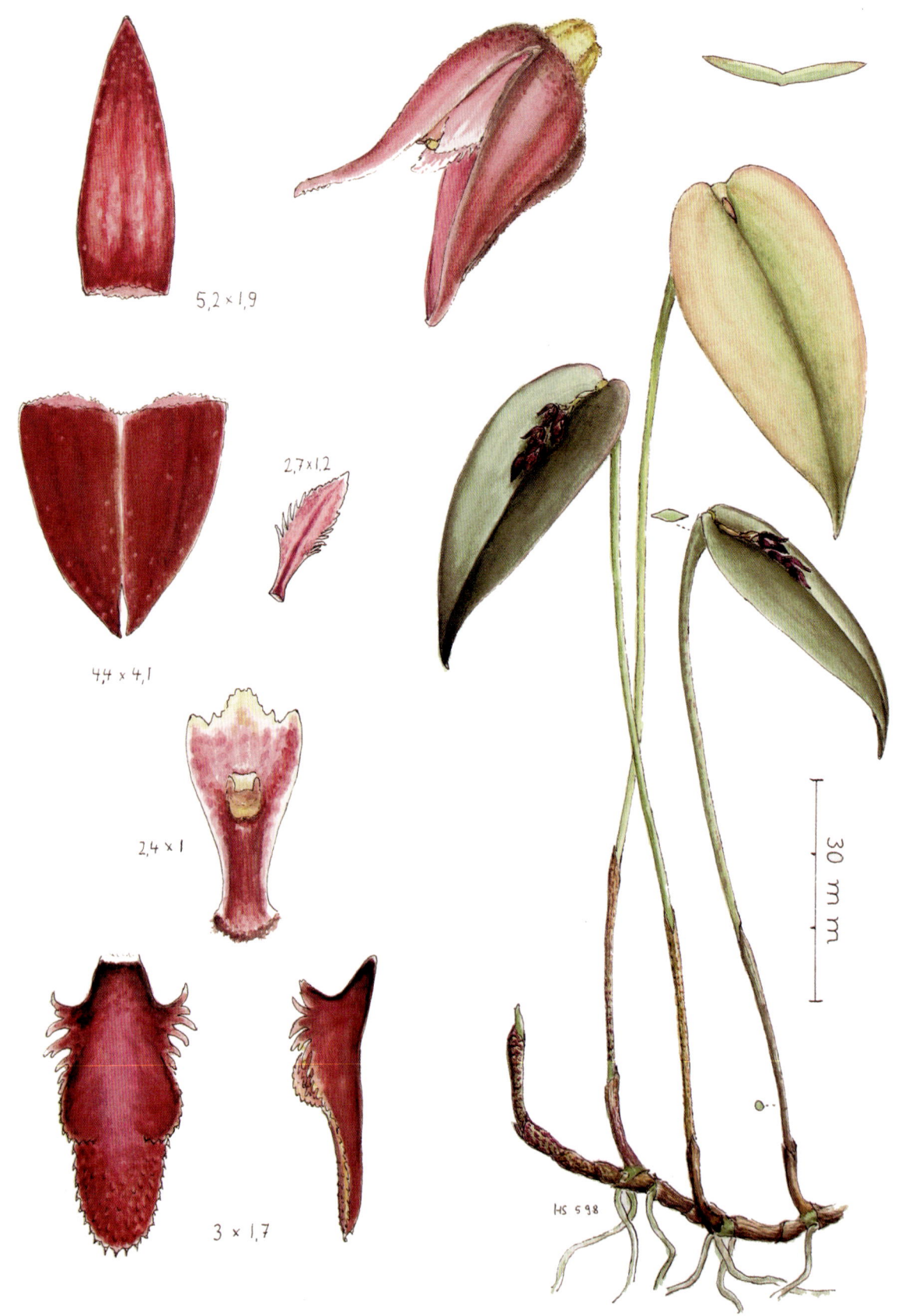

5,2 × 1,9

2,7 × 1,2

**P
L
A
T
E**

XIV

4,4 × 4,1

2,4 × 1

30 m m

3 × 1,7

HS 5 98

Pleurothallis prolifera

1241

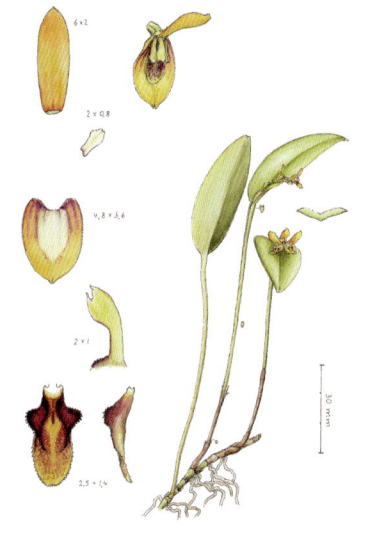

Pleurothallis exarticulata 1242

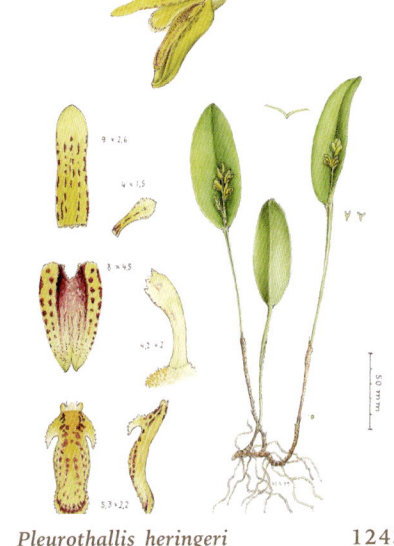

Pleurothallis heringeri 1243

Pleurothallis luteola 1244

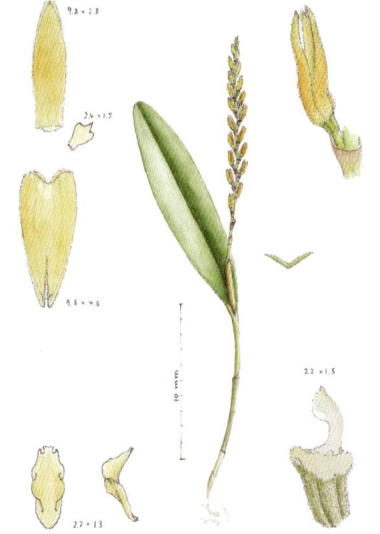

Pleurothallis acuminatipetala 1245

Pleurothallis binotii 1247

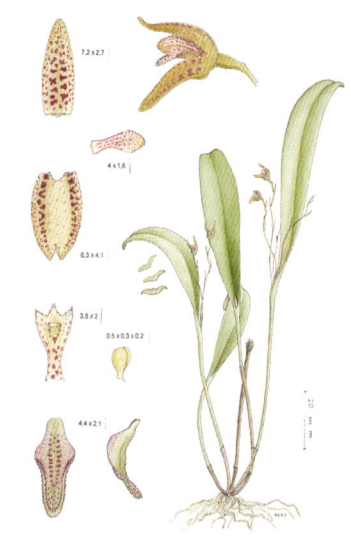

Pleurothallis pantherina 1251b

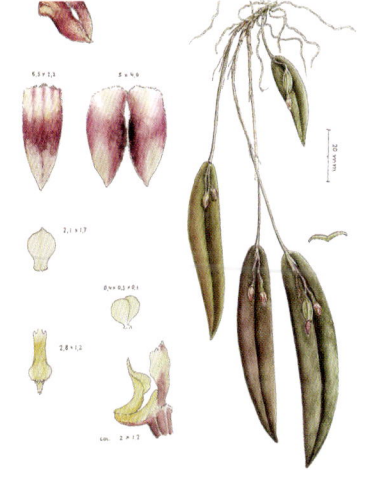

Pleurothallis cryptophoranthoides 1252

Pleurothallis dracula 1252b

Pleurothallis klotzschiana 1255

P
L
A
T
E

XV

| 513 |

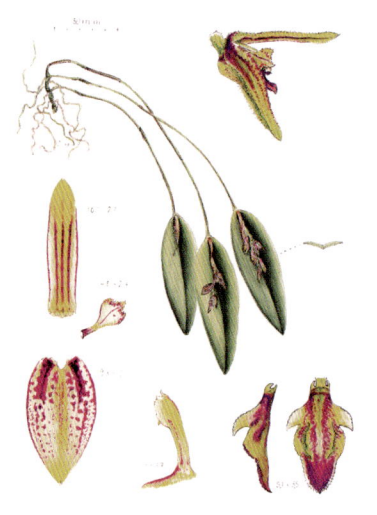

Pleurothallis porphyrantha   1257

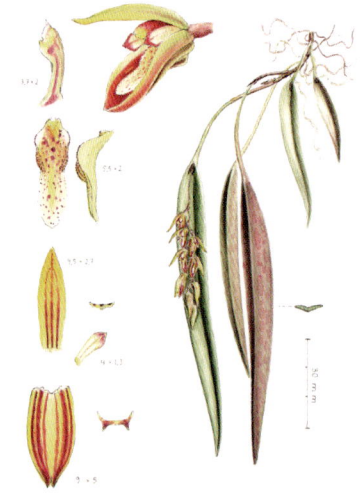

Pleurothallis albo-rosea   1260

Pleurothallis macropoda   1262

P L A T E

XVI

Pleurothallis octophrys   1273

Pleurothallis strictophylla   1275

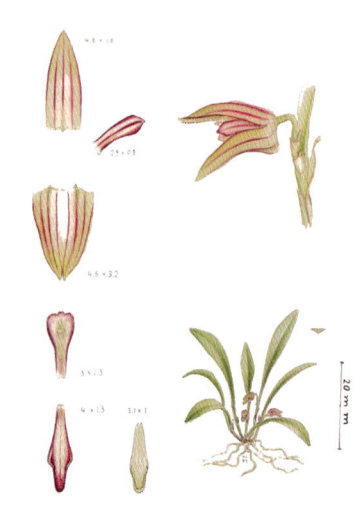

Pleurothallis lingua   1279

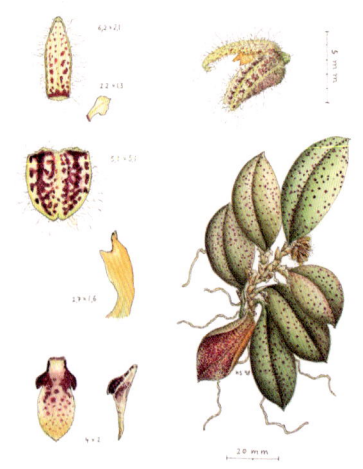

Pleurothallis crinita   1280

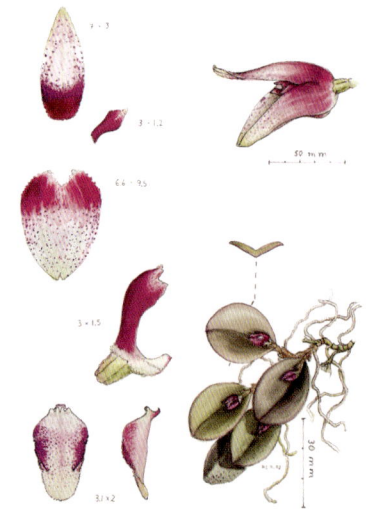

Pleurothallis karlii   1281

Pleurothallis recurva   1286

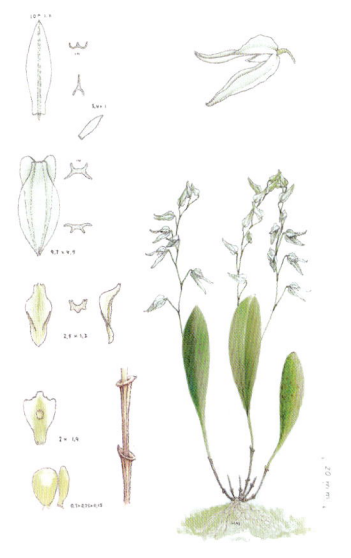

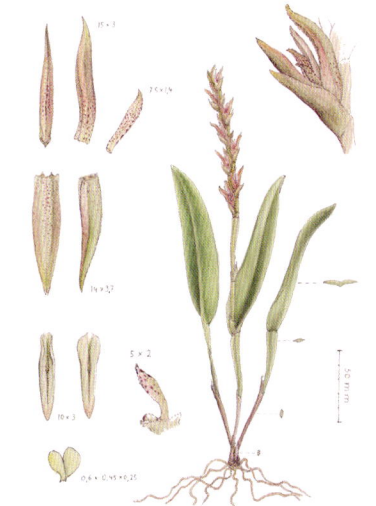

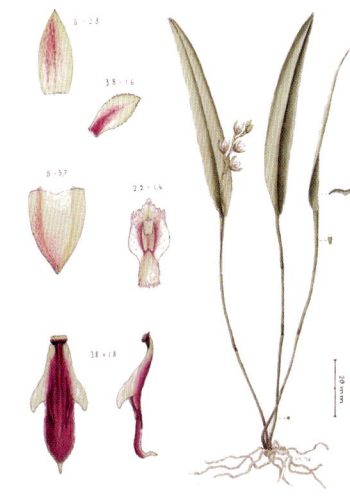
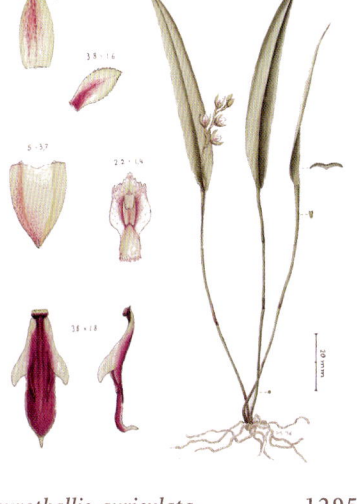

Pleurothallis leucopyramis   1289

Pleurothallis platystachya   1293

Pleurothallis auriculata   1295

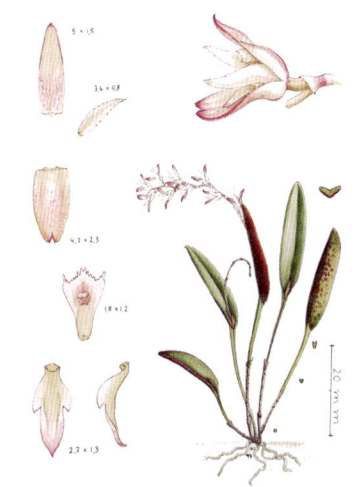

Pleurothallis capillaris   1298

Pleurothallis hygrophila   1301

Pleurothallis longicaulis   1303

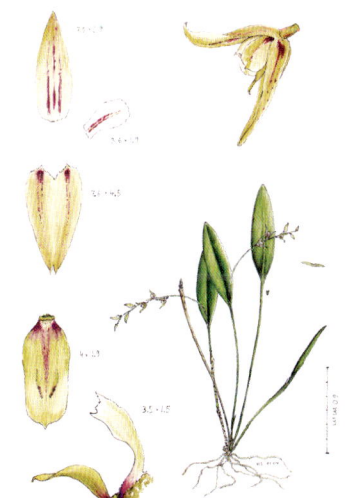

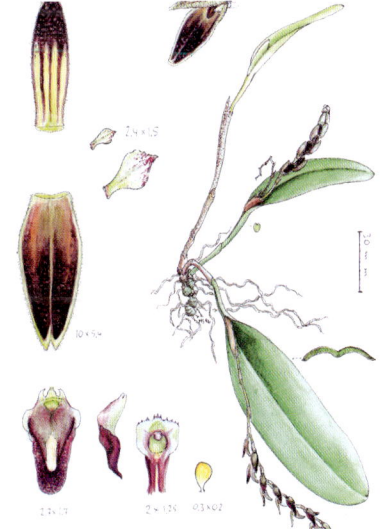

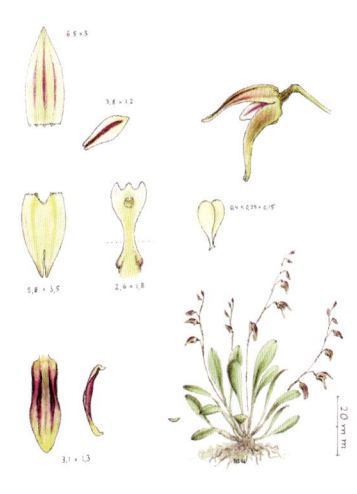

Pleurothallis malachantha   1304

Pleurothallis saurocephala   1314

Pleurothallis marginalis   1319

P
L
A
T
E

XVII

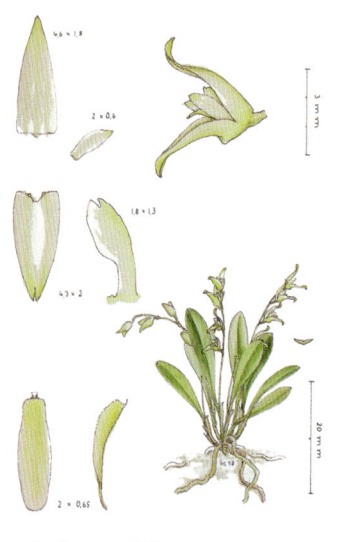

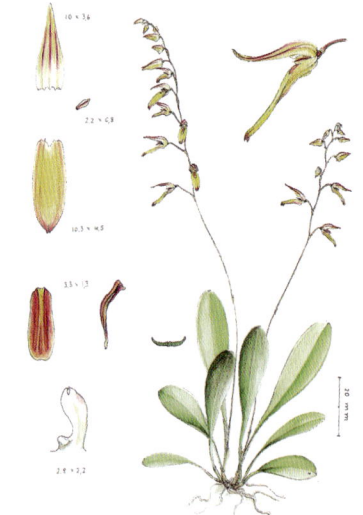

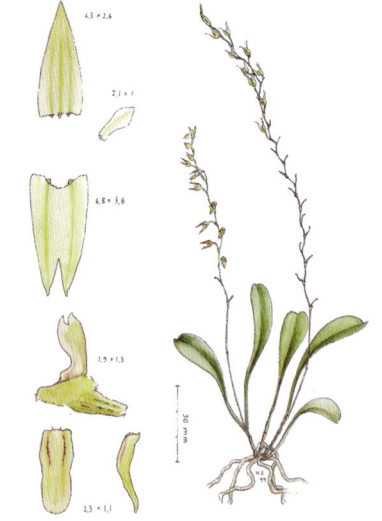

Pleurothallis viridiflora   1321b   *Pleurothallis grobyi*   1323   *Pleurothallis aff. grobyi*   1323b

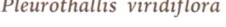

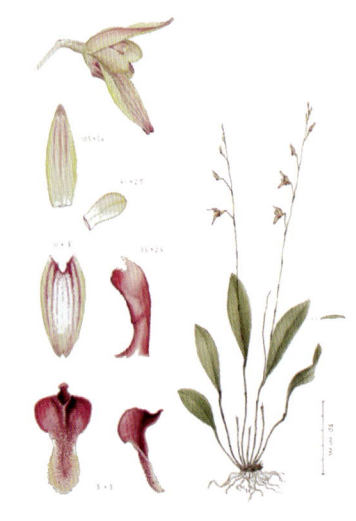

Pleurothallis subpicta   1325   *Pleurothallis edwallii* var. *major* 1329a
Pleurothallis edwallii var. *palida* 1329b   *Pleurothallis bocainensis*   1335

Pleurothallis depauperata   1338   *Pleurothallis eugenii*   1339   *Pleurothallis linearifolia*   1341

Pleurothallis malmeana   1342

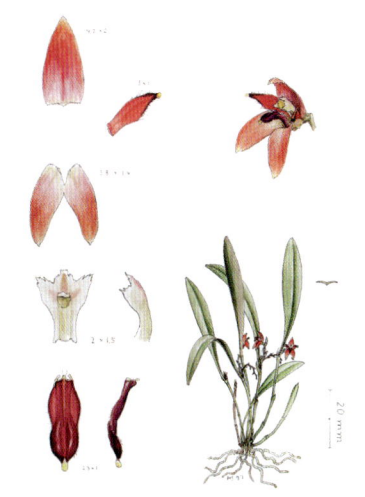

Pleurothallis adenochila   1348

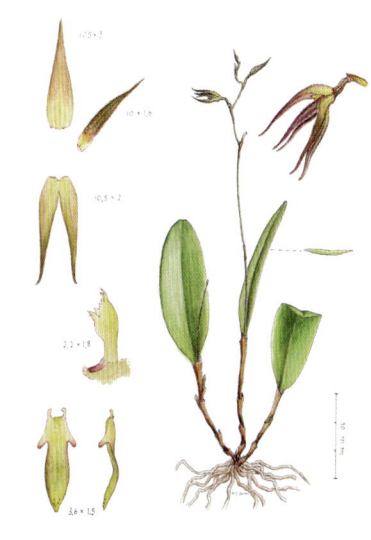

Pleurothallis tigridens   1357

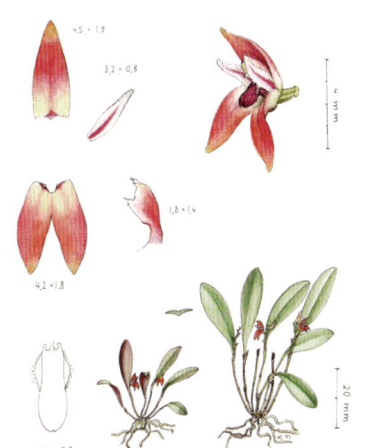

Pleurothallis microgemma   1362

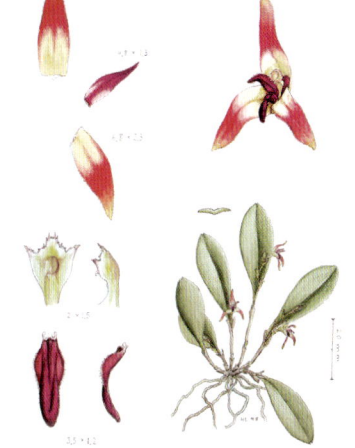

Pleurothallis modesta   1363

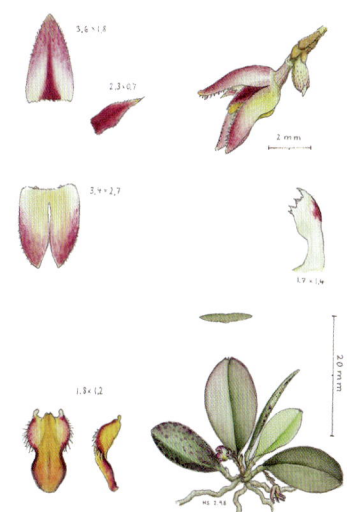

Pleurothallis recurvipetala   1365

Pleurothallis aff. laciniata   1371b

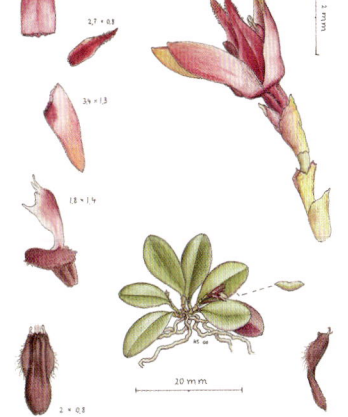
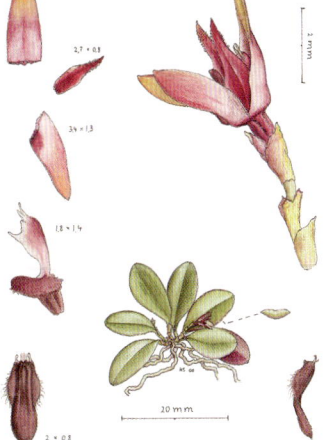

Pleurothallis microphyta   1374

Pleurothallis aff. microphyta   1374b

P
L
A
T
E

XIX

| 517 |

4,0 × 1,9

2,8 × 1

3,7 × 1,8

3 m m

P
L
A
T
E

XX

2 × 1,2

20 m m

2,5 × 1,2

Pleurothallis rubro-limbata

1375

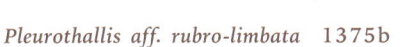

Pleurothallis aff. rubro-limbata   1375b

Pleurothallis corticicola   1376

Pleurothallis aff. corticicola   1376b

Pleurothallis foliata   1379

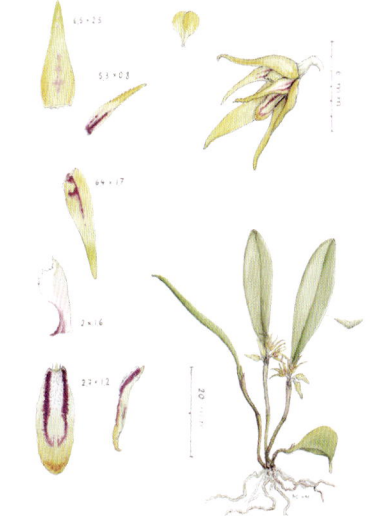

Pleurothallis spannageliana   1384

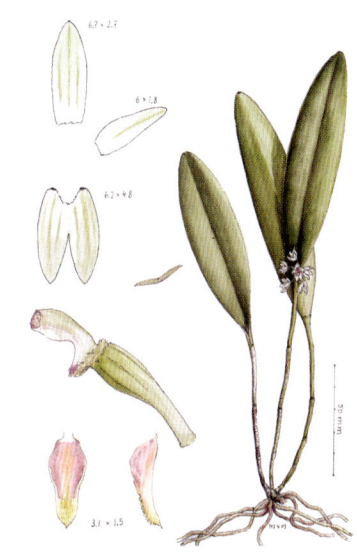

Pleurothallis pulvinata   1385

P
L
A
T
E

XXI

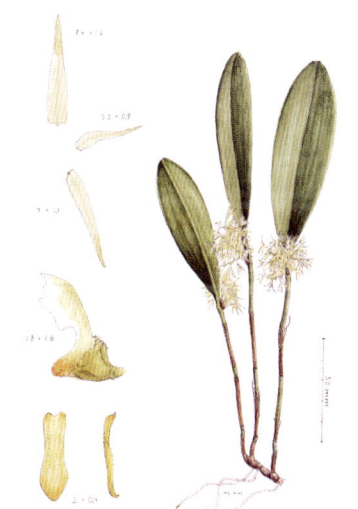

Pleurothallis fasciculata   1386

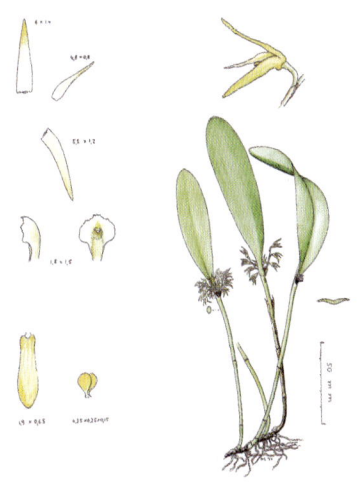

Pleurothallis fasciculata   1386b
var. densiflora

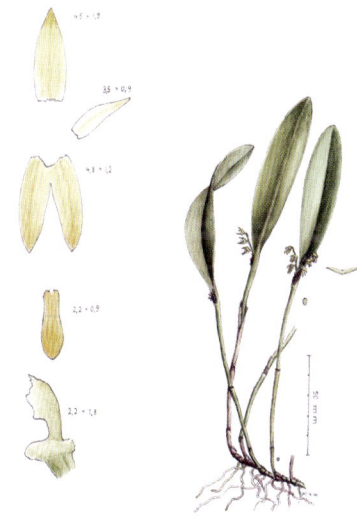

Pleurothallis obovata   1387

| 519 |

Pleurothallis dryadum   1388

Pleurothallis ferdinandiana   1389

Pleurothallis montipelladensis   1394

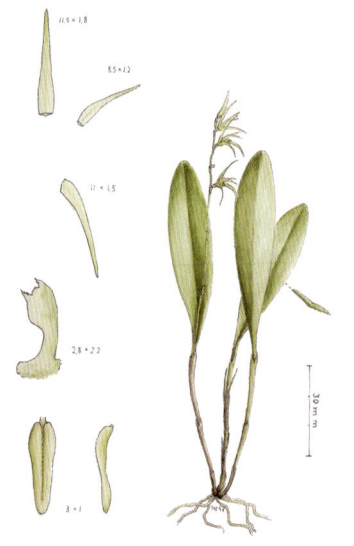

Pleurothallis pabstii   1395

Pleurothallis simpliciglossa   1397

Pleurothallis sclerophylla   1399

Pleurothallis radialis   1404

Pleurothallis rubens   1405

Pleurothallis heterophylla   1407

Barbosella australis   1427

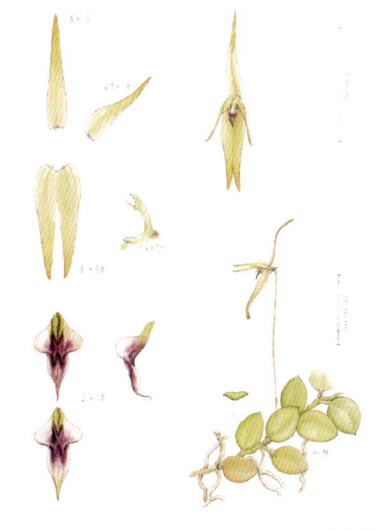

Barbosella gardneri   1429

Barbosella gardneri
var. microphylla   1432

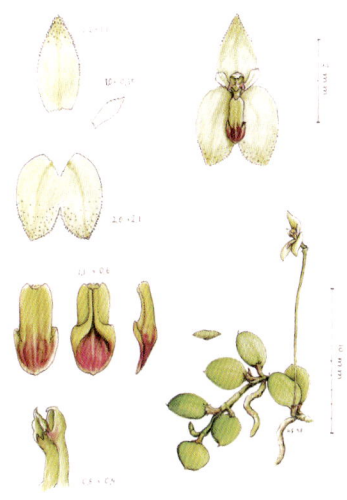

Barbrodia miersii   1433

Brachionidium dungsii   1437

Octomeria alpina   1440

P
L
A
T
E

XXIII

Octomeria crassifolia   1443

Octomeria densiflora   1445

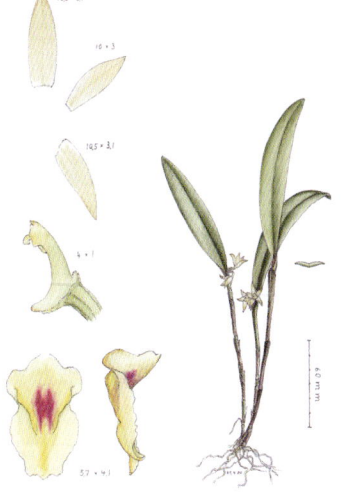

Octomeria ementosa   1446

P
L
A
T
E

XXIV

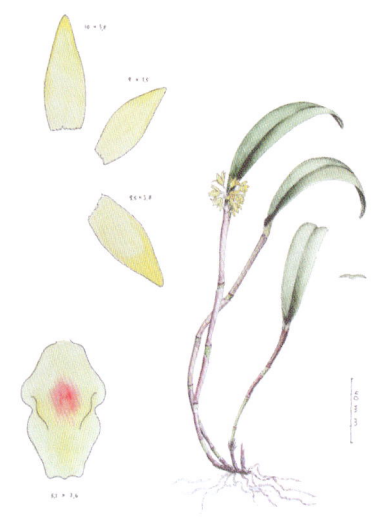

Octomeria fasciculata　　　　1448

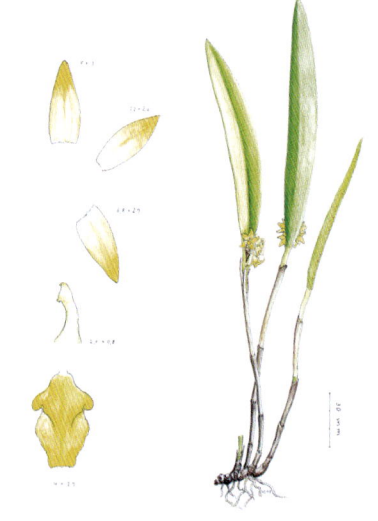

Octomeria robusta　　　　1454

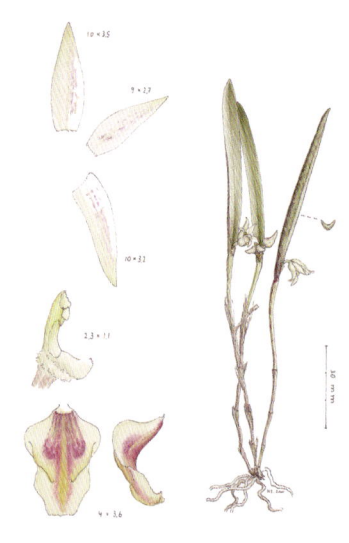

Octomeria sarcophylla　　　　1455

Octomeria serrana　　　　1457

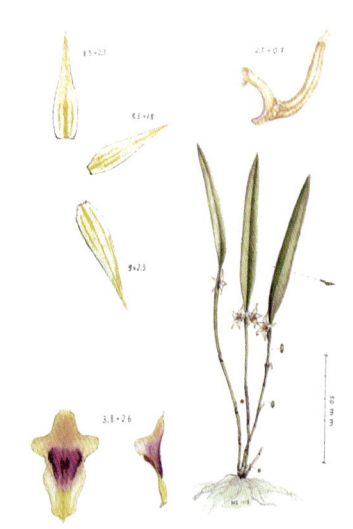

Octomeria anceps　　　　1458

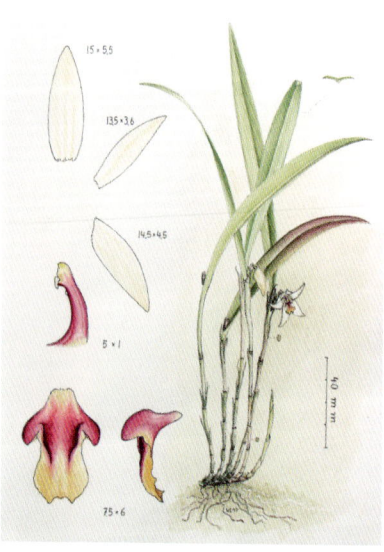

Octomeria grandiflora　　　　1460

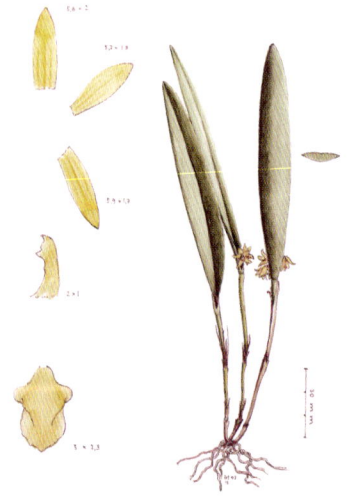

Octomeria gracilicaulis　　　　1462

Octomeria oxichela　　　　1465

Octomeria concolor　　　　1469

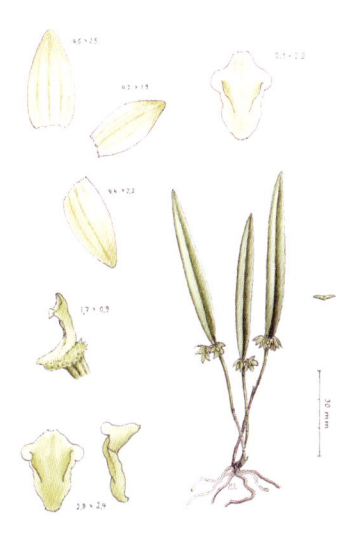

Octomeria micrantha 1471

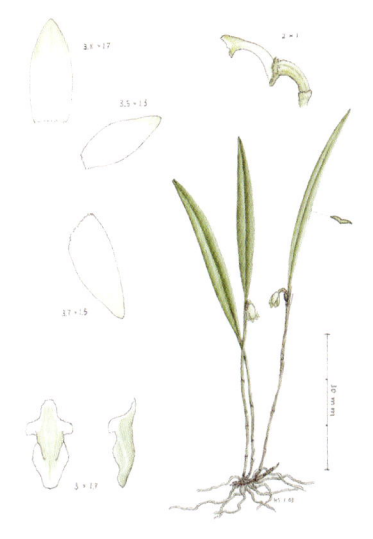

Octomeria aff. micrantha 1471b

Octomeria bradei 1476

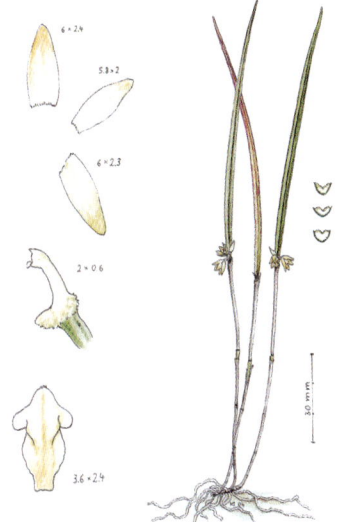

Octomeria linearifolia 1478

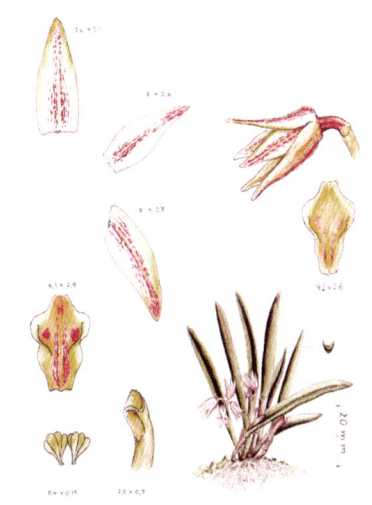

Octomeria aff. linearifolia 1478b

Octomeria aff. linearifolia 1478c

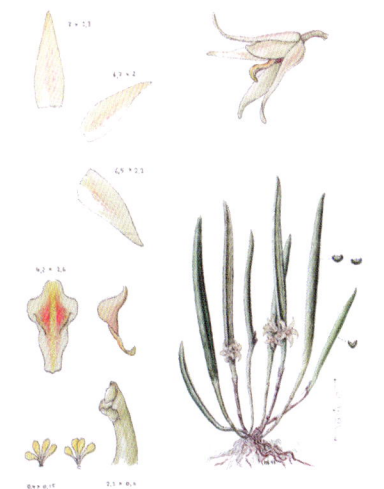

Octomeria aff. linearifolia 1478d

Octomeria albopurpurea 1480

Octomeria crassilabia 1483

P
L
A
T
E

XXV

P
L
A
T
E

XXVI

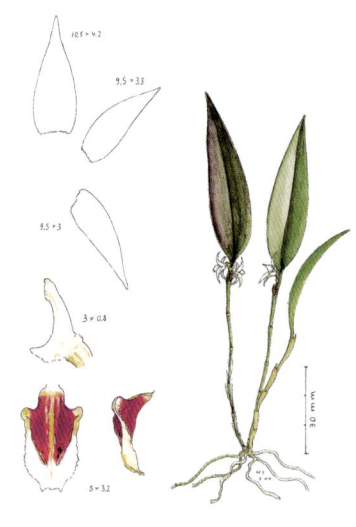

Octomeria diaphana   1484

Octomeria glazioviana   1486

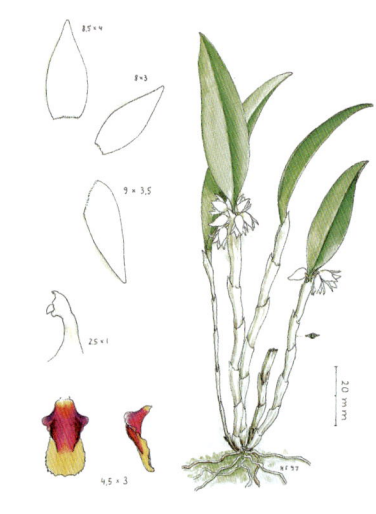

Octomeria aff. glazioviana   1486b

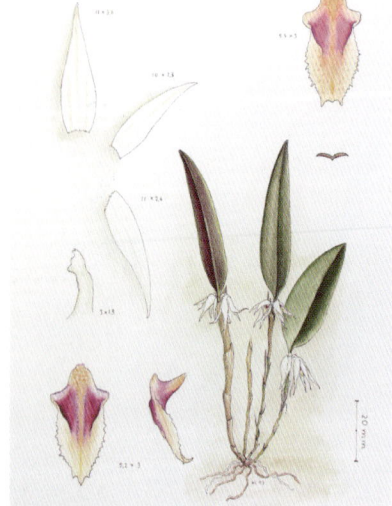

Octomeria rechiana   1487

Octomeria reitzii   1488

Octomeria rodriguesii   1489

Octomeria rotundiglossa   1490

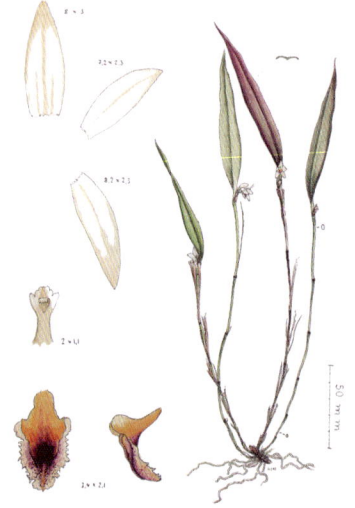

Octomeria tricolor   1491

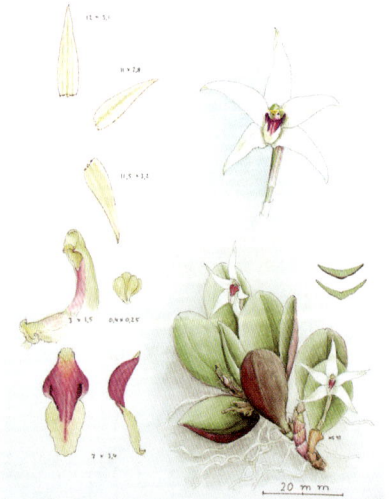

Octomeria estrellensis   1493

Octomeria minuta   1495

Octomeria aff. ochroleuca   1497b

Octomeria rhodoglossa   1498

Octomeria aloefolia   1506

Octomeria fimbriata   1508

Octomeria chamaeleptotes   1515

Octomeria riograndensis   1517

Octomeria alexandri   1518

Octomeria sancti-angeli   1519

P
L
A
T
E

XXVII

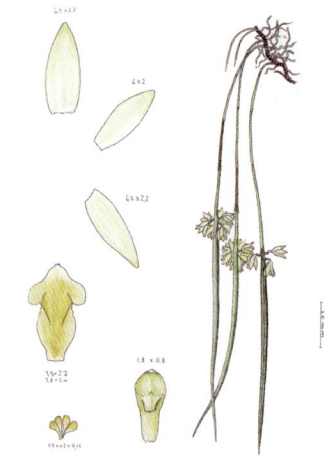

Octomeria decumbens   1521   *Octomeria geraensis*   1522   *Octomeria aff. geraensis*   1522b

**P
L
A
T
E**

XXVIII

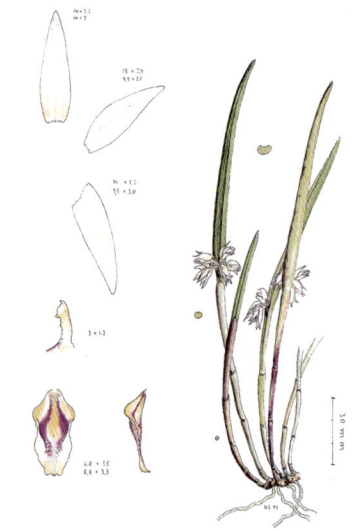

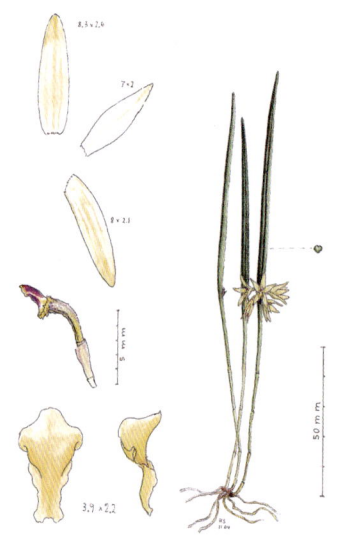

Octomeria juncifolia   1523   *Octomeria praestans*   1527   *Octomeria truncicola*   1528

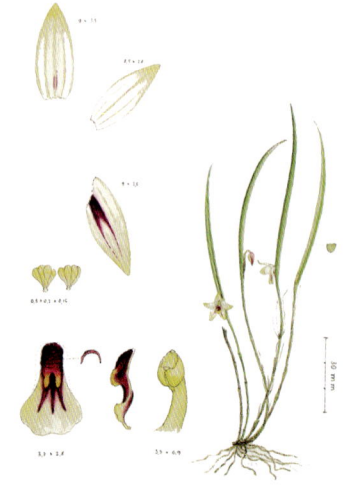

Octomeria wilsoniana   1529   *Octomeria gracilis*   1530   *Octomeria campos-portoi*   1536

BIBLIOGRAPHIC REFERENCES

ANDERSON, A. B. **Alternative to Deforestation** (Columbia University Press – NY – 1990).

ARQUIVOS do Jardim Botânico do Rio de Janeiro.

BARBOSA RODRIGUES, J. **Iconographie dês Orchidees du Brésil** (Vol. 1 e 2 – Friedrich Reinhardt Verlag, Basle – 1996).

BECHTEL, H., Cribb P., e Launart, E. **Manual of Cultivated Orchid Species** (3ª edição, Blandford Press – 1992).

BOCK, I. **Particular**, and German Orchid Society.

BOLETIM **CAOB** – 1996 – 2005 São Paulo.

BRASIL ORQUÍDEAS Editora Brasil Orquídeas Ltda (No. 01 – 13 São Paulo).

COGNIAUX. **Orchidacea Brasiliensis**; Flora Brasiliensis Martii.

CORREA, M. Pio. **Diccionário das plantas úteis do Brasil** – Vols. 1 – 6 (Imprensa Nacional – RJ – 1926).

DARWIN, C. **The Voyage of the Beagle** (Bantam Books Inc. – 1958).

DARWIN, C. **On the Various Contrivances by which British and Foreign Orchids are Fertilized by Insects** (J. Murray, London – 1862).

DEAN, W. A Ferro e Fogo – **A História e a Devastação da Mata Atlântica Brasileira** (Companhia das Letras – 1996).

DEPTº GEOGRÁFICO DO ESTADO DO RIO DE JANEIRO (**Anuário Geográfico do Estado do Rio de Janeiro**).

DRESSLER, R. **The Orchids, Natural History and Classification** (Harvard U.P, Cambridge, Mass. USA. – 1981).

DRESSLER, R. e Dobson, C.H. **Phylogeny and Classification in the Orchidaceae** (Annals of the Missouri Botanical Garden 47: 25-68 – 1960).

DRUMMOND, J.A. **Devastação e Preservação Ambiental no Rio de Janeiro** (Editora da Universidade Federal Fluminense – 1997).

DUNNING, R.S. **South American Birds** (Harrowood Books Pennsylvania – U.S.A. – 1998).

DUNSTERVILLE, G. & Dunsterville, E. Orchid Hunting In The Lost World (And Elsewhere in Venezuela). (American Orchid Society, Florida – 1988).

EMMONS, L.H. **Neotropical Rain Forest Animals** (University of Chicago Press – 1990).

FREYRE, G. **Ingleses no Brasil** (Top Books Editora e Distribuidora – 1977).

FREYRE, G. **Casa Grande e Senzala** (José Olimpio Editora – 1946).

GARAY, L.A. **On the origins of the Orchidaceae** (In Botanical Museum Leaflets of Harvard University 19(3) 57-95 – 1960).

GARDNER, G. **Travels in the Interior of Brazil** (Reeve Brothers, King Street – London – 1846).

GARDNER, G. **Viagem ao interior do Brasil** (Editora Itatiaia – Belo Horizonte – 1975).

GARAY, L.A. **Noticia Orchidologia II**.

GOVERNO DO ESTADO DO RIO DE JANEIRO (**Centro de Informações e Dados do Estado do Rio de Janeiro**).

HEYWOOD, V. **Flowering Plants of the world** (B.T. Batsford Ltd – London – 1993).

HOEHNE, F. C. **Iconografia das Gesneriaceas do Brasil** (Secretaria de Cultura, Instituto de Botânica – SP – 1970).

HOEHNE, F.C. **Cinquenta e uma Nova Espécies da Flora do Brasil** (Arquivos de Botânica do Estado de São Paulo).

HOEHNE, F.C. **Iconografia de Orchidaceaes do Brasil** (São Paulo – 1949).

HOLST, A.W. **The World of Catasetums** (Timber Press – 1999).

JACCOUND, R.L. de S. – **História, contos e lendas da velha Nova Friburgo** (Múltipla Cultura – 1999).

JUDZIEWICZ, E.J., Clark,L.G., Londoño,X. & Stern M.J. **American Bamboos** (The Smithsonian Institution Press Washington and London – 1999).

KOEHLER, S. et al. **Phylogeny of the Bifrenaria complex based on morphology and sequence data from nucler + DNA internal transcribed spacers(ITS) and cloroplast trnl-trnf region** (UNICAMP, Depto. de Biologia).

KOEHLER, S. & Amaral, M.C.E. **A taxonomic study of the South American Genus Bifrenaria Lindl.** (Brittonia 56(4) 2004 pp 314-345).

LIMA, P.M. e Guedes-Bruni, R.R. **Reserva Ecológica de Macaé de Cima Nova Friburgo-RJ** – Aspectos Florísticos das Espécies Vasculares – Vol.1 (Jardim Botânico do Rio de Janeiro – 1994).

LIMA, P.M. e Guedes-Bruni, R.R. **Reserva Ecológica de Macaé de Cima Nova Friburgo** – RJ – Aspectos Florísticos das Espécies Vasculares – Vol. 2 (Jardim Botânico do Rio de Janeiro – 1996).

LIMA, H,C. e Guedes-Bruni, R.R. **Serra de Macaé de Cima:** Diversidade Florística e Conservação em Mata Atlântica (Instituto de Pesquisas Jardim Botânico do Rio de Janeiro – 1997).

LINDLEY, J. **Folia Orchidaceae**.

LOREZI, H. **Árvores Brasileiras** – Vols. 01 e 02 (Instituto Plantarum de Estudos da Flora Ltda – Nova Odessa – SP – 1992 e 1998).

LUER, C. **Icones Pleurothallidinarium I – XVI**.

MARQUES Otavio.A.V., Eterovic,A. e Sazima, I. **Serpentes da Mata Atlântica** (Holos, Editora Ltda-ME – SP – 2001).

MARTINELLI, G. **Campos de Altitude** (Editora Index, Rio de Janeiro – 1989).

MARTINELLI G., Cavalcanti, M.J. e Schuback, P. **Conservação da Biodiversidade** (Jardim Botânico do Rio de Janeiro – 1990).

MARTIUS, Von. **Flora Brasiliensis**.

MENEZES, L.C. Genus Cyrtopodium, **Espécies Brasileiras** (IBAMA – Brasília – 2000).

MILLER, D. and Warren, R.C. **Orquídeas do Alto da Serra** (Salamandra Consultoria Ltda – 1998).

NICOULIN, M. **A Gênese de Nova Friburgo** (Fundação Biblioteca Nacional – 1995).

PABST, G. e Dungs, F. **Orchidaceae Brasiliensis** (2 volumes. Brücke Verlag, Hildeshein –1975-77).

PANSARIN, E. **Biologia Floreal de Cleistes Macrantha** (UNICAMP, Depto. de Biologia – 2003).

PRIDGEON, Cribb, Chase & Rasmussen **Gênera Orchidacearum** (vol. 1 Oxford University Press).

RAPOSO, Pe. J.G. **Etimologia das Orquídeas do Brasil** (Ave Maria Editora – 1995).

RAUH, W. **Bromeliads** (Blandford Press Ltd. – U.K. – 1979).

REICHENBACH, H. **Otia Botânica Hamburgenia, Kenntnis der Orchideen, Kranzlin, Fr. Orchideen Flora Südamerikas.**

RUSCHI, A. **Beija Flores do Estado do Espírito Santo** (Editora Rios Ltda. – S.P. – 1982).

SILVA, J.J.da (**Arquivos da Associação dos Engenheiros da Estrada de Ferro Leopoldina** – R.J. – 2002).

SINGER, R. e Koehler, S. **Pollinarium Morphology & Floral Rewards in Brazilian Maxillariinae** (UNICAMP, Depto. de Biologia – 2004).

SINGER, R. **The Pollination Mechanism in Trigonidium Obtusum Lindl.:** sexual mimicry and trap flowers (UNI-CAMP, Depto. de Biologia – 2002).

SINGER, R. & Koehler, S. **Notes on Pollination of Notylia Nemorosa** (Orchidaceae) UNICAMP, Depto. de Biologia – 2003).

SINGER, R. & Koehler, S. **Toward a Phylogeny of Maxillariinae Orchids** (UNICAMP, Depto. de Biologia).

THOMAS, K. **O Homem e o Mundo Natural** (Companhia das Letras – 1988).

TUKEY, H.B. Jr., Wittwer, S.H. e Tukey, H.B. **Am. Soc.. Hort. Sci.** 71.496 – 1958.

WATSON,W. e Chapman, H.J. **Orchids:** Their Culture and Management (Charles Scribener's Sons, New York – 1903).

WETTSTEIN, R. Bot. **Zeitschrift.**

WITHNER, C.L. **The Cattleyas and Their Relatives** (Volume I – VI Timber Press, Oregon, USA – 1990).

WITHNER & HARDING **The Debatable Epidendrums** (Timber Press Oregon, USA – 2004).

THE AUTHORS

All are naturalists, conservationists and orchidists from way back. Miller, Warrer and Moura published the book "Orchids of the High Mountain Rain Forest in South Eastern Brazil" (1994). In 1996, a Portuguese language edition amplified and up dated was published "Orquídeas do Alto da Serra". Both versions were received in Brazil and in the rest of the world. 5.000 copies were printed and sold.

Due to the success of these books, the authors decided to produced a book with a much larger geographic scope - "The Organ Mountain range, its History and its Orchids".

Richard Warren, PhD in botany, and one of the most important growers of orchid species "in vitro".

Helmut Seehawer over the last ten years has concentrated his forces in studying, describing and illustratng the orchids of the Pleurothallidinae subtribe.

Izabel Moura Miller during the last 20 years has photographed Brazilian native orchid species. Her work has been published in various countries.

David Miller is a consultant in the area of sustainable development and conservation.

David and Izabel Moura Miller have bought and managed substantial areas of original forest with the intention of their permanent conservation.

INDEX

I

L

M

N

(CIP)
BIBLIOTECA PÚBLICA MUNICIPAL DE NOVA FRIBURGO - PUBLIC LIBRAY OF NOVA FRIBURGO (RJ/BRASIL)

M647o	Miller, David et al
	The Organ Mountain Range, its history and its orchids / David Miller, Izabel Moura Miller, Richard Warren and Helmut Se-ehawer; ilustrated by Alvaro Pessanha and organized by Inghra Ursula Scart. - Nova Friburgo (RJ): Editora Scart, 2008. 544 p.: il. ; 20,5x27,5 (Brochure)
	ISBN 978-85-60217-01-4
	1. Orchids. 2. Atlantic Rain Forest - Rio de Janeiro. 3. Botanic. I.Miller, David. II.Miller, Izabel Moura. III.Warren, Richard. IV.Seehawer, Helmut. V.Pessanha, Álvaro, ilust. VI.Scart, Inghra, org. VI. Title
	CDD 584.158153

19th Edition
Elaborated by Alcíria Araújo - Librarian/CRB7 RJ 4007 - Public Library of Nova Friburgo

For questions or suggestions about
this book please contact us at
riotrust@hotmail.com or
mountainrange@terra.com.br